Praise for

THE WILDERNESS WARRIOR

"Although Roosevelt's presidency ended a hundred years ago, Mr. Brinkley finds ways to make his presidential portrait a timely one. . . . *The Wilderness Warrior* describes a vigorously hands-on president, eager to fight more than one battle at a time. . . . Brinkley's fervent enthusiasm for his material eventually prevails. . . . He conveys the great vigor with which Roosevelt approached his conservation mission."

—Janet Maslin, *The New York Times*

"A fascinating, immensely readable book regarding Teddy Roosevelt's intense love for nature and how, during his presidency, he essentially became one of the very first eco-warriors."

—Johnny Depp, *Entertainment Weekly*

"An excellent book! It's a work that surely will rank as the most comprehensive and thoughtful study of Roosevelt's endless love affair with nature." —James E. McWilliams, *Austin American-Statesman*

"A magnificent and magisterial biography."

—Glenn Altschuler, *Baltimore Sun*

"To understand America, you need to appreciate Teddy Roosevelt. Doug Brinkley brilliantly uses the lens of Roosevelt's love of nature to show why he is so influential, fascinating, and relevant to our own times. This wonderful book is as vibrant as he was."

—Walter Isaacson, author of *Einstein*

"A compelling rendition of a grand American life."

—Susan Larson, *New Orleans Times-Picayune*

"Brinkley fully inhabits Roosevelt's mind." —*The New Yorker*

"A masterful look at T.R. as environmental crusader and ultimate outdoorsman. . . . An engrossing, compellingly written book. . . . Roosevelt is the very man we need today as the number of species in America continues to decline. . . . Brinkley's Roosevelt is P.T. Barnum, Walt Whitman, and Captain Ahab all rolled into one."

—Chris Irmscher, *The Los Angeles Times*

"Brinkley writes easy and lively prose. . . . The wider orbit of what might be called 'proto-conservation' in America receives excellent treatment in Brinkley's book, which is populated by a host of colorful minor characters, some of whom were major figures of the time. . . . Getting to know this cast of characters is one of the many pleasures of Brinkley's book." —Robert P. Harrison, *The New York Review of Books*

"Douglas Brinkley has brought us an important, deeply researched, compellingly readable, and inspiring story. No earlier historian and biographer has done such a splendid job of showing how much we all owe to T.R.'s activism as wilderness warrior. Exactly a century after his presidency, there could not be a better time to revisit and celebrate T.R.'s unfinished environmental legacy."

—Michael Beschloss, author of *Presidential Courage*

"Brinkley gives us the most insightful account yet of Roosevelt's evolution from sickly, bird-nest-collecting schoolboy to the biggest, baddest conservationist of the twentieth century. . . . Readers will close this book with a better appreciation for Roosevelt's forward-thinking genius—and, just as satisfying, the history of the American conservationism movement in its formative years."

—Dianna Delling, *Outside.*

"A stirring account of the man who turned our attention to conservation and the many glories of our American landscape." —Ken Burns

"Brinkley, a professional historian who ranks as one of the most prolific academics writing books for general readerships, has performed superb research at archives across the nation to fill the book with compelling details." —Steve Weinberg, *The Minneapolis Star Tribune*

"In *The Wilderness Warrior*, Douglas Brinkley brings into relief the biography, cultural influences, and political record of the most effective conservationist in history. . . . Like the Grand Canyon that as president he more or less rescued from development and mining interests in one fell swoop, Roosevelt is one of those American treasures that can make you wonder how you missed getting around to for so long. . . . Interesting and thorough. . . . Brinkley is a veteran author on twentieth-century Americana." —Bob Blaisdell, *The San Francisco Chronicle*

"Douglas Brinkley's *The Wilderness Warrior* proves conclusively that, by at least one measure, Theodore Roosevelt was our greatest president: he was the most effective conservationist who ever occupied the Oval Office. Brinkley paints Roosevelt in a compelling shade of green. His narrative is brisk and compelling, and it features an engaging cast of characters. *The Wilderness Warrior* is a brilliant exploration of this fascinating life. And for anyone tired of seeing conservationism mocked as a fringe movement, lying outside the American political mainstream, Douglas Brinkley's book is a persuasive and uplifting rebuttal."

—Peter Rowe, *The San Diego Union-Tribune*

"Groundbreaking. . . . No other author has done the excellent job Brinkley does in bringing Roosevelt into focus."

—*Orange County Register*

"*The Wilderness Warrior* is magnificent. Brinkley catches the full-blown spirit of the manic, insomniac president who left us with many splendid gifts—national parks, state parks, wildlife refuges, park rangers, national forests and monuments, a wildlife conservation ethic, and the idea of clean water." —Wendy Williams, *The Providence Journal*

"Brilliant portrait of a great president." —*Anchorage Daily News*

"Our greatest conservation president, Roosevelt lived much of his life in, around, and for nature—a fact that biographers have tended to ignore until Brinkley stepped forth. This new portrait of T.R. is one of the best we have. It is a grand portrait of a man and his era, rich in detail and infectious in enthusiasm." —Donald Worster, author of *A Passion for Nature: The Life of John Muir*

"What a grand tale of the nexus of political intrigue and natural preservation. . . . *The Wilderness Warrior* does justice to the magnitude of the man and his environmental legacy. . . . Mr. Brinkley is at the top of his game. . . . Brinkley's love of natural history coupled with a nuanced understanding of presidential power makes this a unique take on T.R.'s years as president." —Claude R. Marx, *The Washington Times*

"A brilliant, exhilarating portrait of a president and his unquenchable passion for the natural world." —*St. Petersburg Times*

"This monumental work explores how a Harvard Phi Beta Kappa who was a member of Manhattan's Knickerbocker elite became the champion of America's forests and wildlife. Author Douglas Brinkley presents Roosevelt as a man in full, complete with contradictions."

—Bill Millsaps, *Richmond Times-Dispatch*

"Riveting. . . . There is much to admire in Roosevelt, and the book is a lively read about this larger-than-life personality."

—Jim Prentice, Canada's Minister of the Environment, in *National Post*

Danny Turner

About the Author

DOUGLAS BRINKLEY is a professor of history at Rice University and a contributing editor at *Vanity Fair*. The *Chicago Tribune* has dubbed him "America's new past master." Seven of his books have been selected as *New York Times* Notable Books of the Year. *The Great Deluge*, his account of Hurricane Katrina and its aftermath, won the Robert F. Kennedy Book Award. He lives in Texas with his wife and three children.

Also by Douglas Brinkley

The Reagan Diaries (editor)

The Great Deluge: Hurricane Katrina, New Orleans, and the Mississippi Gulf Coast

The Boys of Pointe du Hoc: Ronald Reagan, D-Day, and the U.S. Army 2nd Ranger Battalion

Tour of Duty: John Kerry and the Vietnam War

Windblown World: The Journals of Jack Kerouac, 1947–1954 (editor)

Wheels for the World: Henry Ford, His Company, and a Century of Progress, 1903–2003

The Mississippi and the Making of a Nation (with Stephen E. Ambrose)

American Heritage History of the United States

The Western Paradox: Bernard DeVoto Conservation Reader (editor, with Patricia Nelson Limerick)

Rosa Parks

The Unfinished Presidency: Jimmy Carter's Journey beyond the White House

John F. Kennedy and Europe (editor)

Rise to Globalism: American Foreign Policy since 1939, Eighth Edition (with Stephen E. Ambrose)

The Majic Bus: An American Odyssey

Dean Acheson: The Cold War Years, 1953–1971

Driven Patriot: The Life and Times of James Forrestal (with Townsend Hoopes)

FDR and the Creation of the U.N.

THE WILDERNESS WARRIOR

THEODORE ROOSEVELT AND THE CRUSADE FOR AMERICA

DOUGLAS BRINKLEY

HARPER PERENNIAL

NEW YORK • LONDON • TORONTO • SYDNEY • NEW DELHI • AUCKLAND

HARPER PERENNIAL

An extension of this copyright page appears on pages 903–904.

A hardcover edition of this book was published in 2009 by HarperCollins Publishers.

FIRST HARPER PERENNIAL EDITION PUBLISHED 2010.

Maps by Nick Springer

The Library of Congress has catalogued the hardcover edition as follows:

Brinkley, Douglas.
The wilderness warrior : Theodore Roosevelt and the crusade for America / Douglas Brinkley—1st ed.
xv, 940 p. : ill., maps ; 24 cm.
Includes bibliographical references (p. [831]–896) and index.
ISBN 978-0-06-056528-2
1. Roosevelt, Theodore, 1858–1919. 2. Conservation of natural resources—United States—History. 3. Nature conservation—United States—History—20th century. 4. Conservationists—United States—Biography. 5. Presidents—United States—Biography. I. Title.
E757 .B856 2009
973.91'1092B 22 2009291902

ISBN 978-0-06-056531-2 (pbk.)

17 18 DIX/RRD 10 9 8

Dedicated to the memory of
Dr. John A. Gable (1943–2005),
executive director of the Theodore Roosevelt Association;
and
Sheila Schafer of Medora, North Dakota,
whom I love with all my heart;
and

Robert M. Utley (aka "Old Bison")
Historian of the American West

Defenders of the short-sighted men who in their greed and selfishness will, if permitted, rob our country of half its charm by their reckless extermination of all useful and beautiful wild things sometimes seek to champion them by saying that "the game belongs to the people." So it does; and not merely to the people now alive, but to the unborn people. The "greatest good for the greatest number" applies to the number within the womb of time, compared to which those now alive form but an insignificant fraction. Our duty to the whole, including the unborn generations, bids us to restrain an unprincipled present-day minority from wasting the heritage of these unborn generations. The movement for the conservation of wild life and the larger movement for the conservation of all our natural resources are essentially democratic in spirit, purpose, and method.

—THEODORE ROOSEVELT, *A Book-Lover's Holidays in the Open* (1916)

And learn power, however sweet they call you, learn power, the smash of the holy once more, and signed by its name. Be victim to abruptness and seizures, events intercalated, swellings of heart. You'll climb trees. You won't be able to sleep, or need to, for the joy of it.

—ANNIE DILLARD, *Holy the Firm* (1984)

CONTENTS

Former president Theodore Roosevelt visiting Arizona in 1913. From his White House bully pulpit, Roosevelt had saved such magnificent Arizona landscapes as the Grand Canyon and the Petrified Forest. His conservation policies, in general, became the template future presidents followed.

PROLOGUE

"I So Declare It": Pelican Island, Florida

(February–March 1903)

I

On a wintry morning in 1903 President Theodore Roosevelt arrived at a White House cabinet meeting unexpectedly and with great exuberance. Something of genuine importance had obviously just happened. All eyes were fixated on Roosevelt, who was quaking like a dervish with either excitement or agitation—it was unclear which. Having endured the assassinations of three Republican presidents—Abraham Lincoln, James Garfield, and William McKinley—Roosevelt's so-called kitchen cabinet at least had the consolation of knowing that their boss, at the moment, was out of harm's way. Still, they leaned forward, bracing for the worst. "Gentlemen, do you know what has happened this morning?" Roosevelt breathlessly asked, as everybody leaned forward with bated breath for the bad news. "Just now I saw a chestnut-sided warbler—and this is only February!"[1]

The collective sigh of relief was palpable. His cabinet probably should have known that T.R.—an ardent Audubonist—had a bird epiphany. With a greenish-yellow cap, a white breast, and maroon streaks down their sides, these warblers usually wintered in Central America; his spotting one in Washington, D.C., truly was an aberration. By February 1903, after his seventeen months as president of the United States following the murder of McKinley in Buffalo by a crazed anarchist, it was common talk that Theodore Roosevelt was a strenuous preservationist when it came to saving American wilderness and wildlife. His track record was in this regard peerless among the nation's political class. "I need hardly say how heartily I sympathize with the purposes of the Audubon Society," Roosevelt had written to Frank M. Chapman, curator of ornithology and mammalogy at the American Museum of Natural History in New York, just two years before becoming president. "I would like to see all harmless wild things, but especially all birds, protected in every way. I do not understand how any man or woman who really loves nature can fail to try to exert influence in support of such objects as those of the Audubon Society. Spring would not be spring without bird songs, any more than it would be spring without buds and flowers, and I only wish that besides protecting the songsters, the birds of the grove, the orchard, the garden and the meadow, we could also protect the birds of the sea-shore and of the wilderness."[2]

By the time Roosevelt wrote that letter, Chapman—a droll, hard-working, unshowy activist spearheading the Audubon movement—was a legend in ornithological circles, considered by many the father of modern bird-watching. Back in February 1886 Chapman had stirred up a serious commotion in a letter to the editor of *Forest and Stream* titled "Birds and Bonnets" in which he lamented the fact that in New York City alone three-quarters of all women's hats sold were capped by an exotic feather from a gun-shot bird. A devotee of comprehensive assessments and long-range planning for protecting aviaries, Chapman deemed the mutilation of birds for fashion "vulgar" and "unconscionable."[3]

Raised on an estate in New Jersey just across the Hudson River from New York City, Chapman had a love of birds from an early age. Although his rich parents had pushed him into the financial world, his passion was ornithology. Still, he went to work on Wall Street, without going to college first—a university degree wasn't required for a nineteenth-century gentleman banker. But the financial rewards of his brokerage work didn't satisfy him, so the dapper Chapman walked away from wealth to pursue a career in ornithology. He began volunteering at the American Museum of Natural History and worked his way up to become the preeminent expert in the Department of Birds. Even without an academic degree, Chapman, with his cleft chin, pursed mouth, and perfectly groomed mustache, became something of a dandyish town crier for his adopted profession, as well as a pioneer in using a camera to study the nesting habits and egg hatching of birds. He believed the modern ornithologist needed to take behavior, psychology, breeding, biology, migration, locomotion, and ecology into consideration during fieldwork.[4]

Theodore Roosevelt, whose father was a founder of the American Museum of Natural History, not only followed Chapman's rising career but cheered on his pro-bird activities every step of the way. Thrilled by Chapman's autonomy from academia, Roosevelt embraced his "public service" work aimed at helping everyday citizens to better understand the wild creatures flittering about in their own backyards. Before Chapman, for example, ornithologists practiced taxidermy on birds, stuffing them with cotton and lining them up on museum shelves. For every specimen on display, there were many others in storage. Bored by this strictly "study skins" approach, Chapman developed innovative dioramas in which habitat was also included as part of the educational experience.[5] A profound, inexplicable infatuation with birds was simply part of Chapman's curious chemistry, and he shared his zeal with Roosevelt and other outdoor enthusiasts. As a protector of "Citizen Bird," Chapman insisted that ornithologists needed to teach fellow hunters that often "a bird in the bush is worth two in the hand."[6]

As the editor of *Bird-Lore* magazine (precursor to *Audubon*)—and author of numerous popular bird guides, including *Bird-Life: A Guide to the Study of Our Most Common Birds* in 1897—Chapman was *the* bird authority of his generation. Roosevelt enjoyed being his enthusiastic sponsor. Chapman insisted on saving not just birds but their habitat—particularly breeding and nesting grounds in Florida. It was *the* essential condition, he insisted, for dozens of migratory species' survival. To Chapman—and Roosevelt—creating "federal reserves" for wildlife and forests wasn't debatable; it was an urgent imperative.

Roosevelt and Chapman weren't unique in their promotion of vast reserves. They were, in fact, reviving conservationist convictions that had been stalled by shortsighted politicians. Since the American Revolution the idea of game bird laws and habitat conservation had struck a responsive chord. In 1828 President John Quincy Adams set aside more than 1,378 acres of live oaks on Santa Rosa Island in Pensacola Bay.[7] Although Adams's personal journals did, at times, show an abiding interest in birds, his motivation for saving Santa Rosa Island was ultimately utilitarian: its durable wood could be used to construct future U.S. naval vessels. But even such a low-grade conservationist effort as Adams's tree preserve drew a fierce backlash. Running for president in 1832, Andrew Jackson denounced Adams's tree farm as an un-American federal land grab, an unlawful attempt to deny Floridians timber to use as they saw fit. "Old Hickory," as Jackson was nicknamed, believed God made hardwood hammock to cut and birds to eat. He ridiculed New England swells like Adams as effete, anachronistic sportsmen overflowing with ridiculous notions of "fair chase" rules and regulations for simply killing critters.[8]

While Jackson clearly lacked the conservationists' foresight, he was correct in labeling Adams and others who applied etiquette to hunting as aristocrats. Because New England had such strong cultural ties to Great Britain—where the idea of wildlife preserves (hunting) for aristocrats was an accepted part of the society since the reign of King William IV (1830–1837)—it's little surprise that America's first true conservationists came from the northeast. Starting in 1783 there were dozens of "sportsman" companion books, which promoted strict guidelines for upper-class gentleman hunters in places like New York, Boston, and Philadelphia. Furthermore, in 1832 the painter and sportsman George Catlin, returning from a sketching trip in the Dakotas, lobbied the U.S. government to establish "a magnificent park" in that region, to be populated by buffalo, elk, and Indians and marketed as a world-class tourist attraction. Filling his western reports with exclamatory prose, Catlin envisioned a "nation's park" that would contain "man and beast, in all the wildness and freshness of their nature's beauty!"[9]

That same year John James Audubon hinted at the need for aviaries when he intrepidly journeyed around Florida, paint box and gun in hand, traveling from Saint Augustine to Ponce de Leon Springs and the Saint Johns River to Indian Key to Cape Sable to Sardes Key and finally to Key West and the Dry Tortugas.[10] Yet he still wrote enthusiastically about massacring brown pelicans and legions of other shorebirds in the Florida Keys. "Over those enormous mud-flats, a foot or two of water is quite sufficient to drive all the birds ashore, even the tallest Heron or Flamingo, and the tide seems to flow at once over the whole expanse," he wrote. "Each of us, provided with a gun, posted himself behind a bush, and no sooner had the water forced the winged creatures to approach the shore than the work of destruction commenced. When it at length ceased, the collected mass of birds of different kinds looked not unlike a small haycock."[11]

Even though the vast majority of nineteenth-century U.S. conservationists enthralled by the "great Audubon" were elite hunters and anglers, there was also a slow-burning idiosyncratic group of naturalists, epitomized by Henry David Thoreau. Thoreau was a careful student of the New England ecosystem and was deeply influenced by William Bartram's *Travels through North and South Carolina, Georgia, East and West Florida, the Cherokee Country, the Extensive Territories of the Muscogulges or Creek Confederacy, and the Country of Chactaws* (1794). His own long, sulking sojourns at Walden Pond and lonely hikes in the dank woodlands of Maine had transformed this onetime Harvard man of letters into a semi-hermitic Concord naturalist. It was Thoreau, in a seminal article in the *Atlantic Monthly* in 1858, who most passionately articulated a need to save wilderness for wilderness's sake. "Why should not we," Thoreau asked with mounting enthusiasm, "have our national preserves . . . in which the bear and panther, and some even of the hunter race [Indians], may still exist, and not be 'civilized off the face of the earth'—[and] our forests [saved] . . . not for idle sport or food, but for inspiration and our own true recreation?"[12]

Although this prescient article was added as a last chapter to Thoreau's classic *The Maine Woods* after his death, our great national hermit, in truth, was an anomaly in pre–Civil War America. His condemnation of the "war on wilderness" was, as the conservation scholar Doug Stewart put it, "a mere whisper in the popular conscience."[13] Instead, the pilot-light credit for galvanizing what the conservationist Aldo Leopold, in *A Sand County Almanac* (1949), called "the land ethic" belonged to well-to-do Eastern Seaboard hunters who loomed over the early campaigns to create wilderness preserves. In other words, Thoreau the poet *contemplated* nature preserves in the *Atlantic Monthly* while hunting clubs like the Adirondack Club and the Bisby Club circa 1870 started actually *creating* preserves in the Adirondacks.[14]

Long before Theodore Roosevelt, John Muir, and Gifford Pinchot were born, in fact, New York's aristocratic hunters, using sportsmen's newspapers and circulars to deliver their message, challenged loggers and sawmill operators and every other kind of forest exploiter to abandon their reckless clear-cutting. They wanted places like the Adirondacks saved for aesthetic and recreational pleasures. The precedent these pioneering gentlemen hunters started needed an indefatigable champion like Theodore Roosevelt to put the U.S. government fully on the side of the bird and game and forest preserves. "When the story of the national government's part in wild-life protection is finally written, it will be found that while he was president, Theodore Roosevelt made a record in that field that is indeed enough to make a reign illustrious," William T. Hornaday, the legendary director of the New York Zoological Park, wrote in *Our Vanishing Wild Life* (1913). "He aided every wild-life cause that lay within the bounds of possibility, and he gave the vanishing birds and mammals the benefit of every doubt."[15]

Even though Roosevelt's alliance with Chapman (and other visionary naturalists like Hornaday) launched the modern conservation movement between 1901 and 1909, Roosevelt's preservationist vein, first developed in 1887, has been unfairly minimized by scholars. Partly that's due to a bias against aristocratic hunters. In addition, historians studying the progressive era have been confused by, or failed even to recognize, the distinction between hunting game birds and helping save song birds that are unfit to eat. Crowds of scholars have unfairly rounded on Roosevelt for having a bloodlust. Nevertheless, to Roosevelt, gentleman hunters were the true front line in the nature preservation movement. Over the years, however, historians have usually deemed Roosevelt first and foremost a "conservationist"—a term first seriously coined in 1865 by George

Perkins Marsh in *Man and Nature* but not popularized until the publication, in 1910, of Gifford Pinchot's manifesto *The Fight for Conservation* (to which ex-President Roosevelt provided an introduction). "Conservation," Pinchot famously wrote, "means the greatest good to the greatest number for the longest time."[16]

A wildlife enthusiast since childhood, Roosevelt in 1887 cofounded the Boone and Crockett Club with George Bird Grinnell in order to create bison, elk, and antelope preserves for future generations of Americans to enjoy. Smitten with "the chase," he had also written a fine trilogy of books largely about his hunting experiences in the Dakota Territory: *Hunting Trips of a Ranchman* (1885), *Ranch Life and the Hunting-Trail* (1888), and *The Wilderness Hunter* (1893). While living at the Elkhorn Ranch thirty-five miles north of Medora, North Dakota, for extended periods between 1883 and 1892 (and shorter ones thereafter), Roosevelt developed a highly original theory about land management and wildlife protection. As president he promoted the pro-wildlife approach with revolutionary zeal. The immortal beauty of America's rivers and its vast prairies, rugged mountains, and lonely deserts stirred him to nearly religious fervor. Yet he remained a proud hunter to his dying day. In fact, Sagamore Hill, Roosevelt's home in Oyster Bay, New York, had walls lined with trophy heads and skins of birds and mammals. Boom (an elk), Pow-Pow (a buffalo head stuffed for library display), and Pop-Pop-Pop (a massive 28-point blacktail buck head spanning more than fifty inches) were his to showcase.[17] They represented Roosevelt's enthusiasm for big game hunting.

On the other hand, President Roosevelt, with scholar's fortitude, kept detailed lists of birds he saw grace the White House lawn. An avid birder, he spied on Baltimore orioles as they flicked their orange-edged tails and on crimson cardinals building sturdy nests. Dutifully he would record their numbers and habits in notebooks. Paradoxically, even though Roosevelt hunted game birds, when songbirds were the issue he agreed with the naturalist John Burroughs, who wrote in *Signs and Seasons* (1886) that the "true ornithologist leaves his gun at home."[18] He understood the clear distinction between game birds (like ducks and ruffed grouse), which were hard to drop, and songbirds (like robins and mockingbirds), which were easy to shoot on the wing but not dinner table fare.

Certain bird species—herons, terns, and ibises, for example—mesmerized Roosevelt. As president, he insisted that killing one of these Florida exotics was a federal crime. And although he wasn't an expert on brown pelicans, he had carefully studied the freshwater white pelicans of North Dakota and Minnesota, who left their lakes near the Canadian border and migrated to the Indian River region in Florida like clockwork every autumn. Although T.R. had never been to Pelican Island, a teeming bird rookery, he had read a great deal about the place, thanks to Chapman. Situated in a narrow lagoon located near Vero Beach on the Atlantic coast of Florida, Pelican Island, a five-and-a-half-acre dollop of shells and mangrove hammocks, was abundant with flocks of wading birds, something akin to the Galápagos Islands (in miniature) when Charles Darwin began his evolutionary studies in 1835. If Roosevelt paddled around the island he would have heard the loud murmur of bird chatter, a dozen species all singing in different keys, yet all somehow in unison, giving the Indian River rookery the distinct feel of a God-ordained sanctuary. Exuberant streams of birds actually congregated on Pelican Island like figures in a timeless dream. Great blue herons, for example, lingered for long and often hot hours, statue-still while somehow still managing to groom their breeding plumage, including ornate onyx head feathers that seductively lured a mate. Reading about the calls, stillnesses, and hesitations of these long-legged birds fascinated Roosevelt no end.

The most prominent resident of Pelican Island, however, was its namesake—the brown pelican (*Pelecanus occidentalis*). Chapman had taken dozens of photographs of brown pelicans congregating there, often carrying silver-colored fish in their elongated beaks as they flew contentedly over the tumbling waters of the Indian River. Studying Chapman's photographs in his 1900 book *Bird Studies with a Camera*, Roosevelt knew these funny-looking birds were of incalculably greater value alive than dead; if the brown pelican passed into extinction, Florida, he believed, would lose one of its most enchanting charms.

Clearly, Roosevelt understood that wildlife had a sacred order and pelicans were part of this grand design or teleology. For more than 2 million years, by adapting to changed circumstances, prowling for fish by turning downwind, half-folding their wings and then almost belly-flopping into brackish or saline water, they had avoided extinction. With their huge heads submerged, the brown pelicans' narrow beaks—the attached pouches serving as a dip net—scooped fish amid swarms of mosquitoes and midges in Florida's glassy lagoons.[19] For all their innate awkwardness, these playful birds were actually very efficient hunters. By dive-bombing for mullets from as high as fifty or sixty feet in the air, a healthy brown pelican could consume up to seven pounds of fish per day. Their daily hunting range was a radius of about fifty to sixty miles. And it wasn't just the frenetic avian activity of pelicans, egrets, ibises, and roseate spoonbills on Pelican Island that Roosevelt embraced as a biological hymnal. He studied the state's weather and its terrain, and kept records of its climate. He loved every little thing that grew in wild Florida, studying the beach mice, the green anoles, the gopher tortoises, the ants, the sea turtles, and the osprey, all with biological sympathy.

Ornithologists like Chapman who journeyed to wild Florida in the 1880s learned to love the shimmering wild egrets and elegant spoonbills that populated the rookeries, but only the brown pelicans made them laugh out loud. These were the clowns of the bird world. Their combination of short legs, long necks, and four webbed toes (which enhanced their swimming ability) made them seem clumsy. Because their bodies were so heavy, takeoff was something of a burlesque act. More than a few bird students (like Roosevelt) noted that when a pelican flew solo—which was often—it left an indelible impression. At times the pelicans resembled helium balloons with bricks attached to their feet, frantically flapping to get airborne, seemingly feverish with fatigue, desperate to defy the law of gravity. Nevertheless, they always managed to lift off.

Underlying President Roosevelt's love of pelicans and other birds was a staunch belief in the healing powers of nature. That he had a mighty strong Thoreauvian "back to nature" aesthetic strain coursing through his veins becomes evident when we read his voluminous correspondence with Chapman, Hornaday, and other leading naturalists of his day, including John Burroughs, William Dutcher, George Bird Grinnell, John Muir, and Fairfield Osborn. Through a combination of book learning and field observations, Roosevelt had a keen sense of the importance of what would come to be known as biological diversity and deep ecology. His appreciation of the beauty of nongame birds like pelicans imbued him with a stout resoluteness to protect these endangered avians. To him the destruction of pelicans—and other nongame birds—was emblematic of industrialization run amok. In fact, with the exception of his family, birds probably touched him more deeply than anything else in his life.

Starting after the Civil War, Americans were faced with the revolutionary impact of Darwinism: everybody, it seemed, weighed in for or against evolutionary theory. To Roosevelt, who read the revolutionary *On the Origin of Species* as a young teenager, Charles Darwin was practically a god, the Isaac Newton of biology. Besides being an excellent scientist, Darwin was a fantastic imaginative writer who had wandered the world far and wide. Because of his intense interest in Darwin, naturalist studies became Roosevelt's guiding principle. Only the Hebrew scriptures had a more profound impact on human societies than *On the Origin of Species*.[20] Although there was a Creator, Roosevelt believed, the natural world was a series of accidents. Yet he also held a romantic view of the planet, a belief that *Homo sapiens* had a sacred obligation to protect its natural wonders and diverse species. He believed every American needed to get acquainted with mountains, deserts, rivers, and seas. One ethereal experience with nature, he believed, made the world whole and God's omnipotence indisputable. "Roosevelt," the historian John Morton Blum concluded, accepted the Darwinian belief in "evolution through struggle as an axiom in all his thinking. Life, for him, was strife."[21]

II

After the Civil War, a new "gold rush" throughout America fomented the massacring of wildlife for profit and sport. Game laws were practically nonexistent in much of the interior west and south of the Mason-Dixon line up until the 1890s. Roosevelt was repulsed by firsthand dispatches he received about the abominable eradication of species throughout America. The glorious bison (once somewhere around thirty million to forty million strong) were nearly exterminated from the Great Plains, and jaguars along the Rio Grande simply disappeared into the Sierra Madre of Mexico. Pronghorn antelope could no longer outrun the market hunters and ranchers. The colorful Carolina parakeet (*Conuropsis carolinensis*) and the ubiquitous passenger pigeon (*Ectopistes migratorius*) were about to vanish forever. So was the ivory-billed woodpecker (*Campephilus principalis*). It was already too late for the great auk (*Pinguinus impennis*) and the Labrador duck (*Camptorhynchus labradorius*)—both species had been permanently eliminated from the planet.[22] Using satire to open resistant minds to the conservation crusade, William T. Hornaday's prophetic *Our Vanishing Wild Life* featured an illustration of a tombstone with "Sacred" carved on top and "Exterminated by Civilized Man 1840–1910" on the bottom. Scrolling downward on the tombstone, Hornaday listed birds made extinct by the epic brutality of humans—the Eskimo curlew, Gosse's macaw, and purple Guadalupe parakeet among them.[23]

By the turn of the twentieth century the situation in Florida was particularly acute. Once deemed a vast swamp of little value, the state was experiencing a boom due to the fashion trendiness of its birds—especially their feathers. Ironically, the Florida birds' splendid display of colorful plumes—nature's design to draw female birds into a mating ritual—had done its job too well: upper-class women of the "gilded age" were drawn to the male bird's fanciful plumage and it became the rage to adorn their hats with the beguiling feathers. As a result plume hunters poured into the state, guns in hand, determined to bag wading birds for the exotic feathers then in high demand. A pound of roseate spoonbill or great white heron wings, for example, was worth more than a pound of gold. For unrepentant old Confederates and lowlifes on the lam, wild Florida's vast thickets and tangled vegetation offered not only a haven but also a source of easy income. Along the banks of Florida's coastal rivers, the pallid shine of oil-wick lamps was a common sight. It emanated from plumer camps, where hunters were poised to gun down nongame birds for the New York millinery industry, which paid handsome sums for pallets of feathers.[24]

Most Florida plume-hunters were uneducated country bumpkins hired as day laborers. A lone plumer working the shallow pools along the Atlantic Ocean could collect 10,000 skins in a single season. A full-sized egret could yield fifty suitable ornamental feathers. Besides skinning the curlews, plovers, and turnstones, the hunters would put the carcasses on ice and ship them to New York by the barrel, where they were considered delicious "bird dishes" in some fine Manhattan restaurants.[25] Still, the real dollars came from the fashion industry. White feathers, particularly those of the American egret (known today as the great egret) and the snowy egret, were the most coveted plumage of all. Although the pink feathers of flamingos (stragglers from the Bahamas) and roseate spoonbills were in high demand as trimming, their plumage started to fade away to an anemic pink after a year or two. The egrets' white feathers epitomized decorative elegance and high status. The demand for beautifully adorned hats fueled an entire industry. By 1900 millinery companies employed around 83,000 Americans, mainly women, to trim bonnets and make sprays of feathers known as aigrettes.[26]

Although feathers had been used to adorn men and women for centuries, both in the courts of Europe and among indigenous peoples around the world, the garish gilded age took them to a new level of popularity.[27] The demand was advanced, in large part, by the proliferation of women's fashion magazines, where exotic feathers were shown adorning gowns, capes, and parasols. "The desire to be fashionable led scores of thousands of women to milliners for something eye-catching and elegant," the historian Robin Doughty wrote in *Feathers and Bird Protection.* "If plumes were costly looking, then ladies demanded them by the crateload, and the elegant trimmings pictured regularly in journals meant that bird populations all over the world fell under the gun."[28] Low-gauge shotguns were the weapon of choice. But starting around 1880, the introduction of semiautomatic rifles—although these were only sporadically used—made wholesale slaughter of wading birds much easier.[29]

By 1886, when George Bird Grinnell founded the Audubon Society, more than 5 million birds were being massacred yearly to satisfy the booming North American millinery trade. Along Manhattan's Ladies' Mile—the principal shopping district, centered on Broadway and Twenty-Third Street—retail stores sold the feathers of snowy egrets, white ibises, and great blue herons. Dense bird colonies were being wiped out in Florida so that women of the "private carriage crowd" could make a fashion statement by shopping for aigrettes. Some women even wanted a stuffed owl head on their bonnets and a full hummingbird wrapped in bejeweled vegetation as a brooch. However, others were aghast at ostentatious displays of feathered hats and jewelry. Led by many of the same women who were agitating for the right to vote, a backlash movement to banish ornamental feathers was under way. The fashion pendulum was slowly swinging away from using birds for exhibitionism. Extravagant birds' rights tenets and oaths were being advocated by many leading U.S. women suffragists, who took their lead from Queen Victoria, who had issued a public proclamation denouncing ornamental feathers in Great Britain.[30]

Terrified by the genocide of birds, Frank Chapman, the leading popular ornithologist in America, began delivering a lecture titled "Woman as a Bird Enemy" around New York. He hoped to shame women into abandoning their cruel fashion statements.[31] Convinced that an inventory of Florida's birds was of paramount importance, he also organized the first Christmas bird count; it quickly grew into the largest volunteer wildlife census in the world. Before long more than 2,000 Floridians began participating in bird counts during an annual three-week period around Christmas.[32]

Early in 1903, the tireless Chapman knew that Theodore Roosevelt, now the president of the United States, remained a "born bird-lover."[33] As governor of New York, T.R. delivered a bold speech on avian rights and cheered on the Lacey Act (landmark legislation passed by Congress on May 25, 1900, to protect birds from illegal interstate commerce). As President William McKinley's vice president, Roosevelt issued an unequivocal statement endorsing the eighteen state Audubon societies in the United States: "The Audubon Society, which has done far more than any other single agency in creating and fostering an enlightened public sentiment for the preservation of our useful and attractive birds, is [an organization] consisting of men and women who in these matters look further ahead than their fellows, and who have the precious gift of sympathetic imagination, so that they are able to see, and wish to preserve for their children's children, the beauty and wonder of nature."[34]

Once Roosevelt became president, under his initiative, the U.S. Department of Agriculture had already publicly supported the various

Audubon societies, and in its *Yearbook 1902* it pleaded with farmers and hunters to leave nongame birds alone.[35] With the future of Pelican Island in the balance, the bird population about to be wiped out, Chapman understood that the time to seek President Roosevelt's support on banning the bird slaughter there was *now*. If the dollop of land was not declared a USDA reservation, it would soon be a dead zone like the ground-down New Jersey Flats.

III

In early March 1903 President Roosevelt was mired at Capitol Hill in Washington, D.C., trying to push forward an anti-anarchy bill and was meeting with newly elected U.S. senators (from Idaho, Kentucky, Washington, and Utah) at the White House. Nevertheless, he still made time for his ornithologist friends. William Dutcher updated T.R. on the status of lighthouse keepers employed by the American Ornithologists Union (AOU) in Key West and the Dry Tortugas (seven islands located seventy miles off the mainland in the Straits of Florida) to protect nesting roosts. The bird-lovers also swapped stories about the health and well-being of their various friends in the Florida Audubon Society, an organization of which Roosevelt happened to be an honorary founder.[36] (Dutcher himself would soon become the first president of the new National Association of Audubon Societies.*)

The gregarious president liked showing off his extensive knowledge about the state's ecosystem, which included varied habitats like sea grass beds, salt marshes, and tree hammocks. Roosevelt's library had a half-shelf of books about Florida's wildlife. During the Spanish-American War he had been stationed at Tampa Bay waiting to be dispatched to Cuba. His uncle Robert Barnwell Roosevelt, his father's brother, a famous mid-nineteenth-century naturalist, had written a landmark ornithological book in 1884, *Florida and the Game Water-Birds of the Atlantic Coast and the Lakes of the United States*. (It was Uncle Rob who taught Theodore about the importance of what is now called ecology.) At the time T.R. was forty-four years old. He was stocky, with piercing blue eyes. His rimless spectacles and robust mustache dominated a remarkably unlined face. He spoke in clipped sentences, often making hand gestures and grimaces to underscore a point. This was followed by a hearty chuckle that bellowed up from his very depths. Emphatic and worldly in manner, a tireless op-

*Although the term Audubon Society is commonly used, many state Audubons are separate entities in fierce competition with the "National Audubon."

timist with thousands of enthusiasms to juggle, in Rudyard Kipling's terminology Roosevelt—who liked to be called the Colonel, in recognition of his service in the Spanish-American War—was quite simply a "first-class fighting man." The journalist William Allen White perhaps summed up Roosevelt's gregarious personality best: "There was no twilight and evening star for him," White wrote. "He plunged headlong snorting into the breakers of the tide that swept him to another bourne, full armed breasting the waves, a strong swimmer undaunted."[37]

As expected, Roosevelt assured both visitors that *of course* he cared a great deal about the fate of Florida's brown pelican, egrets, ibises, and roseate spoonbills. He always had, since childhood. He had, in fact, recently read Chapman's *Bird Studies with a Camera* and loved the vivid chapter on Pelican Island. Chapman and Dutcher couldn't have had a more receptive audience that March afternoon in Washington.

The American Ornithologists Union had been trying to purchase Pelican Island outright from the federal government for three years, to no avail. That winter, members of the AOU finally had a constructive meeting with William A. Richards, the Department of the Interior's new General Land Office (GLO) commissioner.[38] Dutcher, acting as chair of the AOU's committee on bird protection, along with Frank Bond, explained their quandary to Richards (a no-nonsense former governor of Wyoming). For years AOU had demanded that Pelican Island be surveyed—a prerequisite for placing a purchase bid on it. Now, with the official 1902 survey about to be filed, AOU felt boxed in. Legally, homesteaders' applications had to be given preference when GLO land was sold. With homestead filings imminent, the AOU's application would be shunned or given a low priority. And that meant the brown pelicans might not survive as a species on the Atlantic coast.

A hunter and conservationist himself, Richards wanted to help AOU, and he summoned Charles L. DuBois, his chief of the Public Surveys Division, into the meeting. Was there an ingenious way to circumvent the homesteaders-first provision? DuBois, a jurist who always dotted the i's and crossed the t's, at first said no. But he offered Dutcher and Bond one long-shot alternative. President Roosevelt *could* make Pelican Island a bird refuge by issuing an Executive Order. Worried that a firestorm would ensue if the U.S. Department of the Interior seemed to be in collusion with AOU, DuBois instead suggested pushing the Executive Order through the USDA, where it would go virtually unnoticed in the Biological Survey Division headed by Dr. C. Hart Merriam.

Now that the AOU had a credible, legal way to protect Pelican Island,

Dutcher wrote to Secretary of Agriculture James Wilson asking that a federal bird reservation be created. A stamp on the top corner shows that the secretary received it on February 27. Immediately using Frank M. Chapman as his conduit, Dutcher pushed for a meeting with the president about Pelican Island. Time was of the essence. With minimal difficulty Chapman procured a White House meeting that March.[39]

After listening attentively to their description of Pelican Island's quandary, and sickened by the update on the plumers' slaughter for millinery ornaments, Roosevelt asked, "Is there any law that will prevent me from declaring Pelican Island a Federal Bird Reservation?" The answer was a decided "No"; the island, after all, *was* federal property. "Very well then," Roosevelt said with marvelous quickness. "I So Declare It."[40]

For the first time in history the U.S. government had set aside hallowed, timeless land for what became the first unit of the present U.S. Fish and Wildlife Service's National Refuge System. History teaches that a zeitgeist sometimes develops around a fountainhead figure, that sometimes a transforming agent—in this case President Theodore Roosevelt—serves as an uplifting impetus for a new wave of collective thinking. Building on a growing ardor for federal intervention into the regulation of the private sector, Roosevelt's "I So Declare It" was in line with the legacies of all the Republican presidents since Lincoln. The Union victory in the Civil War, in fact, meant that the U.S. federal government had emerged as the principal proponent of national reform movements like conservation.

Recognizing the need for scientific wildlife and land management, every U.S. president in the gilded age considered himself conservationist-minded to some limited degree: certainly Benjamin Harrison, Grover Cleveland, and William McKinley did. Each, in fact, had landmark "forest reserves" accomplishments in his portfolio to showcase for history. Yet they all lacked long-term vision, concerned instead with only the forestry issues and water-shortage emergencies of the moment. But Roosevelt was vastly different; nature was his rock and salvation. Refusing to be hemmed in by the orthodoxies of his time, he burst onto the national stage—first as civil service commissioner and governor and vice president and then as president—promoting the Gospel of Wilderness. Bridging the gap as a naturalist-hunter, he deemed songbirds liberators of the soul and bison herds incalculably valuable to the collective psyche of the nation. Even though local communities across the American West complained about federal land grabs, Roosevelt insisted he was preserving wilderness for their own good, for the sake of the American heritage.

With nationalistic optimism, Roosevelt's patriotic summons essentially called for deranking the Louvre, Westminster Abbey, and the Taj Mahal as world heritage sites. The United States had far more spectacular natural wonders than these worn and tired man-made spectacles: it had the Grand Canyon, Crater Lake, the Petrified Forest, Key West, the Farralon Islands, the Tongass, Devils Tower, and the Bighorns. American bird flocks, he insisted, were far more glorious than those found in the steppes and forests of staid Old Europe. The implicit assumption was that Roosevelt's utter love of "American Wilderness" always had a heavy component of raw nationalism. When asked as ex-president in 1918 why he loved wildlife so much, Roosevelt had a characteristically direct yet unreflective answer: "I can no more explain why I like natural history," he said, "than why I like California canned peaches."[41]

Now, with this imperious decree of March 1903, the irrepressible naturalist was saying that a part of wild Florida should be saved for the sake of imperiled birds and endangered animals. President Roosevelt's guiding eco-philosophy was that habitat preservation for animals mattered, completely. Any reasonable person, he believed, should understand this. In the new century, market hunters had an obligation to stop their rampage and bow to the forces of biological conservationism and utilitarian progressivism as far as land and wildlife management were concerned. Forests needed to be treasured as if life-giving shrines. Citizens had to rally to save remnant populations of wildlife *everywhere* before species extinction became epidemic. Biodiversity was apparent and essential in nature, Roosevelt believed, wherever open minds looked. A huge cornucopia of wild creatures and plants, diverse in purpose and structure, with beauty and utilitarianism beyond the most fertile imagination, was an omnipotent God's blessed gift to America.

A relieved Chapman rejoiced when he heard Roosevelt's verdict—"I So Declare It"—realizing this was a new precedent for wildlife protection. He vowed to convey to future generations that March 1903 was *the* turning point in the birds' rights movement. True to his word, Chapman would laud Roosevelt in *Camps and Cruises of an Ornithologist* (1908) and *Autobiography of a Bird-Lover* (1933). Filed away in Chapman's personal papers on the fifth floor of the American Museum of Natural History, in fact, is the letter he wrote to Roosevelt in 1908, claiming that "The Naturalist President" had, "more than any other person," inspired him to write *Camps and Cruises of an Ornithologist.*[42] In that long memoir, Chapman credited the "characteristic directness" of President Roosevelt with guaranteeing the "future safety of pelicans" for perpetuity.[43] "Not only shall

I enjoy the book, but what is more important, I feel the keenest pride in your having written it," Roosevelt wrote to Chapman in gratitude. "I like to have an American do a piece of work really worth doing."[44]

With that one sweeping "I So Declare It," President Roosevelt, the big game hunter, had entered John Muir's aesthetic preservation domain. And Pelican Island wasn't a passing whim of a president showing off to ornithologist colleagues. It was an opening salvo on behalf of the natural environment. No longer would slackness prevail with regard to conservationism, for Roosevelt—the wilderness warrior—would coordinate the disparate elements in the U.S. government around a common "great wildlife crusade." Perhaps the historian Kathleen Dalton in *Theodore Roosevelt: A Strenuous Life* summed up Roosevelt's evolved attitude toward biota circa 1903 best: "Despite his official commitment to the policy of conservation of natural resources for use by humans he held preservationist and romantic attachments to nature and animals far stronger than the average conservationist."[45]

On March 14, 1903 President Roosevelt officially signed the Executive Order saving Pelican Island. By slipping the federal bird reservation into the domain of the U.S. Department of Agriculture, as Charles DuBois of the Interior Department had suggested, T.R. was hoping to avert the notice or controversy that keeping it in the Interior Department would have generated. Whenever he was faced with an obstacle, Roosevelt liked figuring out a way to circumvent it. Remarkably, T.R.'s Executive Order sailed through the bureaucracy in just two weeks. Legally it had to be approved by both Agriculture and Interior before the president could sign it.[46] Without any note of toughness—and only fifty words long—the order was a seminal moment in U.S. Wildlife History: "It is hereby ordered that Pelican Island in Indian River in section nine, township thirty-one south, range thirty-nine east, State of Florida, be, and it is hereby, reserved and set apart for the use of the Department of Agriculture as a preserve and breeding ground for native birds."[47]

The first unit of the U.S. National Wildlife Refuge System was now a reality. And Sebastian, Florida—the hamlet closest to Pelican Island—was its birthplace. Eighteen months later, Roosevelt created the second federal bird reservation, at Breton Island, Louisiana. By 2003, when Pelican Island celebrated its centennial, the U.S. National Wildlife Refuge System comprised more than 540 wildlife refuges on more than 95 million acres. Taken together, these woodlands, bayous, desert scapes, bird rocks, tundra, prairie, and marshland make up 4 percent of all United States territory.[48] At the time, however, the Pelican Island declaration garnered very little national attention—the *New York Times* never mentioned it, nor did the Jacksonville *Florida Times-Union*.[49] But future generations took serious notice; the impetus for a National Wildlife System had sprung to life. Saving Pelican Island initiated Theodore Roosevelt's evolving idea of creating greenbelts of federal wildlife refuges everywhere the American flag flew. Very quickly these refuges grew exponentially in numbers under Roosevelt's influence until the map of the lower forty-eight states was vastly altered. From this single small island in Florida's Indian River Lagoon grew the world's greatest system of land for wildlife. In the remaining six years he was in office, Roosevelt created fifty more wildlife refuges. Writing in his well-received *An Autobiography* (1913), Roosevelt explained how his ambition hardened to create these refuges without his ever making an on-site inspection trip:

> The establishment by Executive Order between March 14, 1903, and March 4, 1909, of fifty-one National Bird Reservations distributed in seventeen States and Territories from Puerto Rico to Hawaii and Alaska. The creation of these reservations at once placed the United States in the front rank in the world work of bird protection. Among these reservations are the celebrated Pelican Island rookery in Indian River, Florida; The Mosquito Inlet Reservation, Florida, the northernmost home of the manatee; the extensive marshes bor-

Michael McCurdy, well-known illustrator of John Muir reprint books, pays homage to President Roosevelt's saving of Florida's brown and white pelican rookeries.

> dering Klamath and Malheur Lakes in Oregon, formerly the scene of slaughter of ducks for market and ruthless destruction of plume birds for the millinery trade; the Tortugas Key, Florida, where, in connection with the Carnegie Institute, experiments have been made on the homing instinct of birds; and the great bird colonies on Laysan and sister islets in Hawaii, some of the greatest colonies of sea birds in the world.[50]

What Roosevelt doesn't mention in *An Autobiography* was the backlash against his creation of bird refuges. The plumers and the millinery industry fought back, appealing to public opinion, lobbying Congress, and, in the most extreme cases, shooting at bird wardens. A battle royal ensued between powerful exploiters of nature versus beleaguered preservationists. Determined to win the so-called Feather Wars against plumers and market hunters—not to give an inch and to use the full force of the U.S. federal government as his arsenal—Roosevelt declared Passage Key, another brown pelican nesting area in Florida, the third federal refuge in October 1905.* This sixty-three-acre island was located offshore from Saint Petersburg, Florida, at the entrance to Tampa Bay. Roosevelt had studied it in 1898, when his legendary Rough Riders were waiting to transfer to Cuba to fight in the Spanish-American War, so he knew firsthand the high quantity of both migratory and year-round birds using it.[51]

Now as president, with another "I So Declare It" decree on Florida's behalf along the Gulf Coast, Roosevelt had helped every bird at Passage Key to continue to survive and thrive in its marine habitat. Slowly but steadily, the federal bird reservations grew. Many of his first reserves were in Florida—Indian Key, Mosquito Inlet, Tortugas Keys, Key West, Pine Island, Matlacha Pass, Palma Sole, and Island Bay. Roosevelt's "Great Wildlife Crusade" also protected colonies of white-rumped sandpipers, black-bellied plovers, and piping plovers on the East Timbalier Island preserve in Louisiana; provided safe nesting grounds for herring gulls on the Huron Islands Reservation in Lake Superior three miles off the shore of Michigan; and offered sanctuary to the sooty and noddy terns on the Dry Tortugas Reservation in the Gulf of Mexico. At the Pathfinder Federal Bird Reservation in Wyoming—created on February 25, 1909, just before T.R. left office—the president not only saved an essential waterfowl mi-

*He created two federal bird reservations at the same time in Michigan: Siskiwit Islands and Huron Islands.

gration stopover place on the western edge of the Central Flyway but also preserved herds of pronghorns, the fastest mammal in North America.

IV

When writing or lecturing about American birds Roosevelt often became lyrical, sometimes even songlike. His sparkling writings are often good enough to put him in the company of such first-rate naturalist writers as John Muir, Rachel Carson, Annie Dillard, Edward Abbey, Louise Erdrich, Peter Matthiessen, and John Burroughs. In 2008 the nature writer Bill McKibben included Roosevelt's 1904 speech at the Grand Canyon in Library of America's *American Earth: Environmental Writing Since Thoreau.*[52] "To lose the chance to see frigate-birds soaring in circles above the storm," Roosevelt wrote in *A Book Lover's Holidays in the Open* (1916), "or a file of pelicans winging their way homeward across the crimson afterglow of the sunset, or myriad terns flashing in the bright light of midday as they hover in the shifting maze above the beach—why, the loss is like the loss of a gallery of the masterpieces of the artists of old time."[53]

During his presidency, Roosevelt also instituted the first federal irrigation projects, national monuments, and conservation commissions. He quadrupled America's forest reserves and, recognizing the need to save the buffalo from extinction, he made Oklahoma's Wichita Forest and Montana's National Bison Range big game preserves. Others were created to protect moose and elk. To cap it off he established five national parks, protecting such "heirlooms" as Oregon's iridescent blue Crater Lake, South Dakota's subterranean wonder Wind Cave, and the Anasazi cliff dwellings at Mesa Verde in Colorado. Courtesy of an executive decree, Roosevelt saved the Grand Canyon—a 1,900-square-mile hallowed site in Arizona—from destructive zinc and copper mining interests. The scrawl of his signature, a conservationist weapon, helped set aside for posterity (or for "the people unborn" as he put it) a legacy of over 234 million acres, almost the size of the Atlantic coast states from Maine to Florida (or equal to one out of every ten acres in the United States, including Alaska*).[54] All told, Roosevelt's legacy was almost half the landmass Thomas Jefferson had acquired from France in the Louisiana Purchase of 1803.[55]

Full of environmental rectitude, Roosevelt turned saving certain species into a crusade. Unafraid of opposition, always watchful of political timing, and constantly ready with a riposte, Roosevelt acted with the

*The 234 million acres is a legacy figure, based on the wild sites that T.R. saved, many of which were later enlarged.

prowling boldness of a mountain lion on the hunt. Suddenly, before people knew what hit them, strange-sounding place-names like Snoqualmie, Nebo, and Kootenai were national forest reserves. Because Florida was known as a bird haven, perhaps turning Pelican Island and Passage Key into Federal Bird Reserves wasn't too shocking. But imagine how perplexed people were when Roosevelt ventured into the supposedly arid desert territories of Arizona and New Mexico, establishing federal bird reservations at Salt River and Carlsbad. Roosevelt believed there was no type of American topography that posterity wouldn't enjoy for recreational purposes and spiritual uplift: extinct volcanoes, limestone caverns, oyster bars, tropical rain forests, arctic tundra, pine woods—the list goes on and on. As Roosevelt noted when dedicating a Yellowstone Park gateway in 1903, the "essential feature" of federal parks was their "essential democracy," to be shared for "people as a whole."[56]

With the power of the bully pulpit, Roosevelt—repeatedly befuddling both market hunters and insatiable developers—issued "I So Declare It" orders over and over again. Refusing to poke at the edges of the conservation movement like his Republican presidential successors, Roosevelt entered the fray double-barreled, determined to save the American wilderness from deforestation and unnecessary duress. The limited (though significant) forest reserve acts of the Harrison, Cleveland, and McKinley administrations were magnified 100 times over once Roosevelt entered the White House. From the beginning to the end of his presidency, Roosevelt, in fact, did far more for the long-term protection of wilderness than all of his White House predecessors combined. In a fundamental way, Roosevelt was a conservation visionary, aware of the pitfalls of hyper-industrialization, fearful that speed-logging, blast-rock mining, overgrazing, reckless hunting, oil drilling, population growth, and all types of pollution would leave the planet in biological peril. "The natural resources of our country," President Roosevelt warned Congress, the Supreme Court, and the state governors at a conservation conference he had called to session, "are in danger of exhaustion if we permit the old wasteful methods of exploiting them longer to continue."[57]

Wildlife protection and forest conservation, Roosevelt insisted, were a moral imperative and represented the high-water mark of his entire tenure at the White House. In an age when industrialism and corporatism were running largely unregulated, and dollar determinism was holding favor, Roosevelt, the famous Wall Street trustbuster, went after the "unintelligent butchers" of his day with a ferocity unheard of in a U.S. president. As if recruiting soldiers for battle, Roosevelt embraced rangers and wardens far and wide—even in Hawaii, Alaska, and Puerto Rico—insisting that the time was ripe to protect American wildlife from destructive insouciance. By reorienting and redirecting Washington, D.C., bureaucracy toward conservation, Roosevelt's crusade to save the American wilderness can now be viewed as one of the greatest presidential initiatives between Abraham Lincoln's Emancipation Proclamation and Woodrow Wilson's decision to enter World War I. It was Roosevelt—not Muir or Pinchot—who set the nation's environmental mechanisms in place and turned conservationism into a universalist endeavor.

"Surely our people do not understand even yet the rich heritage that is theirs," Roosevelt said in his book *Outdoor Pastimes of an American Hunter*, in 1905. "There can be nothing in the world more beautiful than the Yosemite, the groves of giant sequoias and redwoods, the Canyon of the Colorado, the Canyon in the Yellowstone, the Three Tetons; and our people should see to it that they are preserved for their children and their children's children forever with their majestic beauty unmarred."[58]

Theodore Roosevelt spent his adult life crusading on behalf of the national parks and monuments. Here he is at Yosemite in 1903.

CHAPTER ONE

THE EDUCATION OF A DARWINIAN NATURALIST

I

At a very early age, Theodore Roosevelt started studying the anatomy of more than 600 species of birds in North America. You might say that his natural affinity for ornithology was part of his metabolism. As a child Roosevelt became a skilled field birder, acquiring a fine taxidermy collection while recording firsthand observations in notebooks now housed at Harvard University.[1] And it wasn't just birds. When Roosevelt was four or five years old he saw a fox in a book and declared it "the face of God."[2] This wasn't merely a flight of fancy. The mere sight of a jackrabbit, flying squirrel, or box turtle caused Roosevelt to light up with glee. The multilayered puzzle that was Roosevelt, in fact, was titillated by the very sound of species names in both English and Latin—wolverines (*Gulo gulo*), red-bellied salamanders (*Taricha rivularis*), snapping mackerel (*Pomatomus saltatrix*), and so on. At times revealing a Saint Francis complex regarding animals, Roosevelt wanted to understand all living matter. "Roosevelt began his life as a naturalist, and he ended his life as a naturalist," wrote the historian Paul Russell Cutright. "Throughout a half century of strenuous activity his interest in wildlife, though subject to ebb and flow, was never abandoned at any time."[3]

Roosevelt was born on October 27, 1858, in the noisy hubbub of gaslit New York City. He struggled with ill health well into adulthood. "I was," Roosevelt later wrote, "a wretched mite suffering acutely with asthma."[4] Often wheezing, the young Roosevelt found physical relief by simply observing creatures' habits and breathing fresh air. Nature served as a curative agent for Roosevelt, as it's been known to do for millions afflicted with respiratory illness. "There are no words that can tell the hidden spirit of the wilderness," he wrote, "that can reveal its mystery, its melancholy and its charm."[5] Literally from childhood until his death in January 1919—following his arduous journey down the River of Doubt through Brazil's uncharted Amazon jungle—he epitomized Ralph Waldo Emerson's criterion for being an incurable naturalist. "The lover of nature is he whose inward and outward senses are still truly adjusted to each other," Emerson explained, "who has retained the spirit of infancy even into the era of manhood."[6]

As a precocious child Theodore Roosevelt became an amateur ornithologist. Trained by one of John James Audubon's student taxidermists, Roosevelt started his own natural history museum in New York City to show off his specimens. Later in life Roosevelt argued that parents had a moral obligation to make sure their children didn't suffer from nature deficiency.

Just over a year after Roosevelt's birth, the British naturalist Charles Darwin published his great scientific treatise *On the Origin of Species by Means of Natural Selection or the Preservation of Favoured Races in the Struggle for Life.* Deeply personal in tone, without esoteric graphs or undecipherable tables, *On the Origin of Species* set off heated global debate over religious beliefs that underlay the then current theories of biology. Darwin's great idea was evolution by natural selection, a death knell to the ancien régime of rudimentary biology. Although the Darwinian catchphrase "survival of the fittest"—which was first coined by the economist Herbert Spencer and which Roosevelt adopted practically as a creed—didn't appear until a revised 1869 edition, by the time Theodore was ten or eleven the biologist was his touchstone, a Noah-like hero. He was enthralled by the idea of collecting species in faraway places, and in his youthful imagination the Garden of Eden was replaced by Darwin's Galápagos Islands. (Almost twenty years before *On the Origin of Species* raised tantalizing questions about the Creation, Darwin had written about circumnavigating the world collecting specimens of animals and plants—both alive and dead—in *The Voyage of the Beagle*). Once Roosevelt grasped the concept of natural selection his bird-watching instincts went into overdrive. Suddenly he understood that the biological world wasn't static. Observed similarities between living creatures were often a product of shared evolutionary history. With Darwinian eyes he now studied every bird beak and eye stripe, hoping to reconcile anomalies in the natural world.[7] As an adult he would often carry *On the Origin of Species* with him in his saddlebag or cartridge case while on hunts.[8]

There was nothing unusual about a nineteenth-century child being enamored of animals and wildlife. Whether it's *Aesop's Fables* or *Mother Goose*, the most enduring children's literature often features lovable, talking animals. But the young Roosevelt was different from most other children: from an early age he liked to learn about wildlife scientifically, by firsthand observation. The cuteness of anthropomorphized animals in the popular press annoyed Theodore; Darwinian wildlife biology, on the other hand, captured his imagination and had the effect of smelling salts. Roosevelt loved the way the British naturalist had gone beyond physical similarities of anatomy and physiology to include behavioral similarities in his extended analysis *The Expression of the Emotions in Man and Animals* (1872). This devotion to Darwin, a real sense of awe, continued long after Roosevelt was an adult. Even when he was president, grappling with showcasing the Great White Fleet and building the Panama Canal, stories abounded about Roosevelt hurrying across the White House lawn exclaiming "Very early for a fox sparrow!" and then suddenly stopping to pick up a feather for closer coloration inspection.[9] Believing that evolution was factual, President Roosevelt nevertheless conceded that the concept of natural selection needed to undergo constant scientific experimentation, and the more data the better.[10]

But before Roosevelt discovered Darwin there were the picture books and outdoors narratives aimed at the boys' market. Every parent recognizes the moment when a child displays a special aptitude or precocity for learning, when hopes arise that it's a harbinger of great educational accomplishment to come. Such sudden bursts of enthusiasm from a toddler indicate both personality and preference. When Theodore Roosevelt obsessed over the lavish illustrations in David Livingstone's *Missionary Travels and Researches in South Africa* and asked questions about Darwin's theory of evolution, his parents, Theodore Roosevelt, Sr., and Martha Bulloch Roosevelt, realized their son was an aspiring naturalist.[11]

A Scottish physician and African missionary, Livingstone always had a high-minded scientific purpose for his jungle explorations—for instance, to discover the headwaters of a river. Even though the elegantly bound *Missionary Travels* was almost too heavy for young Theodore to carry, he would stare at the photographs of zebras, lions, and hippopotamuses for hours on end, thirsting for Africa. His early fancy for animals was the most appealing and tenderest part of his adolescence. "When I cast around for a starting-point," his friend Jacob A. Riis wrote in *Theodore Roosevelt: The Citizen* (1904), "there rises up before me the picture of a little lad, in stiff white petticoats, with a curl right on top of his head, toiling laboriously along with a big fat volume under his arm, 'David Livingstone's Travels and Researches in South Africa.' "[12]

Nearly coinciding with the publication of *On the Origin of Species*, a Neanderthal skullcap was found three years earlier in Neander Valley, Germany. For anybody even remotely interested in the relationship between animals and man the discovery of the first pre-*sapiens* fossil was stunning news. Suddenly Thomas Huxley, a discerning British biologist with long, wild sideburns, began saying in his lectures that the skull was proof that man was a primate, a direct descendant of apes. Just as exciting was Huxley's work on fossil fish, which he collected and classified with gusto. Although Huxley had been skeptical of Darwin's evolutionary theory, the publication of *On the Origin of Species* changed that. As the introverted Darwin retreated to a more private life, spending time with family and friends, Huxley became *the* leading interpreter of Darwin, explaining the master's theories and articles to rapt audiences all over the world. Determined to defend evolution to the hilt, Huxley declared himself "Darwin's bulldog," ably drawing gorillas on blackboards to explain to the old-school scientists how man evolved from them. Whereas Darwin was a field naturalist, his advocate Huxley practiced anatomy; together they constituted a nearly lethal one-two punch on behalf of modern biology.

Although Theodore couldn't possibly have understood the intricacies of evolutionary theory as a young boy, the explicit fact that man had evolved from apes appealed mightily to him. As a naturalist Darwin was unafraid to cut into the tissue of a cadaver looking for clues to creation. Merely having the temerity to write that man, for all his nobility, still bore "in his bodily frame the indelible stamp of his lowly origin" made Darwin heroic to Roosevelt.[13] "Thank Heaven," Roosevelt wrote about his childhood a year before his death, "I sat at the feet of Darwin and Huxley."[14]

Besides reading heavily illustrated wildlife picture books and hearing about the evolutionary theories of Darwin and Huxley from his family, Theodore gravitated to the Irish adventure writer Captain Mayne Reid. Generally speaking Captain Reid—a school tutor turned frontiersman on the Missouri and Platte rivers—wrote about the "Wilderness Out There" in a highly romantic way, as in a cowboy western.[15] His seventy-five adventure novels and oodles of short stories are full of backwoods contrivance. In *The Scalp Hunters* (1851), for example, Captain Reid, with an air of superior wisdom, went so far as to declare that the Rocky Mountains region was a sacred place where "every object wears the impress of God's image."[16] But Reid also appreciated evolution, filling his writings with sophomoric Darwinian analysis. He never missed a chance to describe birds, animals, and plants in vivid and apposite detail.[17] Although Captain Reid never made much money with his hair-raising tales, he consistently milked his Mexican-American War military service for a string of successful plays. Strange wild locales were among Captain Reid's specialties; for example, in *The Boy Hunters* (1853) he made the Texas plains, Louisiana canebrakes, and Mississippi River flyover seem like teeming paradises for any youngster interested in birds. There was, in fact, an able naturalist lurking underneath his often racist (even by mid-nineteenth-century standards) dime-novel prose.

Anybody wanting to understand Roosevelt as an outdoors writer must turn to *The Boy Hunters*. The plot is fairly straightforward—a former colonel in Napoléon's army moves to Louisiana with his three sons and a servant, determined to be at one with nature—but by Chapter 2 the narrative takes a strange twist. One afternoon a letter arrives from Napoléon's hunter-naturalist brother asking the old colonel to procure a white buffalo skin for France. Feeling too arthritic to tramp the Louisiana Territory in search of the rare buffalo, the colonel sends his sons—the "boy hunters"—in pursuit of the rare beast. Accompanied by the faithful servant, the adventurous boys head into the dangerous wilderness, determined to find a white buffalo, thought to be a sacred symbol in many Native American religions.* (Starting in 1917 the white buffalo also became a featured image in the state flag of Wyoming.)

Swooning over such chapter titles as "A Fox Squirrel in a Fix," "The Prong-Horns," and "Besieged by Grizzly Bears," Roosevelt loved every page of *The Boy Hunters*. Much of the novel's action took place in the Big Thicket of Texas, where wild pigs and horses roamed freely. After exposure to the American West the boys were no longer content shooting at birds: they coveted big game. Their ambition, Reid wrote, was not "satisfied with anything less exciting than a panther, bear, or buffalo hunt."[18] Like a trio of well-armed Eagle Scouts, the boy hunters grew to be com-

* *The Boy Hunters* was dedicated to "The Boy Readers of England and America." Reid hoped that the novel would "Interest Them So as to Rival in Their Affections the Top, the Ball, and the Kite—That It May Impress Them, So as to Create a Taste for that Most Refining Study, the Study of Nature." As for the hunting of white buffalo, Reid got the idea from the true story of the Cheyenne killing one in 1833.

pletely self-reliant, able to ride horseback, dive into rivers, lasso cattle, and climb huge trees like black bears. They scaled a steep cliff and shot birds on the wing with bow and arrow. They were taught to "sleep in the open air—in the dark forest—on the unsheltered prairie—along the white snow-wreath—anywhere—with but a blanket or a buffalo-robe for their beds." Drawing on the legends of the mountain men like Jim Bridger, Reid, preaching the strenuous life for boys, created little Natty Bumppos who could "kindle a fire without either flint, steel, or detonating powder."[19] These boys didn't need a compass for direction. They could read all the rocks and trees of that "vast wilderness that stretched from their own home to the far shores of the Pacific Ocean."[20]

The Boy Hunters included marvelous Hogarth-like woodcut illustrations of ferocious cougars and a bear wrestling an alligator, which spiced up the narrative. Lovingly Captain Reid described in credible naturalistic detail tulip trees and the fanlike leaves of palmettos, weird yuccas, and lofty sugar maples. Most ambitiously, however, he anticipated Darwinian theory in anecdotes about the food chain.[21] Academics have (accurately) criticized Captain Reid for harboring racist views of Indians and black slaves, but they've traditionally overlooked his essentially Marxist analogies about how the rich preyed on the poor in the mid-nineteenth century. He was a radical Republican, and to him the mega-capitalists were "king vultures" who didn't have a single positive trait and who abused the common vultures (*aura* and *atratus*) without mercy. Later in life, when Roosevelt fought in the Spanish-American War, he borrowed much of Captain Reid's observations of vultures for color in his own memoir *The Rough Riders*.[22]

Unlike James Fenimore Cooper or the *Crockett Almanacs*, Reid's half-fictions awkwardly offered up the proper Latin names for wildlife and plants he encountered. "About noon, as they were riding through a thicket of the wild sage (*artemisia tridentata*)," he wrote in *The Boy Hunters*, "a brace of those singular birds, sage cocks or prairie grouse (*tetrao urophasianus*), the largest of all the grouse family, whirred up before the heads of their horses."[23] Passages like this occur in dozens of Reid's books, often with the Latin binomials slowing down the otherwise fast-paced prose. Clearly, Reid wasn't a local-color writer like Bret Harte or Alfred Henry Lewis recounting desperado hijinks and cowboy yarns from the Wild West. Roosevelt, like many American and English boys, made no bones about his idolatry of Captain Reid and his risky romps throughout the West.

Later in life, when Roosevelt moved to the Dakota Territory for intermittent spells to ranch and research a trilogy of books on the outdoors and hunting, Reid remained his literary model. Reid's prose exploits of an American "cibolero" (buffalo hunter) in *The White Chief: A Legend of North Mexico* (1855), for example, motivated Roosevelt to lustily track down a nonwhite bison himself. In following its protagonist on a wild, woolly hunt for trophies in the Rockies, the book celebrated manifest destiny while offering Roosevelt a "manly" lesson in natural history. "To [the cibolero] the open plain or the mountain was alike a home," Captain Reid wrote. "He needed no roof. The starry canopy was as welcome as the gilded ceiling of a palace."[24]

Whenever Roosevelt wrote about the wilderness or birds, even as a boy, traces of Reid's hyper-romantic style are easily detectable.[25] Although other naturalist writers also captivated young Theodore's imagination—for example, John James Audubon and Spencer Fullerton Baird, the foremost naturalist of the era*—Captain Reid, whom Edgar Allan Poe witheringly described as "a colossal but most picturesque liar," remained his role model.[26] (Captain Reid seldom purposely lied about nature, but he always embellished his own supposed rough-and-ready heroics.) An impressive five times in his *An Autobiography* of 1913, Roosevelt wrote about how "dearly loved" Reid was.[27] By contrast, other outdoors books for boys didn't much appeal to Roosevelt. For example, he dismissed Johann D. Wyss's *The Swiss Family Robinson* for its clumsy zoology and was bored by its tropical escapism and dragfoot juvenility. According to young T.R., Wyss had assembled a "wholly impossible collection of animals" to pad his plot about survival after a shipwreck.[28]

One naturalist Roosevelt did enjoy was the British scholar Reverend J. G. Wood, whose *Illustrated Natural History* was published to wide acclaim in 1851. Wood made studying wildlife fun for nonscientific minds as well as specialists, without dumbing it down. Ignoring the complications of Darwinism, he simply took the view that natural history was "far better than a play" and "one gets the fresh air besides."[29] In particular, Roosevelt loved Wood's *Homes without Hands: Being a Description of the Habitations of Animals, Classed According to Their Principle of Construction* (1866). Roosevelt marveled at passages in this weighty 632-page tome about ant nests and tunnels, deciding on the spot to emulate the author.[30] The eight-year-old Roosevelt sat down and wrote a short essay, his first known written work, titled "The Foraging Ant." There were more than 10,000 ant spe-

*Baird's scientific naturalist works—*Catalogue of North America Birds*, *Review of American Birds*, *North American Reptiles*, and *Catalogue of North America Mammals*—were reference bibles to Roosevelt.

cies in the world and Roosevelt wanted to understand their differences.* Proud of "The Foraging Ant," which was about the workaholism of ants, Roosevelt read the essay out loud to his parents, who were complimentary about the pseudoscientific earnestness of his naturalist prose.[31]

More impressive than the essay itself, however, was the mere fact that young T.R. had read *Homes without Hands.* Although the book was replete with engravings of numerous species ranging from ospreys to wasps, the prose was fairly dense. Ostensibly, Wood was describing the varied hearths wild creatures built to live in—moles' burrows, prairie dogs' tunnels, rabbits' warrens, beavers' dams, spiders' nests, and so on. But *Homes without Hands* was not aimed at the children's book marketplace; it was serious adult fare, the kind of comprehensive text required in mid-nineteenth-century university biology or zoology introductory courses. Full of Latin identifications, choking with minutiae about the abdomen colors of bees and conchologistical (conchology is the study of mollusks) insights into *Saxicava*, this book was a far cry from the elegant musings of Thoreau about the shifting light at Walden Pond or a *National Geographic* article on white-water runs down the Shenandoah River. That Roosevelt gravitated to its heft pointed to his future avocation as one of the most astute wildlife observers our country has ever produced. One object from his Twentieth Street home impressed him more than any other: a Swiss wood-carving that showed a hunter stalking up a mountain after a herd of chamois, including a kid. As Roosevelt recalled in *An Autobiography*, he fretted regularly for the tiny chamois, fearful that "the hunter might come on it and kill it."[32]

Stifled by city life, Roosevelt educated himself as best he could about zoology on the streets of Manhattan. As if cramming for a final exam, he grew determined to learn the song of every fast-fluttering bird in New York and the nesting habits of every small mammal in Central Park. Studying marine species in the nearby Atlantic Ocean was another interest; he actually enjoyed analyzing the radula (mouthparts) of mollusks. One afternoon he spied a dead seal at a fish stand among the piles of maritime products available for sale on the wharf—scallops, tuna, and other food fish; dried sea horses and pipefishes hawked for their "medicinal" qualities; and much more. Roosevelt fixated on the dead seal's bulk and whiskers.[33] The fact that the species had been netted in New York

*By 2009 the naturalist E. O. Wilson of Harvard University had documented 14,001 ant species. See Nicholas Wade, "Scientist At Work: Taking a Cue From Ants on Evolution of Humans," *New York Times*, July 15, 2008.

Harbor, where it is a rare visitor, flabbergasted him. Day after day, as if drawn to a talisman, he kept begging to be brought to the pier to study the seal's anatomy more closely.[34] As a budding naturalist, acquainted with *The Voyage of the Beagle*, the seven- or eight-year-old noted that seals had no external ears and that their back flippers didn't bend forward to help their bodies when they were on dry land. Zoology books taught him that there were nineteen seal species in the world, most living around Antarctica and the Arctic Circle.

"That seal filled me with every possible feeling of romance and adventure," he later recalled in his autobiography. "I had already begun to read some of Mayne Reid's books and other boys' books of adventure, and I felt that this seal brought all these adventures in realistic fashion before me. As long as that seal remained there I haunted the neighborhood of the market day after day. I measured it, and I recall that, not having a tape measure, I had to do my best to get its girth with a folding pocket foot-rule, a difficult undertaking. I carefully made a record of the utterly useless measurements, and at once began to write a natural history of my own, on the strength of that seal."[35]

Eventually, once the seal's body was sold for blubber, Roosevelt was given the head as a souvenir. With this in hand, the ten-year-old created his "Roosevelt Museum of Natural History." Its purpose was to help train him to become a natural history professional like Darwin. Bookshelf space was made in the upstairs hall at 28 East Twentieth Street for "Tag #1," and before long he had rows of bird nests, dead insects, and mouse skeletons. He considered himself a "general collector," with a particular interest in salamanders and squirrels. After a few months, however, he turned his focus primarily to birds. Before long, he had numerous specimens to call his own. There is, in fact, a precious historical document housed at Harvard University, handwritten and five pages long, that illuminates the sheer earnestness with which young Theodore maintained his natural history collection. Titled "Record of the Roosevelt Museum," it begins with a proclamation of professional accomplishment. "At the commencement of the year 1867 Mr. T. Roosevelt, Jr. started the Museum with 12 specimins [*sic*]: at the close of the same year Mr. J. W. Roosevelt [West Roosevelt, his cousin] joined him but each kept his own specimens, these amounting to hardly 100. During 1868 they accumulated 150 specimens, making a total of 250 specimens."[36]

Roosevelt scrambled for new specimens of mammals, birds, reptiles, amphibians, and insects wherever he could. Conch shells, larvae, hollowed eggs, and even a common cockroach were worth inclusion into his growing windowsill and bookshelf collection. In an article titled "My Life as a Naturalist," written in 1918 when he was an ex-president, Roosevelt recalled that he collected specimens the way other boys collected stamps.[37] Jars of tadpoles and minnows were particular favorites in his museum. Heading up Third Avenue toward the Harlem River, Roosevelt would wander the fields looking for rabbit holes and woodcock feathers. Leaves found in Gramercy Park were pressed into books and preserved. Young Theodore prided himself on being able to distinguish a hardwood from a conifer. Sometimes he would study the differences between leaves from the same tree, like a perplexed botanist.

As a boy, Roosevelt developed a keen sense of hearing, perhaps as compensation for his extreme nearsightedness. When he was twelve, his parents finally bought him spectacles, and a whole new world of avian color blossomed in front of him: for example, the bright plumage of indigo buntings and red cardinals shone for the first time as though dipped in Day-Glo. Roosevelt was interrogative about birds, believing they were messengers from prehistoric times. "It was the world of birds—birds, above all—that burst upon him now, upstaging all else in his eyes," the historian David McCullough explained of the adolescent Roosevelt, "now that he could actually see them in colors and in numbers beyond anything he had ever imagined."[38]

As a budding ornithologist Roosevelt also mastered the identification of a bird by the way it flew. If he saw a moderate rise and fall, the bird could be a woodpecker or northern flicker. Other birds, however—like hawks, egrets, herons, and crows—flew in a straight line, flapping their wings in constant rhythm. He learned that studying a bird's feather shape, color, and pattern was another way to properly identify it. And then, of course, there were the differences in bird's beaks, which from a utilitarian standpoint were better than Swiss army knives: they caught and held food, preened feathers, and built nests. There were thousands of different types of birds in the world and Roosevelt wanted to identify them all: a lofty goal that became a lifelong pursuit, which he never abandoned. "Roosevelt was part of a generation of men," the naturalist Jonathan Rosen has explained in *The Life of the Skies*, "who helped mark the transition from the nature-collecting frenzy that followed the Civil War to what we today recognize as birdwatching."[39]

II

Roosevelt's mother greatly encouraged her son's fascination with birds. Martha Bulloch Roosevelt (or Mittie, as she was usually called) was born on July 8, 1835, in Connecticut but grew up in Savannah and then Roswell, Georgia, north of Atlanta. She was something of a southern belle, with a gentle air about her; her two brothers had fought for the Confederate army in the Civil War. Even though Mittie married a Yankee—Theodore Roosevelt, Sr., of New York—and moved to Manhattan, she never abandoned her Johnny Rebel sympathies. After the surrender at Appomattox she tended to romanticize the "lost cause," always championing the Gray over the Blue, furious over the way captured Georgian soldiers had been mistreated in prison camps in Illinois and Maryland. Most of Mittie's reflections on the South, however, were idyllic. Her yarns about Georgia red clay, black bears, and beige panthers always kept her children wide-eyed. Because of Mittie, in fact, the piney woods of Georgia—the shortleaf, longleaf, loblolly, and slash pine—had an enduring fascination for Theodore. Encouraged by Mittie, Theodore became a die-hard enthusiast of natural history, a collector enraptured by birds. She even allowed into the Roosevelt household exotic live creatures like flying squirrels and newts. "My triumphs," Roosevelt recalled of his childhood, "consisted in such things as bringing home and raising—by the aid of milk and a syringe—a family of very young gray squirrels, in fruitlessly endeavoring to tame an excessively unamiable woodchuck, and in making friends with a gentle, pretty, trustful white-footed mouse which reared her family in an empty flower pot."[40]

In the spring of 1868, Theodore's parents traveled to Roswell to visit Martha's relatives. A fascinating correspondence between mother and son has survived, illuminating young Theodore's love of collecting all things wild. "I have just received your letter," he wrote to his mother on April 28. "My mouth opened wide in astonishment when I heard how many flowers were sent to you. I could revel in the buggie ones. I jumped with delight when I found you had heard a mockingbird. Get some of the feathers if you can. . . . I am sorry the trees have been cut down. . . . In the letters you write do tell me how many curiosities and living things you have got for me. . . . I wish I were with you . . . for I could hunt for myself."[41]

Two days later the precocious nine-year-old responded to a letter from his father, angling for hard-to-find plants. "I have a request to make of you, will you do it?" Roosevelt asked his father drily. "I hope you will. If you will it will figure greatly in my museum. You know what supple jacks are, do you not? Please get two for Ellie [Elliott, his younger brother] and one for me. Ask your friend to let you cut off the tiger-cat's tail, and get some long moos and have it mated together. One of the supple jacks (I am talking of mine now) must be about as thick as your finger and thumb must be four feet long and the other must be three feet long. One of my mice got crushed. It was the mouse I liked best though it was a common mouse. Its name was Brownie."[42]

As a young boy Roosevelt not only collected live mice and a woodchuck but also sketched them in notepads, a few of which have survived undamaged. Eastern moles were another favorite of the young Roosevelt, and in a series of drawings housed at Harvard he captured all the salient details of these blind underground tunnelers: flat fur; eyes covered with skin; the absence of ears; and side-facing feet ideal for digging. Then there are his fine sketches of robins and wrens. Clearly, Roosevelt was blessed with a mind that could absorb physical detail. His bird and animal drawings were quite accomplished for a boy of his age. To most children one field mouse was indistinguishable from another, but Roosevelt knew that a canyon mouse (*Peromyscus crinitus*) or a pinyon mouse (*P. truei*) was different from a cotton mouse (*P. gossypinus*). And he didn't recoil from drawing them.[43]

III

An astonishing fact about young Roosevelt's prodigious interest in nature was that he kept semi-regular diaries. Studying his unaffected scrawl and chronic misspellings brings crucial understanding and clarity to his later achievements as a conservationist and wildlife protectionist. Being a naturalist in the 1850s and 1860s was fashionable, and the young Roosevelt had caught the bug. Darwin had sounded a clarion call for a new generation of biological collectors, and Roosevelt had heard it loud and clear. Roosevelt called his first diary *My Life* and kept it between August 10 and September 5, 1868. That summer the Roosevelt family spent their holiday in Barrytown, a busy railroad depot and steamboat landing across the Hudson River. Leasing the country home of John Aspinwall, the Roosevelts immediately began exploring the Hudson River valley, which had recently been celebrated on canvas by such remarkable landscape painters as Thomas Cole, Jasper Cropsey, and Albert Bierstadt. Sitting on the riverbank, they could watch steamers, canal boats towed by tugs, and tall sails drifting by in silence.[44]

Barrytown provided Roosevelt his first opportunity to hear the earth inhale and exhale without interference from the noise and stench of the industrial revolution. In virtually every diary entry throughout the late summer of 1868, Roosevelt wrote enthusiastically about animals he had either encountered or heard of in the dales of Barrytown. There were deer, big and small, to be studied; he learned how to analyze their tracks, like

Influenced by Audubon and Darwin, Roosevelt started sketching birds and mammals in his notebooks. In these pencil drawings, he marks the sub-species variations of shrews.

an Iroquois or Algonquin scout in training. There were encounters with packs of wild dogs; mid-afternoon pony rides; times for feeding horses sweet grass, green meat, and bran mash; lazy hours reeling in bluegills at fishing ponds; walks along fast-running brooks teeming with crayfish, eels, salamanders, and water bugs (all easily trapped from the low bank); and strange spiderweb filigrees to analyze. Then there were the nests of wasps to fear and weasel holes to avoid when riding a pony. And, of course, bears (or at least rumors of bears)—young Theodore was ready to discuss bears at any time, night or day.[45]

Throughout his life, bears mesmerized Roosevelt. In *An Autobiography*, in fact, he wrote wide-eyed of Bear Bob—a former slave once owned by his mother's family in Georgia—who had been scalped by an angry black bear's claw.[46] (That violent, degrading spectacle haunted Roosevelt's imagination.) Even at nine, Roosevelt saw bears—lumbering giants that can run as fast as a racehorse for short distances—as a symbol of wilderness. Enamored of their cunning, curiosity, and ferocity (when provoked), Roosevelt marveled that their acute sense of smell enabled them to detect game from nearly one mile away. He also noted that bears belong to a mammal group referred to as plantigrade (i.e., flat-footed, in contrast to digitigrade animals, which walk on their toes, like cats and dogs). Young Roosevelt rejected the human-like Thomas Nast cartoons of these omnivores in *Harper's Weekly* (popular in the 1860s) in favor of cutting-edge zoology books with exact details on nonretractile claws' agility (among black bears only) in climbing trees. With scientific detachment, young Theodore would correct people who casually claimed that all bears hibernate in winter; they don't, and such sloppy misconceptions annoyed him greatly.[47]

For example, he believed bears were "wicked smart," able to backtrack in their own footprints to shake hunting dogs off their easily detectable scent. Yet popular culture usually portrayed bears as slothful, honey-loving buffoons hiding in a den. Often black bears were written about as small cousins of grizzlies. To a degree, that assessment was fair. But it bothered Roosevelt, who knew that there was nothing small about black bears. Hunters had recently killed adult black bear males weighing well over 300 pounds, with short, stubby claws used to scale trees quickly. In Native American lore, so-called spirit bears—black, brown, grizzly, and polar—were all endowed with sacred characteristics; but in European culture they often had a menacing reputation, as in *Goldilocks and the Three Bears*. As a boy Roosevelt was largely unimpressed with both caricatures of bears. He wanted hard, factual, empirical data about their instinctive habits. "My own experience with bears," he later wrote, "tends to make me lay special emphasis upon their variation of temper."[48]

The Hudson River valley was the first American wilderness area Roosevelt fell in love with. He tossed fist-sized rocks into boggy creeks and used miniature nets to scoop up minnows from local muddy bottoms. Such was his enthusiasm for the region that Roosevelt pouted the following spring when his family left for a year in Europe (from May 12, 1869, to May 25, 1870). He preferred studying North American wildlife in the woods of New York to what his parents thought would be a sweeping "continental" educational experience for their four children. Capitulating to his father's will, Theodore put on a cheery face and grew excited about the prospect of inspecting the zoos and natural sites of old Europe. While their steamship, the *Scotia*, headed across the Atlantic, Roosevelt's journal commented briefly on staterooms and seasickness, then focused on wildlife. Numerous entries in his log for 1869–1870—the spelling atrocious—mention some encounter or another with wildlife or domestic animals.

May 19th, 1869 [at sea]

Warmer. Read and played. We saw two ships one shark some fish severel gulls and the boatswain (sea bird) so named because its tail feathers are supposed to resemble the warlike spike with which a boatswain (man) is usually represented.

June 23rd, 1869 (London)

We went agoin to the zoological gardens. We saw some more kinds of animals not common to most menageries. We saw some ixhrinumens, little earthdog queer wolves and foxes, badgers, and racoons and rattels with queer antics. We saw two she boars and a wildcat and a caracal fight.[49]

Everywhere the Roosevelts traveled in Europe, young Theodore cast a friendly eye toward living creatures. Whether it was hauling in English lake fishes with his sister Corinne or listening to the roar of a waterfall in Hastings where herons congregated, Roosevelt observed nature like a budding biologist. Even when he was touring museums or cathedrals, his eyes immediately darted toward heraldic shields with bears and bulls, statues of wooden dogs, art showing centaurs, and any scrawny stray cats that prowled territorially around the grounds as if they owned the property. Vivid descriptions of ornery mules and trotting horses populated the diaries. In "My Journal in Switzerland," for example, Roosevelt wrote of climbing the Alps around Mont Blanc and trying to trap animals to learn "natural history from nature."[50] Although he was bored by the manicured European botanical gardens, Roosevelt marveled on seeing his first glacier, which he anointed "Mother of Ice."[51] While in Lugano, he was given two chameleons by the family's carriage driver to keep as pets; he was intrigued by their uncanny ability to change colors with such ease.[52]

By September 1869 the Roosevelts had made their way to southern Italy. Unfortunately, his asthmatic condition was making Theodore miserable, discoloring his face, forcing him to wheeze for air, his chest heaving and filled with mucus. Clearing his throat was necessary so often that it became a tic. Often he had to sleep sitting up in order to breathe. In his diary he cast horrific spells as awful attacks of "the asmer." There was no effective treatment available in the mid-nineteenth century, so he suffered, gasping for air whenever flare-ups occurred. After one sick, asthmatic afternoon, bored and bedridden, Theodore suddenly perked up at the sight of a performing canine. "In the evening a Lady came with a little dog who is the cuningest litleest fellow in the world," he said; "he had a great many tricke letting you kiss its hand whipping you standing on its hind legs and having a dress on."[53]

Although the Roosevelts traveled through Europe in high style, young Theodore missed his tiny museum back on Twentieth Street in New York City. Discovering English birds' nests and French snakeskins on walks, he pocketed them to haul back home for his expansive collection. Not wanting to let his natural history training lapse, he read Wood and Reid and Baird every chance he got. In Dresden, Germany, he ventured by himself to the Royal Zoological Museum, a venerable institution founded in 1728. Housing more than 100 animals, the museum had a particularly fine collection of reptiles and fishes. The main draw at the museum, however, was the intricate glass models of sea anemones precisely blown by the artist Leopold Blaschka.[54]

But it was the aquarium in Berlin that really got Roosevelt's blood running, bringing him up close to live perched ravens, rare ducks, and feisty cormorants. "We saw birds in there nests on trees and anemonese and snakes and lizards," he recorded on October 27. "We had a walk and played chamois and after goats which I a grizzely bear tried to eat."[55] Tellingly, even the Berlin aquarium left him missing the United States. "Perhaps when I'mm 14 I'll go to Minnesota," he wrote, dreaming of bird-watching in the Red River valley; "hip, hip, hurrah'hhh!"[56] In these diaries are the beginnings of Roosevelt's imagining America as a wilderness, a mosaic of longleaf pine woods and forested wetlands, of large floodplains and swamps and lakes. Europe may have had prairies, but America had dry *and* wet prairies. To Roosevelt America had now risen in stature as a true wilderness laboratory—Europe, he sensed, was spent.

Whenever young Theodore saw animals neglected or abused, his diaries reflect his outrage. The sight of a puppy run over in a Parisian gutter sent him into a deep funk. He lamented how cows and goats were mistreated by peasants, and was upset at seeing horses suffer from whips, bits, epilepsy, frostbite, and heat exhaustion. On December 7, in Nice, he recorded how atrociously a French peasant treated his livestock. "We saw a brutal conveyance of lambs," he wrote. "They were tied by their legs and swung across a donkeys back. We saw also a young vilain who swung a poor animal around."[57]

Three weeks later, young Theodore was horrified by the way Italians treated their animals. "It seemed as if we would never get out of Naples which I was very anxius to do," he wrote, "to avoid seeing the cruelty to the poor donkeys in making them draw heavey loads and nearly starving them."[58] In Rome he bragged that "we saw a beautiful big huge dog and I was the only one he shook hands with and I am proud."[59] Paradoxically, other entries from the same European trip extol the thrill of what hunters call the chase. Sometimes Roosevelt would patrol the streets of Rome pursuing wild dogs, cornering them so he could watch them snap and snarl up close, taunting them with his gun muzzle to provoke them to show their wolflike teeth. In prose he used dramatic imagery (indeed, clichés) about "trapping wild dogs" that were "growling furiously," stray mongrels that "thrust into his face."[60]

At other times during his European tour, Roosevelt sketched animals in the zoos he visited. This practice may have been inspired by his meeting Professor Daniel G. Elliot, one of the world's foremost naturalists, in Florence on March 1. Well-to-do, Elliot came from New York but had been embraced in European scientific circles as an ornithologist and mammalogist of rare talent. Some claimed, in fact, that Elliot had been put on earth to study a wide range of species including Californian salmon and American bison. Writing and often illustrating numerous books, Elliot was considered the leading expert on cats, both wild and domestic. But what most captivated Roosevelt was that Professor Elliot had undertaken to produce two volumes on North American birds in 1869, covering the numerous species that the great Audubon had missed.[61] An easy touch, the critic Roosevelt deemed Elliot's *The New and Heretofore Unfigured Species of the Birds of North America* a "beautiful" accomplishment of lasting integrity."[62]

Warmed by the glow of spending time with Elliot, T.R. sought out more books by naturalists. He started carrying around illustrated volumes by Alexander Wilson and Thomas Nuttall. At the zoological garden in Florence, he got to feed the bears and sketch them in a notebook. Even when Roosevelt toured the drafty stone museums of Europe, he would try to find a link to animals, noting, for example, that an ornate sleigh resembled a "crouching leopard." The last entry in young Theodore's European diary was made on their westbound ship before the family disembarked in New York. After complaining that he hadn't seen any birds or fish, his luck turned. "In the afternoon we saw some young whales," he wrote, "one of whom came so near the boat that when it spouted some of its water came on me."[63]

IV

Upon returning from Europe, fatigued from touring Catholic cathedrals and German dance halls, Roosevelt was flushed with the ambition of seeing the American West with his own eyes. Smitten with the whole idea of "Go West, young man!" once promulgated by Horace Greeley, he wanted to straddle the Continental Divide clad in buckskin, riding off into the wild Rockies. Roosevelt's parents, however, preferred the thick woodlands of northernmost New York and Vermont instead of places with names like Dead Man Gulch or Hellville. Proximity, one supposes, played a large role in his parents' decision about the itinerary. All T.R. could do was nod in acquiescence; at least the Hudson River valley was better than being seasick on another voyage to Europe. During the summers of 1870, 1871, and 1872 the Roosevelt family rented country houses along the Hudson River valley—in Dobbs Ferry (George Washington's headquarters before the march to Yorktown in 1781) and Spuyten Duyvil (near Riverdale).[64] The Hudson enchanted Roosevelt, even though the constant houseboat traffic and train whistles in these commuter towns not far from the city meant he wasn't having the kind of wilderness experience his imagination craved.

Conscious that he hadn't toured his homeland west of the Atlantic coast, Roosevelt titled his post-Europe diary "Now My Journal in the United States" (May 25 to September 10, 1870). Spending much of that summer in Spuyten Duyvil, Roosevelt quaintly dated his diary with "country" at the top of each entry. Sounding like a modern-day summer camper, he wrote of swimming, hiking, and shooting a bow and arrow. "We began to build a hut," he wrote on June 6, "and had a nice time and found a bird's nest with 3 eggs (but we did not take them)."[65]

When that particular diary ends abruptly, in September, there is a gap for the next eleven months. In August 1871, when the next journal begins, the awesome natural sites in the Adirondack Park—Mount Marcy, Blake Peak, and Lake Tear-of-the-Clouds (the source of the Hudson River)—animate his writing. Bursting with excitement about the Adirondacks (and the White Mountains), he tramped around, imagining himself in the footsteps of frontiersmen like Ethan Allen and George Rogers Clark. The Adirondacks at this time were very much in vogue. Two years prior to the Roosevelt family's visit, the Congregationalist minister W. H. H. Murray had published the best-selling *Adventures in the Wilderness; Or Camp-Life in the Adirondacks*, in which he claimed that the upstate New York "wilderness" helped cure consumption and other lung ailments.[66]

Owing to the unexpected success of Murray's book, tourists and health seekers alike came pouring into the Adirondacks, fishing in Lake Placid, hiking up Whiteface Mountain, and simply inhaling the air along the MacIntyre Range.[67]

Young Theodore counted himself in the front line of the new enthusiasts for the Adirondacks. Like Thomas Jefferson—who in 1791 deemed the thirty-two-mile-long Lake George "without comparison, the most beautiful water I ever saw"[68]—Roosevelt wanted to explore the shoreline and islands of this natural wonder. In fact, the "Queen of American Lakes" always held a special place in Roosevelt's heart. "We started on the Minnehaha up Lake George," he wrote. "We passed innumerable islands on the way up it. At the head or rather *tail* of this lake, where it is connected with Lake Champlain the mountains were very abrupt and the lake very narrow. The scenery at this point is so wild that you would think that no man had ever set foot there."[69]

In these diary entries from the Adirondacks Roosevelt uses a vivid exactness in describing the the piney woods and incomparable lakes he played in. Satisfied and comfortable, he watched a blue-gray bird with a shaggy crest dive into the lake and quickly identified it as a kingfisher. Careful distinctions were made between *coveys* of quail and *runs* of common loons. Strange as it seems, Roosevelt spent an hour just observing a little blue heron with a daggerlike bill that sat on the lake edge sunning itself; the sight was worth 100 Lord's Prayers. Sometimes it almost seemed that when Roosevelt saw a bird he became the bird for a short spell.

In the Adirondack Park and the White Mountains, Roosevelt also discovered the fiction of James Fenimore Cooper. His father felt that reading a novel like *The Last of the Mohicans* (1826) while staying in the area where the French and Indian War took place would enrich the literary experience. Cooper's hero Natty Bumppo (also called Leatherstocking or Hawkeye) was a bold white scout, paddling across Lake Champlain, climbing lofty summits, and wearing a bearskin to gain entrance to a Huron village. Later in life Roosevelt remembered that he read all five of *The Leatherstocking Tales* in the order in which Cooper had written them. This means that in addition to absorbing *The Last of the Mohicans*, he read what in retrospect are the two most important American conservationist novels of the nineteenth century, narratives that dealt, in part, with imperative calls to create forest reserves through visionary natural resource management: *The Pioneers* (1823) and *The Prairie* (1827). "I put Cooper higher than you do," Roosevelt would write to the novelist Josephine Dodge Daskam when he was vice president of the United States. "I do not care very much for his Indians, but Leatherstocking and Long Tom Coffin are, as Thackery somewhere said, among the great men in fiction."[70]

Before Cooper, forests were viewed as dark, satanic thickets, a regrettable natural obstacle to homesteaders and frontiersmen, something to be clear-cut; streams were similarly considered dangerous, unpredictable torrents. Cooper overturned this concept of the "haunted" wilderness. To him trees were "jewels" and fishes "treasures." In *The Pioneers*, for example, he railed against despoilers of nature for their "wasteful extravagance." Cooper's alter ego, Natty Bumppo, firmly believed that the unnecessary slaughter of wildlife was a crime against God. Cooper even anticipated the extermination of the ubiquitous passenger pigeon. "It's wicked to be shooting into flocks in this wastey manner," Cooper wrote in *The Pioneers*. "If a body has a craving for pigeon's flesh, why, it's made the same as all other creatures, for man's eating; but not to kill twenty and eat one. When I want such a thing, I go into the woods till I find one to my liking, and then I shoot him off the branches without touching the feather of another, though there might be a hundred on the same tree."[71]

On that summer's holiday, stimulated by his evening campfire readings in *The Last of the Mohicans*, young Theodore turned explorer, stalking chipmunks burrowed in fallen trees, and spying on woodpeckers making a racket with their beaks like plier claws. Pretending to be Natty Bumppo, he carefully studied salamander markings, finding them hidden under water-soaked logs. To the bafflement of his parents he gathered more than 100 different species of lichens and fungi under rocks and in dense undergrowth. He brought out from caves unusual samplings of moss to scrutinize back home under a magnifying glass. And, of course, there was daily talk of bears. "There is a tame bear here who eats cake like a Christian," he recorded at White River Junction, "and appears very anxious to come to close quarters with us."[72]

V

As these diaries attest, the future president's father encouraged T.R.'s love of nature. Straitlaced, slightly pious, and a dutiful husband, Theodore Roosevelt, Sr., preferred philanthropy over business and was always ready to aid the poor and needy. An abolitionist before the Civil War, he now—during the Reconstruction era—had an "emancipationist memory" (as the historian David W. Blight called the belief that the federal government should intervene to help the poor and disenfranchised), and a determination to be part of the progressive movements of his day.[73] Theodore Roosevelt, Sr., worried that in the thirst for post–Civil War reconcilia-

Theodore Roosevelt, Sr., was a founder of the American Museum of Natural History in New York.

tion, African-Americans were going to be discriminated against. But the main thrust of his philanthropy was to extol natural history and animal protection. Promoting the humane treatment of horses, making sure they weren't abused, became one of his civic concerns. In *An Autobiography*, T.R., in fact, goes on for a couple of pages about his father's prowess with horse reins, bragging that he was a fine four-in-hand carriage driver. A man of medium height, Theodore Sr. was so well proportioned and carried himself with such perfect posture that he seemed much larger. "He was a big, powerful man, with a leonine face, and his heart filled with gentleness for those who needed help or protection," Roosevelt recalled, "and with the possibility of much wrath against a bully or an oppressor."[74]

Driven to succeed, Theodore Sr. usually dressed in a white neckcloth and a dark suit, forgoing the broad-brimmed high hats favored by his fellow aristocrats. He gave the impression of a groomed man on top of his life's game. His nephew Emlen Roosevelt, in fact, claimed that he personified "abundant strength and power" in every pursuit he decided to tackle.[75] The wellspring of Theodore Sr.'s enthusiasm for living creatures—both wild and domestic—came from his father, Cornelius Van Schaack Roosevelt (T.R.'s grandfather), who was born in 1794. Cornelius dropped out of Columbia College and devoted his full attention to Roosevelt and Son hardware. Before long he had a near-monopoly on American plate glass imports. With money to invest, Cornelius founded Chemical Bank of New York. Basically, the Roosevelt family business, as befitting the urban gentry, was now financial investment.

The son of the exceedingly rich Cornelius Roosevelt, Theodore Sr., a leisured gentleman, born into the Dutch aristocracy of New York, never experienced poverty. Still, he had compassion for the underprivileged. (So did his son Theodore Jr., the future president.) Theodore Sr. mixed easily with both rich and poor. Among his many civic-minded good deeds was his cofounding of the Newsboys' Lodging Home and New York's Children's Aid Society. Calm, cheerful, deeply thoughtful, and a devoted Dutch Reform Protestant, Roosevelt was determined that his children—Anna (nicknamed Bamie), Theodore Jr. (Teedie), Elliott (Nell or Ellie), and Corinne (Conie)—would have an ideal childhood. Studying the natural world, he insisted, would be a big part of their education. Summers spent in the fresh air of upstate New York and New England helped fulfill this desire. Trips abroad were always structured for maximum educational uplift. Although not a naturalist, the elder Theodore Roosevelt was known throughout Manhattan as a devotee of the Hudson River valley. It's little wonder that he became a founder and trustee of the American Museum of Natural History. To be a civilized person, Theodore Sr. believed, meant honoring the biological history of *On the Origin of Species*, the most revolutionary book of the nineteenth century.

Now a New York landmark and a national treasure, the museum began when Dr. Albert S. Bickmore, who had been a student of the zoologist Louis Agassiz at Harvard, arrived in Manhattan. Fresh from collecting species throughout the East Indian archipelago, Bickmore began telling colleagues of a dream he had of founding a museum of natural history. They all agreed it would take a lot of money. So Bickmore was directed to 28 East Twentieth Street, where Theodore Roosevelt, Sr., resided. The elder Roosevelt had an abiding philanthropic interest in promoting nature education and was friendly with the leading Darwinian scientists in America. Although his reputation for good works was renowned, he wasn't a pushover. If Theodore Sr. didn't like a sales pitch, he'd give a firm, definitive no, even if his directness bruised sensibilities. Theodore Jr. wrote in *An Autobiography* that his father was the "best man" he ever knew but "the only man of whom I was ever really afraid."[76]

The courteous and knowledgeable Bickmore, however, received a warm welcome. Hours of fine conversation ensued, with Bickmore's dream coming into realistic focus by the time the teapot was emptied. "Professor," the elder Roosevelt told his guest, "New York wants a museum of natural history and it shall have one, and if you will stay here and cooperate with us, you shall be its first head."[77]

This was not idle encouragement. Applying his Dutch fortitude and beneficence, the elder Roosevelt helped spearhead the nascent effort to build a world-class nature museum in Manhattan. Vouching for Bickmore with his rich friends, he envisioned the museum as an educational place to offer schoolchildren and working people a chance to study stuffed elephants, desert dioramas, birds' nests, and porcupine quills. A few months later, on April 8, 1869, the charter for the American Museum of Natural History was approved in the front parlor of the Roosevelt brownstone. Two days later, Governor John Thompson Hoffman of New York signed a bill establishing the museum. The elder Roosevelt's partners in this grand philanthropic endeavor included the future U.S. ambassador to the United Kingdom Joseph Choate, the finance magnate J. Pierpont Morgan, and the U.S. congressman William E. Dodge, Jr.[78]

Theodore Roosevelt, Sr., was one of eighteen Manhattanites who signed a letter to the commissioner of Central Park requesting a building site on the Upper West Side for the new museum. The spot chosen was Manhattan Square, which ran along Central Park West from Seventy-Seventh to Eighty-First Street. The *New York Times* singled out the elder Roosevelt as being "deeply interested in the enterprise" from the moment Bickmore had knocked on his door.[79] President Ulysses S. Grant laid the museum's cornerstone in 1874. As was to be expected, ideas regarding the mandate and magnitude of the new museum collided. Early bickering, for example, centered on whether dinosaur fossils should be displayed. The elder Roosevelt—a trustee—was for displaying them; paleontology, after all, as the French scientist Georges Cuvier had dramatically proved, was an exhilarating part of natural history. (Critics of fossils, however, believed they were a Darwinist ploy to somehow discredit what would become Creationism.) At Theodore Sr.'s request a delegation was sent to France to study Cuvier's dinosaur bones and acquire outstanding taxidermy by Jules Verreaux.[80]

From the time it opened its doors in 1877, the American Museum of Natural History was a marvelous ticket-taking success. Visitors streamed en masse to see the new architectural wonder looming over a largely underdeveloped neighborhood. The original neo-Gothic building anchored the area, then home to squatters and goats. Huge turrets, conical roofs with heraldic eagles, and granite walls all added to its baronial elegance. At nighttime the illuminated museum could be seen from New Jersey, across the Hudson River.[81] The interior decor was heavy with black walnut and ash. The museum attracted nearly as much press attention as the opening of the Empire State Building eventually would; even President Rutherford B. Hayes was on hand for the formal opening late in 1877.[82] The inaugural displays focused on the glory of Darwinism (even though *On the Origin of*

Species had been published eighteen years earlier, its impact in America was just being felt), scientific achievement, geographical discovery, and the natural sciences. An extraordinary collection of fossil invertebrates—purchased for $65,000 from the official New York state geologist—attracted the longest lines.[83]

Young Theodore came home from Harvard University to attend the museum's opening and marveled at the seashells, beetles, and bird skins. He had donated twelve mice, a turtle, four bird eggs, and a red squirrel skull to the collection.[84] Respectfully, he listened to President C. W. Eliot of Harvard, a dull blade on the podium, extol the "divinity" of the natural sciences. He was particularly drawn to the second floor of the main hall, which showcased Professor Daniel Elliot's collection of North American birds. A stuffed dodo, a mummified crocodile, and a huge badger all apparently fascinated him no end.[85]

Nobody could have guessed that Theodore Jr., running around the museum excited about a mammoth tooth and a badger claw, would decades later have a wing of the museum dedicated in his honor for his efforts on behalf of U.S. conservation. Other great presidents—like Washington, Lincoln, and Jefferson—have beautiful memorials in Washington, D.C., along the Potomac River tidal basin, to celebrate their statesmanship. The New York museum, by contrast, fittingly became the New York state shrine to Roosevelt (along with a nature preserve called Roosevelt Island in the Potomac River in Washington, D.C., and Theodore Roosevelt National Park in the badlands of North Dakota). The Upper West Side museum was rededicated after Roosevelt's death to honor the "scientific, educational, outdoor and exploration aspects" of his life—not the political aspects. And while it's true that sagacious conservationist sayings of his were carved on the marble walls—"The nation behaves well if it treats the natural resources as assets which it must turn over to the next generation increased; and not impaired in value" might be the best—they were undercut by some of his most strident imperialist maxims.[86]

But the museum was a repository for dead wildlife only, whereas the elder Roosevelt—like young Theodore—also cared deeply about live animals. Wandering down the corridors of the museum lacked the domestic intimacy of mousing with a cat, let alone the thrill of hearing a cougar snort. So Theodore Sr., teaching his four children another lesson, brought the family into the animal protection movement.

CHAPTER TWO

ANIMAL RIGHTS AND EVOLUTION

I

Cruelty to animals infuriated Theodore Roosevelt even when he was a child. The mere sight of a horse being flogged or dog kicked made him sick at heart. Overcrowded poultry cages and bounties for shooting cats for no clear-cut scientific reason bothered him. After the Civil War, companion animals like dogs and cats were seen by society at large as a quirk of sophisticates. Working-class people, worried about rabies and flea infestation, thought of such animals mainly as disease vectors.[1] (Roosevelt himself, as a college student, shot a mad dog that was menacing his horse on Long Island.[2]) To the Roosevelt family, however, a domesticated dog or cat was part of the family. During the Victorian era one's obligation to a pet included making sure it didn't suffer undue pain. No less a person than Darwin decided not to be a physician (even though his father and grandfather were physicians) because he couldn't stomach watching hideous premodern surgical procedures (such as amputations) performed without anesthetics. Morphine and ether would not come into medical practice until the 1850s. "When the concern about pain was combined with the changing understanding of similarities between humans and animals, it became obvious that animals were also capable of suffering and feeling pain," the historian Stephen Zawistowski has explained in *Companion Animals in Society*. "Thinking on this had come full circle from the Cartesian assertion that animals were automatons incapable of feeling pain."[3]

The Darwinian naturalists—including young Roosevelt—believed all animals and birds could feel pain; therefore, its deliberate infliction had to be stopped. Wild turkeys, lizards in pet shops, passenger pigeons, overworked donkeys—all needed to be treated fairly, in a humane, pain-free way. Roosevelt even believed some animals had thought processes and emotional lives similar to those of humans. During his presidency, he wrote to Mark Sullivan, the editor of *Collier's*, "I believe that the higher mammals and birds have reasoning powers, which differ in degree rather than kind from the lower reasoning powers of, for instance, the lower savages."[4]

As for hunters, Roosevelt, like his father, insisted that they follow an ethical code that would protect "wild creatures" from "destruction" by "greed and wantonness."[5] True gentleman sportsmen, Roosevelt learned as a child, needed to seize the initiative in protecting both domestic and wild animals from abusive treatment. For example, gravely wounded animals, in all circumstances, had to be *immediately* put out of their misery by hunters. For his entire life Roosevelt disdained steel-jaw traps. Animal shelters and horse ambulance fleets, the Roosevelt family believed, needed to be established in major cities. A form of sterilization (spaying) was needed to stop dogs and cats from overbreeding. Drawing on British notions that cockfighting and bullbaiting were intolerable, the Roosevelt clan insisted that mercy be extended to all of God's creatures.[6]

As young T.R. saw it, Darwin's *Descent of Man* and *The Expression of the Emotions in Man and Animals* broke down the egocentric notion that humans were godlike and all other creatures—even chimpanzees, baboons, orangutans, and gorillas—were inconsequential vermin or a food source. In 1641 the French philosopher Réne Descartes had actually promoted the notion that animals had no feelings and were essentially machines; that same year, however, the Massachusetts Bay Colony passed the first legal code to protect domestic animals in North America.[7] Influenced by his grandfather Cornelius V. Roosevelt and his father, Theodore Roosevelt Sr., throughout his life T.R. held that animal protection groups needed to be established and maintained. (At first he had domestic animals in mind; later in life, he included wild animals.) As he put it, "harmless life" deserved to be treated with respect, not as "waste products without feelings."[8]

Certainly, Roosevelt wasn't a vegetarian or what would later be called a vegan. Although he was only five-foot-eight, he ate enough beef and wild game for a football squad. Nobody of his generation promoted hunting—and eating game—more assiduously than Roosevelt. As Darwin's idea of survival of the fittest implied, the natural world was violent. Deer, elk, rabbits—and many other species—usually died violently, torn to pieces by predators. Hunting, if done correctly, was the *least* violent way for an animal to die. A principle with universal application, according to Roosevelt, was that living creatures, even cattle or lambs on their way to slaughter, needed to be handled with dignity. (Roosevelt was prescient in this regard. In 1958 the federal Humane Methods of Slaughter Act mandated that a cow had to be knocked out as painlessly as possible before an incision was made.)[9]

II

One of the Roosevelt family's friends, in fact, was Henry Bergh, a founder of the modern animal protection movement. The son of a wealthy shipbuilder, Bergh had grown up in New York City and then drifted through his early adulthood, spending the Civil War years traveling around Europe. After a brief stint during the Lincoln Administration as a secretary to the American legation in Russia, where he had befriended the czar, Bergh and his wife, Matilda, settled down in London. There, he had a religious epiphany, a humane vision pertaining to animals.[10] Since childhood, Bergh had always been concerned about their well-being. In London he met Lord Harrowby of the Royal Society for the Prevention of Cruelty to Animals to explore how he could take the group's anticruelty creed to the United States and create a similar organization in New York City. An incensed Bergh explained to Lord Harrowby how sickened he had been watching Russian and Caucasian peasants whip donkeys with sticks until their legs buckled and their backsides were full of festering welts. Such mistreatment, he believed, needed to be stopped. A prohibition movement had to be launched at once.[11] His new motto for his life emanated from the pages of Thomas Paine's *The Age of Reason*: "Everything of persecution and revenge between man and man, and everything of cruelty to animals is a violation of moral duty."[12]

Bergh's father had died, leaving Henry and his brother and sister an enormous inheritance. Once Bergh returned to Manhattan from abroad, immediately claiming his money, he embarked on changing the way Americans looked on animals, pushing for numerous new anticruelty laws. Many of his early supporters had become active in both the abolition of slavery and public health advocacy.[13] Whether it was designing ambulances for horses or sponsoring "clay pigeons" for shootists, Bergh and his followers were on a tub-thumping mission to reverse barbaric behavior. Stray dogs in the city pound, for example, were being captured and crammed into large iron cages, and then lowered into the Hudson and East rivers and drowned. Aghast at such rank cruelty, Bergh wanted to find more humane ways to euthanize dogs. His efforts in this regard were successful, as a progression of more ethical killing techniques—for example, administering gas into decompression chambers and sodium pentobarbital injections—soon prevailed.[14]

A serial pamphleteer, Bergh printed a circular trumpeting a society that would prohibit "thoughtless and inhuman persons" from hurting "dumb animals." He mailed this broadside throughout the state, but

Puck *magazine showed Henry Bergh as "The Only Mourner" to follow the dog cart to the New York City pound.*

his search for people to sign a petition encountered a lot of resistance. The pervading sentiment in New York was that an owner of an animal could treat it however he or she wanted. To start legislating whether a farmer should slaughter a billy goat or whether a barkeeper could shoot a garbage-eating tabby was tantamount, most people thought, to interfering with constitutional rights of ownership. "Without animals you would have no meat, no milk, no eggs," Bergh snapped at his critics in a public statement. "There would be fewer vegetables and little grain, because the farmer would have to pull his own plow. You would have to walk everywhere you go instead of riding. Your shoes, your coat, that beaver hat, your gloves, the silk scarf you were wearing—all of these things and many more you have only because of the world's dumb creatures. Since we are so dependent on them, I consider it morally wrong to be needlessly cruel to them."[15]

While some other New Yorkers scoffed at Bergh's plea for the humane treatment of animals, the Roosevelt family embraced his animal rights program (as did the legendary editor Horace Greeley of the *New York Tribune* and former president Millard Fillmore). A massive lobbying effort began in 1865, with Bergh heading the grassroots effort in Albany. A constant force at his side was Cornelius V. S. Roosevelt, who insisted that the humane treatment of animals had to become law. On April 10, 1866, the American Society for the Prevention of Cruelty to Animals (ASPCA) was officially incorporated in New York state and laws were passed prohibiting abuse of animals.[16] A man who beat a mule or horse could now be charged with a misdemeanor. When the ASPCA was officially incorporated at a public meeting at Clinton Hall in Manhattan later in April—with Mayor John T. Hoffman in attendance—Bergh was unanimously elected president. Today the ASPCA's first annual report (1867) is on display at the organization's facility on Ninety-Second Street in Manhattan. Both Cornelius V. S. Roosevelt and John J. Roosevelt (T.R.'s granduncle, who was twice elected to the New York state legislature) had chartered the new organization.[17]

Knowing that Cornelius V. S. Roosevelt, John J. Roosevelt, Theodore Roosevelt, Sr., and other powerful New York philanthropists were on his side, an empowered Bergh now began patrolling downtown like a cop twirling a billy club, looking for animal abusers to arrest.* For cruel attitudes and practices to end, a few animal abusers would have to be handcuffed and carted off to jail. Or maybe—as the extreme believers in animal rights advocated—these abusers could be put in the stocks or dunked in a barrel. Somehow, *an example* had to be made of the animal abusers if the laws were to have an impact. It didn't take Bergh long to act. Just outside the ASPCA's tiny office at the corner of Broadway and Twelfth Street, he spied a butcher with a cart full of hog-tied calves, crammed so tightly together that they were bleeding from hoof lacerations and bellowing loudly in agony. Bergh, who was on horseback, chased the butcher's cart, and finally caught up with it at the Williamsburg Ferry slip. Hopping off his horse, Bergh implored the butcher to release the calves or else go to jail. "Yah," the butcher yelled, as if being accosted by a lunatic. "You're crazy."[18]

Tempers flared and a shouting match ensued. The police were called in to settle the rancorous dispute. The result was a lawsuit filed by the ASPCA against the butcher. It was the first of many suits Bergh would file in the coming years. Sounding as impassioned as the captured John Brown after his Harper's Ferry raid, Bergh resorted to courtroom theatrics, hoping to persuade the magistrate to side with the humane movement. His passionate pleas, however, fell on deaf ears. Half of New York's farmers and hay hands would be imprisoned, the judge reasoned, if cattle mistreatment were prosecuted to the letter of the new law. So the judge imposed only a minuscule fine on the relieved butcher and asked the

*Other allies of Bergh included the poet William Cullen Bryant, the industrialist Peter Cooper (who formed Cooper Union), and the politician John A. Dix (whom Fort Dix, New Jersey, was named after).

ASPCA not to bring such frivolous cases before his court anymore. "Ridicule was regularly cast on the efforts of the society to enforce the law," the historian Roswell McCrea recalled in *The Humane Movement*, "and many of its supporters became discouraged."[19]

The determined Bergh, however, continued pushing his ASPCA agenda forward. The rumor soon spread that he was a fanatic, an unhinged fool who thought feral cats, kitchen pigs, and leopard frogs had rights. Bergh, for example, challenged the poultry industry, insisting it was inhumane to drop live chickens into a scalding bath. Old horses, he also insisted, should never be turned out in the streets to die. The great impresario P. T. Barnum, livid because Bergh had the temerity to claim he treated his prize circus animals in an "abominable" fashion, fired back a volley, denouncing Bergh as a "despot."[20] Quite simply, Bergh became a citywide laughingstock, accused of caring more about hens than about battered women. Lies circulated that he had pet dragonflies and caterpillars. "It hurt him when people were amused at a picture of him with donkey ears, surrounded by animals laughing at him," his biographer Mildred Mastin Pace said, "or a caricature of him wearing a horse blanket instead of a coat."[21]

A repetitive cycle soon developed that stymied the ASPCA. Every time Bergh succeeded in arresting an abuser, the judge would throw the case out, usually ridiculing it. Even his friends said he had a crusader complex. Recognizing that for the ASPCA to become effective he would have to forgo street rumbles and win a landmark legal decision, Bergh tempered his impetuous patrols. Matter-of-factly, he now started searching for a litmus test, a case he could win. It was hard to make above-the-fold news in the *New York Times* or *Herald-Examiner* in cases about whipping mules. There wasn't much inherent drama in a stubborn mule that wouldn't budge; it boiled down to a semantic argument over what constituted a heavy hand. To succeed, Bergh needed a ruse that would grab everybody by the lapels and say, "Wake up! New laws have been passed!" Seeking controversy, Bergh made a strange but inspired tactical leap. In 1866 he decided to bet the ASPCA franchise on defending green sea turtles that were being systematically tortured on the East River wharves without so much as a murmur of public protest.[22]

For starters, Bergh felt pity for the sea turtles, which had been shipped to New York from the Mosquito Lagoon area in Florida (and various Caribbean islands) to be sold for soup and stews. They were valued for their meat and calipee (the cartilaginous part of the shell), and their eggs were also being collected from beaches to be used in cooking and as curatives. The East River fishermen had pierced the huge fins of their captured greens with a screw-like device and then tied them up with straitjacket thongs, giving them no food or water. The 400-pound turtles lay on the dock, writhing in pain, piled up like cordwood, one on top of another. Often they were left upside down, struggling to right themselves. Marching up to Captain Nehemiah Calhoun, the fisherman responsible for the turtles' mistreatment, Bergh informed him that he was under arrest for cruelty to animals. As Bergh and Calhoun argued back and forth a mob of spectators arrived hoping for a pistol duel or a fistfight. Before violence ensued, the police intervened and Captain Calhoun was escorted to the nearby police precinct. Waving a copy of the anticruelty bill, Bergh had the police arrest the captain. After Calhoun endured fingerprinting, a court date was set for later in the week.

Bergh was defeated in the courtroom, but this loss turned out to be a boon for the ASPCA. After the hearing, a counter-sentiment in Bergh's favor started to swell. The courtroom loss jacked up ASPCA's fund-raising a hundredfold or more. Thousands started writing contribution checks or offering their pocket money to the ASPCA. Others just sent in coins as if dropping them into a collection basket. Donations arrived in amounts ranging from one cent to $1,000. Meanwhile, owing to the media attention, Bergh had become an A-list celebrity in New York, appearing at Broadway openings so frequently that he was mistaken for a stagehand, imploring autograph seekers to demonstrate moral elasticity when it came to animal rights. Everybody had now heard of the ASPCA and the ubiquitous "turtle man." That was a good thing. Daily, Bergh's office was besieged with reports of starving horses, mauled dogs, cats set on fire, and illegal cockfights. Determined to win in the long run, full of resilience, refusing to let the humane movement be ghettoized or demoted into the kooky slipstream of the time, Bergh started hiring investigators to look into complaints of abuse, paying them ten to sixteen dollars a week.

Getting hard-earned traction for the ASPCA—laws against abuse of animals were starting to be seriously enforced—in 1873 Bergh once again upped the ante. With the legendary attorney Elbridge T. Gerry—a great-grandson of a signer of the Declaration of Independence—at his side, Bergh rescued a young girl named Mary Ellen Wilson from an abusive home. Owing to a series of family deaths, the ten-year-old Mary Ellen had ended up in a deplorable tenement house, and her body was bruised and scarred when a social-worker type discovered her.[23] The Bergh-Gerry team successfully used an innovative interpretation of the writ of habeas corpus as their legal weapon. Acting as a private citizen, feeding his friends at the *New York Times* details of Mary Ellen's plight, Bergh took the child's abusers to court. Deeply saddened by the way children were regularly mistreated, Bergh, with Theodore Roosevelt, Sr., as a principal ally, formed the Society for the Prevention of Cruelty to Children; the modern-day humane movement for child protection was born.[24]

III

Young Roosevelt admired his family for embracing the humane movement of the late 1860s and 1870s in all its forms. Constantly throughout his life, he would place both his grandfather Cornelius and Theodore Sr. on pedestals, pleased by their association with high-minded men like Henry Bergh. But he was also ashamed of his father for having avoided service in the Union Army during the Civil War. Family obligations, Theodore Sr. said, prevented him from fighting. When Fort Sumter was attacked, Theodore Sr. was supporting, at the Twentieth Street house, his wife, Mittie; her mother, Martha; her sister Annie; and his and Mittie's own three children (and they had a fourth coming). These were ample reasons for not volunteering in the Union Army. Instead of marching off to war, Theodore Sr., as rich men were apt to do, hired a surrogate soldier. Although it is true that Theodore Sr. helped create an Allotment Commission, which eased the financial burdens for Union soldiers fighting at places like Shiloh and Antietam, he himself nevertheless had avoided combat. (The commission saved for the New York families of 80,000 soldiers more than $5 million of their wages.[25]) Some 620,000 men had died in the war. American women endured losing sons, husbands, and fathers over high-minded principles like abolition. Theodore Sr., however, never even smelled gunpowder from the time of Bull Run to Appomattox. Mortified by this, T.R. spent his entire life waging policy and battlefield wars, anxious to prove that cowardice didn't run in the family's bloodline.[26]

Theodore Sr. routinely supported nonprofit activists like Bickmore and Bergh, not to mention grappling with family financial matters, and he was nothing if not intrepid. As Emerson had been fond of saying—and Theodore Sr. believed wholeheartedly—there was only truth in transit. Having squired his children around Europe, the elder Roosevelt decided that having his brood see the Holy Land and Egypt was an essential component of their education. A few days before T.R.'s thirteenth birthday, the family set sail from New York on the S.S. *Russia*, bound for Liverpool. As was to be expected, en route T.R. kept a diary and drew images of animals he saw, many of them resembling cave paintings but others looking like thoughtless doodles. Even with the tumultuous Atlantic affecting his stomach, he dutifully reported seeing gulls, terns, and kittiwakes. While the ship was in the Irish Sea, he grew excited because a snow bunting was on board and was captured by the crew.[27]

But T.R. had something more in mind than sketchbook drawings on this journey to the Nile. Recently he had taken taxidermy lessons from John Bell, a wizard at the art who had trekked with John James Audubon up the Missouri River in 1843. According to the *Ornithological Dictionary of the United States*, the sage sparrow (*Amphispiza belli*), in fact, was named after Bell by the great Audubon himself.[28] Decades later, Roosevelt would recall the lanky, white-haired Bell fondly as being "straight as an Indian," with a "smooth-shaven clean-cut face" and a "dignified figure always in a black frock."[29] It was at Bell's crammed shop at the corner of Broadway and Worth that T.R. eagerly learned the "art of preparing" wildlife. The high odor of microminerals, in fact, became almost a perfume to Roosevelt. In particular, he was attracted to cleaning skulls and other bones with the larvae of dermestid beetles. To Roosevelt, Bell's office was straight from Mr. Venus's shop in Charles Dickens's *Our Mutual Friend.* In his "Notebook on Natural History," he recorded a story Bell had told him about a death match between a rooster and a field mouse. "The field mouse fought long and valiantly," Roosevelt recounted, "but at last was overcome, although not until after a protracted battle."[30]

So now, when the *Russia* dropped anchor, T.R., coughing unmercifully, his stomach weak from seasickness, went wandering through the shopping district of Liverpool looking for arsenic, an odorless ingredient used by ornithologists for skinning and cotton-stuffing birds. "In the streets I was much annoyed by the street boys who immediately knew me for a Yankee and pestered me fearfully," he wrote on October 26, 1872. "Requiring to buy a pound of Arsenic (for skinning purposes) I was informed that I must bring a witness to prove I was not going to commit murder, suicide or any such dreadful thing, before I could have it!"[31]

By the 1860s, taxidermy—a craft practically as old as Egypt itself—had become a high art, with specialists the world over vying to be the very best mounters. The etymology of the word derived from the Greek *taxis* (arrangement) and *derma* (skin). Taxidermy took a steady surgeon's hands adept at making incisions carefully and treating skins delicately. Always the loyalist, young Theodore bragged that his prized teacher, John Bell, one of Audubon's right-hand men, was *the* top taxidermist, the world champion.[32] But critics instead preferred Jules Verreaux, whose innovative exhibit "Arab Courier Attacked by Lions" at the Paris Exposition of 1867 was novel in that the skinned and mounted animals were displayed in a harrowing, gruesome diorama. The innovative Verreaux actually had a Barbary lion (now an extinct subspecies) ripping a helpless camel. Authentic skulls and teeth were used in the diorama, and a human figure wrapped in plaster of paris was added for increased dramatic effect. In 1869 the American Museum of Natural History acquired this pioneering work of taxidermy, which initiated a new way of displaying wildlife worldwide. Young Theodore was enthralled with Verreaux, because he opened up new possibilities for how naturalists and taxidermists could expand science and art.[33] But his loyalty was with Bell for one overriding reason: Bell had the great Audubon on his résumé.

After procuring the arsenic, T.R. raced to see Liverpool's natural history museum. He didn't like its displays, concluding that Americans did taxidermy better. Perhaps because he was being teased by working-class European boys, a pronounced nationalism and family chauvinism began seeping into T.R.'s diaries. "The specimens," he wrote in his diary, "are neither so well mounted or so rare as those in our own American Museum of Natural History in New York."[34]

To practice his taxidermy in England, Roosevelt would purchase snipe and partridges at marketplace stands. Meanwhile, John James Audubon had entered Roosevelt's life in a profound way. Studiously, Roosevelt pored over leatherbound volumes of Audubon in a succession of hotel suites and railway compartments, clutching an aunt's copy of *Birds of America* like a fumbled but recovered football. And he also started to become enamored with Audubon's three-volume study *The Viviparous Quadrupeds of North America*. While he was studying Audubon's works, species extinction and animal abuse started to worry Roosevelt. In Turin, for example, he lamented the "miserable" treatment of horses. And his Berghian sentiments about how European carriage horses were treated in paddocks were once again noted in his diary. "The cabhorses are worse [off] than those of New York," he complained at one juncture, "and the animals in general are treated much more badly."[35]

Young Theodore was happy to be done with Europe when the Roosevelts finally arrived in Egypt a few days after Thanksgiving 1872. The very sight of the ancient city of Alexandria was like a tonic to him. "How I gazed on it!" Roosevelt wrote. "It was Egypt, the land of my dreams; Egypt the most ancient of all countries! A land that was old when Rome was bright, was old when Babylon was in its glory, was old when Troy was taken! It was a sight to awaken a thousand thoughts, and it did."[36]

No sooner did Roosevelt arrive at the dock than he started filling his diaries with descriptions of braying donkeys, foxlike yellow dogs with

A group photo of the Roosevelt family, taken in Egypt in 1872. Theodore Roosevelt, Sr., is at center back row, seated; Mrs. Theodore Roosevelt is at his left. Seated on floor: Anna, Corinne, Theodore Roosevelt, and Elliott.

perked ears, leashed baboons doing circus tricks on a street corner, and camels smaller than he expected. He hurried to a fowl stand and purchased a quail—after haggling with the Arab vendor over the price—for taxidermy. As the family continued on to Cairo by train, T.R. was enthusiastic about the wildlife he saw from his compartment window. "We passed through the Delta of the Nile and on all sides numerous birds of various spieces areose, while heards of buffaloes and zebus grazed quietly in the marshy fields," he wrote. "Among the birds were snipe, plover, quail, hawks, and great black vultures."[37]

The historian David McCullough has written that young Theodore had a deep aversion to attending church. He reached this conclusion by reading the boyhood diaries. He even ventured to claim that T.R.'s asthma attacks were often precipitated by not wanting to go to Sunday services. However, McCullough only hints at something that becomes abundantly clear in the diary entries of 1872; T.R. wrote religiously about the animals he saw *before* and *after* services. As a "natural theologist" of sorts, Roosevelt had an image of God that wasn't a gray-bearded patriarch hurling thunderbolts down to earth but an omnipotent guardian who devised the exquisite intricacies of all living things. Here is a sample entry from Egypt on December 1:

> We went to the English Church at the New Hotel. The sermon was quite good. Afterwards we visited the gardens where we wandered about till lunch time. There were many beautiful trees and at one place an artificial cave with brooks, and cascades, and winding passages and stone cut stairs, and *so* cool and refreshing! The ground moreover was covered with a creeping sort of vine, which looked like grass, and on this tame ravens and crows hopped about, while warblers sang in the bushes. It was altogether very beautiful and we passed a very happay time there. Coming home we were followed by a man who showed off a funny little longeared hedgshog and wanted us to let him charm a cobra which he had in his shirt.
>
> In the evening Father and Mother dined at Mr. and Mrs. Blodged. I went to the gardens before dinner and wandered about for some time. The birds had all gone to their nests but innumerable bats were flitting over the surface of the water.[38]

Once the Roosevelt family left Cairo for a houseboat journey up the Nile,* the amount of writing T.R. produced about birds was staggering. He later stated that his "first real collecting as a student of natural history" began on this Egyptian holiday.[39] Yet he never collected a truly rare bird of scientific value.[40] The journals from this trip have titles like "Remarks on Birds," "Ornithological Observations," "Ornithological Record," "Catalogue of Birds," and "Zoological Record." Now that they were in the Middle East, T.R.'s father gave him a French pin-fire double-barreled shotgun to use. That changed everything. Everywhere he went, his gun was close at hand. "The mechanism of the pin-fire gun was without springs and therefore could not get out of order," Roosevelt recalled, "an important point, as my mechanical ability was nil."[41]

Although the Pyramid of Cheops enthralled him, he got a bigger thrill shooting for specimens along the riverbanks. When it came to identifying Egyptian wildlife, he put Darwin aside and instead listened carefully to the observations of local guides. Hunting with his father on December 13, T.R. shot his first bird—a warbler-like species. It was the first of thousands of birds he would kill in the name of taxidermy, science, survival, and sport in the coming decades. Never before in his young life had he felt as vigorous and vital as in Egypt with shotgun in hand.

* According to the historian Patricia O'Toole, Henry Adams was touring the Nile River at the same time as the Roosevelt family.

Perhaps the dry desert climate, so rehabilitating for asthmatics, helped. Every day, it seemed his spirits kept lifting. "In the morning, we passed a large flock of about sixty Egyptian geese," Roosevelt wrote on December 29. "They were wading in the shallows, but swam out into deep water, where they arranged themselves in an irregular long line and as we approached, divided themselves into several squads and flew off in various directions. At about 12 oclock we stopped and took a walk, during which I observed no less than seven species of hawks crows, stercho finches, and small waders in easy shot."*[42]

Even though young Theodore flourished in the desert climate, his American chauvinism didn't dissipate as the family moved on to the Holy Land. Everything was bigger and better back home. When he arrived in Jerusalem, his first instinct was to declare it "remarkably small."[43] Bathing in the Jordan River produced the lament that it was only "what we should call a small creek in America."[44] To the manger of Bethlehem in Palestine, where Jesus Christ was born, T.R.'s irreverent reaction was to shoot two "very pretty little finches" for his collection.[45] Upon arriving in Damascus, the sacred ground where Paul became a disciple of Christ, Roosevelt went jackal hunting. "On we went over hills, and through gulleys, where none but a Syrian horse could go," Roosevelt wrote. "I gained rapidly on him and was within a few yards of him when [he] leaped over a cliff some fifteen feet high, and while I made a detour around he got in among some rocky hills where I could not get at him. I killed a large vulture afterwards."[46]

The historian Edmund Morris pointed out the jackal hunt was the future president's first attempt to hunt wildlife for "sport, rather than science."[47] Nothing has baffled Roosevelt scholars over the decades more than how Theodore, who vehemently opposed cruelty to animals, could nevertheless kill wildlife with such ease. Although T.R. often hunted for science (in the Middle East he was collecting for his Roosevelt Museum), one can't escape the conclusion that he relished the thrill of the chase as a sport. Basically Roosevelt took his cues from Captain Reid in *The Boy Hunters*. The chapter "About Alligators" defended old-time naturalists like Alexander Wilson and John James Audubon against the overclassification of the laboratory school. By hunting their own specimens down, Audubon and Wilson had truly learned to understand color variations (and the eating and breeding habits) of species better than stationary

---

*Theodore noted that these "seven species" were "in easy shot" but did not raise his rifle to them, as December 29 was a Sunday and his father forbade shooting.

bio-lab technicians in Cambridge or New Haven. According to Captain Reid, a real naturalist lived outdoors while the "old mummy-hunters of museums" sat around like shriveled prunes making divisions and subdivisions of *Crocodilida* ad nauseam.[48]

So hunting, to young Roosevelt, was a prerequisite for being a *real* faunal naturalist of the old Audubon school. But mistreating beasts of burden, who often suffered and died in streets, had nothing to do with hunting. All persons with a moral compass—as the ASPCA claimed—knew that. Also, the slaughterhouses of the world, Roosevelt complained, weren't regulated in any way, shape, or form. Rancid meat and salmonella were commonplace. As a budding sportsman and an advocate of the humane movement, Roosevelt simply wanted hunting and the treatment of domestic animals *regulated.* Species extinction, torture of animals, overhunting, lack of seasonal bag limits, cock and bull fighting—such activities were anathema to his gentlemanly outlook on life. Killing game like cougars or bears with a knife was fine—but tormenting or teasing the animal was deemed unforgivable. Like his father and paternal grandfather, Theodore believed that animals had feelings and perhaps even communicated with one another in ways undecipherable by humans, and that they needed to be treated mercifully. By shooting finches in Egypt, for example, carefully studying their eye bands and plumage, taking careful notes of their demeanor, and lovingly stuffing them so as not to damage their plumage, Roosevelt believed he was *honoring* the species. Most other men would simply shoot birds. Roosevelt, by contrast, shot and collected them for scientific scrutiny. Only by learning everything about a species could you eventually save it from the maw of industrial man. If Roosevelt's views pertaining to animals seem contradictory, consider this: they are essentially the hunting and animal rights codes American society abides by in the twenty-first century.

IV

When Roosevelt boarded a ship for Greece, leaving the Middle East, his asthma flared up again—perhaps returning to "civilization" made him ill. Wherever Theodore went in Europe he pouted and wheezed, believing that Greek ruins and Turkish mosques were a waste of time for an ornithologist like himself. Arriving in Vienna on April 19, 1873, he bemoaned the fact that his father needed to spend *months* in Austria on business. Young Theodore's letters from Vienna attested to his depression: "I bought a black cock and used up all my arsenic on him." "The last few weeks have been spent in the most dreary monotony."[49] "If I stayed here much longer I should spend all my money on books and birds *pour passer le temps*."[50]

While Theodore Sr. stayed in Vienna, he sent three of his children—Theodore, Corinne, and Elliott—to Dresden (known then as the Florence of the Elbe) to live with a German family. His cousins John and Maud Elliott were also there. The idea was for the American youngsters to become fully immersed in German culture. What interested the young Roosevelt most about Germany, however, were the romantic painters who had studied in Düsseldorf during the 1860s—Albert Bierstadt and George Caleb Bingham among them—and had then turned toward the American West for inspiration. And there was also his utter fascination with the white storks of Dresden, which nested in chimneys and could be found around the nearby pond. Instead of immersing himself in German history or language, he continued to play Audubon, studying the storks for variations in color and size. "My scientific pursuits cause the family a good deal of consternation," he wrote to an aunt from Dresden. "My arsenic was confiscated and my mice thrown (with the tongs) out of the window."[51]

But something else had occurred on this Old World trip. For the first time Roosevelt had carefully read *On the Origin of Species* himself instead of being spoon-fed Darwin's theories by an uncle, a cousin, or an adult friend. (Darwin at this time was said to be *talked* about more than *read*.) Somewhat pretentiously, catching the rising wind, Theodore now began imitating the great evolutionary theorist, talking in Darwinspeak about animal variation and natural selection, one of the basic mechanisms of evolution along with genetic drift, migration, and mutation. What was exciting about Darwin was that he was a scientist *and* an explorer; he thereby met Captain Reid's criterion for greatness while epitomizing modernity in science. Next, Theodore read *The Descent of Man*, in which he learned that *Homo sapiens* had evolved from apes, shrews, and birds. The effect of reading these books was that Roosevelt began to sound like the character in Henry James's *The Madonna of the Future* who breathlessly said, "Cats and monkeys—monkeys and cats—all human life is there!"[52] Roosevelt, in fact, held to Darwin's belief that men were biological relatives of apes until his dying day.[53]

More than anything else, Darwin offered the young Roosevelt the philosophy of biology. Darwin was part of the first generation ever to revolt against Aristotle's concept of *scala naturae*, the story of man's march to perfection.[54] What Roosevelt grew to appreciate about Darwin was that he described geological events and natural selection in historical terms.

Evolutionists also embraced the notion that nothing was predetermined; everything was adaptive. In one swoop Darwin erased determinism from the blackboard of human collective experience. Roosevelt was not very good at physics but had a fine grasp of history. He took comfort in mathematical facts, not supernaturalism. Basically, he saw Darwin as explaining the history of the world in an orchard, a finch, a tortoise, or a desert. Darwin even offered possible answers for the extinction of the dinosaurs at the end of the Cretaceous.

Writing to Oliver Wendell Holmes, Jr., just before the 1904 presidential election (which he won), Roosevelt explained why Darwin was *the* force in the "tremendous intellectual revolution" of their time. Beaming himself thousands of years into the future, Roosevelt predicted that Darwin's work would have an unparalleled "position in history" and that it would have been "superseded by the work of the very men to whom it pointed out the way."[55]

Roosevelt in 1873—nearly thirty years before becoming president—had already decided to become a foot soldier in the Darwinian "revolution of natural history." Darwin had looked "with confidence to the future, to young and rising naturalists" who could understand creation from an evolutionary perspective—and he found a willing volunteer in Roosevelt. As Roosevelt later wrote, Darwin had "originality" going for him, unlike those "well meaning little creatures at universities" who were "only fit for microscopic work in the laboratory."[56] The time had come, Darwin had said, for modern biology to lead the way toward enlightenment. "When we no longer look at an organic being as a savage looks at a ship, as something wholly beyond his comprehension," Darwin wrote, "when we regard every production of nature as one which has had a long history; when we contemplate every complex structure and instinct as the summing up of many contrivances, each useful to the possessor, in the same way as any great mechanical invention is the summing up of the labour, the experience, the reason, and even the blunders of numerous workmen; when we thus view each organic being, how far more interesting—I speak from experience—does the study of natural history become!"[57]

Careful to avoid taxonomic errors, Roosevelt began tagging the yellow wagtails and pelicans he shot. He tried to record each bird's most minute distinguishing traits. He was particularly proud of an Egyptian plover he collected and mounted.[58] Whenever he struggled to identify and classify the birds he killed, the Reverend Alfred Charles Smith's *The Attractions of the Nile and Its Banks, a Journal of Travel in Egypt and Nubia* would help. In fact, young Theodore's Middle East diary marks the professionalization

As I have no news to write I will close up with some illustrations on the Darwinian Theory.

Theodore Elliott Johnny

Stork Bull monkey

Theodore Roosevelt drew Darwinian evolutionary ideas using his family as natural selection case studies. This illustration—one of a series—was done on September 21, 1873, while he was in Dresden, Germany. He was fourteen years old.

of his youthful enthusiasm for wildlife and domestic animals.[59] No longer was it enough to record *seeing* "snipes"; homologies were now equally important to him. The variations between the great snipe (*Gallinago media*) and the common snipe (*G. gallinago*), for example, made all the difference in the world. The time had arrived, Roosevelt believed, for him to understand the biological reasons that some birds had nonfunctioning wings whereas hummingbirds couldn't stop flapping theirs. "We behold the faces of nature bright with greatness," Darwin had written, but "we forget the birds which are idly singing around us mostly live on insects or seeds, and are thus constantly destroying life; or we forget how largely these songsters, or their eggs, or their nestlings, are destroyed by birds and beasts of prey."[60]

In *On the Origin of Species*, unlike *Missionary Travels*, there were no lavish illustrations, no photographs of great zebra herds or wallowing hippopotamuses. Just carefully reading the text, however, became something of a personal benediction to Roosevelt. Enhancing Darwin's allure were profiles of the British explorer-naturalist that appeared in popular boys' magazines. It was also exciting that Charles Darwin and Abraham Lincoln—young Roosevelt's two idols—had both been born on February 12, 1809; this is the kind of coincidence children love. Like young Theodore studying under John Bell, Darwin once had a taxidermy apprenticeship with John Edmonston, an escaped West Indian slave who moved to Scotland. And, just as Theodore Roosevelt, Sr., had helped create the American Museum of Natural History, Darwin's paternal grandfather had written *Zoonomia* (1794–1796), which dealt with animal transmutation. It was as if zoology was in the bloodlines of both Roosevelt and Darwin. An admonition Darwin's father had once shouted at young Charles could very well have been blurted out in the Roosevelt household: "You care for nothing but shooting, dogs, and rat catching, and you will be a disgrace to yourself and all your family."[61] Maybe someday, Roosevelt hoped, he too would be lucky enough to catch rats on a ship around the world like the *Beagle*, all in the name of natural history.*

Taxidermist, illustrator, diarist, voracious reader, hunter, ornithologist, mammalogist, animal rights advocate, naturalist, and now Darwinian evolutionist, Roosevelt—all of fourteen years old—was über-precocious. Taken together, all these sides indicated a deep appreciation of wildlife, and an understanding of how little biologists understood about the living world. "When I was young I fell into the usual fashion of those days and collected 'specimens' industriously, thereby committing an entirely needless butchery of our ordinary birds," Roosevelt wrote to his hunting friend Philip Stewart as vice president of the United States in 1901. "I am happy to say that there has been a great change for the better since then in our ways of looking at these things."[62]

When we go through the ruck of evidence about Roosevelt's childhood, in fact, one document stands out. In whimsical letters written from Dresden to his mother and sister Bamie, T.R. drew charts showing Darwin's evolutionary theory in terms of the Roosevelt family's genealogy. Done in a young person's hand, the illustration resembled what today would be called "outsider art" or a doodle on the back of an envelope. It was presented to mother as "Some illustrations on the Darwinian theory," broken down into four stages. T.R. demonstrated how, as a close reader of Darwin's work, he *personally* could have actually evolved from a Dresden stork—the kind that the Danish author Hans Christian Andersen popularized in a fairy tale. Theodore, in fact, was so enamored of these long-legged birds living in the chimneys of Dresden that he imagined his own descent from them. Elliott, on the other hand, came from a bull. And his

*The voyage of the *Beagle* lasted from December 1831 to October 1836. Its mission was a hydrographical survey of South America and the South Pacific. Besides the Galápagos, Darwin got to visit the Falkland Islands, Cape de Verde Islands, Brazil, Tahiti, New Zealand, and dozens of other nirvanas for an aspiring naturalist.

cousin Johnny—Darwin would have undoubtedly approved of this—had evolved from a monkey.[63]

Indeed, Roosevelt's illustration most assuredly was modeled after the frontpiece of the eminent naturalist Thomas Huxley's 1863 *Evidence of Man's Place in Nature*, which showed, in sequence, a gibbon slowly evolving into man. Huxley had worked tirelessly to help decipher the 500-page *On the Origin of Species* for a mass audience, distilling complicated scientific facts for the comprehension of the general public.[64]

Humorous aspects of the drawing aside, as of 1873 Roosevelt was dead serious about spending his life as a faunal naturalist (or biological explorer). After all, there were naturalist mysteries to be solved on this little-known planet earth. Religious leaders had long argued over the origin and development of life. Now Darwin and Huxley had provided an answer. Bursting with the enthusiasm of a convert, Roosevelt swallowed natural selection hook, line, and sinker. For the rest of his life, in fact, he used evolutionary theory as his guiding light; it illuminated his views on everything from politics to geography to fatherhood.[65]

CHAPTER THREE

Of Science, Fish, and Robert B. Roosevelt

I

No photographic images exist of the fifteen-year-old Theodore Roosevelt with his trunkful of bird booty, on board the S.S. *Russia* and anxious to arrive at New York Harbor. Shortly before sailing home, however, he described in a letter to his mother (who had returned early) how the combination of asthma and mumps had made his face as puffy as "an antiquated woodchuck with his cheeks filled with nuts."[1] Besides homesickness for his Roosevelt Museum and his friends, there was another reason the sickly Theodore was eager to return to New York. While the Roosevelts were in the Middle East and Europe, a new family mansion had been constructed at 6 West Fifty-Seventh Street—near Central Park and much closer to the American Museum of Natural History. Young Theodore couldn't wait to see the new house and unpack his bee-eaters, wagtails, bullfinches, and other specimens in the space especially assigned to his Roosevelt Museum. To Roosevelt the new home meant bigger and better display space for his painstakingly acquired wildlife specimens. His Roosevelt Museum was now three or four times larger than before. Each new bird added was assigned an inventory tag, including its Latin binomial.[2] "My first knowledge of Latin," Roosevelt later recalled of his childhood, "was obtained by learning the scientific names of the birds and mammals which I collected."[3]

Roosevelt studied Carolus Linnaeus's *Species Plantarum* and *Systemae Naturae* to help himself learn the binomial system. Naturalists revered Linnaeus as their patron saint because in the mid-eighteenth century he had created this simple, universal two-part Latin nomenclature, which revolutionized taxonomy. Thanks to Linnaeus, botanists and zoologists everywhere could use the same language when discussing species. Rejecting previous assumptions about the "subjective value of perfection of animals," Linnaeus based his sytem on "objective, observed similarities in anatomical structure."[4] His system of biological classification—still used—was developed with a sense of cataloging God's creatures. Honoring all life-forms with both a genus and a species name, Linnaeus also placed humans in his binomial system, as *Homo sapiens* (the term for man beginning in the Pleistocene epoch 1.8 million years ago). By the time

Theodore Roosevelt was growing up, scientists and explorers seeking glory ranged far and wide in the remote wilderness, racing to discover organisms that could be named after themselves. "The Linnaean system eliminated the confusion of having, for example, a butterfly called the mourning cloak in the United States, the yellow edge in Canada, and the Camberwell beauty in Britain," Nancy Pick explains in *The Rarest of the Rare*. "People all over the world, whatever their language, can understand *Nymphalis antiopa*."[5]

In considering the Roosevelt Museum it's important to remember that Theodore was always trying to emulate both Linnaeus and his father. Just as Theodore Sr. had planning meetings, fund-raising drives, and specimen hunts, so too did his son. Ecstatic about the Fifty-Seventh Street house, particularly because it had more room for his collectibles and curiosities—not to mention a modern gymnasium on the top floor with free weights and parallel bars—Theodore decided to hold a spontaneous Roosevelt Museum "directors' meeting" on December 26, 1873. Playing the adult, he met with his cousins James West Roosevelt and William Emlen Roosevelt behind closed doors. Big issues needed to be decided, now that he had acquired all these strange Old World birds to add to the New World collection. A professionalization process—imitating the Museum of Comparative Zoology at Harvard University—was under way. The document that T.R. drafted illustrates just how businesslike he had become about being a naturalist and scientist:

> There has been no meeting for two years owing to the absence of a majority of the members in foreign countries collecting specimens.
>
> Whereas the size of the Museum requires entire reorganization it is resolved that a new constitution be adopted.
>
> Said new constitution having been read and signed by directors. It is also resolved that Mrs. James K. Gracie, Miss Elizabeth Lewis, Mr. Elliott Roosevelt and Mr. John Elliott be constituted members.
>
> It is also resolved that in consideration of the great services rendered by Messrs. Elliott Roosevelt and John Roosevelt that they be not obliged to pay any initiation fee.
>
> It is also resolved that any of the directors be authorized to sell or exchange duplicate specimens of the Museum the proceeds to be given to the Museum.
>
> It is also resolved that Mr. Theodore Roosevelt, Jr. President of the Roosevelt Library present at the next meeting written proposition for the incorporation of the Library and Museum.

Present at the meeting:

T. ROOSEVELT, JR.
J. W. [JAMES WEST] ROOSEVELT
W. E. [WILLIAM EMLEN] ROOSEVELT.[6]

Clearly there was a lot of faux erudition going on here. Although Theodore did keep minutes of the Christmastime meeting, the so-called board didn't convene again for five months, and then met only to agree on letting Theodore spend $8.50 to purchase new ornithological specimens.[7] That modest financial request is the last surviving document pertaining to the Roosevelt Museum. West Roosevelt eventually became chief resident physician at Roosevelt Hospital in New York City and Emlen Roosevelt became the senior partner of the Wall Street firm Roosevelt and Son. But Theodore stepped up his naturalist pursuits through a private tutor instead of a board of directors. According to biographer Edmund Morris, acceptance at Harvard "floated ever nearer" and if the "grail eluded his reach, he might not have the strength to grasp it again."[8]

Theodore Sr. was elated with his son's advances toward maturity and with young Theodore's steady attempt to streamline his naturalist hobby into a serious scientific pursuit. In *An Autobiography*, in fact, Roosevelt reflected on his father's sage advice that "if I wished to become a scientific man I could do so," but as a corollary he had to be dead certain "that I really intensely desired to do scientific work."[9] As of the early 1870s biological studies such as botany and zoology had become truly scientific disciplines. The German scientist Alexander von Humboldt, the founder of modern geography—who died when Roosevelt was one year old—had pioneered in studying the interchange between organisms, habitat, and geography.[10] Then, in the early 1880s, the microscope had been invented to study germs. Theodore enjoyed *On the Origin of Species*, but Theodore Sr. worried that his boy wasn't going to buckle down in chemistry and physics classes; his son liked bear variations, not cell theory. "I am a belated member of the generation that regarded Audubon with veneration, that accepted [Charles] Waterton—Audubon's violent critic—as the ideal of the wandering naturalist," T.R. later reflected, "and that looked upon [Nikolaus] Brahm as a delightful but rather awesomely erudite example of advanced scientific thought."[11]* Clearly, Roosevelt was

*Charles Waterton (1856–1912) was a British naturalist and explorer who influenced Charles Darwin, publishing a number of books recounting his specimen collecting missions. Nikolaus Brahm (1751–1812) was an eccentric German

the kind of natural history student who would travel great distances to shoot specimens in forests. But would he sit still and do the semi-stifled indoor laboratory work that came after great hunts? His father had doubts.

Because as an adult T.R. preached the strenuous life, recording his rugged outdoor adventures and battlefield prowess with such dramatic flair, it's often forgotten that his aptitude for professions other than naturalist or author was minimal. His abysmal health ruled out an appointment to West Point or Annapolis. He had no predilection for the world of commerce. Managing the family's fortune didn't interest him—in fact, he brushed off the chance to become a partner in Roosevelt and Sons. He felt antipathy toward bankers, accountants, and financiers. And every time Theodore gazed at a law book, his eyes glazed over in boredom. Even though his uncle Robert B. Roosevelt tried to nudge him toward the law, even a squad of top-notch defense or prosecuting attorneys could never have convinced him that habeas corpus was a more interesting concept than the migratory habits of *Ectopistes migratorius*—the passenger pigeon.

Therefore, Theodore's career path seemed as plain as day: to become a Harvard-trained biologist or naturalist.[12] Someday his specimens could become part of the Harvard Museum of Comparative Zoology Collection, Roosevelt hoped, displayed beside the natural history artifacts donated by George Washington, Meriwether Lewis, Charles Darwin, Henry David Thoreau, and John James Audubon.[13] For most of his life, Theodore had been homeschooled. He was never enrolled in public school, and he attended a private school in Manhattan for only a brief time. Theodore Sr. hired a highly regarded tutor, Arthur Cutler, to help Theodore brush up for Harvard's notoriously difficult entrance exam. Years later Cutler was asked to recall his former student's inherent strengths and weaknesses. "[Theodore] never seemed to know what idleness was," he said. "Every leisure moment would find the last novel, some English classic, or some abstruse book on Natural History in his hand."[14]

In 1874, continuing the family tradition of fleeing Manhattan every

zoologist who was obsessed with naming larvae. Roosevelt liked the stories about Waterton—another eccentric, to put it mildly—who pretended to be a dog biting dinner guests on their legs. Waterton also invented a way to preserve animal skins, molding them to create caricatures of his opponents. A fierce opponent of the British soap factories that polluted a lake near his house, Thomas Waterton eventually had a national park named after him in Alberta, Canada. Waterton turned his own estate into the world's first waterfowl and nature reserve. Roosevelt may have later taken his ideas about federal bird reservations from Waterton.

summer, Theodore Sr. decided to rent a house in Oyster Bay. A jutting curve of land on the North Shore of Long Island, less than an hour by train from midtown Manhattan, Oyster Bay had been founded in 1653. By the time of the American Revolution, it was a bustling seaport. Although it later boasted that "George Washington slept here,"[15] Oyster Bay had actually been a hotbed of loyalist sentiment once the port was occupied by British troops after the Battle of Long Island. By the time the Roosevelts started summering in Oyster Bay, wealthy New Yorkers, attracted by the breezes coming off Long Island Sound, had built blocks of Victorian and colonial homes. Just after the Civil War, Oyster Bay had grown into a popular summertime getaway location for New Yorkers desperate to escape the clamor and congestion of urban life. Appropriately, the town seal of Oyster Bay eventually became a seagull in flight.[16]

Theodore's mother, Mittie, named the waterfront estate her husband leased "Tranquility." With its columns, attractive veranda, and huge parlor, it was like something out of the antebellum South. The name, however, was a misnomer, for the house constantly sang with activity. As the old adage goes, life begins at the water's edge. And Oyster Bay was no exception. On the North Shore the quaint meadows, low hills, and dense woodlands were a bird lover's paradise. Just rocking on Tranquility's veranda allowed Theodore to possibly see out-of-season surf scoters, old squaws, herring gulls, red-throated loons, catbirds, and chickadees. The ornithological diversity of the open bay was conveyed by these preening birds, and all young Theodore had to do when the four o'clock shadows arrived was sit back and watch and listen with pure elation.

By 1874 the bard of Long Island was Walt Whitman, who wrote poetically about the clouds drifting over sand dunes and eagles dallying in the sky. In his old age Whitman, in fact, would claim that he had "incorporated" Long Island into himself.[17] Now, at every chance possible, Roosevelt would tramp around rural Glen Cove and Lloyd Neck as if he were Whitman on the prowl, jotting down wildlife sightings with scientific exactitude. His ear was always cocked to catch a vireo's robinlike warble or a pesky gull's squawk. His notebooks were no longer travel diaries: the two he kept in Oyster Bay between 1874 and 1876 were labeled "Journal of Natural History" and "Remarks on the Zoology of Oyster Bay."[18] Inside, amid long lists of birds he observed in Oyster Bay, were symbols used to identify the avians as male or female. Struggling to be a professional, he jotted down sightings of prairie warblers perched in cedars and yellowthroats eating caterpillars. "My contributions to original research were of minimum worth," Roosevelt recalled of his fieldwork;

"they were limited to occasional records of such birds as the dominica warbler at Oyster Bay, or to seeing a duck hawk work havoc in a loose gang of night herons, or to noting the bloodthirsty conduct of a captive mole shrew."[19]

Although Oyster Bay appealed to Roosevelt's fancy for birds, the forested environment of the Adirondacks tugged at him in a more primordial way. Long Island had woods. The Adirondacks had wilderness. What exactly did or didn't constitute wilderness has long been debated, with no definitive verdict. Certainly, just hearing somebody utter the word "wilderness" conjures up remote landscapes—that is, sparsely populated or untraveled areas undisturbed by too many footprints or other obtrusions of human beings. One imagines wild—not domestic—animals in a wilderness area. Yet, just when you start homing in on a definition, you have to contend with the hard reality that both outer space and oceans are often referred to as wilderness. To many people, in fact, wilderness is nothing more or less than a state of mind. Basically, wilderness is a subjective concept, as the historian Roderick Frazier Nash noted in his landmark study *Wilderness and the American Mind*, first published in 1967. Trying to arrive at a "universally acceptable definition of wilderness," Nash wrote, is nothing short of impossible. "One man's wilderness may be another man's roadside picnic ground," he believed. "The Yukon trapper would consider a trip to northern Minnesota a return to civilization while for the vacationer from Chicago it is a wilderness adventure indeed."[20]

Given the fact that Roosevelt was a Manhattanite, it's easy to assume that the Adirondacks, where he headed next, were his idea of a genuine wilderness. The Adirondacks embodied what the ecologist and philosopher Aldo Leopold would define as "wilderness" in 1921: the ability of a certain geographical terrain to "absorb a two weeks' pack trip" away from human-centered activities.[21] The historian Patricia Nelson Limerick in *Something in the Soil* described the western wilderness as a place where "mass society's regimentation and standardization" find a "restorative alternative."[22] (The Wilderness Act of 1964 officially defined wilderness as being "untrammeled by man and where man himself is a visitor who does not remain."[23])

So when young Theodore arrived in the Adirondacks in August 1874, he wasn't looking for the silken comfort of Paul Smith's hotel, perched on a bank of the Lower Saint Regis Lake. Only by "roughing it" in the backcountry could he encounter bears, deer, and raccoons up close in a wilderness setting. Rhapsodically and steadfastly, he began keeping a diary called "Journal of a Trip to the Adirondacks" about the millions of acres west of Lake Champlain and north of the Mohawk River. Hiring a rugged Canadian backwoods guide, Mose Sawyer (from Keese's Mill, New York), as his hiking companion, Roosevelt intended to hike the highest mountains—Algonquin, Marcy, and Whiteface. (Even with Sawyer, he only got about halfway up them.) These were summits his parents had previously prohibited him from climbing, worried about his health and well-being. Roosevelt and Sawyer set up a camp along the south branch of the Saint Regis River near McDonald's Pond. Occasionally visitors from New York City would accompany them on day outings in the wooded ranges.

Sawyer claimed that Roosevelt—who preferred venison to brook trout for dinner—had no interest in hunting deer or fishing. It was all birds, birds, birds. Once, in fact, Theodore quietly glided past fifty grazing deer, not lifting his rifle. Arm in arm through the beech woods, Sawyer and Roosevelt traveled; a lifelong bond was forged. As partial payment for taking him into the remote hardwood and softwood forests of the Adirondacks, Roosevelt presented Sawyer with four of the birds he had mounted in Oyster Bay (unfortunately, Sawyer's cat ate them and died of arsenic poisoning). "He was a queer looker," Sawyer recalled of Roosevelt in a newspaper, *The Adirondack Experience*, decades later, "but smart." According to Sawyer, the oddest aspect of Roosevelt was that he didn't "smoke, drink or swear." Sawyer testified that Roosevelt had a puritanical streak lurking in everything he did and said. For example, two prominent New York City doctors—Dr. Wright and Dr. Loomis—spent an evening with Roosevelt and Sawyer at Blue Mountain Lake. After listening to the budding ornithologist pontificate, one of them concluded that Roosevelt was "the smartest idiot I ever saw."[24]

What jumps out from "Journal of a Trip to the Adirondacks" is the way Roosevelt was now injecting the botanical history of trees and fauna into his wildlife narrative. No longer was it enough to record seeing a purple finch or a barn swallow from his canoe; now he also described the geographical surroundings in which he observed a blue jay or common sparrow, specifying whether the tree it was perched in was a white pine, balsam fir, or red spruce. Roosevelt enjoyed having the sweat shine on his face as he hiked the alpine terrain to see hapland, rosebay, diaspensia, mountain sandwort, and other vegetation native to the Adirondacks. His hope, as he took topographical notes about Saint Regis Lake that August and on follow-up visits in 1875 and 1877, was to write a booklet about the birds of the Adirondacks.[25] But since Dr. C. Hart Merriam and other top-notch naturalists had written seriously on Adirondacks wildlife, Roosevelt pragmatically realized that it was hard to add much to "the sum of human knowledge."[26]

On each of these trips Roosevelt's irreplaceable guide was Sawyer: a real outdoorsman, what loggers called a seven-sided son of a bitch, who knew every bit of the Adirondacks' mid-mountain and cloud forests like the back of his hand. "The region more particularly dealt with is at an elevation of over 2000 feet, is strongly hilly, or one might almost say mountainous, and is thickly studded with small lakes and ponds whose outlets are narrow and very crooked streams," Roosevelt recorded in his diary. "It is densely wooded; the few open places being the clearing around houses and the 'slashes' where the woods have been burned. The forests are chiefly evergreen (largely intermixed with birches, beeches, and maples, however); the bulk being composed of pines, balsams and spruces, with numbers of hemlocks on the ridges. In the low level lands there are frequently extensive tamarack swamps. . . . The characteristic features of the fauna of the region are, among mammals, the presence in small numbers of the larger carnivora and furbearing spieces; among birds the abundance of woodpeckers, the breeding of numerous warblers, and the presence of a number of northern species, as *Picoides arctus* [black-backed woodpecker], *Parus hudsonicus* [boreal chickadee], *Perisoreus canadensis* [gray jay], *Contopus borealis* [olive-sided flycatcher] and *Falcipennis canadensis* [spruce grouse]."[27]

Unable to find a black bear and uninterested in shooting white-tailed deer (which had been almost wiped out in upstate New York), Roosevelt instead ended up hunting muskrats and squirrels. Each specimen he bagged was studied for coloration and markings back at what Sawyer called the "Great Camp"; Darwin would have approved of the young man's detailed approach to his wildlife studies. Minor revelations about mammalogy were plentiful for Roosevelt. For the first time he understood why farmers considered chipmunks such a menace; he pulled eighteen wild peas out of a single pouch of one he shot. A highlight that first summer was Roosevelt's hearing the far-off howl of an eastern timber wolf. (According to a study by the University of Toronto, wolves howl more frequently in August than any other month; Roosevelt was the beneficiary of this zoological fact.[28]) After hearing the wolf's lonesome contact and reunion call, Roosevelt kept a special eye on underground burrows and rock caves where he knew *Canis lupus* packs lived in dens. He wanted to see one of the wolves with his own eyes—a feat he never accomplished.

But that was about it. Roosevelt's most detailed, lively, and interesting passages from the Adirondacks journals were his bird entries. Noting that there was a plethora of woodpeckers, Roosevelt successfully collected five species for his Roosevelt Museum. To Roosevelt, woodpeckers were an evolutionary link to a prehistoric era when giganotosauruses and stegosauruses roamed the prairies of North America. Each hammer of their beaks was interconnected to the world of trees and insects, and to time immemorial. The Adirondacks without an echo of a woodpecker sharpening its beak, he reasoned, would be deafening in their silence. "There is a grandeur in this view of life," Darwin had written to end *On the Origin of Species*, "that, whilst this planet had gone cycling on according to the fixed law of gravity, from so simple a beginning endless forms most beautiful and most wonderful have been, and are being evolved."[29]

Many of the birds Roosevelt collected and observed in the Adirondacks later became much less common and were placed on the National Audubon Society's "Watch List."[30] There was, for example, the rare Bicknell's thrush, a shy, buff-breasted, long-distance migrant whose call—"vee-ah"—Roosevelt enjoyed immensely. Even the wood thrush, common in the 1870s, found itself on the watch lists by late 2008, a casualty of the disappearance of extensive deciduous tracts of woodlands in upstate New York. To Roosevelt the wood thrush's song at twilight—a rising and falling "ee-o-lee, ee-o-lay" as heard in the Adirondacks—was more beautiful than any other sound on earth.[31]

The world of the Adirondacks as presented by Sawyer and as heard with his own finely tuned ears caused Roosevelt to dream about heading out to Hawk's Ridge in Duluth, to see great flocks of raptors in their autumnal glory. Like Audubon and Darwin, he would refuse to become an indoor naturalist. From Minnesota he could bundle up in fur and sheepskin like Zebulon Pike and go southward to rendezvous with the headwaters of the Arkansas and Red rivers. The West was tugging on Roosevelt's imagination like a mighty altar. "A naturalist can find employment any where—can gather both instruction and amusement where others would die of *ennui* and idleness," Captain Mayne Reid had said in *The Boy Hunters*. "Remember! There are 'sermons in stones, and books in running brooks.' "[32]

II

When young Theodore was growing up, the interior American West was still a raw wilderness of snow-choked mountains, pristine forest, black lava rock, unknown canyons, and a buffalo-trodden prairie larger than Europe. Because Roosevelt had never traveled west of the Erie Canal, his understanding of the American West in the 1870s derived primarily from newspaper accounts and photographs (with the invention of the portable wet-plate camera, images of Pikes Peak, Old Faithful, and the Three Sisters of the Tetons were now regularly appearing). Although the westward expansion of the 1870s had no artist to equal Mathew Brady during the Civil War, the photographer W. H. Jackson beautifully documented Yellowstone. Also, Timothy O'Sullivan did a fantastic job of capturing the raw natural essence of Utah's Wasatch Mountains. Snow squalls, scorching desert sun, hailstorms, dense fog, and steep cliffs were among the many natural obstacles these photographers faced. Emulating the landscape paintings of Albert Bierstadt and Thomas Moran with his camera, Jackson took sweeping panoramic portraits of the cliff dwellings of Mesa Verde, Colorado; O'Sullivan did the same with ancient ruins at Canyon de Chelly, Arizona. Both places were later saved as national sites by President Theodore Roosevelt.[33]

But something else exciting was occurring in western photography that elated the young Roosevelt. Photographs started to appear in periodicals celebrating the explorers of the late 1860s and 1870s. There was Lieutenant Colonel George Armstrong Custer standing over his first hunted grizzly bear (taken by Wild Illingworth), and Major John Wesley Powell on horseback talking to an Apache guide (courtesy of Jack Hillers). Like many other teenage boys, Roosevelt swooned over these photographic images of the frontier, as he had over Audubon's birds and Catlin's Indians. The Cascades and the Bitterroots had now been opened up to him as stunning visual experiences. Certainly many of these post–Civil War explorers—Custer and Powell leap first to mind—used photography as self-advertisements for their loftiest exploits. Young Theodore, however, saw them through rose-tinted glasses, in a blur of romance, as explorers braver than brave.

At the end of the Civil War new geographical revelations were appearing almost daily in the public press, along with pictures, because the U.S. Congress was eager to inventory the mineral wealth west of the Mississippi River.[34] The U.S. government held title to more than 1.2 billion acres—mostly west of Kansas City—but had surveyed only about one-sixth of this land. In March 1867 Congress approved sweeping geological studies of the western lands by the General Land Office and Corps of Engineers. Suddenly, engineers educated at West Point were the new trailblazers in the Rocky Mountains and beyond. Every square yard of foothills, badlands, thin forests, or drainage ditches would be mapped. On March 3, 1879, Congress created the U.S. Geological Survey to inventory the national domain acreage that President Jefferson had acquired from the Louisiana Purchase, and much more.[35]

With so much mineral-rich land open for the taking, speculators were scheming to grab stakes, mine gold, discover oil, dig coal, control wheat markets, fish out rivers, and clear-cut virgin forests. This entrepreneurial fever also provided a commercial opening for trained civil engineers, geologists, and biologists (to a limited degree). Railroad workers and miners were greatly in demand, as were cartographers and surveyors. Clarence King, first director of the U.S. Geological Survey, noted that 1867 marked a pivotal point when "science ceased to be dragged in the dust of rapid exploration and took a commanding position in the professional work of the country."[36]

The historian William H. Goetzmann, in his Pulitzer Prize–winning *Exploration and Empire*, broke down the nineteenth-century exploration of the American West into three distinct phases: (1) the Lewis and Clark era (1803 to 1840), when the desire was to gather practical information and discover likely trade routes; (2) the manifest destiny years (1840 to 1865), when families (and railroad companies) headed west looking for fertile land, natural resources, and big bonanzas; and (3) the post–Civil War "frontier laboratory" phase (1860 to 1900), when botanists, paleontologists, ethnographers, and engineers sought scientific information. This third period was the time of the "Great Surveys," when a wave of scientists headed west for reconnaissance and inventories. "It was also a time for sober second thoughts as to the proper nature, purpose, and future directions of Western Settlement," Goetzmann wrote. "Incipient conservation and planning in the national interest became in vogue, signifying the way that the West had come of age and its future had become securely wedded to the fortunes of the nation."[37] But it wasn't until March 3, 1885 that a Biological Survey—a U.S. Department of Agriculture unit within the Division of Entomology that Roosevelt championed like a fight promoter—was established.*

When Roosevelt was growing up, the federal government did show tiny signs of a burgeoning new awareness that protecting wildlife and animals mattered. On June 30, 1864, for example, when Roosevelt was

*When it was created in 1885 by Congress the Biological Survey was for "economic ornithology." The following year mammals were incorporated. In 1896 the unit's name was changed to the Division of Biological Survey. On March 3, 1905 T.R. officially upgraded the outfit to the Bureau of Biological Survey. Throughout the text I use simply Biological Survey.

only five years old, in what is widely considered the initial federal intervention on behalf of wildlife resources, Congress transferred the Yosemite Valley from the public domain to the state of California. Even though Sacramento would eventually return the Yosemite Valley to the U.S. government, in 1906, an important precedent was established, for in the transfer agreement California agreed to "provide against the wanton destruction of fish and game found within said park, and against their capture or destruction for the purposes of merchandise or profit."[38]

A timeless hero to Roosevelt was President Ulysses S. Grant, particularly for his strategic skill in securing a Union victory in the Civil War. Just as T.R. turned fourteen in 1872, and was taking an interest in the American West, President Grant signed the bill creating, at least on paper, Yellowstone National Park. Enthralled by stories of grizzly bears, time-clock geysers, petrified logs, and massive elk herds—all of which appeared in his favorite boys' magazine *Our Young Folks*—Roosevelt vowed to visit the new national park someday. Two great interests—General Grant and American wildlife—had come together in the Yellowstone story. Unfortunately, President Grant didn't fully comprehend the thuggishness of western poachers. The 2.2-million-acre park had no uniformed police force (wardens) to enforce the law, and wildlife was killed there indiscriminately until 1894, when the Yellowstone Park Protection Act clamped down on the criminals.[39]

Accounts of Alaska's brute beauty were also extremely popular as Roosevelt was coming of age. Virtually no important biologists had inventoried this American acquisition. Back in 1867, Secretary of State William Seward had acquired more than 650,000 square miles of Alaska from Russia for a song—$7.2 million (less than 2 cents an acre).[40] Antiexpansionists called the purchase "Seward's Folly" and considered it a frozen wasteland not worth a trillionth of a dollar, but other Americans cheered it. Determined to prevent both the Russians and the Japanese from killing what were now American seals in the Bering Sea rookeries, President Grant set aside the Pribilof Islands to protect them in 1869; this measure was approved by Congress the following year.

Living on the five tiny Pribilof Islands—only two of which, Saint Paul and Saint George, were habitable by humans—were the finest seal herds in the world, tens of thousands of bulls with harems swimming in the frigid fogbound waters and coming onto rocky beaches. The Pribilof Islands were, in essence, cities of seals. But Russian, American, and Japanese vessels were slaughtering these mammals in the Bering Sea at a rapid rate. To these hunters and harpooners, the Pribilofs were a killing ground.

Rudyard Kipling included in his second *Jungle Book* the short story "The White Seal," about fierce battles between nations over seal fur from the Bering Sea. Now that the United States owned these islands—considered the most lucrative rookeries of fur mammals anywhere—President Grant wanted to make sure he could protect the hundreds of thousands of fur seals.[41]

Like Grant, President Benjamin Harrison also cast an eye on Alaska. By means of an Executive Order on March 30, 1891, he created the Afognak Island Forest and Fish Culture Reserve to protect another part of Alaska's "bays and rocks and territorial waters, including among others the sea lion and sea otter islands."[42] (The island is today part of Katmai National Monument, the second-largest area in the National Park System.[43]) A side effect of Harrison's executive action on behalf of Alaskan wildlife was that the thick-billed murres and red-legged kittiwakes were saved from extinction. Flocks of auklets were able to flourish, and the blue-gray arctic fox could continue devouring their eggs. The food chain had been preserved.[44]

Neither Grant nor Harrison has been properly credited for small steps in protecting Alaskan wildlife from harpooners and market hunters. As Roosevelt later noted, the foresight of his two fellow Republican presidents regarding Alaskan wildlife had mattered. It proved that the federal government could, when necessary, intervene effectively to help mammals survive as species. Also, it showed that the government would protect the intricate tapestry of nature if the reason for doing so was economically compelling—in the case of Yellowstone, wildlife was being protected for tourism; in Alaska, the motivation was the long-term benefit of the seal-hunting industry.

This economic conservationism was a postulation that T.R.'s uncle Robert Barnwell Roosevelt—who lived in a brownstone adjacent to T.R.'s birthplace on Twentieth Street in New York City—put to the test regarding the fished-out streams and lakes of North America.

III

As a boy Theodore Roosevelt always had a soft spot for Robert Barnwell Roosevelt, his vaguely bohemian "Uncle Rob." A creative contrarian and lover of animals, birds, and especially fish, Uncle Rob was always caught up in the maelstrom of his times. He simply never sat upright in his carriage. With a bristle-cropped beard, receding hairline, bad eyesight, and frameless spectacles, R.B.R. (as the family called him) would be easy to mistake for any of the bewhiskered American presidents elected be-

Robert B. Roosevelt was considered the great U.S. conservationist during the years from the Civil War to the Spanish-American War. "Uncle Rob," as his nephew Theodore Roosevelt called him, was a many-sided Victorian reformer. In recent years the New York Historical Society (Manuscript Department) has opened R.B.R.'s personal scrapbooks for scholarly viewing.

tween Grant and McKinley. Imbued with an outlandish catholicity of interests, the conspicuous Robert B. Roosevelt epitomized the many-sided Victorian-era bon vivant. For decades the Roosevelt family—owing to his philandering with chorus girls and trollops—treated him as a black sheep. Somehow he just couldn't control his streak of lechery. Embarrassed by his wayward morals and intellectual incongruities the family, to some extent, banished R.B.R. from the fold. Nevertheless, his shiftiness aside, he was greatly admired in his day by the public at large.[45]

While not quite a bigamist, Robert B. Roosevelt came awfully close to being one. The simple fact was that he led a double life, married to Elizabeth Ellis Roosevelt (T.R.'s Aunt Lizzie) while having a long-standing affair with a neighbor, Minnie O'Shea (Mrs. Robert F. Fortescue), whom he hired at the *New York Citizen*.[46] Secretly R.B.R. fathered four children with this mistress. Later, after Elizabeth died, he married Minnie.[47] Unfortunately, this affair was so widely frowned upon that Robert B. Roosevelt has not been fully appreciated by recent environmental historians. His diaries and papers were largely hidden by the Roosevelt family for 100 years. Usually R.B.R. sported a porkpie hat and a gray jacket as he squired women around town. As a kind of territorial marker he gave elegant green gloves—purchased in bulk at A.T. Stewart's department store—to the women he slept with. Society types used to laugh whenever they walked down Fifth Avenue or rode a carriage in Central Park and spied a woman wearing these gloves, which might as well have been the Scarlet Letter.[48] Apparently, these women didn't realize the negative connotation of the gloves.

But R.B.R.'s womanizing was, in truth, the least interesting aspect of him. To many it seemed that he had distinct parts and switched roles dexterously. Casting off and putting on personas, he could be a barroom enthusiast, a romantic adventurer, a barrister, a rousing orator, a husband, an adulterer, a sage, an animal protection advocate, a gourmet cook, a humble farmer—R.B.R. could even write memorable prose as a novelist and satirist. As was typical of the British upper class, Robert, never in need of money, dedicated himself to public service.[49] Known for switching jobs with kaleidoscopic quickness, he served as coeditor of the reformist newspaper *New York Citizen* (which rallied its readers against the nefarious William Marcy Tweed's ring).[50] He was one of the Committee of Seventy, which broke the Tweed ring, causing the notorious boss to leave public life in disgrace.

During the Civil War R.B.R. served in a New York volunteer regiment; this experience provided him with enduring friendships. During Reconstruction he took his gusto for reform into the political sphere in 1870 and was elected to Congress; the only reason he ran was to establish federal fish hatcheries. Following a successful two-year stint in Washington, D.C., promoting pisciculture, he wrote plays and became commissioner of the Brooklyn Bridge (1879–1881), overseeing architectural adjustments and maintenance issues. As U.S. minister to the Netherlands under President Grover Cleveland in 1888, Roosevelt was at his smooth-functioning best in promoting a special "Dutch-American" relationship. Extremely adept at fund-raising, he served as treasurer of the Democratic National Committee in 1892, helping elect Cleveland for a second (nonconsecutive) presidential term.[51]

Erudition came easily to Robert B. Roosevelt. A self-styled man of letters, he wrote one mediocre novel—*Love and Luck* (1880)—which didn't sell well. But his comic satire *Five Acres Too Much* (1869), which spoofed the virtues of farming, struck a chord in the immediate post–Civil War years, when laughs were in short supply.[52] His other offbeat novel, *Progressive Petticoats* (1874), a lampoon of women suffragists, was also enthusiastically received. (For a modern-day comparison, he was the David Lodge of his generation.) His pasquinades had ardent fans. As a personal favor, Robert B. Roosevelt also edited the flowery poetry of General Charles G. Halpine (Miles O'Reily), his coeditor at the *New York Citizen*.[53] Sometimes R.B.R. wrote limericks himself—which were topical, like most efforts in this genre, and uniformly bad. On the whole, Roosevelt's literary efforts, read a century after they were written, no longer sparkle; posterity has thrown them overboard. Only occasionally does his wit hold up. But R.B.R.'s furious advocacy of "fish rights," the nonprofit sine qua non that became his sustainable legacy, has enduring historical importance, though it has been undervalued by academic scholars.[54] R.B.R., more than any other direct influence, turned Theodore Roosevelt into a conservationist as a teenager.

Born on August 27, 1829, Robert B. Roosevelt grew up in New York City. Turning his back on the mercantile life of his father, Cornelius V. S. Roosevelt, and the Quakerism of his mother, Margaret Barnhill Roosevelt, R.B.R. became a maverick who gravitated toward the pageantry of wild things.[55] Later in life Hamilton Fish—who had served as governor of New York and Grant's secretary of state—dubbed R.B.R. "Father of all the Fishes."[56] Robert's unconventionality first became clear when he changed his middle name, Barnhill, to Barnwell to avoid jokes about manure. As a teenager, he fished the waters of Long Island Sound for striped bass and bluefish whenever the opportunity presented itself. College, however, wasn't important to the short, portly R.B.R. For all his erudition, he was happiest outdoors, whether at sea or on land. He flouted civil niceties, always speaking his mind candidly. Even his enemies—and there were many—couldn't deny that he was frank.

When Robert turned twenty-one years old, he married Elizabeth Ellis and embarked on a political career as a Democrat. His decision to remain a Democrat, even after Abraham Lincoln walked onto the national stage, cast a lingering haze of suspicion over him in family circles. In aggregate, the reason R.B.R. gave for not following the family line was the galloping tempo, unhampered corruption, and rake-offs that characterized New York politics in the late 1850s and 1860s. Democrats, he believed, while fools, were less beholden to robber barons, and that made all the difference to him.[57]

Like other New York gallants Robert B. Roosevelt belonged to the so-called club set of the late 1850s. With corks popping, often overdrinking and straining his constitution, he enjoyed discussing political, literary, and culinary affairs. R.B.R., in fact, developed a virtual mania for joining elite "societies," perhaps because he was very sociable. Robert (or "Rob" as he was usually called) was very everything, in fact. Very blue-blooded, though he aspired to be a populist. Very mannered, though he used the rake's characteristic tactics of cajolery and the nod and the wink.

As a gourmet chef he proudly founded the Pot Luck Club, whose intellectual members cooked five-star meals, judged truffles, and drank the finest French wines. Every year he also chaired the annual dinner of the Ichthyophagous Club, helping educate aspiring gourmets about the "unsuspected excellence" of many neglected varieties of American fish. Usually, the naturalist Joaquin Miller (whom he called the "sweet singer of the Sierras") was at his side. Fishing stories in a comical vein were told at these dinners. The British, R.B.R. used to say, while always filling the champagne glasses, had to learn there was more to life than cod. The roomful of fish lovers roared with delight. A few specialty dishes were *Darne de saumon, garni d'éperlans à la Roosevelt* and *Filet of striped bass with shrimps à la Baird.*[58]

Perhaps more than anybody else in his era, Robert B. Roosevelt argued for fish conservation because he enjoyed eating perch and trout so much. Turtle soup was another one of his specialties. Writing in his diary of the "turtle trial," in fact, Robert was a contrarian, considering both Henry Bergh and the judge flat wrong. Sea turtles, he believed, weren't insects or reptiles, but fish. Deeming the Pot Luck Club the "very head and centre of gastronomic, ichthyologic, zoo-logic," he declared turtle too delicious a culinary delight to be an insect. "Then we have the sea turtle," he wrote, "the glorious monster who sleeps in the mid-ocean in the amplitude of his thousand pounds of excellence." Just hearing all that talk about sea turtles during the turtle trial, in fact, made R.B.R. want to "luxuriate in the green fat and the yellow fat." Only an imbecile, he scoffed, could possibly eat turtles thinking they were insects or reptiles. "It had been claimed that turtles are snakes developed in Darwinian theory," he recorded. "That a very intelligent and highly gifted snake, feeling sensitively the unprotected nature of a tail that dragged its slow length an unnecessary distance behind by taking much thought added a shell to his body and converted the longest and slimmest of figures into the stoutest and roundest. That a turtle, and above all a snapping turtle, is but a snake in disguise. . . . I, for my part, cannot accept a serpent when I ask for a turtle; I repudiate the 'snaix' even under the head of a terrapin. Why should not a turtle be a fish?"[59]

Even though Robert B. Roosevelt found Manhattan's social life fulfilling, he was continually drawn to the woods of rural Long Island. Trolling in the waters of the Great South Bay became his favorite hobby. Eventually, Robert purchased property along the bay, where the snipe and geese were plentiful, to build a country estate surrounded by pristine nature. But R.B.R.'s primary residence remained the five-story brownstone on

East Twentieth Street in Manhattan, next door to what the National Park Service now oversees as the Theodore Roosevelt Birthplace.* A door on the second floor connected the two residences, making them a single-unit dwelling.[60] (There remains speculation in the Roosevelt family that Theodore Sr. moved to West Fifty-Seventh Street to get his impressionable children away from their colorful uncle Rob.[61])

Animal stories were the mainstay of R.B.R.'s skill as a first-rate raconteur. To Uncle Rob treed opossums and sneaky foxes were more colorful than boring gossip about the sexual affairs of fellow socialites. Whenever young Theodore visited his uncle's home, in fact—as he often did—he encountered a veritable menagerie: guinea pigs, chickens, parrots, and ducks coexisting in total mayhem. A cow even pastured in the parlor while a pony walked in circles around the dining room table.[62] A pet spider monkey dressed in ruffled shirts would often greet visitors at the front door. A German shepherd was allowed to dine at the main table. Somewhere along the line Uncle Rob made it a consuming hobby to collect "Brer Rabbit" stories from African-Americans and dutifully wrote them down as an ethnologist would, mastering the slave-culture diction. Unfortunately, when he published these stories in *Harper's*, readers paid them little mind. "They fell flat," T.R. would later recall, although he was proud that Uncle Rob had been ahead of Joel Chandler Harris in collecting this Georgian folklore. "This was a good many years before a genius arose who in 'Uncle Remus' made the stories immortal."[63]

That R.B.R.'s affinity for animals eventually developed into conservationist activism owes much to the influence of a British-born outdoor writer, Henry William Herbert. The indefatigable Herbert had been educated at Cambridge University and launched a literary career as a romance novelist, influenced by Sir Walter Scott and William Wordsworth. Moving to America in 1831, he adopted the pseudonym Frank Forester for his outdoor writing.[64] Both a fine writer and a pen-and-ink artist, Herbert cofounded the *Atlantic Monthly* in 1833, believing that outdoors prose needed a fresh, clever popular outlet. The magazine flourished. But it was Herbert's conservationist activism in such books as *The Warwick Woodlands* (1845)—denouncing "game hogs" and advocating a measured conservationist approach—that caused alarm in hunter-angler circles.[65] R.B.R., for one, took keen notice of Herbert's grim warning that "the

*Ironically, in 1919 R.B.R.'s brownstone became headquarters for the Woman's Roosevelt Memorial Association.

game that swarmed of yore in all the fields and creeks of its vast territory are in such peril of becoming speedily extinct."[66]

Robert B. Roosevelt joined Herbert's crusade to replenish the trampled fisheries of New York. In 1844 Herbert helped found the New York Sportsmen's Club, whose mission was to have the state legislature place seasonal limits on the hunting of deer, quail, and woodcock. The club's first successes were the passage of model game laws in three counties: New York (Manhattan), Orange, and Rockland. As soon as R.B.R. joined, he championed an important auxiliary mission for the club, declaring that his priority in life was to enact laws protecting trout, shad, and perch. Indiscriminate fishing practices in general, he believed, had to be curtailed at once if the waterways of New York were to maintain even a semblance of their old abundance.[67] Thus R.B.R. became—as his editor friend at the *New York Times* John Bigelow dubbed him—the "Piscicultural High Priest or Pontifex Maximus."[68]

What Robert B. Roosevelt admired about Frank Forester was that he was an able practitioner of "outdoor" literature. Writers in this genre recounted hunting and fishing trips, as opposed to nature writers (who usually refrained from killing wildlife) and latter-day environmental writers (who were more technical in their approach). When Theodore Roosevelt was growing up, most outdoors writing was Anglophile. It was impossible to find a good angling book, for example, about Florida's reefs, Louisiana's bayous, Hawaiian atolls. John Skinner's *American Turf Register and Sporting Magazine* in 1829 was filled with pedestrian prose. Henry's creation of Frank Forester changed all this, demonstrating that American soil was fertile ground for outdoor writing. Single-handedly, he made "hook and bullet books" respectable; this was the genre at which Theodore Roosevelt was determined to excel.[69]

Young Theodore, however, never fully cottoned to Frank Forester's instructive style. To him, outdoor writing meant pushing one's limits, seeking danger, and developing survivalist instincts in the brutal wild. Forester was too tame for him, too concerned with what was proper hunting attire for flushing grouse after noontime tea. T.R. was drawn to the more adventurous stories of Captain Mayne Reid. (Roosevelt also fell under the spell of the founder of *Forest and Stream*,* Charlie Hallock, whose 1877 *Vacation Rambles in Michigan* made him curious about the Great Lakes

*Originally published as *Forest and Stream* in 1873, the magazine changed its name to *Field and Stream* in 1930.

islands and grayling.[70]) Eight-pound trout, snarling cougars, grizzly-bear paws larger than a human's head, virgin forest that even Lewis and Clark hadn't trampled—these were the kind of "manly" pursuits the teenager wanted to read about in the 1860s and early 1870s. He wanted to experience outdoor life before the telegraph wire spoiled it all. "Unfortunately, [Forester] was a true cockney, who cared little for really wild sports," T.R. later wrote, "and he was afflicted with that dreadful pedantry which pays more heed to ceremonial and terminology than to the thing itself."[71]

Following Herbert's suicide in 1858, R.B.R. eased away from his other civic obligations and focused on protecting the fish populations. He wrote a Herbert-like tract on New York and New England angling. *Game Fish of the Northern States of America, and British Provinces* was a sensational critical hit, blending personal experiences and conservationist convictions with recipes for cooking fish.[72]

Young Roosevelt was only four years old in 1862, when the success of *Game Fish* resulted in Uncle Rob's being praised as the "Izaak Walton of America."[73] Although R.B.R.'s main concern was recreational fishing, his book included chapters about how millions of Americans would become deprived of a great foodstuff if U.S. rivers and lakes were fished out. Much of the book was about fish hatching. The editor of London's *Land and Water*, in fact, credited R.B.R. with introducing American fish into English waters, feeling he deserved more public credit for this transatlantic pisciculture.[74] Usually fishing books, even Walton's venerable *The Compleat Angler*, appealed only to a select audience of patrician sportsmen. But *Game Fish* managed to have a profound influence, becoming the mid-nineteenth-century equivalent of Rachel Carson's *Silent Spring*. Only Thaddeus Norris's encyclopedic *American Angler Book* (1864) came close with regard to celebrating the fish found in the United States' waters.

What worried R.B.R. most was the scarcity of fish. The two waterways flanking Manhattan—the East and Hudson rivers—for example, had become cesspools for industrial waste and raw sewage. The situation wasn't much better for the waterways of Long Island, which were serving as garbage dumps. In the Hudson, river men used to netting tons of shad were coming back to land empty-handed. Regarding the Great Lakes Robert worried the walleye and trout might soon go the way of the dodo and the great auk. Ferociously, he corresponded with fellow gentlemen anglers in New Hampshire and Vermont, who had toyed with scientific concepts like "artificial propagation" to replenish their fished-out lakes and streams.[75]

Wildlife management was embryonic in the mid-nineteenth century, but R.B.R. pioneered in introducing scientific concepts relating to fish. Refusing to temporize, in a torrent of lucid (if self-indulgent) books, articles, and harangues, R.B.R. urged fishermen to develop "moderation, humanity, patience, and kindness under all circumstances." He promoted fly-fishing mainly because it's far harder and more of a sporting challenge than bait fishing (it also resulted in fewer ancillary kills), as he simultaneously awakened the nation to the perils of overharvesting lake fish. In his second book, *Superior Fishing; Or, the Striped Bass, Trout, and Black Bass of the Northern States* (1865), which recounted a trip to Lake Superior and its tributaries (plus other freshwater fishing sites), Roosevelt jumped to the then radical conclusion that fish "poachers," those scoundrels who netted whitefish and yellow perch out of season, should receive the "contempt of all good sportsmen" and deserved nothing less than "the felon's doom."[76]

To fully comprehend the importance of R.B.R.'s legacy, it's best to remember that he—like his nephew T.R.—relished slaying dragons, pounding away at adversaries. In political fights Robert B. Roosevelt was always taking on fiendish "rings" with "off with their heads delight." These included the "Rivermen Ring," the "District of Columbia Ring," and the "Tweed Ring"; he wanted to be a pallbearer for them all. Corruption of any kind was anathema to R.B.R.'s code of noblesse oblige. In his third conservation book, *The Game Birds of the Coasts and Lakes of the Northern States of America*, R.B.R. merged the sportsman's ethics with natural history and prosecutorial prose. Basking in his newfound literary celebrity, he spearheaded the preservationist agenda of the New York Sportsmen's Club. In 1874, at R.B.R.'s urging, the club changed its British-sounding name, becoming the New York Association for the Protection of Game (NYAPG). Three years later R.B.R. was elected president (a post he held until his death in 1906).[77] Whether the issue was white-tailed deer, mallard ducks, or brook trout the members of NYAPG were hunters who believed wholeheartedly in conservation. The organization was instrumental in the majority of New York legislation aimed at saving wildlife during the 1870s and 1880s.[78]

What distinguished R.B.R. from other members of NYAPG was that whereas they promoted preservation, he fought for "restoration." In this regard R.B.R. was furthering the teachings of Henry W. Herbert, challenging the so-called big bugs of his day for killing off America's waterways. There were many fine books published on fish culture in the nineteenth century—including the U.S. Fish Commission's annual *Reports*, which began in 1871—but none had the literary flair of R.B.R.'s efforts.[79]

IV

Analyzing contemporary "fish culture" and heading a conservationist club weren't enough for Robert B. Roosevelt. For twenty years, he served as the head of the New York State Fish Commission, an unpaid position. What enraged him most was the fine-mesh nets fishermen draped across the Hudson to catch huge schools of shad, thereby preventing any fish from swimming upstream to their spawning grounds. A firm believer in effecting change through the legislative process, R.B.R. lobbied state lawmakers in Albany, and soon it was decreed that nets could have a mesh no smaller than 4½ inches. Numerous laws followed: fines would be issued to fishermen who operated nets or traps on Sunday; fishing seasons for some species were established; rivers were ordered restocked; and catch limits were enacted to further protect shad during the two-month season (April 15 to June 15).[80]

Robert B. Roosevelt was a workhorse on the fish commission. Because he was independently wealthy, he had time to send his fellow commissioners a barrage of white papers on pisciculture, and his colleagues usually rubber-stamped his decisions. Nobody doubted that R.B.R. was *the* voice of the commission. Still, Robert was too much of an autocratic gentleman to persuade working fishermen who needed to troll long hours just to feed their families. A spokesman was needed who could talk with no hint of refinement about the virtues of artificial propagation of fish. R.B.R. recruited Seth Green—the fashion-defying "Fish King," considered the premier angler in America around 1868—to join the commission and be his mouthpiece in promoting hatcheries.

Raised near Rochester, New York, Green learned to hunt and fish around the Genesee River's lower falls. As an adolescent, he learned "fishing secrets" from the local Seneca Indians. By the time Green turned forty, in 1857, he was the top fish dealer in New York and was considered the ace commercial fisherman in America; he and his crew caught ten to twenty-five tons of fish a month.[81]

Meanwhile, Green developed the rudimentary science of artificial insemination. He stripped the female trout of her eggs, catching them in a tin pan like the ones gold miners used. Next, he milked the male trout's milt to drain into the same pan. A series of other tasks were performed before these pans were placed on the hatching beds. A high level of patience was necessary. Forty or fifty days later, using a magnifying glass, he was able to detect the eyes of a little fry. Around 120 days later eggs hatched.

At first, Green had a success rate of only 25 percent. Not content with that, he experimented, and before long his trial-and-error approach paid off. He discovered that water should never be mixed in with the spawn and milt, because it diluted potency. Using a method he called "dry impregnation," Green had a 97 percent efficiency rate. He secretly produced fish at his compound for a couple of years, cutting brook trout fillets for market and selling his spawn in Buffalo, Niagara, and Rochester.[82]

This anonymity, however, didn't last forever. A series of admiring articles about Seth Green's upstate hatchery appeared in New York City newspapers. His discovery resulted overnight in a rush to create trout ponds everywhere. Such fish farms were seen as a highly profitable investment, an easier way to get fish onto the nation's dinner tables than fishing in streams and lakes with rod and reel or nets. Green was suddenly in high demand as a teacher and demonstrator. In 1867 New England fish commissioners hired Green to start propagating the American shad on the Connecticut River, where stocks had been seriously depleted.[83] Although he was subjected to torrents of antiscientific ridicule from disbelieving fishermen, Green carefully erected dams, waste gates, and hatching boxes in four New England states. Fishermen on the Connecticut River, now deeply resentful of Green's wizardry, vandalized his hatchery equipment and cut holes through all his nets. Undaunted, Green began keeping watch over his homemade boxes, leaping out of bushes in the morning hours with a loaded Winchester to frighten would-be saboteurs away.

For a while, Green had a bigger worry than angry fishermen: it turned out that shad couldn't be hatched through artificial impregnation the same way trout were. From a way station in Holyoke, Massachusetts, he experimented with burying the eggs in gravel placed in the troughs. Every day he made scientific adjustments to his contraptions, but the shad eggs wouldn't hatch. Despite the continued harassment of local fishermen, Green's stoic persistence and surgical repairing eventually paid off. When he checked the boxes one afternoon, the shad had hatched. It was an important moment: he had established that shad eggs could hatch in only thirty-three hours, far less time than trout eggs took. And his success rate was even higher: Green claimed that 999 out of 1,000 shad eggs hatched under his new protocol. By the time he closed his Holyoke shop in 1872, Green had released 40 million shad into the Connecticut River. But the river men weren't wrong. Before Seth Green arrived on the scene, shad had been selling at 100 for forty dollars. By the time he finished replenishing the river, the market price per 100 had plummeted to three dollars.

Although Green received excellent press coverage, due in part to his *Trout Culture*, published in 1870,[84] New England's fish commissioners gave him only a measly stipend of $200 for all his innovative hard work. (By contrast, in 1871 the California Fish and Game Commission introduced hatchery shad into the Sacramento River and paid Green handsomely for inspiring the hatcheries.[85]) Outraged at his shabby treatment by the New Englanders, Green accused the commissioners of not understanding the magnitude of his accomplishment. His goal was to restock America's lakes, rivers, and streams, but he needed a sponsor for such a huge undertaking.

It was at this point that Green joined forces with Robert B. Roosevelt, the nation's richest enthusiast for fish culture. Using his political contacts in Albany, R.B.R. had already petitioned the legislature to launch the state fishing commission. When Green was appointed to the commission's oversight board, the state of New York provided him with a $1,000 grant to inventory the Hudson River shad population and then start a hatchery operation.

At first, the team of Roosevelt and Green made a tactical blunder in initiating public fish hatcheries in New York. The folksy Green went along the banks of the Hudson, telling groups of river men that the hatchery movement was about to make "fish cheap on the open markets." It was poor public relations, and the New Yorkers were soon acting like their Connecticut neighbors. The fishermen used oars, axes, and sledge hammers to smash shad-hatching boxes, destroying all Green's initial work.

An infuriated R.B.R. cursed the "inborn cussedness of human nature" and suggested that wardens were needed on the Hudson River to protect state property. Nevertheless, for a few months R.B.R. also sought a rapprochement with the river men. But when Roosevelt heard that Green had been physically harassed by river men for placing hatchery boxes in the Hudson—cigarette butts were flicked at Green and dead shad were thrown in his face—he headed to river towns such as Beacon and Poughkeepsie, threatening to have the saboteurs clapped into prison. "Furious, Roosevelt went in person and harangued the men," a daughter of R.B.R. wrote in her diary. "Anyone who knew him would realise that this must have been to him not only a relief but a genuine pleasure. He had a surpassing command of irony, sarcasm, and vitriolic incentive combined with a powerfully paternal method of appealing to one's better nature, that would bring a sob to the throat of the most callous and horny-handed son of toil. So long as the latter was unaware that it was merely forensic eloquence . . . the fishermen had no chance."[86]

Over the years a strong friendship developed between R.B.R. and Green. Whenever Roosevelt took his yacht to Newfoundland or Maine, Green went along in search of nature's secrets. Constantly trying to update the general public on scientific improvements, in 1879 they cowrote *Fish Hatching and Fish Catching* and received solid reviews. Their explorations in 1883 of wild Florida's "abundance, beauty and fragrance of flowers" resulted in another book, *Florida and the Game Water Birds*. In it, Green was portrayed as a stubborn foil, continually asking unanswerable questions about local diamond-backed terrapins, stingrays, and sharks.[87] Taken as a whole, Florida was described by R.B.R. as a "floral El Dorado."[88] Never before had he seen so many ducks and waterfowl. For an educational appendix to *Florida and the Game Water Birds*, R.B.R. provided a brief paragraph about each avian species he encountered in Florida. He was building on the traditions of William Bartram and John James Audubon. "There are no dangerous animals in Florida, only a few of Eve's old enemies," R.B.R. innocently wrote, "and the sportsman is safer in the woods at night under the moss-covered trees and on his moss-constructed mattress than in his bed in the family mansion on Fifth Avenue."[89]

The pliable Green, however, wasn't always just a sidekick to R.B.R.'s Huck Finn. He did write Robert B. Roosevelt a letter later that year as his rich friend was fueling speculation about running for mayor of New York City. Green saw that his employer's pigheadedness would make political compromise impossible. "I know you have a big solid head," he wrote to Roosevelt. "But the ware and tare of if you was mayor of New York would be more than it would be on the yacht. There you don't have but a few to conquer & some times you find you are wrong and have to take water. . . . I know you would be always right if you was mayor but there is so many thieves they would keep you awake nights. I know you would get the best of them but it would take a heap of work."[90]

Truth be told, R.B.R. was enjoying rural life on Long Island's Great South Bay (a lagoon) too much to be mayor of any city. In 1873 he had paid $14,000 for a two-acre estate near Sayville. The *Suffolk County News* described the estate as "a comfortable but unpretentious villa."[91] Living in a dream of bliss, R.B.R. named the house Lotus Lake and enjoyed playing a new role, that of gentleman farmer, yachtsman, hard-clammer, and authority on fish. Sailing around Fire Island, an alarmed R.B.R. declared that New York City's waste was killing off the eelgrass. Whenever free time was available, he read up on pirates like Blackbeard and Captain Redeye. Playing at being a farmer gave him material for an ongoing spoof about the vicissitudes of growing cucumbers and black wax yellow-pod beans. He joined the Suffolk County Agriculture Society and kept diaries about his farming triumphs and woes.[92] Regularly, however, while at Lotus Lake, he corresponded with Spencer F. Baird, head of the U.S. Fish and Fisheries Commission, about the need for hatcheries and an Audubon-quality "fish plates" book of all the American species.[93]

R.B.R.'s voluminous diaries about fish, crabs, frogs, and turtles, kept through the mid-1870s, are even more telling of his daily commitment to studying nature than his cucumber and beanstalk logs. Here is an example of his science-laden style:

> March 14, 1877
>
> I visited State Hatching House. Everything in splendid order. Fish eggs clear and bright. Hatched in Hutton boxes till almost ready to come out, then placed on trays in troughs. All that come out head forest die. . . . Young Cal. brook trout and young Cal. salmon quite alike, and former handsomer than our B. trout but with blunter head than salmon; they have no carmine specks on their sides. Kennebeck salmon yearlings have yellow sides, much more so than Cal. salmon. Impregnated some eggs of B. trout Mar. 15, while there were 100,000 of fry with the sack absorbed. The spawning season lasting all winter.[94]

V

Robert B. Roosevelt obsessed over a varied group of wildlife species. And, why not, since he was Dr. Doolittle incarnate? Sometimes he would sit in front of Madison Square with George Francis Train, who published his own quirky newsletter, feeding crumbs to sparrows. Tremendously proud that the Roosevelt coat of arms showed ostrich feathers in plumes, R.B.R. said that they were "always borne with their tops curled over."[95] His personal papers are filled with long, well-written observations on oysters, including their alleged aphrodisiac properties. Minnows were another specialty of his, and he pioneered in studying their spring spawn. He took copious notes on eels, which he collected in tanks to scrutinize. By 1876, in fact, he was considered America's authority on eels, although he admitted to not fully understanding their role in the food chain. Spencer

Fullerton Baird was known to collect snakes in a barrel so R.B.R. decided to one-up him with eels. "Eels—are they kin to snakes?" he once asked in his diary. "We shall leave that question to Darwin and Huxley. You know they are the leaders of modern thought; and it takes a thought leader to find out a thing of that kind. They say eels are a connecting link between the batrachians and the true fishes, and, standing in that position, they are no kin, or, if any, very little, to snakes; though they may be cousin-german to a salamander or mud-puppy. But there is another question: how did the eel get into this position of middle-men? Did he evolute, so to speak, from his cousin catfish? Or did he involute from his cousin mud-puppy? Or did he proceed from that great practical evolutionist, his uncle bull-frog, who used to be a tadpole?"[96]

Robert B. Roosevelt's unpublished notes on species are cheeky and he asks the same kind of Darwinian-era questions as his precocious nephew Theodore. If Darwin could write entire chapters on orchids and beehives, R.B.R. saw no reason not to do a similar study on oysters and eels. Like most naturalists, R.B.R. valued observation more than reasoning, so his notes on fish—shad, pickerel, bass, bream, and sturgeon, in particular—are fiercely detailed. And few alive knew more about frogs—the animal, he claimed, "easiest victimized"—than Roosevelt. Whereas some naturalists dreamed of climbing Clingmans Dome in the Great Smokies or observing timber wolves at Isle Royale, R.B.R. fantasized about visiting a pond in Illinois where 250,000 frogs were believed to live.

For far too long, environmental history has obscured R.B.R.'s influence on his nephew's desire to become a naturalist. You might say the future president was a hybrid—half his father, the other half Uncle Rob. Clearly, Robert B. Roosevelt had taught his nephew that ruinous times would ensue if waterways weren't properly managed. Later in life, T.R. collected live animals exactly the way R.B.R. did. The conservationist books and articles T.R. wrote about the American West were merely more sophisticated versions of *Superior Fishing* and *Florida and the Game Water Birds*. It's not a stretch to believe that T.R. inherited his idea of owning a beautiful Long Island estate surrounded by teeming wildlife—what became his beloved Sagamore Hill in Oyster Bay in 1887—directly from his parents' Tranquility and his Uncle Rob's Lotus Lake. Everybody, it seemed, wanted to visit R.B.R. on Long Island, even Oscar Wilde, who arrived one afternoon with a "wreath of daisies" for a hatband.[97]

To anybody interested in the angler's life, R.B.R. was a true celebrity as an author and activist. His motto—"Remember, no man ever caught a trout in a dirty place"—galvanized anglers to support antipollution laws.[98] Everybody, it seemed, wanted to fish with Robert B. Roosevelt. So as Theodore Roosevelt prepared to attend Harvard University in the fall of 1876, his uncle was already an irrepressible crusader for fish and wildlife.[99] Certainly every natural history professor T.R. took a course with at Harvard would have known him as R.B.R.'s nephew; R.B.R.'s fame was that widespread in biology circles. Robert's books, while quirky, were honored at the Museum of Comparative Zoology. Robert had earned a place in the history of pioneering conservationists. "Robert B. Roosevelt was among the first to understand that our wild species were being decimated," Ernest Schwiebert, the renowned author of *Trout* and *Matching the Hatch*, wrote. "Our cities and factories were already spewing their waste into our waters. Timber was cut with a mindless rapacity, and land poorly suited to agriculture was being stripped for farmsteads. Roosevelt worked tirelessly for conservation."[100]

CHAPTER FOUR

HARVARD AND THE NORTH WOODS OF MAINE

I

At age thirteen, when Theodore was deemed mature enough, his father sent him on a 500-mile excursion by train and stagecoach from Manhattan Island to Moosehead Lake to convalesce in a serene alpine environment after his bouts of asthma. The lake was the largest in Maine, with more than 400 miles of rugged shoreline, most of it untrampled wilderness in the 1870s. All was going well for Roosevelt on the unescorted journey to the lake until he arrived at the Bangor and Piscataquis Railroad depot and station. As he waited for a stagecoach bound for Moosehead Lake, a couple of local youths began taunting him for being a sissy. A nauseated, demoralized feeling rose up behind Roosevelt's breastbone. Timidly, he put up his dukes and in return got pummeled. "They found that I was a foreordained and predestined victim, and industriously proceeded to make life miserable for me," he recalled in *An Autobiography*. "The worst feature was that when I finally tried to fight them I discovered that either one singly could not only handle me with easy contempt, but handle me so as not to hurt me much and yet to prevent my doing any damage whatever in return."[1]

From his lakefront lodge, Roosevelt, piqued by the hazing incident, embarrassed by his feebleness, stared zombie-like at the blue water, which swallowed strands of dark green spruces along its shoreline.[2] Instead of resenting his tormentors, he envied their hardiness, brawn, and condensed force. Following the humiliation at Moosehead Lake, Roosevelt made a pact with himself: he was going to become a man of true physical strength. Boxing, weight lifting, calisthenics, hiking—he would do whatever it took. Someday, he promised, those same bullies would treat him with respect. "The experience taught me what probably no amount of good advice could have taught me," he recalled. "I made up my mind . . . that I would not again be put in such a helpless position; and having become quickly and bitterly conscious that I did not have the natural prowess to hold my own, I decided that I would try to supply its place by training."[3]

In a sense, Roosevelt was determined to evolve from prey to predator. He started pumping weights and doing push-ups regularly, and he also improved his skills as an equestrian and marksman. Meanwhile, his bird collecting in the woodlands of New York continued unabated. Whenever time permitted, Roosevelt had a birder friend accompany him on these tramps. His usual companion was Frederick Osborn, an irresolute ornithologist who, like himself, was in awe of God's wild creatures; no nest, egg, or bird sighting failed to enthrall Osborn. Memories of their bird-watching times together remained with Roosevelt for the remainder of his life, one more vivid than another.

One winter afternoon in the 1870s near Bear Mountain—a breathtaking rise on the west bank of the Hudson River—Roosevelt and Osborn went on a hunt. Full of anticipation, Roosevelt had journeyed upriver from Manhattan to see Osborn, who lived in nearby Garrison, a ferry and railroad depot directly across the river from the U.S. Military Academy at West Point. Much of the forest around Bear Mountain was still pristine, but quarrymen had of late been mining toprock (basalt) to provide building material for eastern metropolises. Clomping down a snowy path, they suddenly stopped dead in their tracks. It was a moment of unprecedented excitement. In front of them was a flock of gorgeous red crossbills. Both Theodore and Frederick had long coveted this species to add to their collections. They were determined to be the very best of the new breed of post-Darwinic ornithologists.

In unison Roosevelt and Osborn rapidly fired their shotguns in a succession of blasts, three or four times each. When the dust settled, finch carcasses lay on the stump-filled field. Red crossbills—with different bills from their brethren in the West—could now be added to the bounty bags of pine siskins, common redpolls, and pine grosbeaks. Working without hunting dogs, Roosevelt and Osborn anxiously sprinted to retrieve their prized birds from the ground. But Roosevelt tripped on a concealed rock or tree root, stumbled forward, and barely recovered his balance. He was smacked in the face by a low-hanging branch or twig and his spectacles went flying. Half-blind, squinting, and shaking off disorientation, he quickly recollected himself and scanned the ground. "But dim though my vision was, I could still make out the red birds lying on the snow; and to me they were treasures of such importance," he wrote years later, "that I abandoned all thought of my glasses and began a nearsighted hunt for my quarry."[4]

Once Roosevelt secured the red crossbills he went searching for his glasses, but in vain. From that moment onward he made a pact with himself always to carry a reserve pair of spectacles—with rims made out of steel—in his breast pocket. (When Roosevelt ran for president as the candidate of the Bull Moose Party in 1912, he was shot in the chest by an anarchist in Milwaukee. The extra bird-watching spectacles absorbed the impact of the bullet and probably saved his life.[5])

Just four months after the day of the red crossbills, Osborn—whose father, William, was president of the Illinois Central—died in a river accident. Roosevelt was emotionally crushed by his friend's death. His mind held a montage of all the wilderness tramps they had gone on together. Osborn, he loyally believed, was a rising prince, as kindhearted as the day was long, the companion of *belle jeunesse*. His death jarred Roosevelt's outlook on life. With a contracted brow and grimace on his face it became the day the fun stopped. "He was drowned, in his gallant youth," Roosevelt mourned decades later. "But he comes as vividly before my eyes now as if he were still alive."[6]

Losing his favorite birding side-kick was emotionally tough on Roosevelt, but his naturalist journals continued without a hitch. Whenever he found a flock of sandpipers or a delicate nest of golden-crowned kinglets—to cite just two examples from that spring—he'd dutifully record the observations in notebooks labeled with pseudoscientific titles like "Remarks on the Zoology of Oyster Bay" or "Field Book of Zoology." Using Elliott Coues's excellent *Key to North American Birds*—first published in 1872—as his principal classification tool, Roosevelt was determined to carry the ornithological torch for both Osborn and himself. Depression, he decided, simply wasn't allowed. March, after the snow melted, left the ground parched. April, as poets often wrote, was a fickle month. But in May—the month when Osborn died, in his prime—there was a sense of rehabilitation in the air. As any naturalist knew, death must always precede rebirth. When the May dandelions arrived, everything turned green, the migratory birds returned to New York, and the fields were filled with thistle and thyme. "I just begin to realize it," he wrote to his mother, who was traveling in Georgia, "the birds even are commencing to come back. While in the Park [Central Park] the other day I saw great numbers of robins, uttering their cheery notes from almost every grove. That winter had only just departed was evident from the number of little snowbirds (clad in black with white waist coats) which were about."*[7]

If any American writer caught Roosevelt's attention in the 1870s it was Coues. Born in Portsmouth, New Hampshire, while John Tyler was president, Coues had lived a storybook life as a western man of science. During

*Probably Roosevelt meant snow buntings, which would have been rare indeed in New York in May, or possibly dark-eyed juncos.

the Civil War he served at Fort Whipple, Arizona, as an assistant surgeon. Later he was appointed to Fort Randall, South Dakota, as a naturalist for the U.S. Northern Boundary Commission. Although he was considered one of the few top-rated surgeons west of the Mississippi River, Coues preferred writing wildlife books to performing tonsillectomies and amputations. His *Key to North American Birds* made him the heir apparent to Audubon. Although he served the Geological Survey of the Territories from 1876 to 1880 as a secretary and naturalist (under the supervision of F. V. Hayden), Coues eventually quit and returned to Arizona to document the amazing variety of migratory birds that wintered at such prime grounds as the Chiricahua Mountains and Patagonia Lake.[8]

Biographers of Theodore Roosevelt have dizzied themselves trying to monitor their subject's travels between Manhattan Island, Moosehead Lake, the Hudson River valley, the New Jersey Palisades, the Adirondacks, and Oyster Bay from 1874 to 1876, before he went to Harvard. Restless in the extreme, Roosevelt switched locales as regularly as a salesman or trail guide. In the two years before he entered Harvard, he tried to get away from West Fifty-Seventh Street on nature excursions as often as possible, usually alone. Instead of Osborn, Coues's *Key* was his new companion, and he religiously followed the books' systematic standards of trinomial nomenclature (taxonomic classification of subspecies).

Often, however, when he was studying foreign languages or arithmetic under the guidance of his tutor, Arthur Cutler, Roosevelt's mind drifted to the Adirondack timberlands and the Long Island meadows, daydreaming about troops of new birds for his collection. Every time he could escape from Cutler's apron strings, he made a beeline for Oyster Bay, part of the Eastern Flyway for migratory birds, eager to hear reedy wails and lovely carols. Whenever Cutler—who now prided himself on being headmaster of the Cutler School of New York—tried regimenting Theodore and getting him to focus on French or Greek, his prize student's attention instead drifted to meditations on plebian robins and exotic waterbirds.[9] "The study of Natural History was his chief recreation then as it continued to be," Cutler recalled later. "He had an unusually large collection of birds and small animals; shot and mounted by himself and ranging in habitat from Egypt to the woods of Pennsylvania. In his excursions outside the city, his rifle [actually a shotgun] was always with him, and the outfit of a taxidermist was in use on every camping trip."[10]

That autumn, Roosevelt's natural history journals document eight visits to the North Shore of Long Island in two months. Even though Roosevelt kept bird counts doggedly, even fiercely, he sometimes panicked over his lack of ornithological expertise. When the Oyster Bay fields turned dark and the nighthawks were no longer doing arabesques, Roosevelt would study the frail skeletons of doves and pigeons in his collection. Unlike those of other vertebrates, many of these birds' bones were hollow tubes. The larger the bird, Roosevelt noticed, the more hollow the bones were. Thanks to Darwin's *On the Origin of Species*, Roosevelt understood this to be a result of evolution—the lighter the bird, the easier it was to fly. Staying up late, long past the shifting dusk, though his parents usually retired early, Roosevelt pondered each avian's aerodynamic design, fascinated that ostriches and penguins had abandoned flight. Throughout these ritualistic midnight inspections, Roosevelt bore down as a scientist, unemotional and sleepless.

Roosevelt sometimes worked eight or nine hours a day on ornithological pursuits when he was in Oyster Bay, and his descriptions of birds increased in vividness and freshness. Each bird Roosevelt saw in a field or grove attracted his serious attention. Even the coarse materials songbirds used to build nests or the exact number of wing beats it took to defy gravity came under his careful scrutiny, in a way Robert B. Roosevelt would have approved. "It becomes very fat in August and is at all times insectivorous," Theodore wrote of the white-throated sparrow. "It has a singularly sweet and plaintive song, uttered with clear, whistling notes; it sings all day long especially if the weather be cloudly, and I have frequently heard it at night, but its favorite time is in the morning when it begins long before daybreak; indeed, excepting the thrushes, it sings earlier than any other bird. The song consists generally of two long notes, the second the highest and with a rising inflection, followed by five or six short ones (as duuduu), but there are many variations. A very common one is to have but two short notes (as uu); sometimes the second note is broken into two (as uuu). It sings all through the summer."[11]

On July 8, 1874, Roosevelt shot, skinned, and mounted a male passenger pigeon in Oyster Bay. What interested him the most was the pigeon's esophagus crop, the sac where food was stored to later regurgitate and feed hatchlings. This bird also had a crop to produce a special milk for its hungry babies. Because these pigeons were still plentiful, Roosevelt wasn't overenthusiastic about them in his journal. During the Jefferson era, there had been millions of these tan-and-burned-orange birds in America, huge flocks migrating regularly from north to south. The ornithologist Alexander Wilson, in fact, once recorded more than 2 million in a single flock flying over Kentucky. Since passenger pigeons were edible and marketable, their slaughter was at full throttle in the middle of the nineteenth century when Roosevelt shot his specimen at Oyster Bay. One New York man, in fact, boasted of killing 10,000 pigeons in a single day. The growth of cities such as Philadelphia and Chicago hastened their destruction by ruining their natural habitat.

Meanwhile, while birding, Roosevelt had started "seeing" Edith Carow of New York. Edith's father, Charles Carow, was perhaps Robert B. Roosevelt's closest fly-fishing friend. "Charles cast the fly simply to perfection," R.B.R. wrote, "and with [him] I have fished many and many a day on the waters of old Long Island and elsewhere as well."[12] At Oyster Bay Theodore and Edith argued over the merits of popular fiction, played board games, and rowed around the Long Island Sound. Everything about Edith—her sharp quick eyes, playful countenance, and air of general smartness—appealed to Theodore.[13] Too young to really understand love, Theodore nevertheless knew this much: Edith, who was destined to become his second wife, was *the* girl he wanted to impress. So they dated. Nothing too serious, just occasional hand-holding or good-night hugs. They were very young, and whatever romance existed between them didn't last long. Once Theodore arrived at Harvard, he quickly found another girlfriend, who came from Chestnut Hill in Boston.[14]

Truth be told, Roosevelt—as of 1874—didn't have the self-confidence or the desire to date with marriage in mind. He was busy just trying to understand his father's expectations. The awkward Roosevelt, who was five feet eight inches tall and weighed 124 pounds, strove to improve himself in every way. He grew bushy side-whiskers, developed compact forearms, and wore pressed clothes fit for a sportsman hunting. He always looked as if somebody had slapped the creases out of the fabric for the sake of upholding the family name. Although his parents often straitjacketed Theodore in formal attire, he preferred dressing down like a muskrat trapper or stable keeper; this fashion attitude would change at Harvard.

Visiting Edith at her summer home at Sea Bright, New Jersey, for a week in July 1874, Roosevelt noted that the sand dunes of the barrier beach were brimming with "ornithological enjoyment and reptilian rapture."[15] Playing a Darwinian biologist, he pickled in jars all the toads, frogs, and salamanders he caught for more careful study of their evolutionary stages. Romance with Edith was put on a back burner in favor of writing about New Jersey's avians in his *Notes on Natural History*. "Whether inland or on the coast the most conspicuous bird was the fishawk," he wrote. "It was most plentiful by ponds, and over one of these several pairs of singular birds could almost always be observed, circling through the air on almost motionless wings, usually far out of gunshot range. On a suitable fish being seen, the hawks swoops down with arrow like swiftness, causing a whistling, booming sound as it descended, and stooping with such force as sometimes to totally immerse itself in water."[16]

With the hungry speed of youth, Roosevelt was dead certain about his career direction. As his father conceded when they returned from the trip to Europe and the Middle East, he was predisposed to be a scientific naturalist or wildlife biologist. But Theodore Sr. warned his son that a life of science meant long hours in a sterile laboratory; field collecting was only a small part of the profession. Either way, the importance of being accepted at Harvard weighed greatly on young Theodore. When word of his acceptance came in the spring of 1876, he was elated, but he also knew that the time had come to professionalize his infatuation with animals. "When I entered college, I was devoted to out-of-doors natural history, and my ambition was to be a scientific man of the Audubon, or Wilson, or Baird, or Coues type," Roosevelt recalled in *An Autobiography*, "a man like Hart Merriam, or Frank Chapman, or Hornaday, to-day." The key phrase here is "out-of-doors"; he didn't want to be an indoor scientist, the kind of scientist Captain Reid had shunned.[17]

Throughout the summer of 1876, as America was in the midst of celebrating its centennial, a Grand Exposition was held in Fairmount Park in Philadelphia. It celebrated the industrial march of progress from Alexander Graham Bell demonstrating the telephone to a massive 650-ton steam

Roosevelt at Harvard University in 1877. Desperate to become masculine he worked-out everyday with free weights.

engine built by Rhode Island's George Corliss (standing seventy feet high, it was the largest engine ever constructed). At one point, Ulysses S. Grant and Frederick Douglass sat side by side onstage; a photograph of them together spoke volumes about the Union victory in the Civil War. The Grand Exposition, in fact, was the first world's fair held in the United States, and more than 10 million people attended. Young Theodore was one of them. The Department of the Interior created a pioneering display showcasing America's original native inhabitants in colorful ethnological detail. When Roosevelt visited the fair, what he marveled at most was the display of U.S. wildlife (assembled with help from the Smithsonian Institution), including a fifteen-foot walrus and an Alaskan polar bear. Huge aquarium tanks displayed the plethora of American fish, such as salmon and perch, an aspect of the exposition that Robert B. Roosevelt had been instrumental in creating.[18]

Also, 1876 was a presidential election year, with his father backing Rutherford B. Hayes of Ohio and Uncle Rob behind Samuel J. Tilden of New York. T.R. didn't get very involved even if, like all good young Republicans, he believed the campaign hoopla: "Hurrah for Hayes and Honest Ways!" Politics was thin gruel to Roosevelt. He complained that both Hayes and Tilden sneered at the teachings of naturalists such as Coues and Darwin. Roosevelt instead pored over zoology books, including the new *Manual of the Vertebrates of the Northern United States*, whose author, David Starr Jordan, was only seven years older than himself.[19] Suddenly, Roosevelt, influenced by Dr. Jordan and Uncle Rob, became interested in game fishes, particularly bass and trout; he was determined to enter Harvard as the best-rounded naturalist of his up-and-coming generation. As promised since the family trip down the Nile, Roosevelt was going to be a Harvard-trained foot soldier in the Darwinian revolution.

II

Before classes started in late September, Roosevelt tried to get his asthma under control. Unfortunately, his breathing had been thick for much of the year. It was as if a fungus had taken hold in his lungs. Because Harvard offered him only first-floor dormitory rooms, Roosevelt sought lodgings elsewhere. (Basements or ground-level rooms developed dampness and mildew, he claimed, which in turn triggered his coughing fits.) Roosevelt—already segregated from the mainstream before the first day of class—rented a second-floor suite in a house at 16 Winthrop Street (since torn down) just blocks from the campus. There were shady elm trees to admire from his four picture windows and, better yet, he could see the Charles River from the rooftop. Worried that he'd turn his living quarters into a taxidermy studio, Roosevelt's overprotective sister Bamie prepared the rooms, fixing up his study and alcove.[20] "Ever since I came here I have been wondering what I should have done if you had not fitted up my room for me," he wrote to Bamie in earnest gratitude. "When I get my pictures and books, I do not think there will be a room in College more handsome."[21]

Within two or three weeks Roosevelt had transformed his Winthrop Street quarters into a virtual vivarium. Mounted birds cluttered his desk and a well-used portfolio of Audubon's *Birds of America* was placed on his shelf like an old friend, soothing and familiar. Stuffed owls, deer antlers, bottles of formaldehyde, arsenic paste, wren's nests, and colorful eggs abounded—and those were just the inanimate objects. Roosevelt was like a golden retriever; you never knew, when he entered 16 Winthrop Street, whether he would be carrying a wounded squirrel or a kicking rabbit. Then there were the live finches and tadpoles, mice litters, and a formicary. One evening, his landlady—a Mrs. Richardson—nearly tripped over one of his escaped turtles; her face turned white with fright. Even though Roosevelt's rent money was good, the notion of having a philotherian tenant made her uneasy.[22]

Having untamed animals running around indoors marked young Theodore as an eccentric at Harvard. This wasn't necessarily a bad thing, since singularity had its time-honored virtues. After all, Cambridge was replete with brilliant characters. Oliver Wendell Holmes scouted the outdoor bookstalls for rare editions, and Henry Wadsworth Longfellow could be seen strolling down Brattle Street window-shopping for elegant canes. The great historian Francis Parkman was sometimes found in the library working on *Montcalm and Wolfe*, and Charles Francis Adams served as university "overseer." Charles Darwin's chief American supporter, the botanist Asa Gray, was no longer teaching at Harvard, but he lived near the campus and was available to answer questions about evolution if any student knocked on his front door. Ralph Waldo Emerson was spied from time to time poking around the campus. (His remark "Nature encourages no looseness, pardons no errors" always appealed to Roosevelt's bare-knuckled Darwinian sensibility.[23]) Being different brooked no condescension at Harvard if you were intellectually astute.[24]

While other freshmen were enthralled that William James and George Santayana were on the Harvard faculty, Roosevelt bemoaned the fact that professor Louis Agassiz—who, like Uncle Rob, had published a landmark book on Lake Superior—had died three years before. During the last year of his life Agassiz—founder of Harvard's Museum of Comparative Zoology, which soon rivaled its counterparts in London and Paris—traveled to Brazil as an ichthyologist, participated in a deep-sea dredging project sponsored by the U.S. Coast Survey, and founded the Anderson School of Natural History at Penikese Island off the southern coast of Massachusetts. "Study nature, not books" was Agassiz's dictum. Clearly Roosevelt would have had a lot to learn from the charismatic zoologist, even though Agassiz had been an antievolutionist to the bitter end. One got the feeling Roosevelt had wanted to challenge Agassiz in class over the accuracy of *On the Origin of Species.* (The old-fashioned Agassiz had fought against having evolution as part of the museum.[25])

Perhaps because Roosevelt was an autodidact, he wasn't inspired by any of the geology or zoology professors at Harvard. This attitude caused some classmates to consider Roosevelt a presumptuous snob, and it caused some of the faculty to write him off as lazy and pedantic. These professors weren't pathfinders or explorers like Audubon or Darwin, and their lectures, without trailblazing deeds, did little to stir the imagination. Microscopes and laboratory garb didn't interest Roosevelt in the least; the mastodon in Boylston Hall certainly did. A slightly above-average student, earning a cumulative seventy-five in his freshman year and an eighty-two as a sophomore, Roosevelt struggled most with foreign languages.[26] Nevertheless, he excelled in German, perhaps because he had lived with a German family in Dresden. Still, he extolled the virtues of Americanism every chance he could, and a slow-burning resentment toward European claims of intellectual superiority was detectable in his letters and diaries. He lampooned the smallness of England and the boorishness of Germany (although the primitive *Volkmoot* of the ancient German forests always interested him). Perhaps because they were predominantly of Dutch ancestry, the Roosevelts often blamed Germany for the worst influences on American life. For example, in his novel *Five Acres Too Much* (1869), Robert B. Roosevelt complained that Staten Island was "overrun by sour-kraut-eating, lager-beer-drinking, and small-bird-shooting Germans, who trespass with Teutonic determination wherever their notions of sportsmanship or the influence of lager leads them."[27]

As for natural history, his major, Roosevelt's excellence became legendary. However, the example of his faculty adviser, Professor Nathaniel Southgate Shaler, turned Roosevelt off even more from the idea of becoming a scientific naturalist. Ostensibly, Roosevelt should have liked Shaler, who was a prized student of Agassiz and who specialized in the study of what were then called earth sciences. Serving as a captain in the Union Army during the Civil War, Shaler fought nobly; he returned to Harvard as a twenty-seven-year-old professor of zoology and geology in 1865 and would stay at Harvard the rest of his life. By the time Roosevelt arrived to take his Introduction to Geology course in 1876, Professor Shaler had made Harvard the center of American geological research, and his *The First Book of Geology* was considered a new classic in the field.[28] Training students in how to biologically observe natural selection was a specialty of Shaler's; he was one of the first Americans to adopt Darwin's specimen collection methods.[29]

But for a young man who'd already shot plovers in the Middle East (as well as the Adirondacks), and was the son of a New York millionaire, Shaler's explanations about shifting tectonic plates and paleontology seemed dull; the professor might be a young Turk, but he had abandoned nature's awesome drama, Roosevelt believed, by an overreliance on arcane theories about ancient dirt and meteorite rock. It was naive of Roosevelt to think that naturalists were buckskin-clad outdoorsmen like Audubon or Wilson, spending their lives in the wild. To Roosevelt, Shaler was a pedant who talked about how the mechanisms of evolution still needed to be ironed out instead of hitting the trail on a treasure hunt. Roosevelt wanted to learn about types of skunks, not faux erudition from a self-conscious Kentuckian struggling to become a prig, poor man. Theodore challenged his professor regularly, until they squared off one afternoon. "Now look here, Roosevelt, let me talk," an exasperated Shaler stated. "I'm running this course."[30]

Roosevelt whizzed his way through a slate of courses in the natural history department: Comparative Anatomy and Physiology of Vertebrates, Elementary Botany, Physical Geography and Meteorology, Geology, and Elementary and Advanced Zoology.[31] And in truth, even though Roosevelt refused to give him any credit, Shaler was an inspirational teacher, known for taking students on geology field trips in the tradition of his mentor, Agassiz. Roosevelt, in fact, went on an excursion with Professor Shaler to study glacially formed cliffs. Nevertheless, Roosevelt, in a swipe at Shaler, lamented that Harvard's president, Charles Eliot, had failed to hire a first-rate "faunal naturalist, the outdoor naturalist, and observer of nature."[32] Yet Eliot believed T.R. and Shaler were two peas-in-a-pod, regardless of intellectual differences.[33] Likewise, Roosevelt would convey to Gifford Pinchot that no matter what their squabbles had been, Shaler, in fact, was "a very dear friend of mine."[34]

Despite his apparent eccentricities, Theodore was popular with and respected by his classmates. Nobody ever questioned his bedrock honesty or inbred decency. Years later, Richard Welling—in an article in *American Legion Monthly*—revealed perhaps the strangest personality trait of Roosevelt at Harvard. Whenever Roosevelt was performing gymnastics indoors—doing somersaults, skipping rope, or doing pull-ups—he acted hesitant and clumsy. His caution and trepidation were evident. As Welling bluntly put it, Roosevelt was "verily a youth in the kindergarten stage of physical development." But when Welling went ice-skating with Roosevelt at Fresh Pond—a kettle-hole lake that served as a reservoir for Cambridge—with the wind-chill factor around zero, the brutal cold would send everybody else home shivering, but Theodore would get an adrenaline rush, shouting, "Isn't this bully!" He had no fear of frostbite or pneumonia.* This was unusual behavior. Welling realized his friend was overcompensating for something. Nature at its most brutal flipped some switch of fortitude in Roosevelt's peculiar makeup. "Roosevelt had neither health or muscle," Welling concluded. "But he had a superabundance of a third quality, vitality, and he seemed to realize that this nervous vitality had been given in order to help him get the other two things."[35]

Detachment might be the best word to characterize Roosevelt's measured indifference to Harvard in 1876 and 1877. Even though his grades were good, he didn't buy into all of the traditional aspects of undergraduate life as one would have expected. Regularly he would send his father updates from Cambridge, describing in detail his asthma flare-ups and his prayer sessions at the local church. There were soirees at Chestnut Hill and Beacon Hill to report, usually with reassurances that his morals were intact, but his reportage from Harvard seemed contrived. It was his two touchstone places—Oyster Bay and the Adirondacks—that continued to arouse his enthusiasm. And he thought a lot about Maine.

III

Ever since Frederick Osborn had drowned in the Hudson River, a void had existed in Roosevelt's life. He needed a close chum to share his enthusiasm for ornithology, someone with whom to prowl his favorite hunting grounds come summer break. Now, as a Harvard freshman, he found such a friend in Henry "Hal" Davis Minot of Roxbury, Massachusetts. Hal, a lanky young man with piercing blue-gray eyes and a thin brown

*Some scholars question whether Roosevelt was using "bully" at this early point in his life. Certainly he was saying it in the early 1880s. Other expressions that he constantly used during his college years were "By Jove" and "My Boy."

beard, had already written a booklet: *The Land and Game Birds of New England* would be published the following spring. Together Theodore and Hal hatched plans to spend the next summer collecting warblers and thrushes. "Our lessons will be over by the twentieth of June," Roosevelt excitedly wrote to his parents from Harvard, "and then Henry Minot and I intend leaving immediately for the Adirondacks, so as to get the birds in as good plumage as possible, and in two or three weeks we will get down to Oyster Bay, where I should like to have him spend a few days with us. He is a very quiet fellow, and would not be the least trouble."[36]

Haunted by his humiliation at Moosehead Lake, Roosevelt—to improve his physique—boxed regularly during his freshman year. He learned how to throw a pretty good one-two punch, but his eyesight was terrible and he was never light on his feet. What made him quite remarkable in the ring, however, was his godawful ability to take a thrashing, to be pummeled unmercifully but still come back for more. This wasn't a recipe for winning matches, but it did win the respect of his classmates. Clearly, Roosevelt had a genius for pushing the limits. Instead of hopping onto a streetcar to downtown Boston as most of his classmates would do, Roosevelt often chose to walk the three or four miles. As an oarsman, Roosevelt preferred rowing when a heavy nor'easter kicked up, seeking the challenge of advancing forward in rivers and lakes when the wind was least favorable.

As planned, during the summer of 1877—between his freshman and sophomore years—Roosevelt spent weeks camping in the Adirondacks with Minot. Because Edith had been excited about spending time with Theodore at Oyster Bay, the news that birds came first may have bruised her feelings. As the historian Edmund Morris joked in *The Rise of Theodore Roosevelt*, Edith "could compete with the belles of Boston, but what were her charms compared with those of the Orange-Throated Warbler, the Red-Bellied Nuthatch, and the Hairy Woodpecker?"[37] As soon as classes ended at Harvard on June 21, Roosevelt and Minot headed straight to Saint Regis Lake "so as to get the birds in as good plumage as possible."[38] Once again Moses Sawyer served as the trusted guide through the rugged Adirondack forests, on what turned out to be the most serious bird collecting trip of Roosevelt's life. Spurred on by his classmate, he made careful notes and pulled together his own first publication on birds: it was modeled on Minot's *The Land and Game Birds of New England*, which had just been published by Estes and Lauriat (in cooperation with the Naturalist Agency of Salem, Massachusetts) to fine peer reviews. In fact, *Harper's New Monthly*, which had just been launched, commended *The Land and Game Birds* to "all who care for out-door sights and sounds."[39]

The resulting booklet by Roosevelt, *The Summer Birds of the Adirondacks*, not much more than a broadside, was instantly the finest list on the subject in print. Although Minot was credited as coauthor, *The Summer Birds* was clearly Roosevelt's accomplishment, the end product of four collecting trips to the Adirondacks. Roosevelt's ample descriptions of nearly 100 species were based on firsthand outdoors observations. *The Summer Birds*, in fact, was impressive in its thoroughness. With an exacting eye, Roosevelt delineated everything from the sprightliness of juncos to the "strikingly" common least flycatchers. With brevity he analyzed the nests of the Swainson's thrush and Wilson's warbler. It was clearly a work aimed at specialists—only a true-blue bird enthusiast would want such a detailed local key.

Sticking to straightforward scientific observation, Roosevelt, perhaps fearful of being trivialized as a populizer, edited his most poetic journal writing out of *The Summer Birds*. Here, for example, was a journal passage from June 23, 1877, that he apparently deemed too fanciful to include in his booklet. The "we" in the following journal entry refers to Minot, Sawyer, and himself:

> Perhaps the sweetest bird music I ever listened to was uttered by a hermit thrush. It was while hunting deer on a small lake, in the heart of the wilderness; the night was dark, for the moon had not yet risen, but there were clouds. . . . I could distinguish dimly the outlines of the gloomy and impenetrable pine forests by which we were surrounded. We had been out for two or three hours but had seen nothing; once we heard a tree fall with a dull, heavy crash, and two or three times the harsh hooting of an owl had been answered by the unholy laughter of a loon from the bosom of the lake, but otherwise nothing had occurred to break the death-like stillness of the night; not even a breath of air stirred among the tops of the tall pine trees.[40]

Theodore spent the rest of his summer rowing around Long Island, discussing birds with his father, and preparing his little book for publication. Roosevelt considered it his highest intellectual achievement to date, proof that he had learned "how to read and enjoy the wonder-book of nature."[41] By the time Roosevelt returned to Harvard in September 1877 for his sophomore year, he was ruddy-cheeked, bronzed, hardened, and strangely handsome, and his asthma was in remission. In October a few hundred copies of *The Summer Birds* were printed. He received accolades from scholars—an important development in enhancing his self-confidence as a naturalist. At the time bird lists for specific locations were considered very important, as national data were just starting to be collected seriously.

No less a personage than Dr. C. Hart Merriam, in fact, a graduate of Yale University's Sheffield Scientific School, embraced Roosevelt with open arms in the April 1878 *Bulletin of the Nuttall Ornithological Club* (the house organ for the first organization in North America devoted to ornithology). It was the list's accuracy that impressed Dr. Merriam, who had just published his own *Birds of Connecticut.* "By far the best of these recent [bird] lists which I have seen is that of *The Summer Birds of the Adirondacks in Franklin County, N.Y.*, by Theodore Roosevelt and H. D. Minot," Dr. Merriam wrote. "Though not redundant with information and mentioning but 97 species, it bears prima facie evidence of reliability—which seems to be a great desideratum in birds lists nowadays."[42]

Although the down-to-earth Merriam was only three years older than Roosevelt, he was already a legend among naturalists. Born in Locust Grove, New York, on December 5, 1855, Merriam spent much of his childhood exploring the Adirondacks and hunting mammals with bow and arrow. Like Roosevelt, Merriam had put together his own wildlife museum as a boy. Hart's mother, objecting to the stench of dead rabbits and squirrels, hired an old army surgeon to teach him taxidermy. Using corrosive sublimate and carbolic acid as his preservatives, young Merriam became something of a prodigy at his craft. In 1871 Hart's father, a U.S. congressman from New York, playing the role of promoter and coach, took his son to meet Spencer F. Baird at the Smithsonian Institution. Once Baird saw the boy's work he was awestruck; every specimen of bird or mammal was perfectly embalmed. Even the mounted butterflies and grasshoppers were first-rate. That same year Merriam struck up a correspondence with John Muir regarding glaciation in Yosemite Valley.

Worried that he couldn't make a living at ornithology, Merriam, whose expertise on the subject of small mammals and fauna led to an assignment by the Hayden Survey classifying animal populations in Yellowstone National Park, was a prodigy who at age seventeen collected 313 bird skins and sixty-seven nests. His resulting fifty-page report was published in 1873 to high praise from zoological circles. Merriam went on to study at Yale University's Sheffield Science School, and then in the medical school at Columbia University. He practiced medicine in Locust Grove from 1879 to 1885, delivering many babies. Nevertheless, if you wanted to understand elk, deer, or grizzly bears, Merriam was the man to turn to.[43]

With Merriam's glowing review in the *Bulletin*, Roosevelt was anointed an up-and-coming naturalist of the Ivy League and was listed in the 1877 *Naturalists' Directory*. He had been accepted by the fraternity of scientists as one of their own.[44] The Nuttall Ornithology Club—founded as recently as 1873—had said so. That was good enough for Roosevelt. Although Merriam hadn't yet reached fame as perhaps the top U.S. government biologist of his generation, the fact that he saluted the Harvard sophomore in such a high-minded fashion won Roosevelt over. Thereafter, he would always hold Merriam in high regard. Nobody working in the Darwinian specimen collecting circuit, Roosevelt believed, knew more about North American wildlife than Merriam. Building on Roosevelt's *Summer Birds*, Merriam, in fact, in 1881 published a better "Preliminary Life of the Birds of the Adirondacks." It was his last foray with birds before his focus shifted to mammals.

Roosevelt followed up on the success of *The Summer Birds* with a set of profiles, drawn from his Oyster Bay notebooks, of the seventeen rarest birds he had encountered on the North Shore. Privately published in March 1879, while Roosevelt was backpacking in Maine, *Notes on Some Birds of Oyster Bay, Long Island* featured an unusual range of shorebirds, all listed with their Latin names. Some of his observations were new. Others were coeval with findings of Spencer F. Baird. Dive-bombing mockingbirds had a southern range, for example, but Roosevelt found one, imitating the vocalizations of half a dozen other species near the family's country home, Tranquility. Four different species of warblers—prairie, golden-winged, pine, and Connecticut—were matter-of-factly included in *Notes*. But he was most clearly proud of having shot and collected two unusual species: a fish-crow, which constantly harassed gulls to relinquish their prey; and an Ipswich sparrow, which had been discovered by a farmer in 1872. "I shot an Ipswich sparrow on a strip of ice-rimmed beach," he later wrote, "where the long coarse grass waved in front of a growth of blue berries, beach plums, and stunted pines."[45]

Meanwhile, all of the rowing in Oyster Bay and hiking in the Adirondacks was starting to pay off for Roosevelt. He no longer looked like the weakling who'd been harassed by the youths of Moosehead Lake. Physical exertion was now part of his daily routine. He rowed on the Charles River, he did sit-ups and jumping-jacks, and he crammed as much activity into each and every hour as possible.

But all wasn't well in the Roosevelt family. Theodore Sr. had been diagnosed with stomach cancer, which the *New York Times* then called a "Hopeless Disease."[46] Carefully, Theodore and his brothers and sisters monitored their father's failing health, concerned about every chill and cough. He was clearly in deep pain, and his prognosis wasn't good. During the Christmas season, Theodore was cautiously optimistic, elated that his father's high temperature had returned to normal. Yet, as Corinne noted, all the "fearful suffering" was turning Theodore Sr. gray when he had "not a white hair before."[47]

On Christmas day, Roosevelt listed all the birds and mammals he'd collected for his museum in 1877. For the first time, he included fish (fifty-two brook trout and 120 Atlantic mackerel). Many bird species were also inventoried; it was a strong year for snipes and herons. As for big game, he'd killed his first deer with a rifle. And he felt that much similar success lay ahead because his sick father had given him a double-barreled shotgun as a Christmas present. Just holding it, feeling its lead weight, made Roosevelt anticipate future hunts.[48]

Later in the winter, once he was back at Harvard, word arrived that his father had died. Theodore took the grim news of February 10, 1878, as a body blow, and his grief was overwhelming. Bravely, he struggled to cope with the loss of "the one I loved dearest on the earth."[49] A curtain had fallen on his life. He had difficulties studying properly and felt haunted by his father's visage. Roosevelt's diary entries that winter and spring are full of lamentations: "Sometimes when I fully realize my loss I feel as if I should go wild." "Oh Father, Father, how bitterly I miss you, mourn you and long for you." The mainspring of Roosevelt's life was gone. He prayed incessantly for his father, feeling terribly inferior to him in every way. Comparative anatomy and biology courses lost all appeal to Roosevelt. Slumping silently around Cambridge, he craved the recuperative powers of the wilderness.

Adding to Roosevelt's depression, Hal decided to leave Harvard midway through his sophomore year, intending to learn law and increase the family fortune. Wasn't it better, his friend hypothesized, to make big money and be a field ornithologist publishing popular books than to waste years in Harvard Square? Or to act like a New York dandy?

Discombobulated by his father's death and Minot's departure, restless beyond words, Roosevelt abandoned the notion of becoming a professional scientist. Years later, in *An Autobiography*, he explained how his disillusion with the professors at Harvard led him to set out on a different course. "I did not, for the simple reason that at the time Harvard, and I suppose our other colleges, utterly ignored the possibilities of the faunal naturalist, the outdoor naturalist and observer of nature," he wrote of his defection, with a trace of contempt. "They treated biology as purely a science of the laboratory and the microscope, a science whose adherents were to spend their time in the study of minute forms of marine life, or else in section-cutting and the study of the tissues of the higher organisms under the microscope. This attitude was, no doubt, in part due to the fact that in most colleges then there was a not always intelligent copying of what was done in the great German universities. The sound revolt against superficiality of study had been carried to an extreme; thoroughness in minutiae as the only end of study had been erected into a fetish."[50]

Like Minot, Roosevelt had come to the conclusion that a naturalist huntsman who lost daily communication with wilderness was inconsequential. Desperately he craved the tonic of the Adirondacks or Maine, where his mourning could be salved and his nerves steadied. Only fresh, unobstructed air would clear his head. Thoreau had left Concord and headed to Maine's Mount Katahdin for a rebirth of vitality. To escape the institutional dourness of Harvard, Roosevelt was looking for his own Katahdin to climb. Even though he didn't consider himself a Transcendentalist, he was thirsting for nature trails and the wide-open wilderness. And he found it in September 1878 near where Thoreau did—in the North Woods of Maine.[51]

As a faunal naturalist, Roosevelt was revolting against the uninspiring idea that Moosehead Lake or Mount Katahdin could best be understood by specialists studying pine cones under a microscope. Decades later, in *Outdoor Pastimes of an American Hunter*, he praised Thoreau for writing *The Maine Woods* but downgraded him for being "slightly anaemic" where hunting was concerned.[52] Essentially, Roosevelt saw himself as a bridge figure between Darwinian naturalists and old-time explorers and big game hunters. A conduit between the often clashing fields of biology and the humanities. Not that he was going to bail out of Harvard as Minot had done. To the contrary. His rebellion was of the inner kind. Meanwhile, T.R. had inherited $125,000 (a fortune back then) from his father. Invested wisely, it would yield $8,000 a year, money he began spending on books, wilderness guides, and expeditions.[53]

On campus Roosevelt started cultivating a lasting reputation for extracurricular achievements such as rowing and boxing. He didn't turn his back on the scientific establishment completely, for he wrote and delivered papers before the Harvard Natural History Society about "Coloration of Birds" and "The Gills of Crustaceans." Whenever possible, he interacted with the Nuttall Ornithological Club, which was just starting to champion "citizen bird." To aged club members, however, Roosevelt seemed a little too self-confident and "cocksure."[54]

IV

At last Roosevelt had the wilderness experience he was craving. The place was Island Falls and it was tucked away in the upper reaches of Maine along the shore of Lake Mattawamkeag, an arduous ninety miles north of Bangor. Ever since Arthur Cutler spent a few weeks there with T.R.'s cousins Emlen and James West during the summer of 1876, the region's vastness, as Cutler reported back to his pupil, had beckoned him. The North Woods of Maine had dramatic storms that rolled in from the Atlantic, crisp air, fast-moving rivers, speckled brook trout, white-tailed deer, herds of caribou, cascading waterfalls, and much more.[55] Here an aspiring naturalist could find moose with huge antlers, beaver kits, and flocks of Canada geese. Here was where a real American hunter could test his mettle in the chill of the new morning. "I was not a boy of any natural prowess and for that very reason," Roosevelt wrote, "the vigorous outdoor life was just what I needed."[56]

As Roosevelt eagerly anticipated hiking in the North Woods and canoeing on Lake Mattawamkeag with his West cousins, Cutler arranged for the guide from the trip two years earlier. Will Sewall was a native of Island Falls, whose mother and sister ran a lodge while he tramped all over Penobscot and Aroostook counties. Standing six feet tall, with kindly, sparkling eyes that said, "I've seen it all, boys," the gravel-voiced Sewall prided himself on being a stoic and on being early to bed and early to rise. Difficult to startle, wise to the ways of the North Woods, he could forage off the forest by sapping maple trees or eating wild berries.[57] Roosevelt immediately enjoyed being in his firm grip.

But there was more than backwoods hardiness to Sewall—he was a virtuous Bible-reading Yankee—who, much like Theodore Sr., frowned upon swearing, drinking, and fornicating. He knew Norse mythology and English literature as well, and he had a penchant for romantic novels like *Ivanhoe* and the rhyming verse of Longfellow, a fellow Mainer. Sometimes on a long hike he recited John Keats's "The Eve of St. Agnes" aloud. It begins with the lines "St. Agnes' Eve—Ah, bitter chill it was! The owl, for all his feathers, was a-cold; The hare limp'd trembling through the frozen grass, And silent was the flock in woolly fold." Unschooled in Darwin, Sewall, as Roosevelt would soon learn, knew as much as Harvard scholars about Nordic figures like the one-eyed god Odin and the hammer-wielding mighty Thor. To Roosevelt, that Sewall romanticized Vikings as seafaring pagans wielding battleaxes was just another component of his charm as a storyteller.

So when Roosevelt headed north from Boston on the Maine Central (accompanied by his cousins Emlen and James West) past the sawmill centers of downstate Maine he was terribly excited both to explore the virgin sprucelands and to meet Sewall. Roosevelt had spent most of the summer of 1878 in Oyster Bay, playing the ornithologist and the swell. Now, for three weeks in September, just before having to start his junior year at Harvard, he was going to "get lost" in the wilderness. Meanwhile, Sewall's job was to make sure he really didn't.

Theodore and his cousins arrived at Sewall's lodge in the first week of September 1878. Island Falls seemed more like a frontier outpost surrounded by howling wilderness than a town. It was scenic in the extreme: the largest peaks of Maine could be seen from its practically nonexistent town center. This was the great wide-open land. Even as late as 1878, glory-hound outdoorsmen could walk in practically any direction and name a creek or mountain ridge after themselves. The entire area was proof positive that you didn't need to light out for the western territories to play Jim Bridger or John C. Frémont. Upper Maine was still raw and pristine and inhabited by Native American tribes who knew where the wild grapes grew.

After a round of greetings, the Roosevelt party was ready to set out into the surrounding wilderness. To help out on the trail, the bearded Sewall asked his clean-shaven nephew Wilmot Dow to join them. Sewall was the brains, and Dow had the sheer muscle. "Wilmot," Roosevelt later wrote, "was from every standpoint one of the best men I ever knew."[58] Both guides were the soul of fidelity and honor. "Theodore was about eighteen when he first came to Maine," Sewall later recalled. "He had an idea that he was going to be a naturalist and used to carry with him a little bottle of arsenic and go around picking up bugs."[59]

They began by canoeing down the Mattawamkeag River until they got to the lake. Then it was another seven miles until they set up camp. As Cutler had expected, Sewall was a bracing antidote to Theodore's Harvard blues; Roosevelt relished every minute of the outdoors strain. Nobody in the Agassiz School of Zoology, of course, would have thought of Sewall as a naturalist, but Roosevelt did. Bestowing that title on Sewall was part of his rebellion against the laboratory. Roosevelt was immediately impressed by Sewall. "I was accepted as part of the household; and the family and friends represented in their lives the kind of Americanism—

self-respecting, duty-performing, life-enjoying," he wrote, "which is the most valuable possession that one generation can hand the next."

During their eighteen days in the woods (September 7 to 26), Roosevelt and Sewall shot grouse, flushed out bats, bathed in the river, read scripture, and doused lanterns to sleep under the stars. Instead of inventorying the behavior of birds, Roosevelt seemed bent on assessing his own survivalist abilities. Sadly, Roosevelt felt that he was falling short of the expectations he had set for himself. As if hiking 110 miles with Sewall and Dow weren't enough, Roosevelt bemoaned how woefully inadequate he found himself as a marksman. "I don't think I ever made as many consecutive bad shots as I have this week," he wrote, in a way all true hunters could sympathize with. "I'm disgusted with myself."[60]

Such bouts of self-criticism aside, Roosevelt excelled during these September days, and he was entranced by the beauty of the North Woods. Although Sewall thought Roosevelt a "different fellow to guide from what I had ever seen," he marveled at the way Theodore was "posted" about the politics and literature of the times.[61] Even though Theodore was struggling with asthmatic attacks—"guffle-ing" was the way Sewall put it—he admired how the young man never complained, never lagged behind, and never asked for sympathy. There wasn't anything about Theodore, in fact, that Sewall or Dow didn't like. "Some folks said that he was headstrong and aggressive," Sewall later wrote, "but I never found him so except when necessary; and I've always thought being headstrong and aggressive, on occasion, was a pretty good thing."[62]

V

When Roosevelt arrived for classes at Harvard at the beginning of his junior year, all he could talk about was the North Woods of Maine: the looming mountains, the hemlocks and elms, the mulberries, the thick pines, the birches and wildflowers. He was now a self-styled outdoorsman. "He would come back with tales of exposure and hardship," his classmate Charles G. Washburn recalled, "and, it seemed to us, which he had enjoyed."[63]

To the walls at 16 Winthrop, Roosevelt added hard-earned trophies of Island Falls—a raccoon skin and stuffed ducks. A photograph taken of his room shows it as fairly neat and tidy, with the mounted birds placed under bell jars. Whether Maine was the inspiration or not, his grades went up during his junior year; they included near-perfect scores in zoology and political economy. His natural history library grew, and he treasured a personally inscribed copy of Coues's *Birds of the Colorado Valley*.[64] Following the lead of his uncle Robert B. Roosevelt, he joined several clubs at Harvard, with Porcellian as his primary focus. His activities as a clubman included serving as vice president of the Natural History Society; he quit this organization during his senior year, however, in an effort to downsize social life at Harvard in favor of time spent with his new love.

That autumn Roosevelt had started dating a conservative Boston girl, Alice Lee, whose parents lived on Chestnut Hill. A beautiful seventeen-year-old, Alice was gregarious and had impeccable manners. Theodore fell for this blond, who was nicknamed "Sunshine." With compassionate eyes, a tennis player's physique, and a flirtatious giggle, Alice was the true first love of Theodore's life. Although there is no evidence to suggest that she shared his enthusiasm for birds, they wandered through meadows together, collected chestnuts, and admired wisteria. Alice, however, had many suitors, and Theodore had to pour on his charm and his gentlemanly ways.

By March 1879 Roosevelt was ready for another adventure in Maine with Sewall and Dow. While other classmates headed south in search of bright weather, Roosevelt took the train north to Island Falls, arriving to find three feet of snow on the ground. Sewall picked Roosevelt up by sleigh at the depot and, with bells jingling, took him to his cabin. For the first time, Roosevelt wore clumsy-looking snowshoes. They were Indian-made with rims of white ash, closed lacing, and the highest-quality raw-

Theodore Roosevelt vacationed regularly in the North Woods of Maine as a Harvard student. In March 1879 he went snowshoeing in the wild. Here (left to right) are Bill Sewall, Wilmot Dow, and Theodore Roosevelt.

hide available. Three times longer than they were wide, they enabled Roosevelt to easily get traction on icy surfaces. Temperatures outside had dropped to ten degrees below zero. Ice coated all the bushes and every trace of road. Huge ten-foot snowdrifts buried low-lying cabins like coffins. At times the wind was so sharp with snow that it froze lips shut. Most sane people were huddled indoors by a fireplace, with plenty of lynx furs and wool blankets, but Roosevelt traipsed around the wintry woods in great spirits.

There was nothing passive about the way Roosevelt listened to lumberjack slang and absorbed North Woods tall tales of the Paul Bunyan variety. "Even then," Bill Sewall recalled, "he was quick to find the real man in the very simple men."[65] Sewall, for example, had two brothers whom T.R. got to know. "Sam was a deacon," Roosevelt recalled of the pair. "Dave was NOT a deacon. It was from Dave that I heard an expression which ever after remained in my mind." Apparently one afternoon Dave Sewall was "speaking of a local personage of shifty character who was very adroit in using fair-sounding words which completely nullified the meaning of other fair-sounding words which proceeded them." Finding such mealy-mouthedness disingenuous, Dave fired off the backwoods insult that they were "weasel words." As Roosevelt recounted, Dave Sewall said, "just like a weasel when he sucks the meat out of an egg and leaves nothing but the shell." Roosevelt always remembered "weasel words" as applicable to disingenuous oratory delivered by sham publications.[66] Later in life, during his Bull Moose years, Roosevelt used the term "weasel words" in both an article in *Outlook* and a major Missouri address.[67]

There was another aspect of Sewall's life that was similar to Roosevelt's—childhood sickness. Although he didn't have asthma, he had been afflicted with hyperthermia, diphtheria, and hearing difficulties. Growing up fragile in such unforgiving country wasn't an option. Sewall either had to get in shape or die. Much like Roosevelt, he began a successful fitness regime and became an inspiration for the philosophy of mind over matter. Treating lumbering as a sport, he practiced day and night with his ax, priding himself at being able to chop a sycamore down faster than anybody else in Aroostook County. As if performing for Buffalo Bill's show, he could toss an ax in the air, catch it, and then split a pine log, all in one motion. Starting at age sixteen, he became a boss man for lumber drives, overseeing pine logs floating down the Mattawamkeag River every harvest season, the months of April to July.

Back at Harvard that spring, a self-confident Theodore acted a bit like a roughneck weaned on frozen rivers. The gloves had come off. Bile stirred in his stomach when he thought about all his anti-outdoors naturalist professors. The wilderness had been so intoxicating. One day he was in Cambridge, feeling hemmed in; the next day he was gazing at an ice-sheeted blue lake looking for moose. The North Woods had toughened him, or so he believed. The convivial ways of Sewall and Dow also lingered in him. No longer was he content behaving like an adolescent rajah collecting bird feathers and speckled eggs. Primitive Maine had knocked some of the tameness out of him, significantly. He now considered his asthma an ugly by-product of civilization's stresses. Reading the Old Testament in Maine by a roaring fire, Roosevelt felt like an American Adam, uncontaminated by the corruptions of Tammany Hall politics or Harvard's pecking orders. Once again Island Falls—even in a blind frenzy of snow—had redeemed his despair and fixed his determination to marry Alice Lee. "I never have passed a pleasanter two weeks than those just gone by," Roosevelt wrote to his mother. "I enjoyed every moment. The first two or three days I had asthma but, funnily enough, this left me entirely as soon as I went into camp. The thermometer was below zero pretty often, but I was not bothered by the cold at all."[68]

Just as bathing in Walden Pond had been a baptism for Thoreau, Roosevelt now felt the cleansing effect of the horizontally blowing snow in the Mooseleuk Range. It was as if Roosevelt's worries about asthma had been stolen away by the blanket of icy whiteness. "I have never seen a grander or more beautiful sight than the northern woods in winter," he wrote. "The evergreens laden with snow make the most beautiful contrast of green and white, and when it freezes after a rain all the trees look as though they were made of crystal. The snow under foot being about three feet deep, and drifting to twice that depth in places, completely changes the aspect of things."[69]

VI

No sooner had the spring semester at Harvard begun than Roosevelt started plotting for a return visit to Maine. The spell of the North Woods had fallen over him. Some of his classmates, however, snickered that all Roosevelt learned in Maine was the art of manly bragging. Even though he would spend June and July in Oyster Bay, tending to family obligations, come August he was going to climb the peak that Thoreau had written about, the 5,268-foot Mount Katahdin. Even with Sewall as principal guide, that would be quite a mountaineering feat for a first-timer. Make no mistake about it—this was a big mountain. When a wall of white clouds hovered in the sky, you couldn't even see the pinnacle from base camp. Mountain climbing, Roosevelt knew, was a far trickier endeavor than hiking in river valleys.

While guests at Oyster Bay were having tea and crumpets, Roosevelt was purchasing camping gear from Greenville Sanders & Sons. His private diaries list a flannel shirt, duck trousers, long underwear, a parka, wool socks, bandannas, a thick blanket, and a bag of "necessaries." The plan was to make camp at the eastern hinge of Mount Katahdin and climb the narrow ridge among the blackflies and ticks. He hoped plenty of bull moose would be encountered around ponds and lakes as they dined on aquatic plants—Roosevelt desperately wanted a moose's palmate antlers for his trophy collection, and he wanted to eat a moose steak. In *The Maine Woods* Thoreau maintained that moose meat was "like tender beef with perhaps more flavor, sometimes like veal."[70] Because Roosevelt had studied the characteristics of bull moose carefully, he understood both their rutting habits and their boreal–mixed deciduous forest habitat.

Roosevelt relished the fact that moose were the largest member of the deer family populating North America, standing six feet tall, with the males weighing up to 1,300 pounds. (A bull bison, however, can be larger, weighing up to a ton.) With their poor eyesight, moose rely on their excellent hearing to escape predators at night. Certainly Roosevelt would have identified with the way the moose turned this deficit into an asset. Their sense of smell was so heightened that a favorite saying of Maine moose hunters was "Keep the wind in your face and the sun at your back" so as not to be detected. Able to run thirty-five miles an hour, neck craning with every sudden noise, the docile-looking bull moose became astonishingly aggressive when feeling threatened. And if a hunter killed a bull moose—which usually took more than a single bullet—two or three grown men were needed to carry the carcass off. Roosevelt was also impressed with the raw strength of the bull moose.

VII

Unfortunately, there is no photograph of Roosevelt on August 23, 1879, when he arrived in Island Falls full of vim and vigor. He was carrying a forty-five-pound pack on his back and held both a shotgun and a rifle in his hands. There was no telling what oddments were inside the pack. You could have recognized him as a greenhorn from as far away as the North Pole. Along with Emlen West and Arthur Cutler, he headed over to the Sewall cabin to plan their climb. Staring at Mount Katahdin, pointing at it with his rifle, Roosevelt spontaneously decided to canoe more than twenty miles to Lake Mattawamkeag (with Emlen) just to loosen up for their big outing.

For eight days the Roosevelt-Sewall party camped and hiked in the Mount Katahdin area. (Today it's part of Baxter State Park.) A huge mound looming out over a sea of woodlands, Katahdin had a soothing Japanese Zen-like aura. Just staring at the peak made you want to write a haiku. On a clear day from the Katahdin summit, Canada poured open on both sides of the mountain. You could see New Brunswick to the east and Quebec to the west. But that serenity was misleading. One misjudged step on the way up Katahdin could mean a broken neck or death.

About three-quarters of the way up Katahdin, both Cutler and Emlen collapsed from fatigue. Roosevelt soldiered on; he bragged in his diary that he had learned to "endure fatigue" as stoically as any lumberjack, even though his throat was burning and his joints were aching. At one juncture, he claimed he was "fagged" but was not stricken with asthma. His mind, however, was on fire. Looking toward the west, Roosevelt probably could see Moosehead Lake, where "the humiliation" had taken place. He'd come a long way since that hazing. As Hudson Stuck, a member of the first team to climb Mount McKinley in Alaska explained, first-time mountaineers like Roosevelt were embarked on a "privileged communion" with the "high places of earth."[71] And, indeed, as he was ascending Thoreau's peak, Roosevelt's heart filled with pure joy; he was double-charged with life, wearing everybody else out.

Once they all returned to Island Falls, Cutler and Emlen decided to call it quits. But Roosevelt wanted another wilderness adventure *now*, while the adrenaline was still coursing through his body like a river of fire. Full of spark and fizz, he said good-bye to his tutor and cousin and turned his sights to the caribou country around the Munsungun Lake region. The idea was for Roosevelt and Sewall to take a pirogue up the Aroostook River on a hunting jag in search of moose and caribou. They loaded up on hardtack, pork jerky, and flour, gearing up for a grueling nine or ten days combating rocks and rapids.

Refusing to take shortcuts, they forded rivers, slipped on stones in fast-moving streams, shot doves for dinner, and hiked until they dropped in clammy exhaustion. With each effort Roosevelt grew merrier. Tellingly, Cutler later wrote to Sewall that the next time T.R. came to Maine for his "semi-annual visit," it would be easier to tether a "tame moose" for him to shoot instead of enduring an endless series of obstacle tests around the Munsungun Lake region.

Book Two of Yagyu Munenori's *The Life Giving Sword* (part of his samurai meditation on martial arts) gives some insight on Roosevelt's euphoria during this sojourn. Writing in the seventeenth century, Munenori explains a state of mind he calls "total removal," a swooping moment when sickness of the mind disappears.[72] Conquering Katahdin was the culmination of something Roosevelt's father had told him: that the body and mind needed to run in tandem, that he had to be whole to succeed. As Roosevelt was going through "total removal," one can only wonder what Bill Sewall thought when his eager client shouted "Bully!" or "By Jove!" every time the dugout flooded or a thunderstorm drenched them to the bone. But when they parted that September, Sewall promised to keep an eye on the Cambridge-bound Roosevelt from the North Woods. Shared experience, after all, is the cement of all friendships. They had forged an alliance that had all the power of a blood bond.

When Roosevelt left the depot at Kingman, Maine, on September 24, he didn't know he was saying good-bye to the North Woods forever (the next summer he would visit the well-heeled Maine coast). Over the previous year, he had spent sixty-nine days with Sewall and traveled more than 1,000 miles of rugged backcountry by wagon, canoe, pirogue, and foot. Everywhere they went was as serene as a forgotten battlefield. Never again would Roosevelt write about debilitating asthma attacks or the disease of puniness. In "Jabberwocky" (in *Alice through the Looking-Glass* and then in *The Hunting of the Shark*), Lewis Carroll used a word he coined: "galumphing," meaning, roughly, galloping triumphantly or marching exultantly with "irregular bounding movements." No word devised before or since then has better described Roosevelt when he conquered Mount Katahdin.[73] "As usual it rained," Roosevelt noted, "but I am enjoying myself exceedingly, am in superb health and as tough as a pine knot." [74]

The fact that Roosevelt left Maine's North Woods didn't mean that the North Woods left him. Ardor for the state stayed with him, as permanent as a ring in a redwood tree. Later in life, Roosevelt adopted the bull moose as his political symbol. He even dubbed his Progressive Party of 1912 the Bull Moose Party. Gleefully, Roosevelt, taking a cue from Dave Sewall, constantly accused his opponents of taking every opportunity to use "weasel words." While serving as president, Roosevelt named his Blue Ridge Mountains retreat—a secluded cabin just outside Charlottesville, Virginia—"Pine Knot." Bill Sewall's cabin, where Roosevelt used to lodge, became the first official historic site in Island Falls; eventually, even the lean-to hunting camp on Mattawamkeag Lake was preserved. Meanwhile, the spot along the Mattawamkeag River, where it's believed

Roosevelt read the Bible on Sundays, now has a historic marker in the ground, detailing Roosevelt's Maine adventures in the late 1870s.

And Roosevelt wasn't done with the straightforwardness of either Sewall or Dow. Because they "hitched well," as Sewall put it, Roosevelt summoned them to the Dakota Territory to operate a cattle business in 1884. But that was six years away. First he had to graduate from Harvard and marry Alice Lee—those were his two primary objectives.

Near the end of his life, Roosevelt presented his wilderness days in the North Woods as the apogee of his happiness during adolescence. He became the adventitious expert on Maine. In a nostalgic essay, "My Debt to Maine," published as part of a volume celebrating the Pine Tree State's centennial in 1919, Colonel Roosevelt (as his byline read in *Maine, My State*) noted that camping out in the North Woods had been transformative for him. His memories ran deep: the soft-needled branches; collecting kindling; shoveling snow; building a rainproof bonfire; stirring up embers with a walking stick; the smell of drifting woodsmoke; the cry of a hawk, loud and clear; roasting grouse, venison, or trout on a spit; concocting new dishes like muskrat and fish-duck (merganser) stew; and ladling out Boston baked beans for breakfast. But more than anything else, those shrill high-wind whistles, all those aromas of pine and hemlock and spruce, never faded away.

Maine, to Roosevelt, was where he first found his authentic self. Men acquired the skills of survival early, in such a rugged terrain, if they wanted to succeed in life. In the North Woods, unlike official Washington or Harvard, there was no gyp game. Nor was Maine benevolent or controllable. Death by avalanche, death by frostbite, death by becoming lost—these were hazards men like Sewall and Dow had learned to overcome. The entire state had a stimulating effect on Roosevelt, almost as if it made him drunk. "I owe a personal debt to Maine because of my association with certain staunch friends in Aroostook County," he wrote, "an association that helped and benefited me throughout my life in more ways than one."[75]

CHAPTER FIVE

MIDWEST TRAMPING AND THE CONQUERING OF THE MATTERHORN

I

The Harvard Athletic Association was sponsoring its spring boxing competition and twenty-year-old Theodore Roosevelt had entered in the lightweight division. The rounds were all held on campus, at the old gymnasium near Memorial Hall. Although once denounced as immoral because of its brutality, boxing in 1879 had become chic, particularly in the Boston area, where the Irish-American John L. Sullivan had brazenly challenged anyone with enough guts to fight him for a $500 wager. Even the *New York Times* and *Harper's Weekly* were pro-boxing, asserting that the sport enhanced physical fitness and manliness. Sullivan was nicknamed the Boston Strongboy; Roosevelt deserved a more Ivy League moniker like the Cambridge Clerk or the Harvard Horticulturist. Even though Roosevelt had been training for months, nobody imagined he'd actually be in a twenty-four-foot ring fighting for a college championship trophy.

That was precisely what happened on March 22, 1879. Besides undergoing intensive training Roosevelt had carefully studied the official Queensberry rules, determined not to lose by default owing to a technical infraction or an illegal maneuver. Grueling as it sounds, the Harvard Athletic Association organized the competition by a process of elimination. You boxed not just one match but two or three in the same day. With the gymnasium packed, his friends and classmates cheering him on, the 130-pound Roosevelt, to the shock of most present, with a couple of good right punches, actually beat a senior, W. W. Coolidge of the class of 1879, in his first square-off. It was considered something of an upset. According to the *Harvard Advocate*, Roosevelt "displayed more coolness and skill than his opponent." Meanwhile, C. S. Hanks of the class of 1879 had defeated his opponent in his semifinal round. Therefore, the championship fight that afternoon would be Roosevelt versus Hanks.[1]

Sitting on a floor seat watching the fisticuffs was Owen Wister, an aristocratic Pennsylvanian who would go on to write the classic western novel *The Virginian.* Two years behind Roosevelt at Harvard, Wister was something of a class clown, famously contributing both the music and the libretto for the Hasty Pudding Club's comic opera *Dido and Aeneas.* As a product of boarding schools in New England and Switzerland, Wister had become extremely erudite by the time he arrived in Cambridge. Just as Roosevelt was an accomplished ornithologist of sorts upon entering Harvard, the easy-tempered Wister had composed songs he thought rivaled the worst of Stephen Foster, which was at least a starting place in show biz. Like Roosevelt, Wister suffered from bad health. Throughout his life he had nervous breakdowns, migraine headaches, sudden tremors, and even prolonged hallucinations; and—again as with Roosevelt—only Mother Nature, it seemed, brought him relief from his physical anguish.[2]

A shrewd judge of character, Wister studied Roosevelt with puzzlement that afternoon in the old gymnasium, figuring he was going to get his block knocked off by Hanks. As a freshman Wister knew Roosevelt only by reputation but was pulling for him as the well-muscled underdog of the bout. According to the *Advocate*, it was a "spirited contest," but Hanks got the "best of his opponent" by his impressive "quickness and power of endurance." Yet something occurred that afternoon that Wister never forgot, and years later he showcased it as the prized anecdote of his memoir *Roosevelt: The Story of a Friendship.* According to Wister, amazingly, near defeat, Roosevelt, by virtue of his good sportsmanship, became the real winner of the Harvard bout. As Wister put it, the packed crowd witnessed that "prophetic flash of the Roosevelt that was to come."

At one point during the bout, in accordance with the Queensberry rules, the referee called time-out, thereby ending a round. However, in the frenzy of the fight, Hanks didn't hear the referee's intervention, and just as Roosevelt relinquished his guard, Hanks smashed him in the face. Blood spurted everywhere. The audience gasped; boos and hisses filled the gym. What a cheap shot! But Roosevelt held a boxing glove up in a theatrical gesture, demanding silence from the crowd. "It's all right," he reassured the crowd. "He didn't hear him."[3]

As Wister recounted in his memoir, he watched mesmerized as the junior walked over to Hanks with extended hand, simply refusing to be victimized. (Some scholars, however, doubt the veracity of Wister's story, feeling that the novelist had probably confabulated the bloody-champ aspect of the spectacle.[4]) With his fine eye for nuance, Wister noticed that Roosevelt's conciliatory gesture combined dash and spirit. "He was his own limelight, and could not help it," Wister surmised; "a creature charged with such voltage as his, became the central presence at once, whether he stepped on a platform or entered a room."[5] One can only imagine how proud Alice Lee must have felt learning that her Theodore won over the crowd's heart by losing the boxing match with such dignity. (Wister intimated that Lee was among the spectators, but it seems unlikely.)

Although Roosevelt kept collecting a multitude of birds, as 1879 turned to 1880 he toyed with the idea of a career in politics for the first time. Business or law brought home income; ornithology clearly didn't. Also, he was itching to be a public servant for the state of New York; politics ran deep in the family gene pool. Upon his engagement to Alice Lee in February, in fact, he wrote Minot a very telling letter about his future plans. "I have made everything subordinate to winning her," he wrote, "so you can perhaps understand a change in my ideas as regards to science." Roosevelt's main goal in life, as he put it, was to "keep up" the family name.[6] He and Alice would marry in October. "Natural History was to remain a genuine avocation," his biographer Carleton Putnam rightly noted in *Theodore Roosevelt: The Formative Years*, "but it never loomed again as a feasible career."[7]

By Roosevelt's senior year at Harvard his classmates respected him for more than just losing boxing bouts and misplacing turtles at 16 Winthrop with a patrician air. Everywhere he walked on campus (or took his dogcart, pulled by his favorite horse, Lightfoot) he was met with "uproarious cordiality." The combination of having his own horse plus his doggedness and vitality had made Roosevelt legendary by his junior and senior years. Everybody had to admit that Roosevelt was sui generis, that he wore no man's collar. Although President Eliot later scoffed that Roosevelt took "soft courses" during his last years, keeping a "very light schedule," he nevertheless received A's and B's. Reading over his journal for the senior year 1879–1880 makes it clear that Roosevelt's worst problem was insomnia. Unable to turn off his mind, he'd spend "night after night" walking by himself in the Cambridge woods near Fresh Pond, sometimes never catching even a couple of hours of shut-eye.[8] It appears that Roosevelt was afflicted with some kind of mania.

In Kay Redfield Jamison's 2004 book about manic depression, *Exuberance*, Roosevelt is exhibit A for this condition. His set of symptoms—propulsive behavior, deep grief, chronic insomnia, and an all-around hyperactive disposition—demonstrate both the manic and the depressive phases of bipolar disorder. Too often, Dr. Jamison argued, people mistakenly thought manic depression meant despondence and withdrawal from human endeavors. Usually it does. But those afflicted with exuberance, she argued, go in the opposite direction; behaving as relentless human blowtorches they're unable to turn down their own flame. Diagnosing Roosevelt's medical condition more than eighty years after his death, Jamison claimed that the highs of the exuberance phase brought many wonderful gifts; but, she warned, there was also a sharp-edged downside. Living by throwing up skyrockets—as P. T. Barnum once put it—wore one down to nothing. No sleep, for example, wasn't good for the heart or other vital organs. Only by exhausting oneself in physical activity—like climbing Mount Katahdin or ice-skating on the Charles River in a winter storm—could an exuberant manic like Roosevelt turn himself off.[9]

Essentially, Roosevelt's exuberance syndrome was both a source of power and a sometimes curse. The poet Robert Lowell once described manics like T.R. as harboring "pathological enthusiasm";[10] Jamison tended to agree. What Jamison admired about Roosevelt, however, was that he channeled his manic-depressive energies in constructive ways, taking what could have been a terrible handicap and using it as an asset. A friend of Roosevelt once colorfully explained T.R.'s ceaseless zest as the "unpacking of endless Christmas stockings," a description Roosevelt wouldn't have minded.[11] Constantly calling life "The Great Adventure," Roosevelt derived "literally delirious joy" from Christmas, never wanting the holiday season to end. The more candles lit and carols sung, the happier he was.[12]

Unfortunately, even though Roosevelt felt fit as a fiddle operating on intermittent sleep, his exultancy was taking a physical toll. On March 26, 1880, Roosevelt went to see Dr. Dudley A. Sargeant, the university physician. On the preexamination form, Roosevelt noted that asthma had bothered him since childhood. As of late, however, he was feeling well and expected a clean bill of health. After thoroughly examining Roosevelt from head to toe, however, Dr. Sargeant pulled his patient aside with troublesome news. There was something wrong with Roosevelt's heart—it was terribly weak. If he kept exerting himself, he would die young. Sternly Dr. Sargeant told Roosevelt to cease all activity that would make his heart rate go up. Mountaineering, twenty-mile hikes, and even climbing staircases would have to stop. All exertion was unhealthy.

Such a bleak diagnosis didn't go over well with Roosevelt. He didn't want to live gently. If he had only a few seasons left to breathe, so be it. Instead of pampering himself or living like a baby he would fight with both fists against the tide of gloomy fatalism. Going back to being a weakling, a runt in the litter of life, was unacceptable to him. Dismissing Dr. Sargeant's verdict out of hand, Roosevelt started planning a six-week hunting trip with his younger brother, Elliott, to the Midwest heartland. They would start from Chicago, go northwest to Iowa, and eventually wind up on the western edge of Minnesota. Inspired by an earful of Elliott's Texas bird-hunting triumphs in Galveston and the Big Thicket, eager to learn more about grouse and prairie chickens, Theodore read Coues's pioneering *Birds of the Colorado* in preparation.

Later in life, as a politician, Roosevelt downplayed his Harvard years, recognizing them as a liability in a country more impressed with log cabins and cowboy mythology. Furthermore, only one in every 5,000 Americans graduated from college in 1880; merely receiving a diploma meant that one was part of an elite.[13] Roosevelt's diaries, however, show that he was elated at finishing twenty-first in a class of 177.[14] And the Phi Beta Kappa graduate had even managed to publish two ornithology chapbooks, a thesis on women's rights, and wrote chapters of *The Naval War of 1812*. "I have certainly lived like a prince for my last two years in college," he recorded in his diary. "I have had just as much money as I could spend; belonged to the Porcellian Club; have had some capital hunting trips; my life has been varied; I have kept a good horse and cart; I have had half a dozen good and true friends in college, and several very pleasant families outside; a lovely home; I have had but little work, only enough to give me an occupation; and to crown all infinitely above everything else put together—I have won the sweetest girl for my wife. No man ever had so pleasant a college course."[15]

After dutifully following all the rituals on commencement day—spending time, for example, with Alice Lee's family in Chestnut Hill—Roosevelt packed up the contents of Winthrop Street and shipped them to the Fifty-Seventh Street house in New York. His plan was to first spend a few weeks in Oyster Bay and then head to the blue-green Maine coast to sun, sail, and explore. Instead of climbing Mount Katahdin, he would enjoy the tumble of the surf on Mount Desert Island, the second-largest island on the Eastern Seaboard. Surrounding the main island were numerous tiny shore islands, each with marine enchantment all its own. After some days with friends Alice would join him.

Partly because Roosevelt was writing *The Naval War of 1812*—an act of genuine hubris for a twenty-two-year-old—he wanted proximity to the boundless Atlantic Ocean to study the cold, buffeting waters around Mount Desert Island.* Two of Roosevelt's favorite painters of the Hudson

*Much of Mount Desert Island became preserved. It was originally established as Sieur de Monts National Monument in 1916, then as Lafayette National Park in 1919, and was renamed Acadia National Park in 1929.

River school, Frederick Church and Thomas Cole, had summered around the Maine island in the 1850s. Recognizing the island as one of God's great galleries, Church had both painted and sketched landscapes of the indented shoreline with great skill. Two of his faithful renderings—*Fog of Mount Desert* and *Newport Mountain*—were beloved by locals for generations. As for Cole, he sketched sixteen natural sites on Mount Desert Island ranging from bold headlands to strands of northern white cedar, red spruce, and black spruce. A particular emphasis was given to blasted pine standing alone on rocky cliffs and to the gorgeous islets that dot Frenchman Bay.[16]

Roosevelt went to Mount Desert Island with his friends Dick Saltonstall and Jack Tebbetts. They lodged four miles from Bar Harbor near Schooner Head, a huge jagged rock very near the Atlantic Ocean. The trio called the bungalow where they slept "bachelors' hall." Right outside their door was the stony beach where crab skeletons, seaweed tangles, and broken shell bits were washed ashore. An immediate favorite locale for Roosevelt was Cadillac Mountain, at 1,530 feet the highest point along the North Atlantic seaboard, and the first place in the United States where you could see the sunrise. Roosevelt also rode horseback over stone bridges and hunted for sea urchins among the shell heaps on Fernald's Point. He was lulled by the murmuring ocean; he picked baskets of cranberries, collected shellfish in the tidal marsh, and gathered wild berries; and when Alice, unchaperoned, arrived, strolled "with my darling in the woods and on the rocky shores."

To an ornithologist the sheer diversity of the marine environment of Mount Desert Island offered merriment. Seabirds such as jaegers, shearwaters, puffins, and razorbills were everywhere, prancing around in the surf then flying away when the shades of twilight fell under the full onset of the sea. At Thunder Hole, Theodore and Alice sat entranced as seawater waves rushed in and out of a perfectly formed cave while debonair black skimmers circled above. Soon, however, Roosevelt was sick again, this time stricken with cholera morbus. Dehydration, diarrhea, and body flux ensued. There were no pills or port or morphine to make him feel better. Not only was he unable to show off for Alice but, as he wrote to his sister Corinne on July 24, the infectious gastroenteritis was "very embarrassing for a lover . . . unromantic . . . suggestive of too much ripe fruit."[17] Refusing to be an invalid, Roosevelt decided to climb Newport Mountain—with Bar Harbor at its base—as a quick cure while he was recuperating. Onward and upward he went for more than 1,000 feet, peering down at the little harbor skiffs, which looked like bathtub toys from that crow's nest vantage point. Given that he was ill, Roosevelt's mountaineering feat at Newport can be attributed only to sheer will—a will ever growing, ever persistent in overcoming obstacles.

II

There was another reason, however, that Roosevelt didn't collect birds on Mount Desert Island—his mind was reeling over his coming Midwest hunting trip with his brother, Elliott. Together they were going to explore the broad expanse of the Great Plains. Ever since Dresden, Theodore had been struggling to keep pace with Elliott, the most troubled of the four Roosevelt children. Handsome, irreverent, and charming, Elliott—a tenderhearted bon vivant—constantly fought against fatigue, dizzy spells, and bouts of depression. He gave no meaningful signs of professional ambition; he simply excused himself from serious work, preferring pleasure. But he was very sweet-natured. According to a well-circulated family story, Elliott, when he was seven years old, took a walk one winter morning only to return without his overcoat. On being interrogated by his parents Elliott explained that he had seen a homeless "street urchin" shivering, so naturally he gave the poor lad his own coat. "I can think of many occasions in his later life when generosity of the same kind actuated him, not, perhaps, to wise giving, for unlike some people he never could learn to control his heart by his head," his daughter Eleanor Roosevelt, first lady from 1933 to 1945, recalled. "With him the heart always dominated."[18]

Theodore Roosevelt and his brother, Elliott, with a big game hunting coach.

A better all-around student than Theodore, Elliott (or "Nell" as the family affectionately called him) was also a tremendous hunter and equestrian who excelled at polo.[19] By 1880 Elliott had already hunted wild turkey in Florida and Bengal tigers in India. "Everything is in an advanced state in Texas," he had written to his father from Fort McKavett, where he was bagging around a dozen birds daily. "By everything I mean all fruits, flowers and vegetables and by Texas I mean the civilized portions thereof."[20] A crack shot and excellent rider, sly as a magpie, Elliott was not overtly proper like his father; he had unfortunately inherited Uncle Rob's libertine ways and was attracted to the bottle.[21]

Leaving Maine on August 6, Theodore visited Alice at Chestnut Hill and then his family at Oyster Bay. In the middle of the month Elliott and Theodore boarded a Chicago-bound train from Manhattan, ready to roll across the prairies Francis Parkman had written about so dramatically in *The Oregon Trail: Sketches of Prairie and Rocky Mountain Life*.[22] Like Roosevelt, Parkman believed that actually visiting American landscapes was essential for gaining impressionistic reportorial material to help make a historical narrative come alive for the reader. It electrified Roosevelt that Parkman, a fellow Harvard alumnus (class of 1846), had used his classical education to honor the western frontier in serious historical prose. Roosevelt adopted Parkman—a devoted naturalist and horticulturist, with expertise in roses and lilies—as his guiding light in history studies. And Parkman knew the forests of America better than anybody else alive. To Roosevelt, Parkman, who also suffered from bad eyesight and was nearing blindness, was quite simply "the greatest historian whom the United States had yet produced."[23] Given a choice between *Walden* and *The Oregon Trail*, Roosevelt would have chosen the latter every time.

On the weekend before the Roosevelt brothers' "Midwest tramp," Theodore was pining for Alice in Chestnut Hill. Although he was quite excited about seeing Chicago and crossing the Mississippi River for the first time, he was already "frightfully homesick" for her. Still, there was a lot of packing and there were many good-byes to make for what he was calling his western trip. And once they were under way to Chicago, he was filled with excitement, behaving like an able-bodied seaman about to discover the world. "Traveled all day through the wooded hills of Pennsylvania and the rolling prairie of Ohio," Roosevelt excitedly recorded in his diary. "It is great fun to be off with old Nell; he and I can do about anything together; we never lose our temper under difficulties and always accomplish what we set about."[24]

Chicago in 1880 was the regional hub for the entire Midwest. All the major grain-producing states—Iowa, Illinois, Indiana, and Nebraska—used the city as their in-transit wholesale distribution center. If Roosevelt had looked at a railroad map of the trans-Allegheny West, in fact, he would have seen clearly that Chicago was the crossroads—or like a fist with all the ubiquitous track lines extending outward like fingers. The novelist Theodore Dreiser referred to late-nineteenth-century Chicago as the "magnet" city of the Midwest and West. Pioneers overlanding to California by covered wagon or railroad, or on foot, usually began their journey in Chicago. The city had been built on the bottom lip of Lake Michigan and rebuilt after the great fire of 1871. Freighters carried timber and iron ore south from Wisconsin and Minnesota to the city's cargo ports, warehouses, and railroad yards. Not only did railroads converge there, but the Illinois and Michigan Canal linked Lake Michigan to the Mississippi River, opening up the agricultural markets of the Midwest. In just four years the first skyscraper—the Home Insurance Building—would be erected on the corner of LaSalle and Adams streets. The city was expanding at an amazing clip.

The eye of Chicago seemed to be looking everywhere across America. As the historian William Cronon pointed out in his landmark study *Nature's Metropolis* (1991), Chicago in the late nineteenth century oversaw an economic domain that stretched from the Sierra Nevada to the Appalachians, and from Duluth, Minnesota, in the north to Cairo, Illinois, in the south. All the varied ecosystems of the Great Plains about which Roosevelt would later be so enthusiastic—the Sandhills of Nebraska and the Wichita Mountains of Oklahoma, the Great Basin of Nevada and the Badlands of the Dakotas—were linked to Chicago in one way or another.[25] Given his predilection for the outdoors, Theodore wasn't elated with Chicago; he itched to leave for the neighboring prairies. Having conquered the Adirondacks, the North Woods, and Mount Desert Island, he craved the fabled pastoral life of the Great Plains, which Washington Irving had written about in *A Tour on the Prairies* (1835).[26] As a fan of Zebulon Pike, he had a headful of frontier myths about discovering the headwaters of the Red and Arkansas rivers (he never quite found them on this Midwest tramp). Over the coming weeks, Theodore and Elliott would hunt grouse, with the euphoria of treasure hunters, in three separate locales: Huntley, Illinois; Carroll, Iowa; and Moorhead, Minnesota, which sat on the border of the Dakota Territories.

The Roosevelt boys' guide in Illinois was "a man named Wilcox"—his first name remains unknown—whom Theodore mentions only perfunctorily in the diaries he kept.* Clearly, to Theodore's mind, Wilcox was no Bill Sewall or Moses Sawyer. But in a letter of August 22 to his sister Anna—posted from the Wilcox farm in Illinois—T.R. did reflect on the hardworking midwesterners he was meeting. Huntley was a tiny village with only one paved street. From the town center there was plenty of cropland to be seen in every direction, but there were virtually no woods—only a few scattered trees. "The farm people are pretty rough but I like them very much," he wrote to Anna. "Like all rural Americans they are intensely independent; and indeed I don't wonder at their thinking us their equals, for we are dressed about as badly as mortals could be, with our cropped heads, unshaven faces, dirty gray shirts, still dirties [dirtier] yellow trousers and cowhide boots; moreover we can shoot as well as they can (or at least Elliott can) and can stand as much fatigue."[27]

In many respects, the Midwest tramp became a hunting competition. Which brother could bag the most game? Because Elliott, who was two years younger, had already worked up his competitive appetite by flushing out prairie chickens in scrub brush in Texas, he was the veteran. To Theodore, by contrast, it was all a new experience. He noted in his diary that it was "great fun to try this open plains shooting to which I am entirely unaccustomed among such vast, almost level fields, with so few trees."[28] After a couple of days in what Theodore referred to as the "fertile grain prairies,"[29] both brothers had bagged many kinds of game birds—doves, ducks, snipes, grouse, plovers. They also collected gophers, impressed by their curved claws, used for tunneling through loose soil. "We had three good days of shooting," he wrote to Anna, "and I feel twice the man for it already."[30]

Yet ultimately the rural folks he encountered in Huntley fascinated Roosevelt more than the prairie chickens in the brush. Being a hunter and bird-watcher had taught him the art of observation, and now he was applying it to studying both rural and transient people. This would become a trademark of his future hunting and wilderness books. "I have been much amused by the people in this house, especially the labourers; a great, strong, jovial, blundering Irish boy; a quiet, intelligent yankee; a reformed desperado (he's very silent but when we can get him to talk

*Throughout the entire Midwest tramp Theodore kept regular diaries, which have (strangely) remained unpublished. They are permanently housed at the Library of Congress in Washington, D.C. A university press should jump at the chance of publishing an edition.

his reminiscences are very interesting—and startling); a good natured German boy who is delighted to find we understand and can speak 'hoch-deutch,' " he wrote to his mother on August 25. "There are but two women; a clumsy, giggling, pretty Irish girl, and a hard-featured back-woods woman who sings methodist hymns and swears like a trapper on occasion."[31]

At times on his Midwest tramp it almost seemed that Roosevelt was an onlooker, observing the styles and fashions and habits of American characters with the eye of a novelist. Although never abandoning his aristocratic bearing and always staying a bit removed, Roosevelt sometimes actually wore the garb and adopted the folkways of the regional people he encountered, in hopes of blending in. You might say he was a method actor of the Stella Adler school, playing in an American Arthuriad while mixing it up with different midwestern types like Iowa wheat farmers or Illinois dairymen. Elliott still dressed like a man of substance, sticking to gold scarves and polished boots. His wardrobe expressed his innate sense of aristocratic entitlement, and he had even taken to smoking a long-stemmed pipe. By contrast, Theodore wore dungarees and cotton work shirts and preferred his boots mud-stained. "I am afraid you would disown me if you could see me," he boasted to his mother from the Illinois prairie. "I am awfully disreputable looking."[32]

The bird collecting in Huntley, however, was a disappointment. "Before the sun was up we started off, tramping in sullen silence through the wet prairie grass," he wrote in his diary, "but we found few birds and shot very badly."[33] Truth be told, the flat Illinois countryside lacked the geographical breadth Roosevelt had hoped to encounter. "I broke both of my guns, Elliott dented his, and the shooting was not as good as we expected," he wrote to his sister Corinne. "I got bitten by a snake and chucked headforemost out of the wagon."[34] On August 27, after a particularly bad day's hunting, a dejected Roosevelt recorded, "The country is shot out."[35]

The Roosevelt brothers returned to Chicago for a few days to clean up and plot the next phase of their expedition. They stayed at the Hotel Sherman, which had burned down in the great fire and been rebuilt. Theodore found it off-putting and dreary. If he had wanted society life or afternoon tea, he could have gone to the Fifth Avenue Hotel. Still, he wasn't much interested in meeting labor activists or radicals of any stripe, either. Nor did he write about the swirl of immigrants—Germans, Irish, Poles, and Swedes—who were pouring into the city. What was exciting to him, however, was Chicago's role as the gateway to the great West. Walking the railyards he could for the first time imagine Lincoln's rise as a populist, and the drama of Bleeding Kansas. How strange it was to think that Mary Todd Lincoln was living downstate in corn-stubbled Springfield, where her husband was buried, housebound after being released from the Bellevue Place insane asylum. It had been fifteen years since the tragedy at Ford's Theatre, but Roosevelt still considered himself a Lincoln man. He could sense the power of Lincoln's presence everywhere in Illinois and the surrounding prairie—the ghost of Lincoln, as the poet Vachel Lindsay wrote, haunted the streets, and "the sick world" still cried.[36]

Once again Wilcox collected the Roosevelts in Chicago—this time with an engineer and a former Confederate soldier in tow—for the second leg of the Midwest tramp. The party boarded the Chicago and North Western—nicknamed the "Pioneer Railroad"—and headed straight toward the Mississippi River as night fell. After crossing the Big Muddy at Davenport, their train rumbled on through Iowa City, where a state university flourished, and onto Grinnell along Rock Creek Lake, where farmers angled for crappies and bluegills in the long Indian summer. Eventually they found themselves in the small town of Carroll in western Iowa. This county seat, surrounded by strands of big bluestem grass, had a steel flour mill, eight general stores, five restaurants, and two grain warehouses, but life ran at a slow pace. People sometimes just sat by the river, a meandering brown stretch of the Des Moines. The previous year, most of the business district of Carroll had been destroyed in a fire, and the townsfolk were slowly starting to rebuild. There was nothing urban about the semi-prosperous community except the railroad depot.[37]

The Roosevelts leased a two-horse wagon and started hunting on the outskirts of Carroll, with a pack of eager dogs, roughly following the North Raccoon River northwest into what was considered one of the best prairie fowl hunting areas in the United States.* Theodore observed how "absolutely treeless" and "sparsely scattered over with settlers" western Iowa was.[38] There was no misunderstanding why locals were called flatlanders—only the occasional low hill, high bluff, or forlorn dale was to be found in Carroll County. The creek grades were so gentle that horses and oxen meandered across them with little or no difficulty. The Homestead Act of 1862 had drawn German, Scandinavian, and Czech immi-

*There are actually three Raccoon rivers in Iowa: North, South, and Middle. They are tributaries of the Des Moines River and part of the Mississippi River watershed.

grants to western Iowa, and Roosevelt enjoyed meeting them, saluting their durable "pioneer stock." Many of these flatlanders, however, were struggling in the 1880s; commodity prices had fallen, and it was difficult to overcome the fixed costs levied by grain elevators, railroad concerns, and butcher-yard operators.

Stricken with both asthma and cholera in town, Roosevelt nevertheless managed to bag the most game birds of his life around a plateau called Wall Lake, frequently referred to by locals as "goose pond."[39] Wildlife was abundant here. Using five Irish setters—"three of which worked well, the other two simple nuisances"—they kicked up covey after covey. Hardly trying, Roosevelt bagged thirty-eight grouse, five quail, one bittern, one grebe, and thirty-six yellowlegs (tall, long-legged shorebirds of freshwater ponds with a white rump, known for announcing themselves with a piercing "tew-tew-tew" siren wail). The sojourn in Illinois, by contrast, had yielded only thirty-five birds. And there were lots of other birds Roosevelt didn't shoot in Iowa, only observing them in his diary. "There was a large flock of pelicans on the lake," he noted on September 8, "and thousands of yellow-headed blackbirds."*[40]

In the Iowa brakes the sodbusters had a keen sense of nature's beauty and bounty. Even their economic problems couldn't get them to abandon the land. There was something about the constrained landscape—whether it was the natural meadows or the planted acres of wheat—that soothed the soul. There was natural beauty everywhere in Iowa if you only knew how to look. Joy seemed to be found by modest Iowa farmers even in the dust of summer. They were what Emily Dickinson had referred to in her poem "A Narrow Fellow in the Grass" as "nature's people."[41] Back east, city dwellers like himself studied nature too much; in Iowa people lived with it as if by the grace of God. There was a subtlety to the Iowa plains that liberated Roosevelt's psyche in a way he hadn't anticipated. "No nation has ever achieved permanent greatness unless this greatness was based on the well-being of the great farmer class, the men who live on the soil," Roosevelt would write of the Midwest as president, greatly influenced by this trip. "For it is upon their welfare, material and moral, that the welfare of the rest of the nation ultimately rests."[42]

---

*Although the game birds of western Iowa seemed plentiful in 1880, that was an illusion. Overhunting would by 1932 cause the extinction of the heath hen, and by 2009 Attwater's prairie chicken would top the endangered species list.

III

Once again, after shooting their fill of birds in Carroll County, muscles sore, the Roosevelts headed back to bustling Chicago to regroup. Although Roosevelt never declared it right out, his brother was grating on him a little. Perceiving himself as an authentic naturalist hunter of the Mayne Reid school, Theodore wrote to his sister Corinne that Elliott, by contrast, "revels in the change to civilization—and epicurean pleasures." Because Roosevelt never overate and never drank much more than a glass or two of alcohol, he mocked his younger brother's gluttonous ways. "As soon as we got here [to Chicago] he took some ale to get the dust out of his throat," Roosevelt continued, "then a milk punch because he was thirsty; a mint julep because he was hot; a brandy smash 'to keep the cold out of his stomach'; and then sherry and bitters to give him an appetite."[43]

Holed up again at the Hotel Sherman, the Roosevelts were particularly glad to be rid of Wilcox's two companions, the prattling engineer and the unreliable ex-Confederate soldier. They visited a gunsmith, who was unable to fix their damaged rifles, so they both bought new ones. They were now more determined than ever to turn their next adventure into even more birds bagged. Leg three of their Midwest tramp would be to the Red River country of Minnesota, part of the vast Hudson Bay watershed and reportedly abounding with wildlife. Newspapers back in 1880 often called the 550-mile tributary the "Red River of the North" to help differentiate it from the southern tributary of the Mississippi River, which formed part of the border between Texas and Oklahoma. The Roosevelt brothers stayed in Moorhead, a small city located on the border of North Dakota (or the Dakotah Territory, as it was known then). That's where their cousin Jack Elliott lived, and the brothers were excited about spending time with him flushing out ruffed grouse from the forestland. Not since Dresden had the three been together.

Unlike Harvey and Carroll, Moorhead was irrigable, and a transportation hub in the Wild West for merchandise and agricultural products, situated conveniently between the Twin Cities of Minneapolis–Saint Paul and Winnipeg, Manitoba. Founded in 1871 by William G. Moorhead, a director of the Northern Pacific Railroad, it was notorious as a "sin city" because of its 100 or more smoky bars (and, one assumes, brothels).[44] There was a certain unhurried stillness in the air around Moorhead that had an enduring appeal for Roosevelt. Here, among the "guns of autumn" (as fall hunting was called in Minnesota), for the first time in his life, Roosevelt felt the lure and tug of the Wild West he had read about. Staying at a "miserable old hotel" surrounded by a strip of bars, he could imagine himself in Tucson or Dodge City, where the saloon doors always swung open. (The outlaw Jesse James had launched his career as a notorious bank robber in 1876, just over 200 miles down the road in Northfield, Minnesota.)

Taking a buggy out to the countryside with his brother and Jack Elliott, Theodore marveled at how easy it was to snare grouse in Minnesota. The great challenge in shooting these birds was that they clustered in dense, prickly thickets and usually flushed dramatically without warning.[45] There was perhaps another reason that Roosevelt wanted to hunt in the lazy Red River region along the border between Minnesota and Dakotah territories. The Red River was in the middle of a crucial migratory route for dozens of bird species. Surely Roosevelt also knew that eagles and owls roosted permanently around the river bottom forests and remnant prairie.[46] With the assistance of a "stub tailed old pointer" (and a Moorhead barkeeper who daylighted as a cart driver) he borrowed along the way in Saint Paul, Theodore's goal was simple: shoot more game birds in Minnesota than he had shot in Illinois and Iowa combined; fill up those tarpaulin sacks.[47]

Unsurprisingly, Roosevelt relished playing a Minnesota-Dakota sportsman with shells in his jacket pocket, wandering through plowed land listening for the whir of wild ducks, and flushing ruffed grouse out of open grasslands and forest thickets. The desolate countryside was crisp and lovely in mid-September. Columns of cumulonimbus clouds ascended and spread in the blue sky. There wasn't a hitching post for miles. Already the aspen and ash were starting to take on fall colors as dramatic as those in Vermont. This was the autumn rutting season, and the young four-point deer were challenging the eight-pointers over females. Roosevelt's most memorable nights in Moorhead were spent camping under the stars, the fall nip adding an aura of romance to the outings. Although Theodore and Elliott were not good cooks, they had mastered the art of cleaning the chicken-sized grouse. Every night they stuffed themselves with fire-roasted bird until they burped.

For ten days in Minnesota, Roosevelt bagged more than 203 game birds, carefully listing them in his diaries: 95 grouse, 51 snipe, seventeen ducks, sixteen plovers, one goose, and so on. He had beaten his own Iowa record. Clearly Theodore was no longer collecting specimens or playing ornithologist. More telling, however, was his side note that Elliott had bagged only 201.[48] That means he had beaten the "old boy" by two birds. Much of the time in western Minnesota Roosevelt had terrible asthma and was forced to sleep sitting up, dreaming intermittently, perhaps, about the prairie potholes of the Dakotas, where the hunting was supposed to be even better than in Minnesota's Red River region.

One morning Theodore and Elliott awoke at dawn and went searching for grouse hot spots. Carefully listening for the birds' drumming noise—which sounded like a stomach growling—the Roosevelts got their game. They also got lost in a "cold driving rain storm"—so lost, in fact, that at dusk they were forced to knock on a Norwegian farmer's "neat but frail little house" asking to "put up for the night."[49] It was a scene straight from the pages of *Giants in the Earth*. According to his diary Theodore lay silently that evening in front of the fireplace while the wind blew hard outside. He was covered by a bison robe but was unable to sleep. This was his first real western experience. To Roosevelt, however, getting lost in Minnesota was "lovely," a softer alternative to "bully." During the coming days the Roosevelt party would camp along a bend of the Buffalo River—an eighty-eight-mile tributary of the Red River, lined with hardwoods, which looked like a poorly maintained Dutch canal.[50]

Roosevelt's fierce competition with Elliott does show that he used the sport as a mechanism to enhance his manly self-worth. His grousing also reflected class-consciousness. Men with an aristocratic mien shot game birds with a retinue of guides; by contrast, the working class went after squirrels and raccoons with a mongrel dog. Game laws made sense when you were rich, looking for sport. Poor people shot game—with or without laws—just to stay alive. Much like being initiated into Porcellian at Harvard, grouse hunting to Theodore was a rite of initiation into manhood with his younger brother. The very act of flushing grouse together was a brotherly bonding experience.[51]

Another outcome of Roosevelt's midwestern tramp was that he grew less interested in birds and more intrigued by the notion of hunting big game. After all, being wrapped in a buffalo robe in Minnesota wasn't the same as either shooting a bison or seeing a thinned-down wild herd with his own eyes. "No man who is not of an adventurous temper, and able to stand rough food and living," he soon wrote, "will penetrate to the haunts of the buffalo. The animal is so tough and tenacious of life that it must be hit in the right spot; and care must be used in approaching it, for its nose is very keen, and though its sight is dull, yet, on the other hand, the plains it frequents are singularly bare of cover; while, finally, there is just a faint spice of danger in the pursuit, for the bison, though the least dangerous of all bovine animals will, on occasions, turn upon the hunter and though its attack is, as a rule, easily avoided, yet in rare cases it manages to charge home."[52]

After so many grouse, Roosevelt wanted to someday travel west of the Red River into buffalo country. (And he would have to do it soon, for the cattle were taking over.) Being in Moorhead was akin to traveling 1,400 miles to see a major attraction—buffalo—and then never entering the admission gate. He had, however, seen plenty of cattle. Between 1877 and 1885 more than 2.5 million head of cattle were slaughtered in Chicago; Roosevelt had been able to see the stockades and corrals for himself.[53] What he hadn't been able to witness on his midwestern tramp, however, were the cattle drovers herding from Fort Worth north by the ninety-ninth meridian to Ogallala, Nebraska, and from there into the northern plains of Montana, Wyoming, and the Dakotas. Because he had not seen the great herds of buffalo and cattle, Roosevelt believed he had missed the main thrust of the western expansion. He vowed to return. Nobody would have to lend him a buffalo robe. He'd earn the next one for himself.

Paradoxically, that odd night with the buffalo robe in a stranger's home was in many respects Roosevelt's good-bye to his rigorous outdoors life for a while. Maturity and responsibility beckoned. Perhaps the recent college graduate wanted to push on westward through the Dakotah Territories to the Bitterroots, where the elk ran in healthy herds and the geysers of Yellowstone were reliable. But he couldn't. He was planning to start studying law in New York and he was getting married the next month. His diary, in fact, abounds with paeans to Alice Lee: "She is so pure and holy that it seems almost profanation to touch her, no matter how gently and tenderly." "My happiness now is almost too great."[54] It was as if he couldn't write her name without bursting a vein. Pausing in Saint Paul about to board an eastbound train, burning with impatience to get home, he sounded tired: "How glad I am it is over," the wanderer wrote from the depot, "and I am to see Alice."[55]

Once he caught up with Alice in Chestnut Hill, everything became a whirlwind. In quick succession he enrolled at Columbia law school, purchased a diamond ring, spent two weekends at Oyster Bay with his mother, rode Lightfoot far and wide, chopped down trees for winter firewood, and prepared himself for his vows at the Unitarian Church in Brookline, Massachusetts. Sometimes he took long walks across green fields, and nobody knows what he thought.

IV

Everything went flawlessly on the wedding day of Theodore and Alice, October 27, 1880. Elliott arrived on cue to serve as best man. "At twelve o'clock on my twenty-second birthday, Alice and I were married," Roosevelt wrote in his diary. "She made an ideally beautiful bride; and it was a lovely wedding. We came on for the night to Springfield [Massachusetts] where I had taken a suite of rooms. . . . Our intense happiness is too sacred to be written about."[56] The newlyweds honeymooned at Oyster Bay, with plans for a long springtime journey to Europe. To T.R. marriage clearly meant that hunting had to come second or third. No longer does he fill his diaries with bird sightings. Although he never slackened his pace, still hurrying on in everything he pursued, a domestication had taken place. Never once did his marriage become imprisoning. His diary entries now mention silk jackets, pajamas, lovely drives, and even chats with Alice while she was sewing. He gloats in his diaries about how he "won" Alice, worshipping the trill of her voice and the shape of her body. Politics also rose to the forefront of his thinking. On November 2, for example, he recounted how he proudly voted for James Garfield, the Republican standard-bearer, for president.

After the New Year Theodore and Alice moved into Manhattan, to the mansion on West Fifty-Seventh Street. Life was quite good. Roosevelt had learned to dam his pent-up tears whenever his father's memory was evoked. Wandering around the Bronx from time to time to get his nature fix amid the clamor of the new elevated railroads, Roosevelt daydreamed about the open spaces of the Far West. New York was getting crowded. Too many of the gadgets of tomorrow first showcased at the 1876 Philadelphia Exposition were now fairly commonplace. Bicycles were all the rave, as were typewriters. Theodore took to neither. (Robert B. Roosevelt, however, was a champion of both.)

Amid the many New York galas and soirees, Roosevelt, however, seemed to have deferred his need for wilderness. Besides attending Columbia law school, he was writing the last chapters of *The Naval War of 1812*. Books about Commodore Oliver Hazard Perry and Admiral John Paul Jones, not Elliott Coues's newest bird key, now caught his eye. As a married man, he found that his diary writing waned, as did his taxidermy. His sketch pads were filled not with mice but with brig sloops. When drafting chapters about the Great Lakes, for example, Roosevelt never mentioned the nesting areas of plovers, gulls, or terns.

There was, however, an exception: he was writing an essay about an ornithological trip with Elliott in December 1880 aboard a twenty-one-foot sailboat around Long Island Sound. The short nautical-naturalist essay was called "Sou'-Sou'-Southerly"—the vernacular name for old-squaws

(*Clangula hyemalis*).* On March 24, in fact, a frustrated Roosevelt wrote in his diary that he was "still working hard at . . . one or two unsuccessful literary projects."[57] The first of these was *The Naval War of 1812;* the second was the Robert B. Rooseveltish "Sou'-Sou'-Southerly."

A strong case can be made that "Sou'-Sou'-Southerly" was the first authentic naturalist essay ever fully realized by Roosevelt.[58] It was written for publication in a sporting magazine but was—for whatever reason—never published. Based on his frozen, white-capped nautical journey with Elliott in a sloop traveling from Oyster Bay to Huntington Bay, the narrative hinged on the perils of duck hunting in the bitter cold of Long Island Sound. The Roosevelts overcame ragged floes, heavy seas, and icicles overtaking their beards, all in the name of bird hunting. "The snow storm had now fairly set in, the hard flakes, mingled with flying spray, driving fiercely into our faces, and (for the short winter day was already becoming even duller and grayer as evening drew on)," Roosevelt wrote, setting the stage, "the land was entirely veiled from our sight thought not far distant. Sometimes there would be a few minutes lull and partial clearing off, and then with redoubled fury the fitful gusts would strike us again, shrouding us from stern to stern in the scudding spoon drift."[59]

If all Roosevelt wrote about in "Sou'-Sou'-Southerly" was ice-laden waters and wind, then history could chalk it up as a solid first effort by an aspiring adventure writer. Many of his nautical references, in fact, gave the impression that he was showing off. And, as usual, Roosevelt also wrote with far too many semicolons. But owing to its Audubon Society overtones Roosevelt's essay offered much more than what one expects from an initial essay—"Sou'-Sou'-Southerly" is filled with his able portraits of such winter birds as coots, dippers, sheldrakes, and bluebills. Checker-back loons wade in shallow coves trying to shelter themselves from arctic-like gusts, and black ducks with white-and-chestnut plumage shake off the curlers breaking over their water-resistant bodies. Only a committed naturalist who cared deeply about shorebirds could have written "Sou'-Sou'-Southerly." Although Roosevelt wrote about hunting black ducks in this piece, he had also captured their magnificence in real-time camera-like prose.[60]

---

*The essay remained unpublished until 1988, when the scholar John Rousmaniere unearthed it from the Library of Congress's T.R. Collection and published it in *Gray's Sporting Journal*. (Rousmaniere had been alerted to the essay's existence by a footnote in *The Rise of Theodore Roosevelt*.)

Even though Roosevelt was writing *The Naval War of 1812* and perfecting "Sou'-Sou'-Southerly," he followed through on his commitment to squire his wife around the grand capitals of Europe that summer. Emulating his father, he created a breakneck itinerary for them and kept it. Even though Alice had horrendous stomach problems they toured Irish castles, steamboated on the Rhine, shopped in Munich, and fished on Lake Como. Surprisingly, he didn't care for much of the European art. Strangely, the sensuous women models of Rubens seemed to him like "handsome animals."[61] But for the most part Europe—as it had when he was just a little boy carting around arsenic paste—made him homesick. While he was in France and Italy, the word "wretchedness" became one of his favorite adjectives. A longing for the American wilderness returned to the forefront of his thinking. Even London's Zoological Gardens disappointed him. Ironically, the British zoologists weren't taking Darwinian advancements into their displays. Then there was the awful news that President Garfield had been shot and was in critical condition. "Frightful calamity for America," he wrote in his diary.[62] It seems the tragedy made him want to return home, but he didn't.

Although they hadn't seen each other for a few years, one of Roosevelt's principal correspondents from that summer of 1881 was old Bill Sewall of Island Falls, who eagerly collected his young friend's special-delivery missives from the little postal box. Clearly Roosevelt was still thinking about the Maine wild—and the exhilaration of climbing Mount Katahdin. Disregarding his Harvard physician's recommendation to watch his heart, Roosevelt prepared to hike up the Matterhorn on his own while Alice rested in a hotel in Zermatt. Remembering all the lessons he had learned from Sewall, he had decided, quite spontaneously, to climb Switzerland's famous peak as a retort to a cabal of snobby English climbers he had accidentally encountered in the hotel lobby. He was determined to prove to them that "a Yankee could climb just as well as they could."[63] Writing to his sister Bamie, he added that conquering the Matterhorn that August would give him at least the credential of being a "subordinate kind of mountaineer."[64]

For serious mountain climbers, conquering the Matterhorn was an initiation ritual; if you could make it to the summit you were accepted as a player. The 14,690-foot Matterhorn was first successfully climbed twenty years before, in 1861, by the Englishman Edward Whymper. His 1880 book *The Ascent of the Matterhorn* was all the rage in Europe. Since Whymper's historic climb, hundreds of others had accomplished the feat (including Lucy Walker, the first woman to make the ascent), and the

Swiss Alpine Club had built a shelter for climbers to rest and sleep in at 12,500 feet, which made a big difference.[65] Here is an excerpt from a long letter of August 5, 1881, that Theodore sent to his sister Anna, proud of his feat: "We left the hut at three-forty, after seeing a most glorious sunrise which crowned the countless snow peaks and billowy, white clouds with a strange crimson iridescence, reached the summit at seven, and were down at the foot of the Matterhorn proper by one. It was like going up and down enormous stairs on your hands and knees for nine hours."[66]

By making it to the top of the Matterhorn, Roosevelt, in essence, felt he had conquered Europe. He also climbed the Jungfrau, a peak only slightly less difficult for experienced mountaineers. Nevertheless, in Roosevelt's mind the wilderness of America was more divine than the tame Alps and Pyrenees. As the historian Louis S. Warren noted in *The Hunter's Game*, Roosevelt—like many of his generation—had come to believe that the United States was "nature's nation," that the pristine landscape represented God's best work.[67]

In retrospect the most amazing part of Roosevelt's European jaunt with Alice Lee—exemplified by climbing the Matterhorn—was his stamina. Zigzagging from city to city, he nevertheless kept assiduously working on *The Naval War of 1812*, preparing his manuscript for publication in May–June 1882. His powers of concentration that summer were amazing. No matter what task Roosevelt undertook, he was like boll weevils eating their way through fields of cotton. Over the years scholars of Roosevelt as a military man have garnered plenty of useful biographical tidbits from reading his diary entries about standing at Napoléon's tomb and contemplating Caesar, Tamerlane, and Genghis Kahn. But for the conservationist-minded, the most interesting aspect of these months abroad was Roosevelt's rejection of European nature in favor of American wilderness. He believed that the Europeans, with the exception of Scandinavians and British, had recklessly shot out all the wildlife. Because this was Roosevelt's first foreign trip since experiencing Maine and Minnesota, he was now touting his glorious homeland as a Garden of Eden. "The summer I have passed traveling through Europe, and though I have enjoyed it greatly," he wrote to Sewall, "yet the more I see, the better satisfied I am that I am an American; free born and free bred, where I acknowledge no man as my superior, except for his own worth, or as my inferior, except for his own demerit."[68]

V

Upon returning from Europe, having squandered part of his inheritance, Roosevelt immediately threw himself back into the urban fray. He was happily married, enjoyed learning law, and, as an impassioned conversationalist, had an easy time making new friends. His asthma and weak heart weren't giving him problems. He was finishing his book, which was scheduled for publication the following spring by G.P. Putnam's Sons. To top the year off, that November he was elected to the New York state assembly from the twenty-first district.

When Roosevelt took office on January 2, 1882, he swore he'd be a steel-fisted reformer like Uncle Rob. He would hunt down thieves, swindlers, polecats, and robber barons; more controversially, he was willing to expose the frauds and shenanigans of the very governing class he was part of. He was itching to earn his spurs in the rough-and-tumble of New York politics. He wanted everything cleaner and better. It was no coincidence that the first bill he embraced would improve street cleaning in the city, and that it had a provision for the better treatment of workhorses.

Vivid stories abound about Roosevelt in Albany, dashing around in his frock coat trying to learn the rules of engagement. He was determined not to run with the wrong crowd, fearing being lampooned in the press as "politics as usual." Fancying himself a change agent or reformer, he refused to see the world in gray, making snap judgments of his fellow legislators' personalities that were often unfair. They were either good or evil, trustworthy or untrustworthy, front-parlor fresh or operators of smoke-filled backrooms. "He would come into that house like a thunderbolt," Isaac Hunt, a fellow Republican legislator and Swiss cattle breeder from Watertown, recalled. "He would swing the door open and he would be half way up the stairs before that door would come together with a bang. Such a super-abundance of animal life was hardly condensed in a human life."[69]

Nobody in the legislature knew quite what to make of Roosevelt. He was like a jaybird on the house roof, loud and sudden. Mocked as a "Squirt," a "Punkin-Lily," and a "Jane-Dandy"—and much worse—Roosevelt was held in contempt on both sides of the aisle.[70] Annoyed by his reformist proclamations, irritated that he seemed above the give-and-take of politics, the longtime Republican speaker of the state house of representatives complained that with Roosevelt now in Albany the Republicans' strength was "sixty and one-half members." Rarely had Albany had an independent-minded legislator so determined to be nonpartisan.

The Republicans, then, saw Roosevelt as an annoyance, and the Democrats loathed him no end; these bad feelings were reciprocated. "There are some twenty-five Irish Democrats in the House," Roosevelt wrote in his diary. "They are a stupid, sodden, vicious lot, most of them being equally deficient in brains and virtue."[71] On another occasion, unafraid of sounding elitist, Roosevelt noted that Tammany Hall Democrats were "totally unable to speak with even an approximation to good grammar; not even one of them can string three intelligible sentences together to save his neck."[72]

Roosevelt's closest friend in the New York legislature was—not surprisingly—Bill O'Neill, who lived in remote Saint Regis Falls in the Adirondacks. Much like Sewall, O'Neill was an honest backwoods type, obedient to existing laws, the owner of a rural general store who also ran a creamery. O'Neill later recalled that Roosevelt—who had published the only bird key of his Franklin County district—had constantly worried him; he was rocking the boat too much in 1882–1883 with his uncompromising reformist zeal. "In all the unimportant things we seemed far apart," Roosevelt wrote fondly about O'Neill in *An Autobiography*, "but in all the important things we were close together. . . . Fortune favored me, whereas her hand was heavy against Billy O'Neill. All his life he had to strive hard to wring his bread from harsh surroundings and a reluctant fate; if fate had been but a little kinder, I believe he would have had a great political career."[73]

A telling sign that Roosevelt was drifting away from being a professional naturalist and toward a career in politics was a letter he wrote to Elliott Coues in April 1882, just three months after taking office. Unsentimentally, T.R. offered to donate the bulk of his "Roosevelt Museum" holdings to the Smithsonian Institution. Coues immediately forwarded Roosevelt's letter to Spencer F. Baird, the secretary of the Smithsonian.[74]

Up to that point only Louis Agassiz of Harvard University had done more than Baird to promote American zoology. Raised in eastern Pennsylvania, Baird had attended Dickinson College, where he was known as the "opossum hunter." Baird's career was helped when, on a collecting trip in Vermont during the summer of 1847, he encountered Congressman George Perkins Marsh, the originator of the term "conservationism in modern usage." Stunned by Baird's self-taught knowledge of American wildlife, Marsh ended up recommending a few years later that the young outdoorsman be hired by the new Smithsonian Institution. Baird embarked on a prestigious career there, and in 1878 became its second leader. Beyond his administrative duties, Baird inventoried North American birds, sponsored wilderness explorations, promoted systematic biology, and of course tirelessly raised funds on behalf of the Smithsonian.[75] If all that wasn't enough, with the possible exception of Robert B. Roosevelt—with whom he corresponded—nobody rallied against the depletion of American fish with as much vim and vigor as Baird, who simultaneously served as the commissioner of the U.S. Commission of Fish and Fisheries.[76]

From young manhood onward, Baird, known for his trademark thoughtful frown, was America's genius at collecting and classifying wildlife. Audubon respected Baird so much that he named his last bird Baird's bunting (*Ammadramus bairdi*). At a time when natural history was an avocation, Baird upgraded specimen collecting to a vocation. He was a "collector of collectors," and Robert B. Roosevelt was one of his finest clients and friends.[77] When Baird was appointed as the first commissioner for the U.S. Commission on Fish and Fisheries in 1871 (by President Grant), tasked with replenishing fish populations and promoting fish culture, R.B.R. cheered. When Commissioner Baird established a salmon fertilization project in California the following year, shipping eggs by train to New York, R.B.R. was one of the first recipients. Together they tried to answer the difficult question of whether ocean fish populations could be restored. And when Baird founded Woods Hole, Massachusetts, as the largest biological laboratory in the world, R.B.R. was his guest at the ribbon cutting.

Even though Baird had never heard of young T.R., the Roosevelt name always rang magically in conservationist circles. That ring was the sound of coins: a cashed donation check from both Theodore Roosevelt, Sr., and Robert B. Roosevelt. "Dr. Coues has sent me your letter offering certain specimens to the Smithsonian Institution," Baird wrote back to T.R. "In reply I beg to say that the same will be very acceptable to use even should there be nothing actually new, for they will give us the opportunity at least of supplying some Museum at home or abroad, and of obtaining in exchange a possible rarity. . . . May I ask what relation you are to my much esteemed friend Robert B. Roosevelt or Mr. Theodore?"[78]

Immediately upon receiving Baird's letter Roosevelt replied. "Dear Sir: I am the son of Theodore Roosevelt and a nephew of Robert B. I am very much obliged for your kind letter, and shall send you the [bird] skins; would your collection include Egyptian skins, as I have some of them? Very truly yours."[79]

Baird responded quickly, and the two men were close for the rest of their lives. "I shall be very happy indeed to have the Egyptian skins, referred to in your letter, as well as others, from different parts of the world," Baird wrote to Roosevelt, "which you may be disposed to contribute to the museum. I am very glad to know something of your personality. I was well-acquainted with your father and, in common with all his other friends, esteemed him most highly."[80]

As Roosevelt crated up his species for the Smithsonian, he did not overlook the American Museum of Natural History, which his late father had been so instrumental in starting. T.R. set aside 125 specimens for his hometown institution even though he sent the lion's share to Washington, D.C. He was not spurning the local institution, but he wanted to contribute to the great cause of building a national wildlife collection. Perhaps the principal reason that Roosevelt gave away his collection, however, was that *The Naval War of 1812* was published in May 1882, to overwhelmingly good reviews. His chapters pertaining to the Great Lakes were praised by military historians all over the world. His prose was lively, filled with brave sailors firing cannons, brigs burning, and creoles fighting to save New Orleans from the huge British armada.[81] An overriding lesson from his study was that in warfare both preparation and training were essential. Roosevelt was thrilled when the U.S. government ordered that copies become assigned reading on every American naval vessel.

Upon receiving Roosevelt's collection Baird—a prolific correspondent, who wrote about 3,500 letters a year—immediately sent an acknowledgment, saying he "was by no means prepared for so admirable or extensive a contribution, and beg to thank you very much for it. There are many specimens in the series which will be a great service to us in extending and completing the collections of the several compartments. I need hardly to say that whatever [else] you can furnish in the way of specimens of natural history will always be gladly received."[82]

An affectionate name for the Smithsonian has long been "America's attic," a fitting designation for our vast depository of national heirlooms. Roosevelt's birds had ornithological value in 1882, and today they are invaluable as a window into our twenty-sixth president's youth. The thorough accession records at the Smithsonian are nothing short of awesome, and the detailed accounting of all aspects of Roosevelt's bird specimens is something that would make even Price Waterhouse proud. Clearly Roosevelt's birds were valued by Baird, for they immediately became an integral part of his natural history collection. In twenty-five pages the Smithsonian provided proper binomials and data on characteristics and colorization for each and every one of the 622 bird skins T.R. had donated. Fifty-three of them came from abroad (specifically, 31 from Egypt, six from Syria, five from Austria, one from Germany, one from France, two from England). As for his U.S. birds, they overwhelmingly came from Oyster Bay, the Adirondacks, and Garrison, New York.[83]

What mattered most to Roosevelt, it seemed, wasn't whether his birds ended up under glass at the Smithsonian Institution—to his mind the greatest museum in the world—but the fact that Baird had actually accepted his taxidermy as being excellent. From then on, whenever Roosevelt went hunting the Smithsonian Institution was the primary beneficiary of his prowess. Three of Roosevelt's favorite Egyptian birds—a crocodile bird, a white-tailed lapwing, and a spur-wing lapwing—were gifted to the American Museum of Natural History. A white snowy owl he had shot near Oyster Bay in 1876 also was deeded to the New York museum. One of the largest birds in the arctic region, the snowy owl often migrated southward in the winter; a few were once discovered in the Caribbean. Covered with velvety, fine-textured, white, downy feathers, this owl epitomized gracefulness, swooping down and using its sharp talons to seize prey in a single elegant motion. Even before Roosevelt was famous, just a twenty-three-year-old assemblyman, this expertly mounted snowy-white became something of a tourist attraction at the American Museum. Over time, as his legend grew, this snowy owl likewise grew in significance. Today it is recognized as the high-water mark of Roosevelt's ornithological career.

CHAPTER SIX

Chasing Buffalo in the Badlands and Grizzlies in the Bighorns

I

Although Theodore Roosevelt had donated his vast natural history collection to the Smithsonian Institution, he nevertheless desperately longed for the head of a free-ranging buffalo to hang on his library wall in New York. Roosevelt preferred to call them by the proper zoological classification "bison." In *Hunting Trips of a Ranchman* (1895) he titled one chapter "The Lordly Buffalo" and was full of reverence for the horned species. With zoological precision he was also careful to note that there were two subspecies of the mammal in North America: *Bison bison bison* (Plains buffalo) and the lesser *Bison bison athabascae* (wood buffalo) found primarily along the Pacific Coast.*[1]

In late 1882, Roosevelt purchased a small brownstone off Fifth Avenue, at 55 West Forty-Fifth Street, hoping to get away from his mother's tight grip and start a family of his own. The new home was, according to a close friend, a "pleasant" hearth where Theodore and Alice entertained guests with "the kind of generous warmth that characterized them both."[2] Roosevelt decided that his heads of indigenous game—buffalo, moose, elk, antelope, grizzly bear, etc.—would be showcased throughout the residence. "Back again in my own lovely little home," Roosevelt wrote in January 1883 following a stint in Albany, "with the sweetest and prettiest of all little wives—my own sunny darling. I can imagine nothing more happy in life than an evening spent in my cozy little sitting room, before a bright fire of soft coal, my books all around me, the playing backgammon with my own dainty mistress."[3]

By summer Roosevelt had set his sights on a country house as well. Craving open-air diversions, he acquired 155 acres of pristine land, half-

*Originally consisting of North Dakota, South Dakota, Montana, and parts of Idaho, Wyoming and Nebraska, some of the Dakota Territories were broken up into the Idaho Territory, Wyoming Territory, and Nebraska Territory. On November 2, 1889, the remaining Dakota Territories split to become the separate states of North Dakota and South Dakota.

wooded, near the family estate on Long Island's north shore, and he started building an eclectic, roomy three-story mansion, with a view from upstairs of Oyster Bay and Cold Spring Harbor. Originally called Leeholm, this mansion would become known as Sagamore Hill (after the Indian Chief Sagamore Mohannis, who had deeded away rights to the property 200 years earlier).[4] The estate became Long Island's great *wunder krammer* (room of wonders) for natural history. Its oak-paneled library would eventually house a first-rate naturalist book collection, and the walls would groan with trophies from the West—including skins and heads of all the North American big game Roosevelt shot—running the gamut from bear to wapiti. (Sagamore Hill would also become the summer White House from 1902 to 1908.[5])

That spring Roosevelt was reading Eugene V. Smalley's *History of the Northern Pacific Railroad*, just released by G.P. Putnam's Sons, the same firm that had published *The Naval War of 1812* the previous year. Roosevelt had grown so enthralled with Putnam that he became a partner of the house, investing $20,000.[6] The timing of Smalley's book was propitious—near the end of the summer the second transcontinental railroad would open, with great hullabaloo. New Yorkers like Roosevelt could now easily travel to the northwest territories that Lewis and Clark had first bravely explored in 1804–1806.[7] Dime novels had popularized past western heroes like Kit Carson and Jim Bridger as updates of the Leatherstocking sagas. Roosevelt wasn't impressed by such hack writers as Ned Buntline or Prentiss Ingraham, but he touted western cowboys as the American equivalent of the British knights popularized by Sir Walter Scott. A mere train ride to the western edge of the Dakota Territory, Roosevelt hoped, would bring an array of Homeric pioneer characters into his fairly aristocratic eastern-centered orbit.

Everything about the Northern Pacific Railroad—a pet project of presidents Lincoln and Grant—enthralled an unabashed expansionist like Roosevelt. Linking Lake Superior to Puget Sound, the Northern Pacific was somehow more romantic than the first transcontinental line, the Union Pacific, which chugged from Omaha to Sacramento through the salt flats of Utah, where Mormon settlements were springing up. With the Northern Pacific, places like the Bighorns, Yellowstone National Park, the Cascade Mountains, and the Olympic rainforests were now more easily reachable from the Atlantic East. As a direct consequence of the connecting spike a greater number of emigrants and fortune seekers now departed for the Minnesota and Dakota territories in droves to grow wheat in the excellent prairie soil. Not only did *History of the Northern*

Pacific Railroad include attractive black-and-white photographs of Pyramid Butte and a panoramic shot of the sediment-laden Little Missouri River (the largest tributary of the Missouri in the region); it also included a shot of a "Ranchman's Log 'Schack' " that exuded unvarnished frontier charm in a classic western landscape.[8]

Having already traveled on the Northern Pacific from Saint Paul to Moorhead, Roosevelt was now eager to take it farther west to a bizarre area that Smalley devoted an entire chapter to: the "Bad Lands." Located in what is now western North Dakota,* the Badlands are a surreal jumble of scoria hills, towering buttes, buffalo wallows, grassy draws, and narrow valleys following the 560-mile Little Missouri River. Taken together the geography resembled a blasted-out Grand Canyon on a small scale. Created as the ancestral Rocky Mountains were being formed 60 million years ago, the Badlands were full of dinosaur bones and fossils, easily found on digs in sandstone beds and soft siltstone.[9] (In 2007 scientists discovered in the Hill Creek Formation a rare "dinosaur mummy" of a 67-million-year-old fossilized duckbilled hadrosaur named Dakota; much of its tissue and bone was preserved in an envelope of skin.[10]) As in the Painted Desert of Arizona, petrified wood was scattered about the Badlands for hundreds of miles, with silica coating the dead tree trunks and old stumps.[11]

That spring of 1883 Theodore (with Alice) spent a lot of time at Tranquility in Oyster Bay, reading Smalley and other books pertaining to western exploration and Dakota wildlife while commissioning the architecture firm of Lamb & Rich to build Leeholm. During the workweek Theodore commuted into Manhattan on the Long Island Rail Road to take care of family business and give political speeches. At a Free Trade Club dinner in May at Clark's Tavern, for example, he delivered an address on "The Tariff in Politics." That evening Roosevelt struck up a conversation with Henry H. Gorringe, a blunt-spoken commodore who'd recently resigned from the U.S. Navy. One can only assume that *The Naval War of 1812* was discussed, for Gorringe (like Roosevelt) was an outspoken advocate for a much larger and more modern U.S. fleet. Gorringe was so blunt, in fact, that Secretary of Navy William Eaton Chandler had found him insubordinate and forced him to resign that February.

*Originally consisting of North Dakota, South Dakota, Montana, and parts of Idaho, Wyoming and Nebraska, some of the Dakota Territories were broken up into the Idaho Territory, Wyoming Territory, and Nebraska Territory. On November 2, 1889, the remaining Dakota Territories split to become the separate states of North Dakota and South Dakota.

Deeply civic-spirited, Gorringe had supported Assemblyman Theodore Roosevelt's reformist initiatives in New York to clean up tenement buildings and improve sanitation, including the clearing of snow.[12] Roosevelt and Gorringe had more in common than their interest in Oliver Hazard Perry and reformist politics: they shared a romanticized view of hunting and ranching along the Northern Pacific Railroad line. Gorringe planned to open a hunting lodge and cattle ranch in the Badlands, taking over a cantonment abandoned by the U.S. government along the languid Little Missouri River. The cantonment was originally built to protect railroad workers from Indian attacks, but Gorringe now envisioned it as a sportsmen's resort. Roosevelt told Gorringe that he was dying to bag a free-ranger "while there were still buffalo left to shoot."[13]

The response from Gorringe was a salesman-like "no problem." Recently, newspapers such as the *Bismarck Tribune* and *Dickinson* (North Dakota) *Press* had boasted that a couple of hunters there had bagged ninety deer and fifteen elk in a few weeks. Gorringe was already part owner of the Pyramid Park Hotel, which he was also hoping to make into a sportsmen's resort, in Little Missouri, a village along the Northern Pacific Railroad route. Since the days of Lewis and Clark the Little Missouri area had been considered excellent hunting country (for bears, elk, antelope, beavers, black-tails, and white-tails) by the Crow, Sioux, Arikara, Mandan, Cheyenne, and Gros Ventre Indians.[14] Now, Gorringe said, it was time for the white man to take advantage of such happy grounds.

Unbeknownst to Roosevelt was that finding buffalo to shoot anywhere—even in the Badlands—was nearly impossible. For example, an outfit in Miles City, Montana, that very September had corralled wagons, bedrolls, horses, tents, pots and pans, playing cards—all the necessary provisions for a high-end Dakota-Montana buffalo hunt. It had signed up numerous eastern clients, promising the head of a 2,000-pound buffalo (plus an immense hump of the delicious muscle that supported the huge skull) for a high fee. Clearly Roosevelt wasn't the only New York hunter craving the ultimate wall trophy, and buffalo steak by campfire under a starry sky. The Miles City outfit, however, couldn't deliver on its sales pitch; disappointed clients, in fact, feeling ripped off, demanded a full refund.

That summer buffalo herds were disappearing from the entire northern range. One of Minneapolis's legendary fur buyers, for example, sent able scouts trudging over the northern plains in buckskins and mackinaws looking for buffalo hides. Finding them proved to be almost impossible. Back in 1881 one Montana dealer had acquired more than 250,000 buffalo hides for his little operation; now, just two years later, he was lucky to get ten. The buffalo hunter himself was becoming extinct. "Almost every wild buffalo had been done away with," the historian Tom McHugh lamented in *The Time of the Buffalo*. "All that remained was the conspicuous leftover of carrion rotting on the prairie."[15]

Weeks after the Free Trade Club dinner where he met Gorringe, Roosevelt fell ill again with both asthma and cholera. Even escaping to an upscale Catskills resort in Richfield Springs didn't help his breathing much. Although he said enthusiastically that the "scenery was superb," being a convalescent made him feel puny. "For the first time in my life, I came within an ace of fainting when I got out of the bath this morning," he wrote to his sister. "I have a bad headache, a general feeling of lassitude, and am bored out of my life by having nothing whatever to do, and being placed in that quintessence of abomination, a large summer hotel at a watering place for underbred and overdressed girls, fat old female scandal mongers, and a select collection of assorted cripples and consumptives."[16]

Following a familiar pattern, Roosevelt started to crave wide-open spaces as a cure-all. A Catskills hotel simply couldn't do the trick. Another month in New York and his entire nervous system would have short-circuited. Gorringe's Badlands beckoned him more than ever. Also gnawing at him was the fact that Elliott had returned from hunting in the dense jungles of India and had brought tiger heads; it wasn't right for an older brother like himself to be trumped like that. Adding insult to injury, Elliott had already been stampeded by frightened buffalo in the Staked Plains of Texas, nearly losing his life for a trophy head. (Later, Theodore would commission Frederic Remington to sketch his brother's brave technique—splitting the herd—as an illustration for his 1888 book *Ranch Life and the Hunting-Trail*.[17])

With a tone of desperation, Roosevelt wrote Gorringe on August 23 to request that plans for their buffalo hunt be completed and the date set.[18] Perhaps the fact that Alice was pregnant put him under additional stress. He was already equipped with two double-barreled shotguns—a No. 10 choke-bore made by Thomas of Chicago and a No. 16 hammerless especially made for him by Kennedy of Saint Paul. He also told Gorringe that he owned a .45-caliber Sharps, considered one of the finest buffalo guns.[19]

Gorringe backed out of going, leaving Roosevelt companionless for the hunt in the Dakota backcountry. Still sick with asthma (but with Alice safely ensconced with her family in Massachusetts), Roosevelt left by himself for Chicago, then switched trains for Saint Paul. Writing his mother a quick letter, he boasted about "feeling like a fighting cock again."[20] Proudly heading west on the Northern Pacific, Roosevelt steamed past the Lake Park region of Minnesota and the wheat fields of the Red River valley across the billowy plains around Jamestown to nearly treeless Bismarck and on to the desolate Badlands of his dreams.

II

At around two o'clock in the morning on September 8, 1883, Roosevelt arrived in the hamlet of Little Missouri (called "Little Misery" by locals) on the western edge of the Dakota Territory. There was no waiting platform or porter to greet him; he was the sole passenger, disembarking in the still darkness. Along the Little Missouri River you could hear a rustle of cottonwoods like waves along a dock. Everything about the scene had an eerie, ethereal cast. Not far from where Roosevelt was standing, George Armstrong Custer had camped with his detachment in 1876 on his way to the fatal battle of Little Bighorn. And just a couple of days prior to Roosevelt's arrival, the former president Ulysses S. Grant had passed through Little Missouri, riding the railroad west to Gold Creek, Montana, where he would celebrate the hammering of the gold spike connecting the Northern Pacific to the Pacific Coast. About 200 miles to the southwest, the pacified Hunkpapa leader Sitting Bull was now living on the Standing Rock Indian Reservation Agency, isolated in a patch of forlorn prairie along the present-day border between North and South Dakota; the U.S. government held him there as a prisoner of war.[21] Either consciously or unconsciously, Roosevelt was about to insert himself into the closing act of the western frontier's historical pageant.

As Gorringe had instructed, Roosevelt made his way in the pitch blackness along the main street to the Pyramid Park Hotel. The gruff manager let him in and ushered him to a cot in a large communal room. Roosevelt collapsed and fell asleep, happy to have made it to the real West at last. In the morning light, as he rose alongside touchy frontiersmen and saddle-sore wranglers, it all looked very primitive. The washbasin where he tried to shave was clogged with dirty water and stubble, and the hotel towel was soiled with alkali dust. Instead of complaining, Roosevelt seemed to relish the lack of amenities. After breakfast, when he sauntered out of the hotel, his jaw dropped at the exquisite scenery. Instead of the flat, rolling prairies he had encountered in Fargo and Bismarck, here were the ill-shaped bluffs of the fabled Badlands. He set off on a hike of six or seven miles, just to get a quick feel of the imposing landscape and the unvarnished little Dakota town. The horizon seemed infinite. He was for once speechless. "There were all kinds of things of which I was afraid at first," Roosevelt later recalled, "ranging from grizzly bears to 'mean' horses and gunfighters; but by acting as if I was not afraid I gradually ceased to be afraid."[22]

Although the distinctive landscape made the Badlands difficult to travel through—that was why French trappers had originally applied the designation "bad"—there were two reasons it was prime cattle country: an abundance of nourishing stem-cured grasses, and the buttes themselves, which offered heifers decent shelter during winter storms. Owing in part to Brisbin's book (its subtitle was *How to Get Rich on the Plains*), cowboys and others who believed that beef was the new cash cow stampeded to the area. "Montana has undoubtedly the best grazing-grounds in America," Brisbin wrote, "and the parts of Dakota stand next."[23]

Theodore Roosevelt himself had been caught up in this cowboy uproar. Even before he read Brisbin or set foot in the Badlands, he gambled on the cattle business. Along with a Harvard classmate, Richard Trimble, he had ponied up $10,000 to be part owner of a ranch north of Cheyenne, Wyoming, called the Teschmaker and Debillion Cattle Company.[24] Many Wyoming ranchers of the early 1880s preferred the label "drover" to "ranching cowboy" (a term that originated in Ireland around AD 1000), but Roosevelt preferred the latter. Somewhat naively he predicted that the spools of barbed wire would never overtake Wyoming as they had overtaken Texas. His romantic vision of himself was quite specific: a hunting cowboy on the open range. Quite correctly Roosevelt understood that cowboy culture was based on three principles: mobility, custom, and survival of the fittest. As a side project he hoped to document cowboy life for magazines such as *Scribner's* and *Collier's*. "It was a frontier institution," the historian David Dary noted of the first generation of cowboys, "and it died when the frontier died."[25]

By 1883, Texan grangers—merchants of fresh beef for military forts, Indian agencies, immigrant communities, and mining outfits—had discovered that longhorns loved the northern range grasses and could survive the blue winters.[26] That year saw the first great Texan cattle drive to the Little Missouri; and as cowboys swarmed up north, the great Western Trail that went from Bandera through Dodge City to Ogallala was bringing cowboys from Texas to the Dakota Territory in search of open-range opportunities. Down in Pecos, Texas, the world's first rodeo had been held (although in 1989 the *New York Times* noted that two Arizona communities—Prescott and Payson—also claimed bragging rights in this regard.)[27] In Omaha, Nebraska, an Iowa showman, William "Buffalo Bill" Cody, premiered his first Wild West Show; the rage for cowboys and Indians was at full throttle. Meanwhile, the completion of the Northern Pacific Railroad provided owners of livestock in the Dakota Territory relatively easy market access to both coasts. It gave outdoors enthusiasts like Roosevelt the opportunity for a quick western trip between boring sessions of the New York state legislature filled with mundane sheaves of legalese, bills, and charts.

As he recuperated from the summer bouts of asthma and cholera, Roosevelt kept focused on the buffalo trophy he wanted for his library wall. He set his sights on shooting an older buffalo, one past its prime. Too old and exhausted from the commotion of rutting to stay in the herd and unable to court cows anymore, these bulls, known as lonesome Georges, straggled hundreds of yards from the rest of the herd, providing an easier target. The sullen, sick lonesome Georges might symbolize a vanishing West, but Roosevelt would be ecstatic to find any. Their mature heads made ideal trophies.

After an initial hike around Little Missouri to get his bearings, impressed by the solemnity of it all, Roosevelt hired Joe Ferris as his Badlands guide. It was said that if anybody could track down a lonesome George it was Ferris, a Canadian (a New Brunswickian, to be exact) who had moved to the region just a year earlier. There were still, in fact, Acadian inflections in his speech. Virtually everybody said that this onetime lumberjack was a self-starter but never arrogant, an individual who always kept his wits about him—and also, most importantly, a puritan of sorts with Spartan instincts, who never bragged. Still, as Louis L'Amour once wrote of a character, if you stepped on one of Joe Ferris's toes, the other nine would light out after you.

Ferris told Roosevelt that finding either a nimble or a dying buffalo was unlikely. The days when George Catlin could recline in a canvas chair and paint great buffalo hunts were over. From the Osage Hills of Oklahoma to the Flint Hills of Kansas all the way north to the billowing grasslands along the Canadian border, a buffalo was hard to find. Earlier that summer the U.S. government had hired a band of Sioux to slaughter around 5,000 buffalo along the Northern Pacific line, so that the grazing beasts would not cause a train wreck. If you followed the tracks across the Badlands in 1883, in fact, you would have found the bleached bones of buffalo scattered and piled high in mounds. Then, as a follow-up to the "golden spike" ceremony in Montana, the federal government—specifically James McLaughlin, superintendent of the Standing Rock

Indian Agency—again dispatched the Sioux (Lakota) tribe to butcher an additional 10,000 bison. A barbarous bloodbath took place on the Great Plains, and back east the newspapers cheered. "Again, the slaughter was carried out with full federal approval," the historian Edmund Morris later observed, offering an additional reason for the extermination of the buffalo. "Washington knew that plains bare of buffalo would soon be bare of Indians too."[28]

Another pernicious enemy of the buffalo was the telegraph companies. Because buffalo were constantly being attacked by flies—black, snipe, and horse—their backs constantly itched. Regularly, buffalo looked for trees to lean into and scratch against, rubbing so hard that they frequently knocked the trees over. After the Civil War, as telegraph lines were strung across the continent, the buffalo took to the poles as scratch posts, causing them to topple. One telegraph company wizard decided that fastening bradawls to the poles might solve the problem, but the opposite happened.[29] "For the first time [buffalo] came to scratch sure of a sensation in their thick hides that thrilled them from horn to tail," the *Kansas Daily Commercial* lamented. "They would go fifteen miles to find a bradawl. They fought huge battles around the poles containing them, and the victor would proudly climb the mountainous heap of rump and hump of the fallen, and scratch himself into bliss until the bradawl broke or the pole came down." With the failure of the bradawl strategy, the telegraph industry also started slaughtering the animals.[30]

Even though Ferris was reluctant to take Roosevelt buffalo hunting, the rich New Yorker's money was enticing. Eyeing Roosevelt with suspicion, Ferris reluctantly agreed to be his guide. He later mocked the chore as "trundling a tenderfoot."[31]

The first service Ferris rendered Roosevelt was to borrow a proper buffalo hunting gun from crotchety old Eldridge Paddock, a local trapper who also dabbled in real estate. Then the pair saddled up and headed seven miles south in a buckboard to the Maltese Cross Ranch (often referred to as the Chimney Butte Ranch), where they planned on meeting up with two other Canadians, William Merrifield and Sylvane Ferris (Joe's brother). For the first time, Roosevelt saw black-tailed prairie dog towns like those Washington Irving had written about in *A Tour on the Prairies* (1835). These burrowing, yellowish-tan ground squirrels were racing about from hole to hole each yip-yip-yipping the "all-clear" sign to others in the colony, popping in and out of multichambered burrows.[32]

Nothing had prepared Roosevelt for the awesome rugged reaches of the Badlands that he encountered on his horse ride with Ferris in the noontime September heat.[33] With delighted murmurs of awe, Roosevelt was essentially following the so-called Custer's Trail he had read about back in New York. The geography was forbiddingly different, a memorial to stark erosion and sculptured sandstone. There was a prehistoric quality to the outcroppings and battlements, and fierce wind had shaped clay in a helter-skelter fashion unique in the world. (The closest geological counterparts to the Badlands were the arroyos of the Gobi and Namib deserts.[34]) General Alfred Sully, an old Indian fighter, had famously called the arid Badlands "hell with the fires out." The Sioux—like the French—called the terrain Mako Shika ("land bad"). Writing in an 1876 edition of the esteemed journal *The American Naturalist*, to which Theodore Roosevelt subscribed, J. A. Allen described the area as a "boundless expanse" that reminded him of a "fierce sea."[35]

More than any other landscape that Roosevelt would ever encounter, the Badlands had an inspiring resilience that swept him away into an almost spiritual state of appreciation. To him the desolate stretch of ridges and bluffs seemed "hardly proper to belong to this earth."[36] The towering buttes and scarred escarpments told geological stories of the prehistoric upheavals, the deposits, the erosion of forgotten times.[37] There was, he said, a sacredness to the Badlands silhouette against the oceanic sky that exuded a cosmic sense of God's Creation as described in Genesis. A cowboy could disappear into the Badlands and never be heard from again. Everything to Roosevelt, in fact, seemed magically contorted. Famously, he joked that the Badlands reminded him of the way Edgar Allan Poe wrote tales and poems.[38] These buff buttes and towering sandstone pinnacles seemed to change shades by the hour, from heliotrope red to horizon blue to nickel gray to a blaze of different oranges. Everywhere bands of brownish yellow formed by shale exposed heavily cut Badlands ravines. "In coloring they are as bizarre as in form," Roosevelt would write. "Among the level, parallel strata which make up the land are some of coal. When a coal vein gets on fire it makes what is called a burning mine, and the clay above it is turned into brick; so that where water wears away the side of a hill sharp streaks of black and red are seen across it, mingled with the grays, purples, and browns."[39]

When Roosevelt and Ferris finally arrived at the Maltese Cross Ranch, they were met with reserve. Roosevelt wore spectacles; he spoke in a falsetto voice, which to the uninitiated could be as irritating as a whistle; and his talk was peppered with "by Joves" and "my boys," dead giveaways that he was an aristocratic swell who never before had been west of the Yellowstone divide. "When I went among strangers I always had to spend twenty-four hours in living down the fact that I wore spectacles," Roosevelt recalled, "remaining as long as I could judiciously deaf to any side remarks about 'four eyes' unless it became evident that my being quiet was misconstrued and that it was better to bring matters to a head at once."[40]

Day after day fanciful hunters like Roosevelt came to kill game in the Dakotas, and enthusiastic comparisons were made to the bush country of British East Africa, where everything was wild. As advertised, wildlife truly did prolifically flourish in the Badlands region. Tree-rich bottomlands, for example, created an ideal habitat for browser animals of all shapes and sizes. Hunters often marveled at the swelling, verdureless red surface extending as far as the eye could see. But then autumn ended and the hunters fled, and the long winter of the Badlands, routinely colder than the bluest days in Maine, hammered down with a numbing thud and all locals were left to eat venison jerky, stay warm, and wait for the springtime thaw.

Although both Ferris and Merrifield—Roosevelt's ranching partners—were cordial, they were clearly lukewarm about hunting down a tired old buffalo with an aristocratic swell. The two of them wore identical expressions, which read, "Not bloody likely." In any event, they were busy raising 150 cattle, hoping for a big payday when heifers were sold.

Luckily for Roosevelt, shortly after their arrival at the Maltese Cross Ranch, a hungry bobcat (weighing approximately twenty-five pounds) got loose in the chicken coop, creating havoc and sending feathers flying. All four men raced out of the ranch cabin—an edifice of a story and a half with a high-pitched shingled roof—to ambush the agile predator. The bobcat got away, but the attendant laughing and cursing broke the ice. The initially cold attitude toward Roosevelt dissipated in favor of cowboy camaraderie.[41] Still, only when Roosevelt offered to pay did Sylvane and Merrifield grudgingly lend him a mare for his buffalo quest.

Although lonesome Georges were more easily hunted than other Great Plains game like antelope or white-tailed deer, the chase still presented a serious challenge. Far from being a "tame amusement,"[42] as Roosevelt put it, a buffalo could turn mean and with a belching snort charge like a mad bull in an unexpected flash. The novelist Thomas Berger accurately noted the inherent danger when he wrote in *Little Big Man* that "buffalo run a mile in one minute and will stampede on a change of wind."[43] Roosevelt, in his conservation-tinged hunting essays, was somewhat defensive about his compulsion for shooting an endangered species. "It is genuine sport," he insisted; "it needs skill, marksmanship, and hardihood in the man who follows it, and if he hunts on horseback, it needs also pluck and good riding. It is in no way akin to various forms of so-called sport in vogue in parts of the East, such as killing deer in a lake or by fire hunting, or even by watching at a runaway."[44]

III

Roosevelt took a real shine to William Merrifield, later writing in his diary that Merrifield was "a good-looking fellow who shoots and rides beautifully, a reckless, self-confident man."[45] The evening of the bobcat's attack, Roosevelt slept on the hard clay floor at the one-room Maltese Cross Ranch cabin, insisting on high-minded principle that he'd never stoop so low as to take another man's bed. Perhaps he wanted to replicate that evening of the buffalo robe three Septembers earlier in the Red River valley. At any rate the gesture was keenly noted by Joe Ferris and Bill Merrifield. Rising at dawn, Roosevelt saddled up his horse (named "Nell" after his brother Elliott), grasped the reins, and trotted south to hunt his buffalo trophy. Jouncing beside him was Joe Ferris, whose horse pulled a wagon full of outback supplies.

For once Roosevelt seemed to be at a loss for words as they followed a creek meandering across a valley tucked between skyscraper rock and curtain wall. The hypnotic landscape was the promised land for anybody afflicted with even a touch of claustrophobia. Nobody has ever visited North Dakota and felt hemmed in. Like all creeks in the Badlands this one ran into the Little Missouri River. If you studied an aerial photograph, the topography looked like random lines on a hand palm or leaf veins squiggling in all directions. There was a trickling creek, it seemed, around every bend. As if living out the dream of an "old regular" or half-breed trapper, Roosevelt experienced in the Badlands the freedom to live without the shackles of the urban world. Windswept plains, unmapped wilds, the howls of hungry coyotes—all this was part of the Badlands experience for Roosevelt.

What Roosevelt had to offer Joe Ferris—and every Dakotan he rode with—was his encyclopedic knowledge of Badlands birds. A particular favorite of his were the nocturnal brown thrashers. Ferris was no doubt impressed that his hunting partner could identify sparrow species or melodic songsters just by tilting his ear. "One of our sweetest, loudest songsters is the meadow-lark," Roosevelt wrote. "This I could hardly get use to at first, for it looks exactly like the eastern meadow-lark, which utters nothing but a harsh, disagreeable chatter. But the plains air seems to give it a voice, and it will perch on the top of a bush or tree and sing for hours in rich, bubbling tones."[46]

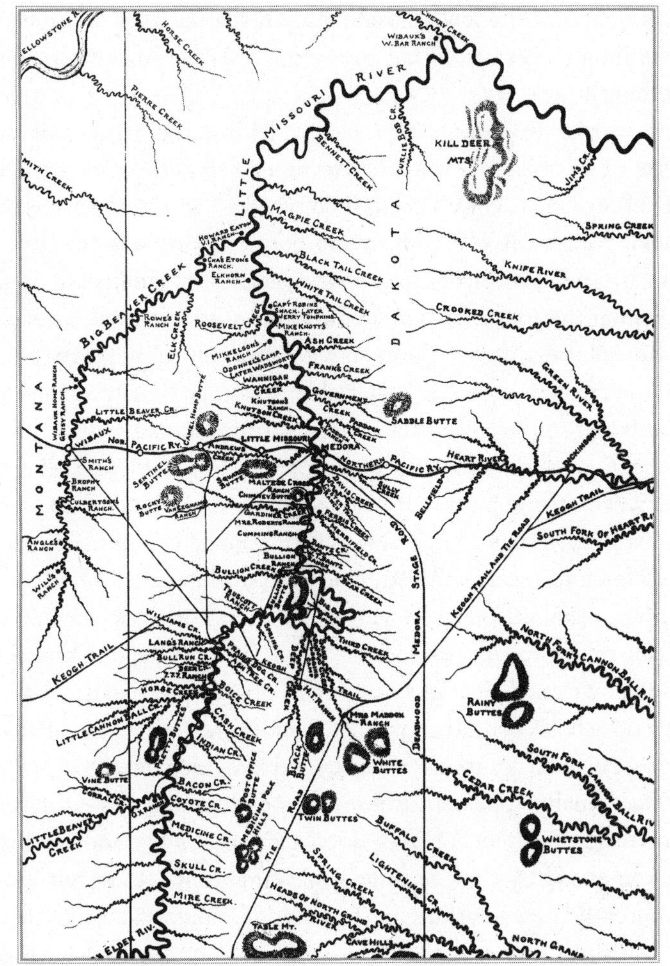

Map of the Little Missouri River in the Dakota Badlands with all its creeks and offshoots.

Roosevelt and Ferris made it by dusk to their destination—a ramshackle, rat-infested cabin in a field situated at the mouth of Little Cannonball Creek. Out to greet them with extended hands were Gregor Lang and his sixteen-year-old son, Lincoln, who were operating the Neimmela Ranch for the rich London capitalist Sir John Pender.[47] According to Lincoln, Roosevelt was full of hearty cheer, saying, "Dee-lighted to meet you!" In his memoir, *Ranching with Roosevelt*, Lincoln recalled the wild-eyed "radio-active" enthusiasm of the future president. "I could make out that he was a young man, who wore large conspicuous-looking glasses, through which I was being regarded with a pair of twinkling eyes," Lang wrote. "Amply supporting them was the expansive grin overspreading his prominent, forceful lower face, plainly revealing a set of larger white teeth. Smiling teeth, yet withal conveying a strong suggestion of hang-and-rattle."[48]

After supper Roosevelt held court, telling the Langs his stories about the world at large. Even after the others fell asleep, Gregor, a sharp-whiskered Scotsman, and Theodore kept going at the big issues of the day, locking horns over literature and politics. This was the beginning of a lifelong friendship Roosevelt would have with the Langs. As fellow plainsmen on the prowl for buffalo during the next week, Roosevelt and his new hired friends grew closer. What Roosevelt relished about buffalo hunting, it seemed, was that social class was temporarily suspended. A man's skill and courage were the criteria for acceptance.

The next morning, Roosevelt's hunt party traveled together in a heavy rain, between conical buttes, anxious to find a single lonesome George or a small band of younger bulls. The soaking made it impossible to track any buffalo, and it made the Badlands clay (often called gumbo) slimy and slippery; this was very dangerous for their horses, who could easily break a leg trying to climb hillocks.[49] With creeks rising quickly and rain pounding down on their backs, Roosevelt's party spent most of the time avoiding ooze holes and slow sand. At the rate they were going Roosevelt would have been lucky to bag a turkey vulture or common skunk. When a mule deer finally appeared, Roosevelt took aim and missed. It was an embarrassingly bad shot. Quickly Joe Ferris took a try and got his kill. Deeply impressed by the Canadian's marksmanship, Roosevelt shouted, "By Godfrey I'd give anything in the world if I could shoot like that."[50]

That smallish deer was the high-water mark of the hunt for the first four days. Ferris hinted that it might be wise to venture back to Little Missouri and dry off for a spell; but Roosevelt insisted they grind on. Lincoln Lang was surprised at how calm Roosevelt seemed to be in the teeth of a downpour. He positively glowed in the deluge. At one point, wallowing in the flash-flood puddles, he applied mud to his face as an emollient, almost like a Lakota Sioux putting on war paint. The other men watched in shocked silence, but Lincoln dubbed him the Great White Chief.[51]

Despite the unrelenting bad weather, Roosevelt continued to be entranced by his surroundings. The winds were as fierce as those along any seashore. The light—when it got a chance to break through the clouds—was often an amazing chartreuse. Like the ocean floor, much of the region was still unknown to cartographers. Every day in the Badlands he encountered some new revelatory feature of intense geological interest. Here he felt like a French-Canadian *voyageur* far away from the stresses of civilization. This, of course, wasn't the first time Roosevelt had succumbed to the lure of a wild place; he had similar bouts of euphoria in the Adirondacks and the North Woods. But this was somewhat different; and he began to entertain the romantic notion of becoming a Dakotan rancher. Even the clumps of box elder and prickly plants appealed to him. "Clearly I recall his wild enthusiasm over the Badlands," Lincoln Lang later wrote. "It had taken root in the congenial soil of his consciousness, like an ineradicable, creeping plant, as it were, to thrive and permeate it thereafter, causing him more and more and more to think in the broad gauge terms of nature—of the real earth."[52]

After days of striking out the Roosevelt party caught a break. They discovered fresh spoor, and off they went in pursuit. Suddenly, there was a buffalo in sight, but upon hearing their clamor it galloped off. For several miles Roosevelt chased the bull through a rough patch of prickly shrubs and eroded gullies, to no avail. Later that afternoon, the men came across three buffalo grazing within fairly easy firing distance. Roosevelt quickly dismounted, took aim, and fired. The bullet penetrated the flesh of one, but the wounded buffalo ran off. Desperate for his big game trophy, Roosevelt chased the buffalo for seven or eight miles, only to miss with his next shot. Once again the buffalo got away. Once again Roosevelt was embarrassed.

Even though Roosevelt loved hunting, he was not a great shot; his poor eyesight prevented that. "Whatever success I have had in game-hunting," Roosevelt later wrote, "has been due, as well as I can make it out, to three causes: first, common sense and good judgment; second, perseverance, which is the only way of allowing one to make good one's own blunders; third the fact that I shoot as well at game as at a target. This did not make me hit difficult shots, but it prevented my missing easy shots, which a good target shot will often do in the field."[53] What he brought to hunting was instead a bookish knowledge of the species' habits and coloration. But Roosevelt was so excited by the windswept panoramas that he didn't comprehend his own clumsiness with a rifle. That evening by the campfire, he remained optimistic about bagging his trophy. It was raining again the next morning when the Roosevelt party stumbled upon a couple of grazing buffalo. Theodore fired and missed his mark again. This time, at least, he could blame the weather.

The hardships Roosevelt endured in pursuit of the buffalo were many. Ants had built huge communities eight or nine feet deep. On one occasion, crawling in the sage to get closer to a bull, Roosevelt stumbled right into a cactus patch, and his hands were suddenly filled with needles as if they were pincushions; they stayed swollen for days. When the hunt party decided to charge at a couple of buffalo, Nell got spooked and tossed its head dramatically backward, causing Roosevelt's rifle to smack against his forehead. According to Roosevelt the blood literally "poured" into his eyes from the stitchable gash.[54] As the blood congealed, however, he spoke excitedly about the prospect of returning home with a purple scar. That evening, his face bruised, forced to sleep in the cold rain, with nothing but dry biscuit in his stomach, Roosevelt glowed with enthusiasm, refusing to engage in tremulous self-pity. It was the experience of freezing while skating at Cambridge all over again. A miserable Joe Ferris, shivering under a wet blanket, marooned in the backcountry darkness, was baffled that evening to hear Roosevelt exclaim, "By Godfrey, but this is fun!"[55]

Doggedly Roosevelt kept hunting through the broken plains and pony paths along Little Cannonball Creek for his buffalo trophy. All a frustrated Joe Ferris could remember thinking was that "bad luck" was following them "like a yellow dog follows a drunkard."[56] On the morning of September 20 Merrifield and Sylvane Ferris left the hunt for an entirely unanticipated reason: Roosevelt, in a fit of exuberance, had handed them a $14,000 check to guarantee his partnership in the Maltese Cross Ranch. Roosevelt was to become a Dakota rancher. As if they had just won the lottery, Merrifield and Sylvane were ecstatic to be trusted with an investment check and tapped to be his highly paid new managers. The two were catching a train to Minnesota to iron out all the business and banking details. Basically, by signing his name once, Roosevelt had bought the boys hundreds of new cattle on spec.

Roosevelt's luck finally changed as, for the first time in his life, he ventured into Montana Territory, hoping to find his buffalo. Noticing that his horse was sniffing something in the air, Roosevelt dismounted, jogged up to a ridge, and peered over. There, grazing on grass, was a buffalo. "His glossy fall coat was in fine trim and shone in the rays of sun," he later wrote, "while his pride of bearing showed him to be in the lusty vigor of his prime." This wasn't a lonesome George in size, but close enough. Stealthily Roosevelt advanced, one quiet foot at a time, to get within range. When he was about fifty yards away he fired a single shot. The bullet entered the buffalo's massive shoulder. "The wound was an almost immediately fatal one," Roosevelt wrote, "yet with surprising agility for so large and heavy an animal, he bounded up the opposite side of the ravine, heedless of two more balls, both of which went into his flank and ranged forwards, and disappeared over the ridge at a lumbering gallop, the blood pouring from his mouth and nostrils."[57]

Sprinting ahead Roosevelt, sweating profusely, followed the blood trail until he found the buffalo "stark dead" in a ditch. All the buttes surrounding Roosevelt now took on a special glow. Hopping from foot to foot, Roosevelt encircled the buffalo, whooping and chanting as if he were White Bull or Two Moons in an effort to pay this "lordly buffalo" due reverence. A perplexed Joe Ferris had never imagined any white man behaving in such a queer fashion, imitating a Sioux or Cheyenne. An exhilarated Roosevelt, in an act of spontaneous generosity, next opened his wallet and handed Ferris $100. "I never saw any one so enthused in my life," Ferris recalled, "and, by golly, I was enthused myself. . . . I was plumb tired out . . . I wanted to see him kill his first one as badly as he wanted to kill it."[58]

That evening the men stuffed themselves on buffalo steak, Roosevelt claiming that the meat from the hump tasted best; this was contrary to George Catlin's promotion of buffalo tongue being the true delicacy. To Roosevelt buffalo meat was barely distinguishable from beef. The hunters didn't sever the head or skin the carcass, however, until the following day. "The flesh of this bull tasted uncommonly good to us," Roosevelt wrote, "for we had been without fresh meat for a week." The *New York World* had caricatured him as a Harvard-educated aristocrat, but from now on he'd be an all-American buffalo hunter.[59]

IV

When Roosevelt returned to Little Missouri on September 23, to spend another night at the Pyramid Park Hotel before heading east, he was a changed man. Francis Parkman had been right: only by living out the western experience could a scholar effectively write about it. Roosevelt's fifteen-day growth of beard in the Dakota wilderness, and his rumble in the West, had strengthened him both mentally and physically. Now, as he slept on a cot at the Pyramid Park Hotel, he felt that he was one of the hardy trappers in the Jim Bridger vein, not Jane Dandy or Lil' Punkin. In *Manliness and Civilization: A Cultural History of Gender and Race in the United States 1880–1917*, the historian Gail Bederman dissects Roosevelt's obsession with becoming a "man's man." Pointing out how his political opponents in New York used to ridicule him as the "exquisite Mr. Roosevelt," Bederman argues that Roosevelt's "cowboy of the Dakotas" persona was an attempt to stamp out any traces of effeminacy. No longer would he be publicly insulted as "given to sucking the knob of an ivory cane," a phallic insult Roosevelt had been forced to endure, for he was now a virile buffalo hunter straight out of "Cowboy Land."[60]

Bederman also explains the two different strains of masculinity that Theodore was juggling at age twenty-four. From his father he had inherited Victorian codes of "moral manliness" including unselfishness, chastity, physical strength, honesty, and altruism. Yet, as noted, his father had rejected serving as a Union soldier in the Civil War, hiring a surrogate in his place. Young T.R. had been humiliated by his father's wartime decision. So perhaps he tried overcompensating in his effort to embrace the ethos of frontier masculinity in which a propensity for violence was often rewarded. Killing an animal, winning a fistfight, and declaring a duel, in other words, were obvious ways to cultivate his deficient "natural man" side. Therefore, Roosevelt felt a need to emulate Indians while simultaneously conquering them, and a need to worship big game like buffalo only to hunt them down. The Victorian mannered man turned to prim Europe whereas the "natural" American always had a westward focus. "On his first trip to the Badlands in 1883, he was giddy with delight and behaved as much like a Mayne Reid hero as possible," Bederman wrote. "He flung himself into battle with nature and hunted the largest and fiercest game he could find. As a child, he had been attracted to natural history as a displacement of his desire to be a Western hero. Now, shooting buffalo and bullying obstreperous cowboys, he could style himself as the real thing."[61]

Roosevelt's solid Victorian morals, however, were never expunged as he became a Great Plains hunter and Dakota rancher. Unlike most men wanting a buffalo head, he actually thought about, and was angered by, the possibility that the great herds might become extinct. His Harvard education in Darwinian biology and naturalist studies gave him a perspective on western wildlife that no ordinary cowboy or hunter could have had. As Lincoln Lang later noted, every day Roosevelt increasingly came to understand the "definite purpose of every natural [object] he saw in the Bad Lands."[62] In the coming decades, his "man's man" side hunted big game while the intellectual Harvard part of his personality would preserve things of great environmental beauty and consequence.

Considered in the light of Bederman's thesis, hunting in the Dakota hills and killing a buffalo were the culmination of a series of masculine initiation rites T.R. had put himself through starting in Maine in 1871 with the incident at Moosehead Lake. Overall, life was going well—he had established himself as a historian, a hunter, and a reform politician;

the very fact that Alice was pregnant proved (to his mind) his virility; his health, while still fickle, was on an upswing. No wonder Roosevelt felt ebullient as he boarded his eastbound train. For in addition to everything else, he no longer believed himself to be a weakling or tenderfoot. Any remaining hints of self-disgust had been vanquished. Although he went to exaggerated extremes to get there and was physically spent by exaltation and fatigue, Roosevelt now saw himself as a western man, not a rich boy whose father had refused to serve in the Civil War.

Once Roosevelt had his buffalo head onboard the train, he was ready to journey back to New York in a Pullman berth. Roosevelt wrapped his prize (weighing approximately twenty-five to thirty pounds) in burlap, loaded it onto a Northern Pacific railroad car, and headed east to Saint Paul.[63] Unlike Texas longhorns, buffalo were singularly unimpressive if you stripped off their horns, just two stubby prongs jutting upward. A rack of deer or elk antlers was, by contrast, far more impressive. But a buffalo head, in all its grandeur, had become coveted all over America for saloon and library walls. The Union Pacific railroad system even acquired buffalo heads to hang in all its scattered depot offices.[64]

Roosevelt returned to Alice as a conquering hunter hero. Of course he proudly hung his buffalo head (a taxidermist in Saint Paul had mounted it) in their Manhattan home. That fall, he talked excessively about the freshness and vigor of the West. After winning a third term in the New York state legislature, he put himself forward for speaker of the assembly at the end of the year, offering a capsule biographical sketch that claimed he was a man of Harvard, Albany, and the Dakota Territory.[65] Yet when he lost the speakership, he characteristically found the silver lining. "The fact that I had fought hard and efficiently . . . and that I had made the fight single-handed, with no machine back of me, assured my standing as floor leader," he wrote. "My defeat in the end materially strengthened my position, and enabled me to accomplish far more than I could have accomplished as Speaker."[66]

Starting in January 1884 Roosevelt found himself working almost full-time in Albany. He wanted nothing less than to break up the political machines of *both* parties in New York City and was also consumed with passing a series of municipal reform laws. Strapped for cash after writing the fat check in the Dakotas for cattle, Roosevelt decided to lease out his brownstone and move back into the house on West Fifty-Seventh Street with his mother. T.R.'s sister Bamie, married to Douglas Robinson, had recently given birth, and she also moved in. Alice suddenly had two family members—Mittie and Bamie—to look after her as her own pregnancy moved into its ninth month. Meanwhile, the construction of Leeholm (Sagamore Hill) continued. "How I did hate to leave my bright, sunny little love yesterday afternoon," Roosevelt wrote to Alice in early February from Albany. "I love you and long for you all the time."[67]

Just days later tragedy struck the Roosevelt clan. On February 13, Theodore received a telegram announcing that Alice had just given birth to a girl. A plethora of hearty congratulations were telegraphed from fellow legislators and friends. Cigars were lit and glasses hoisted in his honor. But then a second telegram arrived. Although it didn't survive, it was probably from Elliott and read something like: "There is a curse on this house. Mother is dying, and Alice is dying too."[68] (For that is what Elliott later told his brother in person at midnight.) Alice was afflicted with Bright's disease (medically termed acute or chronic nephritis), and his mother had typhoid fever. A panic-stricken Theodore raced to board a train for Grand Central Station. The fog through Dutchess County was pea-soup thick as if village after village were floating in clouds. All he could do was sit and pray. It was around midnight when he finally arrived in Manhattan and made his way to West Fifty-Seventh Street. The pall of death hung all about as he entered the parlor and climbed the stairs to see Alice and the baby on the third floor.

Alice was drifting into and out of consciousness. As Roosevelt took her in his arms, his spirit broke down. It was as if the fog had entered his throat. With his surety evaporated, he simply didn't know what to do except clutch her and sob. As he watched her head sinking into the pillow his emotions ran the gamut from contempt of God to guilt for being away in Albany. Her breathing was soft and low, and he berated himself for not being a better husband. Bright's disease ravaged the kidneys—its symptoms included vomiting, high fever, and excruciating back pain. Breathing became difficult, the body became puffy, and the urine turned bloody. It was death by slow torture.[69]

The situation was no better on the second floor. Roosevelt's mother was in utter misery, with a sustained fever of over 104 degrees as well as gastroenteritis and diarrhea. In the 1880s, when there were no antibiotics, death took one out of every ten patients afflicted with typhoid. The situation was beyond bleak. Mother had always been his one-woman support system. Without her he feared being rudderless.

On February 14 (Valentine's Day) Mittie died at two o'clock in the morning. Twelve hours later so did Alice.[70] Two days later a cold spell gripped New York as two hearses made their way to the Presbyterian Church on Fifty-Fifth Street and Fifth Avenue. All the leading philanthropists and politicians in the city—including the Astors and Harrimans—arrived to pay their last respects to the deceased Roosevelt women.[71] The *New York Times* and *New York Sun* covered the double funeral as if it were an important event.[72] After an opening prayer, the old hymn "Rock of Ages" was sung by a chorus of mourners paying their respects. On top of the two rosewood coffins were wreaths of roses and green vines. Following the benedictions, Roosevelt's mother and wife were buried at Greenwood Cemetery next to his father. During these painful days of February, Roosevelt returned to keeping his diaries. "The light has gone out of my life," he wrote, and his words were accompanied by a huge cross on the page. A couple of days later he added, "For joy or for sorrow my life has now been lived out."[73]

Nobody will ever know the depths of the private pain that Roosevelt felt as winter changed into spring. For months afterward, everybody used kid gloves when dealing with him. His former tutor Arthur Cutler wrote to Bill Sewall in Maine that Theodore was stuck in a "dazed, stunned state." Roosevelt himself put on a stoic veneer when writing to Sewall: "It was a grim and evil fate, but I have never believed it did any good to flinch or yield for any blow, nor does it lighten the blow to cease from working."[74]

During sleepless nights that spring, Roosevelt would sit in a rocking chair, silent as smoke, and read natural history books. The world seemed quite diabolical. He wondered whether the Badlands—where even the half-clad buttes had an unstable equilibrium—might be the best place to heal and hatched a plan to light out for the Dakota Territory following the Republican National Convention in Chicago. Perhaps solace could be found in a ranch house with undraped windows surrounded by roping corrals and branching chutes. A saddle horse would probably be his best companion, his true equal and friend.

People always devise their own ways of coping with loss. Roosevelt took the route of bottling up his emotions, seldom mentioning his wife by name, submerging her memory, and never reminiscing about her legacy to his daughter Alice. Oddly, he didn't even invoke her name in his own *An Autobiography*. It was as if Roosevelt believed he could best respect his beloved wife in silence. Nevertheless, upon a return visit to North Dakota, he holed up in the Maltese Cross cabin and edited a volume of memorials about Alice; he had it privately printed by G.P. Putnam's Sons. "She was beautiful in face and form, and lovelier still in spirit; as a flower she grew, and as a fair young flower she died," he wrote in the introduction. "Her life had been always in the sunshine; there had never come to her a single great sorrow; and none ever knew her who did not love and revere her for her bright, sunny temper and her saintly unselfishness."[75]

Despite his grief Roosevelt that spring nevertheless engaged in politics at Albany with full fervor. Even though he loved the notion of General William T. Sherman as the Republican nominee for president, he reluctantly settled on the more pedestrian James G. Blaine. More and more his political coach was Henry Cabot Lodge of Massachusetts. Although Lodge didn't share Roosevelt's enthusiasm for the wilderness he was keenly interested in organized fox hunts. A trusting Roosevelt used Lodge as a confidant and sounding board. Unlike Roosevelt, Lodge was terse, calculating, and unemotional. But he was also deeply honest, loyal, and a gentleman. Both shared a bedrock belief in the virtues of American exceptionalism. Both were dogged in their pursuit of western expansion. That spring Roosevelt and Lodge traveled together by train to Chicago for the smoky bedlam of the Republican National Convention. (Roosevelt left his infant daughter, whom he'd soon nickname Mouseskeins, in the care of his sister Bamie.) Although Lodge knew that Roosevelt had developed a reputation as a reformer, he was surprised at what a folk hero his New York friend had become with the western Republican delegates, merely for shooting a buffalo in the Dakota Territory. A rumor circulated at the convention that when the territory became a state, Roosevelt would probably be its first U.S. senator.

V

For his part, Roosevelt couldn't wait to get out of Chicago. As soon as the convention was wrapped up, with Blaine as the nominee, he took a train to Saint Paul. Near a nervous breakdown, his entire exhausted body in low-grade pain, Roosevelt turned into a semi-recluse, not wanting to read newspapers or receive telegrams from anybody. Arriving in the Little Missouri area on June 9, he went directly to the Maltese Cross Ranch, anxious to begin his life as a cowboy and hunter.[76]

Since September Merrifield and Ferris had tended to Roosevelt's cattle herd of around 440 head; only twenty-five had been killed by wolves or the cold.[77] The coulees and buttes, as hoped, had adequately protected the herd. Riding around the region and seeing the new buildings that had sprung up in the tiny boomtown of Medora bolstered Roosevelt's morale—even if he still couldn't imagine life without Alice and Mittie. In another burst of Rooseveltian enthusiasm, he wrote Gregor and Lincoln Lang a $21,000 check to acquire 1,000 new cattle. His investment in the Badlands was now more than $35,000.

In photographs taken at the time, Roosevelt is often wearing a custom-made buckskin suit. He had commissioned the outfit from a seamstress in Amidon, North Dakota, because its "inconspicuous color" was ideal for hunting antelope; it caused, he wrote, "less rustling" than other fabrics "when passing among projecting twigs." But the show-off in him also wanted the fringed suit and its accompanying hunting shirt because they were "the most picturesque and distinctively national dress ever worn in America," the uniform in which "Daniel Boone was clad when he first passed through the trackless forests of the Alleghanies. . . . It was the dress worn by grim old Davy Crockett when he fell at the Alamo."[78]

Despite his fanciful wardrobe, Roosevelt, as rancher, was a workhorse (not a showhorse). He participated with a vengeance in round-ups and brandings, becoming a decent roper and a cool presence during stampedes. While his horsemanship wasn't exceptional he always had a good rapport with his mount. He learned how to braid a halter and bridle rein as if born on the range. With the crack of dawn he was up, anxious to perform morning chores. Whether it was going to find a stray or fixing a fence or coping with foul weather, Roosevelt always volunteered, at least to the point of showing to Merrifield and Ferris that the elitist in him had disappeared forever.[79]

Intoxicated with the Badlands, Roosevelt decided to ask his North Woods friends Bill Sewall and Wilmot Dow to come jump-start his North Dakota ranch with him at the Maltese Cross. Generously Roosevelt insisted he would share all profits with them. They would all reunite in the West as business partners and kinsmen.

Roosevelt spent only three weeks in the Badlands. On June 30 he headed back east to spend time with his baby daughter in Massachusetts. But his thoughts kept returning to the West, as he thought about building a second ranch, to be called the Elkhorn, about twenty-five miles north of Medora. The Maltese Cross was too close to town, attracting a constant stream of locals eager to shoot the breeze with a newsy New Yorker. If he wanted to write books about the Badlands, he would need solitude. His tentative plan was to divide his time between New York and the Little Missouri River area (the building of his Sagamore Hill estate continued).

No sooner did Theodore arrive back east than he wrote Bill Sewall another letter. To Roosevelt, Sewall was like one of the characters Chekhov wrote about who were the salt of the earth but were so virtuous that they never tasted success; he was now hoping to change this. "If you are afraid of hard work and privation, do not come west," Roosevelt wrote to Sewall. "If you expect to make a fortune in a year or two, do not come west. If you will give up under temporary discouragements, do not come west. If, on the other hand, you are willing to work hard, especially the first year; if you realize that for a couple of years you cannot expect to make much more than you are now making; and if you also know at the end of that time you will be [in] receipt of about a thousand dollars for the third year, with an unlimited rise ahead of you and a future as bright as you yourself choose to make it, then come."[80]

In late July, Roosevelt announced to the New York press that he would indeed, though reluctantly, back James G. Blaine for president—causing a wave of speculation that he was abandoning his reformist independence to embrace the Republican machine. Before he once again headed west a reporter for the *New York Tribune* buttonholed Roosevelt and asked about his sudden advocacy of Blaine. Roosevelt snapped that he was "disinclined to talk about the political situation" yet happy to discuss his newfound "life in the West."[81]

Accompanying Roosevelt on this trip to the Badlands were Sewall and Dow (the straight talk in Roosevelt's letter had worked). They had journeyed down from Maine to join their boss at the New York railroad station and take the Chicago Limited. The *New York Herald* reported that Roosevelt was carting along on the train all sorts of western paraphernalia—a heavy monogrammed saddle, angora chaps, a pearl-handled revolver, and silver-inlaid bits and spurs. He had temporarily purchased only squatters' rights, so the game plan was to start living in an existing dilapidated hut at once and then purchase 1,000 new shorthorn cattle in Minnesota. Timber for a new primary ranch house—a close approximation to the "Ranchman's Log 'Schack' " as featured in the *History of the Northern Pacific Railroad*—would be cut in September, and construction would commence early the next year. Referring to the neophytes from Maine as his "backwoods babies," Roosevelt got a huge kick out of pointing out, from the train window, Wisconsin dells, Minnesota lakes, and emerging rimrock canyons of the Badlands. Writing to Bamie on August 12, Roosevelt said that his lumberjack friends exhibited "absolute astonishment and delight at everything they saw" as they traversed the upper Midwest and that their "very shrewd and yet wonderfully simple remarks were a perfect delight to me."[82]

From the outset, however, Sewall had reservations about the Badlands as cattle country. He understood Roosevelt's infectious excitement about the surreal terrain, but it seemed too arid for cattle. After meeting the Ferris brothers and Merrifield and working long days to get the Elkhorn

Ranch built to Roosevelt's specifications, Sewall wrote to his brother in Maine what he really thought of the whole "range management" enterprise: "Tell all who wish to know that I think this is a good place for a man with plenty of money to make more," he said, "but if I had enough money to start here I never would come."[83]

What Sewall didn't fully comprehend was that Roosevelt was multitasking: ranching at the Maltese Cross and Elkhorn was an excuse to hunt and write a popular book about his adventures. Never thrifty with money, essentially a horrific businessman, Roosevelt was less concerned that he had squandered one-fifth of his fortune in the cattle business than he was with collecting naturalist data and hunting anecdotes for the book he wanted to write about the Badlands. In addition, just being in the Dakota Territory was a balm to his grief, bringing him much-needed clarity. Life was short, he felt, so make the most of it. In fact, no sooner were the first logs split for the Elkhorn Ranch than Roosevelt announced that he was ready for a 1,000-mile hunt on horseback. The very name of the range he wanted to explore—the Bighorn Mountains—had him salivating.

VI

Located in both southern Montana Territory and north-central Wyoming Territory, the Bighorns—a range of the Rockies—had a truly varied ecosystem for more than 200 miles, with everything from sheer mountain walls to tall grasslands, from glacier-cut valleys to alpine meadows populated by an amazing array of raptors. Accounts of the mountain men who first saw Cloud Peak and Black Tooth Peak abruptly rising out of the rolling prairie were legendary. Jim Bridger had floated through the lofty Bighorns on a raft, and Jedediah Smith had been mauled by a grizzly bear not too far away. When escaping General Crook's Army in 1876 the Sioux had taken refuge in the lodgepole pine and spruce forests of the Bighorns as if these were the last haven on earth.[84] Old-time trappers with horseshoe mustaches said this eastern front range of the Rockies had more game animals than the human eye had ever seen—and particularly the thought of hunting a grizzly bear inspired Roosevelt to undertake the most arduous expedition of his life so far.

Most of the Bighorn peaks were rounded on top, with flanks that gently sloped. The glacial lakes throughout the range were as bright blue as those in Alberta or the Yukon. Temperatures during the winter months could unexpectedly drop to forty degrees below zero in a few hours. Yet, compared with other Rocky Mountain zones, the Bighorns received little snow. All the spring rains emanated from general weather systems. Vegetation encountered in the Bighorns depended completely on whether one was below or above the timberline. Lumber companies were eyeing the area as a prime source of timber.[85]

To ride horseback and hike into the Bighorns, Roosevelt took with him Merrifield and Norman Lebo, an old Union soldier and blacksmith from Ohio. Sewall and Dow stayed at the Maltese Cross and Elkhorn ranches to tend the cattle. While Roosevelt and Merrifield rode horses, Lebo followed with the "prairie schooner" supply wagon.[86] From Medora to the foothills of the Bighorns was nearly 300 miles across the chilly flatlands toward the town of Buffalo, Wyoming. With no map to guide them the trio simply mounted and started heading toward the Montana line as if nomadic characters in a Zane Grey novel. "We had no directions as to where the Big Horns were," Merrifield recalled, "except that they lay to the southwest."[87]

No sooner did the trio reach Montana than a storm appeared. Horrific cloudbursts filled the big sky as lightning bolts popped and boomed. Day turned to night. Heavy raindrops fell, and their horses tried to run away. The odor of ozone, stronger than in the East, was almost intoxicating. "The storm rolled down toward us at furious speed and the wind shrieked and moaned as it swept over the prairie," Roosevelt recalled in *Hunting Trips of a Ranchman*. "We spurred hard to get out of the open, riding with loose reins for the creek. . . . The first gust caught us a few hundred yards from the creek, almost taking us from the saddle, and driving the rain and hail in stinging level sheets against us."[88]

The detailed diaries Roosevelt kept of the trek to the Bighorns tell exciting stories of shooting duck at Lake Stanton (filled with cutthroat trout) and hearing coyotes wail all night when the men camped along the Powder River. Nearly every Wild West cliché happened to Roosevelt on the trail: a non-injurious shooting powwow with the Cheyenne; the near-loss of their wagon to quicksand; the hunting of enough grouse to feed a village; magnificent herds of white-tailed deer. At one point Roosevelt miraculously shot two deer with a single bullet. "I elevated the sights (a thing I hardly ever do) to four hundred yards," he wrote, "and waited for the second buck to come out further, which he did immediately and stood still just alongside of the first. I aimed above his shoulder and pulled the trigger. Over went the two bucks! . . . This was much the best shot I ever made."

A full three days before reaching them, Roosevelt could see the Bighorns rise in the distance over the plateau. He couldn't wait to hike up into them—after all, he had climbed both the Matterhorn and Mount

Katahdin. But his enthusiasm was dangerously naive. Already the September nights in the Wyoming mountains were bitter cold. The weather was known to be freaky; it didn't snow much, but that didn't mean three or four feet couldn't be dumped within a couple of hours. Packhorses—even first-rate ones—would have a hard time making it up the sides of the steep ridges. Challenging nature, they would have to leave their horses in the valley. Because of the spruces, of course, there would be plenty of wood to build a bonfire. But this was no guarantee of survival. Many men had perished in the Bighorns mistakenly believing that fire trumped sleet; it never did. Even the most sure-footed mountaineer was no match for the raw natural powers of this Wyoming wilderness. Whether it was wolf packs prowling or wind whistling through the canyons, only a fool wasn't reduced to humbleness in such potentially lethal terrain. At the curl of twilight everything was ghostly and mysterious beyond even the deepest backcountry of the Adirondacks. "If I listened long enough, it would almost seem that I heard thunderous voices laughing and calling to one another," Roosevelt wrote, "and as if at any moment some shape might stalk out of the darkness into the dim light of the embers."[89]

Roosevelt took to calling the elk "lordly," just as he had done with the buffalo. Even though the naturalist in him carefully studied every coloration and trait variation of those he killed, he began seeing the elk as almost holy. Nonhunters might be perplexed by this, but the northern Cheyenne would have completely understood the spiritual aspect of Roosevelt's search for game. "From morning till night I was on foot," he wrote, "in cool, bracing air, now moving silently through the vast, melancholy pine forests, now treading the brink of high, rocky precipices, always amid the most grand and beautiful scenery; and always after as noble and lordly game as is to be found in the Western world."[90]

Since his boyhood, grizzly bears had enthralled Roosevelt. Carefully he studied all their zoological traits, realizing that they were essentially shy and not predatory. The sheer hulk of the omnivorous grizzly—a member of the brown bear family often weighing up to 1,300 pounds—made it the true king of the Rockies. With their astounding senses of hearing and smell, grizzlies were hard to hunt. But as any trapper could testify, they had terrible eyesight, and if you happened to stumble upon one in a refuge like the Bighorns it could lunge without warning or retreat. Come October, all the grizzlies would hibernate in dens until late April. So Roosevelt knew that his best chance for getting a large, full-grown grizzly was in September. But this was also the time of year when the bears had spent months actively digging for rodents and roots, so that their claws were the longest. As Darwin would have appreciated, these fearsome claws had an additional purpose besides warding off predators: they enabled grizzlies to dig winter dens with relative ease.[91]

The damp afternoon when Roosevelt stumbled upon his first grizzly provides one of the classic stories in American outdoor literature. Roosevelt's desire for precision and suspense was urgent, even if the latter wasn't always fulfilled. As recounted in *Hunting Trips of a Ranchman*, the odyssey began at sunset on September 12, 1884, when he happened to encounter bear tracks. It's a testimony to Roosevelt's familiarity with bears that he knew the tracks belonged to a grizzly. A strange "eerie feeling" of expectancy swept over him. Alone, he followed the footprints from tree stand to stand. As darkness neared, however, he could no longer see anymore. It was time to head back to camp. He vowed to pick up the bear's trail in the morning.

Upon waking, Roosevelt and Merrifield checked on the carcass of an elk they had killed. To their surprise a grizzly had gnawed the body during the night. Wearing moccasins so as not to scare away game, the two men began following the bear marks. The mingling odors of pine and sweat filled Roosevelt's nose. He drank it all in. "When in the middle of the thicket we crossed what was almost a breastwork of fallen logs, and Merrifield, who was leading, passed by the upright stem of a great pine," Roosevelt recalled. "As soon as he was by it he sank suddenly on one knee, turning half round, his face fairly aflame with excitement; and as I strode past him, with my rifle at the ready, there, not ten steps off, was the great bear, slowly rising from his bed among the young spruces."[92]

And what a huge grizzly it was, standing about nine feet tall and weighing more than 1,200 pounds. It would have been impossible for Roosevelt to have found a better specimen for his North American mammal collection.[93] "He had heard us but apparently hardly knew exactly where or what we were, for he reared up on his haunches sideways to us," he wrote. "Then he saw us and dropped down again on all fours, the shaggy hair on his neck and shoulders seeming to bristle as he turned toward us. As he sank down to his forefeet I had raised the rifle . . . Half-rising up, the huge beast fell over on his side in the death throes, the ball having gone into his brain, striking as fairly between the eyes as if the distance had been measured by a carpenter's rule."[94]

The killing of this first grizzly bear was cathartic. During the coming days Roosevelt bagged two more: a mother and a cub. Why was he was so compelled to slaughter an animal he loved so deeply? How could he have shot a baby face-to-face? Roosevelt would claim that he needed multiple specimens for scientific study. He would claim that the bear meat went into the evening pot. But both answers were bunk. Quite simply, he enjoyed shooting the birds and animals he loved the most. The brutality of such acts never seemed to bother Roosevelt, for he considered himself privileged as a Darwinian well-schooled in tooth-and-claw violence.

Feeling like a champion hunter, Roosevelt descended with Merrifield and Lebo out of the Bighorns carrying enough trophies to fill the walls of a small Wyoming lodge. They arrived in the town of Buffalo on September 18 full of superlatives, and rented rooms at the Occidental Hotel.[95] That evening Roosevelt, the harried traveler, dined with U.S. Cavalry officers at Fort McKinney, listening to snatches of conversations about the peace settlements with the northern Cheyenne.[96]

Even though Lebo was a blacksmith, the Roosevelt party's horses were going lame from collapsing in creeks and ravines.[97] Although local wisdom dictated that any cowboy needed about a dozen mounts for roundups or 1,000-mile treks, the Roosevelt trio started their journey back to Medora with just a couple of horses.[98] On October 1 they got caught in what Roosevelt called a "furious hurricane" that whirled with "driving rain squalls."[99] For a couple of days they were forced to hide out in butte alcoves, desperate to stay warm and dry. By the time the weather cleared, Roosevelt had had enough. Leaving Lebo behind with the prairie schooner of supplies, he started riding off with Merrifield toward Medora. With winter around the corner and a presidential election just weeks away, Roosevelt was eager to return to New York with his fine trophies.

In *Hunting Trips of a Ranchman*, Roosevelt wrote eloquently of what it was like to be an adventurer in the Bighorns and to see your ranch appear on the open range, promising clean sheets and a library shelf packed with books by Shakespeare and Hawthorne. He bowed to the unassailable beauty of the West. If nothing else, the Badlands had encouraged Roosevelt to be more poetic as a writer. He was inspired by nature, and his writing now took on a more colorful cast. Whatever hardships he endured had been distilled into only postcard memories. Clearly, he had the talent to succeed as a wilderness writer. As the naturalist E. O. Wilson of Harvard once aptly noted, field biologists have a lot more "gee whiz" or "sense of wonder" than other kinds of scientists.[100]

"The rolling plains stretched out on all sides of us, shimmering in the clear moonlight; and occasionally a band of spectral-looking antelope swept silently away from before our path," Roosevelt wrote. "Once we went by a drove of Texan cattle, who stared wildly at the intruders; as we passed they charged down by us, the ground rumbling beneath their tread, while their long horns knocked against each other with a sound like the clattering of a multitude of castanets. We could see clearly enough to keep our general course over the trackless plain, steering by the stars where the prairie was perfectly level and without landmarks; and our ride was timed well, for as we galloped down into the valley of the Little Missouri the sky above the line of the level bluffs in our front was crimson with the glow of the unrisen sun."[101]

CHAPTER SEVEN

CRADLE OF CONSERVATION: THE ELKHORN RANCH OF NORTH DAKOTA

I

At some point in the fall of 1884, Roosevelt began assembling his jottings about the Badlands and the Bighorns into a book. Updating Captain Mayne Reid, he considered how *Hunting Trips of a Ranchman* (as he called the project) could combine vivid natural history with tales of big-game hunting; he wanted to offer an antidote to the artificiality of money-driven urban life, which he felt was hampering the democratic spirit as well as feminizing a generation of American men.[1] Always a romantic, Roosevelt originally intended to write *Hunting Trips* at the Elkhorn and Maltese Cross ranches, even though there was no decent reference library in the entire Dakota Territory. Pragmatism, however, eventually held sway (as it usually did with Roosevelt), and he ended up composing *Hunting Trips* back east.

By October, in fact, Roosevelt was back in Manhattan, having abandoned his plan of getting away to Dakota. With the presidential election looming, he put *Hunting Trips* on hold and threw himself wholeheartedly into the heated political contest. Using Bamie's home at 689 Madison Avenue as his pied-à-terre, Roosevelt stumped incessantly for the Republican, James G. Blaine. The fact that the Democratic nominee, Grover Cleveland, was hounded by charges of adultery and had fathered an illegitimate child increased Roosevelt's zeal to elect Blaine. According to the *New York Sun*, Theodore, forever the puritan, chafed at the unholy notion of a womanizing rogue becoming commander in chief.[2] (Perhaps if Roosevelt had fully known that Cleveland was a true outdoorsman, he would have been less antagonistic.) In private, however, Roosevelt didn't care for the partisan stammerings of either candidate. They were both, he believed, old-school mugwumps while he was a new-school reformer. In any case, Blaine accused the Democrates of representing "rum, Romanism, and rebellion." The negative campaign tactic didn't work. Blaine went down in defeat on Election Day. (Cleveland won by a relatively close margin: 219 electoral votes to 182.[3]) Writing to Henry Cabot Lodge, who had himself lost a congressional race in Massachusetts, Roosevelt carped

Theodore Roosevelt wearing a customized Badlands hunting costume. This photograph was used to promote Hunting Trips of a Ranchman.

that the Republican Party had been done in by so-called "Independents" whom he deemed "pharisaical fools and knaves."[4]

Disappointed by Blaine's loss, Roosevelt headed back to the Badlands just two days after the elections in November, eager to gather more material for *Hunting Trips* and to track bighorn sheep along the Montana line. His spirits rose once he was on a horse. He spent a few days at the cabin at Maltese Cross, catching up with Dakotan friends. Then, with hard snow falling, he trotted north on his horse Manitou. The solitude of the Elkhorn Ranch, he figured, would offer minimum distractions and he could start writing *Hunting Trips* in earnest.

The prairie winds of the Dakota Territory could be ruthless, and Roosevelt, traveling by himself, was nearly blinded when snow squalls started blowing in his face. As he forded the Little Missouri River the ice cracked and fear ran up and down his spine. Then, to use Jack London's term in *The Call of the Wild*,[5] the "dominant primordial beast" welled up in Roosevelt. Undaunted by his precarious predicament, he took the inclement weather as a challenge. By twilight, new snow was falling so heavily that Roosevelt was forced to seek shelter in a lean-to that he luckily stumbled upon. He'd forgotten to bring hard tack with him, so dinner consisted of only tea as snowdrifts layered up against his door.[6] Roosevelt reported in his private diary that he slept, warm and without vexation, by a small fire while wolves—which he deemed "the beast of waste and desolation"—howled nearby.[7] At daybreak a narrow band of light appeared in the east, intimating that the storm had subsided.

Having endured the wintry ordeal, a famished Roosevelt grabbed his shotgun and hunted sharptail grouse in the sparkling white snowdrifts. Pioneers in the Dakota Territory and Minnesota used to claim that the brushland was so filled with sharptails that when they flocked the sun was blocked (although this was a dubious claim, because grouse don't rise that high), and indeed Roosevelt bagged five that day. "The sharptails fly strongly and steadily, springing into the air when they rise, and then going off in a straight line, alternately sailing and giving a succession of rapid wing-beats," Roosevelt wrote. "Sometimes they will sail a long distance with set wings before alighting, and when they are passing overhead with their wings outstretched each of the separate wing feathers can be seen, rigid and distinct."[8]

Immediately, Roosevelt roasted two grouse over a small fire. They were uncommonly tasty. Fortified, he continued on to the frozen trail to the Elkhorn. Upon arriving at the ranch, he was cheerfully greeted by Sewall and Dow. Roosevelt was pleased to learn that his hardy cattle were in relatively fine shape. The idea of going on a hunt was bandied about, but the trio decided to first procure firewood—lots of firewood—for the blustery winter days ahead. For hours they chopped down trees and collected kindling. The jocular Maine lumberjacks teased Roosevelt, saying that he was a rank amateur when it came to felling trees. As Dow mockingly told a rancher after three days of clear-cutting cottonwoods, "Well, Bill cut down fifty-three, I cut forty-nine, and the boss, he beavered down seventeen."[9] Always attuned to animal metaphors, Roosevelt knew he was being good-naturedly mocked. Beavers gnawed down cottonwoods and willows slowly and painstakingly, eating bark while they worked. For a tenderfoot trying to be a bull moose, being perceived by his workers as a "beaver" was a real put-down.

Nevertheless, with temperatures dropping to thirty degrees below zero, Roosevelt wisely retreated back to the Maltese Cross, where the primitive creature comforts of Medora were near at hand. He spent hours reading poetry, shooting mule deer, and lunching with the Marquis de Mores at his château in Medora. Roosevelt loved his winter outfit of coonskin cap, long overcoat, and fur-lined gloves. But most of his free time was spent indoors, writing, and his deep love and appreciation for the wilderness in winter became evident in his prose. With a craftsman's care he began pondering the power of death, the howling prairie, and the bitter cold. New England poetry was, of course, famous for bleakness, and Roosevelt imitated its tone. The deep-seated sentiment of "iron desolation" permeated his writing. (The naturalist John Burroughs had used iron as a poetic metaphor for a forest's forlornness in his 1879 book *Locusts and Wild Honey*, which greatly influenced Roosevelt.[10]) "When the days have dwindled to their shortest, and the nights seem never ending, then all the great northern plains are changed into an abode of iron desolation," Roosevelt wrote. "Sometimes furious gales blow out of the north, driving before them the clouds of blinding snow-dust, wrapping the mantle of death round every unsheltered being that faces their unshackled anger. They roar in a thunderous bass as they sweep across the prairie or whirl through the naked cañons; they shiver the great brittle cottonwoods, and beneath their rough touch the icy limbs of the pines that cluster in the gorges sing like the chords of an aeolian harp."[11]

One winter day Roosevelt was informed that some bighorn sheep were climbing buttes only twenty-five miles from the Maltese Cross. Ever since he had first arrived in the Badlands, he had wanted a ram's head for his trophy collection at Sagamore Hill. The hunt itself would be recorded in Chapter 7 of *Hunting Trips of a Ranchman*, "A Trip after Mountain Sheep." With Merrifield at his side, Roosevelt rode deep into the "fantastic shapes" of the "curiously twisted" Badlands.[12] Because bighorns lived in rocky precipices, they didn't leave detectable footprints, so Roosevelt had only his rifle and luck to guide him.[13]

Stalking bighorn was a difficult proposition requiring mountaineering skills, stamina, and tenacity. Larger than a deer, a bighorn ram weighed around 300 pounds and was swift and sure-footed. "In his movements he is not light and graceful like the pronghorn and other antelopes, his marvellous agility seeming rather to proceed from sturdy strength and wonderful command over iron sinews and muscles," Roosevelt wrote. "The huge horns are carried proudly erect by the massive neck; every motion of the body is made with perfect poise, and there seems to be no ground so difficult that the big-horn cannot cross it. There is probably no animal in the world his superior in climbing, and his only equals are the other species of mountain sheep and the ibexes."[14]

After days of unstable tracking on slippery ledges and knifelike ridges, Roosevelt got his handsome sheep. Tuckered out, he admitted that it was a lucky shot. He also claimed that skill was also a factor. Strapping the ram onto his horse Manitou's back, he brought the prize to the Maltese Cross ranch and feasted on mountain "mutton."[15]

For Roosevelt, his wilderness experiences always got back to his desire for good health *and* bragging rights. "I have just returned from a three day trip in the Badlands after mountain sheep; and after tramping over the most awful country that can be imagined I have finally shot a young ram with a fine head," he wrote to his sister Anna. "I have now killed every kind of plains game."[16] (By the time Roosevelt became president in 1901, the bighorn sheep in the Badlands had been wiped out.)

As Christmas 1884 approached, however, the merciless "iron desolation" and strange landforms of the interior plains were too much for Roosevelt. He was homesick for his daughter Alice (or Baby Lee, as he often called her). The numbing Dakota cold proved unrestful and intellectually unproductive. Scooping up his notes for *Hunting Trips* once again, Roosevelt boarded the eastbound train. He was frustrated because writing about the Badlands while *in* the Badlands had proved elusive. After enduring the sad holiday in New York—the eggnog parties and Christmas packages were not the same without his mother and his wife—Roosevelt hunkered down to write seriously about the Badlands and Bighorns. Nothing could distract him from the arduous chore at hand. Unlike *The Naval War of 1812*, this first-person effort would be a memoir from Dakota, Montana, and Wyoming intermixed with Burroughsian observations on natural history, the sportsman's code, hunting stories, warnings about biological conservation, and cowboy lore.

After Roosevelt settled down to write, and consuming pots of black coffee, he pushed himself relentlessly, usually writing for two or three sessions a day. By February 1885, he had written 95,000 words, and the next month *Hunting Trips*, a collation of wilderness experiences, was finished. The pace had exhausted him. But once Roosevelt's depleted health was restored, after days of almost nonstop sleep, he returned to Medora to spend a few weeks checking up on his ranches.[17] Early on, Roosevelt—a bit out of practice in the saddle—was tossed from Manitou into the frigid Little Missouri River. Chunks of ice kept him from being swept away in the current, and somehow he managed to get a grip on the situation and save himself and his horse. Perversely, he was delighted by the thrill of being near death and by the tingly, numbing cold water. Days later, abruptly, he purposely tossed himself into the river to relive the experience. "I had to strike my own line for twenty miles over broken country before I reached home and could dry myself," he boasted to Bamie. "However it all makes me feel very healthy and strong."[18]

Meanwhile, G.P. Putnam's Sons was preparing to publish *Hunting Trips* (dedicated to Elliott Roosevelt, "That Keenest of Sportsmen and Truest of Friends") in July, as a so-called sporting book.[19] No other well-known politician in America, the advance notices boasted, could have written such a gripping hunting narrative. A photograph of Roosevelt posed in a fringed buckskin suit, Winchester rifle at his side, was used to promote the author as a gentleman-sportsman. Taken in a New York studio, the photo, a contrived combination of Buffalo Bill and John James Audubon, reeked of Broadway hokeyness, right down to the backdrop of ferns and an artificial grass carpet. But the actual book, filled with etchings and woodcuts and published in a first edition of only 500 copies, printed on quarto-size sheets of handwoven paper, remains a true collector's item. Although it had a strong conservationist ethos, *Hunting Trips* was primarily aimed at gentleman-sportsmen like the writer, aristocrats who could afford hunting holidays, chuck-wagon hands, and what was then a hefty retail price of fifteen dollars.[20]

All of Roosevelt's major outdoors adventures between 1880 and 1884 were vividly recounted in *Hunting Trips of a Ranchman* (subtitled *Sketches of Sport on the Northern Cattle Plains*). Putting his college education to good use, he wrote about Minnesota grouse, Montana buffalo, Dakota Territory bighorn sheep, Great Plains antelope, and Bighorns bears. Chronology was abandoned, often to the reader's confusion, in favor of biological and topographical edification. Showcasing his erudition as a naturalist was Roosevelt's first priority; recounting thrilling hunts was a close second. Most chapter titles, in fact, had to do with wildlife: "Water Fowl," "The Grouse of the Northern Cattle Plains," "The Deer of the River-Bottoms," "The Black-Tail Deer." Roosevelt wrote that the American West was a Darwinian laboratory full of amazing wildlife action. "The doctrine seems merciless, and so it is; but it is just and rational for all that," Roosevelt wrote about *On the Origin of Species*. "It does not do to be merciful to a few, at the cost of justice to the many."[21]

The villains of *Hunting Trips* were the "swinish game butchers" who ruthlessly hunted for hides "not for sport or actual food," and who cold-bloodedly murdered the "gravid doe and the spotted fawn with as little hesitation as they would kill a buck of ten points."[22] Whenever T.R. turned polemical on behalf of good sportmanship, he echoed the ethical sentiments and concerns of his uncle Robert B. Roosevelt and the sporting press, such as *Forest and Stream*. Like Uncle Rob pontificating on the essential beauty of shad, trout, and eels, throughout *Hunting Trips* Roosevelt gave loving naturalist observations about the elk, antelope, and buffalo he had hunted. Not all, however, was blood and thunder. There was an "Indian guide" feel to much of the prose. For example, Roosevelt wrote quietly about stumbling upon a white-tailed deer's resting spot with the "blades of grass still slowly rising, after the hasty departure of the weight that has flattened them down."[23] Reading *Hunting Trips* makes it abundantly clear that Roosevelt deeply respected these deer.

Although cherry-picking is required, genuine conservationist beliefs can be excavated from the pages of *Hunting Trips*. For example, true western outdoorsmen, Roosevelt wrote, would have to become citizen-protectors of the wildlife being devastated by bands of destructive rogues. In almost every chapter he feared the day when elk, buffalo, and prairie chickens would vanish forever. "No one who is not himself a sportsman and lover of nature can realize the intense indignation with which a true hunter sees these butchers at their brutal work of slaughtering the game, in season and out," he wrote in *Hunting Trips*, "for the sake of the few dollars they are too lazy to earn in any other and more honest way."[24] By expressing such views in 1885, Roosevelt was pitting himself against the railroad behemoths, telegraph companies, real estate brokers, and even Buffalo Bill, whom he respected as a master horse-breaker.[25] Although there are only a few such passages—in a book that promoted the joys of big game hunting—*Hunting Trips* nevertheless marked the beginning of Roosevelt's great crusade for the conservation of deer, elk, antelope, bighorn sheep, and bears. Wringing a livelihood from the "outdoors" literary marketplace instead of U.S. naval history now became an all-important occupational pursuit for Roosevelt to juggle along with politics, ranching, and managing the family trust. And in wildlife protection he had found his cause.

Hunting Trips received impressive reviews that July. The *New York Times*, for example, said that the book was clear-eyed and would seize "a leading position in the literature of the American sportsman." Although the first part of the review focused on Roosevelt's ethnological delineation of cowboy culture, the *Times* also noted that his naturalist writing on the survivalist tactics of white-tailed deer was exemplary. "The common deer, or whitetailed deer, found in almost any State in the Union, he tells us was not so plentiful five years ago on the northern plains as it is today," the unidentified reviewer wrote. "With this deer its increase seems to be due to its particular habits. It seeks the densest coverts, is fond of wet and swampy places, and is rarely jumped by accident. It demonstrates the survival of the fittest."[26]

Overnight *Hunting Trips* became the seminal study of both the Badlands and the Bighorns. The core message Roosevelt conveyed was that hunting big game was good for the American soul. Bouts of barbarism, Roosevelt believed, reawakened the primitive and the savage in a man, to good effect. It was a theme that pervaded his writings for the rest of his life. "In hunting, the finding and killing of the game is after all but a part of the whole," he later wrote. "The free, self-reliant, adventurous life, with its rugged and stalwart democracy; the wild surroundings, the grand beauty of the scenery, the chance to study the ways and habits of the woodland creatures—all these unite to give to the career of the wilderness hunter its peculiar charm. The chase is among the best of all national pastimes; it cultivates that vigorous manliness for the lack of which in a nation, as in an individual, the possession of no other qualities can possibly atone."[27]

II

One mixed review, however, caught Roosevelt's attention amid the cavalcade of raves.[28] The thirty-five-year-old naturalist George Bird Grinnell, the esteemed editor of *Forest and Stream*, did compare *Hunting Trips* to Francis Parkman's *The California and Oregon Trail* and Lewis H. Garrard's *Wahtoyah and the Taos Trail* because of its "freshness, spontaneity, and enthusiasm"; on the other hand, Grinnell criticized Roosevelt for generalizing too much about the western species he had encountered while hunting, for failing to discuss color variations properly, and for other inaccuracies of zoological detail. The slightly patronizing review observed that although the youthful Roosevelt had studied a particular antelope herd in Montana, the herds in Manitoba or Saskatchewan were not necessarily identical to it. Grinnell believed that Roosevelt was talented, but that to be a real Darwinian zoologist he should have spent more time doing field observations before rushing into print with his first impressions, which were sometimes inaccurate despite being well written. "Mr. Roosevelt is not well known as a sportsman, and his experience of the Western country is quite limited, but this very fact in one way lends an added charm to this book," Grinnell wrote, damning Roosevelt with faint praise. "He has not become accustomed to all the various sights and sounds of the plains and the mountains, and for him all the difference which exists between the East and the West are still sharply defined. . . .

George Bird Grinnell was the editor of Forest and Stream *and a co-founder of the Boone and Crockett Club with Theodore Roosevelt. The* New York Times *deemed him the "father of conservation."*

We are sorry to see that a number of hunting myths are given as fact, but it was after all scarcely to be expected that with the author's limited experience he could sift the wheat from the chaff and distinguish the true from the false."[29]

The review stung Roosevelt, who prided himself on the scientific exactitude of his animal descriptions. Grinnell, it seemed, had taken Roosevelt down a notch. Doubly frustrating was the fact that Grinnell had championed Roosevelt's conservation activism in *Forest and Stream* the previous year, praising his efforts in the New York state assembly to halt the damming of streams that fed the Hudson River. "It is satisfying to see," Grinnell had written, "now and then, in our legislative halls, a man whom neither money, nor influences, nor politics can induce to turn from what he believes to be right to what he knows to be wrong."[30] But now, in 1885, Grinnell had become Roosevelt's enemy.

After reading the review, Roosevelt stormed into the offices of *Forest and Stream* demanding a meeting. Always cordial, Grinnell agreed, and they sat together for hours going through *Hunting Trips* almost page by page. To Roosevelt's surprise, the erudite editor seemed to know more about bighorn sheep and white-tailed deer than he did. Once Roosevelt's bruised ego was salved, the conversation turned to conservation issues, specifically big game protection. "I told him something about game destruction in Montana for the hides, which, so far as small game was concerned, had begun in the West only a few years before that," Grinnell recalled; "though the slaughter of the buffalo for their skins had been going on much longer and their extermination had been substantially completed. Straggling buffalo were occasionally killed for some years after this, but by this time . . . the last of the big herds had disappeared."[31]

By the time Roosevelt left the headquarters of *Forest and Stream* no lingering animosity or grudge would spoil his new friendship with George Bird Grinnell, who fast became as close a friend as Henry Cabot Lodge. When it came to saving wildlife the two men were in sync. And Roosevelt had learned a lesson: never again would he leave himself vulnerable to charges of faking about nature or of mischaracterizing species. Instead of being rivals, Roosevelt and Grinnell united in what would become a lifelong crusade to save the big game animals of the American West from extinction.[32] "Roosevelt called often at my office to discuss the broad country that we both loved, and we came to know each other extremely well," Grinnell recalled decades later. "Though chiefly interested in big game and its hunting, and telling interestingly of events that had occurred on his own hunting trips, Roosevelt enjoyed hearing of the birds, the small mammals, the Indians, and the incidents of travel of early expeditions on which I had gone. He was always fond of natural history, having begun, as so many boys have done, with birds; but as he saw more and more of outdoor life his interest in the subject broadened and later it became a passion with him."[33]

It was easy to see why Roosevelt was captivated by Grinnell, who was nine years his senior. Grinnell, a native of Brooklyn, was an explorer, rancher, hunter, bird-watcher, ethnologist, published author, first-rate editor, and western folklorist. (And as if that weren't enough, the *New York Times* would later call him the "father of American conservation."[34]) Like the Roosevelts, the Grinnells had deep roots in the United States, having arrived in Rhode Island as far back as 1630.[35] A real sophisticate, always impeccably dressed, with a pipe close at hand, he knew more about the American West, and more about North America's 650 mammal species, than any other scholar alive. Puff-puff-puffing, he would discuss why kit foxes were the pygmies of the fox group and why the spotted skunk had the strongest scent of the species.

When Grinnell was seven, his family had moved to Audubon Park in upper Manhattan, the thirty-acre estate that had served as the great ornithologist's last home. The hallways there were cluttered with overhanging elk and deer antlers, "which supported guns, shot pouches, powder flasks, and belts."[36] The grounds were full of wild animals for Audubon to study and draw. The walls of Audubon Park, in fact, groaned with paintings and hunting trophies once belonging to the great Audubon.[37] A close friendship developed between the young Grinnell and Audubon's widow, Madame Lucy. The old barn was filled with ornithologists' collections of skins and specimens, and Grinnell absorbed Audubon's influence. The nearby Hudson River became his bird-watching laboratory. "In winter the river was often very full of ice, and eagles and crows were constantly seen walking about on the ice, no doubt feeding on refuse and the bodies of animals thrown into the stream north," Grinnell wrote in a partially unpublished memoir. "The crows used to roost on a cedar-covered knoll north of the Harlem River in what is now the Bronx, not very far from Highbridge, and each morning they flew low among the tree tops."[38]

As an undergraduate at Yale, Grinnell had the same problems in the classroom that Roosevelt would have at Harvard. He was too enamored with the idea of following in the footsteps of his idol to sit still in the classroom. Grinnell's life mission crystallized when he read Audubon's 1843 account of traversing the Missouri and Yellowstone rivers. A passage Audubon wrote in his journal, a lament about the buffalo slaughter on the Great Plains, was seared into Grinnell's mind and became, in a sense, a mission statement. "What a terrible destruction of life," Audubon wrote, "as it were for nothing . . . as the tongues only were brought in, and the flesh of these fine animals was left to beasts and birds of prey, or to rot on the spots where they fell. The prairies are literally *covered* with the skulls of the victims."[39]

Then Audubon had fired off a verbal challenge that would launch the modern conservation movement. "This cannot last," Audubon said of the buffalo slaughter. "Even now there is a perceptible difference in the size of the herds, and before many years the Buffalo, like the Great Auk, will have disappeared; surely this should not be permitted."[40]

Those last six words—"surely this should not be permitted"—galvanized Grinnell. With the encouragement of Professor Othniel C. Marsh (no relation to George Perkins Marsh), the leading paleontologist in the United States, Grinnell volunteered to work on a Yale-sponsored Great Plains dinosaur dig in 1870, writing that he was "bound for a West that was then really wild and wooly."[41]

Grinnell had read every one of Mayne Reid's books as a boy, so the American West beckoned to him like the star of Bethlehem. While collecting fossils at Antelope Station in Nebraska, Grinnell encountered buckskin scouts, drifters, gold-seekers, Christian farmers, itinerant preachers, and Plains Indians. It was just the first of many trips west, during which he befriended such legendary figures as Charley Reynolds, Buffalo Bill, and Frank and Luther North. By the time he reviewed Roosevelt's *Hunting Trips* in July 1885, he not only had been part of the Marsh Paleontological Expedition but had made scientific discoveries in support of Darwinism, had accompanied Lieutenant Colonel George Armstrong Custer to the Black Hills in 1874 when gold was discovered, and had joined Captain William Ludlow of the Army Corps of Engineers the next year in surveying Yellowstone.[42] And nobody alive wrote about duck hunting with more authority than Grinnell.[43] Believing that Native Americans had been "shamefully robbed" by the U.S. government, Grinnell worked side by side with Plains tribes,[44] inspiring enough trust and confidence that many of the Indian bands gave him a special name: to the Pawnee, he was "White Wolf," an honorary member of the tribe; to the Cheyenne he was *wikis* ("migratory bird"); to the Blackfeet he was "Fisher Hat," in recognition of his ability to find fish in seemingly depleted streams; and to the Gros Ventres he was "Gray Clothes," because of the dull suit he often wore.[45] By 1885 he was known as *the* American expert on the ethnology of the Plains Indians. The anthropologist Margaret Mead saluted Grinnell's pioneering efforts on behalf of saving Indian tribal culture as recently as 1960, using the word "classic"[46] to describe his book about the Cheyenne.[47] (The great western writer Mari Sandoz did the same in 1962.[48])

So when Roosevelt stormed into Grinnell's office at *Forest and Stream* he was dealing with a heavyweight. Starting in 1882, using the magazine as a soapbox, the editor began crusading to save natural resources. Among other environmental causes he promoted seasonal licenses, laws against killing young animals, the need to preserve habitat, and the need for game wardens. Grinnell was small in stature, with large ears, and usually sported a well-trimmed mustache or goatee; his regal personality stood in sharp contrast to that of the bombastic Roosevelt. Grinnell was also soft-spoken, self-effacing, and humble—yet there was nothing timid about his approach. When he spoke about the American West, people listened. He believed strongly that scientists should get mud on their boots, and he dreamed of forest reserves, bison parks, restocked rivers, and greenbelts around western cities. To the scientific-minded Grinnell, there was an interconnectedness to nature. Even if the United States had the best game laws in the world, they meant nothing without forest protection. Last but not least, Grinnell encouraged states to create zoological societies—a lobbying campaign at which he proved successful in New York.

As a self-appointed watchdog for Yellowstone National Park, Grinnell constantly denounced overcommercialization and federal mismanagement. After witnessing hunters slaughter elk and deer in the park in 1875, he had written a scolding letter promoting big game conservation there and included it in the Ludlow Expedition report. Over the years to come, with a great deal of success, Grinnell would lobby the U.S. Senate to preserve the territorial integrity of Yellowstone. "My account of big-game destruction [in Yellowstone] much impressed Roosevelt, and gave him his first direct and detailed information about this slaughter of elk, deer, antelope, and mountain-sheep," Grinnell recalled. "No doubt it had some influence in making him the ardent game protector that he later became, just as my own experiences had started me along the same road."[49]

Early in 1886, a few months after the publication of *Hunting Trips*, Grinnell helped form the Audubon Society to protect birds from extinction. From the get-go he had no stronger ally in those efforts than Theodore Roosevelt. As the historian John Reiger observed in 1972 in *The Passing of the Great West*, "Grinnell, the originator and amalgamator of ideas, *prepared* Roosevelt for Gifford Pinchot, the President's famous environmental administrator."[50] It was the alliance of Roosevelt and Grinnell (not Roosevelt and Pinchot) that launched the modern conservation movement in earnest. To Roosevelt, Grinnell was an American treasure whose likeness should have been cast in granite.

III

By late August 1885, following the publication of *Hunting Trips*, Roosevelt was back in the Badlands. The Elkhorn Ranch was now completely built, with eight rooms, a large stone fireplace, numerous windows, and a center hall—all adorned with taxidermy. Buffalo robes, deer antlers, and bearskins were strewn about the place. There were so many mule deer sheds and elk sheds on the piazza that it looked like an antler museum. Roosevelt converted the cellar into a photography darkroom; taking pictures of nature was yet another of his hobbies. There were two stables, together often housing as many as thirty horses. Roosevelt entertained and wrote at the Elkhorn (most of the real ranching work, however, was done from the Maltese Cross),[51] and especially enjoyed sitting on the porch facing the piazza-like area in front of the ranch house. "Just in front of the ranch veranda is a line of old cottonwoods that shade it during the fierce heats of summer, rendering it always cool and pleasant," Roosevelt wrote. "But a few feet beyond these trees comes the cut-off bank of the river through whose broad sandy bed the shallow stream winds as if lost except when a freshet fills it from brim to brim with foaming yellow water."[52]

The big news around Medora was that Marquis de Mores had been arrested for murdering Riley Luffsey, a buffalo hunter who loathed de Mores's barbed wire. De Mores was now in jail in Bismarck, awaiting trial. Furthermore, there was a rumor that he was furious at Roosevelt over land boundaries, cattle prices, Roosevelt's rude ranchhands, and much else. Because Roosevelt was essentially a squatter, in the last years before fences and deeds transformed the open range into fixed property, enforcing boundaries was constantly a cause of friction. Sharp letters were exchanged between the two rich cattlemen. Talk of a pistol duel was even bandied about, but it proved empty. Eventually de Mores was found not guilty of murdering Luffsey.[53] In any event, his arrest had been, at worst, an unpleasant distraction to Roosevelt, who had founded the Little Missouri Stockmen's Association and prided himself on his western leadership role perhaps more than on being a New York assemblyman. And the stockmen's association did more than settle land and brand issues. Because of a drought, brushfires were common in the Badlands that summer. As head of the association, Roosevelt worked side by side with his neighbors to put out the blazes, many started by Plains Indians angry at white settlement.

By September 16 Roosevelt was headed back east, stopping at the Bismarck jail for a brief visit with de Mores. In the spring, writing *Hunting Trips* had left Roosevelt physically depleted, but now he was in high spirits back home in New York. He attended the state Republican convention in Saratoga Springs and spent time with his little Alice. Friends were impressed by his general happiness and vitality. For the first time since his wife's death, he was open to the idea of a new romance. This changed attitude allowed him to reconnect with his childhood sweetheart, Edith Carow. "For nineteen months (since the deaths of Alice and Mittie) they had successfully avoided each other," Sylvia Jukes Morris wrote in *Edith Kermit Roosevelt*. "But sometime early that fall, either by chance or design, they met."[54]

Deeply refined, quietly attractive, and unmarried, Edith Carow was then twenty-four and still infatuated with Theodore. As Roosevelt had been winning elections and writing critically acclaimed books, his sister Bamie had constantly updated Edith, with whom she'd remained friends. Victorian etiquette called for a long mourning period, so as Theodore and Edith grew closer, they were very discreet. Even after Edith accepted Roosevelt's proposal that November, they behaved in public only as friends for a full year. Somehow three years "in waiting" seemed much more socially appropriate than only two.

That Christmas season Roosevelt circulated at numerous social gatherings with Edith at his side. Nevertheless, he didn't use her full name in his diary, referring to her as only "E" and reserving all his affection in the journal for Alice. No love letters between the two survive—Edith ordered their correspondence destroyed—and, in fact, it's quite possible there weren't any. By all accounts, Edith was an intensely private woman, with a unique ability of tamping down Roosevelt's over-the-top enthusiasms. "Edith was not the sort of person to encourage rhapsodies anyway," according to the historian Edmund Morris. "She disapproved of excess, whether it be in language, behavior, clothes, food, or drink."[55]

Keeping Sagamore Hill running and maintaining the Dakota ranches were expensive propositions for Roosevelt. As marriage plans were privately discussed, he dreamed of having a large family. Keeping bloodlines alive mattered to him a great deal. Although he could live comfortably on his trust fund, he needed book advances to feel financially secure. At Henry Cabot Lodge's suggestion Roosevelt signed a contract with Houghton Mifflin to write a biography of Thomas Hart Benton, the senator from Missouri who from 1821 to 1851 had fervently encouraged westward expansion.[56] It would be part of a new series called American Statesmen. During January and February 1886 Roosevelt started writing *Thomas Hart Benton* in earnest, taking advantage of New York's fine research libraries. Edith's company was welcome, too, but still he kept thinking about the Badlands. A conscientious businessman, he knew, always checked up on his investment. Imagining Sewall and Dow suffering in the cold, worried that his dogies and yearlings wouldn't make it through the winter, he returned to Medora at the end of the winter planning to write *Benton* there while collecting new material for a sequel to *Hunting Trips*. "I got out here all right, and was met at the station by my men," he wrote to Bamie on March 20 from the Elkhorn. "I was really heartily glad to see the great, stalwart, bearded fellows again, and they were honestly pleased to see me. Joe Ferris is married, and his wife made me most comfortable the night I spent in town. Next morning snow covered the ground; but we pushed to this ranch, which we reached long after sunset, the full moon flooding the landscape with light. There has been an ice gorge right in front of the house, the swelling mass of broken fragments having been pushed almost up to our doorstep. . . . No horse could by any chance get across; we men have a boat, and even then it is most laborious carrying it out to the water; we work like Arctic explorers."[57]

Four days later thieves stole Roosevelt's boat (an unusual object in the semiarid Badlands) by using a knife to cut the towline tied to the piazza. As chairman of the Little Missouri Stockmen's Association of Dakota-

Theodore Roosevelt guarding the thieves who stole his boat. The staged photo was taken in September 1884.

Montana, Roosevelt had been given the position of deputy sheriff of what is today Billings County.* He felt it was his duty to catch the scoundrels, so once he calmed down he hatched a plan. They would construct another scow and then light out after the thieves. (You could almost hear the wheels turning: what a good article or essay the catching of the crooks would make in *Century* magazine.) So the new boat was built, and off they went in hot pursuit, like Pat Garrett. Disconcertingly, Roosevelt, always bent on self-improvement, brought along copies of Leo Tolstoy's *Anna Karenina* and Matthew Arnold's collected poetry with him so as not to be bored.

For a few days Roosevelt—with the help of a wagon driver—tracked the thieves in this new boat, sleuthing for clues along the riverbank. Occasionally he found fortuitous footprints as clear as if sealed in wax. Most of his daytime hours, however, were spent navigating around ice floes. For supper he killed deer and rabbits. After pursuing his prey more than eighty miles he captured the three thieves (Burnsted, Pfannenbach, and Finnegan) at Cherry Creek in McKenzie County. By the time he marched back to Dickinson, all six men had blistered feet and frostbitten toes.

*Billings County was created by the 1879 territorial legislature. The first government, however, wasn't organized until May 4, 1886.

Typically Roosevelt boasted that the man-hunt was a "bully affair" (and he got his whopping good story to write about for *Century*). If he felt exhausted he moaned in silence. He even finished *Anna Karenina* along the way for good measure. Typically vainglorious, Roosevelt later had reenactment photographs taken of himself, his rifle keeping his three weary prisoners at bay. Nevertheless everybody in Dickinson was abuzz about their daring new deputy sheriff. "He was all teeth and eyes," the town doctor, Victor Stickney, wrote of first encountering T.R. "His clothes were in rags from forcing his way through the rosebushes that covered the river bottoms."[58]

News of the bravery of Roosevelt, Dow, and Sewall spread from Stark and Billings counties all the way back to New York. With Edith steaming across the Atlantic for the summer to spend time with her family in Europe, Roosevelt basked in his new status in the Badlands. For certain he was no longer a Jane Dandy or Lil' Pumpkin in the Dakota Territory. The combination of writing *Hunting Trips* plus the episode of the boat thieves had transformed him into a minor Wild West legend. Everywhere he went in Medora or Dickinson, people cuffed him on the back in admiration. Locals—even Sewall and Dow—called him "Mr. Roosevelt," and meant it. (Since the death of Alice he bristled if anybody dared call him "Teddy.") The consensus was that the New York politician was a "fearless bugger."[59]

That spring and summer Roosevelt felt "strong as a bear, full of healthiness of mind."[60] Constantly he wrote naturalist riffs about shimmering cottonwoods, low buttes, and prairie grasses. Taken together his prose amounted to a love song to the Badlands. "I have my time fully occupied with work of which I am fond; and so have none of my usual restless, raged wolf feeling," he wrote to his sister Anna on May 15. "I work two days out of three at my book or papers; and I hunt, ride and lead the wild, half adventurous life of a ranchman all through it."[61]

By Independence Day, Roosevelt had finished *Thomas Hart Benton*. Most of it had been written in the cool quiet of mornings at his desk at the Elkhorn Ranch, with its view of the Little Missouri River. When *Benton* was published, reviewers hailed it as workmanlike and a success.[62] Nothing more than that. What Roosevelt most admired about Benton, it seemed, was his belief in the regenerative power of the American West, the fact that he championed frontiersmen with the "tenacity of a snapping turtle."[63] In discussing Benton's support of westward expansion, Roosevelt insisted that the federal government should have acquired even more territory: the Baja peninsula from Mexico; and British Columbia,

Saskatchewan, and Manitoba from Great Britain. (His experience hunting in Minnesota made him want all of Canada to belong to the United States.) Only the United States, he maintained, knew how to properly manage land and rivers. He was acting as the wilderness warden of America. "No foot of soil to which we had any title in the Northwest should have been given up," he wrote; "we were the people who could use it best, and we ought to have taken it all."[64] The United States, he argued, needed to "swallow up the land of all adjoining nations who were too weak to withstand us." (In Roosevelt's obsession with the western lands, conservationists could perhaps see the seeds of his future belief in vast forest reserves.)

Although the only book Roosevelt actually wrote at Elkhorn was *Benton* (and perhaps a couple of chapters of *Hunting Trips of a Ranchman*) the conservationist and scholar Lowell E. Baier—a longtime official of the Boone and Crockett Club—nevertheless called the Badlands cabin the "cradle of conservation" in an important 2007 article in the *Theodore Roosevelt Association Journal*.[65] It was at the Elkhorn that Roosevelt found his voice to caution against careless growth, deforestation, wildlife depletion, and environmental degradation. That July Fourth, Roosevelt traveled from the Elkhorn to a Dickinson's Independence Day ceremony. He boasted about the largeness of the American landscape, its "big prairies, big forests and mountains." Addressing an admiring crowd of ranchers and farmers, Roosevelt warned that the "Far West" might be raped by those who exercised their democratic rights "either wickedly or thoughtlessly."[66]

IV

Between March and August 1886, Roosevelt wrote six articles for *The Outing Magazine*, each time using the Elkhorn Ranch as his lead.[67] This was a coup for the glossy magazine edited by Poultney Bigelow, a Yale-educated outdoors enthusiast.[68] *Outing*, aimed at men, was rolling in advertising revenue because of its popular dog and horse stories. One contributor to *Outing*, trying to describe the magazine's readership, said it was for "plain, uneducated, shrewd minded men of sport."[69] Illustrated with drawings by J. R. Chapin, R. Swain Gifford, and J. B. Sword, among others, the action-packed pieces by Roosevelt bemoaned the depletion of game in the West. One was titled "The Last of the Elk." Although all these articles were tied together by hunting and bravado, Roosevelt nevertheless discussed nongame birds like avocets and stilts.[70]

Although Roosevelt admired the illustrators assigned to his own articles, he was astonished by the exquisite pen-and-ink sketches of an obscure artist named Frederic Remington that he discovered elsewhere in the magazine, accompanying stories about the Apache Wars along the Arizona-Sonora border. Impressed by Remington's clear, honest eye, Roosevelt decided to tap him to illustrate future stories he planned on writing about the Badlands for *Century*.

In the *Outing* article "The Ranch," Roosevelt expressed his environmental concerns in earnest. "To see the rapidity with which larger kinds of game animals are being exterminated throughout the United States is really melancholy," he grumbled. "Fifteen years ago, the Western plains and mountains were places fairly thronged with deer, elk, antelope, and buffalo. . . . All this has now been changed, or else is being changed at a really remarkable rate of speed. The buffalo are already gone; a few straggling individuals, and perhaps here and there a herd so small that it can hardly be called more than a squad, are all that remain. Over four-fifths of their former range the same fate has befallen the elk; and their number . . . is greatly decreased. The shrinkage among deer and antelope has been relatively nearly as serious. There are but few places left now where it is profitable for a man to take to hunting as a profession; the brutal skin-hunters and greasy Nimrods are now themselves sharing the fate of the game that has disappeared from before their rifles."[71]

In August 1886 Roosevelt took another hunting trip, with Merrifield as sidekick, this time to the Coeur d'Alene mountains of Montana and Idaho in search of white goats. He would draw on this trip for two essays about these sure-footed climbers, which he considered the "queerest wild beasts in North America." Roosevelt, in fact, wrote naturalist essays about mountain goats (he sometimes called them white goats) for both his second book on the Badlands (*Ranch Life and the Hunting-Trail*, published in 1888) and *Harper's Round Table* in 1897. They were among his very best prose efforts. Even though Roosevelt thought that the meat of mountain goats was musky (and that trying to compete with them in mountain climbing was a fool's errand), he developed a deep fondness for them. Tracing their lineage back to the Himalayas, Roosevelt described their agility, long tail, and distinctive hump. "If a goat is on its guard, and can get its back to a rock," he enthused, "both wolf and panther [mountain lion] will fight shy of facing the thrust of the dagger-like horns."[72]

The hunt for the white goat marked the beginning of Roosevelt's injecting the "fair chase" doctrine (or code of ethics) into his personal relationships with westerners. Roosevelt employed a market hunter, Joe Willis, as his guide in the Coeur d'Alene mountains.[73] Up to this point

Roosevelt had treated his rugged, unwashed guides almost as equals (although he insisted that they all call him "Mr. Roosevelt," never by his first name). Not anymore—on the white goat hunt, Roosevelt continually lectured Willis about changing his careless hunting habits. Grinnell would later describe how Theodore "made himself agreeable as usual and preached so effectively the doctrine of game preservation that he wholly converted Willis, who up to this time had been a skin and meat hunter, considering game animals valuable only for the dollars they yielded the hunter. Roosevelt was constantly doing such individual useful work in conservation matters."[74]

Once back in Medora with his new white goat trophies, Roosevelt received a jolt. A gossip columnist in New York had broken the news that he was secretly engaged to Edith Carow, who was spending the summer with family in Europe. Angry but embarrassed, he wrote to his sister Bamie, who'd been kept in the dark: "I am engaged to Edith and before Christmas I shall cross the ocean and marry her," he confessed. "You are the first person to whom I have breathed a word on this subject." He flogged himself for not staying devoted to his late wife. "I utterly disbelieve in and disapprove of second marriages; I have always considered that they argued weakness in a man's character. You could not reproach me one half as bitterly for my inconstancy and unfaithfulness as I reproach myself. Were I sure there was a heaven my one prayer would be I might never go there, lest I should meet those I loved on earth who are dead."

Politically, the autumn of 1886 wasn't a good one for Roosevelt, either. Never fond of caution, he had quite impetuously decided to run for mayor of New York City, but he lost to Abram Hewitt, the nominee of Tammany and other Democratic organizations. Before Roosevelt had time to contemplate his rashness, he steamed off to Britain on an ocean liner to marry Edith. Bamie was one of the few people invited to the December wedding in London at St. George's Church along Hanover Square. Just a few British men in top hats and women in silks attended. It was a very low-key event.[75]

While Roosevelt was in Europe for a fifteen-week honeymoon, disaster struck the Badlands. According to Lincoln Lang, starting in November ice-dust particles as sharp as glass fragments swirled about in cyclone-like gusts. There were whiteouts, and Lang later wrote that the rush of frigid air "coldly burns the skin as it strikes. It finds its way into your nostrils and then into your lungs, rapidly chilling you through and paralyzing the senses."[76] By New Year's Day 1887, temperatures in the Dakota Territory had plummeted to forty-one degrees below zero. Beleaguered

Badlanders boarded up windows, hoping to keep the deadly winds from blasting into their homes. Outside, stock literally froze in their tracks or died in snowdrifts, unable to feed on the buried grasses.[77]

At the end of January, another blizzard hit the area. Ice killed off most of the vegetation. Foodstuffs were suddenly in short supply. Snowdrifts were now ten to fifteen feet high against the buttes. Huge cattle herds had no safe place to huddle and stay warm. The Montanan painter Charles M. Russell later drew a series of stark illustrations showing skeletal cattle, all rib cage, in the grip of famine. Near death, some cattle even rammed their heads through cabin doors hoping desperately for warmth. Even the most sinewy cowboy was afraid of the wind chill and snow squalls along the ice-chewed rock formations. It was, as Edmund Morris aptly dubbed it, the "Winter of the Blue Snow."[78]

Roosevelt didn't know how cataclysmic the "Winter of the Blue Snow" had been in the Dakota Territory until he arrived back in New York from his honeymoon. After reading reports of dead cattle across the Great Plains (from Merrifield and Ferris) he promised to travel to Medora soon to survey the devastation. But he had to keep his priorities straight. Family always came ahead of business—his father had taught him that rule. So Roosevelt settled into Sagamore Hill, spent time with little Alice, and reconnected with friends. Edith was now pregnant; that September she would give birth to Theodore Roosevelt, Jr. (his first son). Roosevelt also needed to discuss his future in politics with Henry Cabot Lodge and his other trusted advisers. Worried that he was cursed when it came to both business ventures and electoral politics, Roosevelt now feared that he had lost his entire Dakota herd and could soon be financially insolvent. Feeling stuck, his intuition gone, not sure whether to advance or retreat from North Dakota, Roosevelt brooded over the cruel fickleness of fate.

Once Roosevelt finally arrived in the Badlands, in April, he was shocked by what he saw. Around 60 to 75 percent of the cattle in the northern plains had frozen to death, and the dead cattle were still piled up along buttes and in bottoms. Day in and day out he tried to inventory his losses. "The land was a mere barren waste," Roosevelt wrote; "not a green thing could be seen; the dead grass eaten off till the country looked as if it had been shaved with a razor."[79] America's great range had been ravaged. Less than half of Roosevelt's own herd had survived the series of blizzards. For once he was hard-pressed to find a silver lining. The only optimistic observation he could muster was that at least a thaw was under way. No longer, however, was the spring roundup a glorious, fun event. Instead of branding and roping, dour-faced local ranchers collected rotting carcasses and scattered bones in wooden carts as if bubonic plague had stricken the region. The winter of 1886–1887 had made the Elkhorn and Maltese Cross ranches nearly go bust as a business venture. "I am bluer than indigo about the cattle," Roosevelt wrote to his sister. "It is even worse than I feared; I wish I was sure I would lose not more than half the money ($80,000) I invested out here. I am planning to get out." All told, his net loss would be $23,556.68.[80]

For the first time in his life nature had been cruel to Roosevelt. The magic of the Badlands had turned menacing and gruesome. No longer was he writing prose hymns about the Missouri lark (Sprague's pipit) being the "sweetest singer" or snow geese "nibbling and jerking at the grass."[81] The wildlife seemed to have died or disappeared. Within a few years Medora would become nearly a ghost town, and the open-range cattle business would be steadily diminished. Nevertheless, over the coming decades, Roosevelt continued to boost North Dakota, telling people that in the Badlands he found vigor based on self-assuredness.[82]

Even though Roosevelt would return to Medora in the coming years (his last visit was in October 1918), his days as a serious rancher were over. Although he didn't abandon the cattle business entirely in 1887—keeping, for example, his ranch brand—after the "Winter of the Blue Snow" he was always downsizing. His years of genuine residency—September 7, 1883, to December 5, 1887—were history. Yet Roosevelt was rich in glorious memories. Even the gray alkali dust, layers of sandstone, and heaps of debris, it seemed, took on a romantic cast in his highly selective mind. Years later, after being president, he told Senator Albert Fall that his days in North Dakota had been far and away the best of his entire life. "Do you know what chapter or experience in all my life I would choose to remember, were the alternative forced upon me to recall one portion of it, and to have erased from my memory all the other experiences?" he asked himself and then answered. "I would take the memory of my life on the ranch with its experiences close to Nature and among the men who lived nearest her."[83]

Although North Dakota provided the "romance" of his life, it was also where his worries about the depletion of America's natural resources took root. Nobody championed the conquering of the West by the U.S. Army, mountaineers, homesteaders, trappers, farmers, and ranchers more than Roosevelt. Already in 1887—as he worked on a biography of Gouverneur Morris, author of much of the Constitution and creator of the U.S. decimal coin system—Roosevelt planned on writing a multiple-volume work he called *The Winning of the West*, an epic history in the Parkman tradition tracing American continental expansionism from Daniel Boone in 1774 to the death of Davy Crockett in 1836. Roosevelt even considered the genocide of Native Americans—which was indeed explored when the work first came out in 1894—as heroic. "The most ultimately righteous of all wars is a war with savages, though it is apt to be also the most terrible and inhuman," he wrote. "The rude, fierce settler who drives the savage from the land lays all civilized mankind under a debt to him."[84]

Because Roosevelt had lived on the Dakota frontier, he felt ideally suited to write a paean to westward expansion. In many ways, he saw *The Winning of the West* as merely the logical next step following the publication of *Hunting Trips of a Ranchman* and *Thomas Hart Benton*. Even though the Allegheny upcountry of the 1770s was vastly different from the Badlands of the 1880s, Roosevelt found them deeply connected.[85] "We guarded our herds of branded cattle and shaggy horses, hunted bear, bison, elk, and deer, established civil government, and put down evil-doers, white and red, on the banks of the Little Missouri, and among the wooded, precipitous foot-hills of the Bighorn, exactly as did the pioneers who a hundred years previously built their log-cabins, beside the Kentucky or in the valleys of the Great Smokies," Roosevelt wrote in the preface of Volume 1 of *The Winning of the West*. "The men who have shared in the fast-vanishing frontier life of the present feel a peculiar sympathy with the already long-vanished frontier of the past."[86]

Such triumphalist "white man's burden" sentiments aside, Roosevelt nevertheless worried that the United States' innate sense of opportunity had recently degenerated into exploitation. America knew how to conquer, but it was failing in the art of properly managing its hard-won resources. The West's virgin woodlands were rapidly being logged and its rolling prairies plowed. Wetlands were being drained, and streams were being fished out. Everywhere Roosevelt went in the Dakota Territory, the topsoil had been leached of nutrients and signs of erosion were commonplace. It sickened him to see wild ungulates being poisoned and slaughtered because they supposedly ate the same grasses as cattle and sheep. Even the very wilderness of the West was disappearing in a maze of train tracks, barbed wire, telegraph lines, and meat-processing plants.

As Roosevelt surveyed the Dakota Territory in 1887, finding it nearly impossible to hunt a buffalo, elk, or pronghorn, he understood that the "winning of the West" had been accomplished at the expense of natural resource management, and it made him melancholy. Saving the American West from environmental ruin after the winter of 1886–1887 became a high priority for public policy. Even while he was counting cattle casualties from the "blue snow," he was planning future western trips: to the Selkirks of British Columbia, the Bitterroots of Wyoming and Idaho, Yellowstone National Park, and the Two Ocean Pass of Wyoming. Each sojourn reinforced his newfound belief that the western terrain was a fragile ecosystem.[87]

CHAPTER EIGHT

Wildlife Protection Business: Boone and Crockett Club Meets the U.S. Biological Survey

I

To offset his losses from the "blue snow," Roosevelt wrote yet another book about his Badlands experiences, to be called *Ranch Life and the Hunting-Trail*, illustrated by Frederic Remington. (This book often gets mixed up with *Hunting Trips of a Ranchman* because, even though they were published three years apart, their titles are very similar.) *Ranch Life* would consist largely of articles Roosevelt had been commissioned to write for *Century* starting in late 1887, along with additional previously unpublished essays. An overarching conservationist message now emerged from Roosevelt's hunting experiences: tragically, American big game was verging on extinction throughout the entire West. Consider how difficult it had been for Roosevelt to shoot a lone buffalo, or to find a grizzly bear. Poacher syndicates were even slaughtering elk within the confines of Yellowstone National Park. If law enforcement didn't round up the illegal shooters and trappers, then doomsday, Roosevelt believed, lurked just around the corner for western wildlife.

The overwhelming question weighing on Roosevelt's conscience as he worked on *Ranch Life* was simple: how could he be proactive to save big game animals? Although Roosevelt's exact moment of reckoning remains unclear, in early December 1887 he found a conservationist solution to his quandary. Borrowing from the way his elders tackled societal ills, he would create a hunting club devoted to saving big game and its habitats. High-powered sportsmen like himself, he believed, banding together, had to lead a new wildlife protection movement. Posterity had a claim that couldn't be ignored: saving American mammals was an imperative. A "fair chase" doctrine—hunting rules and regulations—had been created. And as far as Roosevelt was concerned, the time for watered-down measures had passed; his club would fight for true solutions, its goal being the creation of wilderness preserves all over the American West for buffalo, antelope, mountain goats, elk, and deer.

By the time Roosevelt left the Badlands for New York, his conservationist resolve had grown firm. What his Uncle Rob had done for fish, he would do for American big game. A day after arriving back in New York, in early December 1887, Roosevelt convened some of the best and brightest wildlife lovers and naturalists in the New York area to dine at his sister's Madison Avenue home. He was ready to make a hard sell. If his father could found the American Museum of Natural History from a parlor in Manhattan, Theodore saw no reason why this group, meeting in the cramped uptown quarters he shared with Edith, Bamie, and little Alice when not at Sagamore Hill, couldn't save buffalo and elk in the American West.[1] After all, *Hunting Trips of a Ranchman* had established him as *the* authority on big game. Roosevelt now had a sacred responsibility, he believed, to save herds of North American ungulates from extinction.

As his first step, Roosevelt asked George Bird Grinnell to be a cofounder of the Boone and Crockett Club (named after his two favorite, iconic trailblazers). Grinnell had already successfully created the Audubon Society and was editor of the respected periodical *Forest and Stream*, so he knew how to rally public opinion. Roosevelt always valued experienced help. Although Grinnell disdained lobbying, he was good at it. Grinnell fully approved of the project, and his willingness to join forces with Roosevelt to promote the conservation of big game animals and their habitat boded well for the eventual success of the Boone and Crockett Club. Roosevelt and Grinnell then lured a who's who of other conservation-minded "American hunting riflemen" to serve as founders. All of the original twelve members, they insisted, had to espouse the "fair chase" philosophy and believe in the sanctity of national parks.[2]

Roosevelt tapped his brother, Elliott, and his cousin J. West Roosevelt (both childhood veterans of board meetings for Theodore's Roosevelt Museum during the 1870s) to join the Boone and Crockett Club. It was now the turn of their generation, emerging into maturity, to continue the kind of conservation work Robert B. Roosevelt had long championed. Most of the other founders were New York capitalists with deep pockets, like T.R. himself: E. P. Rogers, a yachtsman and financial investor; Archibald Rogers, the rear commodore of the New York Yacht Club; J. Coleman Drayton, who was John Jacob Astor's son-in-law; Thomas Paton, the husband of the heiress Marion Rowle; and Rutherford Stuyvesant, a wealthy real estate investor. From the outset Roosevelt knew that large sums of money would be necessary to lobby effectively in Washington, D.C.[3] Basically, the founders were from the establishment, easily

A drawing of Roosevelt standing next to a trophy worthy of the Boone and Crockett Club.

distinguishable from the plain citizenry of New York even though none was afraid to get mud on his boots.

In early January 1888, the twelve founders of the Boone and Crockett Club had approved a prescient conservationist constitution at Pinnards Restaurant in Manhattan. Ideas had been allowed to percolate freely. The thought of buffalo once again thundering on the plains aroused the founders' enthusiasm during their inaugural deliberations. A decision was made from the outset that the club would have a permanent membership of exactly 100 hunters. The bylaws also stipulated that a limited number of associate members (no more than fifty) would also be allowed. Roosevelt—who would remain the club president until 1894—filled the associate memberships with a galaxy of truly talented writers, including Owen Wister and Henry Cabot Lodge (two outdoorsmen he admired unconditionally). Scientists, military officers, political leaders, explorers, and industrialists were also recruited as members, in hopes that they'd forge innovative solutions to stop the exploitation of America's natural resources. Roosevelt himself brought in the army generals William T. Sherman and Philip H. Sheridan, the artist Albert Bierstadt, the former secretary of the interior (from 1877 to 1881) Carl Schurz, and the geologists Clarence King and Raphael Pulmelly.[4] "The members of the club, so far as it is developed, are all persons of high social standing," an editorial writer in Grinnell's *Forest and Stream* said of the club's founding, "and it would seem that an organization of this description, composed of men of intelligence and education might wield a great influence for good in matters relating to game protections."[5]

Among the Boone and Crockett Club's original objectives, as stipulated in article 3 of its bylaws, were (1) to promote "manly sport with the rifle"; (2) to promote "travel and exploration in the wild and unknown, or but partially known portions of the country"; (3) to "work for the preservation of the large game of this country, and so far as possible to further legislation for that purpose, and to assist in enforcing the existing laws"; (4) to "promote inquiry into and to record observations on the habits and natural history of various wild animals"; and (5) to "bring about among the members interchange of opinion and ideas on hunting, travel and exploration."[6] Bylaw 3 is the area where the club eventually succeeded beyond even Roosevelt's wildest hopes.

The club's members were indeed an aristocratic clique, pampered but not easily fatigued. Their first meetings were relaxed, as if the members were sitting around the lounge-library of the country manor hotel. Roosevelt served as a moderator, keeping the conversation on track, reminding the wildlife preservation gospelers to be pragmatic. In short order they came up with a set of reasonable rules, none more important than their own membership requirements. To be eligible, a hunter had to have successfully shot at least three varieties of North American big game with a rifle. Once inducted, members were sworn to maintain a strict code of honor—in particular, never lying about a kill and always behaving as serious naturalists. Frontier values, the founders agreed, built character and were to be encouraged. On a hunt, nobody was allowed to pull rank or claim privileges because of his social station—members were absolute equals when tramping. As a covenant all members had to be determined to save big game for future generations of Americans to enjoy. Although it wasn't their main priority, club members all seemed to mourn the yearly loss of the "heritage" of American hunting.

Like membership in Skull and Bones or Porcellian, the very act of going on a hunting trip with a fellow club member created a lifelong bond of fellowship; after all, even Natty Bumppo seldom hiked in the Adirondacks or Green Mountains without Uncas or Chingachgook as guides.[7] A general feeling among the members was that research on wildlife—habits, traits, coloration—could never be overdone. To all of the men, America without big game would be like an oak or sycamore tree that never had leaves: totally unacceptable. By saving wild creatures, the Boone and Crockett Club was saving America's outdoors heritage. Some members were given distinguished Native American names such as Pappago, Little Brave, and Running Waters.[8]

On February 29, 1888, the founders drafted a constitution. The club has been described as the "first-ever organization to be formed with the explicit purpose of affecting national legislation on the environment."[9] Its initial goal for wildlife conservation was to add enforcement provisions to the laws governing Yellowstone National Park. When Congress had established America's first national park in 1872, it neglected to enact regulations punishing poachers, sawyers, vandals, or miners. As a result, many westerners treated the park as if it were ripe for plucking. Particularly, local wildlife was under siege from the Northern Pacific lobby (the so-called Railroad Gang) and its associated real estate speculators. If the park police happened on someone poaching or trying to haul out minerals, all they could do was escort the offenders to the boundary and expel them.[10]

Therefore, the Boone and Crockett Club appointed a committee to "promote useful and proper legislation toward the enlargement of the Yellowstone National Park."[11] As the railroad fought for a right-of-way through the park, and its allies pushed legislation reducing Yellowstone's acreage, the Boone and Crockett committee lobbied for regulations with teeth in them.[12] There are about 130 boxes of largely uncataloged material at the club's archives in Missoula that bear witness to how crucial Yellowstone preservation was to the founding members.

The club had real reasons to worry about wildlife depletion in the West. Audubon's bighorn sheep, the eastern and Merriam's elk, the heath hen, the Carolina parakeet, and the great auk had all gone extinct since the 1870s, to name just six examples. The once plentiful beaver was no longer found east of the Mississippi River. (Even encountering a beaver in the Great Plains and Rocky Mountains was a rare event by 1887.) The buffalo population had dwindled dramatically since the start of the century, from 40 million in 1800 to at most a couple of thousand in 1887. White-tailed deer were faring only slightly better: when Europeans first came to America, there had been approximately 24 million white-tailed deer, compared with just 500,000 at the time of the first Boone and Crockett dinner. Even wild turkeys—once nearly ubiquitous and numbering 15 million—were struggling, with current estimates of only 30,000. (Today the number is around 7 million.) "As sportsmen, the club founders were very aware of what the plight of wildlife meant in terms of future hunting prospects," George B. Ward and Richard E. McCabe explained in *Records of North American Big Game.* "As Americans, they saw clearly what it represented in broader resource terms to the national health. As businessmen, industrialists, journalists, and politicians, they recognized that it lay with them, and others of like vision, to attempt 'so far as possible' to relieve and correct the situation."[13] The survival of two game animals Roosevelt had written about so authoritatively in *Hunting Trips* and in his outstanding articles for *Outing*—the pronghorn and elk—was a high priority of the Boone and Crockett Club. Only 150,000 elk and 25,000 pronghorn remained in the West; each species was down 98 percent from counts early in the nineteenth century.[14] Yellowstone became the club's special cause.

"So far from having this Park cut down it should be extended," Roosevelt declared, "and legislation adopted which would enable the military authorities . . . to punish in the most rigorous way people who trespass against it."[15]

Many historians now believe that the Boone and Crockett Club—Roosevelt's brainchild—was the first wildlife conservation group to lobby *effectively* on behalf of big game. The club sprinkled the issue of wildlife protection with kerosene, struck a match, and watched it take off. While antihunters sat on the sidelines gabbing about the extermination of the buffalo, Roosevelt and Grinnell popularized the sportsman's code and called for protection of the buffalo in Yellowstone. In essence, the club became the most important lobbying group to promote *all* national parks in the late 1880s. Audubon's six words—"surely this should not be permitted"—which Grinnell promulgated had now become the club's rallying cry. And with the help of General Sheridan and Senator Vest, in particular, Roosevelt and his friends achieved very good press during their club's first two years of existence.

The club had awakened a national conscience pertaining to the wanton destruction of America's limited natural resources. Like cavalry officers on a mad charge, Roosevelt and Grinnell often went to Washington, D.C., to demand congressional action on behalf of wildlife. Their message was as clear as a bell. Shame fell upon senators who tried to cross words with Roosevelt and company. "Those who used to boast of their slaughter are now ashamed of it," a triumphant Grinnell declared in 1889, referring to the club's success, during its first year, in convincing Americans about the need for regulated hunting and for federal parks; "and it is becoming recognized fact that a man who wastefully destroys big game, whether for the market, or only for heads, has nothing of the true sportsman about him."[16]

II

While Grinnell used *Forest and Stream* to promote the Boone and Crockett Club during 1888, Roosevelt subtly did the same in *Century* magazine.[17] The six long articles T.R. wrote for *Century* about conservation, ranching, and hunting were expertly illustrated by Frederic Remington, who recalled the landscape of the Badlands from a trip there in 1881. If Custer could take his own photographer along with him to the Black Hills in 1874, Roosevelt saw no reason not to have Remington illustrate his Dakota exploits for posterity. Roosevelt wanted his likeness to exude dignity, stoicism, and righteousness. Despite all his frenetic activity and the difficulty he faced daily, pouring out words as quickly as ideas came to him, Roosevelt preferred being portrayed by Remington (and later by others) as coolheaded and manly—a modern fictional counterpart would be Rooster Cogburn in *True Grit.* Reporters enjoyed covering Roosevelt's bombastic side, but he himself felt that his most impressive quality was fundamental decency.

Remington would soon become nearly as famous Roosevelt. By the time Roosevelt had recruited him, Remington, a native of Canton, New York, had sketched cavalrymen, bronco busters, mountain trappers, desert rats, cowboys, and Indians. (In fact, he had just finished illustrating Elizabeth Custer's *Tenting on the Plains*, to modest acclaim.[18]) Like Roosevelt—and Buffalo Bill, who in 1887 was performing in Great Britain to capacity crowds—Remington knew how to strike a mythic note when portraying the settlement of the American West.[19] Yet, to his credit, Remington, not a particularly talkative man, wasn't interested in the bogus myths of the West such as El Dorado or the Northwest Passage. Like Roosevelt, he wanted to present the Rockies, the Great Plains, and Southwest in a factually accurate way, sketching with enterprise and precision whatever he saw. Belief in Darwin and the science surrounding that belief—humans evolving from apes, natural selection, and survival of the fittest—were also an important component of Remington's artistry.

Although they collaborated brilliantly on the articles for *Century*, Remington didn't personally care for Roosevelt, who was known to mock sheep farmers (Remington had once herded sheep in Kansas). "No man," Roosevelt claimed, "can associate with sheep and maintain his self-

Frederic Remington's Buffalo Hunter Spitting a Bullet into a Gun *(ink wash and watercolor on paper, 1892) was a Roosevelt favorite. This illustration was created for an edition of Francis Parkman's* The Oregon Trail.

respect."*[20] This rift over sheep was rather silly, for on the face of it, the two men had much in common: an Ivy League education (Remington had gone to Yale); a belief in the strenuous life and Darwin's and Huxley's biology; and, of course, a shared interest in wildlife, ranching, cowboys, and the American West in general. Neither of them enjoyed fake western stories about jackelopes, bigfoots, or ring-tailed roarers. But Roosevelt's blue blood rankled the scrappy, middle-class Remington, who was often stone broke and begging for freelance assignments.[21]

Inherited privilege, in fact, annoyed Remington no end. And to Roosevelt, Remington was just a gun for hire, a talented illustrator from whom he had commissioned sixty-four illustrations (plus another nineteen for *Ranch Life and the Hunting-Trail*). Not for a second, however, did either man regret their collaboration. Sheep or no sheep, Remington's illustrations couldn't have been finer. But Roosevelt blanched at the idea of Rem-

*In an article in *Harper's Round Table* titled "Ranching" (August 31, 1897) Roosevelt admitted that being a shepherd was hard work: "A good deal of skill must be shown by the shepherd in managing his flock and in handling the sheep-dogs, ordinarily it is appallingly dreary to sit all day long in the sun, or loll about in the saddle, watching the flocks of fleecy idiots. In times of storm he must work like a demon and know exactly what to do, or his whole flock will die before his eyes, sheep being as tender as horses and cattle are tough."

ington as an equal and never once considered him worthy of membership in the Boone and Crockett Club. To Roosevelt, at least before the Spanish-American War, Remington was a plebian, not fit to share a private club's dais with high-caliber naturalists like Grinnell and Parkman.

III

In the late summer of 1888, having finished *Ranch Life and the Hunting-Trail*, Roosevelt once again went on a big game hunt. His destination was the dense coniferous forests of the Pacific Northwest and his prey was the woodland caribou (*Rangifer tarandus caribou*). As a Harvard student, Roosevelt had tried to hunt caribou in the North Woods of Maine and failed miserably. Now, with his library walls at Sagamore Hill filled with North American big game trophies, the lack of a caribou head was palpable. According to Grinnell, Roosevelt's best chance of finding a herd was in the Idaho Territory, high in the Selkirk Mountains along the border of the Washington Territory. So off Roosevelt went on the Northern Pacific Railroad for a stopover in Medora and then on to the Idaho village of Kootenai on the north side of Lake Pend Oreille. Along the way, whenever possible, Roosevelt worked on the first volumes of *The Winning of the West*. No time was ever wasted when Roosevelt was in a railway car, for he always turned his compartment into a rolling library.

Never before, not even in the Bighorns, had Roosevelt encountered mountains such as the Selkirks. At one point he could see the Columbia River bending and twisting through gorges lined with towering pines. Much of the brush-choked forest had never been explored.* It was delightful to feel like a naturalist explorer again. "The frowning and rugged Selkirks came down sheer to the water's edge," Roosevelt recalled. "So straight were the rock walls that it was difficult for us to land with our batteau, save at the places where the rapid mountain torrents entered the lake. As these streams of swift water broke from their narrow gorges they made little deltas of level ground, with beaches of fine white sand; and the streambanks were edged with cottonwood and poplar, their shimmering foliage relieving the sombre coloring of the evergreen forest."[22]

The village of Kootenai was the head of the famous Wild Horse Trail, a pack path that led to mineral-rich mines near Fort Steele, British Columbia. Roosevelt—with old John Willis (who wasn't attentive to hygiene) and a Kootenai Indian named Ammál (who was built like a heavyweight

*Four of Canada's first national parks—Mount Revelstoke, Glacier, Yoho, and Banff—are all part of the Selkirks.

boxer) as guides—traveled up the swift Pack River on the Wild Horse Trail and over the Continental Divide to the Kootenai River. From there they floated down the bone-chilling river in a pirogue, eventually making camp alongside Kootenai Lake. This sheet of crystal-clear water was considered the heart of caribou country. Situated in a long valley between the Selkirk and Purcell mountains, the glassy lake was approximately seventy miles in length.[23] Naturally, they set up camp at a level place. Roosevelt was so anxious for caribou that he bathed his first morning in Idaho before the sun broke.

Idaho was God's country to Roosevelt, even though the hunting started out slowly. One afternoon, while eating frying-pan bread by a brook, Roosevelt spied an ouzel feeding. Suddenly a water shrew swam into a shallow eddy nearby. Roosevelt had read about this rare little mammal—*Sorex palustris*—in zoology books over the years, but this was his first real-life encounter with it. The water shrew's habitat was northern forest streams surrounded by fallen logs and lichen-covered rocks. Roosevelt's first thoughts were of Spencer Fullerton Baird at the Smithsonian Institution and Dr. C. Hart Merriam at the Department of Agriculture. He knew that they would have "coveted it greatly" for their collections.[24] Forgetting about caribou for the moment, Roosevelt captured the feisty little shrew and studied it carefully. The Kootenai guide, Ammál, who had taken to calling Roosevelt "Boston Man," shook his head in disbelief at the glee his client felt about an insectivore.[25] "It was a soft, pretty creature," Roosevelt wrote, "dark above, snow-white below, with a very long tail." After inspecting it alive, Roosevelt killed the shrew, turning the skin inside and letting it dry. Throughout his days in Idaho, he treated the specimen as a prized possession. Too much handling of the skin, Roosevelt believed, owing to chemicals on one's hands, would lead to discoloration. When Roosevelt returned to New York he sent the shrew to Baird in Washington, D.C., where it became part of the Smithsonian's natural history collection.[26]

But Roosevelt hadn't traveled all the way to Idaho so that a solitary water shrew could be put on display at the Smithsonian. Eventually he killed a black bear, one with "two curious brown streaks down its back," and fried its meat for dinner. The caribou, however, remained elusive. Hiking up steep mountain crests and the faces of cliffs, a frustrated Roosevelt couldn't even find a caribou trail. Willis often ran ahead to reconnoiter, but without luck. At dusk Roosevelt's party felt removed from even the back of civilization. "Indeed the night sounds of these great stretches of mountain woodlands were very weird and strange," Roosevelt wrote. "Though I have often and for long periods dwelt and hunted in the wilderness, yet I never before so well understood why the people who live in lonely forest regions are prone to believe in elves, wood spirits, and other beings of an unseen world."[27]

Back in New York, Roosevelt raved about the unsurpassed beauty of Idaho—and it was also pretty good for the hunter's pot. While Grinnell promoted Montana's Flathead Range as the most gorgeous part of the Rocky Mountains, Roosevelt championed the stupendous ranges of Idaho. Perhaps because he had so fully documented wildlife in the Dakota Territory, he was proud to add the topmost peaks of the Idaho Territory to his area of expertise. Before long he was writing about Idaho's "hoary woodchucks [marmots]," "timid conies [pikas]," and "troops of noisy, parti-colored Clark's crows."[28]

When Roosevelt became president thirteen years after the caribou hunt, saving wild Idaho—which had become a state in 1889—ranked high on his agenda. On January 15, 1907, he created Caribou National Forest, about 200 miles east of Boise. The following year, on July 1, he virtually turned the state into one vast wildlife preserve, setting aside millions of acres in seventeen new national forests with his presidential pen: Pocatello, Cache, Challis, Salmon, Clearwater, Coeur d'Alene, Pend Orielle, Kaniksu, Weiser, Nez Perce, Idaho, Payette, Boise, Sawtooth, Lemhi, Targhee, and Bitterroot.[29] Seldom, if ever, had a hunt resulted in such a momentous conservationist gesture on behalf of wild creatures.

IV

That fall, Roosevelt agreed to campaign for the Republican presidential nominee, Benjamin Harrison. Huge crowds came out to hear Theodore as he and Edith traversed Illinois, Michigan, Wisconsin, and Minnesota. Perhaps because of the fame he'd achieved through his connections with the West, the farmers and ranchers of the upper Midwest and Great Plains loved him. And he was just as popular all the way south to the Rio Grande and Rio Colorado. To Roosevelt the "hurly-burly of a political campaign,"[30] as he once put it, was an enthralling blood sport, the supreme test of personal combat for a genuine warrior.[31] While Harrison ran a "front porch" campaign, delivering speeches from his home in Indianapolis and avoiding perspiration, Roosevelt hit the trail with a vengeance, orating at every crossroads town and village junction. The whole experience, he wrote in a letter to Lodge, was "immense fun."[32]

As Election Day neared, Roosevelt celebrated his thirtieth birthday. It was a rare time for self-reflection. Despite all his exciting work with the Boone and Crockett Club and as a writer, his political career had stagnated. After all, it was hard to build momentum after losing the mayoralty. He hoped that stumping for Harrison, giving the Republican Party his everything as a surrogate, would earn him a major (or even minor) government post in the administration. Tactically, Roosevelt was on track. On November 6, Harrison defeated the incumbent president, Grover Cleveland, by 233 electoral votes to 168. "I am as happy as a king," Roosevelt wrote to his British diplomat friend Cecil Spring-Rice, "—to use a Republican simile."[33] And sure enough, Roosevelt was rewarded with an appointment as U.S. civil service commissioner starting on May 7, 1889. It was a post he would hold for almost six years.[34] His primary task would be to clean up corruption in the federal government.

That Christmas season was a joyous one for Roosevelt. When *Ranch Life and the Hunting-Trail** was published in December, most reviewers applauded his well-crafted prose and Frederic Remington's fine engravings and pen-and-ink line drawings. Whether T.R. was writing about cattle branding at Elkhorn, cowboys' rope tricks, or goat hunting in the canyon of the Coeur d'Alene, Remington delivered spot-on illustrations that electrified the narrative. Although *Hunting Trips of a Ranchman* had been a better book than *Ranch Life and the Hunting-Trail*, Remington's sketches upgraded the latter into an enduring western classic. Nobody alive could draw mountain men, French-Canadian trappers, or timber wolves with the realistic precision of Remington. (For some reason, however, Remington could never properly depict mountain lions, a deficit that annoyed Roosevelt.)

Like Remington, Roosevelt had romantic sentiments about the American West, finding it an almost inexhaustible source of material for books and articles. "Civilization seems as remote as if we were living in an age long past," Roosevelt wrote approvingly about being a westerner in *Ranch Life*. "Ranching is an occupation like those of vigorous, primitive pastoral peoples, having little in common with the humdrum, workaday business world of the nineteenth century; and the free ranchman in his manner of life shows more kinship to an Arab sheik than to a sleek city merchant or tradesman."[35]

Remington wanted merely to draw the reality in *Ranch Life*. "I don't consider that there is any place in the world that offers the subjects that

---

*Amazingly, that same year Roosevelt published two other books: *Life of Gouverneur Morris* and *Essays on Practical Politics*.

the West offers," he once observed. "Everything in the West is life, and you want life in art. . . . The field to me is almost inexhaustible."[36] By contrast, Roosevelt was on an accelerated mission to save its wilderness areas and big game. Underneath his name on the title page of *Ranch Life*, Roosevelt identified himself as president of the Boone and Crockett Club of New York: the club had become a "bully pulpit" for his ideas about wildlife management and forestry reserves. In the text, he mourned for the "fast vanishing" elk and told readers that when hunting deer they should shoot "only the bucks."[37] The descriptions in *Ranch Life* of shifting weather and natural wonders were both precise and poetic. Roosevelt's zoological descriptions rose to the high standard Grinnell had inspired at *Forest and Stream*. Of antelope, for instance, he wrote: "Antelope see much better than deer, their great bulging eyes, placed at the roots of the horns, being as strong as twin telescopes. Extreme care must be taken not to let them catch a glimpse of the intruder, for it is then hopeless to attempt approaching them. On the other hand, there is never the least difficulty about seeing them."[38]

Predictably, Roosevelt ended *Ranch Life* with a boast about the natural grandeur of the American West. With jingoistic but good-natured pride he tried to one-up both Asia and his brother, Elliott. (The fact that he had turned thirty didn't mean the sibling rivalry had dissipated.) Elliott, at this juncture, however, was on an alcoholic downslope, desperately struggling with depression and contemplating suicide. "My brother has done a good deal of ibex, mountain sheep, and markhoor shooting in Cashmere and Thibet [Tibet], and I suppose the sport to be had among the tremendous mountain masses of the Himalayas must stand above all other kinds of hill shooting," Theodore wrote. "Yet, after all, it is hard to believe that it can yield much more pleasure than that felt by the American hunter when he follows the lordly elk and the grizzly among the timbered slopes of the Rockies, or the big-horn and the white-fleeced, jet-horned antelope-goat over their towering and barren peaks."[39]

Unfortunately, T.R. didn't get to bask in the acclaim that *Ranch Life* received. That winter, to meet his publisher's deadline, he was working overtime on *The Winning of the West*. Puffy-eyed, he burned the midnight oil nightly until three or four in the morning. Somewhat naively, he had promised G.P. Putnam's Sons the first two volumes by the spring of 1889. Always tottering toward a physical breakdown, pushing himself beyond the usual human limits, whenever possible Roosevelt locked himself up at Sagamore Hill and wrote. His entire nervous system was strained. Desperately he tried blocking out both good and bad news. With two children to raise, Edith constantly worried that bills had to be paid. (She made sure all financial obligations were met.) Whenever the issue of financial insolvency was raised, however, Roosevelt's blue eyes would darken in disapproval. His exuberance would be temporarily extinguished. Life for Theodore had become a pressure cooker of deadlines, commitments, responsibilities, and financial insecurity. Only the finances, however, made him irritable. His great consolation was that at least he hadn't become a leech, or one of those fellows who always looked for other people to carry the load.

V

Although Roosevelt doesn't mention it in *An Autobiography*, there may have been another impetus for creating the Boone and Crockett Club in 1887. The previous year, anxious to professionalize mammalogy, Dr. C. Hart Merriam of the Department of Agriculture elevated the Economic Ornithology section at the Department of Agriculture into the Division of Economic Ornithology and Mammalogy.[40] In principle, the new division was intended to help farmers cope with pests. But in practice, Dr. Merriam was interested in the distribution of mammals across the United States. Owing to the creation of the Audubon Society and the American Ornithologists Union, birds were starting to be properly studied. Two exhaustive "bulletins," in fact, were published in the late 1880s: W. W. Cooke's "Bird Migration in the Mississippi Valley" and Walter B. Barrows's "The English Sparrow in America."[41] But nobody was publishing similar high-quality bulletins about chipmunks, skunks, squirrels, gophers, ferrets, groundhogs, or dozens of other American mammals.

Merriam—short in stature, with a mustache that made him look like an otter—soon changed that. Merriam's agriculture division began conducting general surveys of mammals (as well as birds), with a keen eye toward biotic community distributions. Using field reports and scientific results, he constructed life-zone maps. For the first time in American history a biological understanding of cougars', wolves', or bears' *ranges* became available to the general public. Nobody had ever inventoried American wildlife quite like Merriam. A crucial component of his success was his uncanny ability to reach out to untrained, backyard naturalists and mammal collectors throughout America. In any given state or territory there was bound to be a local hunter who had preserved the skins of such diurnal species as rabbits and squirrels. "It was from such sources that many of his specimens came and he carried on a large correspondence," the naturalist historian Wilfred H. Osgood recalled of Merriam in the *Journal of Mammalogy*, "promoting interest in mammals by purchasing specimens and, in some cases, by employing collectors or at least by placing standing orders."[42]

A truly practical biologist, Merriam also pioneered in using a new trap for small mammals: called the Cyclone, it was made of tin and wire springs, had an area of only about two square inches when collapsed, and was easily portable in quantity. Merriam sent Cyclones to Roosevelt with the idea that when he was in the Rockies he could set up such traps around cabins and creeks, then carefully perform taxidermy on the little mammals and ship the specimens back to Washington, D.C. (Although this is speculative, Merriam may have sent these traps because he was jealous—since Roosevelt had sent the Idaho water shrew to Baird, Merriam's friendly competitor in collecting.)

Although Merriam's shop was officially the Division of Economic Ornithology and Mammalogy, Roosevelt called it the Bureau of Biological Survey (which is what it officially became on March 3, 1905, the day Roosevelt was inaugurated as president in his own right). The Geological Survey mapped the topography of America after the Civil War, and Roosevelt envisioned the Biological Survey doing the same for the classifications of plants, birds, and animals. The Biological Survey, with Merriam at the helm, proved that temperature extremes were partially responsible for how wildlife was distributed. Although only about five or six crackerjack naturalists were ever on his official payroll, Merriam established a grassroots network of specimen collectors from all over America, a particularly notable achievement in an era when there were no telephones, let alone the Internet. He published a pathbreaking study of the San Francisco Mountain Region and Desert of the Little Colorado, Arizona establishing from summit to base six life zones in the flora of the mountains: Lower Sonoran (Sonoran Desert plants); Upper Sonoran (Colorado pinyon and juniper woods); Transition (Ponderosa pine forest); Canadian (Rocky Mountain Douglas fir and white fir forest); Hudsonian (tree line forests of Rocky Mountain bristlecone pine and Engelmann spruce); and Arctic-Alpine (alpine tundra).*

A founder of the National Geographic Society, Merriam joined the American-British fur seal commission determined to save the great herds

---

*Dr. C. Hart Merriam, *Results of a Biological Survey of the San Francisco Mountain Region and Desert of the Little Colorado, Arizona*, U.S. Department of Agriculture, Division of Ornithology and Mammology, No. 3 (Washington Government Printing Office, 1890).

of Alaska from extinction, in the same way that Roosevelt and Grinnell were protecting the buffalo. By the time Merriam became president of the Biological Society of the Cosmos Club in Washington, D.C. in 1891, he was considered the greatest authority on applying Darwinian theory to American species and topography.[43]

Writing reports from Arizona and saving seals constituted the fun part of Merriam's job at the Biological Survey. As an administrator he had to find ways for wildlife and humans to coexist in America. Everything from poisoning ground squirrels with barley and strychnine to promoting lime and sulfur wash to prevent rabbits from attacking orchards fell under his job description. Anxious to save the beaver from being over-hunted, Merriam became a supersalesman on behalf of muskrat fur as an alternative. Starting in 1886 he also oversaw publication of the *Annual Reports of the Biological Survey*. He also made sure U.S. Department of Agriculture "bulletins" were printed and disseminated all over the country.[44] If you were a farmer in Mississippi or Arkansas, for example, constantly shooting raccoons as varmints, Merriam was the national voice that would say, "Not a good idea." Raccoons fed on crayfish, which infested leveled embankments; so the raccoons were actually providing a pest-control service by destroying the potentially destructive crustaceans. Such pragmatic solutions to wildlife control are why Roosevelt respected Merriam so much. When it came to wildlife protection, Merriam was a living instruction manual.

CHAPTER NINE

Laying the Groundwork with John Burroughs and Benjamin Harrison

I

It is strange now to picture them meeting for the first time in March 1889 at the Fellowcraft Club at 32 West Twenty-Eighth Street in New York. With a membership of around 200 sophisticated newspapermen, writers, and artists, the Fellowcraft had the kind of leather-chair ambience preferred by the literary-minded aristocrats of the gilded age. The thirty-year-old Theodore Roosevelt was one of the club's younger members. Although he was preparing to move to Washington, D.C., for his new post as U.S. civil service commissioner, he nevertheless continued fulminating against the deplorable conditions in New York City's notorious cordon of louse-infected tenement slums. No job could ever rein in his multifarious reformist interests and instincts. But Roosevelt dropped his anti-poverty crusade and his history writing on March 7 for a long-coveted opportunity to meet the naturalist John Burroughs. The date, in fact, should be noted in the annals of U.S. conservation history as the cementing of an extremely significant alliance that would last for nearly three decades. "I thought him very vigorous, alive all over, with a great variety of interests; and it was surprising how well he knew the birds and animals," Burroughs recalled. "He's a rare combination of the sportsman and the naturalist."[1]

Roosevelt and Burroughs's lunch reportedly came about through an intermediary. As an assemblyman, Roosevelt had gotten to know Jacob A. Riis, a Danish-born newspaperman whose *How the Other Half Lives* would soon awake the nation to the suffering of New York City's immigrants. Roosevelt regularly visited a boys' club run by the reformer and polemicist John Jay Chapman in Hell's Kitchen with Riis, and would often give away copies of Burroughs's compilations of essays, such as *Wake-Robin* and *Winter Sunshine*. Chapman, who was a neighbor of Burroughs, arranged to have the two birders meet for lunch at the Fellowcraft Club.[2] Joining them was Elizabeth Custer, the widow of Lieutenant Colonel George Armstrong Custer. Since her husband died at the battle of Little Bighorn

Theodore Roosevelt with John Burroughs in Yellowstone National Park in 1903. The photograph was taken by Illustrated Sporting News.

in 1876, she'd published two western memoirs: *"Boots and Saddles" or Life in Dakota with General Custer* (1885) and *Tenting on the Plains or General Custer in Kansas and Texas* (1887). She, too, would have plenty to converse about with Roosevelt.[3]

T.R. was always trying to engulf people up in the tidal wave of his erudition, which some interpreted as monomaniacal self-regard. As one close friend kindly put it, "He was the prism through which the light of day took on more colors than could be seen in anybody else's company."[4] But Burroughs was a household name—his books were mandatory reading in schoolhouses all over America—and he made Roosevelt feel inferior.[5] Burroughs didn't just tramp around the wooded countryside keeping field notes or hunting game—he *was* the wooded countryside personified. Unlike Roosevelt, who was constantly showing off his credentials as a naturalist, Burroughs, as if half divine, believed there was sanctity in every fallen leaf or grain of sand. (Or, as Charles Dickens wrote of a favorite character in his 1854 novel *Hard Times*, he was a "man who was the Bully of humility."[6])

Concerned over the post–Civil War abandonment of agrarian communities in favor of overcrowded cities, the usually benevolent Burroughs wrote stingingly against "scientific barbarism," even calling humming factories "the devil's laboratory."[7] Raised in the Catskill mountains, Burroughs would rub his eyes in disbelief as he surveyed the environmental degradation of New York City's waterways. Although Burroughs never postulated a social philosophy per se, there was a sympathy for the Transcendentalists in everything he did or said. But he nevertheless admired captains of industry like Henry Ford and Thomas Edison who rose from the tinker's bench to change the world. Roosevelt, by contrast, never fully condemned industrialization—he wanted only to break up trusts and to create great American parks, forest reserves, and bird rookeries to help invigorate city dwellers and uplift their urban spirits from chronic factory smoke and industrial disease.

There were other differences between the two naturalists. Burroughs was contemplative whereas Roosevelt preferred direct action, always ready to rumble. Although both men hunted, Burroughs's pockets didn't always bulge with bullets. While in *Wake-Robin* Burroughs wrote about a deer hunt in the Adirondacks and about shooting rabbits for sport, his literary talent was better suited to promoting the joy of watching wildlife. The older Burroughs got, in fact, the more hunting bored him. To Burroughs, for example, the American Museum of Natural History was a morgue, an abominable mockery of the great web of vibrant life in the animal kingdom. "A bird shot and stuffed and botanized is no bird at all," he later told a group of children who came to visit the museum. "And a bird described by another in cold paint is something less than you deserve. Do not go to museums but find Nature. Do not rely on schoolbooks. Have your mothers and fathers take you to the park or the seashore. Watch the sparrows circle over you, hear the gulls screech, follow the squirrel to his nest in the hollow of an old oak. Nature is nothing at all when it is twice removed. It is only real when you reach out and touch it with your hands."[8]

One can be reasonably sure Roosevelt knew Burroughs's biography from Catskills childhood to *Signs and Seasons* publication practically by heart at the time of their first meeting. Born in Roxbury, New York, on April 3, 1837, Burroughs was six years younger than Theodore Roosevelt, Sr. In his memoir *My Boyhood*, Burroughs wrote of growing up poor on his family's farm yet being enchanted by juniper trees, gurgling streams, apple orchards, and picturesque dairy farms. "I deem it good luck, too, that my birth fell in April, a month in which so many other things find it good to begin life," Burroughs wrote in *My Boyhood*. "Father probably tapped the sugar bush about this time or a little earlier; the blue-bird and the robin and song sparrow may have arrived that very day."[9]

While he was growing up, Burroughs's life revolved around the Roxbury harvest cycle. Enthralled by the cool sweep of the Catskills, he adopted the upstate woodlands as his own "open-air panorama."[10]

By the time he was nine or ten years old, birds—of all kinds—became Burroughs's fixation. One spring day, he later recalled, a cloud of passenger pigeons descended on a grove of beeches. The birds' collective noise in his pasture sounded like a gust at sea or a tornado.[11] Not long afterward—while visiting the U.S. Military Academy at West Point—Burroughs, who had started to style himself as a backwoods ornithologist, dipped into a copy of Audubon's *Birds of America* in the library and couldn't put it down.[12] The beauty of Audubon's flamingos and wild ganders was beyond stimulating. (Years later, in 1902, while Theodore Roosevelt was president, Burroughs wrote a biography of Audubon. Its purpose was to restore Audubon's place as the premier American literary naturalist by virtue of his voluminous journals.[13]) Yet Burroughs noticed that Audubon had also written a ghastly essay on how farmers were slaughtering passenger pigeons by setting tree traps and filling water pots with sulfur. How much more beautiful was the cooing of live passenger pigeons than the heaps of dead birds local farmers used to feed hogs!

To earn a living, Burroughs decided to be a rural schoolmaster. He worked first in New Jersey, not far from the Atlantic coast. When he was twenty years old he traveled to Chicago and had his daguerreotype taken; it shows a rare handsomeness. His longish hair was slicked back, and his straightforward gaze suggested deep wisdom. But Burroughs was always most comfortable in rural settings. Whenever he found himself in New York or Chicago, he would clutch his wallet, worried that pickpockets might spot him as an easy mark. Struggling to find steady work, Burroughs moved to Washington, D.C., in 1862, during the height of the Civil War, and clerked in the Currency Bureau of the Treasury Department. Before long he was befriended by the poet Walt Whitman, who was also living in Washington.

It's unclear whether Whitman fell in love with Burroughs's physical beauty or with his rural innocence (or both) in the fall of 1863, but he did fall in love. (There is, however, no evidence of a sexual encounter between them.) To Whitman, his twenty-six-year-old protégé was like an only son; also, Burroughs was an open-hearted romantic with an unjaded face like "a field of wheat." Mentoring young men like Burroughs came easily to Whitman. He had come to Washington to help nurse the 70,000 Union and Confederate soldiers wounded in action and recuperating in poorly run, unsanitary hospitals. Under Whitman's tutelage, Burroughs started writing about nature in a more intimate way, perfecting his literary craft as a prose stylist when he was not working at his desk in the Treasury Department. There was a refreshing hominess to Burroughs's essays about bullfrogs, maple syrup, and trout spawning. As Burroughs explained, his privileged education under Whitman—who was meticulously working on his cycle of war poems *Drum Taps* when they met—led him to try to "liberate the birds from the scientists." Unlike Roosevelt, Burroughs wasn't overly interested in John James Audubon, the hunter and taxidermist extraordinaire, carting around arsenic paste and a shotgun. But Burroughs *did* want to become a writer about green spaces who would be celebrated as the "Audubon of prose."[14]

For the most part Burroughs spent 1863 to 1873 in Washington, D.C., writing outdoors prose. His friendship with Whitman grew and grew; the gray-bearded poet calling the Catskills writer his own personal "naturalist-in-residence" who felt empathy even for quick-breeding insects. To Burroughs, Whitman's controversial *Leaves of Grass* was "an utterance from Nature, and opposite to modern literature, which is an utterance from Art."[15] When Abraham Lincoln was shot, Whitman—who used to *cher confrère* with Lincoln whenever they passed on the street—mourned like a grieving widow. The great Lincoln, Whitman moaned, had been stolen away in his prime.[16]

Just before the assassination, Burroughs had hiked up Batavia Mountain in the Catskills seeking an afternoon of solitude. Pausing for a moment to catch his breath, he was suddenly mesmerized by the long, ethereal, flutelike song of a hermit thrush. The sounds of this bird held Burroughs transfixed, as if in a dream, for ten or fifteen minutes. Craning his neck high and low, looking up in tree branches and along the ground where the songster might have been foraging, Burroughs struck out. There would be no sighting of a hermit thrush that afternoon. All he took away was a comforting memory of the delicate ringing melody. Upon hearing Burroughs talk effusively about the hermit thrush, Whitman wrote his celebrated eulogy to Lincoln, "When Lilacs Last in the Dooryard Bloom'd." One verse went as follows:

> Solitary the thrush,
> The hermit withdrawn to himself, avoiding the settlements,
> Sings by himself a song.
> Song of the bleeding throat.[17]

While Whitman worked on *Leaves of Grass*, Burroughs became his cheerleader, comparing him to Thoreau and Emerson. Encouraged by

Whitman, confident that *Leaves of Grass* was *the* great American masterpiece, the very embodiment of democracy in verse, Burroughs wrote his own first book, *Notes on Walt Whitman as Poet and Person.*[18] Whitman himself helped edit the manuscript, rearranging quotations and sentences. Walks through Washington's parks now became commonplace for the two nature lovers, who often strolled through the White House gardens or Rock Creek Park in search of veery or ruby-crowned kinglet.[19] And Whitman was the one who chose *Wake-Robin*—a trillium found in eastern American woods—as a title for Burroughs's second book. "He thinks natural history, to be true to life, must be inspired, as well as poetry," Burroughs wrote of Whitman, following one of their hikes. "The true poet and true scientist are close akin. They go forth into nature like friends. . . . The interests of the two in nature are widely different, yet in no true sense are they hostile."[20] Whitman's belief in the special connection between science and poetry was shared by Roosevelt.

As an ice-breaker at the Fellowcraft lunch, Roosevelt had told Burroughs about his European honeymoon with Edith and how Burroughs's nature books—especially *Birds and Poets* and *Locusts and Wild Honey*—made him long terribly for the United States. Roosevelt added that Burroughs's prose was "thoroughly American." Their talk shifted to the reform work Roosevelt was doing to help impoverished New York City boys; Roosevelt told Burroughs that, like Santa Claus, he regularly handed out mint copies of *Wake-Robin* as gifts, instructing the recipients to read every page carefully, for it embodied "all that was good and important in life."[21]

Perhaps somewhat embarrassed, Burroughs feigned mild disbelief at the anecdote, certain that his musings about plovers and blackbirds couldn't uplift ghetto boys who panhandled for stale bread and rotten fruit in the Bowery. But he was pleased that the nephew of Robert B. Roosevelt thought so highly of his work. Totally uncynical, seldom if ever putting anybody down with a jolt of criticism, Burroughs decided that he liked the cut of Roosevelt's jib. (Essentially, Burroughs felt about Roosevelt as he did about a Catskills neighbor: "That man hasn't a lazy bone in his body. But I have lots of 'em—lots of 'em."[22]) After lunch, on the train ride back to the Hudson River valley, as Burroughs passed stops in Tarrytown, Cold Spring, Beacon, Poughkeepsie, and Hyde Park,* he

*Burroughs usually took the train to the Hyde Park station and then caught a little ferry across to his farm at West Park, in Esopus. In the winter, he'd simply walk across the ice.

wrote about his luncheon with Roosevelt, musing on how much luckier rural children like his own son, Julian, were than the urban poor. "How different is the life of Julian," he wrote, "in the country with fresh air, good books, and parents with a measure of leisure—from that of the boys that Chapman and Roosevelt want so much to help."[23]

While Burroughs simply thought of the lunch as enjoyable, Roosevelt had been deeply impressed. Burroughs, he was now certain, was the Thoreau of his time, perhaps the finest literary naturalist America had ever produced. Frequently when fans meet a writer or artist they admire, encountering the celebrity in person is a terrible disappointment. The exact opposite occurred at the Fellowcraft Club; the upshot of the lunch was that Roosevelt was now indissolubly linked to Burroughs. Not since his father died, in fact, had Roosevelt seen such Homeric dimensions in anyone as in John Burroughs. When Roosevelt started writing the last book of his North Dakota trilogy, *The Wilderness Hunter*, his style became infused with Burroughs's naturalist writing. Unlike his previous two Dakota volumes, *The Wilderness Hunter* would emphasize the wildlife-sportsman ethos over even the best-honed hunting yarns.

II

Another factor that contributed to this change of emphasis in Roosevelt's writings was his move to Washington, D.C., to assume his new duties as a member of the Civil Service Commission. Theodore and Edith had decided that she and the children (a second son, Kermit, was born in October 1889) would at first remain at Sagamore Hill while he lived rent-free at Cabot and Nannie Lodge's residence in Washington.[24] An insomniac Roosevelt usually slept only four or five hours a night, so he figured that, after his desk job, there would be plenty of spare time to continue writing *The Wilderness Hunter* and start in earnest to write *The Winning of the West* (his history as mural). Everything was either handwritten or dictated to a stenographer—typing never appealed to him. Every waking hour was a whirlwind of activity. "He was a live wire," Burroughs noted about T.R. in his journal, "if there ever was one in human form."[25] (On another occasion Burroughs said Roosevelt was "a many-sided man and every side was like an electric battery."[26]) Roosevelt himself wrote in *Ranch Life*, "Black care rarely sits behind a rider whose pace is fast enough"—a fitting observation that David McCullough used as the epigraph of *Mornings on Horseback*.[27]

Certainly overworking was preferable to behaving like his brother Elliott. Following a riding accident, the restless, ill-adjusted, but charming

Nell turned to alcohol and opiates to deal with a broken leg and with inner anguish that modern psychiatrists would have probably diagnosed as a form of debilitating depression. A chronic misery had fallen over him. He drifted across the Atlantic, fumbled about London and Rome, occasionally sneaked in some serious hunting, but mostly just squandered opportunities to succeed at anything. For a while he sought rehabilitation in Illinois and worked in Virginia. But spiritual destitution followed him every step of the way. Despite being married to a wonderful wife, Anna Hall, Elliott was nevertheless a serial adulterer, like Uncle Rob. By including Elliott as a founder of the Boone and Crockett Club, Theodore hoped to get his brother refocused on the outdoors life—hunting being the one activity that stabilized Elliott's tormented spirit.[28] However, once Elliott became embroiled in a paternity suit, Theodore lost all patience with his brother.[29] Bringing shame upon the family name, he believed, was never acceptable. "He is evidently a maniac," an agitated Theodore wrote Bamie about Elliott, "morally no less than mentally."[30]

In early 1889, besides worrying about Elliott, Roosevelt wondered whether the Boone and Crockett Club had a staunch ally in Benjamin Harrison. Because Harrison was a Republican—as were most early conservationists—Roosevelt was hopeful. Would the new president fight for forest reserves, fish hatcheries, and big game preservation? The stoop-shouldered Harrison, in fact, was an "aesthetic conservationist" who loved the outdoors almost as much as Roosevelt did.[31] Growing up on a farm along the banks of the Ohio River (near Cincinnati), Harrison hunted duck, fished for smallmouth bass, and hiked around the North Bend woods looking for arrowheads. He had a sharp eye for birds. Harrison appreciated the redemptive quality of wild places and their contribution to building character. Twice before being elected president, he visited Yellowstone National Park. While serving as a U.S. senator from Indiana, Harrison had been instrumental in halting commercial development in Yellowstone, pushing for prohibitive legislation that allowed only ten park acres to be leased for hotel use. Harrison had also introduced a bill in early 1882 that would have set aside land along the Colorado River of Arizona for government preservation. (The legislation failed, but T.R. eventually saved the Grand Canyon via an executive order under the authority of the Antiquities Act of 1906.)

Despite these legislative setbacks, Harrison's conservationist convictions grew. His new secretary of the interior, John W. Noble, was a college friend of his at Yale who'd risen through the Third Iowa Cavalry to become a brigadier general during the Civil War. Following Lee's surrender, Noble moved to Saint Louis, practiced law, and was eventually made a U.S. district attorney.[32] Perhaps because he had seen so much killing in the Civil War, Noble didn't cotton to the slaughtering of bison by market hunters, which had become widespread owing to the demand for the hides. And he worried about a timber famine in the Missouri Ozarks and elsewhere, seeing it as an impending national danger. In 1910 George Bird Grinnell, reflecting on the early history of the conservation movement in his partially unpublished "Brief History of the Boone and Crockett Club," praised Noble in no uncertain terms, writing that he was "a man of the loftiest and broadest views and heartily in sympathy with the efforts to protect the forests."[33]

As an intellectual, Roosevelt spent much of the 1890s competing with Edward Coues for preeminence as the top frontier historian. Each man, in particular, was vying to be considered most knowledgeable about the American West. While Roosevelt was writing *The Winning of the West*, Coues was editing an impressive string of journals and frontier reports about western exploration. The years Coues spent as an army surgeon in forlorn outposts along the Mexican border had allowed him to gather valuable insights for his books. Coues's firsthand knowledge of the West clearly informed his reliable annotations of the classic accounts of exploration he edited in the 1890s: *History of the Expedition of Lewis and Clark* (1893), *Expeditions of Zebulon Montgomery Pike* (1895), *Journals of Alexander Henry and David Thompson* (1897), *Journal of Major Jacob Fowler* (1898), *Forty Years a Fur Trader on the Upper Missouri by Charles Larpenteur* (1898), and *Diary of Francisco Garces* (1900).

Coues's major books were edited, but even so, only Roosevelt himself (and perhaps a few others) could publish at that book-a-year pace. Roosevelt felt Coues, along with Burroughs, was doing the most important work of any U.S. writer or intellectual in the 1890s by editing these six treasured classics of western expansion for future generations to appreciate. When Coues died in 1899 at age fifty-seven, Roosevelt considered it a terrible loss to ornithology, zoology, and frontier history. Coues, he believed, had awakened the popular consciousness to the epic of American exploration.[34] (And then there were Coues's ornithological works.) For the remainder of his life Roosevelt used Coues's *Key to North American Birds* (which went through six editions) as his central reference work regarding classification.

At the Civil Service Commission, Roosevelt continued his war against the entrenched spoils system, a war he'd been waging since he joined the New York state assembly in 1883. Now he had a national platform from which to preach against the epidemic of corruption. Almost as much as

"Bad Lands Cowboy," the label "Civil Service Reformer" soon became attached to Roosevelt in the minds of the American people. Over the next six years, serving both presidents Harrison and Cleveland (the latter won the 1892 presidential election, returning to the White House for a second nonconsecutive term), Roosevelt prosecuted dishonest government officials from coast to coast. Fraud at the U.S. Post Office was his particular focus. He also tried to help the nongovernmental Indian Rights Association (IRA) improve living conditions on territorial reservations in the Dakotas, Nebraska, and Oklahoma.[35] "The spoils system was more fruitful of degradation in our political life than any other that could have possibly been invented," he would write late in his tenure. "The spoil monger, the man who peddled patronage, inevitably breeds the vote-buyer, the vote-seller, and the man guilty of malfeasance in office."[36]

No nook or cranny was off-limits when it came to Roosevelt's determination to eradicate illegal profiteering from the federal government. Fellow Republicans were aghast that Roosevelt doggedly investigated his own party's members, but he believed both parties were unacceptably full of money skimmers. His targets included not only customs officials in New York City but even William Henry Harrison Miller (President Harrison's former law partner in Indianapolis). From that moment, the taciturn president disliked the flamboyant Roosevelt, barely listening when his commissioner pontificated about crooked Wyoming developers determined to carve up poor Yellowstone National Park, or about a new investigation of the U.S. Post Office, or about graft in the Indian Agency. The old general—the grandson of America's ninth president, William Henry Harrison—would tap his finger, bite his lip, and stare straight ahead with a marble face. Roosevelt wasn't oblivious of the icy treatment, writing to his daughter Alice that the five-foot-six-inch Harrison was a "little runt of a President."[37] Often, Roosevelt called the president "Little Ben" behind his back.

Although Roosevelt used Sagamore Hill as his home base that summer of 1889, he often traveled to historic U.S. sites of western expansionism. Greatly encouraged by G.P. Putnam's positive reaction to the first two volumes of *The Winning of the West*, Roosevelt pressed on, doing research in archives in Canada (Ontario), Tennessee, Virginia, and Kentucky. To Roosevelt entering each archive was like entering a mine—he never knew what gem or nugget it might contain.[38] "If nothing else, *The Winning of the West* stands as another monument to Roosevelt's preternatural energy and powers of concentration," the historian John Milton Cooper, Jr., observed. "No other active statesman in the English-speaking world, not even Winston Churchill, produced such a solidly scholarly work of history while he was, as the Romans said, *in medias res*."[39]

III

Whether Roosevelt was hunting bears or attacking spoilsmen, his level of activity wasn't without critics. Ironically, he now got along splendidly with toothless trappers and cattle ropers, but was no longer as comfortable with the refined intelligentsia of the East Coast. Critics like John Hay and Henry Adams, to name the most prominent, belittled his talk of the "strenuous life" as counterfeit and self-aggrandizing (though they both liked his wife, Edith, tremendously). Whenever Roosevelt spoke about humans needing to have "healthy animalism" instilled into their lives, patricians rolled their eyes. Hadn't he learned anything in Porcellian? Whenever he claimed that great knowledge could be gleaned from backwoods types like Hell Roaring Bill Jones or Yellowstone Kelly, they rebuked him for being a literary nationalist at best and folk-obsessed and jingoistic at worst. Hadn't he traveled extensively through Europe and understood the great art of Leonardo and Michelangelo? Equating the beauty of Pike's Peak with *The Last Supper*, they believed, was Wild Wolf macho nonsense.

With his trademark teeth and eyeglasses moving in unison as he spoke, Roosevelt countered that his critics were part of a stifled class, deaf to the clarion call of nation-building, unable to see that the United States' frontier values made the nation vastly superior to Europe's effete culture. His opponents could die in their Washington parlors, but he preferred to go out like a wild animal shot at dusk in an untrampled forest. The whole Hay-Adams circle viewed Roosevelt, in the words of Kathleen Dalton, as "an entertaining but dangerous man to have in a drawing room: he had spilled coffee all over the dress of one governor's wife and bumptiously ripped another woman's hem with a clumsy step."[40]

The poet James Russell Lowell notably bucked this patrician crowd assessment, praising T.R. in the 1890s for being "so energetic, so full of zeal, and, still more, so full of fight."[41] (It didn't hurt that Roosevelt had quoted from Lowell's poem "A Fable for Critics" to open the first volume of *The Winning of the West*.) As a conversationalist, Lowell would say, Roosevelt was in a league of his own. Clearly Roosevelt was a force of nature, a rare phenomenon, a well-rounded intellectual unafraid to enter the fray of national politics, conservation, military affairs, and academic scholarship. With the exception of Henry Cabot Lodge, however, Roosevelt was no longer fully comfortable with the Brahmins of mannered society. He consciously cultivated the manners of a background different from his own, eating with his fingers, reading books at the dinner table, waving off blessings, and carrying a loaded pistol for its shock value. Essentially five generations of etiquette had been abandoned in favor of half-primitive insolence.

Although Roosevelt was impressed with the Hay-Adams crowd, wanted their airy approval, and admired them as perspicacious people who had personally known Lincoln, his *respect* went more to scientists. In the presence of biologists, naturalists, and surveyors like Grinnell, Baird, Coues, or Merriam, for example, Roosevelt was much more modest, soft-spoken, and open to criticism. He listened as much as he spoke. It was as if he had determined that politicians were corrupt and intellectuals fey, whereas U.S. government scientists (that is, those who knew how to write well) and members of the army or navy were the true pillars of American integrity. As for the pioneers themselves, Roosevelt proudly characterized them as "grim, stern people, strong and simple, powerful for good and evil, swayed by gusts of stormy passion."[42]

In July 1889 the first two volumes of *The Winning of the West* were published to another round of critical acclaim. Best read as a *bildungsroman* about how "Young America," as the country was called by Great Britain, had succeeded in its westward expansion, *The Winning of the West* was Roosevelt at his nationalistic apogee. "His Americanism," Burroughs wrote, "charged the very marrow of his bone."[43] Frederick Jackson Turner, still a relatively obscure historian at the University of Wisconsin–Madison, praised the volumes in *The Dial*, saying that Roosevelt had dealt "impartially and sensibly with the relations of the pioneers and Indians whom they disposed."[44] The *Atlantic Monthly* commended Roosevelt for his "natural, simple, picturesque" style.[45] According to the *New York Times*, the volumes were admirable in their "thoroughness" and written by a "man who knew the subject."[46] Great Britain's finest review publications—including the *Saturday Review* and the *Spectator*—all gave thumbs up.[47] What none of these glowing reviews pointed out was that Roosevelt had pioneered in writing a new kind of popular scientific history, melding Parkmanism with Darwinian thinking and a full jigger of Mayne Reid to boot. Some passages directly echoed *The Oregon Trail* and *On the Origin of Species*, and even *The Scalp-Hunter*.[48] The consensus was clear: the historian Roosevelt had a knack for not putting the reader to sleep.

For his own part, Roosevelt was proud that *The Winning of the West* was more in the tradition of Francis Parkman than Henry Adams's *History of the United States in America during the Administrations of Thomas Jefferson and James Madison*, whose first volume also appeared that year. Besides writing about westward expansion as Parkman had done, Roosevelt had infused his narrative history with scientific explanations. "Now I am willing that history shall be treated as a branch of science, but only on condition that it also remains a branch of literature," Roosevelt wrote; "and, furthermore, I believe that as the field of science encroaches on the field of literature there should be a corresponding encroachment of literature upon science; and I hold that one of the great needs, which can only be met by very able men whose culture is broad enough to include literature as well as science, is the need of books for scientific laymen. We need a literature of science which shall be readable."[49]

Yet, as Roosevelt was apt to do, he felt the sting of criticism more than the high-minded accolades. Accusations abounded that the *The Winning of the West* had been a rush job. Typos and minor mistakes could be found. The *Atlantic Monthly*, for example, pointed out that Roosevelt had misidentified John Randolph of Roanoke, and the *New York Sun* charged him with unethically paraphrasing a book by the scholar James R. Gilmore.[50] An exclamation of anger broke from Roosevelt's pen, for he knew his public reputation was under assault. Roosevelt dealt with each charge differently: he befriended the *Atlantic Monthly*'s editor but put the kibosh on the envious Gilmore in a very public rebuttal to the charge of quasi-plagiarism. By confronting his tormentors, Roosevelt escaped the turbulent waters of bad publicity unscathed. By emulating Parkman, Roosevelt prided himself in having written the "history of the American forest."[51]

Basking in the sunshine of literary fame, Roosevelt wrote to Francis Parkman himself—who the previous year had written an important conservation-oriented article, "The Forests of the White Mountains," for *Garden and Forest*[52]—and told Parkman about his future plans as an author. Although not an environmentalist in the modern sense of the term, Parkman was a premier naturalist and horticulturist of his day, running a nursery in Massachusetts to supplement his career as a historian. Clearly Roosevelt wanted to show Parkman that, he too, used wilderness and fauna as his background for historical events.[53] "I am pleased that you like the book," he wrote on July 13, 1889. "I have always had a special admiration for you as the only one—and I may very sincerely say, the greatest—of our two or three first class historians who devoted himself to American history; and made a classic work. . . . I have always intended to devote myself to essential American work; and literature must be my mistress perforce, for though I really enjoy politics I appreciate the exceedingly short nature of my tenure."[54]

IV

In the fall of 1889, Edith moved the three Roosevelt children—Alice, Ted, and Kermit—from Sagamore Hill to a rented house at 1820 Jefferson Street in Washington. (The house, just off Connecticut Avenue, was one-tenth the size of Sagamore Hill.) Theodore, who called his children "bunnies," hoped his family would grow even more.[55] (He would soon get his wish: Edith gave birth to Ethel in 1891 and to Archibald in 1894.) Considering the constraints on his time, Roosevelt was a good, loving father to all five children. Enjoying the hurly-burly of the household, he instructed his brood at a young age how to identify songbirds and insects. In the nation's capital, Theodore was usually more mannered, acting like his own father, determined to teach his children the Ten Commandments, Wordsworth, Keats, and Shakespeare. At Sagamore Hill, however, he encouraged mayhem, coaxing them to swim in the bay and play in the Long Island woods. Dull moments were frowned upon.[56] "Every evening I have a wild romp with them," Roosevelt wrote to his mother-in-law, Gertrude Elizabeth Carow, "usually assuming the role of 'a very big bear' while they are either little bears or a 'racoon and a badger, papa.' "[57]

Over Thanksgiving 1889 Roosevelt began planning a "grand holiday" during which he would bring Edith, Bamie, Robert Munro Ferguson, Corinne (and her husband, Douglas Robinson), and Henry Cabot Lodge's sixteen-year-old son George (nicknamed Bay) to the Badlands and Yellowstone. They would travel by pack train to pristine parts of the upper Rocky Mountains. Because Theodore talked incessantly about the Elkhorn Ranch, it made sense for his wife to see the Medora magic firsthand and then head to Yellowstone. Over the next nine months, as he prepared for this expedition, he devoured every aged calf-bound book ever written about exploration in Yellowstone. Theodore was thrilled to learn that all the Rocky Mountain big game he loved, except the mountain goat and caribou, were to be found in Yellowstone National Park. According to Arnold Hague, a member of the Boone and Crockett Club, deer—both mule and white-tailed—populated the Gallatin Range valleys in high numbers, dashing up hillsides and grazing in meadows. An impatient Roosevelt could barely wait to see the enchanted herds for himself.[58]

And he was likewise eager to show off his scientific knowledge to his family—to explain why some owls nested in prairie dog holes and to describe the mating rituals of elk. Playing geologist, he could explain to Edith the significance of the 2-million-year-old lava on Huckleberry Ridge tuff and how Specimen Ridge had one of the world's largest petrified forests. More and more, he saw himself as an interpreter of both American triumphalism and Darwinian species variation as they related to western U.S. history. In fact, after delivering an address on westward expansion at the American Historical Association's year-end meeting, Roosevelt was acclaimed by his colleagues as the leading proponent of the "new school" of western historians.[59] "I know of no one in the East, besides yourself, who has any conception of Western history," William Frederick Poole, the association's president, wrote to Roosevelt. "You have entered a fresh and most interesting field of research, and I predict you great success."[60]

In January 1890, encouraged by the positive response from the American Historical Association, Roosevelt once again fantasized about quitting the U.S. government so he could be a full-time western historian. He wondered how best to blend his historical research with his conservationist beliefs. Realizing that the myth of American abundance was a national curse, Roosevelt set about to change attitudes about saving wildlife and preserving habitat. There was in America what his friend William T. Hornaday, chief taxidermist of the National Museum (the Smithsonian), called an "army of destruction" that had to be stopped.[61] Roosevelt intended the family trip to Yellowstone, now slated for September, to be a fact-finding mission as well; it would help him better understand the poaching and plundering before he started testifying, as he hoped, before congressional committees on the sanctity of the park.

Perhaps there was another motivation for visiting Yellowstone in 1890. Roosevelt might have felt embarrassed that both President Harrison and George Bird Grinnell—his superiors in national politics and North American big game conservation, respectively—had already toured the national park whereas he had seen Old Faithful and the Tetons only in picture books. He would now be able to even the score. Polishing up his Civil Service badge, Roosevelt would probe into why Wyoming poachers and Montana lumbermen and railway-tie cutters were being permitted in what the law of 1872 deemed a "public park of pleasure-ground for the benefit and enjoyment of the people."[62] Why weren't these intrusive business concerns being collared by local law enforcement or the U.S. Army? How could the U.S. government make sure Yellowstone wasn't "shot out" by horn and hide hunters? His "grand holiday," doubling as a fact-finding mission on behalf of the Boone and Crockett Club, he believed, was an integral part of this journey to the park that the novelist Thomas Wolfe later called "the one place where miracles not only happen, but where they happen all the time."[63]

Even before leaving for the West in late summer, Roosevelt chafed at the loopholes in the original Yellowstone Act, which didn't properly preserve big game. He wanted protective amendments, and fast. In 1872 there had been only a single transcontinental railroad spanning the Rocky Mountains—the Union and Central Pacific, which rumbled across Wyoming far to the south of Yellowstone. Roosevelt was fine with that. But in 1890, influenced by Grinnell, Roosevelt, after deep consideration of the issue, opposed a proposed new Montana Mineral Railroad line aimed at "segregating" the park. Under the sway of the *Forest and Stream* crowd, Roosevelt now fancied himself as the conservationist point man in upbraiding Montana Mineral on the Yellowstone issue. "I am glad to hear that Roosevelt is going to stand back on the question of railways in the Park," Grinnell wrote to a fellow member of Boone and Crockett, "and not to work against us."[64]

The "grand holiday" started out splendidly—a first-class train ride from New York through Chicago and Saint Paul, until the steaming locomotive eventually rolled into the western edge of the Dakotas on September 2. For seven or eight days they mixed it up with the sharp-faced Ferris brothers and T.R.'s hardy Elkhorn ranchhands, such as Bill Merrifield, who lived among the abrupt escarpments like nonconformist characters in a Bret Harte story. Roosevelt's elation with Medora was evident as he pointed out Custer's 1876 campsite, the Marquis de Mores's defunct meatpacking plant, and the innumerable rock formations that gave the Badlands their peculiar charm. One afternoon, coping with washouts and quicksand, they forded the Little Missouri River twenty-three times. Absent-eyed antelope could be seen grazing along a ridge, with muscles suddenly tensed upon the realization of human encroachment. At dusk they watched timid white-tails in bushy gullies and big-eared mules on sage-spangled buttes. "Nothing could be more lonely and nothing more beautiful than the view at nightfall across the prairies to these huge hill masses, when the lengthening shadows had at last merged into one and the faint glow of the red sun filled the west," Roosevelt wrote about these rock landmarks in a publication of the Boone and Crockett Club. "The rolling prairie, sweeping in endless waves to the feet of the great hills, grew purple as the evening darkened, and the buttes loomed into vague, mysterious beauty as their sharp outlines softened in the twilight."[65]

Corinne marveled at how Theodore spoke of Dakota cowboys as if they were heroic knights on horseback and their low-lying cabins splendid castles. Despite the fact that she looked like a Dresden china figurine, Corinne proved to be the real trouper of the holiday, not complaining while trudging across mud holes and calf-high streams, smeared with pine pitch and achy from saddle sores. Just watching Theodore use an iron brand and rope steer yearlings like the other cowboys, in fact, made her proud. "We lunched at midday with round-up wagon," she recalled in her memoir *My Brother Theodore Roosevelt*, "rough life, indeed, but wonderfully invigorating, and as we returned in the evening, galloping over the grassy plateaus of the high buttes, I realized fully that the bridle-path would never again have for me the charm it once had had."[66]

Meanwhile, having heard so many stories from Theodore about Medora, Edith was now pleased to put a face on things. The Badlands stillness seemed unbreakable, eternal, and primeval. Nature, she understood anew, was tonic for her husband; serene solitude of the sagebrush calmed this act down. He simply was more relaxed without gaslights. And she undoubtedly discerned from reading the preface to volume one of *The Winning of the West* that her husband, in an imaginative leap of romantic fancy, equated his Dakota ranching days with those of the late-eighteenth-century pioneers clearing brush through the Allegheny upcountry and Great Smokies valleys. "The men who have shared in the fast-vanishing frontier life of the present," he had written, "feel a peculiar sympathy with the already long-vanished frontier of the past."[67]

On September 9, the Roosevelt party started making its way to Yellowstone National Park. For the next week, everybody's eyes were fixed on wildlife around Inspiration Point and on the condensed force of Yellowstone's Lower Falls as it roared downward 308 feet. Everything about Yellowstone was exhilarating to Roosevelt—although he was disappointed that game wasn't found around the hundreds of geyser basins where the tourists congregated. One afternoon he and Ferguson fished in the Yellowstone River within close view of Tower Falls, bringing strings of brook trout back to camp for supper. According to Corinne's diary, throughout their stay in Yellowstone National Park, Theodore kept copious notes about the wildlife they spotted. She marveled at her brother's ability to distinguish birds at a glance or from merely hearing their thin cries. During just the first four days in Yellowstone they encountered the peregrine falcon, red-tailed hawk, Canada jay, raven, mallard duck, teal duck, nuthatch, dwarf-thrush, robin, water ouzel, sunbird, long-spur, grass finch, bittern, yellow-crowned warbler, Rocky Mountain white-throated sparrow, song sparrow, wren, and pigeon hawk. As a bird-watcher, Roosevelt was most stirred by the golden eagle, which put on an aerial show: these dark-brown raptors glided at fifty miles an hour and then swooped downward for direct strikes on chipmunks and ground squirrels. "Each one of the above I saw with the eyes of Theodore Roosevelt," Corrine recalled, "and can still hear the tones of his voice as he described to me their habits of life and the differences between them and others of their kind."[68]

Although Theodore occasionally sulked about not being able to "rough it"—the cost of having his family in tow—being in the fresh air brought ample reward. "He loved wild places and wild companions, hard tramps and thrilling adventure," Corinne wrote, "and to be a part of the type of trip which women who were not accustomed to actual hunting could take, was really an act of unselfishness on his part." On most days the Yellowstone sky was cloudless; the nights were cold, with frost chilling the eyeballs and causing sinuses to ache. Instead of eating elk venison, as T.R. would have liked, the party's diet usually consisted of cutthroat trout plus canned ham and tomatoes. At night a theatrical Theodore hammed it up, pretending to be a bear on the prowl outside their tents, swollen with laughter until thoroughly spent. As Corinne put it, they were all enjoying the "pretense of roughing it."[69]

As Edith and Corinne soon learned, however, for all his scientific knowledge, Theodore was a reckless escort in the wilderness. For starters, the professional guide he had hired, Ira Dodge, got them terribly lost. Acting as if it were still midsummer, one evening the Roosevelt party camped at an altitude of 7,500 feet, shivering all night under flimsy blankets; even the camp's drinking water, in a pail, froze.[70] Disregarding safety, Theodore thrust people ill equipped for outdoors rigors to push themselves to the point of breakdown or exhaustion. Worse yet, Roosevelt had leased a string of horses unaccustomed to being ridden sidesaddle. On a pack trail ride along stretches of the Continental Divide, which separates waters flowing west from those flowing east, Edith was thrown off her horse, which had reared suddenly, spooked by an erupting geyser. The pain in her back was excruciating, but no doctor was brought in. Her recovery was slow. Soon thereafter, Theodore himself was injured when hunting with Ferguson outside the park. He had "rather strained" his groin and was uncomfortable on horseback for a few days. After visiting the Mammoth Hot Springs in the northwest corner of Yellowstone, where the hot water rose through limestone instead of lava, the Roosevelt party was back at the Elkhorn Ranch on September 23, bruised but all smiles.[71]

The whole Medora-Yellowstone trip was hailed by T.R. as an unsurpassed bonding experience for his family. Only going to a great European spa like Baden-Baden, he believed, had the same rejuvenating effect on citified people as a week in America's great park. (It didn't hurt that he had an office of civil service clerks to mind the store back in Washington,

D.C., during his six-week grand holiday.) "I have rarely seen Edith enjoy anything more than she did the six [weeks] at my ranch, and the trip through the Yellowstone Park," he wrote to his mother-in-law. "And she looks just as well and young and pretty and happy as she did four years ago when I married her—indeed I sometimes think she looks if possible even sweeter and prettier. . . . Edith particularly enjoyed the riding at the ranch, where she had an excellent little horse, named Wire Fence, and the strange, wild beautiful scenery, and the loneliness and freedom of the life fascinated and appealed to her as it did to me."[72]

After the vacation at Yellowstone, Theodore threw himself into his conservation work for the Boone and Crockett Club harder than ever. Arming himself with scientific data, he was determined that his children could someday bring their children to experience Wyoming's Garden of Wonders. Using the newest wildlife science available, Roosevelt wanted the old-time wildlife abundance back. Yellowstone needed to be expanded as a zoological reservation (George Catlin had once called for this), where big game like elk and buffalo could thunder around unmolested by the intrusions of civilization. After all, Roosevelt argued, the West couldn't have been won without buffalo and elk to provide the pathfinders with meat. The time had come to create reserves so that the populations of both species could increase again and be safe. If Robert B. Roosevelt and his amiable helper Seth Green could repopulate Hudson River spawning shad through artificial propagation, then surely a similar repopulation project could be undertaken on behalf of buffalo. Essentially, the visionary Roosevelt was calling for what in the 1980s became the American Prairie Foundation, a nonprofit organization to create a 3.5-million-acre reserve in central Montana for studying North American game, birdwatching, hunting, and hiking.[73]

By 1890 the conservationist movement was no longer embryonic. A new leader had appeared on the West Coast, a man who spoke on behalf of pristine nature with the grace of a literary angel. The California naturalist John Muir's two articles in *Century* magazine (both illustrated by Thomas Moran), "The Treasures of the Yosemite" and "Features of the Proposed Yosemite National Park,"[74] had created a literary sensation. Worried that overgrazing by sheep was denuding the Sierra high country and threatening the groves of old-growth sequoias, Muir wanted to preserve the complete watersheds of the Merced and Tuolumne rivers inside a new national park. Immediately, Roosevelt recognized that California had found its John Burroughs. It was helpful that Robert Underwood Johnson, an editor of *Century*, was himself a strong proponent of national parks in California. Congress created three of them that fall: Sequoia, Yosemite, and General Grant, which is now part of Kings Canyon National Park.

Bolstered by Muir, Roosevelt now argued that wildlife preserves like Yellowstone, Yosemite, Sequoia, and General Grant were among the best ideas of the Gilded Age. Using the Boone and Crockett Club as his pulpit, he argued for tougher antidevelopment and antipoaching laws at Yellowstone. "Through his effort with Grinnell, Roosevelt began to envision the park as a sanctuary and breeding ground for wildlife," the historian Jeremy Johnston explained in *Yellowstone Science*. "Roosevelt hoped that if the park's wildlife were protected, their populations would dramatically increase and spread to the surrounding regions. This would ensure the continuation of hunting, his favorite pastime, outside the park's boundaries. It would also alleviate his fear that as settlement increased, the West would become a series of private game reserves creating a situation where only the rich could hunt."[75]

As Roosevelt touted Boone and Crockett's conservation agenda throughout official Washington, there was talk about conflict of interest. But a sharp (and convenient) distinction had been drawn in Roosevelt's own mind: his club was a watchdog agency guarding against incursions in Yellowstone National Park (federal property). Still, the noisiest of Montana Mineral Railroad's lawyers and Wyoming's developers weren't afraid to publicly smear T.R. as a hypocrite attacking the spoils system from the Civil Service Commission while exploiting his government connections to lobby for conservation. Still, Roosevelt had *rightness* on his side. There was a palpable urgency to what the Boone and Crockett Club was trying to accomplish in terms of saving big game. A new public consciousness was needed to save the untamed beasts of the west. Roosevelt thought that promoting species survival via educational outreach in zoos and museums was an important way to wake up America's youngsters to the plight of animals. He also championed the sculptures of Edward Kerneys (considered America's first animalier), who made anatomically correct bronzes. He collected them like mad.[76] The indifference of big business toward habitat saving annoyed Roosevelt mightily. Instead of thinking of forests as a finite resource and offering to replant as they logged, the railroads preferred the slash-and-burn approach. And the problem of deforestation wasn't only in the West. The soil runoff from speed-logging in the Adirondacks was being blamed by scientists for ruining navigation (by creating sandbars) on the Hudson River. Following John Muir's preservationist tactics as delineated in *Century* with regard to California's world-class forests, Roosevelt started floating the idea of creating an Adirondack National Park in New York.

Roosevelt remained determined, and in January 1891 he ran a very important board meeting of the Boone and Crockett Club in Washington. Roosevelt and Grinnell appealed to the room of dark-suited worthies—most notably Secretary of the Interior John W. Noble—about the importance of protecting wildlife and creating forest reserves.[77] The latter issue was taking on particular urgency, since deforestation was an ever-increasing problem. Western wildfires were epidemic. Railroads had an insatiable appetite for timber, needing wood for railway carriages, stations, platforms, fences, and, of course, the ties for their expanding network of tracks. (In 1887, *Scientific Monthly* estimated that the railroads needed 73 million new ties each year.[78]) Loggers thought of forests as an infinite resource, so no replanting was done. The denuded land was vulnerable to erosion and so, for instance, the soils of Roosevelt's beloved Adirondacks were already clogging the navigable water of the nearby Hudson River. "Roosevelt . . . asked me to say something of the way in which game had disappeared in my time," Grinnell joked to a fellow member of Boone and Crockett, Archibald Rogers, "and I told them a few 'lies' about buffalo, elk, and other large game in the old days."[79]

The board meeting led to White House action to protect the nation's forests. A few days later two members of the Boone and Crockett Club—William Hallett Phillips (a lawyer and diehard angler who accidentally drowned in the Potomac River in 1897, moving Rudyard Kipling to dedicate a poem in *Scribner's* to his memory) and Arnold Hague (a geologist-conservationist with the U.S. Geological Survey who had written an influential report on Yellowstone)—briefed Secretary Noble on how the new science of forestry could prevent deforestation. The Harrison administration quickly pushed legislation through Congress to protect forests on public lands. The Forest Reserve Act of 1891, which the president signed that March, put an end to the virtual giveaway of public land to the railroads and enshrined the government's role in protecting the wildlife in American forests. Most important, its final provision, Section 24, gave the president the right to convert public land into forest reserves. It stated: "That the President of the United States may, from time to time, set apart and reserve, in any state or territory having public land, bearing forests, in any part of the public lands wholly or in part covered with timber or undergrowth, whether of commercial value or not, as public reservations, and the President shall, by public proclamation, declare the establishment of such reservations and the limits thereof."[80] The language of Section 24 would prove crucial to Roosevelt's future conservationist efforts as president.

As soon as the first forest reserve—Yellowstone National Park Timberland Reserve—was established, it was clear that YIC and other would-be developers had suffered a huge, irreversible defeat. The Boone and Crockett Club issued a resolution praising Noble, and Grinnell published a glowing tribute to his efforts in *Forest and Stream.*[81] President Harrison quickly bestowed protection on 13 million acres of American woods, creating eleven forest reserves,* where absolutely no tree cutting was allowed; and six timberland areas, where limited logging was permitted under close supervision. As the conservationist Gifford Pinchot later noted in his memoir *Breaking New Ground*, this was "the most important legislation in the history of Forestry in America," and it "slipped through Congress without question, without debate."[82]

Before this act, land in the American West was being sold by the U.S. federal government to private enterprises. Nearly a quarter of the Montana Territory, for example, had been deeded or sold to the railroads. But President Harrison's act put a wrinkle in that habitual practice. Recognizing that Europe's natural resources were being depleted and its lands deforested and eroded, President Harrison had behaved like a champion for George Perkins Marsh and the Boone and Crockett Club. Working in the administration's favor was the fact that starting in 1876 the Department of Agriculture had created an activist U.S. Division of Forestry. Meanwhile, at the Department of Interior, the U.S. Geological Survey had formed the Irrigation Survey, in which scientists worked to find solutions to America's resource management problems. Like Roosevelt, both

*The reserves would be renamed national forests in 1907, during Roosevelt's presidency. Besides the Yellowstone Park Timberland Reserve in Wyoming these new federal lands included the White River Plateau timberland reserve (Colorado), Pecos River forest reserve (New Mexico), Sierra forest reserve (California), Pacific forest reserve (Washington), Pike's Peak timberland reserve (Colorado), Bull Run timberland reserve (Oregon), Plum Creek timberland reserve (Colorado), South Platte forest reserve (Colorado), San Gabriel timberland reserve (California), Battlement Mesa forest reserve (California), Afognak Forest and Fish Culture reserve (Alaska), Grand Canyon forest reserve (Arizona), Trabuco Canyon forest reserve (California), San Bernardino forest reserve (California), Ashland forest reserve (Oregon), and Cascade Range forest reserve (Oregon). The difference in designations was fairly simple: trees weren't allowed to be cut in forest reserves, but in timberlands limited government-supervised logging was allowed.

government divisions considered themselves enemies of the railroad and mining industries.

Despite the enormous victory, Roosevelt wasn't satisfied. There were still no laws to properly police these public lands; and he was worried that market hunters, loggers, and miners would not be deterred by "No Trespassing" signs if ignoring them had no consequences—no jail time and no heavy fines. In fact, implementing the Forest Reserve Act wasn't easy. Congress grappled over the legal specifics until in 1897 it passed the National Forest Management Act (Organic Act), which clarified "the purposes for which the national forests could be created to preserve and protect the trees in a reservation; secure good water conditions; and furnish timber for the use of the American people."[83] Or, put more simply, to make sure every generation of Americans had healthy forests. As the historian Harold K. Steen explained regarding the 1897 provisions, "Not until the 1960s and 1970s would Congress, and the courts, take another look at those purposes."[84]

The 1897 provision authorized U.S. presidents to set apart and reserve, whenever they chose, government land wholly or in part covered with timber or underwood growth. This executive branch prerogative, in fact, was used by T.R., who founded the Forest Service in 1905, saving 151 million forested acres between 1901 to 1909 as president, mostly in the West (an increase of forest reserves by 300 percent). As Stewart Udall, secretary of the interior under both John F. Kennedy and Lyndon Johnson, perceptively noted, "The Boone and Crockett wildlife creed . . . became national policy when Theodore Roosevelt became president."[85]

Of course, Roosevelt was thrilled that cedars along the Pecos River in New Mexico were now protected and that the ponderosa pine around Los Angeles's San Bernardino Mountains would tower unmarred for decades to come. However, he was most gratified that President Harrison had placed 1.2 million acres of Wyoming forest (an estimated area of 1,936 square miles) adjacent to Yellowstone National Park under federal protection as part of the new Yellowstone National Park Timberland Reserve. This was a crucial component for his idea of a big game preserve to grow properly. Not only did Yellowstone deserve recognition as the first national park, but courtesy of the Harrison administration the Yellowstone National Park Timber Reserve was now also where the national forest system was born. This wasn't quite the same as the enlargement of the park that Roosevelt had lobbied for, but it nevertheless was a huge victory for the Boone and Crockett Club.

Once Noble understood how forests protected watersheds, he never hesitated in his advocacy of the reserve system. "Your associates, Mr. Phillips and Mr. Hague, brought the business to my attention," Noble wrote to Roosevelt on April 16, 1891, describing how the Forest Reserve Act was consummated. "Having been familiar with the subject, I had no hesitation in immediately advising the President favorably as to the proclamation, and I am glad to see that he has promptly appreciated the situation and acted as he did."[86]

V

The early 1890s were the halcyon days of the American conservationist movement. Groups like the American Forestry Association and the American Association for the Advancement of Science were starting to be heard on Capitol Hill. In 1891, when the first Irrigation Congress met in Salt Lake City, a future senator—Francis G. Newlands of Nevada—stated matter-of-factly that "unless the mountains and the hillsides are kept covered with timber the snows which now practically impound the water and hold it until needed will melt the quicker in summer and thus make artificial storage more expensive."[87] The lobbying efforts of leaders like John Muir, George Grinnell, and Theodore Roosevelt were paying off. Few congressmen—except some in the West—wanted be remembered for contributing to the deforestation of America. Senators George Vest of Missouri (D) and Charles Manderson of Nebraska (R) bravely fought for forest reserves every year and eventually influenced the Senate with their pro-conservationist views. If the National Wildlife Federation Conservation Hall of Fame were on the job, both men would have been inducted long ago,* as should Noble and Harrison. "The Executive and its representative, the Department of the Interior," Roosevelt and Grinnell wrote following the act of 1891, "have at all times been most sympathetic and helpful in the movement for forest and game preservation."[88]

But the stunning success of the Forest Reserve Act of 1891 raises a question: why then? Was it a coincidence that these events followed on the heels of the supposed close of the western frontier? In a historic paper delivered at the Chicago World's Fair on July 12, 1893—"The Significance of the Frontier in American History"—Frederick Jackson Turner argued that the frontier closed in 1890. Influenced by Roosevelt's *The Winning of the West*, Turner, a professor at the University of Wisconsin–Madison, claimed that the United States' westward expansion had created a new

*The National Wildlife Federation Conservation Hall of Fame's first inductee was Theodore Roosevelt, in 1964.

sort of citizen: the frontier-spirited outdoorsman (e.g., Carson, Bridger, and Pike). On top of that, pointing to census figures on population destiny in the West, he stated that western expansion was now a *fait accompli.*

Many historians consider Roosevelt the "progenitor" of the frontier thesis because on January 24, 1893, more than six months before Turner delivered his paper at the Chicago World's Fair, Roosevelt had delivered the biennial address before the State Historical Society of Wisconsin in Madison declaring the Old Northwest the "heart of the country."[89] Turner was sitting in the audience dutifully taking notes. The historian Michael L. Collins noted that at the very least Turner owed a huge debt to Roosevelt.[90] Graciously, Roosevelt claimed that Turner had "put into definite shape a good deal of thought which . . . [had] been floating around rather loosely." Essentially, an alliance was formed in promoting the frontier hypothesis, with T.R. as the popular oracle and Professor Turner influencing fellow academics. Even though they developed only what the historian Ray Billington deemed a "corresponding" relationship, Turner, in what is widely interpreted as honoring a debt, quoted T.R.'s Wisconsin address in his own 1920 book *The Frontier in American History.*[91]

Roosevelt and Turner's frontier thesis was clever. In 1890 settlers were no longer riding Conestoga wagons up the Oregon Trail and trailblazers like Kit Carson were no longer tangling with the Navajo at Canyon de Chelly. Thirty years earlier Abraham Lincoln had called for a transcontinental railroad. By 1890, with Chicago as the main terminus, a web of tracks now ran out of Illinois in every direction from coast to coast. Cities all through the American West, such as Albuquerque, Omaha, Sacramento, Seattle, Tucson, Denver, and Portland, were rapidly growing in population. Although Geronimo was making a little noise in the Arizona Territory, the Native American population had by and large been pacified, and reservations were being set up in dozens of states and territories. Indians had now been relocated to Oklahoma reservations (and other locales), the buffalo were nearly gone, and the Great Plains–Rocky Mountains landscape was being developed and mined. Even the stubborn Mormons of Utah had renounced polygamy in the Woodnuff Manifesto.[92] "Literally innumerable short stories and sketches of cowboys, Indians, and soldiers had been, and will be written," Roosevelt wrote Frederic Remington. "Even if very good they will die like mushrooms, unless they are the very best; but the very best will live and will make the cantos in the last Epic of the Western Wilderness, before it ceased being a wilderness."[93]

Now that the "West was won" and the Rocky Mountain wilderness "ceased to be wild," Roosevelt and his fellow members of the Boone and Crockett Club saw the Forest Reserve Act of 1891 as the responsible national starting point for creating a sustainable trans–Mississippi River environment from which *all* Americans could benefit. *The Winning of the West*—plus, equally important, his naturalist writings—made clear Roosevelt's advanced belief in the benefits of timber resource management and regulated hunting and fishing. Shortly after the passage of the act of 1891 Roosevelt and Grinnell cowrote an essay, "Our Forest Reservations," lambasting "corporate greed" and fretting over the lack of game wardens in the American West. "We now have these forest reservations, refuges where the timber and its wilds denizens should be safe from destruction," they wrote. "What are we going to do with them? The mere formal declaration that they have been set aside will contribute but little toward this safety. It will prevent the settlement of the regions, but will not of itself preserve either the timber or the game on them. . . . The forest reservations are absolutely unprotected. Although set aside by presidential proclamation, they are without government and without guards. Timber-thieves may still strip the mountain-sides of the growing trees, and poachers may still kill the game without fear of punishment."[94]

What Roosevelt and Grinnell (and John Muir, for that matter) were arguing for was, in fact, closing much of the western frontier to settlement and development. What good, they asserted, were forest reservations and national parks if these were left unprotected and not administered properly? With his characteristic law-and-order attitude, Roosevelt believed all poachers and despoilers should be imprisoned. Their actions, he felt, were unpardonable. All the wildlife he loved that flourished on federal property in 1891—including walruses on Alaska's Amak Island, sea lions in California's Farallones, and bald eagles in Colorado's White River plateau—needed police protection. The national park and forest movement, both Roosevelt and Grinnell understood, was going to hinge on the federal government's protecting its assets. "The game and timber on a reservation should be regarded as government property, just as are the mules and the cordwood at an army post," they wrote. "If it is a crime to take the latter, it should be a crime to plunder a forest reserve." With strict law enforcement, they believed, no big game species would "become absolutely extinct."[95]

In recent decades the "new western history"—presented by Patricia Nelson Limerick and Richard White—postulated that declaring the "frontier closed," as Turner (and Roosevelt, in a sense) did in 1890 had (in post-modern terms) deeply racist connotations.* They chastised Roosevelt and Turner for believing, as Limerick put it, that the frontier was "where white people got scarce, or alternatively, where white people got scared."[96] The White-Limerick new western history arguments, in hindsight and from the vantage point of multiculturalism, were fundamentally sound. As the premier champion of Anglo-American settlement of North America, Roosevelt treated Native tribes, Spanish settlers, and even French Canadians as riffraff who needed to be cleared away like so many weeds. Caucasian outdoorsmen on the western frontier were, to Roosevelt, almost infallible. For example, Daniel Boone and his cohorts in the Cumberland valley, Roosevelt believed, had been "ordained of God to settle the wilderness."[97] Every chance Roosevelt got, he championed George Rogers Clark and Zebulon Pike. No matter how cruel white backwoodsmen were to "red Indians," the savagery was blamed on the Indians. Although Roosevelt sometimes wrote glowingly of the Indians' wilderness prowess—as he did in *The Winning of the West*—he still seemed to be using them as foils in order to elevate the frontiersmen into first-class guerrilla fighters. At its best, *The Winning of the West* treated Native Americans as Rousseauesque noble savages—a popular concept of the time.

Yet, it's important to remember that although Roosevelt's ethnocentrism and his notion of the white man's burden are repugnant today, they were the accepted tenets of his own time. Nationalistic boasting was in fashion. Ever since Polk won his war in 1848 and the United States acquired parts of California, Nevada, Arizona, and New Mexico, western expansion had been touted as an accomplishment to be celebrated, along with the Declaration of Independence and the Gettysburg Address. So, with half-shut eyes, Roosevelt wrote only the winner's history of the West. Western triumphalism—called "new history" because it dealt with a part of the continent about which very little had been written—became the scholarly norm in the 1890s, with Roosevelt leading the way. Despite all its shortcomings, Roosevelt and Turner deserve kudos for helping create this genre of U.S. western history, on which both Lim-

*The "new western history" of 1990 was different from Roosevelt's "new history" of 1890. All Roosevelt wanted was for the West to be brought into the larger national narrative. Limerick and White had no problem with this, per se. What they objected to was Roosevelt's one-dimensional characterization of the West as all manifest destiny and triumphalism. Their new western history correctly brought in Native Americans, Hispanics, women, and others who had been cut out of the Rooseveltian Anglo-American "new history."

erick and White would build their careers 100 years later. Limerick, in particular, challenged the 1890 frontier-is-closed thesis time and again. "In the American West, too many 'frontier-like' events happened *after* 1890—homesteading continued, short-term extraction even accelerated as the western oil, timber, and uranium booms took off, and contrary to myths of a vanished West, neither Indians nor cowboys disappeared," Limerick noted in 1991, in a speech celebrating the National Forest Service's centennial. "In extractive industries, the familiar boom/bust cycle continued, while Indian, Hispanic, Anglo, and Asian people continued to search for ways to live together. The westward movement didn't stop at 1890; millions more people moved into the West in the twentieth century. If one went by numbers, one would have to call the nineteenth century westward movement the frail prelude to the much more significant twentieth century westward movement. It would be easier to sell me a used car, or a vacuum cleaner, or an encyclopedia set, than it would be to sell me on the idea that the creation of forest reserves was another sign and symbol of the end of the frontier."[98]

While Limerick rightfully threw cold water on Roosevelt and Turner's thesis, a caveat must be added. Three new national parks *were* created in late 1890 out of huge parcels of pristine California wilderness. Approximately 13 million acres of the West *had* been set aside in 1891 by the Forest Reserve Act. (That acreage is more than twice the size of Massachusetts.) Limerick was correct in saying that millions of settlers kept coming west, but they weren't allowed into the sequestered government-owned prime forestlands. The Interior Department was, by 1890, closing off large swaths of the West to future development. By 1898, 40 million acres had been saved as reserves. Therefore, perhaps the appropriate resolution to the dispute between Roosevelt-Turner and Limerick-White can be found by considering the role of forestry science during the gilded age. Roosevelt, as the *New York Times* would note, was a leader in a new post–Civil War generation trying to redefine Americanism in the 1890s. Roosevelt may have named his pony Grant, admired John Hay's ring made of hair from Lincoln's beard, and applauded Sherman and Sheridan for protecting Yellowstone, but he had never personally experienced war—and neither had his wealthy father. As for the western American "frontier," Roosevelt wasn't part of its settlement. Hopping off the Northern Pacific Railroad with thousands of dollars to lavish on wilderness guides and equipment in Medora was hardly Jim Bridger stuff. But the obverse of that reality was also true: Roosevelt never killed a Confederate, an Indian, a Mexican, or any other human on American soil.

What Roosevelt did in the Dakotas (and Grinnell did in Nebraska, Baird in Arizona, and Merriam in California) was collect samples of western wildlife, as ambulating Ivy League scientists were apt to do. For all their Wild West notions, these men—and a dozen like them who graduated from Harvard and Yale between 1870 and 1890—were the children of Charles Darwin. After the surrender at Appomattox in 1865, an entire generation of Ivy League graduates, for the first time, had all studied Darwin. Science was the rage. And to those who—like Roosevelt, Baird, Merriam, and Grinnell—were predisposed to biology, the father of evolutionary theory continued to be a secular saint as they entered their thirties. Once *On the Origin of Species* had been published in 1859, it was virtually impossible for educated Americans like Roosevelt to look at flora or fauna in the same way. In other words, the Forest Reserve Act of 1891 came about not because the frontier closed but because during the 1870s Harvard and Yale had started taking biology, naturalist studies, and forestry seriously in the aftermath of Darwin (and George Perkins Marsh). For the purposes of inventory and study, America's outdoor laboratories (wildlife included) needed to be preserved. That was a scientific imperative. Just as Copernicus realized that the earth wasn't the center of the solar system and Newton discovered laws for the movement of the stars, Darwin made it clear that man must be considered as merely a part of the natural world.

What made Roosevelt different from Grinnell, Baird, or Merriam was that while he fully embraced Darwinism and Marshism, he wouldn't throw away Mayne Reid's potboilers or the notion of the Alamo as a heroic line in the Texas sand. Roosevelt stubbornly refused (or was intellectually unable) to become part of the "dry as dust" world of science. "I know these scientists pretty well, and their limitations are extraordinary, especially when they get to talking of science with a capital S," Roosevelt wrote to Grinnell. "They do good work; but, after all, it is only the very best of them who are more than bricklayers, who laboriously get together bricks out of which other men must build houses. When they think they are architects they are simply a nuisance."[99]

Although when it came to studying bears, elks, deer, and antelope Roosevelt too was something of a bricklayer, he had appointed himself as the architect of the burgeoning scientific conservation movement. It was Roosevelt, for example, who dramatically testified before the Public Lands Committee of the House of Representatives against railroad expansion and the YIC's development schemes. During a question-and-answer session Roosevelt acted like a conservationist hit man, ready to take out any un-American corporations or individuals undermining the new wildlife protection ethos and forestry. Unlike Baird, for example, Roosevelt never demanded data. For both better and worse, Roosevelt believed in the cumulative power of firsthand observations over empirical laboratory results. All the biological conservation theory and forestry science in the world, he insisted, wouldn't add up to much if the American people didn't *believe* the findings.

Although Roosevelt worried that Merriam, at the U.S. Biological Survey, was overdoing classification, he insisted that big game, songbirds, and even reptiles should be saved like rare, precious gems. To protect animals as endangered species, moreover, Roosevelt believed you had to make people *care* about their survival. Roosevelt, susceptible to the ideas of naturalist-inclined poets like Whitman and Burroughs, was the kind of polymath not usually admired by serious scientists. To make nature dull like the "little half-baked scientists," Roosevelt believed, was fraught with peril. Darwinism needed to be communicated directly to people in simple ways that they could understand and that wouldn't dethrone God as the creator. Like Darwin himself, Roosevelt was a "nature theologist," holding that nature was proof positive of the genius of God, who had masterminded everything from sparrows' eyes to Pike's Peak.[100]

Thus, by 1891, serving as civil service commissioner and as president of the Boone and Crockett Club, Roosevelt was already the heart and soul of the burgeoning conservation movement. One reason for this was that he was the only Darwinian-trained biologist who wore a shapeless broad-rimmed rancher's hat, carried guns, and knew how to attract a large audience both inside and outside official Washington. When Roosevelt offered homilies about grizzly bears and elk herds, the general public listened. As a Washington-based politician, he had clout: for example, he dined regularly with Secretary of the Interior Noble at the Metropolitan Club. He could also glad-hand with backwoods types in the West on his hunting trips. It's one thing to set aside timber tracts with Section 24 as President Harrison did; it's quite another to change an American mind-set about wildlife and timber management—a mind-set committed to plowed land and sawmills—as Roosevelt was attempting to do, seemingly overnight.

In the early 1890s people needed familiar points of reference to make the leap from Creationism (God put mammals, fish, minerals, and trees on earth to be used) to Darwinism-Marshism (varied species need lots of protective habitat and enforceable laws to survive). Roosevelt was there as America's conservationist trail guide. Because most subscribers to *Forest and Stream* imagined "Dakota Teedie" as a wilderness hunter in buckskins, he had the credibility to explain to them why game laws and forest reserves were necessary. No other easterner was perceived by so many Americans as embodying the western spirit. Only by living in the log cabin at Elkhorn and writing about it in *Hunting Trips* and *Ranch Life* did Roosevelt earn the right to explain why California's old-growth timber needed saving and why for every tree felled in Wyoming another should be planted.

By mixing Darwinian-Marshian analysis with cowboy campfire yarns, and by applying his inbred prosecutorial disposition, inherited from Uncle Rob, Roosevelt was able to help sell the U.S. Congress, the departments of Agriculture and Interior, and eventually western Americans on the notion that saving natural wonders, wildlife species, timberlands, and diverse habitats was a patriotic endeavor. From his boyhood (when he drew Egyptian storks to demonstrate evolution) until his death in 1919 at age sixty (after an arduous river trek to the Amazon of Brazil), Roosevelt served as the American spokesperson for mainstreaming evolutionary theories. This was something neither Francis Parkman, Henry Adams, nor John Hay had an inclination to do—nor, for that matter, did John Burroughs, Elliott Coues, or John Muir. "He who would fully treat of man must know at least something of biology," Roosevelt would write later in life, "and especially of that science of evolution that is inseparably connected with the great name of Darwin."[101]

VI

Especially after the Forest Reserve Act and the three new national parks in California, it was natural for Roosevelt to support President Harrison for reelection in 1892. Despite their personal differences, the two men were philosophically similar. Roosevelt cheered his fellow Republican's achievements, such as the bold appointment of Frederick Douglass as ambassador to Haiti. When Harrison resolutely confronted Britain and Canada about their overharvesting of fur seals in the Bering Sea, Roosevelt was honored to be part of his administration. When Harrison's wife, Caroline, died of tuberculosis a few weeks before the 1892 presidential election, Roosevelt sympathized with his boss's deep grief and distracted mind. So when Grover Cleveland routed Harrison in the election, Roosevelt, too, had a sense of loss.

As president of the Boone and Crockett Club, Roosevelt hoped that President Cleveland would build on the Forest Reserve Act of 1891. Although the overweight Cleveland—who was a cartoonist's delight because of his girth and his walrus mustache—couldn't be accused of being a typical outdoorsman, he was known to care deeply about the fate of big game. Therefore, Roosevelt planned to engage Cleveland, a fellow New Yorker, in saving Great Plains buffalo from extinction. Recognizing that preserving the territorial integrity of Yellowstone was the initial step if the national park movement was to succeed, Roosevelt refused to reduce the political heat just because Harrison had been rejected by the electorate. He believed that the unflinching Grover Cleveland, who had gone after Tammany Hall's notorious Roscoe Conkling,[102] could be won over by the Boone and Crockett Club through a combination of diplomacy and arm-twisting. After all, most of Roosevelt's fellow club members were extremely rich and were, like Cleveland, from New York state.

On December 5, 1892, as Harrison's term was winding down, Roosevelt wrote a letter to the editor, attacking the villains—mining interests and real estate grabbers—of Cooke City, Montana, located northeast of Yellowstone National Park. To Roosevelt, this mining-camp town seemed to be frying in greed. Through unethical quid pro quos and bribes, local developers in Cooke City had, Roosevelt feared, chipped away at the territorial integrity of President Grant's idea for a park (dating from 1872); President Harrison's wise amendments regarding forestry and timberlands (1891) were simply being ignored. Grinnell had published a series of articles in *Forest and Stream* criticizing the contraband mentality of Cooke City and even disseminated a pamphlet all over Montana aimed at stopping the pilferers by threatening to have the U.S. Army arrest them. The Boone and Crockett Club's hard-line approach was "If you poach in Yellowstone, you will go to jail for two years."

Roosevelt's letter, written on U.S. Civil Service commissioner stationery (and thus implying that the federal government was on his side), didn't mince words. "It is of the utmost importance that the Park shall be kept in its present form as a great forestry preserve and a National pleasure ground, the like of which is not to be found on any other continent than ours; and all public-spirited Americans should join with *Forest and Stream* in the effort to prevent the greed of a little group of speculators, careless of everything save their own selfish interests, from doing the damage they threaten to the whole people of the United States, by wrecking the Yellowstone National Park," he wrote. "So far from having this Park cut down it should be extended, and legislation adopted which would enable the military authorities who now have charge of it to administer it solely in the interests of the whole public, and to punish in the most rigorous way people who trespass upon it. The Yellowstone Park is a park for the people and the representatives of the people should see that it is molested in no way."[103]

In *The Winning of the West*, Roosevelt had promoted manifest destiny and the westward march of U.S. capitalism with the zeal of Horace Greeley, so his new position baffled the developers in Montana. Roosevelt, in fact, had once speculated that Duluth would soon rival Chicago as the citadel of the West and that the Red River valley of the Dakotas would harvest grain for the world. As if he were a bond salesman for Jay Cooke, he had written that Montana would supply the most beef and that the Cascade Mountains of Washington Territory had enough potential timber to construct endless homes for America's growing population.[104] Now, suddenly, Roosevelt was smashing the utilitarian paradigm on behalf of preserving Douglas firs, petrified logs, and elk herds. To the Cooke City folks, Roosevelt's new demands were nothing more than atheistic excuses for a federal land grab.

Such were the deeply anti-Roosevelt protestations of Montanans (and the organized syndicate YIC) in the early 1890s. How were they to know that Roosevelt had developed his preservationist insights by reading sportsman literature and studying Darwinian biology at Harvard? Who knew he had memorized every detail of Audubon's *Birds of America* as if it were a sacred text? How could railroad titans have understood that he took pride in his association with John Burroughs and George Bird Grinnell, who had lured him into the preservationist camp? How were western cattlemen to fathom Roosevelt's preference for open-range grazing because his humane, Berghian side didn't like seeing wild game get tangled up in barbed wire? Could anybody really imagine that his Uncle Rob used to have monkeys leaping around in a New York brownstone and a German shepherd sitting at the dinner table? To T.R.'s thinking, his letter in *Forest and Stream* was just straight talk. Like Muir, he thought the idea of national parks should be adopted, honored, and celebrated by mainstream Americans. Some areas of the American landscape and some types of wildlife, he believed, were simply too magnificent for mankind to destroy for the quick financial profits of scoundrels like the Cooke City crowd.

So when Roosevelt went elk hunting in western Wyoming in September 1891, for the first time since the Forest Reserve Act of the past spring, he was considered by many locals as a bizarre, land-grabbing preservationist zealot. (And that was even before his blistering open letter in *Forest and Stream.*) Accompanied by his friend Robert B. Ferguson, the frustrated forty-niner Tazewell Woody, and the campfire cook Elwood Hofer, Roosevelt wanted to see the elk herds of the Tetons, which the Shoshone spoke about with such reverence, with his own eyes.[105]

Two Ocean Pass was a scenic wonder that left Roosevelt breathless. It was located on the Continental Divide (in what became Bridger-Teton National Forest in 1908). All around him were evergreen forests and eternal rock peaks with "grand domes and lofty spires." Craggy ramparts pierced the sky in this vast mountainous region. Here was a sacred spot for sure. Some streams flowed westward into the Snake River and then the Columbia River, eventually emptying into the Pacific Ocean. Others descended eastward toward the Yellowstone River, which drained into the Missouri River before merging with the Mississippi River at the confluence north of Saint Louis; from there the Mississippi went straight to the Gulf of Mexico.[106]

To an American outdoors romantic like Roosevelt, the forlorn, wild valley of Two Ocean Pass epitomized the miraculous West. He was walled in by the raw, rugged Teton mountain chains, their flanks blasted and slashed by precipice and chasm. Carefully Roosevelt studied the fork of a stream where one branch headed toward the Oregon coast while the other flowed in the direction of Louisiana's bayous. Clad in a buckskin tunic with leggings, Roosevelt was living out his fantasy of a voyage of discovery. Everything around him—mountain valleys; fields of goldenrod, purple aster, bluebells, and white immortelles—was unmarred by mankind. There were no surveyors' stakes, mining shacks, or cattle trails to break the spell. Two Ocean Pass and the Tetons—the Grand Tetons—were becoming known as national treasures as surely as Yellowstone and Yosemite. A poet like Whitman could have written a hymn just by breathing in the crisp Wyoming air. "In the park-country, on the edges of the evergreen forest, were groves of delicate quaking-aspen, the trees often growing to quite a height; their tremulous leaves were already changing to bright green and yellow, occasionally with a reddish blush," Roosevelt wrote in the essay "An Elk-Hunt at Two-Ocean Pass," which appeared in *The Wilderness Hunter.* "In the Rocky Mountains the aspens are almost the only deciduous trees, their foliage offering a pleasant relief to the eye after the monotony of the unending pine and spruce woods, which afford so striking a contrast to the hardwood forest east of the Mississippi."[107]

CHAPTER TEN

The Wilderness Hunter in the Electric Age

I

Ever since Roosevelt arrived in the Dakota Territory in 1883 to ranch cattle, the very idea of Texas enthralled him. Many of the Badlands cowboys he encountered spoke of the Hill Country as a hunter's paradise teeming with big-bodied deer. So in the spring of 1892, as U.S. Civil Service Commissioner, thirty-three-year-old Roosevelt hatched a plan. Officially, he was going to Texas to investigate the dismissal of a few U.S. postal employees solely for partisan political reasons. But he also arranged for a six-day collared peccary hunt in the South Texas Coastal Plain, which would enliven *The Wilderness Hunter*, the outdoors memoir he was writing. Furthermore Roosevelt was hoping to anchor future installments of *The Winning of the West* on Lone Star history. "The next volumes I take up I hope will be the Texan struggle and the Mexican War," Roosevelt would write his friend Madison Grant. "I quite agree with your estimate of these conflicts, and am surprised that they have not received more attention."[1]

Killing a peccary (or "javelina," the term preferred in Texas) during the Gilded Age wasn't easy. In addition to being elusive, peccaries were fierce fighters who traveled in packs, known to slash horses' legs with their daggerlike tusks and stampede over dogs in dense thickets of chaparral and scrub oak. "They were subject to freaks of stupidity, and were pugnacious to a degree," Roosevelt wrote. "Not only would they fight if molested, but they would often attack entirely without provocation."[2]

Roosevelt had imagined Uvalde, Texas—where his friend John Moore ranched—to be a temperate prairie like North Dakota. But the area's proximity to the Gulf of Mexico meant there was a wide range of varied habitat to study. Three different types of rail—King, Clapper, and Virginia—were found in the brushland marshes. Around giant cypress trees or pecan groves Roosevelt discovered uncommon species such as greater pewee and Rufous-capped warbler. Bustling insectivorous redbirds and flycatchers, moving together in concert, abounded. Around wild fruit fields were frugivorous birds, including many whose genera Roosevelt was uncertain about.[3]

After discovering no peccaries along the Frio River, the Roosevelt party headed south along the Nueces River toward the oak-motte prairies of the Gulf Coast near Corpus Christi. The spring air was mild at Choke Canyon, and Roosevelt was delighted to see so much unexpected greenery. Little chimney swifts dashed in front of his horse at regular intervals as they moved seaward avoiding the stinging ants. The horseflies were the biggest he had ever seen. The insects were such a serious problem for Texas settlers that screens covered house windows and smoking coils were lit to ward off the swarms. Those in shacks smoked fern roll-ups to ward them off. Roosevet copiously noted the lilac-colored flowers and wide bands of purplish wildflowers that carpeted the unobstructed Texas prairie. "Great blue herons," he wrote, "were stalking beside these pools, and from one we flushed a white ibis."[4]

Once Roosevelt had absorbed the Nueces River area in exacting detail, the expedition went onward with trophy-hungry determination. At sunrise the hunt party was greeted by the Texas nightingale (the mockingbird) and at sunset by the howls of coyotes. But no javelinas. Just when the hunting looked bleakest of all, however, Roosevelt suddenly stumbled upon his mark. A sow and a long-tusked boar turned their huge heads toward the Roosevelt party, grinding their teeth so loudly it produced a sound like Mexican castanets. Their needle-sharp eyes had that dark, calmly menacing look of a great white shark as it circles prey. Roosevelt shot them both at point-blank range.[5]

That evening the hunt party feasted on peccary and Roosevelt shipped his trophy heads back to New York. More than anything else, it seemed, Roosevelt thoroughly enjoyed the tough-talk style of his Texas compadres. Like a mynah bird, Roosevelt had picked up a lot of sayings and brags which he now constantly repeated back in Washington. He admired, for example, the story of a Texan who carefully studied a tenderfoot's 32-caliber pistol and said: "Stranger, if you ever shot me with that, and I *know'd it*, I would kick you all over Texas." As a corollary, Roosevelt decided that when it came to peccary hunting, guns weren't the armament of choice. "They ought to be killed with a spear," Roosevelt wrote his British friend Cecil Arthur Spring Rice. "The country is so thick, with huge cactus and thorny mesquite trees, that the riding is hard; but they are small and it would be safe to go at them on foot—at any rate for two men."[6]

Texas put a ruddy color back in Roosevelt's cheeks; and his brow, though creased, now showed few traces of stress. The fresh air had once again purged his bureaucratic fatigue, and the open country had given him time to relax and think. The spare campfire meals had thinned him down quite a bit. Once back in Washington, he remained so enchanted with Wild West cowboy lore, in fact, that he made plans to visit Deadwood in August to see with his own eyes where Wild Bill Hickok died. (He went and deemed it "a golden town."[7])

That journey to the Black Hills of South Dakota, however, had a civil service objective: to investigate graft and inhumane conditions on various Sioux reservations following the massacre at Wounded Knee. Roosevelt rode to the Pine Ridge Reservation, where more than 7,000 Sioux lived, largely in squalor, to investigate what had happened twenty months earlier when the Hunkpapa Sioux leader Sitting Bull was murdered by U.S. troops while under house arrest.[8] That killing had triggered the massacre of December 29, 1890, when 500 cavalrymen surrounded an encampment of Lakota Sioux at Wounded Knee Creek. Four rapid artillery-fire Hotchkiss guns were brought in and—after a sharp disagreement with a deaf tribesman who refused to surrender his rifle—more than 300 Lakota Sioux men, women, and children lay murdered in the bloodied snow. "They gathered up the frozen dead in wagons at Wounded Knee," *The American Heritage Book of Indians* later lamented, "and buried them all together in a communal pit."[9]

Naturally, at the time of Roosevelt's inspection, tension between the Sioux residents and white guards at Pine Ridge remained high. Complaints that the Sioux were now being given poisoned food had traveled back to Washington, D.C., and landed on Roosevelt's desk;[10] he was looking into allegations that U.S. officials were diluting and stealing foodstuffs directed toward the Great Plains reservations. (As president, Roosevelt, after reading Upton Sinclair's *The Jungle*, famously took on the Chicago meatpacking industry for selling rancid beef and pork. Now, fourteen years before the Meat Inspection Act was passed, he sided with the discontented tribes who claimed they were being sold poison pork at commissary stores on the reservations.)

As it turned out, Commissioner Roosevelt sided with the Indians on most of the issues. No American, he maintained, should be deliberately served rotting meat and given poor medical attention. Roosevelt's host, Captain Hugh C. Brown, boldly issued a meat recall at Pine Ridge, defying his military orders. The stealing of U.S. supplies directed for the reservations, Roosevelt thundered, had to stop at once. Breaking with General William T. Sherman's philosophy that all Native Americans had to "be killed" or else "maintained as a species of paupers," Roosevelt wanted the tribespeople fully integrated into the fabric of American life.[11] To Roosevelt, the properly maintained reservations were merely a way station to fuller integration, which could be accorded in due time.

At the time of Roosevelt's reservation tour, the number of Indians in the United States was only 250,000, drastically decreased from estimates of the population in 1492, which were in the millions. The surviving Native Americans had overcome disease, conquest, genocide, and assimilation, but Roosevelt worried that the spoils system could do them in. "The Indian problem is difficult enough, heaven only knows," Roosevelt wrote in January 1891 to a friend who advocated Indian rights, "and it is cruel to complicate it by having the Indian service administered on patronage principles."[12]

From Pine Ridge Roosevelt headed south to meet the humanitarian Herbert Welsh of the Indian Rights Association (IRA). Organized in 1882, the IRA believed in the immediate and direct acculturation of Native Americans into the mainstream of U.S. society. The energetic Welsh knew how to lobby effectively on behalf of Indian welfare (or, at least the IRA's vision of it).[13] The IRA believed serious changes needed to be made in state and federal government to create a pathway to full citizenship for all Native Americans.[14]

Roosevelt deemed Welsh the most effective advocate fighting on behalf of Indians in America. Together, they toured the Cheyenne River Reservation in South Dakota. They also visited George Bird Grinnell's old stomping grounds in Nebraska (where he had once befriended the Blackfoot and North Cheyenne while working on Buffalo Bill's ranch near North Platte). In his capacity as civil service commissioner, Roosevelt inspected the Missouri River Indian agencies in South Dakota and Nebraska—Yankton, Santee, Omaha, and Winnebago—pausing at all the old Lewis and Clark campsites for curiosity's sake. Although he stumped for President Harrison's reelection along the way, he also denounced the abuses Native Americans were suffering in Nebraska's reservations at the hands of a delinquent U.S. government. During this inspection trip Roosevelt didn't keep a South Dakota–Nebraska diary, but he did write an official report as civil service commissioner, one that was considered too inflammatory to be published in family newspapers. It was pure Roosevelt, playing the role of muckraker in the style of Lincoln Steffens or Jacob Riis. Point by point he analyzed why Sioux, Cheyenne, and other tribespeople weren't getting a fair shake. The federal government didn't disseminate Roosevelt's final report, but Welsh printed 3,000 copies and distributed them to leading legislators and philanthropists. "By the time of Roosevelt's departure from the Civil Service Commission in 1895," the historian William T. Hagan has noted, "he had earned the admiration of many friends of the Indian."[15]

Although Roosevelt was enough of a social Darwinist to write that the Pawnee and Cherokee were far superior to the Sioux, he was more fair-minded in his assessment of the U.S. government's failings in its Indian policies than most other leading politicians of the era. Exactly *why* Roosevelt behaved so decently to Indians is paradoxical but simple. Never one to romanticize Sitting Bull or Geronimo (deeming both dangerous rabble-rousers), he had invested so mightily in the U.S. Army's western triumphs that he wanted to make sure the defeated Indians were not treated badly or as inferiors. This basic moral premise put him in the IRA camp. Just as presidents Lincoln, Johnson, and Grant forgave the Confederates after the Civil War, welcoming them back into the Union fold, Roosevelt believed that now that the West was won, the vanquished Indians should be brought into the constitutional democracy with the same God-given rights as everybody else. Roosevelt's Americanism—that is, the need for the country to act as one—far outweighed his mistaken interest in armchair eugenics.

Yet there was another factor in play. Tickled to be called the Great White Chief by some Native Americans, Roosevelt truly respected the central role bison continued to play in the culture and religion of the Sioux (and other tribes). Unlike Euro-Americans, the pragmatic Sioux tribes used every part of the buffalo: hides were made into clothing and tepees; horns were eating utensils and cups; and muscles provided glue and bowstrings. After killing a buffalo, the Sioux would first eat the fresh meat and then preserve the rest as sun-dried jerky strips. A positive auxiliary effect of repopulating the Great Plains and Rocky Mountain regions with buffalo, Roosevelt believed, was to properly honor the folkways of the Plains Indians. George Bird Grinnell was in full agreement on this point. Like the Plains Indians, the Boone and Crockett Club hoped, as an Indian once told Grinnell, that someday the prairie lands would once again be "One Robe."[16]

In a kingmaking mood following his successful appearances at the Deadwood Opera House and the Dakota-Nebraska-Kansas reservations, Roosevelt continued to give last-minute speeches back East championing President Harrison's reelection whenever the Republican National Committee asked. His voice, however, wasn't persuasive enough. On November 8, 1892, the Democrat Grover Cleveland easily defeated Harrison by 277 electoral votes to 145.[17] As Cleveland took office on March 4, 1893, Roosevelt offered his resignation from the civil service. The incoming president refused, deciding that having a high-profile Republican reformer like Roosevelt in his administration was a good thing.

His job secure, Roosevelt forged ahead with more outside activity. He wanted the Boone and Crockett Club to have a log cabin exhibit at the World's Columbian Exposition in Chicago, opening in May 1893 to celebrate the 400th anniversary of Christopher Columbus's voyage to America (as well as Chicago's moment to present itself as a world-class city). There were more than 200 European-designed buildings going up on the fairgrounds alongside Lake Michigan, so Roosevelt had focused on displaying the vernacular frontier home, the log cabin, as a point of national pride. The humble birthplaces of Lincoln and Grant were far more impressive, he believed, than Buckingham Palace or the Vatican. (Both replica presidential cabins were exhibited in Chicago under the slogan that the ingenuous American "cuts his coat according to his cloth."[18]) Roosevelt assumed the role of exhibit designer, and his archetypal western log cabin was packed with Davy Crockett relics and old-time hunting and trapping equipment.[19] It was situated on a man-made island called the Wooded Island, in a man-made lagoon, and was next to the Japanese pavilion. The "cabin" staff hired a long-haired hunter as host. Schoolchildren could watch him perform public demonstrations that included curing venison jerky and constructing a box trap.[20] Unfortunately, Roosevelt's Boone and Crockett cabin had to compete with an exact replica of the Old Times Distillery of Kentucky, which gave out free whiskey samples. A New England cabin was also on the grounds, providing "good old-fashioned" seafood stews for the tasting. Still, owing to the Crockett memorabilia, Roosevelt's cabin was a popular tourist destination.

The very announcement that Roosevelt was erecting a frontier cabin at the Chicago Exposition brought him a lot of mail. One was from an old hunting friend and guide on his trip to the Bighorns. In spite of its clearly anti-Indian, illiterate tone Roosevelt relished in the letter's colorful slang:

> Feb 16th 1893; Der Sir: I see in the newspapers that your club the Daniel Boon and Davey Crockit you Intend to erect a fruntier Cabin at the world's Far at Chicago to represent the erley Pianears of our country I would like to see you maik a success I have all my life been a fruntiersman and feel interested in your undertaking and I hoap you wile get a good assortment of relicks I want to maik one suggestion to you that is in regard to getting a good man and a genuine Mauntanner to take charg of our haus at Chicago I want to recommend a man for you to get it is Liver-eating Johnson that is the naim he is generally called he is an olde mauntneer and large and fine look-

ing and one of the Best Story Tellers in the country and Very Polight genteel to every one he meets I wil tel you how he got that naim Livereating in a hard Fight with the Black Feet Indians thay Faught all day Johnson and a few Whites Faught a large Body of Indians all day after the fight . . . Johnson was aut of ammunition and thay faught it out with thar Knives and Johnson got away with the Indian and in the fight cut the livver out of the Indian and said to the Boys did thay want any Liver to eat that is the way he got the naim of Livereating Johnson.

"YOURS TRULY" ETC., ETC.[21]

Another Rooseveltian scheme for the Chicago World's Fair was to commission the artist Alexander Proctor to design and erect life-size sculptures of American wildlife on the bridges connecting the fairgrounds and lagoons. At Roosevelt's behest Proctor, who had been raised in Denver, created life-size polar and grizzly bears, elks, cougars, and moose for public display. To Roosevelt, Proctor's naturalist work, influenced by Darwin, was the highlight of the entire exposition. Both Roosevelt and Proctor insisted on *exactness* of animal composition. Wanting to honor his sculptor friend for a job well done, Roosevelt held a salutatory dinner for him at the Boone and Crockett cabin display in Chicago. In between toasts declaring Proctor the greatest sculptor of the American West, a man who understood the intersection of nature, wildlife, and science, the wildlife artist was asked to join the club. In coming years Proctor achieved some degree of renown for his bas-relief *Moose Family*, commissioned by the forester Gifford Pinchot in 1907 after Roosevelt became president.[22] "For the men of the Boone and Crockett Club," the historians Jesse Donahue and Erik Trump wrote in *Political Animals*, "Proctor was representative of the vanishing West, both through his work and in his person."[23]

Roosevelt also got into the fair's futuristic spirit by offering advice on the Forestry Building interpretive center, constructed with a rustic wraparound veranda made solely out of indigenous wood. He also found the Idaho pavilion impressive: this three-story western cabin was made of basaltic rock, volcanic lava, and stripped cedar logs; with large chimneys, an arched stone entranceway, and a reception room fitted out like a trapper's den, it became a prototype to be emulated in future national parks for information centers or lodges.[24] What Roosevelt appreciated in the Idaho Pavilion was a new type of western architecture, which easily blended into the natural setting. (Little did Roosevelt know that in 1893, the architect Frank Lloyd Wright constructed Winslow House in

River Forest, Illinois, considered the first of his free-flowing prairie-style homes, which brought the natural world directly into the hearth instead of blocking it out.[25])

Although 27 million people streamed into the exposition between May and October 1893, *the* two biggest attractions in Chicago—Buffalo Bill's Wild West Show and the Ferris wheel—operated outside the gates. It has long been speculated that Roosevelt adopted the name Rough Riders for his Spanish-American War outfit from watching William "Buffalo Bill" Cody perform there, with live buffalo herds and cowboy-and-Indian re-creations.*[26] Most of the *Forest and Stream* crowd disdained Buffalo Bill for his "skinning" career—he slaughtered bison for the railroads—but Roosevelt admired the "steel-thewed and iron-nerved" showman for his "daring progress [to open] the Great West to settlement and civilization. His name, like that of Kit Carson, will always be associated with old adventure and pioneer days of hazard and hardship when the great plains and the Rocky Mountains were won for our race."[27]

The 1893 World's Columbian Exposition looked to the future, even as it celebrated the past. In fact, it became a showcase for the revolutionary marvels of harnessed electricity. Everything from the first phosphorus lamps to the first neon lights was on display. Virtual shrines to the wonders of alternating-current power were opened to the public courtesy of Brush, Thomas Edison, Western Electric, and Westinghouse.[28] And there, in the shadow of the electrical exhibit, was Roosevelt's Boone and Crockett Club log cabin, a throwback to a distant era, lit up only on a few chilly autumn nights by a newly trimmed fire. While America was abuzz about the electrical wonders of tomorrow, Roosevelt, with retro satisfaction, busied himself promoting the gospel of rustic renewal. Still, he was extremely proud that American ingenuity—from the log cabin to the electric mansion—was being showcased to the world. "Indeed Chicago *was* worth while," he wrote in June 1893. "The buildings make, I verily believe, the most beautiful architectural exhibit the world's ever seen."[29]

For the history of U.S. wildlife conservation, something else occurred at the fair—something far more important than electricity on parade,

---

*William F. Cody first called the troupe (in 1893) "Buffalo Bill Cody's Congress of Rough Riders of the World." But, in truth, Roosevelt also had fair claim to the term. In August 1883 he had written to a friend from the Elkhorn Ranch, "I think there is some good fighting stuff among these harum-scarum roughriders out here."

or an obscure history professor at the University of Wisconsin-Madison eulogizing the American frontier, or schoolchildren touring log cabins. The National Game, Bird, and Fish Protection Association (NGBFPA) was created that year in Chicago. Going forward, the Boone and Crockett Club, the Audubon Society, and other wildlife preservation organizations would work together, sharing lobbyists and coordinating strategies. By January 1895, the NGBFPA had adopted resolutions to encourage federal propagation of game birds and federal interdiction of interstate game traffic. Even though wildlife didn't have the economic importance of timber or water, more and more Americans were starting to care about species survival.[30]

II

As president of the Boone and Crockett Club, Roosevelt edited *American Big-Game Hunting*, a volume of essays about hunting and conservation, to be published in the fall in time for the fair's last gasp. As fate would have it, this was not a propitious time for selling an expensive book. The Panic of 1893 had brought hard times to most Americans. Unemployment was high; wages were low; money was tight. Speculative finance and laissez-faire capitalism squeezed the wallets of ordinary Americans, from immigrants to workers in urban sweatshops to Midwesterners desperate to redeem silver notes for gold. Many banks failed, as did railroad companies like the Northern Pacific, the Union Pacific, and the Atchinson, Topeka and Santa Fe. According to the journalist J. Anthony Lukas in *Big Trouble*, in Colorado alone 435 mines and 377 related businesses closed because of the panic.[31] Western cities like Denver—known as the Queen City of the Plains—which had relied on the silver mining boom were particularly hard hit. Bitter and broke, many settlers in Pueblo and Durango, Colorado, deemed the uncut Rocky Mountain forests now designated as "federal reserves" a serious insult to their inbred sense of manifest destiny economics.[32]

As the Panic of 1893 gripped the Midwest and West, there were clamorous demands that the Forest Reserve Act of 1891 be revoked. The only people the reserves benefited, their opponents said, were "nature cranks" and the "athletic rich."[33] In this uncertain financial climate, the high price of *Big-Game Hunting*—ten dollars—meant it would appeal only to the well-off or antiquarians. "We thought," Grinnell recalled, "that perhaps there were enough big game hunters in the country to make it possible to publish the book without too great a loss." The idea of the Boone and Crockett Club's publishing venture had originated with Roosevelt.

Grinnell recalled that Roosevelt financed the first printing of 1,000 copies with a personal check of $1,250. "He never said anything about this," Grinnell recalled, "and I never asked about it."[34]

Personally immune to the Panic of 1893, Roosevelt and Grinnell recruited stories for *American Big-Game Hunting* from founding club members. Roosevelt carefully edited and pruned the prose of the submissions, proving to be well suited for the task. Always ready with red pencil, Roosevelt actually asked Grinnell to send an overwritten submission his way so he could "slash it up" by a third.[35] This presented a delicate problem in the case of one submission.

Back in 1887, Roosevelt had tapped the famous landscape painter Albert Bierstadt to become a member of the Boone and Crockett. Bierstadt's sublime paintings of the Rockies and the Mojave Desert, in which settlers were shown (if at all) as dots dwarfed by the vast American West, promoted the inherent value of wilderness. In addition, Bierstadt, whose notebooks are filled with sketches of American wildlife, had killed a huge moose along the New Brunswick–Maine line.[36] With a rack sixty-four and a half inches wide, it was determined to be the eighth-largest set of antlers ever recorded.*

Bierstadt had realistically painted this triumph in *Moose Hunter's Camp*, so Roosevelt was eager to get a first-person account from him for the book. "At the last meeting of the Boone and Crockett Club it was decided, subject to the approval of the rest of the members or of a majority of them, to see if we could not produce a volume to be known by some such title as that of the Boone and Crockett Club, and to consist of various articles on big game hunting, etc. by members of the Club," Roosevelt wrote to Bierstadt in February. "We intend to issue it annually if we find it reasonably successful. To do this would need some money, probably five or ten dollars annual dues for each member being sufficient. I hope you approve of the scheme and that if we decide to get out the book you

*Starting in 1932, the Boone and Crockett Club would distinguish itself as publicly promoting "record book" trophies based on a scoring system for big game. The club took as a model the English Rowland Ward records system. For moose the best score was 224–418 with an antler span of sixty-seven inches and a weight of over 1,800 pounds. A host of categories were soon developed. Take, for example, goats. By the time of World War II, the members of Boone and Crockett were encouraged to be "grand slammers" (i.e., to hunt one sheep from each of four individual subspecies: desert bighorn, Rocky Mountain bighorn, Dall's sheep, and Stone's sheep). The club also would disqualify submissions for a host of reasons: for example, hunting with a telescopic 4X scope was considered not "fair chase."

will give us an article on moose hunting. I should greatly like to have in permanent form one or two of your experiences. Have you ever published an account of the way in which you got your big head? If not, do write it out for us at once."[37]

Of course, the fact that Bierstadt was an excellent painter didn't necessarily mean he wrote well. The artist's mother tongue was German, and his English was only passable. Nevertheless, as requested, Bierstadt wrote an accurate, lively account of his Maine–New Brunswick moose hunt. In reading the essay, Roosevelt discovered a bigger problem than atrocious spelling or awkward syntax: Bierstadt hadn't actually shot the moose; his Indian guide had. This ran afoul of the Boone and Crockett Club's eligibility rules—its members had to have killed a big-game animal personally, in a "fair chase." Worse yet, Bierstadt's submission expressed his disdain for the violence associated with hunting; he wrote that it was wrenching to pull the trigger on such a lovely North Woods moose: "I took the rifle then and ended his misery; he reeled, staggered, and tried to lean against a smaller tree which bent over as he gently breathed his last. My sketch book was in use at once. I have as you will see one big head; but I have made up my mind that I don't want to kill any more moose, but to go and see them in their own haunts is a pleasure."[38]

The situation seemed clear: either the Boone and Crockett's constitution would have to be rewritten or the sixty-nine-year-old Bierstadt would have to be expelled from the club. But Roosevelt found a third way to handle the problem. He adeptly edited the story to make it seem as if Bierstadt had, in fact, bagged the animal himself. This "benign deception," as two scholars later called it in the *New England Quarterly*, was uncharacteristic of the usually up-front Roosevelt. By recasting the death scene in the passive voice—"This bull was killed"—he excised the Indian guide's marksmanship.[39] Initially, Roosevelt's creative edit achieved his overriding goal of preserving Bierstadt's integrity by allowing this story to be published in *American Big-Game Hunting* while also adhering to the club's constitution. However, Bierstadt wouldn't agree to these artfully truthful but misleading edits. If he accepted Roosevelt's solution, his article would, in fact, have degenerated from nonfiction to fiction. The painter suggested a compromise—the Boone and Crockett Club could publish his essay *without* using his byline or signature.

Unwilling to compromise any farther, Roosevelt now balked. As civil service commissioner, he was busting lying scoundrels right and left. If the press discovered his cover-up of Bierstadt's story, it would have a field day at his expense. He wasn't going to risk what reporters call a blind item for the sake of somebody else's problem. "Grinnell and I both feel that it would not do to put in any non-editorial article unsigned, and moreover that when we get a piece of yours it ought to be purely yours, and without emendations from us," Roosevelt wrote to Bierstadt on June 8 from Washington, D.C., unburdening himself of the whole ordeal. "So I shall have to trust to the hope that for our second volume we may persuade you to write a piece needing no emendation, over your own signature."[40]

There were some wonderful pieces in *American Big-Game Hunting*. Grinnell's one contribution, "In Buffalo Days," is arguably the most elegant meditation on buffalo ever written. Truly worried that the species was headed toward extinction, Grinnell expressed his love for the animal by describing every twitch and tail flap he had ever noticed as a naturalist. "It was in spring, when its coat was being shed, that the buffalo, odd-looking enough at any time, presented its most grotesque appearance," he lovingly wrote. "The matted hair and wool of the shoulders and sides began to peel off in great sheets, and these sheets, clinging to the skin and flapping in the wind, gave it the appearance of being clad in rags."[41]

Among the other prominent contributors to *American Big-Game Hunting* were T.R.'s old Harvard friend Owen Wister ("The White Goat and His Country"). Living in Philadelphia but writing about the West, Wister had become a member of Boone and Crockett Club at Roosevelt's invitation. After Harvard the two enthusiasts of the West grew close, frequently discussing the Rockies, buffalo repopulation, and the frontier cowboys. In 1893, Wister had visited Yellowstone and met Frederic Remington there. He and Remington decided to help in Roosevelt's crusade to protect wildlife at Yellowstone. Known for encouraging rows, launching into diatribes, and harboring a sycophantic admiration for Ulysses S. Grant (of whom he published a biography in 1900), Wister was the kind of Harvard man Roosevelt could call a true brother in arms. The two struck up an informal pact in 1893 or 1894: Roosevelt would continue writing about the West as a historian, while Wister would write a great novel about a Rocky Mountains rancher, using the working title *The Virginian*. Along with Remington they would constitute a club of three amigos determined to popularize the American West along the Atlantic seaboard. They even wore the exact same outdoors clothes.[42]

Others whom Roosevelt and Grinnell asked to participate in *American Big-Game Hunting* were Winthrop Chanler ("A Day with Elk"), Archibald Rogers ("Big Game in the Rockies"), and F. C. Crocker ("After Wapiti in Wyoming"). As coeditor of the book, Roosevelt had diligently corresponded with contributors, making suggestions about how to improve their manuscripts and struggling to create an overall unity of effect. Roosevelt, in fact, oversaw the physical look of the volume, choosing deep red cloth and a big-game head for the cover. The first fifth of *American Big-Game Hunting* explained the conservationist objectives of the Boone and Crockett Club to readers. The West Pointer Captain George S. Anderson, superintendent of Yellowstone, for example, led off with the hunter's lament "A Buffalo Story." Picking up the saga in the 1870s, Anderson traced the demise of the buffalo as a species in a longing, heartfelt way. Nevertheless, he gloated over killing a "lonesome George" just for its tongue. A veteran of the Indian Wars with the Cheyenne, Kiowa, and Commanche, Captain Anderson was known as the premier "saddle officer" in the West. That was in the 1870s and 1880s. As of 1890, in a complete reversal, Captain Anderson's enemies were no longer Native Americans in war paint but white settlers around Yellowstone engaged in poaching, vandalizing, and overgrazing.[43] His conservationist evolution from buffalo skinner to buffalo protector was indicative of a new consciousness developing in the American West of the 1890s, one which Roosevelt was instrumental in promoting.

Not wanting to be left out, Roosevelt took up the Great Plains in *American Big-Game Hunting.* His "Coursing the Prongbuck" (lifted, in tone and emphasis, from *Hunting Trips* and *Ranch Life*), again expressed his enthusiasm for the Badlands. He explained that pronghorn herds were thinning out from the Dakotas to Texas because of the pronghorn's own curiosity; they were always investigating prairie schooners or human camps too closely. All a westerner had to do was wave a colored rag from behind a rock or sage, and the antelope would slowly head toward it—and, invariably, be shot as a result. "The pronghorn is the most characteristic and distinctive of American game animals," Roosevelt wrote. "Zoologically speaking, its position is unique. It is the only hollow-horned ruminant which sheds its horns. We speak of it as an antelope, and it does of course represent on our prairies the antelopes of the Old World, and is a distant relative of theirs; but it stands apart from all other horned animals. Its position in the natural world is almost as lonely as that of the giraffe."[44]

Roosevelt also contributed "Literature of American Big-Game Hunting" to the volume, enthusiastically touting his all-time favorite naturalists and sportsmen. "The faunal natural histories, from the days of Audubon and Bachman to those of Hart Merriam, must likewise be included," Roosevelt advised fellow hunters, "and, in addition, no lover of nature would willingly be without the works of those masters of American literature who have written concerning their wanderings in the wilderness, as Parkman did in his *Oregon Trail*, and Irving in his *Tour on the Prairies*; while the volumes of Burroughs and Thoreau have of course a unique literary value for every man who cares for outdoor life in the woods and fields and among the mountains."[45]

One conservationist from whom Theodore Roosevelt didn't solicit an essay was Robert B. Roosevelt. Every time T.R. and his freewheeling, unpredictable uncle tried to collaborate on anything there was a clash of wills. The group R.B.R. had founded, the New York Association for the Protection of Game (NYAPG), had shifted its focus from an early triumph—saving quail—to a new craze for trapshooting.[46] Ever since the Interstate Trapshooting Association was formed in 1890,[47] NYAPG's members were using the association more or less as a rod-and-gun club for blasting plates on manicured fairways. Even the *New York Times*, not immune to the fad, started covering trap tournaments as premier sporting events.[48] But plates were the good part of the fad. Unfortunately, some clubs started using live pigeons instead of clay ones. Although not an animal rights advocate, Grinnell vehemently denounced the shooting of birds as both cruel and unsportsmanlike. In an editorial in *Forest and Stream* he lambasted not only the shooters but also the commercial netters who sold boxes of the pigeons to trapshooting clubs.[49]

There is no record of whether R.B.R. felt hurt about not being tapped for the Boone and Crockett. Spending the late 1880s abroad as ambassador to the Netherlands during President Cleveland's first term, R.B.R. let the conservationist mission of NYAPG slip away; preoccupied in The Hague, he had scant time to seriously challenge poachers in upstate New York. Seizing the opening, T.R. and Grinnell had created the Boone and Crockett at an appropriate time, entering the void left by the NYAPG's slack course. Most important, instead of being a statewide organization like NYAPG, the Boone and Crockett Club was national in scope. Owing to his celebrity Roosevelt was able to attract the best people to his club. With apparently no sense of betrayal NYAPG's lawyer, Charles E. Whitehead, for example, instead began doing pro bono work for T.R.'s Boone and Crockett Club.* "It was, in other words, an opportune time for a new organization," the historian John F. Reiger has explained in *American Sportsmen and the Origins of Conservation*, "one that would have a scope, as well as the self-discipline to stay focused on what was important."[50]

---

*In 1875 Whitehead had won *Phelps v. Racey*, an important case against a Manhattan game dealer who was wholesaling quail shot out of season. It was considered a "landmark decision supporting state authority to limit the sale of game."

Following the publication of *American Big-Game Hunting*, it was becoming clear that Roosevelt was among those Americans best equipped to exert, as his sister Corinne put it, a "potent influence for good in Western affairs."[51] Coincidentally, Roosevelt's *The Wilderness Hunter* reached bookstores around the same time; it included twenty-four full-page engravings (some drawn by Remington). To a modern-day reader, the book smacks of the influence of John Burroughs, starting with epigraphs from Walt Whitman and Joaquin Miller. Boasting like Moses in the Old Testament, Roosevelt declared in his preface that for a "number of years much of my life was spent either in the wilderness or on the borders of the settled country."[52]

Roosevelt's preface goes on to explain that his firsthand outdoors experiences taught him that besides the thrill of the fair chase there was an aesthetic value in nature. Merely soaking in grand scenery and studying woodland creatures united the naturalist's body and soul. "In after years there shall come forever to his mind the memory of endless prairies shimmering in the bright sun," Roosevelt wrote, sounding like Katherine Lee Bates's 1895 patriotic anthem "America the Beautiful," "of vast snow-clad wastes lying desolate under gray skies; of the melancholy marshes; of the rush of mighty rivers; of the breath of the evergreen forest in summer; of the crooning of ice-armored pines at the touch of the winds of winter; of cataracts roaring between hoary mountain masses; of all the innumerable sights and sounds of the wilderness; of its immensity and mystery; and of the silences that brood in its still depths."[53]

Carefully studying Burroughs had taught Roosevelt to make the most minute and detailed observations of nature, from a blade of bunch-grass to a nagging gnat, from a dead fly on a windowsill to the highest branch of a towering redwood tree. Even when Roosevelt was stalking a wapiti (round-horned elk) in the Bitterroots, he now paused to note fallen timber, scolding chickadees, slippery pine needles, and loose gravel. One chapter of *The Wilderness Hunter*, ostensibly about hunting elk, evolved into an informed field study of Rocky Mountain birds modeled on Burroughs's beliefs about the backyard as a universe.

In *The Wilderness Hunter* Roosevelt no longer focused on "manly" bird sounds like eagles' screams, loons' cries, or owls' hoots. There was a softening of presentation in this new book that harked back to the surging memories of his boyhood diaries. He was downright pastoral about celebrating land where the horses didn't boss the streets. "The remarkable and almost amphibious little water wren, with its sweet song, its familiarity, and its very curious habit of running on the bottom of the stream, several feet beneath the surface of the race of rapid water, is the most noticeable of the small birds of the Rocky Mountains," he wrote. "It sometimes sings loudly while floating with half-spread wings on the surface of a little pool."[54]

On August 6, 1893, a red-letter day, the *New York Times* hailed *The Wilderness Hunter* as a five-star delight. Roosevelt's western hunting stories, filled with picaresque and sometimes gory detail, were written, the anonymous reviewer said, from a genuine love of the outdoors, told "without romance and with admirable clearness."

The Hay-Adams circle may have scoffed at Roosevelt's obsession with wildlife, but to the *Times* this biophilic enthusiasm sprang from the same American grain as the Transcendentalists. The reviewer even applauded the influence of Whitman and Miller on the book (unaware that Burroughs was the secret lurking muse). Being a cheerleader about bison and beaver in an age of species eradication, the *Times* implied, was a good thing. "The Americanism of Theodore Roosevelt is not that of the old-fashioned Fourth of July orators," the review noted. "He is a sound-hearted, sound-minded patriot who has realized that in the present day there is no lack in his country of men of learning and influence always too keenly alive to the most trivial faults of our social and governmental systems and ever ready publicly to deplore them, and has wisely set for himself the opposite task of stimulating a love of country in the rising generation. Americanism is not a good-looking word, and it is one that has been sadly misused. Yet we can think of none better to apply to Mr. Roosevelt's creed and practice."[55]

That the review concluded with this approving recognition of Roosevelt's vision—equating the western wilderness with nationalism—must have bolstered his self-confidence immeasurably. Without an iota of equivocation, Roosevelt instructed the outdoors community in *The Wilderness Hunter* that the killing of a female moose or deer was reprehensible. Over and over again, Roosevelt maintained that *real* hunters honored the game they shot, and that the opposite attitude ("butcher spirit") was evil incarnate. Empty-headed hunters, Roosevelt insisted, those who shot wildlife just to kill, were to be rejected by their communities as pariahs. *Forest and Stream*, in a largely positive review, likewise pointed out that Roosevelt was unique among big-game hunters because the "blood and the killing" weren't central to his wilderness reportage. "We can get enough of that," the magazine sniffed, "by interviewing an employee at a slaughter house."[56] Even *The Youth's Companion*, a widely popular boy's magazine, praised Roosevelt's book for lashing out at cold-blooded market hunters who shot moose stuck in snowdrifts, taking the fair chase out of the hunt and making it one-sided.[57]

Besides overlooking his inherent conservationist attitude in *The Wilderness Hunter*, recent environmental historians have mocked Roosevelt as a weekend warrior, an urbanite with money to burn who bought himself a ticket to the wilderness for a few weeks and then returned home. At face value this analysis is true. But from the perspective of 2009 Roosevelt's desire to connect with nature to rejuvenate himself has proved ahead of its time. Today only 1.9 percent of Americans are living in rural areas, compared with 40 percent when *The Wilderness Hunter* was published, so Roosevelt was anticipating a modern trend. As of 2008 Jefferson's agrarian vision and the homesteading of the West were kaput. Even Thoreau's back-to-nature ethos, based on self-reliance, which had a revisionist run in the 1960s, had become cultish at best, a matter of a few survivalists holed up in forlorn mountain cabins in the Sierra Nevada or Appalachians. But Roosevelt's notion of extreme wilderness experiences in short fixes has become widespread. Shooting the rapids, mountain climbing, rappelling—Americans crave an extreme fix from nature in hundreds of different ways. Whole cities such as Boulder, Eugene, and Asheville cater to consumers of nature like Roosevelt: claustrophobic city dwellers and suburbanites desperate to encounter a rare bird or cypress grove or desert ecosystem before it all vanished.

III

With the commercial success of *The Wilderness Hunter*, Roosevelt had enhanced conservationist stature. Using a civil-service issue as a pretext to open a dialogue with Secretary of the Interior Hoke Smith, he tried in April 1894 to influence U.S. government policy on law enforcement in parklands protection. "I am very glad of the position the Interior Department has taken in reference to the Yellowstone Park," Roosevelt wrote to Smith. "The next time we give a dinner I shall ask you to be our guest, as we much appreciate the stand you have taken in forestry matters and in the preservation of these parks. It will be an outrage if this government does not keep the big Sequoia Park, the Yosemite, and such like places under touch."[58]

Roosevelt respected Smith, who was then thirty-eight years old and the publisher of the *Atlanta Evening Journal*. A tall, pickle-nosed, big-bellied Cleveland Democrat, Smith was a shrewd anticorporation lawyer bent on promoting the "New South." President Cleveland believed Smith would "stand fast against land grabbers and exploiters of the public domain." As a corollary, Smith was unafraid to lambaste the Populist Party (or People's Party, founded in 1891 to lobby for free silver coinage), scoffing at its membership as essentially a hatful of charlatan hayseeds.[59] Although Roosevelt was from the opposition party, he respected Smith as a tough, honest, no-nonsense yellow-dog Democrat with the admirable glare of a battle-tested Confederate veteran. Smith was a snappy dresser, always seen wearing a frock coat and slouch hat; there were no wrinkles in his wardrobe. Often he would sport a handkerchief in his jacket pocket as an affectation. Roosevelt, who denounced the Populists as the type of men who didn't wear "under shirts," admired Smith's sense of style. Furthermore, Smith's wife was the daughter of the legendary General Howell Cobb—a Confederate so gray his image should have been chiseled onto Stone Mountain, and also a former secretary of the treasury, having served under President Franklin Pierce. Cobb, it was said, calculated every waking hour in the service of Georgia's greatness.[60]

Shrewdly, Roosevelt used the fact that his own mother had lived in Georgia—as a Bulloch from Roswell—as an ice-breaker with Smith. Peaches, pine trees, and red clay *were* part of his heritage, and many of his first-prize items assigned to his boyhood Roosevelt Museum had come from just north of Atlanta. For his part, Smith, who would go on to serve as a U.S. senator from Georgia, told Roosevelt that he welcomed any strategic wisdom from the Boone and Crockett on how to deal with policing Yellowstone and the California national parks.

Roosevelt's friendship with Smith became extremely important in the bipartisan effort to vex the relentless lobbying of western anticonservation legislators. From the outset of his second term, President Cleveland, using Smith as his megaphone, made it clear he was on the side of the forestry movement. Only weeks after his inauguration Cleveland, in fact, threw down the gauntlet: he "deplored" the grim fact that the western timberlands, which his predecessors Grant and Harrison had saved, were being destroyed by "timber depredators."[61] Scolding Congress, particularly the senators from Colorado and South Dakota, President Cleveland, under Hoke Smith's sway, called for immediate protective legislation.[62]

With the Panic of 1893 giving economic reasons to fight the new forest preserve system, the timber lobby in the West was pushing back against the conservationists. Led by Senator Henry M. Teller of Colorado (himself the secretary of the interior in Chester Arthur's administration, a man who now advertised himself as the "defender of the West"), this group wanted the reserves reduced in size if not abolished completely. To repeat, however: from the outset of his second term, President Cleveland had made it clear that he was on the side of the forestry movement.

In April 1894 Roosevelt was preparing to testify before the Senate Committee on Territories, chaired by Charles Faulkner of West Virginia, in favor of giving Yellowstone Park's U.S. Army soldiers real authority to deal with trespassers; he also testified against the redrawn, smaller boundaries of the park that the timber lobby was championing. An incident in the park the previous month added fuel to Roosevelt's cause.

Captain George S. Anderson—the superintendent of Yellowstone and a member of the Boone and Crockett Club—tracked a suspected poacher, Edgar Howell, for a few days in the Pelican Valley region of the park. At one abandoned campsite Captain Anderson found six buffalo scalps and skulls. On March 13, he stumbled on Howell along Astringent Creek skinning a buffalo that had just been shot. Nearby were the bodies of five other kills. The superintendent arrested the poacher red-handed and immediately wrote to Secretary of the Interior Smith from Cooke City, Montana, recommending that this arrest "be made the occasion for a *direct* appeal to Congress for the passage of an act making it an offense . . . for any one to kill, capture, or injure any wild animal in the Park."[63]

By chance, Emerson Hough was on assignment for *Forest and Stream* in Yellowstone at the time of the arrest, and the official Yellowstone photographer, Jay Haynes, using his new portable Eastman Kodak camera, was able to take pictures of the dead buffalo. This firsthand evidence was the basis for a stinging editorial in *Forest and Stream*: George Bird Grinnell urged "every reader who is interested in the Park" to write to his "Senator and Representative . . . asking them to take an active interest in the protection of the Park" before America's last great buffalo herd was gone forever. Meanwhile, with righteous indignation, Roosevelt publicly lashed out at Howell, claiming that the sleazy marauder should, at the very least, be "sent up for half a dozens years"; Roosevelt personally preferred a stiff rope and a short drop.[64] In his testimony before the Senate Committee on Territories, Roosevelt milked the Howell case for every possible drop of sympathy, resorting to both shaming and tongue-lashing.

Meanwhile, Roosevelt acquired a new Yellowstone ally on the other side of the U.S. Capitol. Representative John F. Lacey of Iowa was the principal sponsor in the House of the administration's Yellowstone Game

Protection Act; it was designed "to protect the birds and animals in Yellowstone National Park, and to punish crimes in said park, and for other purposes."[65]

Lacey was born in Virginia in 1841, and his family moved to Iowa when he was a teenager. The Laceys homesteaded along the Des Moines River, where John was immediately mesmerized by the open prairie and amazed by the endless sea of tall grass and the abundance of songbirds.[66] He enrolled at college in nearby Marshalltown, but after the Civil War broke out he enlisted in the Iowa Volunteer Infantry. The combat he saw, against Confederate forces in northern Missouri at the Battle of Liberty on September 17, 1861, turned him against war forever.[67] He returned to college, studied law, and by 1870 was elected to Iowa's House of Representatives, where he became known as an avid advocate of wildlife conservation.

By the spring of 1894, Lacey was in his second of eight terms representing Iowa in the U.S. House of Representatives. He took fact-finding trips around the West (unusual for a congressman representing the Midwest at the time), assessing timberlands that might well be considered future forest reserves and growing angry at the smoking lumber mills and the stump-dotted slopes that he passed. He always harbored a primal urge, a yearning, to be around nature. According to his daughter Berenice, he was "pained" to see the "wanton destruction" of forests and wildlife.[68]

On May 7, 1894, the president's team of Hoke Smith, John Lacey, and Theodore Roosevelt secured the passage of the Yellowstone Game Protection Act (otherwise known as the Lacey Act of 1894). At long last the federal government could take poachers like Howell to nearby courthouses in Cooke City and Livingston for legal prosecution, instead of merely having them escorted off the park premises.[69] The Act also ensured protection by the U.S. Army against timber harvesting, mineral extraction, and the defacing of geysers or rock formations for the foreseeable future. If you were caught carving your initials in rock—the way William Clark had done in 1806 at Pompey's Pillar in Montana—you could end up in jail. The U.S. Army, which had first started administering the park in 1886, would continue doing so until 1918. Entrepreneurs trying to make a quick buck out of Yellowstone were frowned on by Roosevelt. For example, Edward Waters of Fond du Lac, Wisconsin, operated a buffalo-elk zoo (even getting a permit to exhibit Crow Indians) but was eventually forced to shut down his roadside operation. As purists, Roosevelt and Grinnell even wanted to prohibit steamboat tours of Yellowstone Lake. "In protecting the beautiful wonders of the Park from vandalism," Captain Anderson noted, "the main things to be contended against were the propensities of women to gather specimens, and of men to advertise their folly by writing their names on everything beautiful within their reach."[70]

Overnight Roosevelt's beloved bears had a sanctuary, and they were on their way to becoming a great tourist attraction.[71] At the Fountain Hotel in the park, black bears started showing up regularly at the kitchen garbage dump, begging for leftovers. Their panhandling became almost as reliable as Old Faithful, and a new tourist attraction. "The preservation of the game in the Park has unexpectedly resulted in turning a great many of the bears into scavengers for the hotels within the Park limits," Roosevelt wrote with self-evident glee of the Yellowstone Game Protection Act a few years later. "Their tameness and familiarity are astonishing; they act much more like hogs than beasts of prey. Naturalists now have a chance of studying their character from an entirely new standpoint, and under entirely new conditions. It would be well worth the while of any student of nature to devote an entire season in the Park simply to study of bear life; never before has such an opportunity been afforded."[72]

The passage of the Yellowstone Game Protection Act of 1894 impelled Roosevelt to push for more U.S. government protection over the national parks. There was scant chance wildlife would last, he believed, if U.S. Army guards didn't patrol California parks like Sequoia, General Grant, and Yosemite as protectors. In Yosemite, grazing sheep—domestic animals that John Muir denigrated as "hoofed locusts"—were ruining the integrity of the park's valleys. Instead of merely chasing the flocks off the U.S. government property, Roosevelt wanted illegal shepherds arrested, handcuffed, and marched to jail. For the wondrous California parks to survive, Roosevelt believed, the U.S. Army also needed to be trained in stocking fish, fighting forest fires, and planting trees.

Although Roosevelt still grumbled that the Northern Pacific was trying to "segregate" Yellowstone, in truth the railroad industry was becoming a fierce proponent of establishing national parks throughout the West. The Southern Pacific Railroad company had even helped push the Yosemite bill through Congress. The railroads saw big tourist dollars in luring easterners to see the wonders of Yosemite as well as the Grand Canyon, Mount Rainier, and so on.[73] John Muir himself embraced the notion of tourists coming to visit Yosemite by passenger train. "Even the scenery habit in its most artificial forms," he wrote, "mixed with spectacles, silliness, and kodaks; its devotees arranged more gorgeously than scarlet tanagers, frightening the wild game with red umbrellas—even this is encouraging, and may well be regarded as a hopeful sign of our times."[74]

Whether or not Muir influenced Roosevelt's thinking remains unclear; what we do know, however, is that Roosevelt soon dialed back on pounding the railroad industry. Nevertheless, he continued to assail the avarice of timber barons, illegal game hunters, metallurgical fiends, real estate dealers, and souvenir poachers bent on disregarding the U.S. government's "No Trespassing" postings. With a gleam of white teeth, Roosevelt attacked the "baseness of spirit" of such Coloradan politicians as Senator Teller and Governor Davis Waite with a steady barrage of invective. Tired of Rooseveltian theatrics, they, in turn, wanted to clap a chloroformed bandanna over his mouth once and for all. While John Burroughs preferred fellowship over invective, he knew the "great cause" of wilderness protection needed the bluntness of his new friend's militant temperament. "Roosevelt was the man of the clenched fist," Burroughs wrote years later in *The Last Harvest*, "not one to stir up strife, but a merciless hitter in what he believed a just cause."[75]

Roosevelt continued to push his conservationist agenda forward. He soon wrote to Captain Anderson that the new legislation could be improved. President Cleveland's protection act was "by no means as good," Roosevelt maintained, as what the Boone and Crockett Club demanded. The bill, he feared, had ambiguous overtones. With unabated ardor, Roosevelt wanted opprobrium names and public humiliation hurled at dishonorable men like Howell through Wyoming's newspapers. W. Hallett Phillips of the Boone and Crockett Club wrote to Captain Anderson, in fact, saying that "Roosevelt says you made the greatest mistake of your life in not accidentally having that scoundrel [Howell] killed and he speaks as if he would have shot him on the spot."[76] Regardless, Roosevelt couldn't deny that the Yellowstone Game Protection Act was a giant leap in the right direction. To cough up $1,000 or spend two years in jail—the harsh penalty suddenly imposed on poachers for merely shooting an elk or deer on U.S. government property—was a serious deterrent in 1894. At the end of this letter to Captain Anderson, Roosevelt admitted that Cleveland's act was "a good deal better than the present systems," adding that "at least [it] gives us a groundwork on which to go."[77]

Another worry for Roosevelt after May 7 was his fear that Yellowstone buffalo—believed to be among the last remnant herds in the United States—would still fall prey to poachers if the number of U.S. Army personnel in the park wasn't dramatically increased more than the act provided. There were fewer than twenty buffalo in Yellowstone—with its high altitude and harsh winters the park wasn't a natural environment for them, but inside the park they were now protected. Roosevelt kept grappling with the larger question of how to save the buffalo in the long term. The Smithsonian Institution was floating a plan that involved fencing them in, which he was lukewarm about. Before long, Roosevelt started touting the notion of breeding buffalo in zoos and then reintroducing them throughout the western forest reserves, particularly in their traditional grounds like the Black Hills, Pine Ridge Reservation, Flint Hills, and Wichita Mountains.[78]

Most important, Roosevelt believed that westerners would have to become good wildlife protectionists themselves in order for buffalo herds, national parks, and forest reserves to remain unmolested. Regular citizens would have to turn in poachers, even if the poachers were friends or neighbors. Emerson used to quote Francis Bacon as saying that humans were the ministers and interpreters of nature; Roosevelt wanted to add a modern point: and *protectors*. Somehow, Roosevelt believed, the people of the West needed to adopt the buffalo permanently as their mascot. "Eastern people, and especially Eastern sportsmen, need to keep steadily in mind the fact that the westerners who live in the neighborhood of the forest preserves are the men who, in the last resort, will determine whether or not those preserves are to be permanent," he wrote. "They cannot . . . be kept . . . as game reservations unless the settlers roundabout believe in them and heartily support them."[79]

IV

A shock wave rippled through Roosevelt's life on August 14, 1894, when news reached him that his brother Elliott had died. The *New York Times* cited heart disease as the cause of death, but Theodore knew it was heartbreak. Elliott's wife, Anna, and son, Elliott Jr., had died of diphtheria in the preceding years. Unable to ward off the demons of the double loss, Elliott had been drinking heavily all summer long. Theodore felt downcast, but found a bit of closure at the funeral. "When dead the poor fellow looked very peaceful," he wrote to his sister Bamie, "and so like his old, generous, gallant self of fifteen years ago. The horror, and the terrible mixture of sadness and grotesque, grim evil continued to the very end; and the dreadful flashes of his old sweetness, which made it all even more hopeless."[80] Elliott left behind his ten-year-old daughter, Eleanor, who was then raised by her grandmother Mary Livingston Ludlow Hall.

No sooner was his brother buried than Theodore headed to North Dakota for a couple of recuperative weeks. Once again he left his wife and children behind. With Bill Merrifield accompanying him, Roosevelt went antelope hunting on the plains around Medora in honor of Elliott. Game was scarce that year, chased away by Plains Indians and Dakota sheepherders. Nevertheless, he managed to shoot, skin, and eat five antelope. Drinking in a bit of Badlands solitude helped bring perspective back into his life. Full of deep thoughts, he spent his nights sleeping in his buffalo-lined bag in the open grassy plains with a tarpaulin to pull over him if the wind squalled. As if following the stations of the cross, he visited the great landmarks of the North Dakota Badlands—Sentinel Butte, Square Butte, and Middle Butte. "Great flocks of sandhill cranes passed overhead from time to time," Roosevelt wrote, "the air resounding with their strange, musical, guttural clangor."[81] The deep-rooted sagebrush of the plains colored long stretches of the parched summer landscape. "The cattle aren't doing particularly well," Roosevelt told Lodge in a letter. "The drought has been very severe on everything. However, except for feeling a little blue, I passed a delightful fortnight all the time in the open; and feel as rugged as a bull-moose."[82]

The year 1894 also saw the publication of Volume 3 of *The Winning of the West*, covering the post–Revolutionary War expansion into the Ohio River Valley, the Tennessee Valley, and the Blue Ridge Mountains of Kentucky. The book is marred by stereotypes of Indians as being "cunning and stealthy," essentially "the tigers of the human race." Roosevelt claimed it was only natural that white Americans would persevere. "The rude, fierce settler who drives the savage from the land lays all civilized mankind under a debt to him," he wrote. "American and Indian, Boer and Zulu, Cossack and Tartan, New Zealander and Maori—in each case the victor, horrible though many of his deeds are, has laid deep the foundations for the future greatness of a mighty people."[83]

Such passages, redolent of the white man's burden, occur throughout Volume 3 of *The Winning of the West*—Roosevelt even promotes his theory of "dominant world races" over "aboriginals." In his landmark study *Theodore Roosevelt and Six Friends of the Indian*, the historian William T. Hagan was generally perplexed at the difference in attitude toward Native Americans in Volume 3 of *The Winning of the West* compared with Roosevelt's humane reports of 1894–1895 as civil service commissioner. Influenced by Indian rights activists among his friends—especially Charles L. Lummis, Herbert Walsh, George Bird Grinnell, Hamlin Garland, C. Hart Merriam, and Francis Leupp—Commissioner Roosevelt (unlike the historian Roosevelt) said that the defeated Indians deserved a "square deal" from the U.S. government. These stark differences within Roosevelt, reminiscent of Dr. Jekyll and Mr. Hyde, would have caused Robert Louis Stevenson himself to scratch his head in puzzlement. Roosevelt number one (of

The Winning of the West) seemed to loathe Indians. Roosevelt number two (of the civil service reports) fought for their human rights.

The rational explanation lies in Roosevelt's belief that it was only proper to treat a defeated people with dignity. A true nineteenth-century gentleman, he put his faith in the hope that education, assimilation, and the example of white Americans would improve Native Americans' lot in the near future. In fact, he envisioned the day when high-quality men like Luther Standing Bear and Few Trails would dine in the White House. Regardless of which side was the *real* Roosevelt, his ideas intersected in a singular way: he *consistently* saw the Indians' future in North America in stark Darwinian terms. Once U.S. federal government graft, skimming, and unconstitutional injustice were removed from the reservations, Roosevelt argued, it would be up to individual Indian tribes to survive. He had high hopes for the Cherokee and Pawnee, less so for the Sioux. "We must turn them loose," Roosevelt wrote in one report, "hardening our hearts to the fact that many will sink, exactly as many will swim."[84]

CHAPTER ELEVEN

THE BRONX ZOO FOUNDER

I

In the autumn of 1894, Roosevelt began collaborating with Madison Grant, a lawyer and explorer, on the creation of the New York Zoological Society. With his waxed handlebar mustache, Yale pride, penchant for bow-ties, and habit of always talking with his hands clasped as if in prayer, Grant was a preeminent figure in the world of zoology, credited with discovering several North American subspecies of mammals (a species of Alaska caribou was named *Rangifer granti* in his honor).[1] That very year he had written for *Century* an article entitled "The Vanishing Moose," which Roosevelt loved.[2] Now both wildlife protectionists would turn their attention to the vanishing buffalo as part of their ambitious scheme for a new American zoo.[3]

This zoo would be situated in the Bronx, then a rural section of New York City. Roosevelt and Grant had been disappointed by the European zoos they visited. Little educational information was disseminated to visitors about species variation or habitat, and most zoological parks emphasized the freakishness and oddity of their collections. Such come-ons as a six-legged deer in Berlin and a two-headed turtle in London sickened Roosevelt. Worse yet, the animals in European zoos paced back and forth in tiny cages, like prisoners waiting for the end of a lifetime sentence. This kind of backward zookeeping had to end. As Roosevelt envisioned it, their modern New York zoo would be built "on lines entirely divergent from the Old World zoological gardens."[4] The animals would have more room, in open-air exhibits where possible, and broadsheets would be created specifically for schoolchildren explaining the principles behind wildlife preservation and Darwinian evolution. And while the Bronx Zoo wasn't as showy as a production by P. T. Barnum or Buffalo Bill, the Chicago Exposition had taught Roosevelt to think outside the box when it came to devising a tourist attraction that would bring throngs to see wildlife up close. A subway stop, in fact, was slated to open at the southeastern entrance to the zoological park.

In planning this zoo, Roosevelt and Grant included a singularly ambitious goal: they would breed buffalo in captivity there and eventually would turn them loose throughout the Great Plains and upper Rocky Mountain region. This so-called New York repopulation plan would reintroduce buffalo in their traditional grounds, such as the Black Hills, Pine Ridge Reservation, Flint Hills, Osage Hills, and Wichita Mountains; even the remaining herd in Yellowstone Park would be augmented with Bronx-bred buffalo. As Roosevelt saw it, buffalo would once again be trampling the luxuriant tall grasses into muddy thoroughfares, as in the days before Christopher Columbus. Unlike the 45 million cattle in the Great Plains, reintroduced buffalo wouldn't overgraze the prairie into a dust bowl. He hoped to create a buffalo common. While there were few dramas as frightening as a bison bull at bay, zoologically schooled members of the Boone and Crockett Club like Roosevelt and Grant nevertheless knew that buffaloes were essentially timid creatures, as easy to corral as cattle. Only when a mother buffalo felt that her young were in jeopardy did they turn frothingly hostile, staving off predators such as wolves and cougars with the threat of a horrific stampede. Then look out! Buffalo might appear slow and lumbering, but they could outrun a racehorse. As domesticated creatures, however, they were fairly benign.[5]

While George Bird Grinnell was supportive of a buffalo common in the West, he thought bison needed lots of roving space to survive and that the New York grasses were completely different from those on the plains. Therefore, he wasn't keen on the Boone and Crockett Club's throwing its weight behind acquiring wildlife for display in New York City; he preferred having the members concentrate on enacting tougher hunting laws. Running a zoo was a headache he simply didn't want. To Grinnell it made more practical sense to have the Department of Agriculture help

The Great American Buffalo *was drawn by Audubon.*

C. J. Jones of Garden City, Kansas—who had captured fifty buffalo on his own and purchased an additional eighty in Manitoba—lead a serious repopulation program right in the heartland of "Buffalo Country."[6]

But, as was often the case, Roosevelt got his way. He insisted that the zoo would teach New Yorkers about the perils western big-game species faced. Also, the zoo allowed Roosevelt, as a New York politician, to found something great for the Empire State, an added political bonus. One of the zoo's most tireless advocates, in fact, was Andrew H. Green, then known as the "father" of greater New York City. When Green concurred that a natural-setting zoo was a fine idea, long overdue, Roosevelt knew his brainchild would take off. A truly creative philanthropist, Green had been a close friend of Roosevelt's father, envisioned Central Park as a recreational center, bankrolled the American Museum of Natural History, investigated the Tweed Ring, and created the Niagara Falls Commission to save the falls from destruction in a bilateral agreement with the Canadian government. Grinnell warned his friends against forming a zoo committee in late 1894 before club members could, at the very least, vote on the idea. They should settle their differences, Grinnell believed, through at least a tip of the hat toward the democratic process.

But Roosevelt, spurred by his idea about buffalo, had another important ally besides Green and Grant. Professor Fairfield Osborn of Columbia University, curator of the American Museum of Natural History, sided with Roosevelt (as he always did), even offering his fund-raising contact list. Starting in the 1890s Osborn had become quite a fixture in zoological circles. The refined, fashionable Osborn would tuck a paisley scarf into his collar instead of a tie, sported corduroy pants, and was constantly jotting notes on legal pads while chain-smoking cigarettes. Green was the first president of the New York Zoological Society, but when he became suddenly ill Osborn took over the obligations. Deep down, Roosevelt probably knew that Grinnell was correct in his skepticism. Roosevelt nevertheless wrote to Madison Grant—whose hand-tailored suits and donnish manners made him a kind of WASP caricature, like a cucumber sandwich at afternoon tea—that regardless of *Forest and Stream*'s viewpoint, "I'll go ahead and do it."[7] Being the impresario of a zoo was simply too irresistible to turn down.

In the spring of 1895, a group from the Boone and Crockett Club officially formed the New York Zoological Society. The leaders included Roosevelt, Grant, Green, and Osborn. Behaving like fraternity brothers, they created a crest for the society, with a ram's head in its center and "Founded 1895" underneath, long before the development plan was officially approved in late 1897.[8] The 261-acre forested zoo, with a topographical range from granite ridges to natural meadows and glades to forest, was ideal for a wide variety of animals to thrive in the open air. The Bronx River wound through the site, and the woods were already home to many birds. It would be relatively easy to create marvelous replicas of diverse habitats here, allowing the wildlife to feel at home.

The zoo officially opened in November 1898, after the blitzkrieg of the Spanish-American War. Grinnell essentially fell into line, now and again flashing an over-the-shoulder look that said "I told you so" when any animal emergency transpired; Darwinism, he believed, was veterinarianism. Roosevelt was especially proud of the imposing Antelope House. But the Lion House soon became the premier tourist attraction. These "houses" were established to tell the biological history of species in a truly detailed, exciting, educational way. A special wooded range had been developed for moose, and a wonderful stream surrounded by plants was devised so children could watch industrious beavers construct dams and lodges. At the time, the new zoological park was the largest in the world, and its open-air displays became the model for its successors, such as the zoo in San Diego, which opened in 1922.

As it entertained tourists and educated schoolchildren, the zoo simultaneously served as a scientific laboratory. For example, the twenty acres reserved for buffalo allowed ample space for mammalogists to study these North American quadrupeds through the rutting cycle.[9] Likewise, elk got fifteen acres in which to thrive and be observed. Moose were brought in from Maine and cougar from somewhere west of Kansas City. Aesthetically, what Roosevelt and Grant were trying to avoid was the zoo seeming like an oversized breeding pen. Bear dens were soon erected with awesome caves and rock precipices, and there were plans to build the world's greatest House of Reptiles, which the foremost herpetologist in America, Raymond L. Ditmars, would curate. "It is extremely desirable that all animals living in the open air should be so installed that their surroundings will suggest," the *New York Times* explained about the new zoo, "as a well-kept and accessible natural wilderness rather than as a conventional city park."[10]

Besides breeding buffalo in captivity, the New York Zoological Society funded an extensive scientific report on the concept of creating wildlife sanctuaries or refuges across the country. Could buffalo reclaim the Black Hills, Red River Valley, or Drift Prairie of South Dakota? Would elk be able to roam freely again along the Yellowstone and Missouri rivers? Someday, would moose be protected in Maine and Minnesota? As Lincoln Lang recalled in *Ranching with Roosevelt*, T.R. "had discussed the possibilities of game protection," in the 1880s and even spoke of someday establishing the North Dakota Badlands as a "national preserve."[11] With the zoological society launched, Roosevelt's vision was no longer an impossible goal. A subtle brotherhood of men—hunters and naturalists—were now in the business of species education.

Already at Black Mesa Forest Reserve in the Arizona Territory two members of the Boone and Crockett Club—Dr. Ed W. Nelson and Alden Sampon—were studying the feasibility of a wildlife reserve in the Navajo lands, a reserve that would be completely off-limits to hunters along the upper border of the Little Colorado River basin.[12] The deep box canyons, yellow pine forests, piñons, cedars, and junipers of the Black Mesa needed protection, as did the three intact cliff dwellings of the Pueblos. Although Roosevelt hadn't been to Black Mesa, he knew from Nelson that creating a big-game reserve there would be ideal for black-tailed deer and silver-tipped bears. And it would mean a great deal to the Native American peoples. But Roosevelt also understood that without irrigation the Arizona rock oasis would become uninhabitable; the scattered pines would wither. The residents of the adjoining sheep and cattle settlements were opposed to a Black Mesa wildlife playground.[13]

Roosevelt—unusually for him—made snide, belittling remarks about the Arizona business types who were unable to comprehend the concept of antiquities. Arizona needed to be maintained, not mined. Stopping growth in Black Mesa, Roosevelt worried, was going to be difficult, but he believed the "fantastic barrenness," "incredible wildness," and "desolate majesty" of the Navajo lands needed to be protected forever. With quiet cheerfulness he began looking for ways to get the job done. "No one could paint or describe it," he later wrote after camping out in the Black Mesa valley, "save one of the great masters of imaginative art or literature—a Turner or Browning or Poe."[14]

II

The recruitment of William Temple Hornaday as the first director and general curator of the New York Zoological Society was a brilliant coup by Roosevelt. (Like George Bird Grinnell, Hornaday regularly used all three of his names.) Born in Plainfield, Indiana, in 1854, four years before Roosevelt, and raised in Iowa starting in 1856, Hornaday grew up earthy and dirt-poor.[15] "I shall always believe," he wrote in his memoir *Two Years in the Jungle*, "I was born under a lucky star as a compensation for not having been born rich."[16] He managed to attend Oskaloosa College and then Iowa State College, where his intuitive genius for handling both domestic and wild animals, added to his excellence in taxidermy, opened the doors to a zoological career upon graduation. While Roosevelt was preparing for Harvard with a private tutor, the irascible Hornaday was traveling the world as a young man searching for exotic species to shoot and stuff in the name of science. His primary employer, the Wards National Science Foundation (of Rochester, New York), sent him to the Bahamas, Cuba, Florida, Brazil, Ceylon, Malaysia, and Borneo.

Obsessive, unbuckling, and stubborn beyond words, Hornaday was a highly sophisticated version of Bill Sewall or Moses Sawyer. Unlike Baird or Merriam, Hornaday had calloused hands. There was always a mischievous twinkle in his eyes, like a child who had suddenly aged overnight. He was the kind of immature prankster who yodeled in church just to hear the echo. There was a cultivated crudeness to his manners, and his certainty about zoology bordered on arrogance. His daily conversation was filled with such bio-trivia as the flesh preferences of wolverines and why hawks were copper-clawed. Hornaday also bristled with statistics on the possible extinction of Delaware swans, Louisiana woodpeckers, and Ohio turtledoves. Vermont was the only state, he believed, which managed its wildlife properly. For all his eccentricities, you had to give Hornaday credit: he walked the walk. Unlike most Harvard-trained scientists, Hornaday had actually wallowed in the mud with alligators, tying their mouths with rope like Jim Bowie working the tip jar in a French Quarter sideshow. And when it came to buffalo, nobody—not even Grinnell—understood their psychology as keenly as Hornaday. The West was full of horse whisperers, but Hornaday—with the notable exception of C. J. Jones—was the only buffalo whisperer around. (Unlike Jones, Hornaday at least didn't try to crossbreed wild buffalo with Hereford cattle.)

By 1879 Hornaday was chief taxidermist and director of the prestigious United States National Museum. When Hornaday created the National Society of American Taxidermists the following year, Roosevelt sat up and paid attention. Stitch for stitch, Hornaday was probably the best American mammal taxidermist of his era. When working on a buffalo, for example, he took the extra step of ridding his specimen of screwworm flies (*Cochliomyia macellaria*), which frequently laid eggs in an open wound or sore. He was an artist skilled enough to prepare an exhibit of a buffalo grazing, stampeding, or simply looking forlorn, bringing out the personality of the animal vividly in any mood or situation. Most famously, Hornaday himself shot a huge 1,800-pound buffalo in Montana; it became the centerpiece of a popular diorama at the National Museum of Natural

History. Upon skinning this buffalo, however, Hornaday made a startling discovery, later claiming that it triggered an effusion of sorrow. "Nearly every adult bull we took carried old bullets in his body, and from this one we took four of various sizes that had been fired into him on various occasions," he recalled. "One was sticking fast in one of the lumbar-vertebrae."[17]

When Roosevelt moved to Washington, D.C., in 1889 for his civil service job, he befriended the bawdy Hornaday, then chief taxidermist at the Smithsonian. Despite his skill at mounting, Hornaday was pushing for the Smithsonian to open a "live animal" department. Hornaday insisted that people preferred to see a real wobbly little buffalo rather than a stiff, old, stuffed one. The global killing of wildlife for science was the hackneyed way, Hornaday came to believe, for truly enlightened men and women to study animals. Newly developed netting techniques made it possible to capture everything from a hippopotamus to a cougar alive. He even wanted to start tagging animals in the wild. Once Hornaday was given permission to show live specimens at the Smithsonian turnstile increased three-fold. The public roared its approval and Hornaday prattled on about the advanced wildlife protective ethos.

Hornaday's new vision led to his founding of the National Zoological Park in 1889, in Washington, D.C.[18] But such brilliant, original thinking (like that of Robert B. Roosevelt) can often go hand in hand with a difficult personality. Since Hornaday loved being out in the field, chief among the targets of his lacerating criticism were those zoologists and ornithologists who didn't get grimy tracking down wildlife for science. You might say he had an outbank Audubon complex, like Theodore Roosevelt. Often smelling of buffalo or bears, Hornaday was far more comfortable in alpine hiking clothes than in a suit. Sometimes his hair was matted with dry grass and mud. There was, as noted above, a farmyard crudeness to his manners, and his rumored atheism did nothing to endear him to the pious Methodists and Episcopalians he worked with at the Smithsonian. In fact, when the people who were financing the National Zoological Garden started telling Hornaday how to run his shop, and radically changing his Darwinian plans for animal displays, he resigned and moved to Buffalo, New York. Once he was settled there, he tried to change the city's name to Bison, New York, to be more zoologically accurate.

Although Roosevelt never approved of Hornaday's vulgarity or imperiousness, he knew that Hornaday was the most knowledgeable expert in the world regarding buffalo. It was said that Hornaday, in a quick glance, could identify the precise home range of a buffalo—for example, Nebraska or Manitoba or Oklahoma—by the constitution of its dung. Furious that these wild creatures were treated so shabbily, he nevertheless remained hopeful that repopulation programs and new game laws might be able to reverse the trend toward extinction. When Hornaday told Congress in April 1896 that national bison ranges should be created to save the vanishing herds, Roosevelt fully agreed. Even though Hornaday wasn't considered refined enough for the Boone and Crockett Club, Roosevelt had no hesitation in asking him to head the Bronx Zoo. To Roosevelt's thinking Hornaday was wasting his talent working in the Erie County real estate business and merely serving as a trustee for the Buffalo Museum of Science.

Roosevelt promised to let Hornaday develop exhibits at the Bronx Zoo any way he wanted. He had faith that Hornaday was the man best able to breed buffalo in captivity as the first step in repopulating the Great Plains. Always speeding up the timetables, Roosevelt was impatient and held the view that if it took six years to destroy all the bison in Kansas, Nebraska, Indian Territory, Oklahoma Territory, and the Texas panhandle, then the Bronx Zoo should be able to repopulate those same expansive grasslands with buffalo in an equivalent time. Only a maladjusted obsessive like Hornaday, with fever in the veins, could possibly pull off a breeding program in such unrealistically short order. Promised a handsome salary, Hornaday agreed, with a little pressure from the Boone and Crockett Club, to accept the zoo directorship—a position he held for the next thirty years. During these decades Hornaday wrote a series of books, articles, and lectures that are the core documents of the wildlife conservation movement in America. His *Our Vanishing Wild Life* is perhaps the single most important (if deeply flawed) book ever published on protecting endangered species. "Give the game the benefit of every doubt!," he would lecture. "If it becomes too thick, your gun can quickly thin it out; but if it is once exterminated, it will be impossible to bring it back. Be wise; and take thought for the morrow. Remember the heath hen."[19]

As Hornaday saw it, the Bronx Zoo would celebrate the whole post–Civil War generation of Darwinian naturalists, including the affluent Roosevelt. The center of Hornaday's zoo was Baird Court—the administrative and library offices—named after the head of the Smithsonian Institution. Then there was Lake Agassiz, named after the great Harvard zoologist. The Boone and Crockett Club contributed hundreds of horns and antlers to put on display. Meanwhile, plans to raise buffalo got off to a rocky start. The native Bronx grass on which the bison grazed was not the same as prairie grass; in fact, it was so different that the first domestically raised herd of American bison died. Grinnell, it seemed, had been right. Bitterly disappointed, Hornaday had to remove all the native grasses and then pay zookeepers to feed the buffalo the proper prairie grasses by hand and keep the water hole always full.[20]

That was quite an embarrassing failure for Roosevelt; he felt a real, if momentary, sense of loss when the buffalo died. But Roosevelt wasn't a quitter. He stuck by his determined zoo director. Since buffalo weren't going to be the zoo's only mission, Hornaday, at Roosevelt's request, hired the field researcher Andrew J. Stone to survey the status of Alaska's caribou herds, polar bears, and seal rookeries.[21] For even if the Bronx buffalo range went away, Roosevelt and Hornaday's plan of breeding buffalo in the Indian Territories (and two or three prairie states) was on track. And Grant, lobbying for financing from the New York state legislature, got the Boone and Crockett Club all the appropriation provisions it had requested. Roosevelt essentially left all the fund-raising and architectural details of the Bronx Zoo up to Madison Grant, whose actorish ways could move mountains. When the Bronx Zoo was officially sanctioned by the Park Board around Thanksgiving of 1897,[22] Roosevelt offered Grant thanks. "I congratulate you with all my heart upon your success with the Zoo bill," he wrote. "Really, you have done more than I hoped. I always count myself lucky if I get one out of three or four measures through."[23]

III

The creation of the Bronx Zoo, with the help of regiments of planners, was the capstone to Roosevelt's tenure as president of the Boone and Crockett Club. Over the years 1888 to 1894, he had achieved a scorecard of extraordinary accomplishments. From the Timberland Act of 1891 to the Yellowstone Protection Act of 1894 to the creation of the New York Zoological Society, the Boone and Crockett Club had become the most effective big-game conservation group in America. Even though Roosevelt was sometimes tasked with club work that he despised—such as preparing accounts for audit[24]—he relished promoting outdoor writers in his edited books. Encouraged by the fine reviews of *American Big-Game Hunting*, Roosevelt and Grinnell began working throughout 1894 on a successor volume, one having a more global perspective than the first. How did North American white-tail deer compare with the fallow deer of the Mediterranean region of Europe and Asia Minor? Why did the Nubian ibex of the Palestine countryside have larger scimitar-shaped horns than the mountain goats Roosevelt had hunted in the Rockies? What caused the nyala of southeastern Africa to be slightly faster than the Badlands antelope? These were the types of questions Roosevelt and his coeditor George Bird Grinnell wanted answered in the new book, called *Hunting in Many Lands*.

When commissioning Madison Grant to write on moose in *Hunting in Many Lands*, for example, Roosevelt preferred that his hunt story take place in Canada. Roosevelt wanted to know everything about moose found farther north than Maine and Minnesota, around the headwaters of the Ottawa River along the Ontario-Quebec border. This, Roosevelt wrote to Grant, would run against the current zoological thinking that the differences between species in tropical and temperate zones were the most scientifically important. By focusing on moose in a specific Canadian habitat, Grant could show (so his editor hoped) the variation among moose populations in North America. "The best zoologists nowadays put North America in with North Asia and Europe as one archetypal province, separate from the South American, Indian, Australasian, and South African provinces, which have equal rank," Roosevelt complained to Grant. "Our moose, wapiti, bear, beaver, wolf, etc., differ more or less from those of the Old World but the difference sinks into insignificance when compared with differences between all these forms, Old World and New, from the tropical forms south of them. The wapiti is undoubtedly entirely distinct from the European red deer; but I don't think the difference is as great as between the black-tail [mule] and white-tail deer."[25]

In 1895 *Hunting in Many Lands* was published, with a sterling article by Madison Grant, "A Canadian Moose Hunt," about the upper Ottawa River, essentially a companion piece to his article in *Century*. Grant also eventually became instrumental in two other conservationist causes besides the Boone and Crockett Club and its offshoot, the Bronx Zoo: these were the American Bison Society (founded by Roosevelt and Hornaday in 1905) and Save the Redwoods League.* Unlike *American Big-Game Hunting*, this second volume, comprising sixteen essays (plus appendices, which included the Yellowstone Protection Act), was, as Roosevelt and Grinnell

*Unfortunately, Grant also fancied himself as a eugenicist, and he believed some deeply racist theories of Nordic superiority. His 1916 book *The Passing of the Great Race* caused widespread fear of a new wave of Italian and Slavic immigration coming to the United States after World War I. In *The Great Gatsby* F. Scott Fitzgerald mentioned Grant by name as a race theorist. Others have claimed that Grant's eugenicist works were studied in Germany during the Third Reich.

planned, international in approach. Fine photographs were interspersed throughout the text. The final product included essays on hunting Russian wolves, Sierra Mountain bears, Mexican rams, East African zebras, Korean leopards, and American antelope-deer (by Roosevelt).

And another contributor, the future U.S. secretary of state Henry L. Stimson, wrote of the wildlife he encountered when he was climbing the turret-shaped mountains of northwestern Montana. W. W. Rockhill, a friend of the Dalai Lama, focused on the big game found in Mongolia and Tibet. There was even an essay on dogsledding in Manitoba by D. M. Barringer, a nineteen-year-old graduate of Princeton University who dedicated his life to the "impact theory" (he became the first geologist to discover that Coon Butte, Arizona, was in fact a meteor crater). In the impressive essay "The Cougar," by Casper W. Whitney (the editor of *Outlook*), Roosevelt himself was praised as being *the* world's expert on the mountain lion's "moods."[26] The global approach to wildlife conservation of *Hunting in Many Lands* was smart and innovative. In the preface Roosevelt and Grinnell called for accelerated mammological research. Color variation, hoof sizes, whisker lengths, mating habits—the more scientific data compiled, the easier it would become for "wild creatures" to be taught "to look upon human beings as friends."[27]

In a piece that Roosevelt and Grinnell wrote together, the Yellowstone Protection Act of 1894 was held up as a model for policing wild refuges around the world. All the contributors to *Hunting in Many Lands* (except Elliott Roosevelt, whose posthumous contribution, "A Hunting Trip in India," was embarrassingly antediluvian in attitude), promoted spirited ideas for global forestry and wildlife protection. Charles E. Whitehead of New York, for example, wrote that the "true hunter" took "more pleasure in watching the natural life around him than in killing the game that he meets." His essay "Game Laws" recommended that the U.S. Army be authorized to try and punish poachers under martial law. "When we reflect how many and valuable races of animals in North America have become extinct or nearly so, as the buffalo and the manatee," Whitehead fumed, "how many varieties of birds that afforded us food, or brightened the autumn sky with their migrations, have been annihilated, as have been the prairie fowl in the Eastern States and the passenger pigeon in all our States, the necessity of these laws appears urgent."[28]

Most of the essays in *Hunting in Many Lands* made at least a passing reference to proper land and wildlife management. An argument was also made that all the other existing national parks besides Yellowstone—Yosemite, Sequoia, and General Grant—should likewise have game protection laws.* As for the coeditors themselves, Roosevelt argued that the reports of various hunters would help naturalists and zoologists better understand species. Grinnell endorsed Roosevelt's participation in the Committee on Measurements exhibition, held at Madison Square Garden in New York City in May 1895, which rated the symmetry and coloration detail of big-game trophies.[29] In the registration book Roosevelt, wanting to be associated with the American West, untruthfully gave his residence as "Medora N.D." Along with other furrists and taxidermists, Roosevelt had submitted to the show numerous specimens, none of which drew more attention than his Texas tusked peccary head.

IV

That month also saw a career development: Roosevelt left the U.S. civil service to become a member of the Board of Police Commissioners of New York City.† He had lived as a bureaucrat in the District of Columbia and was ready to return home. At the time, New York's police department was perceived as untrustworthy and corrupt; the new commissioner would have to make public integrity his first priority. From day one, Roosevelt had a two-pronged approach to running the force: do away with bribery and force officers to abide by the law. Making the city's police department more honest, however, proved difficult. The Bowery, in particular, a mile-long row of brothels, bars, and burlesque clubs, was the most notorious tenderloin district in America; the police patrolling the beat were mostly on the take. Always "prudish as a dowager," as one biographer put it, Roosevelt thought prostitution caused moral debasement and was a menace to health.[30] Roosevelt began making surprise visits to Bowery saloons, firing officers if they were found partaking in the draft beer and revelry.[31] Nothing at the police department pre-Roosevelt, it seemed, had been done on the up-and-up. Roosevelt found that even instituting a standard for the promotion of police officers was fraught with controversy. "The public may rest assured that so far as I am

---

*Mackinac National Park—demoted to Mackinac State Park in the summer of 1895—was an exception. So was Hot Springs National Park in Arkansas, slightly more than 1,000 preserved acres, saved only because Bathhouse Row (man-made facilities around therapeutic thermal springs) was what the U.S. government was intent on preserving. That Arkansas location probably should have been declared a national historic site, *not* a national park.

†Mayor William Strong, a Republican, actually appointed a four-man board of police commissioners. T.R. was selected as board president by his three peers.

concerned," Roosevelt stated on accepting the appointment, "there will be no politics in the department and I know that I voice the sentiment of my colleagues in that respect. We are all activated by the desire to so regulate this department that it will earn the respect and confidence of the community."[32]

Commissioner Roosevelt wasn't just handcuff-happy in the Bowery or on the Lower East Side. Bitten by the temperance bug, he tried to enforce the moribund blue law against allowing saloons to be open on Sunday and went after nefarious dealers of exotic pets like a tyrant. Following the old Henry Bergh–ASPCA line, Roosevelt believed in ethical treatment of captured wildlife for both humane and sanitary reasons. Although many Americans liked to have some of the world's 350 parrot species as pets, a majority of these parrots died in transit from South America or Africa. Roosevelt wanted to have such importation regulated to ensure the birds' safe passage. But exotic birds hardly consumed much of his time. To instill discipline on the force, Roosevelt took to taking "midnight prowls" to investigate the cops' beats. Determined to run a squeaky-clean department, he insisted that public integrity was an essential component of proper law enforcement, that a single bad apple poisoned the nobility of the entire force. "Two years and eight months left to me on this Board," he boasted after just a few weeks in office, referring to his appointed term, "and that is time enough to make matters very unpleasant for policemen who shirk their duty."[33]

But even as Roosevelt succeeded at modernizing the police headquarters on Mulberry Street, introducing bicycle squads and implementing pistol shooting practice, he continued his work for the Boone and Crockett Club. In late 1895, for example, the National Academy of Sciences asked him for his opinions on the condition of America's national forests. Gleaning information from the U.S. Geological Survey and the Department of Agriculture's Division of Economic Ornithology and Mammalogy (Biological Survey), Roosevelt didn't like what he found. In his report, he worried that the 13 million acres of national forests set aside by the Forest Reserve Act of 1891 were being plundered for timber, and that the sheep pasturing in forest reservations would destroy all the herbage. In addition, Roosevelt once again called for the U.S. Army to do more policing of western reserves; urged the hiring of dozens of wardens; and, last, recommended to Secretary of the Interior Hoke Smith that still more forest acreage be set aside by the Cleveland administration.

As New York's police commissioner, Roosevelt appeared in the newspapers daily throughout the spring of 1896. In particular, the press reported his searching the streets and saloons for "slacking cops."[34] Not that Roosevelt, in his zeal to end corruption on the force, ignored the need to fight crime. For instance, when Owen Wister spent some time with Roosevelt on Mulberry Street, he was impressed by how committed his old friend was to stopping gambling and prostitution in the city. After one lunch Wister dutifully noted that Roosevelt's jaw was "acquiring a grimness which his experience of life made inevitable; and beneath the laughter and the courage of the blue eyes, a wistfulness had begun to lurk which I had never seen in college; but the warmth, the eagerness, the boisterous boyish recounting of some anecdote, the explosive expression of some opinion about a person, or a thing, or a state of things—these were unchanged, and even to the end still bubbled up unchanged."[35]

Truth be told, Roosevelt wasn't happy as police commissioner, finding the work "grimy" and "inconceivably arduous, disheartening, and irritating."[36] Firing underlings for gross incompetence was demoralizing, and the job left him hardly any time for the outdoors life. Roosevelt found consolation in the fact that at least he wasn't forgotten as a historian. The final volume of *The Winning of the West* was published in June 1896, to a fourth round of positive reviews. However, this time Roosevelt received no psychic uplift from the publication, and perhaps he was relieved that the long work, based on antiquated ideas he had first paraded before the public seven years ago, was at last over. Instead of reviewing Volume 4, the *New York Times* presented a feature article about how Roosevelt couldn't wait to tramp around the West (where he hadn't been in two years) with a new small-caliber Winchester. Roosevelt commuted back and forth on the Long Island Railroad from Sagamore Hill to his Mulberry Street office, preferring the pastoral Oyster Bay to the hurly-burly of the carriage-choked city. Roosevelt would have preferred to make his way to Yosemite or Alaska to write a book, just as John Burroughs had done—his friend had just published *Birds and Bees and Other Studies of Nature*. Roosevelt would title his book, the article intimated, something like *Bears and Deer of the American West and Beyond*.[37]

That August Roosevelt managed to spend a couple of weeks in the Dakotas and Montana. He was preparing to close the Elkhorn ranch while in the West he campaigned tirelessly on behalf of Republican William McKinley of Ohio. Once Roosevelt returned to New York, in fact, he accused the Democratic candidate, William Jennings Bryan, of being a wild-eyed anarchist willing to usher in dissolution and disunion. Roosevelt never before had so much fun belittling an opponent. He told everybody in the Boone and Crockett Club that Bryan, a Nebraskan who served two terms in Congress, besides being an agrarian radical (and admittedly a first-class orator) was against forestry science, wildlife protection, and national parks. Roosevelt warned that as Election Day neared Bryan would become downright demagogic, turning the worst class of voters into a rabble armed with pitchforks, demanding that the dollar be leveraged on the silver standard instead of gold. If Bryan was elected, Roosevelt worried, the Forest Reserve Act of 1891 (and any other wise federal government initiatives) would be overturned, for his supporters had the kind of peasant mentality that would end up even denuding Pikes Peak and Mount Olympus in what Roosevelt saw as a "Witches Sabbath."[38]

In any event, Roosevelt needn't have worried. On Election Day, McKinley bested Bryan and the existing U.S. government's timberlands—at least on paper—were safe.[39] That holiday season Roosevelt was in high spirits, lunching with Burroughs and plotting with Grinnell.

A great moment in U.S. conservation history occurred a few weeks before William McKinley's inauguration on March 4, 1897. On Washington's birthday, February 22, 1897, ten days before the end of his term, the outgoing president, Grover Cleveland, created thirteen new or expanded forest reserves totaling 21 million acres; much of this land was in the verdant Pacific Northwest. Naturally howls of protest came roaring into Washington, D.C., from lumber companies, grazers, and mine owners.[40] "The rage of the lumber and railroad men," the reporter George B. Leighton later noted in *Five Cities*, "knew no bounds."[41]

Timber barons, in particular, felt that they had been blindsided by the fat, hatless departing president. From their entrepreneurial perspective Cleveland had just served them strychnine in their coffee. Western businessmen couldn't believe the sheer treachery of Cleveland's parting shot. They were on the verge of mutiny. Suddenly *all* the milling of lumber and hauling of river stone had to cease at Cleveland's designated forest areas unless the U.S. government said otherwise. Always predisposed to underrate Cleveland, Roosevelt was surprised and grateful that the outgoing president had unsheathed a sword. "This action," he and Grinnell later bragged, "was directly in the line of recommendations urged in the Boone and Crockett Club books."[42]

Of course, Roosevelt wished that millions more acres had been put aside, particularly in the Arizona and New Mexico territories, where the Painted Desert, Black Mesa Valley, Canyon de Chelly, and Grand Canyon lay vulnerable. Although he had seen the Grand Canyon only in photographs from the rim, he knew—ever since reading John Wesley Powell's *The Exploration of the Colorado River and Its Canyons* as a teenager—that

A rare photograph of Theodore Roosevelt with Grover Cleveland (on left).

it needed to become a national treasure. Nevertheless, he heartily approved of the forests and natural wonders the Cleveland administration had the fortitude to save: the San Jacinto and Stanislaus (California); Uinta (Utah); Washington, Mount Rainier,* and Olympic (Washington); Bitterroot, Lewis and Clark, and Flathead (Montana); Black Hills (South Dakota); Priest River (Idaho and Washington); and Teton and Big Horn (Wyoming).[43] "It was a serious matter taking this great mass of forest reservations away from the settlers," wrote Roosevelt. "That it needed to be done admits of no question, but the great bulk of the people themselves strongly objected to its being done; and a great deal of nerve and a good deal of tact were needed in accomplishing it."[44]

President Cleveland met with immediate blowback from many western senators. Words like traitor, fink, thimblerigger, Judas, blackleg, bamboozler, mountebank, stool pigeon, and patsy were hurled his way. "So hostile and powerful were these forces," the historian Char Miller remarked in *Gifford Pinchot and the Making of Modern Environmentalism*, "that through their representatives in Congress they had managed to suspend Cleveland's action pending congressional hearings."[45] Senator John Lockwood

*The existing Pacific Forest Reserve of Washington state was greatly enlarged and renamed Mount Rainier Forest Reserve on February 22, 1897.

Wilson of North Dakota, for example, excoriated Cleveland for a "dastardly blunder" carried out to please East Coast elitists like the Boone and Crockett Club. Wilson predicted that westerners would ignore the edict and continue to log timber as they saw fit. Senator Richard Franklin Pettigrew of South Dakota called Cleveland "a disgrace to civilization and a disgrace to the Republic." Nearly every western senator, in fact, believed that Cleveland had betrayed America. Cleveland's action in kicking over the hornet's nest, they argued, was in part pathological, a punishment because the Democratic Party had lost the 1896 election. (This didn't make any sense, however, because Bryan was no friend of the forest reserves.)

Meanwhile, the Seattle chamber of commerce was in high dudgeon over President Cleveland's last-minute "sneaky" forest grab. The mere pun on his last name—*Cleave-land*—got its members hopping mad. "The reservations, of no benefit to any legitimate object or policy, are of incalculable damage to the present inhabitants of this state," these northwestern businessmen argued. "If they were allowed to stand, not only will the mining industry be destroyed, but the great railroad trunk lines of the Central West which are now heading for Puget Sound will be prevented from coming here. All the passes in the Cascade mountains by which the railroads can reach the Sound are embraced in these reservations."[46]

But the *New York Times*, in a spate of editorials, applauded President Cleveland's parting proclamation as a historic accomplishment on behalf of the general public and posterity. "To leave [pristine forests] to private enterprise is to make sure within a generation or two of reducing the Western land now wooded to the condition in which countries once well watered and fertile, like Greece and Spain, have been reduced by like improvidence," the *Times* argued. "It is to dry up the streams now stored by the forest and to expose the country the water supply which they protect to an alternation of drought and flood."[47] That August John Muir also vigorously defended Cleveland's public lands act in an article in *Atlantic Monthly* titled "The American Forests"—though he also noted that sometimes "wild trees" had to make way for "orchards and cornfields."[48] To Roosevelt's mind the sworn enemies of the Cleveland reserves were (politically speaking) at the polar opposite ends of the political spectrum: Bryan Populist-Democrats from the Midwest and Rocky Mountain regions and Republican Wall Street types and monopoly-minded captains of industry on both coasts.

Unlike Roosevelt, President Cleveland had too much dignity to call his opponents horrific names in 1897. Nevertheless he ably defended himself nine years later in a book titled *Fishing and Shooting Sketches.* Cleveland wrote that the "criticisms" and "persecutions" from "mendacious" newspapers and "shameless" Western politicians were "nothing more serious than gnat stings suffered on the bank of a stream—vexations to be borne with patience and afterward easily submerged in the memory of abundant delightful accompaniments." For the rest of his life Cleveland gloated that the granite-ribbed San Jacinto Mountains around Palm Springs, California, the Uinta Mountains of Utah one hundred miles east of Salt Lake City, and eleven other wilderness areas had been saved due to his boldness.[49]

That May the U.S. Senate tried to make an immediate amendment to President Cleveland's forest lands act. A Lieu Selection Act (passed on June 4, 1897) was created to offer money to homesteaders booted out of the new forest reserves. Emotions ran high. Northern Pacific Railroad agents throughout Washington state, for example, encouraged residents to simply disobey the federal government. It was up to the new president—William McKinley—to grapple with the fracas the anticonservation politicians and extractioners were making. What added to these senators' fury was that President Cleveland had issued his order without consulting them in any way. If Grover Cleveland had stayed in power, the Sundry Civil bill—called the "Washington's Birthday Reserves" Act (by conservationists) or the "Midnight Reserve" Act (by pro-development westerners)—would probably have been nullified. The continuation of the forest reserves rested squarely on President McKinley's broad shoulders. As Muir noted in *Our National Parks*, promoting his aesthetic view of nature, forest reservations were useful not as "fountains of timber" but as "fountains of life" capable of rejuvenating the human spirit and rescuing it from the "vice of over-industry" and the "deadly apathy of luxury."[50] Avoiding political quicksand, and following the old legal adage about cooling out the client, McKinley adroitly held the "Washington's Birthday Reserves" in abeyance for a year; the act didn't become officially operative until March 1, 1898.[51]

The fact that President McKinley didn't recoil from or play the ostrich on Cleveland's 21-million-acre coup impressed Roosevelt tremendously. McKinley, in fact, got lucky, for the discovery of gold in Yukon-Alaska in 1896 eventually caused many people in the Pacific Northwest to give up on lumber and instead start developing Seattle and Portland as major ports and outfitting centers. "I am exceedingly glad that President Cleveland issued the order," Roosevelt wrote to Grinnell that summer, "but none of the trouble came on him at all. He issued the order at the very end of his administration, practically to take effect in the next administration.

In other words he issued an order which it was easy to issue, but difficult to execute and which had to be executed by his successor. . . . I think that credit should be given the man who issues the order, but I think it should be just as strongly given to the man who enforces it. . . . President McKinley and Secretary [Cornelius] Bliss took the matter up, and by great resolution finally prevented its complete overthrow." The estimable point Roosevelt was trying to drive home to Grinnell was that McKinley and Cornelius Bliss, his secretary of the interior, deserved credit equal to Cleveland's for the creation of these thirteen reserves.[52]

Bliss was a New Yorker, a successful businessman, a member of all the right clubs, and a bit of a dandy. When McKinley nominated him to be secretary of the interior, conservationists like Roosevelt knew that their movement would have an ally in the executive branch. Bliss was easily confirmed by the Senate, over the objections of Senator Henry Teller of Colorado, who claimed that such an "Eastern man" knew "nothing of the great Western matters constantly arising in the Department of the Interior."[53] The real reason for Teller's objections was perhaps that he wouldn't be able to make sweetheart deals with a man of Bliss's moral fiber. Within two months of being confirmed, Bliss got a sort of revenge on Teller by appointing a forester, Gifford Pinchot, as his "confidential special agent" to look into how to both protect and create new western reserves.[54]

There was another reason Senator Teller made a terrible mistake in going after Bliss. As the old Arab proverb goes, "The enemy of my enemy is my friend." Given that there were few easterners whom Senator Teller disliked more than Theodore Roosevelt, once Bliss was confirmed, he joined Henry Cabot Lodge in suggesting that Roosevelt become assistant secretary of the navy. Ostensibly, Roosevelt would be in the Navy Department, but Bliss knew he would interfere in western land issues left and right. The secretary of the interior—a prominent contributor to the American Museum of Natural History—welcomed his fellow New Yorker's interference. Nobody was a better backstop than Roosevelt. As secretary of the interior, Bliss, backed by the Boone and Crockett Club, championed forest reservations in Alaska, the surveying of Yosemite National Park, the saving of prehistoric sites in the Arizona Territory, and the commissioning of the special forest agent, Pinchot, to assess how best to preserve and use vast tracts of public land in the Pacific Northwest.[55] As Roosevelt boasted, Bliss was 100 percent in line with the Boone and Crockett Club's agenda. (In 1900 Roosevelt even promoted the idea of Bliss as the Republican vice presidential nominee, instead of himself.)

A graduate of Yale, Pinchot was tasked with making recommendations about forest management and building public support for the "Washington's Birthday Reserves." Known for giving himself airs, he traveled up and down the West Coast, meeting with newspaper editors, politicians, Rotary clubs, and citizen groups. The assignment required a delicate balancing act. Constantly Pinchot had to pluck up enough nerve to tell lumberjack types about the virtues of forestry and preservationists about the need for paper products. In Seattle, for example, Pinchot got into an ugly dispute with John Muir over sheep grazing in national parks and forest reserves. Even though Pinchot had personally reassured Muir, while they were hiking together in the Cascades, that he was against the "hoofed locusts," in an interview with the *Seattle Post-Intelligencer* he switched stories. An infuriated Muir shouted hypocrite, accusing Pinchot of currying favor with the Wool Growers Association. Spitting mad, his "eyes flashing blue flares," Muir told Pinchot, "I don't want anything more to do with you."[56]

Throughout 1897, as groups like the wool growers fought tooth and nail to overturn the federal "lockup" of forest lands, with their congressmen promoting a spate of amendments and nullification bills, Roosevelt vehemently defended Cleveland and McKinley's policy. At least on paper and in principle, many of his most cherished wilderness places (including Wyoming's Bighorns and Tetons and Montana and Idaho's Bitterroots) had been saved in part for his grandchildren and great-grandchildren to enjoy. But westerners didn't like the government's incursion into their lives. The arrival of a federal land officer, a scientist from the Biological Survey, or an inspector from the Interior Department caused many westerners to reach for their guns. A forest ranger coming to an outback town like Bend, Oregon, or Spokane, Washington, for example, was greeted with all the hospitality that would have been extended to a plague of locusts. In Montana alone, suddenly the Flathead reserve was assigned nine rangers, Lewis and Clark seven, and the Bitterroots nine (though part of this reserve is in Idaho). These mounted rangers formed a "chain of patrol" around each forest preserve, looking for fires, poachers, and outlaws.[57]

Meanwhile, a consensus had started to form in America that big business was insensitive to the environment. The educated class was coming to believe that the federal government needed to intervene before the rivers ran dry and the forests disappeared like the buffalo herds. While working for the Federal Writers' Project during the Great Depression, the poet Kenneth Rexroth reflected on the character of a typical 1890s Californian businessman, for example, willing to destroy natural wonders like Mount Shasta for the sake of mineral exploration. "He is most often a stranger to the country in which he operates, with no interest in its well being and no care for the conservation of its resources," Rexroth wrote in the *WPA Guide to California*. "He is interested in the immediate exploitation of the irreplaceable commodity. The effects of that exploitation on the surrounding country and its population, or on the workers . . . [are] the least of his cares. Former mining areas are littered with abandoned machinery, the streams are polluted, the forests are destroyed, and the aboriginal population murdered or enslaved."[58]

The challenge Roosevelt, more than any other high-profile American, addressed in the 1890s was how to get farmers and backwoods families to hop on the new conservation bandwagon. Thousands of settlers in the Rocky Mountains, California, and Pacific Northwest accepted the arguments of the mining and timber industries and disobeyed federal law—for instance, cutting timber down in the reserves in broad daylight. Federal geological reports collected in 1897 made the feelings of local citizens vividly clear. "Nearly all illicit lumbering and other timber depredations are looked upon by settlers as blameless ventures," the investigator George B. Sudworth wrote after on-site investigations of Colorado's White River Reserve. "Such operations furnish a limited amount of employment to the poorer classes. . . . They are considered to be taking only what rightfully belongs alike to them and all other settlers. The depredator's good name is not thought to be sullied by the veritable theft of timber from the national domain. The spirit of some landless settlers . . . is well illustrated by the following remark made to this writer by a party suspected of selling dead building logs: 'This timber belongs to us settlers and we're going to get it! The Government officials can't prevent us either, with an army! If they attempt to stop us, we'll burn the whole region up.' "[59]

Roosevelt remained convinced that increased law enforcement, in the newer "Washington's Birthday Reserves" as well as Yellowstone, was the answer. He wanted to lock up any and all scoundrels trying to despoil the federal forestlands. This was a continuation of his brag that he would have shot the slimy hide hunter Edgar Howell in the face for killing the Yellowstone buffalo. If anarchic, anti–federal government followers of senators Teller and Pettigrew wanted to challenge the authority of the executive branch over the forest reserves by acts of civil disobedience, Roosevelt's view was "Bring them on." Their conniptions were music to his ears; after all, the federal government was now on *his* side. If these debasers defied federal authority, then off they'd go to Fort Leavenworth Prison or Ship Island in the Gulf of Mexico, where they would rot behind bars whittling driftwood as the sun rose and set. Because Roosevelt had studied the life of Zebulon Pike for Volume 4 of *The Winning of the West*, he became especially incensed that shepherds in Colorado were destroying the grasses in the forest reserve named after the bravest scout of the Jeffersonian era. As Muir squared off against the "hoofed locusts" of Yosemite Valley and the High Sierra, Roosevelt likewise fought to save the Colorado Rockies.

V

Drawing rooms and gentlemen's clubs in New York had been abuzz in early 1897 over what job Theodore Roosevelt would get in the new McKinley administration. He was known to want a post that would be intellectually stimulating, and there were some early murmurs that he might be given Interior. The anti-forestry legislators deserved nothing less. But only a fool took them seriously. The effect of Bliss's expedient confirmation put an immediate wet blanket on that low-burning fire. Roosevelt was much too volatile a pro-conservation figure to deal with the western politicians, so McKinley could not make that appointment. However, watching the Republican boss Mark Hanna hold court at a party in New York made Roosevelt grow ashamed and leery of any kind of connection with the McKinley administration. Hanna was a political operative—a breed of man he disliked. Not wanting to be associated with such immoral types, he developed a mild case of self-revulsion.[60] "I felt," Roosevelt wrote, "as if I was personally realizing all of Brooks Adams's gloomiest anticipations of our gold-ridden, capitalist-bestridden, usurer-mastered future."[61] For his part, the new President McKinley was reluctant to appoint Roosevelt to any meaningful post; he wrote Roosevelt off as "too pugnacious."[62]

Things may have remained at a stalemate had Henry Cabot Lodge not orchestrated a lobbying appeal to have his friend appointed as assistant secretary of the navy. As America's foremost expert on the naval battles of the War of 1812, and having been a great success as police commissioner in New York City (where he increased the force by 1,600 men), Roosevelt seemed an ideal number two administrator for the Navy.[63] McKinley had qualms but soon, as a favor to Henry Cabot Lodge and Secretary of the Interior Bliss, he agreed to appoint Roosevelt to the post. Roosevelt assumed his duties on April 19, 1897.

Although the post of assistant secretary hadn't been created until 1861, Roosevelt now felt that he was part of a club that stretched all the way back to John Paul Jones, the naval hero of the Revolutionary War. Roosevelt remained forever grateful to Bliss for vouching for his character and helping to secure the appointment. But if McKinley and Bliss knew what Roosevelt was revealing in his private correspondence that spring, they would probably have fired him. For example, he told the seapower scholar Alfred Thayer Mahan that the McKinley administration planned to annex the Hawaiian Islands, cut a canal through Nicaragua, construct a modern naval fleet, and kick Spain out of the Caribbean. (Those were all programs *he* wanted implemented.*) In a second letter he suggested that Mahan lobby T.R.'s boss, Secretary of the Navy John D. Long, for the United States to build more battleships.

But even as Roosevelt immersed himself in military affairs, he found time to duel intellectually with Dr. C. Hart Merriam over the nature of species and subspecies. Merriam, the reviewer whose praise for *The Summer Birds of the Adirondacks*, back in 1877, had helped establish Roosevelt's bona fides as a naturalist, was now chief of the U.S. Department of Agriculture's Bureau of Ornithology and Mammalogy (Biological Survey). Nobody admired Merriam more than Roosevelt, who regularly sent notes of appreciation for his steady work on behalf of biological inquiry, mammalogy, and biogeography. Roosevelt always enjoyed seeing Merriam's name in print. There was always an air of collaboration about the two men. Sometimes, for example, Roosevelt sent Merriam sketches and drafts that he was working on. Who else but the overly conscientious Merriam would take the time to examine 27,000 specimens of white-footed mice before issuing a report on their habits?[64] Every time Merriam spoke publicly in science forums, even the people in the back rows whispered in awe at his illustrious erudition. There was something about this government scientist's deportment that demanded respect. He had a knack for making even trifles interesting.

Starting in the mid-1890s Merriam plunged headfirst into the debate over the classification of bears. Boldly he declared there were ten bears to be saved, as well as a new subspecies. Back in 1890, when Merriam, following a trip to the San Francisco Mountains of Arizona, announced his "life zone" theory (i.e., that temperature and humidity were *the* leading factors in species development), Roosevelt applauded the findings. But now these sudden pronouncements about bears left him baffled. Perhaps, he thought, Merriam was just overworked. So in an unusual gesture of

---

*After devouring the book in a single sitting, Roosevelt had reviewed Mahan's *The Influence of Sea Power upon History, 1660–1783* positively in the October 1890 *Atlantic Monthly*. And in 1897 he praised Mahan's biography of the British admiral Horatio Nelson in a review in *The Bookman*.

solidarity Roosevelt tried to disagree only quietly, letting the biologist create new textbook designations.

Although Darwinism had been fully embraced in the Ivy League schools, save for a few recalcitrant professors, its crossover into the mainstream culture was fraught with dissent. Being a pioneering biologist like Merriam, one who insisted that germs were living organisms (like people), was mistakenly interpreted as tantamount to declaring Adam and Eve a farce. For hard-core creationists—who were a large majority of Americans—Merriam was pushing scientific inquiry too far for comfort. When Henry Cabot Lodge, serving as a U.S. senator from Massachusetts, wrote a letter to Roosevelt, intimating that Merriam was getting a little too loopy with his Darwinian claims in the U.S. Biological Survey, Roosevelt objected. "Now, I was a little disturbed at what you said to me about Hart Merriam," Roosevelt wrote. "On most matters I accept your judgment as much better than mine. On this you for the time being accept mine. The only two men in the country who rank with Merriam are [Alexander] Agassiz and [David] Jordan."*[65]

Out of all Roosevelt's naturalist friends, only Merriam (and the botanist Asa Gray) took Darwinian pursuits such as cross-pollinating flowers—anthers and pistils—seriously. Merriam was like Roosevelt in that loafing wasn't part of either man's personality. For Merriam, every waking minute was sacred time for further scientific inquiry into the mysteries of life. He had become the workhorse of the U.S. Biological Survey. Besides publishing the definitive two-volume work *The Mammals of the Adirondack Region, Northeastern New York*, he had visited seal rookeries in Newfoundland, helped found the National Geographic Society, conducted collecting trips in the Mojave and Sonoran deserts, served on the American-British Seal Commission, and written a groundbreaking Darwinian interpretive text, "The Geographic Distribution of Life In North America, with Special Reference to the Mammalia" in *Proceedings of the Biological Society of Washington.*[66]

At the time of his disagreement with Roosevelt about overspecialization of organisms, Merriam was organizing an expedition to 14,179-foot Mount Shasta, a dormant volcano in Siskiyou County, California,

*Roosevelt was referring to the son of the great Louis Agassiz. Alexander Agassiz became famous for his numerous comparative zoology reports, including *Marine Animals of Massachusetts Bay* (1871). David Starr Jordan was a renowned ichthyologist who became president of Indiana University and then Stanford University. He went on to be a peace activist and was an expert witness in the Scopes trial of 1925.

the second-highest peak in the Cascade Range. Groves of conifers on its slopes had already been recklessly denuded by a lumber company.* Raised in New York, Merriam was admired as a highly effective administrator by the East Coast aristocracy who frequented the gentlemen's clubs—Metropolitan, University, Cosmos, and Century. The railroad tycoon E. H. Harriman, for example, hired Merriam in 1899 to head a famous eight-week expedition to Alaska. Harriman's primary personal goal was to hunt a brown bear. Paid a retainer, Merriam organized the travel arrangements, booked the best polar scientists for the voyage, and, most famously, hired John Burroughs and George Bird Grinnell to come along. Once in Alaska, Merriam, for the sake of American natural science, hiked across Howling Valley in Glacier Bay, wrote on the volcanic island of Bogoslof, and pondered the fate of seals. Eventually he compiled "The Merriam Report," a multivolume account of everything learned on the expedition, for E. H. Harriman himself. A deskbound Roosevelt was envious because he hadn't been able to go along on the historic expedition.

Considering the high level of mutual admiration, one suspects that what actually started the feud with Merriam was his encroachment into the study of bears, considered Roosevelt's bread-and-butter area of expertise. Merriam stoked up a controversy regarding bears in 1896 by publishing (for the Biological Society of Washington) the paper "A Preliminary Synopsis of the American Bears." Claiming that for fifteen years the classification of North American bears had been done with imperfections and unscientific contractions, Merriam wanted to challenge orthodoxy. Taxonomic revisions of various genera of bears, he said, owing to field research, were now needed. Having collected 200 to 300 bear skulls and skins as samples, he insisted that the results were crystal clear. There were many more bear species than previously known. That possibility seemed so fantastic, so incomprehensible to Roosevelt that he could barely absorb the assertion calmly. Merriam might as well have claimed to have discovered Sasquatch.

Calling for a comprehensive treatise on bears, Merriam admitted that "much additional material is coveted from North British Columbia, and

*For conservationist-naturalists, visiting Mount Shasta—the sixth-tallest mountain in California—had become something of a rite of passage. John Muir wrote of Shasta that his "blood turned to wine" upon seeing the peak. Because it was a volcano Muir also called it a "fire mountain." To the poet Joaquin Miller the summit was "Lonely as God, and white as a winter moon." Theodore Roosevelt, in a letter of 1908, wrote that he considered "the evening twilight on Mt. Shasta one of the grandest sights I have ever witnessed."

the coast regions of Alaska south of the Alaskan peninsula." Nevertheless, from studying so many skulls Merriam confidently declared in his synopsis that there were ten full bear species (and one subspecies in Canada), not the smaller number that Roosevelt had supposed. To give just two examples, there were now a small black (*Ursus floridanus*) whose range was the Everglades and a huge brown bear (*Ursus dalli*) found in Yakutat Bay and the St. Elias Alps which needed to be added to zoology books.[67] In other words, some of the information in Roosevelt's published essays on bears, while entertaining, was, in Merriam's view, technically wrong.

Roosevelt could hold his tongue no longer. Feeling bruised by an article of 1897 in the *New York Times* proffering Merriam's views about bear species, Roosevelt suddenly saw things Henry Cabot Lodge's way. Merriam, it seemed, had indeed gone Darwin-mad, playing the clairvoyant, turning Linnaeus on his head, and wanting to rewrite zoology books to support his field research, which called for new ways to classify species. As if the bears weren't enough, Merriam was about to publish an article in *Science* claiming that more species breakdowns of many other mammals were needed to cover such factors as color variation, differences in horn size and shape, whiskers, and hoofprints. Regarding coyotes in America, for example, Merriam believed there were actually eleven distinct species. Roosevelt, urging modification of Merriam's theory, balked at the species approach to classification. Merriam's heavy emphasis on species classification, he argued, would merely confuse the general public. "I have been greatly interested in Dr. Merriam's article as to discriminating between species and subspecies," Roosevelt wrote to Henry Fairfield Osborn. "With his main thesis I entirely agree. I think that the word 'species' should express degree of differentiation rather than intergradation. I am not quite at one with Dr. Merriam, however, on the question as to how great the degree of differentiation should be in order to establish specific rank."[68]

Osborn, who would go on to become the preeminent advocate of Darwinism in the early twentieth century, was Roosevelt's ally in the well-mannered dispute of 1897. In a letter to Osborn, Roosevelt admitted his own "conservative instincts," but added that when it came to creating entire new species of bears, wolves, elks, and coyotes, he was sanely skeptical.[69] If Merriam's theory were true, that meant his trophy collection at Sagamore Hill of North American big game would never be complete, and every year he'd have to try to bag newly designated species. Roosevelt saw Merriam's idea as akin to having an "old familiar friend" suddenly "cut up into eleven brand new acquaintances." Although Roosevelt loved Merriam dearly, he thought Merriam's new zoology was off-kilter and not worth expounding in serious periodicals like the *New York Times* and *Science.* Turning a blind eye toward Merriam's research, Roosevelt insisted that varied species—like mule and white-tailed deer—were smart "arbitrary divisions" devised for "convenience's sake." But he didn't find value in suddenly catapulting black-tailed deer, for instance—comfortable as a subspecies—into the species category on a biological whim. At the end of the day they were *deer.* He believed their "essential likenesses" far more important than their "minor differences." While Roosevelt fully supported having Merriam's Biological Survey field collectors record variations in species discovered in different regions of America—and in fact coveted such information himself—he didn't want to "lumber up our zoological works" by adding unnecessary new terminology, thereby overloading the binomial system.

One wonders what Secretary of the Navy Long thought of his underling, whom he didn't know well, being involved in naturalist squabbles throughout the spring and summer of 1897 (Long's diary suggests he had strong reservations about Roosevelt's sanity). Roosevelt's Darwinian-influenced views spread into his public policy pronouncements, including his pro-expansionist sentiments, when he flat-out stated that "the rivalry of natural selection" was "one of the features of progress."[70] On April 30, Roosevelt published a rebuttal to Merriam in *Science.* In "A Layman's View on Specific Nomenclature," Roosevelt started out by praising Merriam as "one of the leading mammalogists and he has laid all men interested in biology under a heavy debt"—but then he attacked. Using examples like coyotes of the Rio Grande Valley, bears of the Bighorns, and cougars everywhere, Roosevelt challenged Merriam's thesis as clumsy. "The excessive multiplication of the species in the books cannot, as it seems to me, serve any useful purpose," Roosevelt wrote, "and may eventually destroy all the good of the Latin binomial nomenclature."[71]

The next month, a taxonomic debate between Merriam and Roosevelt was held at the Cosmos Club in Washington, D.C.—a little mansion on Madison Place that had served as something of a living room for John Wesley Powell, Clarence King, and the Geological Survey community in general—in front of an audience of America's leading naturalists. It had been arranged by L. O. Howard of the Bureau of Entomology.[72] Merriam, who lived nearby on Sixteenth Street, often used the Cosmos library as his own salon, occasionally reading Huxley and Thoreau in an easy chair as a break from arsenic and formaldehyde. There were few other places in Washington, D.C., where you could you simply pluck from the bookshelves classics of exploration without so much as consulting a reference librarian.

Just as *Marbury v. Madison* was carefully studied in law schools for decades after the decision, the rancorous disagreement between Roosevelt and Merriam had a long shelf-life in graduate biology programs. At its core was the question: What constituted a species? On Roosevelt's side were "lumpers," old-fashioned taxonomists uncomfortable with "undue cleavage of the genus." The "splitters" were Merriam's followers, who insisted that wildlife that integrated "must be treated as subspecies and bear trinomial names; forms not known to integrate, no matter how closely related, must be treated as full species and bear binomial names." To Roosevelt these "splitters" were essentially perpetuating a gimcrack theory, smothering in its implications (although he didn't phrase his view in quite such a degrading way). Merriam made plenty of valuable points defending his research. Nevertheless he was not, as a rule, a good speaker. For visual effect he brought with him wolf and coyote skulls, with mixed results.[73]

Roosevelt, by contrast, made the room shake when he spoke. Thrusting his hands out of his shirt sleeves, he lectured on the need for biology not to overcomplicate everything. One point, which Roosevelt essentially conceded, was that ornithology was a relatively "finished science" whereas mammalogy, particularly throughout the American West, was "yet in its infancy." Merriam saw this concession as an opening. Daily the Biological Survey was getting mammals with skull variations and tails different from others in the same genus. Was it really so irresponsible to believe, Merriam wanted to know, that new species were being discovered?[74] Essentially, Roosevelt won the debate on extempore elocution while Merriam did better on specifics; in other words, it was a draw.

Besides his sharp argument with Merriam, Roosevelt's obsession with species bled into his job at the Navy Department in other, unexpected ways. On behalf of entomology, for example, Roosevelt wanted the new class of U.S. torpedo boats to carry names like *Wasp*, *Hornet*, and *Yellow-Jacket*.[75] Under the aegis of a decorator, Roosevelt filled his office with a wide assortment of antlers; it looked like a Wyoming hunting lodge. And even though war with Spain was looming and naval procurement was one of his responsibilities, Roosevelt continued to mercilessly prune and edit articles that hunters were submitting to the Boone and Crockett Club. "Wherever the young idiot speaks of papa, father should of course be substituted, and, if possible, the allusion should be left out all together," Roosevelt wrote to Grinnell after reading an article on deer hunting submitted in August 1897. "It is not advisable to put in nursery prattle. In the next place all of the would-be-funny parts must be cut out ruthlessly.

If there exists any particularly vulgar horror on the face of the globe it is the 'funny' hunting story. This of course means that we shall have to cut down the piece to about half its present length; but if that is done I think it will be good."[76]

VI

As Roosevelt put together the Boone and Crockett Club's third volume, *Trail and Camp-Fire*, he retained a bias toward preservation and the kind of songbirds that the Audubon societies had lobbied to protect, as opposed to the plight of milkweed bugs or the angular-winged katydid.* "The geology and the beetles will remain unchanged for ages," he wrote George Bird Grinnell, his coeditor once more, when going over a manuscript about Africa, "but the big game will vanish, and only the pioneer hunters can tell about it. Hunting books of the best type are often of more permanent value than scientific pamphlets; & I think the B&C should differentiate sharply between worthless hunting stories, & those that are of value." Perhaps not wanting to war with his coeditor as well as Dr. Merriam, Roosevelt ended his letter praising Grinnell's recent article on buffalo as "worth more than any but the very best scientific monographs about the beast."[77]

In Roosevelt's correspondence of 1897, four interrelated conservation issues—increased national park protection, more forest reserves, western water reservoirs, and the diminution of buffalo—concern him above all others. He truly believed that the Boone and Crockett Club had made great inroads in raising public consciousness of buffalo's plight. Hoofed game were on the rebound in North America, and the Bronx Zoo would soon get the word out even more throughout the East Coast. Citizens had started warming up to the idea of game and forest wardens being trained in biology. As Roosevelt told Grinnell, *Trail and Camp-Fire* must hit the same urgent notes: "We have made such a point of Yellowstone Park in our two previous volumes," he noted, "that I think we ought to dwell on it in this one."[78]

By September, with *Trail and Camp-Fire* delivered to its publisher, Roosevelt took a rare three deep breaths, sheepishly worried that he had been too roughshod in his exchanges with Merriam both in *Science* and at the

---

*Remember that George Bird Grinnell's original Audubon was defunct. There were only state Audubon societies in the late 1890s. No real national "Audubon Society" was formed until the 1940s. And even as late as 2008, some state Audubons (like Connecticut's) remain separate and independent of the National Audubon.

Cosmos Club. His defiant mood had tapered off. "I almost broke the heart of my beloved friend Merriam," Roosevelt confided to Henry Fairfield Osborn. "He felt as though he had been betrayed in the house of his friends; but he really goes too far. He just sent me a pamphlet announcing the discovery of two species of mountain lion from Nevada. If he is right I will guarantee to produce fifty-seven new species of red fox from Long Island."[79]

Apparently Merriam harbored his own regrets about the dustup, declaring that Roosevelt was "a writer of the best accounts we have ever had of the habits of our larger mammals."[80] Furthermore, while he was in the Olympic Mountains of far western Washington, a stunning Pacific slope cluster of low-lying peaks surrounded by rain forests and considered the wettest spot in the continental United States, Merriam discovered that the elks there had coloration and antler size different from those found in Yellowstone. Appealing to Roosevelt's ego (and perhaps his own wicked sense of humor), Merriam named this new subspecies *Cervus roosevelti*.[81] These huge elk were magnificent creatures. "It is fitting that the noblest deer of America should perpetuate the name of one who, in the midst of a busy public career," Merriam wrote in the December 17, 1897, issue of the *Proceedings of the Biological Society of Washington*, "has found time to study our larger mammals in their native haunts and has written the best accounts we have ever had of their habits and chase."[82]

What could Roosevelt do? There was hardly an honor in the world he would have preferred more than having a species of elk found in the dusky coastal and Cascade ranges of the Pacific Northwest named after him. However, if he embraced *Cervus roosevelti*, other naturalists would dismiss him as a hypocrite bought off by flattery. Nevertheless, here was a heaven-sent opportunity for Roosevelt to make everything right again with Merriam, and he seized it. There is no record of Roosevelt's thought process, but in any event he accepted the new honor, informing Merriam, "No compliment could be paid me that I would appreciate as much as this—in the first place, because of the fact itself, and in the next place because it comes from you. To have the noblest game animal in America named after me by the foremost of living mammalogists is something that really makes me prouder than I can well say. I deeply appreciate the compliment and I am only sorry that I will never be in my power to do anything except to just merely appreciate it."[83]

The deeply touched Roosevelt now felt he had a debt to repay. He began reading everything he could on the Olympic Mountains and sought photographs of *Cervus roosevelti*. The 800-pound "Roosevelt elk" was brown or dark beige with very dark underparts and a yellowish-brown tail. Much like their namesake, these elk were crepuscular, extremely active at dawn and dusk. Focusing on Washington state wildlife for the first time, he learned that the Olympics contained five distinct landscapes: temperate rain forest, rugged mountain terrain, large lowland lakes, cascading rivers, and saltwater beaches. As an ornithologist, Roosevelt hoped to soon see the black oystercatchers, with their long reddish beaks, crack open mollusks along the rugged Pacific shore. The mere thought of aromatic Sitka spruce and western hemlock appealed to Roosevelt just as much as seeing his namesake elk in their natural habitat.

Starting in 1897, Mount Olympus became Roosevelt's new Matterhorn, another peak he wanted to conquer. The highest point in the Olympics chain, it had eight tumbling glaciers and some of the finest strands of Pacific silver fir in North America. Mount Olympus—the very name enthralled Roosevelt—wasn't going to elude him. The Pacific Ocean here was a sea of boulders, many larger than houses. For a marine biologist there were new universes to explore. While Muir championed the Yosemite Valley, Merriam studied northern Arizona, and Grinnell focused on northwestern Montana, Roosevelt developed an abiding fascination for the Olympics of Washington state and the forest reserves of Colorado; they were two unexplored western places (not counting Alaska) on his future itinerary. Fascinated to learn about Washington state's big-leaf maples in rain forests adorned with epiphytic mosses and ferns, he became determined to save them—a feat he accomplished six years later as president. The only other rain forests as temperate as those stretching from Alaska to Oregon along the Pacific Coast were in Chile, New Zealand, and South Australia. Europe had nothing like them, so Roosevelt, as he educated himself about the Olympics, swelled like a toad with pride.

When *Encyclopaedia of Sport*, a British reference guide, asked Roosevelt to contribute an article on elk that year, he turned his focus to the herds he'd been studying with unremitting interest. "There are several aberrant forms of wapiti, including one that dwells in the great Tule swamps of California," Roosevelt wrote. "There is also an entirely distinct species with its centre of abundance in the Olympic mountains of Washington and in Vancouver Island. This species, which Dr. Hart Merriam has recently done the present writer the honour of naming after him (*Cervus roosevelti*)."[84]

Ironically, in the long run, Roosevelt's position, that Merriam was creating too many species of mammals, triumphed. *Cervus roosevelti* would lose its species status in 1899, becoming a subspecies called *Cervus canaden-*

sis roosevelti.[85] Hearing the news of his demotion, Roosevelt asked Merriam, "By the way, is 'Roosevelti' merely a synonym of 'occidentalis,' for the Olympic Wapiti? My only glory gone!"[86] Regardless of its designation, however, "Roosevelt elk" remains the common name for these gorgeous creatures that ranged throughout the dense redwood and rainforest country of the Pacific Northwest. The combination of *his* elk wandering among 2,000-year-old sequoias became—in his dotage—one of Roosevelt's Edenic fantasies.

Roosevelt also worked side by side with Merriam on abolishing the unsportsmanlike practice of chasing deer to the water's edge with packs of hounds in the Adirondack Park. This was a conservationist project in which they could be brothers in arms. The two men also wanted to ban the new practice of jacklighting (shining a spotlight into flocks of ducks which stunned so they could not fly off, and then firing away). As noted, in 1884 Merriam had written *The Mammals of the Adirondacks Region of Northeastern New York*, a careful examination of that area's fauna. The Boone and Crockett Club recommended the book for sportsmen; and Roosevelt praised it in *Trail and Camp-Fire*. Offering an olive branch to Merriam in the public sphere, throwing out grand praise, Roosevelt called the federal biologist a "field naturalist in the highest sense of the term; the model of what we ought to have for the entire American continent."[87]

But Roosevelt's interest in Adirondack deer wasn't simply a matter of rebonding with Merriam. The declining deer population in those home-state woodlands troubled him. Like Uncle Rob, Theodore wanted to protect the Hudson River watershed by not cutting down too many trees—at times it seemed that most of the topsoil of upstate New York was ending up in Manhattan's harbor. In 1897, after a concerted lobbying effort, the New York legislature enacted a law to protect deer. Proper wildlife management, Roosevelt boasted, truly got extraordinary results. "We set to work ridding the Adirondacks of the [hunting] dogs," the New York conservationist John Burnham recalled of Roosevelt and Merriam's push, "and it was a thrilling, dangerous job."[88]

A year after the bans on dogs and jacklighting went into effect, the Adirondack deer and ducks started to rebound. In an early experiment in wildlife relocation, trapped deer from Maine were taken to upstate New York and let loose. The repopulation commenced as hoped for. (A similar reintroduction with buffalo, moose, and elks, however, failed.) Still, Roosevelt was thrilled that his Adirondacks deer were being revived through a combination of wildlife and forestry law. Former president Benjamin Harrison, in fact, had just built a sporting home in the Adirondacks called Berkeley Lodge. Along with Roosevelt and Merriam, Harrison was a high-level proponent of the "Forever Wild" movement to save the Adirondacks from destruction. And he spoke up on behalf of deer, too. Starting in 1897 Roosevelt once again began exploring the region for bird sightings in hopes of updating his nearly twenty-year-old *Summer Birds of the Adirondacks*; he never, however, found the time.

Extraordinarily fine essays on hunting in Africa and Newfoundland were included in *Trail and Camp-Fire* when it was published in late 1897. On the book's cover was a moose head with record-size antlers; the title page had an amateurish illustration of a mountain goat standing on a rock ledge. The volume's scope was ambitious. On the conservationist front, the editors—Roosevelt and Grinnell—took up the cudgels for saving Adirondack deer by championing many new laws. There was also a self-congratulatory essay on the founding of the Bronx Zoo. Finally, Roosevelt once again provided a list of the essential natural history books all true-blue hunters and serious explorers needed to read. Four essays in this fat book—including Roosevelt's "The Bear's Disposition"—dealt with bear hunting and protection issues.

Grinnell's own contribution, the long essay "Wolves and Wolf Nature," was simply brilliant. It could have been published as a monograph instead of in an anthology. "In discussing wild animals, we are all very disposed to consider the species as a whole, and to deal in general terms, jumping to the conclusion that all the individuals of a kind are exactly alike, and not taking into account the marked variation between different individuals, for we consider only their physical aspect," Grinnell wrote. "We forget that to each individual of the species there is a psychological side; that these animals have intelligence, reason, mind, and that at different times they are governed by varying motives and emotions, which differ in degree only from those which influence us."[89]

Later, when Roosevelt became president, he lashed out at the novelist Jack London for not writing accurately about wolves in *Call of the Wild*. Roosevelt's expertise on this matter stemmed largely from firsthand observation and from reading "Wolves and Wolf Nature." Somewhat incongruously, Grinnell's essay also served as an impetus for Roosevelt to go wolf coursing in Oklahoma a few years later with Captain Jack "Catch 'Em Alive" Abernathy. However, Roosevelt took issue with Grinnell's depiction of how gray wolves brought down prey, insisting in his essay "On the Little Missouri" that they attack prey at the hindquarters, feasting first on the flanks. "It will be noticed that in some points my observations about wolves are in seeming conflict with those of Mr. Grinnell," Roosevelt wrote, "but I think the conflict is more seeming than real; and in any event I have concluded to let the article stand just as it is. The great book of Nature contains many passages which are hard to read, and at times conscientious students may well draw up different interpretations of the obscurer and least known texts. It may not be that either observer is at fault; but what is true of an animal in one locality may not be true of the same animal in another, and even in the same locality two individuals of a species may widely differ in their habits."[90]

Roosevelt was now embracing the very criticism Grinnell made of T.R.'s first book, *Hunting Trips of a Ranchman*, as his own clear-headed scientific statement of purpose. Just why Roosevelt felt compelled to have these frays with Merriam and Grinnell is open to speculation; but one can probably attribute it to a mixture of egotism and his belief that he was correct. As a target of Roosevelt's attacks, Grinnell, unlike Merriam, never let the jabs bother him. Supremely self-confident, Grinnell had, in fact, learned how to *use* Roosevelt for his own conservation cause in *Forest and Stream*, unleashing the feisty reformer's combative personality at his own will.[91]

Grinnell and Roosevelt agreed on the precepts of conservation for the West: repopulating it with the buffalo and the elk, saving its natural wonders, and helping Native Americans there reconstitute their heritage. Two days before Christmas 1897, Roosevelt wrote to John A. Merritt, the third assistant postmaster general in the McKinley administration. All those holiday stamps he had licked for Christmas cards had given him an idea. Why not promote the West by issuing new stamps? When Merritt replied asking for specific recommendations, Roosevelt suggested a Cheyenne warrior with a bonnet of eagle feathers, a prairie schooner, a Remington cowboy illustration, and (if a real person could be used) an image of Kit Carson—Roosevelt always promoted Carson at every chance possible. Those were fairly safe choices. But, Roosevelt wrote, if the U.S. Post Office was truly interested in presenting the American West in a stamp series, it should focus on the region's wildlife and wondrous natural sites. "By all means have one of those postage stamps with a buffalo on it," Roosevelt instructed. "The vanished buffalo is typical of almost all the old-time life on the plains, the life of the wild chase, wild warfare, and wild pioneering. If any bit of scenery were taken I should suggest your going up to the Cosmos Club or to the Geological Survey and examine three or four of their photographs of the boldest [Arizona] canyon walls, or of Pike's Peak."[92]

CHAPTER TWELVE

THE ROUGH RIDER

I

Starting in January 1897 William Randolph Hearst's *New York Journal* and Joseph Pulitzer's *New York World* began reporting zealously on the Cuban insurrection against Spain. Up until then the U.S. Senate was for Cuban independence, but back-burnered the issue. There was little clamor to send the American navy into a firefight. Still, through these jingoistic newspapers, the public was made aware of such heinous acts as a Spanish firing squad executing the Cuban revolutionary Adolfo Rodriguez and the horrific conditions of prisons in Havana. Public sympathy in America was turning against the ogre, Spain. Besides the Cuban insurrection, the Spanish authorities were also trying to squelch the Philippine liberation movement (the Philippine Islands were then a Spanish colony). Hatred for all things Spanish became widespread in the United States, fueled by—in large part—the Hearst and Pulitzer papers.

Influenced by this so-called yellow journalism, Roosevelt had no trouble actively disdaining Spaniards in 1897. To Roosevelt the Spanish government exuded a conceited authoritarianism that he despised. He believed that the United States had an obligation to challenge *any* brutal European monarchy that was thumbing its nose at the Monroe Doctrine. Disregarding the fact that American sugar tariffs enacted in 1894 had disrupted the Cuban economy, Roosevelt blamed Spain for Cuba's impoverished living conditions. He was glad, if anything, that the tariff had helped set the Cubans against Spanish rule.

Roosevelt, as a Mahanian naval strategist, had serious concerns about Spain. In particular, he had carefully read "War with Spain—1896," a national security document written by an astute naval intelligence officer, Lieutenant William Wirt Kimball, for the Naval War College. Roosevelt thought Kimball's analysis was spot-on. He had even written to Mahan that the "Kimball Plan" avoided the politics of imperialism versus anti-imperialism altogether. It was a blueprint for war preparedness. Kimball insisted that if war with Spain occurred, a naval engagement would be preferable to a land war in Cuba and the Philippines. This made great sense to Roosevelt. Boiled down, the Kimball Plan called for an offensive strike against Madrid in three war zones: the Philippines, Cuba, and on the high seas against Spanish shipping. The crucial strategic point was for the United States to cleverly draw the Spanish navy into protecting the far-flung Philippines, leaving Cuba wide open to a U.S. invasion. What Roosevelt as a naval historian most deeply admired about the Kimball Plan was its specificity: every Spanish ship was described and analyzed in minute detail. In many ways the document stylistically resembled his own *Naval War of 1812*. If war came T.R. wanted the U.S. Navy prepared for all the challenges the Kimball Plan presented. He wrote to Kimball, "The war will have to, or at least ought to, come sooner than later."[1]

Around Christmas 1897 Assistant Secretary of the Navy Roosevelt became obsessed by the idea of war with Spain. He worried that if the United States didn't confront Spain in the Western Hemisphere then there would be "disastrous long term consequences for future American security."[2] Overflowing with patriotic spirit, Roosevelt believed that fighting for the independence of Cuba and the Philippines was both a noble cause and a strategic imperative. Spain's ambitions had to be thwarted. There were echoes in Roosevelt's thinking of 1886, when he had promoted armed conflict with Mexico, and of 1891, when he had fantasized about shooting "dagos" in Chile.[3] Although he found time to scold Frederic Remington for having drawn badgers improperly in an illustration in *Harper's Weekly* and continued reading up on the Olympic Mountains, for the most part Roosevelt was consumed that holiday season with promoting an imperialistic pamphlet of quotations he'd assembled under the title *Naval Policy of the Presidents*.[4] He also corresponded with Commodore George Dewey (whose fleet was maneuvering in the Pacific) and impetuously directed the North Atlantic Squadron to join the U.S.S. *Maine* at Key West, Florida, to immediately begin winter exercises.[5]

On January 25, 1898, the *Maine* dropped anchor in Havana harbor in what was essentially an exercise in showing the flag.[6] America was inching closer to war with Spain. When on February 15 the *Maine* exploded in a fireball, killing 262 U.S. sailors, Roosevelt's war fervor increased. He blamed Spain for the explosion. While President McKinley ordered a report to find out whether the *Maine* had been sabotaged by Spain or whether a short-circuit fire had blown up the gunpowder kegs, Roosevelt grew impatient. Believing that the U.S. Navy was in a perfect state of readiness, the best it had been since the Civil War, he wanted the Spanish forts in Cuba reduced to burned wood and rubble. All-out war against Spain, he believed, should be declared at once. "Cuban independence," no longer a remote concept, had become Roosevelt's new rallying cry and his response to those who advocated peace at any price. "Personally I cannot understand how the bulk of our people can tolerate the hideous infamy that has attended the last two years of Spanish rule in Cuba," Roosevelt wrote to William Sheffield Cowles, "and still more how they can tolerate the treacherous destruction of the *Maine* and the murder of our men!"[7]

Pledging to go and fight in Cuba himself—even though he was nearing forty and had six children to help raise—Roosevelt famously declared that the cautious President McKinley had "no more backbone than a chocolate éclair."[8] Warring with his own administration, Roosevelt said that come hell or high water he was going to fight on the front lines in Cuba or Puerto Rico (or even the Philippines if need be). He wrote to Dr. William Sturgis Bigelow, a Bostonian physician with a world-class collection of Japanese art, on March 29, 1898, "A man's usefulness depends on his living up to his ideals in so far as he can. Now, I have consistently preached what our opponents are pleased to call 'jingo doctrines' for a good many years. One of the commonest taunts directed at men like myself is that we are arm-chair and parlor jingoes who wish to see others do what we only advocate doing. I care very little for such a taunt, except as it affects my usefulness, but I cannot afford to disregard the fact that my power for good, whatever it may be, would be gone if I didn't try to live up to the doctrines I have tried to preach."[9]

Less than two weeks later, President McKinley reluctantly asked Congress for a declaration to interfere in Cuban affairs. Roosevelt was all for the declaration but emphatically against the annexation of Cuba, unless Havana wanted it.[10] Congress became engulfed in a heated debate. Was war the right choice? Should America defend its honor in Cuba? These questions became academic when, on April 23, Spain declared war on the United States. President McKinley called for three regiments of volunteers to supplement the depleted army. Then on May Day, out of the clear blue sky, astounding news arrived. The previous day Commodore Dewey—known for his fearless firefights along the Mississippi River as a Union naval lieutenant during the Civil War under Admiral David Farragut's command—had crushed the Spanish fleet in Manila Bay, without losing a single U.S. sailor in battle. A few days later, on May 6, Roosevelt simply resigned as assistant secretary so he could implement the Kimball Plan, defend the Monroe Doctrine, and revenge the *Maine*. In quick order he received an army commission, purchased an appropriate outfit at Brooks Brothers, and departed for drill training in San Antonio, Texas. Nobody in official Washington could believe how childishly he was acting. Bigelow, who shared with Roosevelt a love of jujitsu,*[11] wrote to Henry Cabot Lodge, "If T.R. goes, the country will not trust him again."[12] Seconding that opinion was Henry Adams, who asked mutual friends, "Is he quite mad?"[13]

Roosevelt became so distracted by the prospect of war that for the first time since the founding of the Boone and Crockett Club, he missed its annual dinner that January. The concept of American imperial ambitions consumed Roosevelt to the point of monomania. Pestering everybody he knew who was in a position to help, Roosevelt kept pleading, "Send me." Picking up the old slogan "Remember the Alamo" from the Mexican War, Roosevelt was one of the progenitors of "Remember the Maine." Obviously, hunting and bird-watching took a back seat to war that spring. Between trying to take care of Edith, who had undergone surgery to remove abscesses from her hips, and trying to persuade President McKinley to declare war on Spain, he had lost all sense of proportion. "I really think he is going mad," Winthrop Chanler of the Boone and Crockett Club noted on April 29. "The President has asked him twice as a personal favor to stay in the Navy Department, but Theodore is wild to fight and hack and hew. It really is sad. Of course this ends his political career for good."[14]

When Secretary of War Russell Alger called for volunteer regiments "to be composed exclusively of frontiersmen possessing special qualifications as horsemen and marksmen,"[15] Roosevelt leaped at the opportunity. At the very least, his years as deputy sheriff of Billings County, North Dakota, and his stint in the National Guard of New York provided him with legitimate "frontier" credentials. Eventually, Roosevelt was made second in command of the First U.S. Volunteer Cavalry, behind his friend Colonel Leonard Wood, a former Indian fighter who had won the Medal of Honor for pursuing Geronimo and became McKinley's chief army medical adviser. Because of Roosevelt's highly publicized enlistment, more than 23,000 applicants flooded into the War Department offices. Everybody, it seemed, wanted to serve with Roosevelt at his side, including some of his Harvard classmates.[16]

With the newspapers cheering Lieutenant Colonel Roosevelt on, more than 1,250 men were eventually selected to form a top-notch regiment. They were first called "Teddy's Texas Tarantulas" and went through three

*Roosevelt loved absolutely everything about Japanese culture. As U.S. president he hired one of the premier Japanese jujitsu teachers as his personal trainer. He also introduced jujitsu to Annapolis and West Point.

or four other names until "Roosevelt's Rough Riders" stuck. The Rough Riders were assigned to San Antonio for their mobilization. The regiment's encampment was erected on the dusty International Fair Grounds (later renamed Roosevelt Park). The diverse Rough Riders comprised twelve troops—five from New Mexico Territory, three from Arizona Territory, one from Indian Territory (the southeastern part of present-day Oklahoma), one from Oklahoma Territory, and a smattering of upper-crust men from the East Coast, particularly men who had played Ivy League football and soccer. A recruiting table was set up on the outdoor patio of the Menger Hotel where men could register; some gave pseudonyms so not to be held accountable for past crimes. Horses and equipment were supplied from Fort Sam Houston's quartermaster depot. Livestock men in Stetsons milled about El Mercado Square gossiping with Mexicans about the glorious war. Slouch hats, blue flannel shirts, and bandannas were handed out to Rough Riders as uniforms, giving the regiment a distinctive cowboy look.[17] Eventually they were all also issued brown canvas stable fatigues for field service and given machetes to help them whack through the tropical foliage of Cuba.

Among Roosevelt's favorite haunts in San Antonio was the Buckhorn Saloon. Opened for business in 1881 by Albert Friedrich—whose father made high-quality horn furniture—from day one the bar had a standing offer: "Bring in your deer antlers and you can trade them for a shot of whiskey or a beer." Before long the Buckhorn had the finest collection of horns and trophy mounts in the world. Men would actually collect antlers shed in the Texas Hill Country and then ride into San Antonio for their free drinks. In 1882, in fact, for $100 Friedrich acquired a record-making "78 Point Buck"; it was placed behind the bar, where it still remains. Business was so good that Friedrich moved his operation to larger quarters at Houston and Soledad streets, a few blocks from the Alamo. Expanding on the tradition of free alcohol for deer racks, the Buckhorn, by the time Colonel Roosevelt discovered it, was also giving away shots for rattlesnake rattles. (Later, mounted fish from the Seven Seas were included in the freebie deal.) Even though Roosevelt wasn't much of a drinker, he would wander in with fellow Rough Riders, order a beer, nurse it, and listen to an old guitar-picker sing about being a cowhand along the Brazos River.[18]

San Antonio was a fillip to Roosevelt. He liked being dependent on his own horse and seeing sagebrush. No doctor or pharmacist could have uplifted him better than the opportunity to lead lineal descendents of Andrew Jackson's fighting force in the Battle of New Orleans. Whether a volunteer was a Fort Worth bronco buster, a Newport polo swell, or a Tucson shopkeeper, each of the Rough Riders shared traits with the others: they all shot straight, were in good physical condition, hated Spain, and were willing to mobilize quickly. "I suppose about 95 per cent of the men are of native birth," Roosevelt wrote. "But we have a few from everywhere including a score of Indians, and about as many of Mexican origin from New Mexico."[19]

Many fine firsthand accounts have been written of Roosevelt's arrival in San Antonio, colorful portrayals of him pacing around like a bantam rooster in a new fawn uniform with canary-yellow trim, sweat constantly beading on his forehead from the Texas heat. The regiment's chant soon became, "Rough, tough, we're the stuff. We want to fight and we can't get enough."[20] Throughout San Antonio signs were hung welcoming each state and territory and offering hospitality.[21] The Menger Hotel—built twenty-three years after the fall of the Alamo—housed a replica of the pub inside Great Britain's House of Lords; bartenders used to give out free shots of whiskey, in solidarity with the men. (The hotel later renamed the room the Roosevelt Bar.) However, Colonel Wood upbraided the much younger Roosevelt for purchasing beer kegs for volunteers. "Sir," Roosevelt apologized when reprimanded, "I consider myself the damndest ass within ten miles of this camp."[22]

Before Roosevelt headed down to San Antonio for the training, he gave away his Elkhorn Ranch to Sylvane Ferris and sold his last head of cattle. (He had visited the ranch only infrequently in 1893, 1894, and 1896.[23]) Roosevelt nevertheless differentiated himself in San Antonio from the other Ivy Leaguers in the Rough Riders. Without falsity, he presented himself as both a Knickerbocker and a wilderness hunter to the rank and file training along the San Antonio River. As Owen Wister put it, Roosevelt embodied both the East (as a socialite) and the West (as a cowboy). Regularly, Roosevelt jogged and rode horseback for miles in the lean May sunshine with his regiment, not far from the Alamo. Many of the Rough Riders had fought against the Comanche and Apache, and had won. Roosevelt knew that in cow country, along the wild borderlands with Mexico, men gave each other nicknames like Red Jim, Bear Jones, or Dutchey; he was honored to be called "the Colonel" by everybody.[24]

II

The Rough Riders eventually boarded a slow-moving train to Tampa Bay on May 30, with Arizona providing the regimental colors. Before the departure from San Antonio Roosevelt worried that the warhorses, ears pricked, snorting, and rattling the boards in the railcar stalls, were being

Colonel Theodore Roosevelt in his Rough Riders uniform.

bullied and whipped as they were loaded onto the railcars by supposed horse masters. The harassing shouts of "Yahah!" bothered him. Taking charge of the situation, Roosevelt waved the others away and loaded the ponies properly into their compartments for the journey to Florida.[25] Back in 1894 Owen Wister had written a short story, "Balaam and Pedro," in *Harper's Monthly* about the abuse of a horse he encountered on a western trek; the Wyoming character who stopped the inhumane treatment became the hero of *The Virginian.* (Wister, in fact, praised Henry Bergh's movement to prevent cruelty to animals in his 1905 novel.[26]) Now Roosevelt, like the protagonist of Wister's tale, was protecting horses under his command.

Once the train reached Galveston the dry heat was replaced by muggy humidity. Nobody, however, really seemed to mind. The railway cars were roofless, giving the procession the aura of a parade, with Rough Riders waving at villagers as the train passed slowly by. At every stop in Louisiana and Mississippi folks poured out of the countryside to have a quick look at Colonel Roosevelt and his cowboy-garbed regiment. Moms with baked goods and girls with pitchers of fresh milk greeted them at depots. Watermelon wagons appeared regularly, providing snacks for the soldiers. Roosevelt basked in the limelight at each depot, offering a running commentary on American exceptionalism. Before the fighting in Cuba even began, there was going to be a showbiz side to the Rough

Riders, but they would soon also touch the heartstrings of America. And, from the start, Colonel Roosevelt was the willing leader.

These Rough Riders were the pride of Roosevelt's heart, and his inextinguishable enthusiasm kept their morale high. Good-heartedly Roosevelt gave up his private berth to a trooper with measles, taking a coach seat with his men. He was determined not to treat his privates like indentured household servants. To kill time, some of the men, while waving away hatching flies, wrote a jingle about going down the "dusty pike" with Colonel Roosevelt, ready to "throttle the sons of Spain."[27] Roosevelt was reading Edmond Demoulins's *Supériorité des Anglo-Saxons* (1897), a foray into social Darwinism. Demoulins wrote that France, his native country, lacked the "independence and ability to fight life's battles fearlessly," qualities that the Anglo-Saxons possessed in spades.[28] He also believed that education—at which the British excelled—was *the* key to developing a great country. All this was music to Roosevelt's ears. As the train rumbled eastward Roosevelt consoled himself with Demoulins's notions of natural selection in the human arena.

Once he finished the book, Roosevelt, flaunting his authority, called his men together at a depot stop near the Sabine River. He lectured them on Darwinism, describing how natural selection explained everything, from the size of their noses to the wings on the mosquitoes they were swatting. Many of the Rough Riders had taken to calling the Spaniards "greasers" or "dagoes," and Roosevelt promised to explain shortly why the epithet wasn't entirely unfair.[29] Seizing Darwin's image of a "tangled bank," which ended *On the Origin of Species*, Roosevelt now made it his own. "Through all nature," Roosevelt intoned, "it is a case of the survival of the fittest. Look at the magnificent trees along the river. The ones that started crooked were crowded out and died. The strong and the straight saplings appropriated all the food."[30]

That wasn't the extent of Roosevelt's lecture on Darwinism. Inspired by Demoulins, he took a leap forward, bringing humans and mammals into the mix. "It is the same with wild animals," he continued. "The cripples and the inefficients that cannot support themselves are killed off." Humans, Roosevelt maintained, were just highly developed animals. So just as the Chinese purged themselves of the criminal class and the wicked, the United States, if it wanted to achieve greatness, needed to sanitize its society by getting rid of criminals. (To be fair, he did, however, add the important point that the blind or crippled or chronically infirm of sound mind needed to be cared for by society at large.)

Roosevelt then segued into the unfitness of Spaniards, the weak link in European affairs. Most, he said, were shiftless and of weak moral fiber. As Americans, his men had a newfound responsibility to liberate Cuba, Puerto Rico, and the Philippines, because the Spanish had proved themselves tyrannical. Obviously Roosevelt was trying to arouse his fighting forces. Verbally attacking the enemy in training was a practice older than Rome. Taking the jingoist notion one step farther, Roosevelt, as if reeling off a paragraph of Kipling, boasted that the Rough Riders had a sacred obligation to revenge the *Maine*. "The old vigilantes in Montana did not have a single law," he reminded them, "but they did have a simple, wholesome code which everyone knew. Life and property are secure as a consequence."[31]

Notions of natural selection had long ago been stamped into Roosevelt. *On the Origin of Species* had taught him that evolution had some important implications for human societies. Still, the lecture was overblown and felt wrong. While Roosevelt was not a strict evolutionist in human affairs, he nevertheless was in the clutches of the general Spencerian notion "root, hog, or die." To Roosevelt it was a slogan akin to gospel.[32] As he articulated in his 1895 essay "Kidd's Social Evolution," published in *North American Review* (a scholarly analysis of Benjamin Kidd's just-published reflection on natural selection), Roosevelt believed humans had two sides—one inspired by Darwin ("the rivalry of natural selection") and the other being moral character (essentially the Ten Commandments). There were laws, he wrote, which governed mankind's reproduction in every generation "precisely as they govern the reproduction of the lower animals."[33] But Roosevelt also understood that the rivalry of natural selection was just one way—not all-encompassing—in which *Homo sapiens* judged progress:

> Other things being equal, the species where this rivalry is keenest will make most progress; but then "other things" never are equal. In actual life those species make most progress which are farthest removed from the point where the limits of selection are very wide, the selection itself very rigid, and the rivalry very keen. Of course the selection is most rigid where the fecundity of the animal is greatest; but it is precisely the forms which have most fecundity that have made least progress. Some time in the remote past the guinea pig and the dog had a common ancestor. The fecundity of the guinea pig is much greater than that of the dog. Of a given number of guinea pigs born, a much smaller proportion are able to survive in the keen rivalry, so that the limits of selection are wider, and the selection itself more rigid; nevertheless the progress made by the progenitors

of the dog since eocene days has been much more marked and rapid than the progress made by the progenitors of the guinea pig in the same time.[34]

Probably the best way to understand Roosevelt's thinking on social Darwinism—besides reading "Social Evolution," which was reprinted as a chapter of his 1897 book *American Ideals**—is to study a lecture given by John Burroughs in the 1890s, "The Biological Origin of the Ruling Class." Fulsomely embracing Darwin as a naturalist, Burroughs believed that the law of the strong overcoming the weak offered a valuable viewing of "the drama of human politics and business."[35] To Burroughs, the fittest usually rose to the top in American life. Undoubtedly evil charlatans sometimes tried to rig the reality. Scum often rose in life's pond—but on the whole, every generation of Americans produced its heroic natural elite. These winners rode herd on human affairs, directing their course, helping civilizations and societies to survive cataclysms. Whenever truly bad leaders, imposters, reached a pinnacle of power, eventually they would be destroyed by the natural elite. Competition of all kinds, Burroughs went on in his lecture (which was a set piece), should be supported so that the best-and-the-brightest could earn their rightful positions of societal power. As for the poor and disabled, Burroughs believed the natural elite had a moral responsibility to take care of them.[36]

III

After four days on the train, the Rough Riders arrived in Florida. The unit was assigned to the U.S. transport *Yucatan*, but the departure date from Tampa Bay kept changing. Roosevelt worried that if the boat didn't leave soon, his men's livers weren't going to withstand all the hiatus booze. The first day was incredibly humid, with a hot, glassy atmosphere and scant wind. Luckily, Edith came down from Oyster Bay for a few days' reunion at the Hotel Tampa. Anxious for war, Theodore was unperturbed by the omnipresent swarms of chiggers and sand flies. To kill time he studied Florida's botanical wonderland. At a glance, he could distinguish holy trees from blue beech and ironwood. Yet, Roosevelt found waiting deeply frustrating—ceaseless delays, widespread discomfort, missing cargo, confusion in command. One afternoon a jolt of excitement hit the camp: there was a rumor that Spanish warships were patrolling the Straits of Florida. But, alas, it was just a rumor.[37]

*In this book Benjamin Kidd's name was dropped from the chapter title.

While waiting to be shipped out, Roosevelt studied the waterfowl along the wharf front and marshy inlets—ibis, herons, and double-crested cormorants, among scores of others. These were the species his Uncle Rob had written about so ably in *Florida and Game Water Birds*. Beneath Roosevelt's army boots on the Tampa beaches were sunrise tellin and wide-mouthed purpura and ground coral, bay mud, tiny pebbles mixed with barnacles and periwinkles. Writing to his friend Henry Cabot Lodge, Roosevelt turned quasi-geobiologist, delineating Florida's semitropical sun, palm trees, sharks swimming in the shallows, and sandy beaches much like those on the French Riviera.[38] The Gulf of Mexico, the ninth-largest body of water in the world, interested Roosevelt no end. Its barrier islands from Texas to Florida were home to myriad songbirds and shelled mollusks. Captain Mayne Reid had written about the region as a place of high adventure, where swells were 150 feet long, the throw of the surf was unpredictable, and pirate bands camped on isolated coral islands, eating clams and octopuses, fugitives from all governments.

Spending these days in Tampa Bay, various U.S. Fish and Wildlife historians believe, later influenced Roosevelt's creation of federal bird sanctuaries along both of Florida's coasts. What Roosevelt understood from being stationed on the Gulf Coast was that the market hunters were having a bad effect on Florida's ecosystem, including the Everglades, Indian River, Lake Okeechobee, and the Ten Thousand Islands. The previous year, Roosevelt's New York–based ornithologist friend Frank M. Chapman had warned him once again that tricolored herons and snowy egrets were being slaughtered for feathers. Now, huge mounds could be seen around the port of Tampa, bird carcasses piled twenty or thirty yards high and rotting in the sun. If the slaughter wasn't stopped, the bird roosts of Florida would vanish, species going the extant way of the passenger pigeon, the ivory-billed woodpecker, and the Labrador duck.[39*]

As President McKinley concentrated on the Spanish-American War, the American conservationist agenda in 1898 was left in the hands of Secretary of the Interior Bliss. Realizing that creating new forest reserves was inevitably controversial, Bliss focused on enlarging existing federal reserves, such as Pecos River in New Mexico and Trabuco Canyon in California. In addition, the lands in the Alaskan Territory were protected

*Roosevelt was right to worry about the future of Florida wildlife. As Elizabeth Kolbert noted in *The New Yorker* (May 25, 2009), of the fifty billion species to have ever inhabited Earth, over ninety-nine percent have vanished. At least eighty percent of all marine species have become extinct.

under an experimental program for the Department of Agriculture. President McKinley himself bragged about these forest reserves—and other accomplishments—in his third State of the Union address.[40] Meanwhile, Roosevelt noticed in Tampa Bay that Florida—one of the richest states in the Union in terms of wildlife—was being treated as a worthless swamp, instead of as the amazing array of ecosystems his Uncle Rob and Charlie Hallock had written about. As Chapman, who was spending much of his year in Florida, told Roosevelt, wildlife protection had to be enforced there, or else dozens of species would soon be destroyed forever.

When the *Yucatan* finally set sail for Cuba on June 13, Roosevelt was nearly giddy with joy at escaping from Tampa. As the regimental band played "The Star-Spangled Banner" and "The Girl I Left behind Me," he looked at the other forty-eight vessels in the flotilla, neatly aligned in three columns, steaming to war. As his boat headed southward, he used his descriptive powers in his correspondence, saying that the Florida Keys area was "a sapphire sea, wind-rippled, under an almost cloudless sky."[41] There was no sign of an equinoctial storm that could throw the armada off course; it was, to use the sailors' cliché, clear sailing. But Roosevelt was hard pressed to turn a naturalist phrase in his diary. There was a certain unrest about these tiny islands themselves—waves breaking heavily on their beaches, tides advancing and retreating—which transcended description. When he first caught sight of the shoreline of Cuba's Santiago Bay, waves beating in diagonals, Roosevelt finally turned somewhat poetic. "All day we have steamed close to the Cuban Coast," he told his sister Corinne, "high barren looking mountains rising abruptly from the shore, and at a distance looking much like those of Montana. We are well within the tropics, and at night the Southern Cross shows low above the horizon; it seems strange to see it in the same sky with the Dipper."[42]

On June 23 the Rough Riders landed at the fishing village of Siboney about seven miles west of Daiquirí, behind General Henry Ware Lawton's Second Division and General William Shafter's Fifth Corps. They were ready for action. Their attitude toward the Spanish occupation of Cuba was best summed up by Wister's ultimatum in *The Virginian*: "I'll give you till sundown to leave town."[43] In New York and Washington, D.C., Roosevelt had romanticized the Cuban insurgents who were fighting the Spanish. However, he soon called them "the grasshopper people," for the shabby way they had treated the land.[44] The woods and fields were so dry that Roosevelt feared they would catch on fire. Only the little grasses tossing purplish shadows in the sand seemed irrigated. Everything manmade looked battered and cheap. Ironically, the Rough Riders were under the command of Brigadier General S. B. M. Young, whom Roosevelt called "as fine a type of the American fighting soldier as a man could hope to see." By happenstance General Young had once been in command of Yellowstone National Park, and Roosevelt, as president of the Boone and Crockett Club, had worked with him on wildlife protection and forest preservation issues.[45]

The Rough Riders took ashore blanket rolls, pup tents, mess kits, and weaponry, but no one thought to give them any insect repellent. It was hot. There was no wind, and they felt on fire. The tangled jungles and chaparral of Cuba, particularly in early summer, were breeding grounds for flies that now swarmed over the camps. As it turned out, these insects were as much the enemy in the Cuban heat as the Spaniards. They filled the air with psssing, droning, chirping, and humming; not for a second were they quiet. Sleeping with a mosquito net was a must. There were 100 varieties of ants in Cuba, including strange stinging ants that seemed to come from a different world. (Darwin, in *The Descent of Man*, claimed that "the brain of an ant is one of the most marvellous atoms of matter in the world, perhaps more so than the brain of a man."[46]) The little crouching chameleons with coffin-shaped heads, unafraid of the soldiers, changed color from bright green to dark brown depending on the foliage they rested on. "Here there are lots of funny little lizards that run about in the dusty roads very fast," Roosevelt wrote to his daughter Ethel, "and then stand still with their heads up."[47]

Unfortunately, the military mapmakers had failed to tell Colonel Roosevelt and company that Las Guásimas, the dingiest village imaginable, was, with only modest exaggeration, the world's biggest scorpions' nest. Soldiers soon swelled up from scorpion bites, which also caused dizzyness and arthritic-like aches. Stephen Crane, who was then a war correspondent for the *New York World* (and whom Roosevelt disdained as immoral because of his novella *Maggie: A Girl of the Streets*), wrote nastily that the former New York police commissioner and bird-watcher recognized "the beautiful coo of the Cuban wood-dove" but inexplicibly seemed deaf to the fatal noise of a "Spanish guerrilla wood dove which had presaged the death of gallant marines."[48] It was a Craneian cheap shot; still, Roosevelt may have been the only soldier in Cuba who recorded ornithological observations of cardinals and tanagers.

Just two days after landing, the Rough Riders got their taste of combat at the battle of Las Guásimas. Although they were just one of many U.S. outfits assaulting the Spanish fortifications around the coastal city of Santiago, the Rough Riders deserved the praise they've received for their performance. Just like the army regulars, they truly were a bold, well-disciplined fighting force; and McKinley acknowledged this by the eve of the horrific battle of Santiago, promoting Colonel Wood (to brigadier general) and Roosevelt (to colonel) within the week. Trying to deflect all the press attention being showered on him only, Roosevelt often trumpeted the prowess of his gutsy troops. "They were a splendid set of men, these Southwesterners—tall and sinewy, with resolute, weather-beaten faces," he wrote, "and eyes that looked a man straight in the face without flinching."[49]

One Rough Rider to whom Roosevelt took a real shine was Corporal David E. Warford of Troop B (Arizona Territory), under the leadership of Captain James H. McClintock.[50] Warford came from Globe, Arizona, a one-trough mining town in the heart of Apache country. It seemed to Roosevelt that Warford had been born on horseback. Constantly smoking, the most confident equestrian around, and able to call out a bird by its song, Warford never complained and was full of bounce. Colonel Roosevelt grew even closer to Warford after the young volunteer was shot in both thighs in the battle of Las Guásimas and continued fighting, injured, before repairing to a hospital ship.[51] Warford was not literate, and he bragged that he had "kilt" Spaniards and that a wounded fellow Arizonian was "crow-bait," but Roosevelt admired his western spirit. If the forest reserves had brave outdoorsmen like Warford protecting them, Roosevelt later realized, timber thieves and game poachers could finally be stopped.

What Roosevelt called his "crowded hour" occurred on July 1, 1898, when, on horseback, he led the Rough Riders (plus elements of the Ninth and Tenth regiments of regulars, African-American buffalo soldiers, and other units) up Kettle Hill (near San Juan Hill) in what is known in military history as the battle of San Juan Heights. Once the escarpment was captured Roosevelt, now on foot, killed a Spaniard with a pistol which had been raked up from the sunken *Maine*. Social Darwinism seemed to have played out in Roosevelt's favor that day—he was the fittest pistolero. Roosevelt later said that the charge up Kettle Hill surpassed all the other highlights of his life. Somewhat creepily, it was reported, Roosevelt had beamed through the battlefield depredations and gory deaths, always flashing a wide smile, but with his pistol pointed. Whether he was ordering artillery reinforcements, helping men cope with the prostrating heat, finding canned tomatoes to feed the troops, encouraging Cuban *insurgentes*, or miraculously procuring a huge bag of beans, Roosevelt was always on top of the situation, doing whatever was humanly possible to help his men avoid both yellow fever and unnecessary enemy fire. There was no arguing about it: Colonel Roosevelt had distinguished himself at Las Guásimas, San Juan, and Santiago (although the journalists did inflate his heroics to make better copy). By the Fourth of July, Roosevelt had become a legend in the United States, the most beloved paragon produced in what Secretary of State John Hay called "a splendid little war."[52]

With the capture of San Juan Heights—the villages and vista spots overlooking Santiago—the city itself soon surrendered. The war was practically over. The stirring exploits of Colonel Roosevelt were published all over America, turning him overnight in to the kind of gallant warrior he always dreamed of being. But the hardships Roosevelt had suffered were real. Supplies like eggs, meat, sugar, and jerky were nonexistent. Hardtack biscuits—the soldiers' staple—had attracted hideous little worms. Just to stay alive the Rough Riders began frying mangoes.[53] Worse still, the 100-degree heat caused serious dehydration. Then there was the ghastly toll from tropical diseases. Diarrhea and dysentery struck the outfit like a plague. Fatigue became the norm. So many Rough Riders were dying from yellow fever and malaria that Colonel Roosevelt eventually asked the War Department to bring them home to the Maine coast, hoping to save lives.[54] The request showed Roosevelt at his best, putting the welfare of his men first, not worrying that history might misconstrue it as a way to dodge combat. On August 14, the Rough Riders, following a brief stopover in Miami, arrived at Montauk peninsula at the end of Long Island (not Maine) and were placed in quarantine for six weeks.[55]

IV

An odd feature of Roosevelt's leadership of the Rough Riders was his continued biophilic obsession with animals, even when preparing for combat. In fact, this distinguishes his war memoir *The Rough Riders* from all other accounts of the 1898 Cuban campaign. And in his autobiography, Roosevelt presents his theory about the role of pets in sustaining military morale. Compared with military tactics and the toll of yellow fever, such passages can seem frivolous, but they do offer a valuable perspective on Roosevelt as a war leader and as a person. Largely at Roosevelt's instigation, his First Volunteer Cavalry Regiment had three animal mascots, brought all the way from San Antonio through their stay in Tampa Bay. Most famously, there was a young mountain lion, Josephine, brought in by an Arizona trooper named Charles Green, a gift from a supportive citizen in Prescott.[56]

Roosevelt adored everything about the cougar cub: her sand-colored coat, dark rounded ears, white muzzle, and piercing blue eyes, which would turn brown as she matured. He knew that as an adult, Josephine would be able to run elusively at thirty-five miles per hour and leap from boulder to boulder with breathtaking grace. Eventually Josephine would weigh at least ninety pounds and be able to pull down a 750-pound elk with her powerful jaw.[57] But for now she was domesticated, though at times surly. (Roosevelt wrote in *The Rough Riders* that she had an "infernal temper.") As the *New York Times* wrote of Josephine, she "rejoiced" when her name was uttered. She was, in turn, beloved by all the men.[58] Purrs were commonplace, even though Josephine learned to distrust anybody who wasn't wearing a military uniform. As the reporter Edward Marshall put it, Josephine "hated civilians."[59]

Roosevelt spent as much time with the cougar cub as he could. She became something of a shadow cat. One evening when they were in Montauk, Josephine got loose, climbed into bed with a soldier, and began playfully chewing on his toes. Roosevelt later chuckled in *The Rough Riders* that the volunteer "fled into the darkness with yells, much more unnerved than he would have been by the arrival of any number of Spaniards."[60] Writing to his children from Tampa Bay, Roosevelt told how their mother, Edith, who had visited him for a few days, was stunned to find him with a cougar at his side. "The mountain lion is not much more than a kitten yet," he explained, "but it was very cross and treacherous."[61]

Another steadfast companion in the Rough Riders was a golden eagle, one of the largest bird species in North America and the national emblem of Mexico. The volunteers named it Teddy in Roosevelt's honor. As in N. Scott Momaday's novel *House Made of Dawn*, Roosevelt considered himself a charter member of the Eagle Watchers Society.[62] Roosevelt loved following these raptors as they swooped down to pluck up snakes and darting prey, and he had even managed to learn a little about the art of falconry. Wearing leather gloves in order not to get clawed, he would hold his arm out for Teddy, calling the New Mexican–born eagle back to camp after it had had its fill of lizards and squirrels. "The eagle was let loose and not only walked at will up and down the company streets, but also at times flew wherever he wished," Roosevelt recalled. "He was a young bird, having been taken out of his nest when a fledgling. Josephine hated him and was always trying to make a meal of him, especially when we endeavored to take their photographs together. The eagle, though good-natured, was an entirely competent individual and ready at any moment to beat Josephine off."[63]

Colonel Roosevelt pets Teddy the golden eagle while members of the Rough Riders play with Cuba the dog and Josephine the cougar.

Both Josephine and Teddy were left behind in Tampa, since it would obviously have been nonsensical to bring a cougar and an eagle into battle.[64] The third mascot, however, made it to Cuba. Roosevelt's regiment had a "jolly dog" named Cuba, owned by Corporal Cade C. Jackson of Troop A from Flagstaff, Arizona. The mutt had dirty gray poodle-like fur and the personality of a Yorkie. Little Cuba could be easily scooped up with one hand. Frisky as a dog could be, Cuba actually accompanied the regiment "through all the vicissitudes of the campaign." Aboard the *Yucatan*, Roosevelt had a Pawnee Indian friend draw Cuba—who ran "everywhere round the ship, and now and then howls when the band plays"—for his daughter Ethel.[65]

Every time the Rough Riders went into battle, Cuba would run off and disappear into the jungle, frightened by the noise of the artillery. Once the smoke cleared, however, when the men were bandaging wounds or frying eggs over a wood fire, Cuba would suddenly slink back into camp looking for handouts and back-scratches.[66] Later, after the victory, a reunion photograph of the Rough Riders was taken in Montauk, with all three mascots in the same frame, Cuba begging near Colonel Roosevelt's leg for either a treat or attention.[67] According to Roosevelt, the dog was occasionally "oppressed" by Josephine but was sometimes able to "overawe" the mountain lion "by simple decision of character." Sometimes when Josephine growled, however, Cuba backed off, like a horse hearing the hum of a rattlesnake.[68]

Perhaps because Roosevelt was so comfortable with the trio of animals, knowing how to feed mice to the eagle and scratch Josephine behind the ears, these mascots added a Dr. Doolittle dimension to his character. In both San Antonio and Tampa Bay his two horses—Rain-in-the-Face and Texas—practically never left his side. When Vitagraph motion picture technicians were filming the Rough Riders wading ashore in Cuba off the *Yucatan*, a soldier was ordered to bring Roosevelt's steeds safely onto the beach. Unfortunately, a huge wave broke on Rain-in-the-Face, causing him to drown: he inhaled seawater and could not be released from his harness. For the only time during the war days Roosevelt, his mind unsteadied, went berserk, "snorting like a bull," as Albert Smith of Vitagraph recalled, "split[ing] the air with one blasphemy after another." As the other horses were brought onto shore, Roosevelt kept shouting, "Stop that goddamned animal torture!" every time salt water got in a mare's face.[69]

Skeptics of Rough Riders lore point out that Roosevelt was only seeking glory, always appearing—abracadabra—when a camera came along. Some critics carped that he used friendly reporters, such as Richard Harding Davis and John Fox, as tools. Roosevelt—so the opprobrium went—was thinking only about himself in Cuba, seeking fame amid the parlous carnage. What makes it clear that these are misrepresentations is the fact that *all* the surviving Rough Riders, even those who lost their legs or eyes, testified that he was a phenomenal leader. Never once did Roosevelt expect more from any volunteer than he gave of himself. No matter how dangerous the situation, he was in the thick of the action. The Spanish soldiers, for example, used smokeless powder, which made it impossible to tell where bullets were coming from in the jungle. At all hours and in all circumstances, Colonel Roosevelt, placing fate in the hands of God, refused to duck or run for cover as the bullets whizzed by. Calmly, even under enemy fire, Roosevelt helped wounded men make primitive tourniquets out of tree branches and bandannas. "Yesterday we struck the Spaniards and had a brisk fight for 2½ hours before we drove them out of their position," Roosevelt wrote to Corinne and her husband, Douglas Robinson. "We lost a dozen men killed or mortally wounded, and sixty severely or slightly wounded [out of about 500]. One man was killed as he stood beside a tree with me. Another bullet went through a tree behind which I stood and filled my eyes with bark. The last charge I led on the left using a rifle I took from a wounded man. . . . The fire was very hot at one or two points where the men around me went down like ninepins."[70]

The Spanish snipers in Cuba fired high-speed Mauser bullets and had deadly aim. The U.S. volunteers, including the Rough Riders, in fact, faced some of the worst combat in the history of warfare.[71] As Stephen Crane, embedded with the Rough Riders, noted, "The tropical forests were regularly aglow in fighting."[72] Constant barrages of rifleshots resulted in heavy American losses. "In the period of about four and a half months they were together, 37 percent of those who got to Cuba were casualties," historian Virgil Carrington Jones said of the Rough Riders. "Better than one out of every three were killed, wounded, or stricken by disease. It was the highest casualty rate of any American unit that took part in the Spanish-American campaign."[73]

The letters Roosevelt wrote from Cuba crackle with the kind of martial detail also found in Crane's Civil War novel of 1895, *The Red Badge of Courage.* Yet they're also full of natural history, with observations about the "jungle-lined banks," "great open woods of palms," and "mango trees," "vultures wheeling overhead by hundreds," and even a whole command "so weakened and shattered as to be ripe for dying like rotten sheep." Constantly, Roosevelt tried to conjure up nature as a way to increase personal power. When the director Terrence Malik made *The Thin Red Line* in 1998—a film about the Battle of Guadalcanal, based on a World War II novel by James Jones—he constantly cut away to exotic birds. This device helped illustrate that nature always watched the pageant of human combat from the sidelines, waiting for the artillery to cease before coming back to life and inventorying the new morning.

In Roosevelt's correspondence and war memoir the land crab is omnipresent, almost the central metaphor of his Cuban campaign. Experts noted that the local species, *Gecarcinus lateralis*, commonly known as the blackback, Bermuda, or red crab, leaves the tropical forests each spring to mate in the sea. This made for an eerie spectacle all along Cuba's northern coast as these disfigured creatures, many with only one giant claw, crawled out of the forests across roads and beaches to reach the water. Swollen with eggs, the female red crabs made their journey to the Caribbean Sea, which was their incubator, traveling five to six miles a day over every obstacle. Roosevelt noted that they avoided the sun's glare, often gravitating to shade, just like wounded soldiers. As if in a scene from Borges or García Márquez, these burrowing red crabs—their abalone-like shells marked with gaudy dark rainbow swirls—while living on land, still had gills, so they needed to stay cool and moist. "The woods are full of land crabs, some of which are almost as big as rabbits," Roosevelt wrote to Corinne; "when things grew quiet they slowly gathered in gruesome rings around the fallen."[74]

For the first time as an adult, Roosevelt was in the tropics. The very density of vegetation was daunting, the white herons often standing out against the greenery like tombstones. These red crabs were to him what tortoises or finches were to Darwin; everything about them spoke of evolution. Unlike the stone crabs of Maine, these red crabs, by contrast, weren't particularly good-tasting; from a culinary perspective they were off-putting. Still, with food supplies sparse, the soldiers smashed the red crabs with rocks, discarded the shells, and mixed the meat into their hardtack, calling the dish "deviled crab." Although the crabs were not dangerous, many Rough Riders were jarred awake at night by the formidable pincers. And the crabs were persistent—a soldier would shake them away from his bedroll, but after scurrying away, the crabs would come back a short while later. Sleeping off the ground on a hammock became more coveted than having a can of tobacco or bottle of rum. What disturbed Roosevelt the most about the Cuban crabs, however, was their attraction to carrion, fallen soldiers as well as dead animals. It wasn't pleasant to think that the price of liberating Cuba was to die on a lonesome beach with red crabs and ants crawling all over your body, entering your mouth and eyes and ears.

In *The Rough Riders*, Roosevelt ably described the timeworn, brush-covered flat in the island village of Daiquirí where his volunteer regiment camped one evening, on one side of them the tropical jungle and on the other a stagnant, malarial pool fringed with palm trees. After the sacking of Santiago, many of his Rough Riders, a third of whom had served in the Civil War, lay wounded in ditches with flies buzzing around them. Sometimes, after an American died, local Cubans would strip the corpse of all its equipment. Humans could be scavengers, too. Roosevelt turned to images of avians and crustaceans to explain the horror of death. "No man was allowed to drop out to help the wounded," he lamented. "It was hard to leave them there in the jungle, where they might not be found again until the vultures and the land-crabs came, but war is a grim game and there was no choice."[75]

Roosevelt then went on to tell of U.S. volunteer soldiers, comrades in arms, mortally wounded—perhaps shot through the stomach—dying without uttering a sound or gasping out a last wish. Men lucky enough to crawl propped themselves up against palm trees and expired in the dismal shade, their uniforms drenched in sweat, urine, and blood. A little field hospital was set up, and Roosevelt witnessed the pathos of men heaving for air, their lungs collapsed, broken ribs piercing vital organs. "We found all our dead and all the badly wounded," he wrote. "Around one of the latter the big, hideous land-crabs had gathered in a grewsome ring, waiting for life to be extinct. One of our own men and most of the Spanish dead had been found by the vultures before we got to them; and their bodies were mangled, the eyes and wounds being torn."*[76]

If that ghastly scene wasn't harrowing enough, Roosevelt proceeded to tell another story. After staring at a corpse that had been mutilated by vultures, the blood having coagulated hours before, Rough Rider Bucky O'Neill, who at home was the mayor of Prescott, Arizona, came up to Roosevelt, shook his head, and said, "Colonel, isn't it [Walt] Whitman who says of the vultures that 'they pluck the eyes out of princes and tear the flesh of kings'?" Not wanting to discuss poetry, Roosevelt muttered that he wasn't sure about the proper attribution and walked away. Then, as if to demonstrate how tenuous life really was, Roosevelt matter-of-factly noted in *The Rough Riders* that O'Neill himself soon perished in the trenches of Cuba: "Just a week afterward we were shielding his own body from the birds of prey."[77]

V

In his review essay "Kidd's Social Evolution" for *The North American Review*, Roosevelt offered an example of when the dictates of natural selection superseded a love of wildlife. "Even the most enthusiastic naturalist," he wrote, "if attacked by a man-eating shark, would be much more interested in evading or repelling the attack than in determining the precise specific relations of the shark."[78] By this criterion, Roosevelt was a success in Cuba in two ways. He not only thwarted the Spanish sharks but

---

* *The Osprey*, an illustrated magazine of birds and nature, thought the Rough Riders, in their letters home, were unfairly maligning vultures. "Had the men only known, the birds, instead of being their enemies, were in reality, invaluable allies," the editorial argued. "In the thick growth of vegetation that clothes many of the hillsides and valleys of Cuba, the work of the burial parties is slow and difficult, and bodies are often overlooked in the search. The keen senses of the buzzard lead them unerringly to the spot. In many cases his work, nauseating and disgusting as it must be to contemplate, is the means of preserving the health and strength of many of our soldiers." Walter Adams Johnson and Dr. Elliot Coues, "Editorial Notes," *Osprey*, Vol. 3, No. 1 (September 1898), p. 12.

managed to make detailed diary notes about vultures and crabs, which he planned to use in his memoir *The Rough Riders.*

When the victorious Rough Riders returned to the United States, Roosevelt was the most acclaimed man in America. His homeward journey, in fact, had been treated as major news. In hard, good health, taut and fit, his face coppered and his hair cut short, he was living his boyhood fantasy of being a war hero. Roosevelt had endured the vicissitudes of war with commendable grit, and now it was all bouquets. "His personal view of the war was reported to have been extracted from Social Darwinism," the historians Peggy Samuels and Harold Samuels say in their landmark work *Teddy Roosevelt at San Juan.* "The superior Anglo-Saxon race necessarily won over the decadent Spaniards."[79] As Roosevelt wrote in a new foreword to *The Winning of the West*, the Spanish-American War had completed "the work begun over a century before by backwoodsmen" by booting "the Spaniard outright from the western world."[80]

Anglo-Saxonism was hardly all there was to the victorious battlefield prowess of the Rough Riders. Something in the American wilderness experience, Roosevelt believed, gave his regiment the upper hand over the Spaniards. Not a single Rough Rider got cold feet or shrank back. Something about the mesas of New Mexico and Arizona had taught them to be tough. In an important essay, "The Darwinist Frontier," the historian Patrick Sharp has contended that Roosevelt believed the American fighting spirit would continue only as long as outdoorsmen didn't get lazy and rest on the laurels of modernity.[81] Slowly, Roosevelt was developing a theory about this, which he would call the "strenuous life." The majestic open spaces of America like the Red River Valley, Guadalupe Mountains, Black Mesa, Sangre de Cristo Range, Prescott Valley, and Big Chino Wash had hardened his men, teaching them the kind of self-reliance Emerson promoted. Wouldn't Rough Riders make terrific forest rangers and wildlife wardens? Didn't the wildlife protection movement need no-nonsense men in uniform to stop poaching in federal parks? "In all the world there could be no better material for soldiers than that afforded by these grim hunters of the mountains, these wild rough riders of the plains," Roosevelt said. "They were accustomed to following the chase with the rifle, both for sport and as a means of livelihood."[82]

While the Rough Riders recovered from bodily atrophy at Montauk, where they were watched for signs of yellow fever, New York's Republican Party was urging Roosevelt to run for governor that fall. Two prominent local politicians—Lemuel Ely Quigg (who had backed him for mayor in 1894) and Ben Odall, Jr. (chairman of the Republican state committee), met with him on August 19 to strategize how best to turn a war hero, about whom New Yorkers were currently fanatical, into a sitting governor.

After the hot trenches of Cuba, the cool summer breezes on the Montauk peninsula were a welcome relief to Colonel Roosevelt, even though the makeshift barracks had no charm. There were ocean beaches and dunes, shrublands and tidal flats, brackish wetlands and salt marshes. As Roosevelt contemplated his political future, and as everybody clamored to shake his hand, the raccoons and white-tailed deer of Montauk brought balance to his newfound fame. There was even Nantucket juneberry along the sand plains to meticulously study. One hundred years later, to honor the fact that the famous Rough Rider had lived at Camp Wikoff in 1898, the community of Montauk named a 1,157-acre wilderness area Roosevelt County Park.[83]

Much has been written about Roosevelt's 137 days of service in the army, mostly blandishments in the style of *Heroes of American History*. The whole island of Cuba had been a theater to Roosevelt, and he was the lead actor. For more than fifteen years, Roosevelt had cultivated good relationships with reporters, and they delivered fresh copy of his dramatic charges with gusto in 1898. He even appeared on the cover of *Harper's Weekly*. General Nelson A. Miles—who was famous for his part in the Indian wars of the West and had been in Cuba but never saw much action there—complained that Roosevelt had never actually charged up San Juan Hill. Miles was correct—Roosevelt's skirmish was on Kettle Hill—but the misnomer was widespread, and it stuck. Why let a geographic mistake beset a powerful war story? The immodest Roosevelt even put in for a Medal of Honor for himself, only to be rebuked by Secretary of War Russell Alger. Although it took until 2001, Roosevelt, through the lobbying of his family, eventually won the Medal of Honor posthumously for his bravery during the battle for San Juan Heights.[84]

Every day that Colonel Roosevelt was at Montauk, the New York press, seemingly in concert, covered even his humdrum statements as if they were major news. There was no need to light the fuse of his celebrity, for he had already been hurtled onto the front page of every national newspaper. The mascots, in particular, grabbed a lot of notice. The *New York Times* ran a feature story about Roosevelt's tame lioness, Josephine, reporting that the colonel might raise the big cat at Oyster Bay.[85] Edith staunched that plan, however, and instead, Josephine was carted off to tour the West as an icon of the Spanish-American War and as a big-top attraction. Unfortunately, in Chicago Josephine got loose or was stolen and was never seen again.[86]

The eventual fate of Teddy the golden eagle was just as disappointing. Quite sensibly Roosevelt had donated the eagle to the Central Park Zoo, where he became a popular tourist attraction. Everything went well for Teddy during his first nine months at the zoo. But in May 1899 two bald eagles from Brooklyn—nicknamed the "heavenly twins"—were brought into Teddy's cage to keep him company. Holy hell broke out. The feisty Teddy, presumably in an act of territorial protection, attacked one of the bald eagles, molesting the newcomer with his claws and beak. A few days later, the heavenly twins ganged up on Teddy, battering him severely. Within hours Teddy keeled over, dead. The zoo superintendent, John B. Smith, told the press that Teddy had died of a "broken heart," having lost his "prestige" to the bald eagle. The body of the Rough Riders' mascot was shipped to Frank M. Chapman at the American Museum of Natural History, where he was stuffed and put on display.*[87]

The story of Cuba, at least, had a happy ending. Corporal Jackson, after being quarantined at Montauk, headed back to Flagstaff with Cuba at his side. Because he was a footloose type, unable to take care of a pet, he gave the celebrated dog to a family man, Sam Black, who had been a ranger in Arizona Territory. For sixteen years Cuba lived in the lap of luxury, catered to by the Black family. When Cuba eventually died of natural causes, he was buried along the scenic Verde River fifty miles southwest of Flagstaff, having been given a proper military funeral in recognition of his service to his country.[88] Cuba was also given a special pet cemetery memorial at Sagamore Hill.

On August 20, Colonel Roosevelt was allowed to leave quarantine to

*In 1908 Roosevelt's beloved Bronx Zoo displayed two of the largest American bald or golden eagles ever captured; they shared a cage. Their names were Uncle Sam and Teddy Roosevelt. Both birds were popular attractions. Unfortunately, on one hot June afternoon, a field mouse scampered into their cage, pecking for corn kernels on the floor. Simultaneously, both eagles went for the kill, their bodies, according to the *New York Times* "hurtling down like projectiles driven from a twelve-inch gun." A death grapple ensued over the mouse (which escaped unharmed) and supposedly over who was king of the eagle house. "Uncle Sam is a trifle heavier than Teddy Roosevelt," the *Times* noted, as if reporting a boxing match. "And Uncle Sam got in the first blow. It was a vicious wing blow, and the second joint of his massive wing struck Teddy Roosevelt full in the chest. The smaller eagle recoiled, as he did so he lunged with his hooklike beak at Uncle Sam. The beak tore a gash in Uncle Sam's right shoulder." "Uncle Sam and Teddy in a Death Grapple," *New York Times* (June 22, 1908), p. 16.

return to Oyster Bay for five days. By the time he arrived at Sagamore Hill, there was a groundswell of support for his gubernatorial candidacy. From Buffalo to Brooklyn, Roosevelt had become public property, a war hero celebrated as a favorite son. All around Oyster Bay, he was greeted with shouts of "Teddy!" (which he hated) and "Welcome, Colonel!" (which he loved). Not for a minute did he suffer from the aftereffects of war; it was as if he had psychologically inoculated himself against trauma. "I would rather have led this regiment," Roosevelt wrote to a friend, "than be Governor of New York three times."[89]

Cleverly, Roosevelt had kept diaries in Cuba, jotting down exact dialogue and stream-of-consciousness impressions. His editor at Scribner, Robert Bridges, worried that if Roosevelt ran for governor, the war memoir they'd been discussing would have to be postponed. "Not at all," Roosevelt told him, "you shall have the various chapters at the time promised."[90] And there were always his biophilic notes, sent to his children from Cuba. "There is a funny little lizard that comes into my tent and it's quite tame now," read one, "he jumps about like a frog and puffs his throat out. There are ground doves no bigger than big sparrows and cuckoos almost as a large as crows."[91]

Once back at Camp Wikoff, Roosevelt wandered around Montauk Point, taking care of his golden eagle and leading little Cuba on long walks. (The dog greeted many of the Rough Riders from dockside as they returned to the United States.) Roosevelt seemed like a changed man, disconcertingly calm, studying the undercarriage of wigeon as they flew overhead. Sometimes, particularly when reporters were around, he rode his horse up and down the beach with the fervor of a plainsman.[92] Having "driven the Spaniard from the New World," Roosevelt could relax—he had been relieved of the burden of his father's buying his way out of Civil War service. With nothing more to prove, he could excel as a powerful politician, soapbox expansionist, true-blue reformer, naturalist writer, and conservationist. After his "crowded hours" avoiding whizzing bullets and tropical diseases, he turned to studying the shorebirds of Long Island. As always, Roosevelt wanted to be the master of his own backyard, even as he prepared to run for governor of New York.

Just how much Roosevelt identified himself with the American West was evident at the send-off his regiment gave him on September 13. A bugle called, and all the Rough Riders dutifully assembled into formation. In front of them was a card table with a blanket draped over a bulky object. Roosevelt was inside his tent, writing letters, when the troop requested his presence outside; he immediately concurred. The First Volunteer Cavalry had a parting gift for their humane and courageous colonel. After some moving words, the blanket was lifted to reveal a bronze sculpture of 1895 by Frederic Remington, *Bronco-Buster.* ("Cowboy" was the western term for a cattle driver. A "bronco-buster" broke wild broncos to the saddle.[93]) Tears welled up in Roosevelt's eyes, his voice choked, and he stroked the steed's mane as if it were real. "I would have been most deeply touched if the officers had given me this testimonial, but coming from you, my men, I appreciate it tenfold," Roosevelt said. "It comes to me from you who shared the hardships of the campaign with me; who gave me a piece of your hardtack when I had none; and who shared with me your blankets when I had none to lie upon. To have such a gift come from this peculiarly American regiment touches me more than I can say. This is something I shall hand down to my children, and I shall value it more than I do the weapons I carried through the campaign."[94]

The Rough Riders had given Colonel Roosevelt the best gift possible. Remington's bronze was far superior to a gold-plated pistol or signed group photograph. It summed up Theodore Roosevelt well: a fearless western cowboy, stirrups flying free, determined to tame a wild stallion by putting the spurs to it, a quirt in his right hand and a fistful of reins in the other. Like much of Remington's finest pen-and-ink work, *Bronco-Buster,* his first venture into sculpture, was charged with kinetic movement and free-floating energy. At fast glance the horse, forelegs held high, practically jumps to life.[95] Roosevelt had succeeding in transforming his sickly childhood in New York City into a frontier saga worthy of Captain Mayne Reid. "The men of the West and the men of the Southwest, horsemen, and herders of cattle, have been the backbone of this regiment," Roosevelt wrote, "as they are the backbone of their sections of the country."[96] A Remington casting of *Bronco-Buster* is now permanently housed in the White House Oval Office as a table centerpiece. And in the Roosevelt Room hangs an equestrian portrait of T.R. as Rough Rider by the Polish artist Tade Styka.

After the Spanish-American War, Roosevelt and his commanding officer, Leonard Wood, became close personal friends. Together the two veterans would hike Rock Creek Park discussing everything from immigrants' assimilation into America to the intolerable sanitary conditions in Cuba. Sometimes they brought young people with them on these outings. "Colonel Roosevelt especially made these walks of the greatest interest to the children," Wood recalled. "He transmitted to them something of his own keen interest in nature, his love of birds, his interest in woodcraft, and in a thousand ways attempted to instill in them an interest in and an understanding of God's world as he saw it, to implement healthy tastes, and to build up a love for a wholesome outdoor life. At the same time he was full of stories of men and animals, stories which tended to build up a love for birds and animals and of wholesome outdoor life for the woods and the fields at a time when it was easy to lay the right foundation and to plant seed which would bear good fruit. He had a wonderful fund of information about birds and animals which he was continually passing on to the youngsters in a way they could understand."[97]

Roosevelt's campaign for the New York governorship in the fall of 1898 did, however, face obstacles. Serious concerns were raised over whether the colonel was a resident of New York (he'd been paying taxes in Washington, D.C.).[98] Even within the state Republican Party, there were some who refused to let this new hero of San Juan Heights forget his record as a dyed-in-the-wool reformist. At times, working hard at retail politics made Roosevelt feel like a draft animal. But once he took to the "Roosevelt Special," a whistle-stop train in which he toured towns and cities, often with Rough Riders standing proudly at his side, the election was essentially secured. In Syracuse, for example, Roosevelt orated on his fortieth birthday from the back of a train with the ferocity of William Jennings Bryan, pounding his fist, spittle flying, speaking of the greatness of the American flag, and pronouncing that better days were just around the bend. It was quite a show. And on November 8, 1898, Election Day, Roosevelt rode triumphantly to victory, defeating fifty-two-year-old Judge Augustus Van Wyck by 17,794 votes (out of over 1.3 million cast).[99] His election was a testament to his power as mythmaker of self. Roosevelt was once again prepared to take a city—in this case Albany—by storm just like in Cuba.

CHAPTER THIRTEEN

HIGHER POLITICAL PERCHES

I

January 2, 1899—inauguration day—was bitter cold in Albany. A layer of snow coated the streets, and the temperature hovered around zero. Along the Hudson River the bucking wind had brought all vessels, even the icebreakers, to a halt. As governor-elect, Roosevelt headed to the state capitol as part of a parade, the trombones following him "froze in silence," with only the drums keeping up a Sousa-like beat.[1] Even the photographers refused to take pictures, for fear of damaging their expensive equipment. That afternoon, following swearing of the oath of office at the capitol, Roosevelt moved his family into the state executive mansion on Eagle Street. Typically, he refused to allow the inclement weather to extinguish his euphoria at becoming governor. Come spring, the entire front yard would be blooming with a variety of roses known as New Yorkers, plus hundreds of other indigenous plants from every county in the state. But now, as the seasonal cycle went, the lawn offered only bare elms and some evergreen shrubbery caked in frost.[2]

Becoming governor of New York was quite a historic achievement for Roosevelt. He was following in the larger-than-life footsteps of John Jay, Martin Van Buren, and Grover Cleveland. Back in 1891 Roosevelt had published *History of New York City*, for some quick money, anatomizing past governorships for anecdotal well-springs of raw courage.[3] Indeed, Roosevelt understood that he was part of a very special club—the governorship of New York—whose antecedents ran from the Revolutionary War to the Spanish-American War. Seventeen years previously, he had come to Albany as an assemblyman: now he was running the state with the largest population in America. Roosevelt was no longer simply moving through history. He was poised to make it.[4]

That first evening in Albany, Roosevelt left Edith and his children in the executive mansion with a night watchman on duty and ventured outside into the vicious sleet. It wasn't family-friendly weather. Dinner was being served at a friend's house for society types of his own rank, a five-course meal worthy of Delmonico's in Manhattan. He didn't want to be late. Upon his arrival cocktails were handed out on specially engraved silver trays. At the dinner table the talk centered on current events. In an adjacent room a violinist played movingly, but Roosevelt talked over the

Theodore Roosevelt was governor of New York from 1899 to 1900.

mellifluous music. All the hale-looking gentlemen, most with stylish mustaches, sat riveted as he held court over buttery foods. At eleven o'clock the governor-elect said good night to his hosts and was driven back to the executive mansion. Dismissing the carriage prematurely, Roosevelt walked across the veranda to the front door; it was locked. Three or four times he rang the bell, feeling slightly like a Cossack trying to get into the czar's palace. Nobody came to his rescue—not Edith or the night watchman or an awakened child. Losing patience, worried he was going to turn as blue as a winter corpse, an east wind cutting through his topcoat, Roosevelt clutched his key ring as if it were brass knuckles and smashed his fist through the plate-glass window. Reaching through the hole, he flipped the catch and gained entrance for his first night's sleep as governor. The following day the *New York Times* ran a whimsical headline: "Gov. Roosevelt Shut Out."[5]

That was the only time in Theodore Roosevelt's two-year term as New York's governor that the words "shut out" would appear next to his name. Starting with his first annual address, presented shortly after the inauguration, Roosevelt launched an activist reformist agenda. Sandwiched in between discussion of roads and the economy was a section Roosevelt called "The Forests of the State." The governor declared he would work against both the "depredations of man" and "forest fires" to keep parts of New York forever wild. He wanted to realign state government on behalf of conservation and natural resource management, demanding scientific answers to statewide environmental problems. (If Roosevelt had an overriding conceit in early 1899, it was that he thought in terms of geological time and in biological imperatives, whereas lesser politicians in Albany were part of the pettiness and crudeness of campaign cycles.) The first annual message also declared that fish and game laws would be "more rigidly enforced." The Adirondack Park, under Roosevelt's watchful stewardship, truly would become a monument to "the wisdom of its founders."[6]

What infuriated Governor Roosevelt about conservation and wildlife protection in his home state was its politicization, which led to inveterate inefficiency and callous disregard of nature. The state Fisheries, Game, and Forest Commission, for example, was packed with self-serving politicians who didn't know flickers from juncos, who had never read Darwin's *A Naturalist's Voyage* or marveled at John James Audubon's drawings of elegant egrets. Their uninformed anti-forestry views had spread like dermatitis over the state, he believed, negatively affecting the Catskills and Adirondacks. Whether he was taxing corporations, improving the Erie Canal, revamping mental hospitals, issuing new labor laws, or crusading on behalf of forestry, Roosevelt was well aware that his reformist policies would influence other states. Owing to his leadership of the Rough Riders, his national popularity was sky high. Reporters from Maine to California understood that he made terrific copy. Widespread fame had brought him all its perks and degradations. "Everything he did," the historian G. Wallace Chessman reflected in *Governor Theodore Roosevelt*, "would go into the record, to be used for or against him in the future."[7]

From day one as governor, Roosevelt championed the hiring of biologists and scientific experts for the New York Fisheries, Game, and Forest Commission, preferably experts trained in the Ivy League, to replace the politicians—who were like bloodsucking ticks on the state's resources. Too often they used their positions of power to promote sweetheart deals. An investigation of the commission, Governor Roosevelt threatened, was under way. Warily, Roosevelt sought to fire the politicians on the statewide game commission and replace them with independent-minded biologists, zoologists, entomologists, foresters, sportsman hunters, algae specialists, trail guides, botanists, and activists for clean rivers. In Germany this phenomenon was being called *Darwinismus.* "The state of New York is fortunate at present in having a Governor who is not only deeply interested in all matters of game, fish and forest preservation," George Bird Grinnell noted in *Forest and Stream* in May 1899, "but also has so clear an acquaintance with these subjects that he can always be depended upon to act on them for the public good."[8]

Grinnell believed that for all his Rough Rider's machismo, in private Governor Roosevelt wasn't a know-it-all. Constantly, as his own correspondence bears out, Roosevelt would seek advice from acclaimed zoologists and foresters. Which herpetologist was the expert on snapping turtles? Which forester knew about a new strain of invasive fungi? Shouldn't George Perkins Marsh's *The Earth as Modified by Human Action* be carefully studied before the state allowed a paper company access to Bear Mountain? Basically, Governor Roosevelt wanted the state bureaucracy reduced and cronyism purged from natural resource management. One good silviculturist or pisciculturist, he believed, was worth more than any number of politicians looking to have their palms greased. Governor Roosevelt, in fact, advocated changing the commission altogether: from having five members to having only one scientific forestry leader—an idea also enthusiastically promoted by Gifford Pinchot.[9]

Pinchot was only thirty-four years old in 1899 but had established himself as an independent progressive of some note. Although he had been born in Connecticut, he was uncommonly proud of his French ancestry.[10] His father, James Wallace Pinchot, a broad-ranging intellectual, had been an intimate friend of the elder Theodore Roosevelt.[11] A graduate of Yale University (class of 1889, summa cum laude), Pinchot played on its football team as a reserve and was a member of the campus-based YMCA and a regular volunteer at the Y's Grand Street Mission in New Haven.[12] Strong, handsome, and notable for his porcelain-blue eyes, Pinchot was in such fine pulmonary shape that he could read from *The Compleat Angler* while doing a one-armed push-up. After college Pinchot spent more than a year at the École Nationale Forestière in Nancy, France, studying forest conservation. As part of his education he toured the most ably managed ancient woodlands of France and Germany. Meteorology, botany, and even astronomy came easily to Pinchot; he took up anything that could help him decode the mysterious forests of the world. Pinchot's father had been a mainstay of the American Forestry Association, strongly advocating conservation management. Inspired by his father, Pinchot decided in his early twenties to dedicate his life to forest conservation. As proof of his pro-forestry convictions, he helped transform the family estate, Grey Towers in Milford,

Pennsylvania, into a tree nursery, "the first forest experiment station in the nation to encourage the reforestation of denuded lands."[13]

At the time Pinchot began foresting at Grey Towers, the United States had no university or college forestry program. He wanted to change that unfortunate situation. In 1892 he began the first serious systematic forestry work in American history at the timberlands of George Vanderbilt's mansion, Biltmore, outside of Asheville, North Carolina.[14] At the start nobody knew whether it was an advanced pilot program against catastrophism or a hillbilly boondoggle. Before long, however, this wasn't a question: Pinchot's forestry methods helped Biltmore prosper. In 1892 he opened an office in New York City, marketing himself as a forestry consultant. Pinchot wanted Americans to avoid the kind of horrific deforestation that had taken place in Europe ever since the industrial revolution oversaw the reckless wholesale destruction of the continent's natural resources. Pinchot was a devotee of George Perkins Marsh, the pioneering Vermont conservationist who wrote *Man and Nature*, and he took seriously Marsh's concern that parts of Asia Minor, North Africa, Greece, and Alpine Europe had been deforested to the die-off point of being as barren as a moonscape. "The earth is fast becoming an unfit home for its noblest inhabitant, and another era of equal human crime and human improvidence, and of like duration with that through which traces of that crime and that improvidence extend, would reduce it to such a condition of impoverished productiveness," Marsh had written, "of shattered surface, of climatic excess, as to threaten the depravation, barbarism, and perhaps even extinction of the species."[15]

The label applied to Pinchot time and again, borrowed from the Scottish philosopher John Stuart Mill, was "utilitarian." In conservationist terms, this meant a believer in *wise use* of natural resources. But the label also unfairly minimized (and at times maligned) Pinchot's lifetime effort to *preserve* and *expand* many of America's most magnificent forestlands. He was a tireless crusader for both utilitarian forest preserves and wildlife protection. "The eyes do not look as if they read books," Owen Wister wrote of Pinchot, "but as if they gazed upon a Cause."[16]

Pinchot first came to Governor Roosevelt's serious attention in 1896, when President Cleveland appointed the Yalie chief forester; however, they had dined together once, in May 1894. (Roosevelt, in fact, wrote Pinchot a quick note following the meal: "I did not begin to ask you all the questions I wanted to.")[17] By 1897, Roosevelt thought enough of Pinchot to nominate him for the Boone and Crockett Club.[18] And in 1898, when Pinchot became President McKinley's head of the Division of Forestry (renamed in 1905 the United States Forest Service), Roosevelt roared his approval. Independently wealthy, using the family fortune to help promote western reserves, almost British in demeanor, Pinchot saw himself as the Exeter- and Yale-trained advocate, press agent, and spokesperson of a new forestry movement. His nickname was "the Chief" (or "G.P.") and his forestry associates were "Little G.P.s." In his 1936 book *Just Fishing Talk*, Pinchot said that when he was a teenager, the Adirondacks had been his touchstone place, the woodlands where he learned to catch brook trout and painted turtles. As with Roosevelt, the Adirondacks had given Pinchot, a world-class fly-fisherman, "a new and lasting conception of the wilderness."[19]

In early February 1899 Pinchot finally spent significant time with his idol and family friend. The governor had invited him to spend an evening in the Eagle Street mansion. Roosevelt had become something of a roadside attraction in Albany since inauguration day, with everybody wanting a moment of his time. Cognizant of Roosevelt's new Rough Riders fame, Pinchot was grateful to have been included in Roosevelt's frenetically full calendar. Accompanying Pinchot to the meeting was the architect Grant La Farge, the son of the famous painter and draftsman John La Farge (whose closest friend was Henry Adams).[20] Roosevelt not only liked Grant La Farge—a fellow member of the Boone and Crockett Club whose face had a scrubbed Bostonian intellectual look—but named him the New York state architect during his first year as governor. The firm of Heins and La Farge, in fact, was commissioned by Roosevelt to build the Bronx Zoo. At Roosevelt's recommendation La Farge also received the contract to design the first buildings at the State University of New York–Albany.[21]

As Pinchot and La Farge waited to be called into the governor's office, they grew slightly nervous. Understanding that the wildly popular Governor Roosevelt was the most celebrated outdoorsman alive, they hoped to form a united front with him on the pressing forestry issues of the era. Pinchot's principal concern was that every hour, the United States had fewer trees than an hour before. Deforestation in such places as the Olympics, the Cascades, and the Front Range of the Rockies was now widespread in land tracts not protected by the Cleveland Reserves. (And it was taking place even in some acreage that was supposedly protected.) Too many unscrupulous deals for U.S. government leases were being made in and around the western reserves. Pretty soon all the raw land west of Denver might look defiled like in Haiti, China, and Italy. The dire warnings in *Man and Nature* had to be heeded. The lack of water for irrigation was also a serious problem in places such as California and Nevada. Pinchot essen-

Gifford Pinchot and his forestry team.

tially promoted two remedies: creating more forest reserves and allowing some regulated timbering within their boundaries. Pinchot's scheme was to enlist Governor Roosevelt in the great cause. Roosevelt—a politician who refused to sit on the dais while the band played—was his best hope for developing a new, widespread public awareness of the perpetual benefits of the forest realm. America had to remain a land with luxuriant woods and verdant valleys. "We arrived just as the Executive Mansion was under ferocious attack from a band of invisible Indians," Pinchot recalled in his autobiography *Breaking New Ground*, "and the Governor of the Empire State was helping a houseful of children to escape by lowering them out of a second-story window on a rope."[22]

What a grand time Pinchot and La Farge ended up having in Albany with the famous Rough Rider! They cut up like misbehaving kids and acted as if they were trail mates; and Governor Roosevelt told numerous stories about adventures off the beaten path. All three shared a gratifying intellectual curiosity about the natural world. Roosevelt, in fact, acted not as a governor with authority and power, but as a fellow wilderness enthusiast, a fraternity brother from the world of the Boone and Crockett Club. He was excited by the talk of the Pacific Northwest and the Front Range, and his facial muscles flexed as he spoke, while his knees bounced with boyish enthusiasm, as if he were overcaffeinated. Any moment, it seemed, he would climb out the window on a rope himself then break another plate-glass to get back inside the mansion. Keenly observant, Pinchot noted that when the Adirondacks were mentioned Governor Roosevelt perked up like a border collie eyeing sheep. Lake Tear-of-the-Clouds, Algonquin Peak, Upper Ausable Lake, Lake George—such natural wonders had magical connotations for Roosevelt, as they would later for the painters O'Keeffe and Hartley.

Capitalizing on the governor's love of this natural setting, Pinchot hoped to form an alliance with Roosevelt that afternoon and evening, for preserving the deciduous hardwoods of the Adirondacks—especially the sugar maple, American beech, and yellow birch. What was supposed to be a short chat with the governor on their way to examine forested acreage owned by the Adirondack League Club (renamed the Tawahus Club in 1897), turned into hours of rollicking storytelling about the outdoors. Clearly, Roosevelt was fascinated to hear about the old-growth forests of the Olympics and Cascades, which Pinchot had recently toured (and photographed) as President McKinley's "confidential forest agent." But Roosevelt's immediate concern as governor was the deterioration of the Laurentian mixed forests from Nova Scotia to the bogs of Lake of the Woods in Minnesota, especially in the Adirondacks and Catskills.

That first evening together, after hopscotching from one topic to the next like red-bellied nuthatches scouring for insects at one decayed stump after another, Roosevelt and Pinchot—in an act of primordial male bonding—put on gloves and boxed. Weaving and jabbing, throwing right and left jabs, ducking punches, Roosevelt was able to size Pinchot up as an honest man with a killer instinct. "Pinchot truly believes that in case of certain conditions I am perfectly capable of killing either himself or me," an amused Roosevelt wrote. "If conditions were such that only one could live he knows that I should possibly kill him as the weaker of the two, and he, therefore, worships this in me."[23]

But that evening it was the six foot and two inches tall Pinchot who seemed the stronger—at least at first. "I had the honor," Pinchot wrote in his autobiography, "of knocking the future President of the United States off of his very solid pins." Fellow Boone and Crocketters had been saying that the patrician Pinchot was, surprisingly, a "man's man," who could "outride and outshoot" anybody. Roosevelt put the Yalie's reputation to the test and came out with a favorable impression.[24]

Roosevelt, not to be defeated, shrugging off the boxing loss, immediately challenged Pinchot to a wrestling match, anxious to show off some pinning techniques. This time Roosevelt easily won the match. The score was now even at 1 = 1. Shrewdly Pinchot decided that it was best to safeguard the tie; a split decision, he reckoned, was the best outcome in dealing with a family friend with such a large ego and such a competitive disposition. Sometime during the punches and take-downs, Roosevelt decided to trust Pinchot; he liked the Old Boy's gameness, the way he didn't refuse a challenge, his aristocratic mien, and his abiding sense of noblesse oblige. And, more important, Pinchot wholeheartedly shared Roosevelt's ideals regarding scientific forestry. Pinchot was also impetuous, and the governor liked impetuousness in a man. Patience, Roosevelt believed, was a bent card that the dim and selfish played. Grinnell remained Roosevelt's muse on wildlife protection issues, but the irrepressible Pinchot, Roosevelt's junior by seven years, now was effectively anointed his guru on forest policy. Roosevelt and Pinchot formed an alliance that would have a profound effect on the modern conservation movement. Together, they would promote America's forests with firm confidence and zeal.

Ironically, even though Pinchot advocated forest conservation, he was seen as a sellout by thoroughgoing preservationist friends of Roosevelt, such as John Muir and William Temple Hornaday. The fact that Pinchot wanted to allow regulated tree harvesting in the Western Reserves was nearly anathema to them. Governor Roosevelt knew about this, but he thought the put-downs unfair. After all, Pinchot's family was about to donate $150,000 for Yale University to start a forestry school and was starting a forestry camp at Grey Towers to teach a new generation wise use policy.[25] "Gifford Pinchot is the man to whom the nation owes most for what has been accomplished as regards the preservation of the natural resources of our country," Roosevelt later said. "He led, and indeed during its most vital period embodied, the fight for the preservation through use of our forests."[26]

Pinchot embraced Governor Roosevelt's notion that New York should have one superintendent who could replace the five-man Fisheries, Game, and Forest Commission.* Roosevelt would promote this concept, saying that a "system of forestry" needed to develop "along scientific principles."[27] Roosevelt also implored the chief of the U.S. Forestry Division to help him preserve the Adirondacks as completely as Yellowstone and Yosemite. The East Coast population centers, he believed, needed wilderness parks to help revive city dwellers' spent spirits. As Pinchot and La Farge headed to the Adirondacks to help establish a land management

*The superintendent would have the authority to appoint three deputies: a forester, a fish culturist, and a supervisor of marine fisheries. See "The Fisheries, Game, and Forest Commission," *New York Times* (March 3, 1900), p. 6.

plan, setting up camp at Lake Colden, located midway up the mountain, they had the governor on their side. Roosevelt, in fact, gave them carte blanche to use his name as expedient. Roosevelt was starting to understand that Pinchot wasn't merely a forester but a revelation.

What really sealed the deal between Roosevelt and Pinchot was their shared admiration of 5,344-foot Mount Marcy, the tallest peak in New York. More than thirty years after Roosevelt first saw its summit, Mount Marcy (named in 1837 after Governor William Learned Marcy) still had magnetic appeal to him. (Sometimes Mount Marcy was called Tahawus, Cloudsplitter, or High Peak of Essex by locals.[28]) Compared with four larger eastern summits—Mount Mitchell, Mount Washington, Clingman Dome, and Mount Rogers—Marcy was a "bump"; yet its slopes were still covered by primeval forest.[29] Now, in freezing February temperatures, Pinchot and La Farge were planning on snowshoeing to the top of Mount Marcy with the help of two Indian guides; theirs would be only the second ascent ever attempted in winter. Upon hearing about their planned adventure, Roosevelt lit up like a Christmas tree. Bully! If he hadn't just started his job as governor, he would have joined the Ivy League explorers on the historic climb. Reacting as if they were about to go to the North Pole or Antarctica, Roosevelt demanded that Pinchot and La Farge report to him in Albany after the ascent. Hungrily, like a city editor, he wanted details of the twenty-foot snow drifts and ice squalls. The trip to Mount Marcy, Pinchot recalled, was "exactly in his line."

Yes, yes, both Pinchot and La Farge vowed to Roosevelt, they would visit Albany with firsthand reports of the summit immediately following their ascent. They had suddenly become Roosevelt's pro tem wilderness correspondents. True to their word, Pinchot and La Farge braved the mountain, but because of a blizzard it was rougher than they expected. Even Governor Roosevelt would have deemed them demented for challenging Old Man Winter so brazenly. As in tundra country, all the evergreens were, as Pinchot put it, "a monument of snow." Pecking out footholds wherever possible, Pinchot and La Farge pressed forward, half a step at a time, constantly shivering. Underdressed for the arctic temperatures, they nevertheless progressed incrementally in the squall. Both guides quit: one claimed that his snowshoes were too long, and the other had developed numbness in a leg. Normally, the pragmatic Pinchot would have retreated, recognizing that mountaineering in such brutal weather was like Russian roulette: one Canadian cold front could bring death faster than sleep. But he didn't want to tell Governor Roosevelt he had failed. So Pinchot and La Farge, minus the guides, pressed on.

Grant La Farge later wrote about the climb for *Outing*, saying that the gale-force wind was like a "battery of charging razors."[30] Also, visibility was no better than what might be seen through a sheet. About three-quarters up Mount Marcy, they were reduced to crawling on hands and knees to reach the summit. Pinchot said that he held his "head down in the squalls" and stopped "every minute or two to rub my face against freezing."[31] Even their mustaches and eyelashes froze. Eventually, through sheer willpower, they arrived at the summit's signal pole, but they saw nothing but snow and ice. "Got to the top," Pinchot wrote in his diary. "Foolish."[32]

Worried about contracting grippe or whooping cough, Pinchot and La Farge snapped photographs of each other and then crawled back down Mount Marcy as quickly as possible—dizzy, terrified, suffering from frostbite.[33] The three-day ordeal was the most dangerous of their lives. Once they were warm by a lodge fire, Pinchot and La Farge, to their horror, learned that they had been climbing in the blizzard of 1899—called the "Storm King" by the press. Unprecedented arctic temperatures had socked and crippled the entire Northeast. Water pipes, it was said, had burst throughout every county in New York. The weight of snow had caused house roofs to cave in. They were lucky—very lucky—to be alive. Both men later retold the story of ascending Mount Marcy as if they were characters in a knockabout comedy.

Their harrowing climb, however, produced one positive result. Returning to the executive mansion in Albany as promised, Pinchot and La Farge had wild stories to regale Governor Roosevelt with. Feeling left out, Roosevelt announced that he too would conquer the Adirondacks' tallest summit come August or September when the weather got better. Full of "dee-light," pleased to hear about their mountaineering antics, Governor Roosevelt had come to embrace Gifford Pinchot as a new member of his extended outdoors family. Given how close their fathers had been, they fell into an easy camaraderie as if they were long-lost blood brothers. (La Farge, an early member of Boone and Crockett Club, had long ago received T.R.'s stamp of lifetime approval.)[34]

II

One New York leader Governor Roosevelt was constantly trying to outfox was Thomas "Boss" Platt. Ever since he served as a New York assemblyman from 1882 to 1884, Roosevelt refused to join Platt's Republican rubber stamps. He was unimpressed by Platt and distrusted him—Platt had flunked out of Yale and had worked as a pharmaceutical salesman and, in Michigan, as a lumber operator. Roosevelt, however, was deferential to Platt—his elder by twenty-five years—merely because political expediency demanded it. He didn't want to spar unnecessarily with a slugger. As past president of the Tennessee Coal and Iron Company, Platt knew at least one thing about geology: that the land had glorious treasures which could be extracted. He considered nature sightseeing a fey enterprise. He had been elected to the U.S. Senate from New York in 1896, and when his photograph appeared in the *New York Tribune* on January 21, 1897, it was the first halftone reproduction ever published in a daily newspaper. As of 1899, only Governor Roosevelt was a more recognizable New York personality than the fit, trim, bushy-sideburned Platt. In his dogged, confident, shrewd, relentless way, Platt was a formidable counterpart to Roosevelt; their values, however, were at the opposite ends of the Republican Party's spectrum. Nevertheless, after only a month in office, Roosevelt wrote to John Hay that, to his surprise, he was "getting on well" with Platt.[35]

Almost monthly Governor Roosevelt and Senator Platt engaged in parlor debates at roundtables on topics such as corporate taxes, improved schools, and funding state infrastructure. Sometimes Roosevelt and Platt found themselves in general agreement on foreign policy issues including the annexation of Cuba and the building of an interoceanic canal. But when it came to conservation issues they were like positive and negative jumper cables; when their ideas touched, sparks flew in all directions. Quite simply, Roosevelt refused to water down his conservationist beliefs to curry favor with Senator Platt. They differed on scientific forestry, the Palisades, antipollution laws, and the need for watersheds.[36] Wisely, Boss Platt—who was never bored in Roosevelt's "impulsive" presence—let him have the parklands, preferring to have him champion birds' rights than meddle in antimonopoly fights on Wall Street, where huge sums of Republican money were at stake. Platt was careful not to denounce Roosevelt publicly; instead, he paid Roosevelt a backhanded compliment by saying that at least the Colonel wasn't a slacker. "Politicians found [Roosevelt] a hard customer," John Burroughs recalled of his stormy relationship with men like Boss Platt. "His reproof and refusal came quick and sharp. His mannerism was authoritative and stern. . . . His political enemies in Albany, early in his career, laid traps for him, in hopes of tarnishing his reputation but he was too keen for them."[37]

As of April 1899, Governor Roosevelt's ideas about pushing oneself to the limits were only implied, or only written in letters to friends. Certainly, reading about the virtues of the pioneers in *The Winning of the*

West or about military fortitude in *The Rough Riders* made it clear that Roosevelt believed overcoming hardship built character; oddly, he always seemed happiest with no accoutrements except a horse and rifle. There are even wisps of evidence that he was a devotee of recapitulation theory: a belief, propounded by a professor of pedagogy and psychology, G. Stanley Hall, founder of Clark University in Worcester, Massachusetts, that "overcivilization was endangering American manhood." According to Hall, American boys were becoming effeminate and needed to return to primitivism instead of wallowing in Victorian-era "ideologies of self-restrained manliness." Too many American men, Hall argued, were having neurasthenic breakdowns. Among Hall's many prescriptions for this decline of American manliness were the promotion of the "savage" in boys, the introduction of nature into their workweek, the creation of physical fitness regimens, and the rejection of the bureaucratic-corporate economy. Hall "believed that by applying Darwinism to the study of human development," the historian Gail Bederman says, "he could do for psychology what Darwin had done for biology."[38]

Addressing the Hamilton Club in Chicago on April 10, the thirty-fourth anniversary of Lee's surrender at Appomattox Court House, Roosevelt pulled together all his "up from asthma" thoughts and presented them to the American public preparing to enter the twentieth century as the doctrine of the "strenuous life." He was introduced by William "Buffalo Bill" Cody. "In speaking to you, men of the greatest city of the West, men of the State which gave to the country Lincoln and Grant," Roosevelt began, "men who pre-eminently and distinctly embody all that is most American in the American character, I wish to preach, not the doctrine of ignoble ease, but the doctrine of the strenuous life, the life of toil and effort, of labor and strife; to preach that highest form of success which comes, not to the man who desires mere easy peace, but to the man who does not shrink from danger, from hardship, or from bitter toil, and who out of these wins the splendid ultimate triumph."[39]

What immediately strikes one upon reading about Roosevelt's promotion of the "strenuous life"—besides its overtones of recapitulation theory—was that he was preaching a philosophy of survival of the fittest that echoed Herbert Spencer. Roosevelt had larded his "strenuous life" doctrine with sociobiology, the misguided belief that Darwin's evolutionary principles could best be expressed by humans through imperial expansionism, military hyperpreparedness, free-enterprise economics, and eugenics. Damning the "life of ease" and the hesitating manner, Roosevelt wanted Americans to engage in strenuous endeavors of every kind. Tiredness, he said, wasn't fitting in a country of such natural vitality. Nation building, he believed, was undertaken by a population that shunned soft hands and conquered weakness and was engaged to the fullest in the consciousness of its times. Every healthy American man, if he was lucky enough to have leisure time, Roosevelt believed, should hike, camp, hunt, and fish. Men could find exhilaration in the wild. Rules were already available; just follow the sportsman's code: "Let us, therefore, boldly face the life of strife, resolute to do our duty well and manfully; resolute to uphold righteousness by deed and by word, resolute to be both honest and brave, to serve high ideals, yet to use practical methods."[40]

The mere fact that Governor Roosevelt delivered this inspirational speech in Chicago instead of New York made his words newsworthy. New York's governor was telling Americans in Illinois to *go hard* into whatever they believed in, whether it was farming, football, forestry, or factory work. Interestingly, Roosevelt never mentioned God in "The Strenuous Life"; in many ways, in fact, the doctrine defied most Christian traditions by putting the obligation of personal power on the individual rather than in the otherworldly, mystical, or communal. Roosevelt's doctrine not only smacked of Spencer—and Hall—but also had a heavy dose of Nietzsche's superman. The saving grace of Roosevelt's philosophy—which liberates him from what was later called fascism—was that he was democratic in spirit, believing anybody could rise to greatness in America. And there wasn't an iota of cynicism in his doctrine: it was pure free-range optimism.

The following year, Roosevelt's speech in Chicago had become so popular throughout America that it was published as a chapter in the appropriately titled book *The Strenuous Life*. Remembering how he had wisely disregarded the advice of a Massachusetts heart doctor in 1880 who had told him to never climb mountains, Roosevelt now touted exertion and physical education as national imperatives. As governor he wrestled, boxed, practiced jujitsu, and swam in the Hudson River just for the bracing sensation. Citizens didn't have to be frail. Lean into your ailment, he believed, and defeat it. Urbanization had caused an unnatural deficiency in young people, and the schools needed to reverse this unhealthy trend by teaching Emersonian self-reliance. Implying that imperialism could be justified as part of the "strenuous life," Roosevelt was really applying the basic tenets of Darwinism to a program for *Homo sapiens*, in the spirit of Horatio Alger's fictional stories about self-made men. If Darwin was correct in saying that humans had evolved from apes and were therefore animals, then it made sense, Roosevelt believed, for the strongest and swiftest among the species to rule the human kingdom. That meant, in his mind, the *Americans*. As Bederman aptly put it, "Roosevelt believed that bitter evolutionary conflict allowed the fittest species and races to survive, ultimately moving evolution forward toward its ultimate, civilized perfection."[41]

Besides sharing the "strenuous life" of boxing and mountain climbing, Roosevelt and Pinchot believed that vast forestlands were necessary so that men could develop survivalist qualities not known in the overly civilized cities. It was as if once the forests disappeared, manhood would also vanish. Roosevelt, as governor of New York, was now in a position to act. In 1899 alone, Roosevelt had the state purchase 69,380 acres for forest reserves in the Catskills and Adirondacks. He wanted the iron-ore companies regulated. He began the ultimately successful process of turning Watkins Glen—a Finger Lakes scenic spot—into a state park. On November 28, 1899, echoing his first annual message, he wrote a scathing letter to the Fisheries, Game, and Forest Commission, claiming that New York's wardens were woefully ignorant of proper forest and wildlife management techniques.[42] He demanded a full report from the five commissioners on each warden in the state, and he intended to replace most of them with scientific experts and woodsmen. Furthermore, Roosevelt wanted the Adirondacks protected as if the region were a national park, "both from the standpoint of forestry and from the less important but still very important standpoint of game and fish protection."[43]

III

You didn't have to be an investigative reporter or an intellectual to realize that Governor Roosevelt was crazy about birds. Regularly, he invited ornithologists to visit the executive mansion to discuss bird protection issues. His son Theodore "Ted" Roosevelt, Jr., at age twelve, would go nest-gathering with his father each week, amassing a fine collection.[44] (As a matter of ethics, however, they refused to collect the eggs of wild birds.) Using his political clout to promote the Audubon Society (New York), he wrote to Frank M. Chapman, associate curator of the American Museum of Natural History, on February 16, 1899, delineating how to dramatically increase the avian presence in the state. "The loon ought to be, and under wise legislation could be, a feature of every Adirondack lake," Roosevelt said. "Ospreys, as everyone knows, can be the tamest of the tame; and terns should be as plentiful along our shores as swallows around our barns. A tanager or a cardinal makes a point of glowing beauty in the green woods, and the cardinal among the white snows. When the bluebirds were so nearly destroyed by the severe winter a few seasons ago, the loss was like the loss of an old friend, or at least like the burning down of a familiar and dearly loved house. How immensely it would add to our forests if only the great logcock were still found among them!"[45]

What disturbed Roosevelt most was that many bird species, because of human recklessness, were becoming either rare or extinct. As a boy he had shot at passenger pigeons for his Roosevelt Museum and was proud of having done so. But he no longer saw it as an achievement. The lessons of John Burroughs had taught him better. "The destruction of the wild pigeon and the Carolina parakeet has meant a loss as severe as if the Catskills or the Palisades were taken away," Roosevelt wrote to Chapman. "When I hear of the destruction of a species I feel just as if all the works of some great writer had perished; as if we had lost all instead of only part of Polybius or Livy."[46]

And Roosevelt considered the Palisades Park between New York and New Jersey a landscape masterpiece. Getting Andrew H. Green (his go-to guy at the Bronx Zoo) to help him establish a 700-acre refuge from Fort Lee (New Jersey) to Piedmont (New York), in order to preserve the sill cliffs (commonly called "Palisades sill") on the west bank of the Hudson River, had become a priority for Roosevelt. He sought to halt the unsightly mining that was ravaging the local scenery of the world's greatest city. As Roosevelt envisioned it, a thirteen-mile stretch of the preserved Palisades along the Hudson River would become a forerunner of other interstate parks nationwide. To damage the cliffs was sacrilegious. Every month Roosevelt grew more and more disquieted, knowing that New Jersey's quarry operations were destroying the scenic backdrop to the city.[47] The view from Riverside Drive, for example, would be forever ruined if the Jersey side of the Hudson was marred with only factories, storefronts, and houses.

Besides getting the right men onto the New York State Fisheries, Game, and Forest Commission (which today is the New York State Department of Environmental Conservation) and establishing Palisades Park, Roosevelt wanted to find ways to educate New York citizens about nature. On May 2, 1899, when he had been governor for only four months, Roosevelt signed into law an educational initiative very dear to his heart. After looking into the curricula of the public schools, Roosevelt was horrified to learn that natural history and geography weren't being taught. Immediately, he sought appropriation funds so that classes presenting men like Audubon, Darwin, Burroughs, and Marsh could be offered in every county.[48] Young citizens, he believed, needed to understand the evolutionary process and learn why dumping sewage and refuse into the Great Lakes and Hudson River was unacceptable. In a sense, promoting Earth Day seventy years ahead of time, Roosevelt believed that humans couldn't afford to recklessly poison their own environment without incurring a heavy toll in ill health, environmental ugliness, and corrosion of the spirit.

A quick look at Governor Roosevelt's time line for 1899 clearly shows that he wasn't a stationary executive. Even mundane talks about tax law became moments for impassioned theater. Although forestry and wildlife issues didn't monopolize his engagements, these topics were a high priority for him on the speaking trail. Refusing to weaken the conservationist plank in his first annual message, Roosevelt overcame a drubbing from the timber industry for extending state forest reserves in Delaware, Green, Sullivan, and Ulster counties. In May he hiked around the Adirondacks, preaching Pinchot's gospel of forestry science to people living around the McIntyre Iron Works. Seizing the initiative from Robert B. Roosevelt, T.R. got the New York legislature to pass Amendment (Ch 729) to the Fisheries Law, which forbade the pollution of any rivers, lakes, or streams used by the state fish hatcheries.

At one juncture Roosevelt went to inspect Niagara Falls to see if it could become a national park. The Transcendentalist philosopher Margaret Fuller had once stated that the great falls were "the one object in the world that would not disappoint."[49] Roosevelt disagreed. Doing a good amount of fast walking, he spent days surveying the cataract, exploring the cliffs of Goat Island, refusing to ride the new electric streetcar, and upset that a suspension bridge promoted by Boss Platt was going to mar the natural view of the thundering falls. (Wasn't the steamer *Maid of the Mist* enough?) The saga of Niagara Falls had begun 600 million years ago, Roosevelt lamented, and now a group of transportation hotshots, after dollars, wanted to demote the natural wonder into a tourist trap. The garishness of their plans sickened him. In the spirit of Ripley's Believe It or Not, there were already five-legged calves and two-headed goats on display near the falls. Almost nothing irritated Roosevelt more than the use of deformed animals in freak shows. As Governor Roosevelt envisioned the situation, Niagara Falls needed to become an intercountry national park administered by both Washington, D.C., and Ottawa. But Roosevelt dropped the issue—the concessionaires had already seized Niagara Falls, and there was no turning back. Instead, Governor Roosevelt headed southward to camp out for a few days in the Peekskill woods (better known as John Burroughs Country).[50]

Not all of Roosevelt's travels were within New York. In June 1899 he headed by train to Las Vegas, New Mexico, for the first annual Rough Rider's reunion. Las Vegas was situated along a stop on the Santa Fe Trail in New Mexico Territory, and Roosevelt had wanted to see the town for years, particularly the Spanish colonial–style plaza where Stephen W. Kearny once delivered a cracker-jack speech on manifest destiny during the Mexican War.* Western figures like Wyatt Earp, Doc Holliday, and Mysterious Dave Mather had spent so much time in the decorative adobes of Las Vegas that Roosevelt found every block intriguing. Billy the Kid had even once called the town home. By coming to the Rough Riders' reunion, Governor Roosevelt was weaving himself into southwestern lore. (In 1940, Las Vegas, New Mexico, was chosen as the Rough Riders' official reunion headquarters, with a museum dedicated to them.)

Besides making terrific press, Governor Roosevelt was able to see for himself the beauty of the rugged Sangre de Cristo Mountains he had heard about so frequently in Cuba from the Rough Riders. This was where the western edge of the Great Plains met with the southern edge of the Rockies. Las Vegas was in the heart of the Central Flyway, one of the four major migration routes in North America—this flyway followed the Great Plains and extended from Central Canada to the Gulf of Mexico. Tall ponderosa pines rose along the canyon rims, and some of the finest piñon, pine-juniper, and groves of gambel oak could be easily enjoyed. Over 270 bird species spent time in this ecosystem.[51] Just seeing Swainson's hawks—which often congregated in the short-grass prairie that later became the Las Vegas National Wildlife Refuge—was enough of an attraction to induce Roosevelt to travel more than 2,000 miles.[52]

Upon returning home from Las Vegas, as Roosevelt's train went though Kansas, huge crowds greeted him at depots. "I cannot tell you how much impressed I was by the rugged look of power in Kansas men whom I met along the line of the railroad," Roosevelt wrote to William Allen White about the famed journalist's home state. "What a splendid type it is! I can see their faces now. Our country is pretty good after all!"[53]

Only a portion of Governor Roosevelt's energy was given to forestry, birds' rights, and wildlife protection in 1899. For one thing, starting in January, *Scribner's* magazine began serializing Roosevelt's *The Rough Riders*, about his exploits in the Spanish-American War; in May it was published as a book and became a runaway best seller. With a vengeance,

---

*New Mexico and Arizona didn't achieve statehood until 1912. They were the forty-seventh and forty-eighth states, respectively.

Roosevelt also wanted to start taxing corporations in New York to help finance conservation programs. Many disgruntled Republicans, including Boss Platt, insinuated that he was a traitor to his class. Such insults were music to Roosevelt's ears. In May 1899, for example, he addressed the City Club of New York and didn't mince words. "A corporation is simply a collection of men, who may do well or who may do ill," he said. "The thing to do is to make them understand that if they do well you are with them, but if they do ill you are ever and always against them."[54]

Such anticorporation speeches angered some Republican bigwigs, who worried that Roosevelt might paralyze the party. But such rhetoric put independent voters strongly on the side of Roosevelt's "sock it to the rich" spiels. Throughout the summer of 1899, while Roosevelt worked hard on writing a biography of Oliver Cromwell (Lord Protector of England from 1653 to 1658), newspaper editorials began to pop up, suggesting that he should become President McKinley's vice presidential nominee in the coming election of 1900. The word was that Vice President Garret Augustus "Gus" Hobart—who had been raised in luxury in Long Branch, New Jersey—was seriously ill and that Roosevelt would be a logical replacement. Henry Cabot Lodge, always Roosevelt's sponsor, began promoting the nomination wherever he went. Others thought Roosevelt should become secretary of war or of the interior. Ironically, Boss Platt and the corporate financiers wanted Roosevelt out of Albany, and hope was that they could elevate him to Washington, D.C., to get him out of their hair.[55]

On November 21, 1899, Vice President Hobart died. His heart had given out. "No one outside of this home," President McKinley said in Paterson, New Jersey, "feels this loss more deeply than I do."[56] Expectations ran high during the holiday season that Roosevelt would be McKinley's replacement, at least for the reelection ticket in 1900. (Mark Hanna would serve as an interim number two in the chain of command.) The other names bandied about—for example, John D. Long and Timothy Woodruff—by contrast to Roosevelt, seemed stale. Disregarding all this speculation, Roosevelt unambiguously said forget it; he had no interest in the number two spot. Even though President McKinley had treated Hobart almost as a copresident, bringing the vice presidency up from its low estate, Roosevelt really wanted nothing to do with any kind of candidacy.[57] Santa Claus was preparing to visit Oyster Bay, the nineteenth century was coming to a close, and he didn't have time for guessing games or parlor politics. Much of what Roosevelt had to be grateful for—his colonelcy, his status as a war hero, his governorship—had come his way within the past two years. His strenuous life was on the upswing. He was considered an icon. Only Buffalo Bill took the oxygen out of a room as easily as Governor Roosevelt. While Roosevelt's restless urge to make a mark on American politics persisted, he didn't think riding on President McKinley's coattails was the proper way for advancement.

Instead of corresponding about Vice President Hobart's death on November 21, Roosevelt preferred to write about the newest addition to the household menagerie: a baby opossum, which he had requested from his friend Bradley Tyler Johnson. "The opossum arrived all right and Archie received it with such loving admiration that I gave it to him, a little to the jealousy of Ted and Kermit," the governor wrote. "In spite of your assurance that it is tame, Archie does not venture to pet it, and its only intimate acquaintance with him is when I take it up by the nape of the neck, and on such occasions it always opens its jaws like an alligator. Archie still regards it with unqualified respect. It has utterly unsettled the nerves of the terrier who sits in front of its cage for hours, showing the most eager, but I fear unfriendly desire to get in."[58] About ten to fourteen days later Roosevelt, worried that Ted's feelings were hurt because the opossum wasn't his, got Ted a guinea pig to even the score.[59]

Roosevelt was always a fine correspondent, and that December the letters poured out of him. In many he expressed thanks to his close friends, such as George Bird Grinnell and Henry Cabot Lodge, for always being at his side. In others he reflected on how parochial Albany was compared with Washington, D.C. Already Pinchot was soliciting Roosevelt to side with him in a campaign to transfer the forest reserves from the Department of the Interior to the Department of Agriculture.[60] But mainly they were a form of cheerleading; he was trying to unite his allies in his reform crusade through graciousness. "Oh Lord!" Roosevelt wrote Elihu Root, "I wish there were more of you. I think I have made a pretty good Governor, but I am quite honest in saying that I think you would have made a better one; for in just such matters as trusts and the like you have the ideas to work out whereas I have to try to work out what I get from you and men like you."[61]

IV

Governor Roosevelt's second annual message to the New York state legislature on January 3, 1900, was the most important speech about conservation ever delivered by a serious American politician up until that time. Everything from illegal hunting to forest fire protection and watersheds was covered. The governor tried hard to persuade the legislators that forest preservation was urgent to help combat hyper-industrialization. Disposing of land abusers as parasites, Roosevelt stated—insisted, really—that the "Adirondacks and Catskills should be great parks kept in perpetuity for the benefit and enjoyment of our people."[62] His speech led to what *Conservation Biology* later called a "revival of democracy" through the nature movement.[63] Like a sentry standing watch, Roosevelt was going to protect New York's wilderness from despoilers of every stripe. "As railroads tend to encroach on the wilderness," Roosevelt warned, "the temptation to illegal hunting becomes greater, and the danger of forest fires increases."[64]

Although Governor Roosevelt gave a small compliment to the Fisheries, Game, and Forest Commission for the propagation of hatcheries producing valuable food, his address was essentially a litany of woes he wanted corrected. Lumbering in state forests, Roosevelt declared, had to be placed "on strictly scientific principles no less than upon principles of the strictest honesty toward the state." Both lakes and rivers, he said, needed to be protected from the indiscriminate effects of hyper-industrialization. Game wardens, he claimed, weren't doing their jobs correctly. He wanted "woodsmen" with a background in science to take over these posts. State forests had to be consistently treated with the utmost respect by lumber companies. "The subject of forest preservation," he said, "is of the utmost importance to the State." And then he took up specific issues of birds' rights:

> The State should not permit within its limits factories to make bird skins or bird feathers into articles of ornament or wearing apparel. Ordinary birds, and especially song birds, should be rigidly protected. Game birds should never be shot to a greater extent than will offset the natural rate of increase. All Spring shooting should be prohibited and efforts made by correspondence with the neighboring States to secure its prohibition within their borders. Care should be taken not to encourage the use of cold storage or other market systems which are a benefit to no one but the wealthy epicure who can afford to pay a heavy price for luxuries. These systems tend to the destruction of the game: which would bear most severely upon the very men whose rapacity has been appealed to in order to secure its extermination.
>
> The open season for the different species of game and fish should be made uniform throughout the entire State, save that it should be

> shorter on Long Island for certain species which are not plentiful, and which are pursued by a greater number of people than in other game portions of the State.[65]

Never before in U.S. history had a governor championed forest preservation and bird rights with such forthrightness.[66] As Roosevelt wrote to Grant La Farge, he had done this without a "particle of popular backing of the effective kind."[67] Pinchot, who had been somewhere out West at the time the speech was delivered, would later memorize passages as if it were the Gettysburg Address. Overnight, the ornithologists praised Roosevelt's second annual message to the skies; it was a fulfillment of a dream. Frank M. Chapman of the American Museum of Natural History, for example, considered January 3 one of the greatest days of his life—Roosevelt's hard-hitting, visionary defense of birds in his second annual message was, Chapman believed, the tipping point for the Audubon Movement, the wave which crashed down on an entire new generation anxious for preservation to triumph over annihilation. A governor of New York, for the first time, took on both the lumber and the cold storage lobbies.

Not that Chapman was surprised. Starting in early 1900, Governor Roosevelt began promoting the virtues of "citizen bird" with a new zeal. Public awareness, the governor and his followers believed, was always the first step in winning a political battle in the United States. The previous year, the magazine *Audubon* had come into existence; its editor, Frank M. Chapman, was hoping to lead an effort to create bird reservations throughout the United States (particularly in New York and Florida). There was a movement afoot, encouraging individual participation in field research projects, surveys, censuses, and polls. The central idea was that every American community could have its own bird sanctuary. People were encouraged to have binoculars or field glasses ready at home. For the first time backyard bird feeders and ceramic birdbaths were erected by everyday citizens hoping to attract crossbills and grosbeaks. Sunflower seed feeders, for example, attracted jays, finches, and chickadees. One company started manufacturing nectar feeders—tubes filled with sugar water—to attract hummingbirds. Vacations were planned around simply spying on a new bird with alert eyes. Hard-core enthusiasts, those rich enough to travel, could be found looking for the greater prairie chicken in the sand hills of Nebraska or sighting the Chihuahuan raven in the borderlands of Texas.[68]

It is hard for some people to understand what made Roosevelt love birds so deeply when other influential contemporaries paid them so little mind. Its safe to say, for example, that Roosevelt was the only serious bird-watcher to ever become president of the United States. In the final analysis, virtually all ornithologists—Burroughs, Chapman, and Grinnell included—were people who just started counting the birds they saw and got carried away. Blessed with some sixth sense, birders like Roosevelt believed avians could be key to the biblical drama of Genesis. And the pastime of bird-watching wasn't exclusively for the rich. Those actively predisposed to birding—27 million strong in the United States by 2009, making this the nation's single most popular hobby 100 years after Roosevelt's presidency ended—take joy in seeing the sudden movement of a warbler or in hearing a veery pierce the afternoon silence with its song.[69] Despite some differences in temperament, Roosevelt, Burroughs, and other leading naturalists shared a desire to personally witness as many as possible of the 452 species that spent time east of the 100th meridian.

Besides Sagamore Hill, one of Roosevelt's favorite places to go birding in 1900 was the woodlands in and around the Bronx Zoo. Throughout his life Roosevelt would traverse bogs, prairie potholes, and wetlands just to see a particularly rare bird. He knew that the best birding occurred in transition areas where two or three habitats met—what modern ecologists call ecotones. Just wandering around the Bronx grounds reconfirmed his belief that, as Pinchot held, forests were a necessary precondition for species survival. Waterfowl—with the exception of the African pygmy goose—always nested in high places, usually trees. Roosevelt fretted that New York's migratory birds faced triple jeopardy: the fragmentations of northern breeding grounds; the disappearance of nesting and feeding areas along migratory routes; and deforestation of the Adirondacks and wintering grounds in Florida, Cuba, Puerto Rico, and Mexico. He didn't worry about some adaptable species, such as the northern cardinal (*Cardinalis cardinalis*), whose year-round *purty-purty-purty* was a soothing antidote to the industrial noise of the nearby Bronx. Mourning doves and blue jays were also easily found in backyards. But other species he encountered using the zoo as a refuge—for example, the black-capped chickadee (*Poecile atricapilla*) and northern mockingbird (*Mimus polyglottos*)—might need human rehabilitation efforts to help them survive the impact of industrialization. Too much of the New York City habitat had been cut into pieces for roads and buildings, but Roosevelt's zoo would be a thickly forested oasis for some bird flocks being decimated by modern conditions.

Just eight months before the second annual message, at the urging of

Chapman, Roosevelt had signed the Hallock Bird Protection Bill. As of May 2, 1900, it was illegal to kill and sell nongame birds for commercial purposes in New York. For the first time a state government was earnestly in the business of birds' rights.[70] The bill—named after the naturalist author Charles Hallock, a former Yalie who had founded *Forest and Stream*, and sponsored by the Audubon Society—regulated the hunting of birds in New York.[71] Hallock was a hero to both Roosevelt and Grinnell and had a catholicity of interests to equal Thomas Jefferson's. Besides founding *Forest and Stream* he was the leading expert on sunflowers (using the seeds to make clean fuel), established a game reserve in Minnesota, and originated the uniform code of game laws in America. His *Camp Life in Florida* had a huge impact on Roosevelt's eco-sensibility. Hallock wrote fine books about angling, including *The Salmon Fisher* (1890). It was Hallock's *Vacation Rambles in Michigan* (1877), in fact, that taught Roosevelt much of what he knew about the Great Lakes. Hallock maintained that America had four main flyways—Atlantic, Mississippi, Central and Pacific—and his bill would make sure they stayed as they were. Roosevelt was upbeat that America's recklessness toward migratory birds could be rolled back.

Chapman, at the governor's request, toured millinary factories after the law was enacted, threatening state-sanctioned shutdowns if any illegal plumage was found. Roosevelt let the Millinary Merchants Protective Association—an organization he despised—know that undercover "Audubonists" would be inspecting facilities at his request. His threat was unambiguous: the Hallock Bill had to be adhered to, and lawbreakers would be arrested.[72] As John Burroughs noted, when "upholding laws like the Hallock Bill," Roosevelt was "scrupulous in morals," and "unflinching in what he thought to be his duty."[73]

After signing the Hallock Bill, Roosevelt wrote Frank M. Chapman a note, praising the Audubon Society for its mission. "It would be hard to overestimate the importance of its educational effects," Roosevelt said. "Half, and more than half the beauty of the woods and fields is gone when they lose the harmless wild things, while if we could only ever get our people to the point of taking a universal and thoroughly intelligent interest in the preservation of game birds and fish, the result would be an important addition to our food supply. Ultimately people are sure to realize that to kill off all game birds and net out all fish streams is not much more sensible than it would be to kill off all the milch cows and brood mares. As for the birds that are the special object of the preservation of your Society we should keep them just as we keep trees. They add immeasurably to the wholesome beauty of life."[74]

Keep in mind, however, that Governor Roosevelt wasn't working in a vacuum when he signed the Hallock Bird Protection Bill. By the time he had been sworn in as governor in January 1899, studying the typology of birds had become a popular movement in America, in large part because of the success of the Audubon Society's first national promotion of bird-watching. When Roosevelt said that *birds mattered*, millions of people listened because they were already predisposed to the Audubon movement and admired its new celebrity spokesperson. Not that Roosevelt was opposed to killing birds for science—far from it. Only by collecting specimens could a naturalist like himself properly study eye lines for superciliary stripes, eye rings, spectacles, mustache marks, malar marks, and ear patches.[75] What infuriated Roosevelt and aroused his righteous indignation were the market hunters who were harming not just New York but the entire Florida ecosystem. When he invited ornithologists to the executive mansion in Albany, Roosevelt would hold court, floating various ideas on how to derail the millinary industry.

Governor Roosevelt used his new political authority and his status as a war hero to lash out at what John Burroughs (in *Signs and Seasons*) called "bird highwaymen." After the Hallock Bill, these millinary plunderers' destruction, both naturalists believed, had to end in criminal suits. (Unfortunately for Roosevelt and the conservation movement, the modern-day concept of class-action suits against despoilers of the environment did not come to fruition until the Federal Rule of Civil Procedure 23 in 1938.) As far back as 1886, with venom pouring from his pen, Burroughs had gone after the "bird highwaymen" and even science-minded men who overcollected. "The professional nest-robber and skin-collector should be put down," Burroughs wrote, using the kind of fierce language Roosevelt admired, "either by legislation or with dogs and shotguns."[76]

Burroughs was first and foremost a bird lover. He knew how to tread quietly in a woods or marsh so as not to scare the birds away. Like Roosevelt, he used his ears as much as his eyes. He looked for particularly rare species at dawn and dusk, when they were most active. Seldom did he disturb birds that were courting or nesting. He trained himself to detect small movements in the woods, usually by looking out of the corner of his eyes. Sometimes Burroughs would make a squeaking or pishing noise to attract curious songbirds. This seemed to work like a charm with chickadees and kinglets. Burroughs marveled that there were 5 billion wild birds in North America. But extinctions of species like the Labrador duck and the great auk were far too frequent. Every town, Burroughs believed, needed an Audubon Society so that birdsong could seep into people's consciousness.

V

Among all these birds' rights activists who gravitated around Burroughs, Grinnell, and Roosevelt, none were as politically effective at reducing market hunting as William Dutcher of New York. Dutcher—who had a grayish beard like Andrew Carnegie's, wore rimless spectacles, and kept his hair neatly parted—dedicated thirty years of his life to the "citizen bird" movement. At first glance, Dutcher's face suggested a buttoned-down "old chap," a man dutiful about fulfilling obligations and the handshake agreements like those made in the days before the telegraph. Born in New Jersey during the Mexican War, Dutcher was raised to become an apprentice Wall Street banker. But, for whatever reasons, his health faltered in the city. Coughing fits, headaches, bronchitis, sinusitis, although not seriously debilitating, flared up chronically, making him miserable. Repairing to a farm near Springfield, Massachusetts, filling his lungs with fresh air, and hiking through the unfenced woods along the Connecticut River revived Dutcher, and he had an inspiration. Nature, he came to believe, had curative powers more potent than the homeopathic nostrums being peddled in his local pharmacy.

Returning to New York to earn a living, and to make something of himself, Dutcher joined the Brooklyn Life Insurance Company, where he worked his way up from cashier to secretary to top agent. Not unhandsome or overly sophisticated, Dutcher was what the sociologist William H. Whyte, in the 1950s, would call an "organization man," dressed for success and unflaggingly loyal to his boss. Living in Manhattan, however, once again took a toll on his precarious health. All the nagging symptoms he had experienced as a teenager came back, causing him to feel like a voodoo doll being pinpricked by every allergen known to mankind. Relief came only when he escaped for weekend trips to hunt snipes, ducks, and geese. One afternoon at Shinnecock Bay on Long Island he shot a beautiful Wilson's plover. Intrigued by this shorebird's white forehead and distinctive eye stripe, he decided to have the taxidermist John Bell—Audubon's student, who had taught Theodore Roosevelt as a boy—mount it. Bell's museum-quality product, he figured, would bring some much-needed flair to his rather drab office at the insurance company.

From that single Wilson's plover grew one of the best bird collections in New York. Dutcher learned that this plover was named after the pioneering ornithologist Alexander Wilson, who shot a specimen in Cape May, New Jersey in 1813. Dutcher may perhaps have thought that someday a bird would be named after him. Certainly he could have imagined few greater honors. If he opened up new windows in ornithology he was sure others would follow him. With the zeal of a smitten hobbyist, Dutcher became infatuated with all the birds of Long Island. That was his niche as a collector. Many weekends, even in winter, he could be found shooting double-crested cormorants, American golden plovers, and harlequin ducks with his .410-gauge along the North Shore and even at the far-off tip of the island.[77] With great steadiness, he would carefully skin his birds by first making an incision in the breast and belly. Then he would peel the skin off the carcass and remove the meat, replacing it with cotton. An arsenic paste was then rubbed all over the feathers to deter insects while the specimen was drying. Writing out insurance policies by day, and reading John Burroughs by night, and applying the arsenic on weekends, Dutcher, the taxidermist-ornithologist, stepped out of the corporate shadows to become the world's authority on Long Island birds, with his only real rival being Theodore Roosevelt of Oyster Bay.

In September 1883, after a summer of collecting specimens on Long Island, Dutcher was elected an associate member of the American Ornithologists' Union (AOU), the new nonprofit organization promoting avian rights. (He would also participate in the first Audubon movement in 1886.) From his cluttered office at 51 Liberty Street—when he wasn't wrestling with insurance claims—Dutcher carefully crafted position papers for the AOU to disseminate and squeezed annual dues from new members. Refusing to let the AOU become ineffectual, he was fervent about stopping the slaughter of birds. He was no martyr, but it didn't take his colleagues at AOU long to realize that he was a workhorse.

Undoubtedly the most tireless member of AOU's Protection of North American Birds Committee, Dutcher evolved into a strong conservationist, determined to win the battle, never concealing his emotions, drafting model laws to protect nongame birds, and coordinating activities between the various state Audubon societies that were springing up along the Atlantic coast. Determined to thwart the plumers, Dutcher, with the financial assistance of Abbot H. Thayer, created a fund to protect colonies of U.S. seabirds from Maine to Florida. The Thayer Fund, as it was called, had the distinction of being the first conservation effort solely dedicated to saving herons, egrets, pelicans, and hundreds of other seabirds from extinction. A simple philanthropic rule of thumb for William

Dutcher was, "If John James Audubon painted it, the Thayer Fund would protect it."

What made Dutcher an effective lobbyist was his single-minded devotion to his winged clients. Like a determined town crier, he was impossible to silence. He saw setbacks not as defeats, but only as retrenchments. Always, at any hour or minute, when it came to birds' rights he had skin in the game. (In other words, he was a fanatic.) As much as he admired Burroughs's poetic musing about hermit thrushes and the common sparrow, birds' rights, he believed, would be won through the legislative process. Laws were elastic, and he planned to take full advantage of that fact. If the Boone and Crockett Club could lobby successfully for timberland reserves and for the protection of Yellowstone Park, then there was no reason he couldn't achieve model bird laws. Cordial, determined, and always armed with data, Dutcher headed to Albany in an effort to convince the New York legislature that the gulls and terns of the Empire State deserved protection. With Governor Roosevelt and John Burroughs cheering him on, Dutcher persuaded the legislators to approve the assigning of a few wardens all around Long Island to safeguard seabirds' breeding grounds. If Albany agreed to this conservation plan, the AOU, through the Thayer Fund, would foot the employment costs. Dutcher won, and immediately the fund paid for the new wardens to keep what he called "brutalists" in check.

The AOU's victory in New York was just the beginning. The indefatigable Dutcher (a devout Episcopalian) traveled up and down the Atlantic coast like an itinerant preacher offering revival meetings on behalf of birds. Promoting the gospel of birds, actively selling what he called the AOU model law (or Audubon law), Dutcher scored legislative victories in Boston, Trenton, Hartford, and Augusta, Maine. But those were all Yankee capitals, the home turf of legislators who were apt to accept the conservationist arguments of the day. As Governor Roosevelt understood, with characteristic realism, his real challenge would be in Florida, part of the old Confederacy, where the avian slaughter had become big business. What good would it do to protect birds in New York, only to have them slaughtered when they migrated to Florida? To truly protect the sheer diversity of shorebirds, he would have to lobby successfully in Tallahassee.

On March 2, 1900, at the L. F. Dommerich estate in Maitland, Florida—an elegant resort town where presidents Grover Cleveland and Chester Arthur had both wintered—an inaugural meeting of the Florida Audubon Society (FAS) was held. The participants were largely central Floridians, but the new governor of New York—Theodore Roosevelt—was asked to be an officer. (This meeting took place two months before the Hallock Bird Protection Bill was passed in New York.) Concerned about "citizen bird," and eager to save Florida's wildlife from human predators, Roosevelt gladly accepted. Most of the other founders, by contrast, lived in Florida, including Governor W. D. Bloxham and G. M. Ward, the president of Rollins College. Recognizing that Governor Roosevelt was the single most popular advocate for birds' rights in the nation, a friend of pelicans (both brown and white), Dutcher embraced him, you might say, as a parvenu embraces an heiress. Even though the relationship was initially based on Roosevelt's providing political muscle for Dutcher's cause, an abiding affection developed between these two bird lovers.[78]

That afternoon in Maitland, when the FAS joined the existing twenty-four state chapters of the Audubon Society, was the day of salvation for Florida's wildlife. The creation of FAS meant that bird lovers no longer felt discouraged or inept. In unity there was power. And with the popular Governor Roosevelt on board, it was harder for the opposition to dismiss the protectors of pelicans and terns as cranks holding conch shells to their ears to hear plumers' distant gunfire. Unfortunately, Clara Dommerich, a fan of T.R.'s who was known for her bullish stubbornness and was the real driving force of FAS, became ill and died just eight months later. Her funeral, however, was the occasion of a rallying cry for birds. Women in Florida not only started boycotting plumers but created Audubon clubs in town after town to keep Dommerich's and Roosevelt's message alive. Of FAS's founders, ten were women and five were men. Katherine Tippetts, for example, opened a branch office in Saint Petersburg and ran it for thirty-three years, going on to serve as statewide president from 1920 to 1924. During the progressive era she became known in conservation circles as the "Florida bird woman."[79] (Her lobbying led to the creation of the entire state park system.)

Besides the FAS and Governor Roosevelt, Dutcher now had the U.S. government on his side. And once again, the "velvet hammer" of the U.S. conservation movement, John F. Lacey, congressman from Iowa's Sixth District, entered the drama. Lacey was committed to the fate of avian species in his home state, such as blue-gray gnatcatchers and Henslow's sparrows. The Lacey Act of 1900—which he sponsored six years after the Yellowstone Protection Act—made it illegal to transport protected birds across state lines. It was the first federal law protecting game.[80] According to the *Federal Wildlife Laws Handbook*, the Lacey Act authorized the secretary of the interior to "adopt measures to aid in restoring game and

Congressman John F. Lacey of Iowa's Sixth District did more to protect migratory birds than any other politician in American history besides Theodore Roosevelt.

other birds in parts of the U.S. where they have become scarce or extinct and to regulate the introduction of birds and animals in areas where they had not existed."[81]

Under consideration since 1897, the Lacey Act finally passed the House on April 18, 1900, and then, three weeks later, on May 25, passed the Senate. President McKinley signed the Lacey Act into law with no regrets.[82] A new conservationism had arrived in America. No longer could you kill a snowy white heron along the Indian River, for example, and sell the feathers to Macy's department store in New York. As Governor Roosevelt saw it, as a consequence of this federal legislation (plus the Hallock Bird Protection law), Dutcher could now make a citizen's arrest if he saw somebody illegally shoot a bird.[83] The landmark legislation, however, didn't accomplish its immediate goal, for the financially stingy President McKinley initially employed very few game wardens. Worse yet, the state game laws weren't really enforced. Still, the law of the land was now on the side of the conservation movement.[84] It would be up to the Audubon societies and AOU to figure out how to raise funds for wardens.

High-minded speeches like Roosevelt's second annual message and legislation like the Hallock Bill and the Lacey Act were, of course, only the first steps in this reformist movement. Enforcing change—in this case, closing up the millinary industry's loopholes and confronting market hunters and fishermen bent on exterminating pelicans—proved to be difficult. Suddenly, the heavily armed plume hunters in Florida were pitted against newly organized conservationist groups like the FAS and AOU in what were known as the "feather wars." This protracted feud flared up and became deadly between 1900 and 1920. Under the Lacey Act, however, a poacher could no longer bribe or intimidate a local judge to turn a blind eye toward the slaughtering of birds. Now, for the first time, accused poachers operating in Florida would have to go before a federal judge, who would be fully cognizant that the federal government wanted the act vigorously enforced.

Refusing to capitulate, and antagonistic toward what they perceived as Yankee hunting restrictions, the plumers in Florida stood their ground and defied the law. It was reminiscent of the blowback in response to the Washington's Birthday Reserves or Cleveland Reserves. Abetted by the millinary industry, many intransigent Florida market hunters simply refused to pay the new laws any mind. They believed in their God-given right to exterminate species; Washington, D.C., could shove its bird laws as far as they were concerned: the Second Amendment gave them the right to shoot whatever they damn well pleased. Roosevelt—the " born bird-lover"[85]—scoffed at such backward, neo-Confederate thinking. Some plumers, however, decided that it was legally safer to hunt in the North, in the "prairie potholes" of Canada and the upper Midwest, rather than in the southern states, which encompassed the cross-continental bird migration route known as the Mississippi Flyway; in Minnesota and Wisconsin, unlike Florida, wardens weren't patrolling the "Father of Waters."

Nevertheless, a huge pro–Audubon Society change had occurred. And to the consternation of the millinary industry, its worst enemy—Theodore Roosevelt—seemed to be a short-fused time bomb. Within four months after the legislation in Tallahassee, he would be president of the United States.

CHAPTER FOURTEEN

The Advocate of the Strenuous Life

I

By the time the Lacey Act had passed in 1900 Governor Theodore Roosevelt and John Burroughs had solidified the friendship that began at their first meeting at the Fellowcraft Club. Burroughs now had a face full of hard-earned wrinkles, and he had grown a long, bushy beard as white as beach flax; he resembled Charles Darwin as an old man (the British naturalist had died in 1882), or Father Time. There was about him an outdoors aura of a solitary holy man or troglodyte. From 1889 to 1900 Roosevelt and Burroughs had regularly met for lunch, discussed bird laws, and swapped books. The two men shared the admirable traits of enjoying a well-turned phrase and never keeping the inkhorn dry. During his tenure as governor Roosevelt and Burroughs each managed to publish three books. Together they were proud Audubonists. "I send you a copy of my *Big Game Hunting*," Roosevelt wrote to Burroughs on May Day 1900. "It is really a combination of two volumes. In the second part, I wish you would look on p. 65 and before and after, where I speak about some birds. Again and again as I have listened to those plains birds I have longed to have you hear them, and I have longed even more to have you hear the bull elk and the great wolves."[1]

In 1899 James Bryce deemed Roosevelt "the hope of American politics"; Burroughs, however, saw the Colonel as the great public naturalist of the future.[2] An alliance of great importance to the wildlife protection movement was forged between Roosevelt and Burroughs. Burroughs, in fact, invited Roosevelt's twelve-year-old son, Ted—who had straight jet-black hair and a face full of freckles—to spend a long weekend with him, hiking around the clear streams near his rustic cabin, Slabsides (built in 1895 on a bog that was once a celery swamp).*[3] According to Roosevelt, little Ted "grinned with delight" when he heard of Burroughs's offer of hospitality. "How I wish I could be with you!", Roosevelt wrote to Burroughs. "As I have written you, when out on the Ranch in the old days I

* Slabsides was at West Park, Esopus, New York, in the Hudson Valley (Ulster County), a mile and a half inland from the river. Burroughs's main farmhouse, Riverby, as of 2009 was still owned by his descendants.

cannot say how many times I longed to have you there. It was only while I was in the West and on my ranch that I ever had much opportunity of really hearing bird songs. Of course I know our common birds of the East—the thrushes, bobolinks, etc., but it was only while I was on my ranch that I ever lived out of doors."[4]

Having Burroughs provide a guided tour around the Hudson River Valley was akin to having Thoreau describe the salient features of Walden Pond in a walk-around. Burroughs, Roosevelt believed, was the top-drawer nature writer of all time. His voice and memory were those of a master. Burroughs's book of nature essays, *Far and Near*, included "Babes in the Woods," about his fine time with Ted exploring Black Creek.[5] Eastern bluebirds—nesting in dead tree stumps—became the species du jour on their fine tramp. Burroughs explained to Ted the difference between the plaintive female note and the more ardent note of the male. The red, white, and blue eastern bluebird had become signature birds to Burroughs.[6] "Never in your life have you given more happiness than to the small boy who spent last Saturday and Sunday with you," Governor Roosevelt wrote to Burroughs on May 21. "I thank you most sincerely for your kindness to him. Ted is a good little fellow and he appreciated every moment of his stay. He has really been very interesting over some of his experiences, notably the conduct of the two parent bluebirds after you by accident broke down the stump containing their nest and then put it up again."[7]

With Burroughs as his primary muse, Governor Roosevelt had entered the "citizen bird" movement full-throttle. Along with Frank M. Chapman, William Dutcher, and a few others, Roosevelt truly thought of himself as part of the guild of professional naturalists. And Burroughs was their éminence grise. Just as Theodore Roosevelt, Sr., had gotten immersed in the humane movement, his son now believed in the moral impact of Darwin's theory: that benevolence toward other species was compulsory, that society had a sacred obligation to take care of lower species like birds.[8] This "moral" Darwinian impulse had first started fomenting when T.R. had read *Wake-Robin* at Harvard University. (In 1859, when Roosevelt was one year old, Burroughs had written in his *Notebook* the evolutionary sentiment that "from a single atom, by infinite modification, Nature builds the universe.") To Burroughs, *The Descent of Man* was a miraculous "model of patient, timeless, sincere inquiry; such candor, such love of truth, such keen insight into the methods of Nature, such singleness of purpose and such nobility of mind, could not be easily matched."[9]

Roosevelt's heartfelt appreciation of Whitman, Emerson, and Darwin grew as he read more and more of Burroughs's books. If Captain Reid was too juvenile and Charles Darwin too scientific, Burroughs fell into the middle zone; he wrote in a way Roosevelt could emulate. Burroughs, you might say, further stoked *the* key ingredient to Roosevelt's penchant for faunal naturalism: compassion for all wild creatures, especially birds. Once Burroughs became the first vice president of the New York Audubon Society, in fact, Roosevelt would do whatever he could to help the organization (of which he was also a member) flourish. "I know your Society will frown upon the milliner's use of bird skins," Burroughs wrote in 1897, accepting the vice presidency. "I hope it will also discourage the senseless collecting of eggs and nests which so many young people take up as a mere fad, and which results in the destruction of so many of our rarer birds."[10]

To Burroughs, *On the Origin of Species* was a "true wonderbook"—the exact sentiment Roosevelt had toward it. The debt both naturalists felt they owed Darwin could never be repaid. Both men learned to scoff at Darwin's critics as dimwits who couldn't comprehend basic scientific laws of nature. Only Shakespeare and Emerson, they believed, had a comparable grip on the universal condition. Darwin "is the father of a new generation of naturalists," Burroughs enthused in his journal. "He is the first to open the door into Nature's secret senate chambers. His theory confronts and even demands the incalculable geological ages. It is as ample as the earth, and as deep as time. It mates with and matches, and is as grand as, the nebular hypothesis, and is the same line of creative energy."[11]

What about the role of God in all this orthodox Darwinian celebration? Both Roosevelt's and Burroughs's views can be summed up in a single, often quoted line: natural selection may "account for the survival of the fittest, but not for the arrival of the fittest."[12] God was the one who created Darwin's world order. Independently, Roosevelt and Burroughs both understood that Darwin believed natural selection was a *process*, not a cause.[13] "The influence of Darwinian thought on Roosevelt's generation," the historian John Morton Blum noted in *The Republican Roosevelt*, "was profound."[14]

Realizing that calling Burroughs "John" was too pedestrian (and "Mr. Burroughs" too formal) Roosevelt settled on "Oom John"—*oom* being Dutch for uncle. Obviously, this was meant not in the sense of a "Dutch uncle" but as a salutation expressing deep and abiding love and kinship for time immemorial. Even though Burroughs continued to call Roosevelt "Mr." or "Governor" or "Mr. President"—old-school propriety to the maximum degree—to T.R. he was Oom John, the wisest mentor of them all. Oom John, in fact, approved of the Boone and Crockett Club books and thought his politically powerful young friend's book *The Wilderness Hunter*, in particular, excellent. There was, Burroughs recognized about Roosevelt, a fine naturalist buried under the yarns about twelve-point antler trophies and the Rough Rider's braggadocio. Not that Burroughs minded hunting per se. Like Roosevelt, Grinnell, and the others in the Boone and Crockett Club, Burroughs believed seasonal hunting was an imperative for thinning out herds and game-bird flocks. Roosevelt was in awe of Oom John's ability to notice field marks on birds—color, feather pattern, eye-catching markings, gender, and shape. But Burroughs preferred Roosevelt's more poetic side, admiring the way he wrote about nighthawks flying over a canyon or the coloration of Old World chats. To Roosevelt there was no higher compliment than Oom John's writing to him to say that he admired Roosevelt's hunting books because they were infused with such "good sound naturalist writing."[15]

II

Coinciding with his second annual address as governor of New York was Roosevelt's publication of *The Strenuous Life*, his essays expounding his view of the hardy American character being replenished by the outdoors life. Overcrowded and unsanitary big cities, where the air was foul and pestilence was bred, where whole blocks were nothing more than slums, were incompatible with health. In speeches in New York City that month—one given at the Boone and Crockett Club's annual banquet—Roosevelt stated that the time had arrived for wildlife preservation, clean rivers, antipollution laws, and wise use of forests. There was an intensity to Roosevelt during those opening months of 1900 that was almost electric. Writing to Henry L. Sprague, for example, Roosevelt mentioned for the first time that he was fond of the West African proverb "Speak softly and carry a big stick; you will go far."[16] To historians this proverb has become emblematic of Roosevelt's imperialistic belief that the United States should maintain a robust army and navy while using diplomatic channels in foreign affairs, dangling the threat of war over adversaries' heads. "Roosevelt loads his gun too heavy," Burroughs wrote in his diary. "The recoil hurts him more than the shot does his enemy. He is bound to make a big noise but the kick of the gun is so much power taken from the force of the bullet. People react vigorously against him as they always do to his surplus verbal energy."[17]

Like in foreign affairs, there was little soft speaking from Governor Roosevelt regarding forestry and wildlife protection; there was only the big stick. Such intimidation tactics on behalf of wildlife protection, particularly against the millinary industry, often worked. On February 10, for example, a few weeks after his second annual message, the state legislature revised Chapter 31 of the General Laws and approved Act (Ch 20) for the protection of forests, fish, and game. New York state now had the most progressive conservation protection laws in the United States (with the possible exception of Vermont). As the historian G. Wallace Chessman noted, Roosevelt squared off against the powerful Utica Electric Light Company when it tried to purchase private lands in the Adirondack Park; protecting birds, he said, came first.[18] By March 1900 Roosevelt had also won approval for the land, funds, and management for Palisades Park, which he had deeply wanted. Largely because of Roosevelt's opposition to quarrying, the ruination of the cliffs had stopped overnight. That success spurred Roosevelt onward. If New York and New Jersey could create a joint park at the Palisades, he didn't see why Wisconsin and Michigan—to give just one example—couldn't do the same with the islets surrounding Beaver Island in Lake Michigan. Meanwhile, Roosevelt began promoting Gifford Pinchot as "the best authority on forestry in the country." When some people raised the objection that Pinchot was "too political" in his work at the Forest Division of the U.S. Department of Agriculture, Roosevelt squawked. Blithe irresponsibility had guided America's forest policy for far too long; Pinchot was the intellectual antidote. "Pinchot has no more to do with politics than the astronomers of the Harvard observatory have," Roosevelt said in Pinchot's defense. "All he is interested in is his forestry work."[19]

As for the oyster beds of New York, Roosevelt was searching for their protector. Uncle Rob had been conducting all sorts of experiments in oyster farming at Lotus Lake, and his nephew was probably intrigued. Basically, Roosevelt wanted to recruit someone like Seth Green to come work as the "oyster protector" at the Fisheries, Game, and Forest Commission. Writing to the conservationist George McAneny, Roosevelt said, "The man appointed to this position should of course have some literary knowledge, some scholarly attainment, but he must know practically about oysters and be able to row and sail and handle himself on mud flats."[20] Connected to Roosevelt's concern about oyster beds were his hyperactive efforts to promote scientific water resource management and to stop reckless water pollution by the timber companies.[21]

Obviously, as New York's governor Roosevelt engaged in other reformist measures not connected to forestry or oyster conservation. He established a state hospital to care for crippled and deformed children, promoted consistent pharmaceutical standards, fought to end racial segregation in public schools, and demanded antiracism efforts in schools. Compulsory seating areas for employees in factories, he declared, were mandatory. Disapprovingly, he toured sweatshops with Jacob Riis, furious that such sickening squalor existed in America, stopping just long enough to ask the most pertinent questions about city services in the down-and-out neighborhoods. And his interest in integrating Native Americans into the main fabric of national life continued. For example, he pushed for compulsory education on the Allegany and Cattaraugus reservations in New York. The time was long overdue, Roosevelt believed, for Native Americans to be given a fair shake. Long before the term "affirmative action" was in use, Roosevelt was promoting the concept on behalf of Indians. And he started corresponding with three African-American intellectuals whom he deeply admired: Booker T. Washington, William Henry Lewis, and Paul Laurence Dunbar.[22]

By June 1900 Governor Roosevelt was described by the *New York Sun* as the greatest reformer in American politics. That month, as testimony to his meteoric rise, the Republican National Convention in Philadelphia nominated him to run as President William McKinley's vice president, despite Mark Hanna's fervent objections. (Hanna actually had a heart attack shortly after T.R.'s nomination, but lived.) Wearing a huge black top hat, Roosevelt had stood out at the convention amid a sea of straw boaters like, as Edmund Morris put it, "a tent in a wheat-field."[23] He remained coy, however, about whether he wanted the official vice presidential nomination in the first place. Boss Platt understood that in any case Roosevelt would get the nod. Famously, he quipped, "Roosevelt might just as well stand under Niagara Falls and try to spit water back as to stop his nomination by the convention."[24]

Amid the political theater of June 1900, Roosevelt nevertheless found time to read books by fellow naturalists. Back in 1882, he had read *Camps in the Rockies* by the British sportsman William Adolph Baillie-Grohman and was floored by its accurate description of elks. Now, while the Republican convention was going on, Roosevelt read Baillie-Grohman's newest effort, *Sport in the Alps*, and was again pleased. "Ever since I read your *Camps in the Rockies* I have felt tantalized because you had nothing about bear and bison hunting," Roosevelt wrote to him. "I know you look down on the latter, but after all it was something peculiar which has passed away forever and I do wish you had written about it. I shall always feel defrauded until you write a couple of chapters on Wapiti and Big Horn in your *Camps in the Rockies*. By the way, your *Sport in the Alps* was exactly the book which I had long been hoping to see written. Again here I wish you could have extended your researches to take in the records of bison and aurochs shooting in Lithuania and Poland in former centuries." Realizing, perhaps, that he was being too critical of a writer he admired, Roosevelt invited Baillie-Grohman to Sagamore Hill to see his favorite black-tailed buck head, an award-winning Boone and Crockett Club trophy.[25]

Once Roosevelt became the vice presidential nominee in June, he traveled throughout New York, speaking to huge crowds in Minneola, Brooklyn, Newburgh, Auburn, Syracuse, and Niagara Falls, plus numerous hamlets in between. A radiant atmosphere seemed to accompany his every step. "I am as strong as a bull-moose," he told Hanna, who was running McKinley's reelection campaign, "and you can use me to the limit."[26] Once again Roosevelt went to Chicago, on September 3, this time to discuss the "labor question" instead of the "strenuous life." Hordes of frenzied citizens followed his train on horses shouting, "We Want Teddy!" Wherever he went in the West, a great fuss occurred. As Roosevelt prepared to deliver a stem-winder in Kansas, the showman William "Buffalo Bill" Cody hopped onto the bandwagon, endorsing him by saying, "A cyclone from the West had come, no wonder the rats hunted their cellars!"[27] Both Cody and Roosevelt believed that the three great U.S. presidents—Washington, Jackson, and Lincoln—were all outdoorsmen in their youth. They both exuded the Wild West mythology in demeanor, and Americans loved them for it.[28]

Hundreds of veterans of the Rough Riders followed Roosevelt through the Rocky Mountain states, acting as both bodyguards and essential eyewitnesses of his valor in Cuba. Buffalo Bill had signed up sixteen of the veterans to reenact the Battle of San Juan Hill in Cody's Wild West extravaganza.[29] When one populist editor mocked Roosevelt in Cripple Creek, Colorado, a well-armed Rough Rider shot the critic with a revolver; Roosevelt wouldn't denounce this act, perpetrated in defense of his besmirched character. The fact that the Democrats had nominated the pro-silver ticket of William Jennings Bryan and Adlai E. Stevenson (the grandfather of the Democratic presidential nominee of 1952 and 1956) egged on Roosevelt's quarrelsome nature. These were repugnant types, he believed, afraid of Darwinism, the strenuous life, fierce expansionism, vehement nationalism, modern science, and old-fashioned hard work. His opinion of Bryan, in fact, was very low—and the Scopes trial was still twenty years in the future.

The most memorable moment of the American West tour occurred when Roosevelt visited Medora. Suddenly there was a proud luster to his

President William McKinley and Governor Theodore Roosevelt ran together as a ticket in the 1900 presidential election. Because they never were together, this photo was purposely doctored to give the appearance of a policy pow-wow.

gait. The Badlands lay before him, the essence of eternity found in the fossils of ancient fish and odd-shaped buttes. He wanted nothing more than to disappear over the horizon with a fine horse, saddle, and bridle. "The romance of my life," Roosevelt said, "began here."[30] That simple phrase was soon adopted by North Dakotans as something akin to the state motto. Having traveled more than 1,000 miles by rail, delivered more than 1,000 speeches, and met 3 million folks, he found the absolute stillness of the Badlands mighty impressive. His day in Medora was his only moment of sustained reflection that fall. "Tis Teddy alone that's runnin'," Mr. Dooley reported, "an' he ain't a-runnin, he's gallopin'."[31]

For the first time in his storied Democratic career, Robert B. Roosevelt broke from his party to vote for his nephew. Blood was thicker than politics. Spending time at Lotus Lake oyster farming, continuing to perfect fish hatching techniques and breed eels, R.B.R. had retired his bicycle in favor of an automobile (the first Roosevelt to do so). Indeed, he was now fancying himself as a race car driver. Although he also maintained a stable of the best horses in Long Island. Old Charlie Hallock, founder of *Field and Forest*, fresh from his success in obtaining birds' rights, wrote to R.B.R., curious about how he planned on voting in the 1900 election, as a

Democrat with a nephew on the Republican ticket. "I'm glad . . . that you are still fishing and shooting," R.B.R. wrote back. "I have been a Democrat the last year [but] the wild extravagances of Bryan and his populist associates forced me to McKinley. I can go there easier now for Bryan is just as crazy about 16 to 1 as ever. Besides Theodore is half a Democrat and will keep the administration right. The tail will wag the dog."[32]

On November 6, 1900, McKinley and Roosevelt won in an electoral landslide. Boss Platt was perhaps the happiest man in America, having foisted Roosevelt on Washington, D.C., and gotten him out of New York. On the other hand, Mark Hanna openly wept, shaking his head in dismay and mumbling, "That crazy cowboy."[33] The Rough Rider was now vice president (or would be as of his swearing in on March 4 on the U.S. Capitol steps). Roosevelt's Democratic critics immediately scoffed that being vice president was a boot-polisher's job (anticipating John Nance Gardner's remark that it wasn't worth "a bucket of warm spit"). The unconventional Roosevelt, who had made bold reformist improvements in New York, they said, fearlessly standing up to Boss Platt on conservation issues, among others, had been relegated to a lifeless position. Poor Teddy Roosevelt had become the "fifth wheel of the executive coach."[34] Typically, Roosevelt would have none of such talk. "If I have been put on the shelf," he said, "my enemies will find that I can make it a cheerful place of abode."[35] This quip was, of course, touchy defensiveness and smart public relations.

Days after McKinley's reelection, Roosevelt, showing how easily he could shift from being governor of New York to being a national office holder, wrote a long open letter to the National Irrigation Congress about the "vital necessity" of "storing the floods and preserving the forests." No longer was he pontificating just about the Catskills and Adirondacks. Wanting to bring western life to the national forefront, he had arid places like the Great Basin and the San Joaquin Valley in mind. Refusing to mince words he laid out a blueprint that the federal government would soon adopt. Dams and reservoirs would be constructed to help irrigate even the "vast stretches of so-called desert in the West." Herein lay the seeds for what would soon become Roosevelt's reclamation of the American West as U.S. president. Not pausing to think if it was smart to build cities in the Mojave or Sonoran deserts, Roosevelt's open letter made some fine points about how deforestation of the arid West must be prevented. Certainly, Roosevelt had his heart in the right place—even if he wasn't foresighted enough regarding the potential menace of dams.[36]

Besides writing to the National Irrigation Commission, Roosevelt took time that December to settle scores with politicians who had mocked, obfuscated, or taken advantage of loopholes in New York's game and fish laws. Reports from wardens that one of his commissioners was abusive toward them, scoffing at fish hatcheries, really set him off. Roosevelt's ire toward one of his five Fisheries, Game, and Forests commissioners in particular—Percy S. Lansdowne, former secretary of the Erie County Fish and Game Association, nominated by Roosevelt that March—was triggered by the illegal loan of a state wildlife boat as a donor's perk. To Roosevelt this was simply stealing from the state. Furthermore, Lansdowne had been known to mock Roosevelt's idea that the ticky-tacky tourism around Niagara Falls wasn't good for nature. Now, as governor, preparing to be vice president, Roosevelt lit into Landsdowne in a letter, calling him an untrustworthy, lying, thieving scoundrel and part of the "patronage machine."[37]

That December Roosevelt also took stock of his own duty and destiny. As of New Year's Day his tenure as governor would be finished. After delivering the most speeches ever in a U.S. presidential campaign, he felt as if he had jumped off a twenty-story ledge, had landed in a fire department net, and was now strolling around Oyster Bay with his hands in his pockets, glowing and in a "what next" trance. He worried about falling into an inward desolation of spirit. The open letter on western irrigation was for public consumption, but Roosevelt also wrote beautiful, serene letters to fellow naturalists about the disappearance of wildlife and forests in the American West. He lamented that the "great mountain forests" he encountered in 1891 in Idaho and Montana and Wyoming were "growing bare of life."[38] A meditation poured out of his pen on how wilderness had cured him of asthma.[39]

There was, however, a strange melancholy in Roosevelt's voice, perhaps indicating delayed depression after finally leaving the adrenaline-fueled campaign. All he could think about that December were the Colorado Rockies. He needed them as a fix for his declining spirits. He decided to head to Colorado come January and hunt cougars for the Biological Survey of Dr. Merriam. Instead of debating at the Cosmos Club how many types of cougars existed in the West, he would collect specimens to help in the inventorying. "Now I am hard at work endeavoring to assume the vice-president poise," he wrote Elihu Root. "Incidentally I may mention that I am getting altogether too much of it as regards habit of body, and have become so fat and stiff that after the first of January when I'm a private citizen, I shall take two months' holiday in Colorado and hunt mountain lions, if the fates are willing."[40]

Ironically, being elected vice president brought a jolt of jitters to Roosevelt. Had his political career hit an impasse? Had he really become just a "dignified nonentity"? Instead of satisfying his ambition, the vice presidency became distressing. Intoxication could be found, he believed, in the Colorado Rockies. Blocking out six weeks of his calendar in January–February 1901, Roosevelt put himself on assignment for the U.S. Biological Survey. He would collect cougar specimens and then write about his hunts for a magazine. That was, from his perspective, honest work. He wouldn't have to be in Washington, D.C., until the inaugural ceremony on March 4. "How I wish I could wait and make the hunt in March and April, so as to get after bear," Roosevelt wrote to Philip Stewart, "but, of course, I have to be back in time for the Inauguration."[41]

III

Ever since Roosevelt read Winthrop Chanler's essay on forest-clad northeastern Colorado in *American Big-Game Hunting* he had been hankering to hike and hunt in the White River region. Right after New Year's Day 1901, he headed to Colorado Springs—nicknamed "Little London" because so many English tourists came to see the Rockies by train—for a combination holiday and cougar hunt. At long last Roosevelt would get to see snow-tipped Pikes Peak, the parts not desecrated by logging, even though the awful January weather made the summit impossible to ascend. From a Colorado outfitter Roosevelt acquired a large leather coat, sweaters, a corduroy jacket, a buckskin shirt, and loads of other appropriate winter wardrobe accessories. With his clothing secure he would travel by train to Colorado Springs to connect with friends and then head north to Meeker.

Roosevelt's host was Philip B. Stewart of Vermont, Yale class of 1886, a former football captain, now a utility executive. He lived off and on in the resort town. (Stewart's Republican father had served as governor of and congressman from Vermont.) Gladly, Stewart pointed out such landmarks as the Antlers Hotel and Garden of the Gods upon Roosevelt's arrival. And there were others involved in the hunt. A surgeon in the Rough Riders, Dr. Frank Donaldson, chief of the throat and chest clinic at the University of Maryland, who lived part-time in Manitou Springs, Colorado, where he ran the Red Crags wellness clinic and lodge (advertised as "the Saratoga of the Rockies"), was also there to greet Roosevelt.[42] As the nurse Clara Barton noted in her memoir *The Red Cross*, Donaldson had been able to find the best medical supplies in Cuba for his Rough Riders because he *demanded* them.[43] Overflowing with excitement, however, Roosevelt could do nothing but talk to Stewart and Donaldson about the beauty of Pikes Peak—which he climbed a quarter of the way up in a failed attempt to find bighorn sheep.[44] Donaldson, who believed that thin mountain air was healing for asthmatics, enjoyed hearing Roosevelt talk about his personal medical history and how nature helped him breathe. Two other Rough Riders from Colorado Springs—Walter Cash and Ben Daniels—were also on hand for Roosevelt's visit.

The primary purpose of this trip was to collect cougars (and to a lesser extent lynx) from the White River area for Merriam. His job, in fact, was to shoot as many of the predators as possible for the Biological Survey to analyze. This was the new arrangement between Roosevelt and Merriam. Somehow, perhaps as a payback for Roosevelt's Cosmos Club challenge, Merriam now had Roosevelt collecting specimens for him in the Rocky Mountains, a pretty nifty trick. For six weeks Roosevelt hunted north of the White River on horseback—mainly between Coyote Basin and Colorow Mountain—enjoying the high, dry country with the cutting air full of shimmering frost particles. These pearly peaks were a fine diversion from politics. The heavily wooded slopes were wilder than he had imagined, untouched by axes. Meeker, named after a U.S. government Indian agent who had been killed by a band of Utes in 1878, had become a regional center for hunting. The White River Plateau Timberland Reserve (the precursor of what became White River National Forest) had been the favorite hunting grounds of many members of the Boone and Crockett Club.

Roosevelt had written (with some nostalgia for his days in the Badlands around 1886), that he had "no more hesitation in sleeping out in a woods where there were cougars, or walking through it after nightfall, than I should have if the cougars were tomcats."*[45] Nevertheless, he spent most nights sleeping in a forest hut or a rancher's house chosen at random, or at the Hotel Meeker across from the local courthouse. (It was hypothermia he was worried about, not cougars.) As at every hotel in the American West where Roosevelt ever slept—that is, among those surviving the wrecking ball—a bronze plaque would soon be erected at the Meeker Hotel on Main Street bragging that he had once stayed there. Over the years Roosevelt would tell people that the Meeker Hotel was better than any lodge in the Swiss Alps. The rock-hewn splendor of Colorado, he'd

*Published to capitalize on Roosevelt's fame as a Rough Rider, this volume comprised *Hunting Trips of a Ranchman* and *The Wilderness Hunter*, with factual natural history corrections. They were packaged together.

say, was by his estimate a world-class attraction. The tourist board of Colorado loved him for that, even though the timber barons and mining companies wanted him buried in an avalanche of snow. "The sage-brush grows everywhere upon the flats and hillsides," he wrote in what would become a chapter in *Outdoor Pastimes of an American Hunter*. "Large open groves of pinyon and cedar are scattered over the peaks, ridges and table-lands. Tall spruces cluster in the cold ravines."[46]

Having adopted Josephine the cougar as the mascot of the Rough Riders, feeding her milk from a bottle and watching her grow, Roosevelt had become extremely interested in mountain lions (or cougars, as he preferred to call the species). Only Merriam and Winthrop Chanler, he believed, had written proficiently about cougars. (And he even doubted a few of their scientific claims.) False reports about how cougars seized prey and about their size variation annoyed him no end. Some zoologists actually believed bobcats were small cougars, a proposition that T.R. knew was hokum. Even though Roosevelt was a fan of the famous hunter Richard Irving Dodge's *The Plains of the Great West*, he believed that Dodge had misidentified cougars as two separate species. Roosevelt—sounding a bit like Merriam—believed there were also five subspecies such as the Florida panther. "No American beast has been the subject of so much loose writing or of such wild fables as the cougar," Roosevelt complained. "Even its name is unsettled."*

While tramping about the Maroon Bells, a gorgeous group of Paleozoic sandstone and mudstone peaks near Aspen, Roosevelt recognized the wisdom of the Harrison administration in having protected this piece of the Colorado wilderness for all time. On every scientific "relief map" of the continental states that Roosevelt had ever seen, those which reproduced nature exactly, showing the peaks and valleys and other geographic details on a small scale, Colorado was the most intriguing, with its hilly ribs and mountain ranges. This was an ancestral elk range, a place where the Ute once hunted the great herds, a wild evergreen country which, thanks to federal intervention, would stay wild.[47] The elemental and the fundamental were honored in the Rockies. Far away from the thunder of applause and prodigious fame, Roosevelt, advocate of the strenuous life, found peace shaving stubble from his face in a nearly frozen stream. In

*Roosevelt found that in Texas cougars were often called panthers. In California they were mistakenly called catamounts, a term that properly refers to the wildcat or lynx. By 2009 the best place to find cougars in the United States was along the Middle Prong of the Gila River in New Mexico.

Colorado he dreamed. He plotted. He slept and breathed well. He found his scattered wits by cupping his hands, then shouting *Hello*, and not getting a reply. "Some thirty miles to the east and north the mountains rise higher, the evergreen forest becomes continuous, the snow lies deep all through the winter, and such Northern animals as the wolverene, lucivee, and snow-shoe rabbit are found," Roosevelt wrote. "This high country is the summer home of the Colorado elk, now woefully diminished in numbers, and of the Colorado blacktail deer, which are still very plentiful, but which, unless better protected, will follow the elk in the next few decades."[48]

Not that he didn't work hard throughout the six weeks. Daily he visited Colorado homesteads to record reports of cougar sightings, as if he were collecting census data for the federal government. Replaying his days collecting grizzly bear stories in Montana in 1889, he now performed the same oral history task in the winter loneliness of Colorado. What could be better than dusk on horseback in the Rockies looking for cougars and hearing lore from old-timers in the snow? Or watching every night as a crescent moon hung over the mountain peaks. Black-tailed deer abounded, but Roosevelt conscientiously refused to shoot one. He took his Biological Survey job too seriously for that. After all, deer weren't predators. Also, Stewart—who took photographs of the Colorado wilderness with his new Kodak camera—didn't want to take down a deer.[49] "The bucks had not lost their antlers, and were generally, but not always, found in small troops by themselves," Roosevelt wrote, "the does, yearlings, and fawns—now almost yearlings themselves—went in bands. They seemed tame, and we often passed close to them before they took alarm. Of course at that season it was against the law to kill them; and even had this not been so none of our party would have dreamed of molesting them."[50]

In Roosevelt's "With the Cougar Hounds"—which appeared in his 1905 book *Outdoor Pastimes of an American Hunter*—Roosevelt mentions Merriam by name three or four times, with interesting elaborations.[51] Clearly, Roosevelt was determined to make readers understand that his hunt wasn't just for sport; it was a scientific expedition. Fourteen cougars both male and female were shot between January 19 and February 14, and Roosevelt recorded detailed data about each one. The largest was eight feet long, and the shortest was under five feet. Zoological tables were compiled by Roosevelt to help Merriam and others at the survey understand the precise circumstances in which the cougars were killed. His field reports from Colorado were the work of a professional, recording whether the cougars' coats were tawny golden or gray-brown. "The fourteen cougar we killed showed the widest variation not only in size but in color, as shown by the following table," Roosevelt wrote. "Some were as slaty-gray as deer when in the so-called 'blue'; others, rufous, almost as bright as deer in the 'red.' I use these two terms to describe the color phases; though in some instances the tint was very undecided. The color phase evidently has nothing to do with age, sex, season, or locality."[52]

Roosevelt's relationship with cougars was complex. In *The Wilderness Hunter*, as a western rancher, he portrayed them as "bloodthirsty" killers attacking cattle with vicious abandon.[53] There was even an overdramatized illustration by J. Carter Beard of a rabid-looking cougar Roosevelt had shot in September 1889. But Roosevelt's respect for the cougar had grown over the years. He liked the fact that a male cougar would defend between 50 and 400 square miles on its own. But he worried that the cougars' prey were elk and deer. Cougars were obligate carnivores that depended on deer and elk as part of their primary diet. If the cougars weren't controlled in the Rockies, then the big game couldn't come back. So Roosevelt saw himself on a four-pronged mission in Colorado: helping the Biological Survey better understand *Puma concolor*; working to eradicate the cougar from the White River region to enhance the elk and deer populations; claiming his place as the North American authority on these big cats; and selling a couple of articles (illustrated with Stewart's photographs) for the October and November issues of *Scribner's Magazine.* For a sportsman, cougars, blessed with diurnal and nocturnal vision, were extremely difficult to hunt. In the days before radio telemetry devices it took a truly gifted outdoorsman to track them at all.

Accompanying Roosevelt, Stewart, and Webb on the hunt was John B. Goff, considered the finest tracker of the "ghost cats" in Colorado. This was always Roosevelt's secret as an outdoorsman; he had a genius (and the money) for finding the best hunt guides available for every expedition. Roosevelt's deft writing about the hunt—when published in 1905—contained the most anatomically correct descriptions of Colorado cougars, from their white muzzles to their huge paws, ever written up until that time. Carefully crating his kills in Denver, Roosevelt had shipped his cougars' heads, paws, and skins directly to C. G. Gunther's Sons on Fifth Avenue in New York City for preparing.

The trip to Colorado whetted Roosevelt's appetite for more cougar collecting. Word from Yellowstone National Park was that cougars were wreaking havoc on the elk herds. Encouraged by Merriam, Roosevelt planned on heading up to the park within the year to find more specimens for the Biological Survey and help out the bands of elk. "Many conservationists of the day, including Roosevelt, believed limiting predation would increase ungulate populations," the historian Jeremy M. Johnston explained in *Yellowstone Science*, "allowing them to recover from the results of the intensive market hunting that occurred in the park before the ban on hunting."[54]

IV

Roosevelt returned to Washington tremendously improved in appearance. In Colorado, he had written a dozen letters detailing his hunt for the cougar (and lynx). What he seemed to admire most about cougars was that they ate meat only fresh and clean—and, of course, the way they mastered topography; they were able to live in isolated cliffs or remote alpine valleys far removed from civilization. Armed with all his detailed measurements and field notes regarding cougars and lynx, Roosevelt denounced William Henry Hudson, best known in history as the author of *Green Mansions*, for his "preposterous fables" about cougars in his recent book *The Naturalist on the Plateau.*[55] Roosevelt kept score on those whom he considered "nature fakers," preparing for a frontal assault in the near future. Unlike bears—which were omnivores with meal alternatives like berries, roots, shoots, and pine nuts—cougars were able to eat only meat; Roosevelt believed this was the reason they were overdramatized as bloodthirsty killers.

Famously, Mark Hanna once quipped that Roosevelt was a "damn cowboy," now only "one heart beat away from the presidency." But the word "cowboy" would imply that Roosevelt was a rubber stamp for the stockmen's associations of the Rocky Mountains, which he clearly wasn't. The reality, in fact, was far worse than Hanna contemplated. Roosevelt was a pro-forest, pro-buffalo, cougar-infatuated, socialistic land conservationist who had been trained at Harvard as a Darwinian-Huxleyite zoologist and now believed that the moral implications of *On the Origin of Species* needed to be embraced by public policy. The GOP was in trouble.

While Roosevelt had been in Colorado a great oil boom was under way. It was big enough to have befuddled John D. Rockefeller himself. Roosevelt didn't know whether it was a cause for celebration or woe. Near Beaumont, Texas, at a place known as Spindletop, black gold spouted 200 feet in the air. The western plains of Texas and Oklahoma, he knew, would never be the same now that oil had been found. Somehow or other Roosevelt found a way to be at war with the Standard Oil Company for his entire life. Meanwhile, on the eve of Roosevelt's inauguration as vice president J. P. Morgan and Company announced the formation of a humongous corporation, U.S. Steel, with a capitalization of $1.4 billion. That development didn't much impress Roosevelt's usually outsized curiosity. Talk of automobiles replacing horses—which R.B.R. envisioned as the wave of the future—irritated him further. The mere thought of Duryeas in Yellowstone or Wintons in Yosemite was anathema to Roosevelt. Society didn't let Studebakers drive into cathedrals or art museums, did it? A "stern moral code" dictated all aspects of his life, causing him to reject what John Morton Blum called in *The Republican Roosevelt* "the amorality of business."[56]

For all of his cutting-edge talk of science, Roosevelt was really an old-fashioned camper type, a rustic, enamored with the very notion of log cabins or hunters' and naturalists' shacks. As he took his oath of office his primary concern wasn't Spindletop or J. P. Morgan; it was conservation. As for foreign policy, Roosevelt promised he would backstop President McKinley's policies. When it came to building a new great naval fleet and administering the Philippines, Roosevelt believed, President McKinley, to his credit, was an expansionist. Roosevelt's only real complaint (and it was a big one) was that McKinley was a slow-moving, incremental expansionist. As an admirer of Mahan, Roosevelt wanted the United States to make permanent naval bases in Cuba, Panama, Guam, and Puerto Rico. In February 1900 he had written a sharp protest letter to Secretary of State John Hay over the Hay-Pauncefote Treaty; he felt U.S. rights to build an isthmus canal across Panama or Nicaragua hadn't been properly protected under the terms.[57]

The challenge for Vice President Roosevelt—particularly when it came to conservation—was to be a team player in the McKinley administration. The Senate session Roosevelt presided over lasted for only five days—March 4 to 8—and then adjourned until the late fall. Roosevelt's big accomplishment as vice president was, as the historian H. W. Brands succinctly put it, having "gaveled the session open and closed."[58] Roosevelt was forced to console himself by considering that his real work took place during the campaign, so it didn't now matter whether he fell into a life of "unwarrantable idleness."[59]

Fame has an ugly dark side in America: the sniping by the tabloid press. Because Roosevelt was the vice president–elect while he was in Colorado, hunting cougars, he was a particularly tempting target for irresponsible journalism. A story was propagated by Senator Thomas MacDonald Patterson of Colorado—the Democratic, free-silver populist editor of the *Rocky Mountain News*—that Roosevelt had been drunk on a train with Senator Henry Cabot Lodge and a few others. It was pure rubbish. The Associated Press also went after Roosevelt for "grossly dissipated conduct." In defense, Roosevelt claimed that the wire service was "controlled" by Bryan, who wanted him bruised.[60] A barrage of belittling stories appeared in the Colorado press about Roosevelt's White River hunts, alleging that he was afraid of bears and that he shot treed cougars chased down by other men. For the first time ever, Roosevelt had journalists turn on him in packs. "To go mountain lion hunting sounds much more ferocious, but it really is not," Roosevelt wrote to Winthrop Chanler of the Boone and Crockett Club. "The only danger I run is from the infuriated yellow press, and this is moral, not physical. It is very exasperating to have humiliating adventures which never occurred attributed to me in connection with bears and wolves (neither of which animals did I so much as see) and then to have the very same papers that have invented the lies state that they were sent out by my press agent with a view to my own glorification. However, I suppose it is all in a day's work of a public man in our free and enlightened country."[61]

That March Vice President Roosevelt—after reading a draft of "The Merriam Report" from the Harriman expedition—had become obsessed with protecting Alaskan wildlife. A member of the Boone and Crockett Club, Casper Whitney, had been quoted in *Outing* as suggesting that Alaska adopt a single commissioner for forests, fish, and game—an idea that Roosevelt had promoted in New York. Roosevelt agreed that this was the "ideal" solution to stop the slaughter of caribou, elks, and seals. What Alaska needed, Roosevelt believed, was one first-rate advocate of protecting wildlife and forests (like the Boone and Crockett Club's president, W. A. Wadsworth), who would be in charge of managing the territory's natural resources. Any time three, four, or five men were on a playing board, Roosevelt told Whitney, game laws tended to be watered down. "I wish to Heaven it were possible to get Congress to act about Alaska," Roosevelt wrote. "As far as I know they simply provide for free rum, by voting to prohibit the sale of liquor there, and I do not know that they know anything about the game laws. Well, things are a little discouraging at times."[62]

Being back at Oyster Bay gave Roosevelt time to read. Two of his favorite new titles were Thomas Huxley's *Autobiography* and Graham Balfour's *The Life of Robert Louis Stevenson*.[63] The novelist Hamlin Garland had sent Roosevelt his recent collection of outdoors stories, *Her Mountain Lover*, and found a truly receptive audience in the vice president. "Your account of the Alaskan trail appealed to me very strongly," Roosevelt wrote to Garland on April 4. "I suppose I am utterly illogical, but it always gives me a pang to think of the fate that befalls the pack horses under such conditions. I am very glad you brought your pony home and rode it. I find it just as you say—that is, about three days restores me to my case in the saddle; though I am sorry to say I have grown both fat and stiff so I should now hate most bitterly to try to manage what we used to call on the range a 'mean horse.' "

Then, changing the subject Roosevelt told Garland about his recent hunt for cougars, noting that he hadn't shot deer or elk. His tone was that of a hunting addict, pleased that he had found a cure, able to restrain from shooting even when big game was smack in front of him. As a fellow writer Roosevelt knew that Garland had singular gifts, and considered his novel *A Son of the Middle Border* a masterpiece. "As I grow older I find myself uncomfortable in killing things without a complete justification," Roosevelt continued to Garland, "and it was a real relief this year to kill only 'varmits,' and to be able to enjoy myself in looking at the deer, of which I saw scores of hundreds every day and never molested them."[64]

There was also a loose end to take care of—getting Merriam the cougar and lynx skulls and skins, which were still being prepared at C. G. Gunther's Sons; an impatient Roosevelt resented their taking so much time. He also compiled with great exactitude the relevant naturalist data collected in Colorado. Unfortunately, Roosevelt got a blast of bad publicity because the owners of C. G. Gunther's Sons invited the press into their Manhattan shop to see Roosevelt's kills. Jokes were already widespread about Roosevelt disappearing into the wilds of Colorado before his inauguration as vice president, and being more interested in cougars than foreign policy. "Gentlemen," Roosevelt began his curt note to C. G. Gunther's Sons. "I am exceedingly sorry you have written to the press asking them to visit the collection. I had no objection to anyone seeing it who wanted to; but the one thing I was especially anxious to avoid was advertising, or seeming to advertise, it in any shape or way. It is most annoying to have had papers like *Life*, the *Journal* etc. notified. Will you please send on the skulls at once to Dr. Hart Merriam, together with the two largest lynx skins? & begin to make up the other skins; and show them to no one from this time on, unless he had my written authority."[65]

When Merriam received the specimens, from C. G. Gunther's Sons, he sent a congratulatory message to Roosevelt, saying that "your series of skulls from Colorado is incomparably the largest, most complete, and most valuable series ever brought together from any single locality, and will be of inestimable value in determining the amount of individual variation."[66] Two cougars, however, stayed with Roosevelt, serving as rugs for his Sagamore Hill library.

Because Roosevelt kept the cougar heads on these rugs, visitors to Sagamore Hill could be forgiven for thinking that he was showing off his hunting skills. Roosevelt even acquired two Alexander Proctor sculptures of cougars, which he used as props to stimulate conversation about his Colorado hunts. Essentially, he had fallen into the same trap as all persona manipulators. For years he promoted himself as a big game hunter extraordinaire—for example, having photographs taken of himself in buckskin holding a rifle. Although he clearly was the leading light of the wildlife protection movement, many average Americans knew him merely as a hunter. After the bad publicity Roosevelt received over the Colorado hunt, in the future he had naturalist-inclined friends at his side to offer testimonials that he was a scientifically-minded hunter, not a bloodthirsty rogue. Confusion over this issue caused Roosevelt deep anguish throughout his years as vice president, president, and ex-president. The sad reality was that most newspaper readers preferred hearing the details of Roosevelt's hunts, not the biological minutiae about the variation of rings on a lynx's tail. Roosevelt could be a grave, serious man when it came to studying wildlife genera, but hardly anybody knew it.

V

Shortly after Roosevelt became vice president, he began casting a wide net in hopes of bringing first-class men into the Forest Service and the Biological and Geological Survey. Writing to Gifford Pinchot, for example, Roosevelt tried to get Jacob Riis's eighteen-year-old son Edward attached to an "outdoor government trip."[67] He also lobbied to get his old friend from Maine, Bill Sewall, a job as postmaster of Island Falls. "He is a true American type of the best sort," Roosevelt wrote in his letter of recommendation, "as strong as a bull moose, fearless, shrewd, honest and kindly."[68] Worried that animals weren't being properly represented with biological facts in various popular books—especially the short stories of Ernest Seton Thompson—Roosevelt started promoting true animal experts like William Temple Hornaday, not literary imposters.[69] As for hunting, Roosevelt wrote a series of letters touting the use of knives rather than guns; a knife at least improved the odds for the animal being pursued. Roosevelt and William Wells of *Forest and Stream* believed they could, once and for all, get cougars written about in a truthful, detailed, zoological fashion. "That cougar of yours which measured eight feet four inches is the longest of which I have any authentic record," Roosevelt wrote to Wells. "My biggest one measured eight feet and weighed two hundred and twenty-seven pounds. I sent its skull on to Dr. Hart Merriam, the naturalist, at Washington and he writes me that it is the biggest skull that he has ever seen. From this it is easy to see what perfect nonsense is written by those who speak of ten and eleven-foot cougars."[70]

That June the executive committee of the Boone and Crockett Club appointed a special committee to propose ideas for establishing big game refuges throughout America. All of Roosevelt's closest conservationist allies were on the committee, including Caspar Whitney, Gifford Pinchot, George Bird Grinnell, Archibald Rogers, and D. M. Baringer. In 1896 the Supreme Court, in the case of *Geer v. Connecticut*, held that the state owned the wildlife even in a national forest—a verdict Boone and Crockett didn't like. The club found a convenient way around the impasse in *United States v. Blassingame*, in which the court ruled that forest reserves were the private property of the federal government and that it could protect acreage from trespassers in the same manner as any private landowner. The ruling made it legal to create refuges in the national forests—the objective of the club.[71] Alden Sampson served as chairman of the committee and issued the following mission statement. "The general idea of the proposed plan for the creation of Game refuges is that the President shall be empowered to designate certain tracts wherein there shall be no hunting at all, to be set aside as refuges and breeding grounds, and the Biological Survey is accumulating information to be a service in selecting such areas, when the time for creating them shall arrive."[72]

Furthermore, that June Roosevelt started grappling with the whole Darwinian concept of man as descended from apes in a serious way. With the enthusiasm of a cheerleader, he hoped Arthur Erwin Brown, vice president and curator of the Academy of Natural Sciences of Philadelphia, would put all his anthropological articles and pamphlets on the subject of human evolution into a book. What could be more exciting work, Roosevelt thought, than tracing the *real* origins of man? Not for a minute, however, did he suggest that natural selection determined human advancement. He leaned toward eugenics but never accepted it. The species with the greatest fecundity—such as rats and mice—had hardly advanced up the biological pecking order. "It has always seemed to me that we should ultimately have to put the branching off of man's direct ancestors from the mass of the other primates to a remote tertiary period," Roosevelt wrote to Brown, "and I am interested in your view that the parent stem branched off directly from the early lemuroid forms, instead of from some monkeylike form after the latter had itself branched off. As I understand it, the belief now is that the existing species even of the sharply defined and small rhinoceros family represent three stems which have remained wholly distinct since eocene times."[73]

Besides birds and books at Sagamore Hill there were, of course, trips to be taken. Determined to get road dust on his shoes, he escorted Edith that May to the Pan-American Exposition in Buffalo and went back to Colorado on his own to deliver a speech and gather more information about cougars for Merriam. There was also a side trip to give a lecture in Minnesota, where he sneaked some Mississippi River bird-watching into the itinerary. Basically, Roosevelt was trying to stay out of President McKinley's way, avoiding the front page, presenting himself as a loyal lieutenant, not a usurper. He was cognizant that he might be able to run for president in 1904, and it was extremely important that he didn't appear hungry for power. Roosevelt's muse throughout these months was the always blunt Henry Cabot Lodge, who wanted him to cool down the cougar-hunting heroics; they had the deleterious potential of making President McKinley think he was grandstanding in the Rockies for future western votes.

Roosevelt spent some of the summer writing essays on deer, cougar, and other North American big game. And then, time permitting, there would even be a Minnesota-Wisconsin series on wood animals like wolverines or badgers with sharp claws and flesh-eaters' teeth that were difficult to tame. Having time on his hands, Roosevelt made arrangements to study law after the Senate reconvened in the fall, and he started collecting walking sticks as a hobby. "I have very ugly feelings now and then," Roosevelt wrote to William Howard Taft that April, with a straight face, "that I am leading a life of unwarrantable idleness."[74]

That summer Roosevelt also started boning up on Vermont's enlightened conservationist laws. Although President McKinley was only moderately interested in conservation, Roosevelt—with the loyal Secretary Bliss at his side—believed he could increase America's forest reserve acreage at the rate of something like three Delawares a year. Reserves aside, Roosevelt also wanted America to have the same strict game laws as Vermont. In early September, just when the apple orchards were bearing fruit, Roosevelt went up to the Green Mountains of Vermont on a fact-finding trip. He had never before visited Vermont in an official capacity. Basically he wanted to get places like Alaska, Colorado, Montana, and

Wyoming to adopt Vermont's admirable standards of natural resource management. Ostensibly speaking for his dinners, Roosevelt was also on a fact-finding mission.

Much of Roosevelt's first day in Vermont was consumed with making speeches on topics ranging from the Civil War to naval policy. On September 6, however, Roosevelt attended a big-tent luncheon of the Vermont Fish and Game League on Isle LaMotte in Lake Champlain.[75] The league had been highly successful in reintroducing Adirondack white-tailed deer to the state, issuing fishing licenses (as a kind of taxation), and opening game reserves.[76] Pleased that as governor of New York Roosevelt had adopted various of Vermont's conservation ideas for his own state, the conservation league had elected him an honorary member.[77] His primary host in Vermont was Frank Lester Greene, an ardent champion of forestry science who had served valiantly in the Spanish-American War. Harvard may have been America's Darwinian laboratory and Yale the institution where forestry science took hold, but Vermont, owing to George Perkins Marsh's legacy, was the birthplace of early conservationist thinking. There was a down-to-earth pragmatism about the way Vermonters like Greene were wise stewards of the lands. Dairy farmers and town merchants in Vermont understood that mangling woodlands was injurious to every aspect of good living. Roosevelt wanted to learn how Vermonters applied conservation and then implement it on a large scale.

From the second Roosevelt was deposited on the lakefront dock from Seward Webb's yacht *Elfrida*, he explored Isle LaMotte like a tourist. At his side was ex–Rough Rider Guy Murchie. A book had informed Roosevelt that the world's oldest coral reefs existed around Isle LaMotte. The Chazy Reef was buried under the southernmost part of the island; its limestone was more than 450 million years old.[78] Roosevelt couldn't help marveling that Vermont—of all places—had once been under a tropical sea. There were other geological facts that probably intrigued Roosevelt and appealed to his insatiable curiosity about the island. Somehow, the few hundred residents of Isle LaMotte were able to quarry black marble limestone from the reefs without polluting the harbor. Some of the finest marble blocks ever discovered, in fact, came from Lake Champlain and were used as construction materials for the U.S. Capitol and Radio City Music Hall. There were other attractions on the island including a fish culture station said to be spawning more than 1 million eggs a week, incubated in a series of tanks, but he never got around to inspecting them, owing to a tragedy.[79]

That evening of September 6, much of Roosevelt's talk to his fellow conservationists centered on his recent cougar hunt in Colorado. Sitting in the audience was Philip Stewart, his hunting protégé and photographer, who gave him somebody to bounce his anecdotes off in a slightly humorous way. "Stewart took the hunt a shade less seriously than I did," Roosevelt joked. "I wanted to shoot the lions but he wanted to Kodak them. He had a large and Catholic taste and wanted to *Kodak everything.* When the dogs treed the first lion I was riding ahead and had got within fifty yards of the tree and could see the animal in the tree snarling and spitting. I was immensely interested. Suddenly Stewart halted me in a tone almost agonizing in its earnestness, as though a pack of mountain lions was upon us when he proceeded with the air of a villain in melodrama to take a picture of a rabbit on a stump."[80]

After the speech Roosevelt repaired to the grand estate of Lieutenant Governor Nelson Fisk, next door to where the Fish and Game League's banquet had been held. He was going to talk off the record to various conservationists in a little while. Back in 1897 President McKinley had stayed at the mansion, where he claimed a cane chair as his own. Now Roosevelt, as a courtesy, was given the same chair. Roosevelt had anticipated having a delightful evening because the novelist Winston Churchill (the author of *Richard Carvel*) was on hand, along with Senator Redfield Proctor of Vermont. Together they were sure to provide lively debates on politics and literature. At five-thirty that evening Roosevelt was called away from the veranda to the telephone, one of the few on the island. He was informed that President McKinley had been shot twice at point-blank range while visiting the Pan American Exposition in Buffalo. Burying his head in his hands, Roosevelt was heard to gasp "My God!"

To think that a wretched little anarchist, Leon Czolgosz, had tried to snuff out America's twenty-fifth president set Roosevelt's teeth on edge. Securing the only telephone line on the island for hours, Roosevelt was able to get a message to the hospital in Buffalo. Word of McKinley's dire plight spread throughout the Fish and Game League crowd and everybody was aghast. Senator Redfield Proctor made a formal announcement which had the effect of nauseating the shocked conservationists even more. "Friends," he said, "a cloud has fallen over this happy event."[81] A short while later another call arrived, telling Roosevelt that the president was "resting quietly" and that recovery was likely. "Good!" Roosevelt exclaimed, his face relaxing.[82]

Announcing the positive news to the guests at the banquet, who listened with bated breath to Roosevelt's every word, the vice president asked to be excused from the event. Roosevelt's friend Dr. Webb, who owned the *Elfrida*, was going to take him to Arrow Point on the mainland near Burlington, where he could take a special train (Engine 108), which pulled the private car of the president of Rutland Railroad to Buffalo. When Roosevelt was asked by a local reporter about the attempted assassination, his face became impassive. Staring straight ahead, with utter stillness as if he were a statue, he said, "I am so inexpressively grieved, shocked, and horrified that I can say nothing."[83]

Nobody knew what Roosevelt thought as his night train traveled across New York state. There are no accounts describing him as serene or anxious or melancholy. Oddly, this is one of the few historic moments in his life that he himself never recounted. Probably his heart had sunk into his boots, and his mind was reeling like the discord of untuned fiddles, for a collective ominousness held sway over America that evening. Had yet another president been killed in his prime? Not since a bullet struck President Garfield twenty years earlier had the nation's nervous system been given such a jolt. President McKinley's wound was quite serious; the bullets had penetrated his abdomen, damaging his stomach and pancreas. McKinley had been rushed to Exposition Hospital for immediate surgery. He was then moved to the home of John G. Milburn on Delaware Avenue to rest.

Arriving in Buffalo in the hush of dawn, catching a morning chill, Roosevelt hurried to McKinley's bedside where, more or less, he stayed for the next three days. By September 10, President McKinley's health had vastly improved. The situation didn't look fatal; it seemed that the president was going to pull through. A very relieved Roosevelt didn't want to hang around Buffalo any longer (like one of those Cuban land crabs or vultures circling the dead) so he said good-bye and headed for Oyster Bay and then the Adirondacks to reconvene with Edith. Everybody deals with tragedy differently, and Roosevelt now felt the urge to climb Mount Marcy. (Probably the ascent had already been planned and he was getting his itinerary back on track.) Roosevelt knew life was short and he didn't want to miss his home state's glorious peak, which had tugged at him since those long-ago campfire readings of *Last of the Mohicans*. To Roosevelt it was wrong for a governor of New York not to have climbed the great summit. Even though Burroughs had written extensively about his 1863 trek to Mount Marcy, the "sage of Slabsides" had never made it to the top; Roosevelt would do it for him.

For the first time since Yellowstone Edith agreed to climb a mountain with her husband, excited to see the virgin groves of birch, pine, spruce, and fir. Two of Roosevelt's eager children—Kermit and Ethel—

were also going to make the climb and go swimming in Lake Tear-of-the-Clouds, the source of the Hudson River. (Alice and Quentin were ill and couldn't join the family outing.) Roosevelt's conservationist friend James McNaughton had a hunt club near Mount Marcy in the hamlet of Tahawus which was going to serve as the Roosevelts' base camp. His family was waiting for Roosevelt at the Tahawus Club when he arrived wet from the rain; the forest was cast in a blue gloom. Clearly, the hike wasn't going to be a picnic. "The Adirondacks," Edith complained, "is probably the wettest place in the world."[84]

Two ranger guides had volunteered to lead the Roosevelt family from the Tahawus Club up to the summit dome, where there was a spectacular view of the divide between the Hudson and Saint Lawrence rivers. Also accompanying the Roosevelt party were McNaughton, a governess, and two law students from Harvard. Their first day on the Calamity Brook trail started out golden but turned to slate-gray as they boarded canoes and paddled toward Lake Colden at an elevation of 3,500 feet. There, at lakeside, they lodged in two cabins with, as Edith put it, "miserable little cots." The next morning, September 13, a pall of fog hung in the air so thick that it was hard to see five yards ahead. The ledges were getting slippery. At this juncture Edith and the children bailed out of the expedition. Theodore had a ranger take them back down to the Tahawus Club. With his family out of harm's way, Roosevelt, clutching a walking stick, waving the other men forward, grew more determined than ever to reach the summit. If Pinchot could pull it off in a blizzard, surely he could make it in a damp September rain.

As his correspondence bears out, Roosevelt was in a deeply reflective mood during those grim days since the attempted assassination of President McKinley. Like a falling barometer or a dropping temperature, Roosevelt's mood had sunk low in Buffalo. He had written Jacob Riis a cryptic letter about losing his youth, saying that at age forty-two he felt a "shadow" coming over him like a dark shroud. Perhaps reaching the summit of Mount Marcy with his companions would help renew his optimistic spirit. After all, how could he not feel uplifted by the sight of wild New York unfurled beneath him, a blanket of green forestlands and long valleys and a pattern of blue lakes for as far as the eye could see. At that high an elevation, where only balsam fir and a few stunted spruce thrived, he could think in an unmuddled way.

The mountain climber Jon Krakauer, in *Into Thin Air*, wrote about the out-of-body sensation encountered when one is rubbing up against the "enigma of mortality," finally reaching a summit after days of difficult climbing. Krakauer's reward was a glimpse across the "forbidden frontier" of death.[85] Somehow 360-degree views from mountaintops, staring at the horizon in a cyclorama, remained the closest humans could get to comprehending the afterlife before the advent of modern aviation. Some climbers have called it a "rush." To others it's a "little taste of heaven." To Roosevelt it was another moment of perfect clarity like the one he had on Mount Katahdin as a young man. For hours he basked in his own romantic profile; he was the explorer hero, at one with the backwoodsman on the trail in a Leatherstocking story, at one with Audubon and Thoreau, Boone and Crockett, breathing fast, wondering what had happened to McKinley in Buffalo. He made a personal pact to become a habitué of Mount Marcy—the summit was that inspiring. "Beautiful country!" Roosevelt kept repeating, while standing on a great gray rock at the edge of an anorthosite cliff. "Beautiful country!"

Once the spell lifted, Roosevelt, his head cleared, started making his way down the mountain with the others. Unbeknownst to him, meanwhile, President McKinley had taken a sharp turn for the worse. Quite suddenly the vice president was desperately needed in Buffalo; the odds were high that he'd be sworn in as the next American president. The only problem was that nobody knew how to find Roosevelt. The press reported that the vice president was "lost" in nature. The *New York Times*, for example, headlined its story that day "Hunt over Mountains for Mr. Roosevelt."[86] Only the park ranger who had escorted Edith and the children down the mountain had a true idea of his whereabouts. At one-twenty-five on Friday, September 13, Roosevelt—eating a sandwich while sitting at Tear-of-the-Clouds—was met by a hyperventilating Harrison Hall. He appeared to be waving urgent dispatches from Buffalo. Roosevelt intuited what the message said.[87]

Racing down the mountain to rendezvous with his family, his combustible spirit restored, Roosevelt packed his belongings and then headed to the North Creek station. His drafts were now open and his chimney was drawing new air. Like a young giant he had sneaked Mount Marcy in just under the wire.[88] Although McKinley had eminent physicians at his bedside, they had failed to detect a gangrenous infection. "For more than twelve suspenseful hours, the nation had no President," the historian Margaret Leech noted in *In the Days of McKinley*. "Theodore Roosevelt was speeding on Saturday morning across the breadth of New York State."[89]

At every railroad depot—Albany, Amsterdam, Utica, Rome, and Syracuse—reporters mobbed Roosevelt's train in search of a quote. He stayed mum. Roosevelt was soon going to be the new president. He arrived in Buffalo at one-thirty PM on September 13, lodging at his friend Ansley Wilcox's colonial mansion on Delaware Avenue. Every labored last minute had been an hour of agony for poor William McKinley. Roosevelt's usual good nature and high spirits weren't on display. He had a distracted look on his face and seemed self-contained. At two-fourteen AM on the morning of September 14, eight days after being shot, McKinley died. For the third time in thirty-six years an American president had been assassinated. A solemn Roosevelt, dressed in a frock coat, a thin gold watch chain hanging out of a pocket, was immediately sworn in as America's twenty-sixth president; the time was three-thirty PM.

At only forty-two years old Roosevelt had become the youngest president in American history. Oddly, when taking the oath, Roosevelt didn't swear on a Bible; owing to the constitutional separation of church and state, nobody thought it was necessary. Perhaps his recent moments on top of Mount Marcy had brought him as close to God as he was going to get. And he waved off the military escort, claiming that a couple of mounted policemen were quite enough. An American president, he insisted, didn't cower when something dreadful happened. Roosevelt, however, understood that the public needed to be reassured that the government was in stable and experienced hands. Immediately, he announced that all of McKinley's cabinet officers—John Hay, Lynam Gage, Elihu Root, Philander Knox, and Ethan Hitchcock among them—would be retained. The old McKinley administration would continue to provide all the springs to the government. "I wish to say," Roosevelt told the press, "that it shall be my aim to continue, absolutely unbroken, the policy laid down by President McKinley for the peace, the prosperity, and the honor of our beloved country."[90]

CHAPTER FIFTEEN

The Conservationist President and the Bully Pulpit for Forestry

I

Roosevelt's wide, toothy smile took on a glint of extra assurance in the last days of September 1901 after he was sworn in as America's twenty-sixth president. Indeed, the forces of destiny seemed to have had a governing hand in his triumphant storybook career. All over Washington, the phrase being bandied about was "Roosevelt luck." Nobody, it seemed, *enjoyed* being president more than T.R., even though he had reached the mountaintop of American politics because of an assassin's gun. In a conversation with the diplomat William vanden Heuvel during the 1970s, Alice Longsworth, Roosevelt's daughter, was asked about the circumstances of her father's suddenly learning, in the desolate Adirondacks, about President McKinley's imminent death. "That must have been a terrible moment of sadness," vanden Heuvel said. Alice, knowing her father all too well, answered, "Are you kidding?"[1]

Overnight, from the relative obscurity of the vice presidency, Roosevelt was now in a governmental position where his every action could be a thunderbolt. The subtle hazards and perils of being president never dawned on him. He was all forward motion, ready to rule by righteousness and a bit of the belt. A friend once famously quipped that Roosevelt was "the meteor of the age,"[2] imbued with unshakable self-confidence in his own ideas. In power of mind and disposition Roosevelt was an old-school military disciplinarian type always shouting "Charge!" and galloping up policy hills with flamboyant bravado. There was nothing fussy about him. Largely disdainful of automobiles and the telephone, Roosevelt remained a twentieth-century saddle-horse man who favored written correspondence as his primary mode of communication.[3] Untiring at his desk, vigorous and direct in his opinions, he was a virtual writing machine. It's been estimated that he wrote more than 150,000 letters in his lifetime. (Harvard University Press published eight thick volumes of them between 1951 and 1954; these are, in effect, an epistolary biography of Roosevelt.) These letters were like chess moves to Roosevelt, helping keep his hyperactive mind fresh and his fighting instinct well honed.

"Now talking with Roosevelt often does no good because he does all the talking," William Allen White noted in December 1901. "But when you write him and he can't talk back you get a chance to put in more."[4]

After only a few days in office Roosevelt started bombarding western friends with exhortations to gear up for a fight to protect America's heritage. According to the Forest Management Act of 1897, timber and water were the only natural resources that the U.S. government was officially sanctioned to protect inside forest reserves. Therefore, this stop-gap act had opened forest reserves to mining claims. And it ignored foraging because allowing livestock to graze in the reserves was still being hotly debated in Congress. Within a few weeks Roosevelt made it clear that his administration would keep voracious sheep—those goddamn hoofed locusts—out of the forest reserves. "Intellectually a sheep is about on the lowest level of the brute creation," Roosevelt had written; "why the early Christians admired it, whether young or old is . . . always a profound mystery."[5] Roosevelt made it clear that sheepherders, or *borregueros* in the west, would be arrested if they trespassed on federal property. Instead, Roosevelt wanted to promote big game in its old, pre-Columbian range. Roosevelt insisted that white-tailed deer rather than livestock of any sort should populate places like the northern Arizona plateau. Eventually, after seven and a half years as president, he won that specific debate. There are literally hundreds of instances in which Rooseveltian wildlife protection trumped grazing. For example, when Roosevelt became president the Kaibab Forest of Arizona had only about 2,000 deer. Owing to Rooseveltian game management principles, this number swelled to 100,000 within a decade.[6] And not a single sheep's bleat could be heard in the Kaibab Forest.

Furthermore, Roosevelt built on postulations promoted by Charles D. Walcott (director of the U.S. Geological Survey), who insisted that there were "sentimental" reasons to save forests; they had recreational as well as commercial value. But before Roosevelt's "new conservationism" could soar, the obligations of continuity and stability were met as protocol dictated. Roosevelt retained stalwarts from the McKinley administration—notably Secretary of State John Hay and Secretary of War Elihu Root—to reassure the nation regarding foreign affairs. The American government would not abandon Hawaii, Guam, Puerto Rico, or the Philippines on his imperialistic watch. There was only one clear deviation in foreign policy. A more energetic push for an isthmus canal would be prioritized; it was a pet issue for Roosevelt. And, of course, Roosevelt would continue to oversee the building of a first-rate U.S. Navy fleet, as dictated by Mahan's security ideas.

On the domestic front, however, immediately deviating from McKinley's policies, Roosevelt began professionalizing forestry and wildlife protection in both Interior and Agriculture. A battle royal for the future of the West had erupted, and he didn't plan on letting his pro-conservationist side lose. "This immense idea (of conservation) Roosevelt, with high statesmanship, dinned into the ears of the Nation," Robert La Follette of Wisconsin recalled of the months following McKinley's assassination, "until the Nation heeded."[7]

Influenced by Pinchot, Roosevelt believed that the United States was in the Dark Ages when it came to proper scientific management of the reserves. In late 1901 showdowns between preservationists and developers over forest reserves had become common in the West. For example, a spur track had been laid from Williams, Arizona, to the South Rim of the Grand Canyon, a distance of sixty miles, slashing its way through forest. Besides increased tourism, eastern fortune seekers were pouring into the Grand Canyon region seeking minerals and timber rights. Other companies wanted to construct buildings at scenic sites. Plans were under way by the Atchison, Topeka, and Santa Fe railway to erect the luxurious El Tovar Hotel on the canyon brink. Was the Grand Canyon going to be ruined? Roosevelt determined that forest rangers and wildlife protectors should be hired as a police force around the Grand Canyon to deal with the increased tourist, timber, and mining development. Not that Roosevelt was opposed to limited *wise* development.[8]

In actuality, Roosevelt wanted more Americans to spend holidays in the West rather than waste time in Europe. What concerned him was that the U.S. government didn't have enough army troops protecting, for example, California's sequoias or Yellowstone's petrified wood. Roosevelt hoped old-breed mountain men, husbandry experts, and Rough Rider–types could be employed in national parks and forest reserves as rangers. In the Sequoia National Park's *Superintendent's Annual Report of 1901* the term "park ranger" was used for the first time by the U.S. Army. Roosevelt liked the ring of it. According to the historian Charles R. Farabee, Jr., in 1902 Roosevelt's Interior Department created three classifications of rangers: Class A1 (deeply familiar with forestlands and able to survey and inventory) and Classes 2 and 3 (no complex requirements, "but they must be able-bodied, sober, and industrious men fully capable of comprehending and following instructions").[9] Class A1 was paid ninety dollars per month; Class 2 got seventy-five dollars; Class 3 got sixty dollars.[10]

Like a diligent ROTC recruiter, Roosevelt now went after one of the ablest Class A1 westerners he knew—David E. Warford of Arizona, from the Troop B Rough Riders, who had taken two Spanish bullets in his thighs near Santiago—to watch over white-tailed deer and forestlands in the Kaibab Forest north of the Grand Canyon. To Roosevelt, Warford—who had returned to Arizona a war hero—was a "new prototype" of the forest ranger conservationist. Warford would protect the yellow pine groves of central-eastern Arizona in what is today the Apache, Coronado, and Tonto national forests. Because the nearly 4.15 million acres* of noncontiguous reserve lands had irregular boundaries—on a map the land looked like jigsaw pieces—it needed somebody who knew every swath of the entire Great Colorado Plateau like the back of his hand to protect the forests from exploitation. Such a vast territory needed a "ranger"—a term first popularized in America during the French and Indian War but appropriated by the Confederate Army during the Civil War. "You have been appointed a Forest Ranger," Roosevelt wrote to Warford. "Now, I want to write to you very seriously to impress upon you that you have got to do your duty well, not only for your own sake, but for the sake of the honor of the regiment. I recommended you because under me you showed yourself gallant, efficient and obedient. You must continue to show these qualities in the government service exactly as you did in the regiment. You must let no consideration of any kind interfere with the performance of your duty. You are to protect the government's property and the forests and to uphold the interests of the department in every way. Now, remember that I expect you to show yourself an official of far above the average type; and you are to stand or fall strictly on your merits."

It was signed, "*Your old Colonel.*"[11]

Bureaucratic confusion reigned supreme in late 1901 regarding the protection of national parks and western forest reserves. From 1886 to 1918, for example, Yellowstone, General Grant, Sequoia, and a couple of other national parks remained protected by the U.S. Army. Hence when Roosevelt became president the acting superintendent of Yellowstone was Major John Pitcher of the Sixth Cavalry, known for his antipoaching zeal and hyperefficiency. Selected in 1902 as superintendent, a job he held for five years, Pitcher made great improvements in Yellowstone, establishing a fish hatchery, buffalo alfalfa fields, and trout bag limits in Yellowstone Lake. But owing to the Federal Reserve Act of 1891, the supervision of forest reserves adjacent to the park became the responsibility of the Department of the Interior (not the army). Within Interior the reserves fell

*The total acreage of the three reserves effective June 30, 1915, was 4,147,682. From the Reports of the Forest Service, Department of Agriculture.

under the jurisdiction first of the General Land Office (GLO) from 1891 to 1901 and then, during Roosevelt's administration, the Forestry Division. But—and herein lies one of the many confusions—the Department of Agriculture (USDA) also had a Bureau of Forestry. Interior and Agriculture divided the responsibilities of managing reserves. So in 1901 Yellowstone National Park, for example, was run by the U.S. Army, while the abutting Yellowstone Park Timber Land Reserve was overseen by the Secretary of the Interior. The United States desperately needed a streamlining of its natural resource policy. As President Roosevelt saw it, forest management and national greatness were one and the same.

What President Roosevelt recognized in the fall of 1901 was that, ironically, the key to making forest science work depended on Class A1 district rangers—men like Warford whom locals would respect as federal law enforcement officers. It was nearly worthless, Roosevelt believed, to appoint out-of-state Ivy Leaguers as rangers. Communities needed to respect the local ranger, who ideally would have a "shared heritage" with them. The new federal forestry rules and regulations had to be explained, because citizens in the West were accustomed to taking timber and foraging livestock at will. Limits had to be taught. The threatening "No Trespassing" needed to come from the mouth of a homeboy. One of the things Roosevelt liked about Warford, for example, was that he spoke Spanish, which allowed him to communicate with many locals in New Mexico and Arizona. Once this pillar of police ruggedness was in place, then the food scientists and biologists could come in as backup. Another innovation of Roosevelt's was having student assistants—that is, Yale- or Biltmore-trained scientific forestry experts—spend time working side by side with the western-born rangers.* Roosevelt wanted to give the local men the "undivided responsibility" to oversee their respective forest reserve site.

Besides the Arizona reserves, Roosevelt turned to the Black Hills in South Dakota, where his old friend Seth Bullock, sheriff extraordinaire, had been employed as forest supervisor to protect the federal reserve as a result of Roosevelt's lobbying as vice president. "As soon as I was appointed," recalled Bullock, "Washington commenced to send a lot of Dudes out here as Forest Rangers. I didn't want them. I wanted Forest

* Anybody reading an old copy of the *Biographical Record of the Graduates and Former Students of the Yale Forestry School* (published in 1913) can see dozens of Yale Forestry School graduates being sent all over the country to assist seasoned rangers, eventually becoming rangers or forestry scientists themselves.

Rangers who could sleep out in the open with or without a blanket and put out a fire and catch a horse thief. I wrote the Colonel [Roosevelt] about it."[12]

Upon receiving Bullock's letter, President Roosevelt instructed Secretary of the Interior Ethan Allen Hitchcock to give Bullock a "free hand" in administering the 1,211,680-acre South Dakota reserve. The sixty-six-year-old Hitchcock wasn't used to having a president run roughshod over him in such a brazen, unremitting fashion. An old-style southern diplomat from Mobile, Alabama, Hitchcock had been McKinley's minister to Russia before accepting the secretaryship. He was a lineal descendant of the Revolutionary War hero Ethan Allen.[13] A snappy dresser and a calm presence, Hitchcock epitomized Mark Twain's belief that truly civilized men never rushed. Hitchcock dealt with everybody in formal niceties, allergic to conflict, gracious to the point of caricature. A low-grade conservationist himself, Hitchcock was deeply concerned that some of America's richest timberlands had been recklessly destroyed and others were on the verge of destruction. Certainly Hitchcock understood that the combination of intensive industrial production, the application of science and technology to manufacturing, and the encouragement of land developers and urban growth was destroying natural habitats forever. Yet the optimistic Hitchcock knew that even burned-out forests could be reborn, eventually producing new yields. Nature could heal itself. For the most part Roosevelt and Hitchcock were in sync. Yet between 1901 and 1907 they feuded like obdurate brothers. Hitchcock resented the bullish way the president was going about things, acting like Cassandra, exaggerating the long-term societal dangers because America was consuming forests three times faster than they were being reproduced. Roosevelt and Hitchcock's goals and vision concerning conservation had nearly identical implications for policy—where they differed was in the matter of *pace*. It was zoom versus incrementalism.

Roosevelt, apparently sensing that Hitchcock was only three-quarters on board, never fully trusted this cabinet officer. Concerned that pro-development senators from Wyoming and South Dakota wouldn't like the conservationist-cowboy Bullock being given carte blanche in the Black Hills, President Roosevelt disseminated Bullock's letter to every legislator on Capitol Hill. No consultation was going on; Roosevelt was informing the legislators that the legendary lawman Bullock was in charge of South Dakota's rimland management. Hitchcock, contemplating resignation, decided instead to buckle up and join Roosevelt's progressive crusade, even if it meant absorbing all the president's doomsday histrionics.

Seth Bullock and Teddy Roosevelt. This photo was taken early in the twentieth century. Roosevelt first met Bullock on a cattle range in 1884, and they became very good friends. Soon after assuming the presidency, Roosevelt appointed Bullock U.S. marshal for South Dakota and Black Hills forest ranger. Bullock had an open invitation to stay overnight at the White House whenever he pleased.

Balding, with a huge gray mustache, Hitchcock acted around Roosevelt and Pinchot like a wise butler tolerating abhorrent behavior from youngsters because he had no other choice. What Bullock tried to communicate to Hitchcock was that the Black Hills could survive only if timber and water were conserved. "If both are destroyed," Bullock warned, "the richest 100 miles square will become a desert."[14]

Having the vital Bullock—a forerunner of Shane—on his side was a great relief to Roosevelt. Everybody needs a few bad-weather friends. "I hope to see you in Washington this winter," Roosevelt wrote to Bullock on September 24. "I want to have you at dinner at the White House, and we will talk over past events. I have been peculiarly pleased to have a man of your type to execute the forest laws, for I know you will see to it that they are enforced absolutely without regard to anything but the law itself. Above all I hope you will see that any Government official who is guilty of laxity or inefficiency is held to a strict account."[15]

What President Roosevelt was trying to accomplish by circulating his correspondence with Seth Bullock to congressional offices was to demonstrate the kind of top-drawer fellow needed to protect the western reserves. Roosevelt's vision of law enforcement was hundreds of Bullocks reined up like an Interior Department cavalry on parade, ready to protect the forestlands as a backup for the U.S. Army. Since 1872 Bullock had been perhaps the most vociferous promoter of Yellowstone living in the West. After all, it had been Bullock, as a renegade member of the territorial legislature, who introduced the bill to preserve northwest Wyoming as a "great national park." Nobody in the West loathed poachers—and arrested them—with the earnest fervor of Bullock. As sheriff in Deadwood, Bullock—who entered the pop culture kingdom in 2004 as the leading character in the HBO television series *Deadwood*—relished protecting the Black Hills from rogue mining outfits operating without proper claims and from rank outlaws illegally prospecting for gold on federal land. And he expressed a sweeping damnation of all lawbreakers. A veteran of Grisby's Cowboy Regiment in the Spanish-American War and a fierce ally of the Rooseveltian conservation movement, Bullock wanted the upside-down county around Deadwood—specially Devils Tower toward the west and Wind Cave to the east—preserved as national parks.

President Roosevelt modeled his administration's conservation policy after his own governorship in New York—where he had tried to whittle down the Fish, Game, and Forest Commission to one nonpolitical appointee. Thus a new era in forest conservation and wildlife management policy was under way. As governor of New York from 1899 to 1900 Roosevelt had led an effort to measure every stream and brook in the state. He now wanted to apply that idea nationally.[16] With an air of reasonableness, he warned Secretary Hitchcock that at all costs *nobody* should be employed in Interior solely for political reasons. Instead, Roosevelt wanted "good plainsmen and mountain men, able to walk and ride and lie out at night, as any first-class men must be able to do." He wanted wilderness warriors who understood forest reserves to lead to overall social betterment in America. His exhibits A and B were Warford and Bullock: outdoorsmen without an ounce of haughtiness or of susceptibility to greed. Remembering how Yellowstone had been hampered by a lack of law enforcement before the protection act of 1894, Roosevelt basically wanted the western territories of Arizona, New Mexico, Indian, and Oklahoma protected by Rough Riders. "In other words," Roosevelt instructed Secretary Hitchcock, "they are to be rangers in fact and not in name, and no excuse will be tolerated for inability to perform the vigorous bodily work of the position any more than lack of courage and honesty would be excused."[17]

II

President Roosevelt, now living in the White House, became extremely controversial on a number of fronts besides recruiting rangers and running roughshod over Interior in the late fall of 1901. Roosevelt couldn't help showing his thornier side and his streak of independence, scoffing at both the GOP party line and concerns over states' rights. Roosevelt had planned to head to Alabama that fall to meet with the "negro leader" Booker T. Washington at Tuskegee Institute. However, suddenly thrust into the presidency, Roosevelt had to cancel this trip. "I write you at once to say that to my deep regret my visit south must now be given up," Roosevelt informed Washington. "When are you coming North? I must see you as soon as possible. I want to talk over the question of possible appointments in the South exactly on the lines of our last conversation together. I hope that my visit to Tuskegee is merely deferred for a short season."[18]

Everything about Booker T. Washington impressed Roosevelt. Born a slave in Virginia, too poor to attend school, the self-taught Washington nevertheless founded, in 1881, the Tuskegee Institute, a technical school for African-Americans. A believer in Washington's "accommodationist" view toward whites, Roosevelt declared Washington "the most useful, as well as the most distinguished, member of his race in the world." In office only a month, Roosevelt invited Washington to the White House as a courtesy on October 4 to discuss a federal judgeship in Alabama. A productive dialogue on race relations ensued; the two leaders stood four-square on many important national issues. Washington was invited back to the White House in mid-October to brainstorm about ways to enhance educational possibilities for African-Americans in the South. Eventually, Washington and Roosevelt, enjoying each other's breathless enthusiasm, broke for dinner. Joining them were the first lady, Edith Roosevelt (she hadn't gotten used to being called that), and a professional friend, Philip B. Stewart of Colorado cougar hunting fame. Everybody had a most enjoyable time.

Holy hell broke out in the South the next day, however, when an AP wire story simply stated that Roosevelt had dined with Washington in the White House. The ground rumbled and the southern press went berserk. A segregationist code had been shattered. Headlines like "Our Coon-Flavored President" and "Roosevelt Dines a Darkie" appeared throughout the former Confederacy. The *New Orleans Statesman* grumbled that the meal was "little less than a studied insult to the South."[19] The *Memphis*

Theodore Roosevelt and Booker T. Washington became great friends. Besides inviting Washington to dine at the White House in 1901, Roosevelt also visited Tuskegee Institute in 1904.

Scimitar ran an editorial stating, "It is only very recently that President Roosevelt boasted that his mother was a Southern woman, and that he is half Southern by reason of fact. But inviting a nigger to his table he pays his mother small duty. No Southern woman with a proper self-respect would now accept an invitation to the White House, nor would President Roosevelt be welcomed today in Southern homes. He has not inflamed the anger of the Southern people; he has excited their disgust."[20]

Southern segregationists ripped into Roosevelt with a fusillade of cruel, bigoted, ugly language. James K. Vardaman—publisher of the *Greenwood Commonwealth*, who was a veteran of the Spanish-American War and a Mississippi Democrat and would become governor and then a U.S. senator—surmised that the "White House was so saturated with the odor of the nigger that the rats have taken refuge in the stable." Roosevelt became nauseated by these insults—such hatred in America was cancerous. The southerners, Roosevelt lamented, had indicted him for trying to encourage literacy and help fight poverty among African-Americans. The *Richmond* (Virginia) *Times* was aghast at Roosevelt's tolerating the idea that "negroes shall mingle freely with whites in the social circle—that white women may receive attentions from negro men."[21] Never one to cower in the face of threats, Roosevelt decided to go hunting in Mississippi sometime in 1902—to go into the belly of the beast, almost as an act of defiance. Not for a split second was he going to let ex-Confederates—of all people—assault his character. In coming months he would continue consulting with his friend Booker T. Washington; only he back-pedaled away from dinners in favor of meetings at ten o'clock in the morning. Roosevelt, however, *did* live up to his pledge to tour Tuskegee, though not until after the 1904 presidential election when being photographed with a "negro" wouldn't cost him votes.[22]

On December 3, Roosevelt was to deliver his First Annual Message to Congress, so he surveyed various friends (including Chapman, Grinnell, and Merriam) about what he might say regarding conservation and wildlife protection. A key to President Roosevelt's vision of the American West was a vast increase in forest reserves and western irrigation. These tenets would be a fundamental part of his annual message. At the time of McKinley's death in September 1901 the number of U.S. forest reserves had increased from twenty-eight to forty: a total of more than 50 million acres. Not bad. Building on that adequate legacy, President Roosevelt strategized about how to triple McKinley's effort. He succeeded in increasing the number of national forests from forty to 159, with a total of more than 150 million new acres.[23] Put another way, the U.S. forest reserves went from about 43 million acres in 1901 to 194 million acres in 1909 under Roosevelt's leadership. As the historian John Allen Gable computed in the *Theodore Roosevelt Association Journal*, Roosevelt's new forestlands constituted an area larger than France, Belgium, and the Netherlands combined.[24]

Forest reserves aside, President Roosevelt looked back in bafflement over why McKinley had rejected stringent wildlife protection laws. Didn't McKinley want elk and antelope to populate the Great Plains? Was he really opposed to a moose reserve for Maine? The fact of the matter was that McKinley simply hadn't wanted to squander political capital with powerful western senators over what he considered fringe issues, such as protecting ungulates. That indifference immediately changed with Roosevelt in power. From the get-go Pinchot, in fact, at Roosevelt's behest, had brought into the forefront of U.S. conservation policy initiatives which the Boone and Crockett Club had formulated: mainly, having game reserves *inside* national forests. Anxious for his administration to make these bold leaps toward wildlife protection, Roosevelt asked Gifford Pinchot to push the ideas about game reserves through Congress. It didn't prove easy going for Pinchot. Western development interests didn't give a damn about buffalo, deer, and elk. Working with the Boone and Crockett Club and the New York Zoological Society, Roosevelt, with the essential help of Congressman Lacey, nevertheless soon made historic strides in getting wildlife protection legislation.

Deciding unilaterally to change the name of the Executive Mansion to the White House that October, Roosevelt asked Pinchot to head the Division of Forestry in Interior. He promised him "an absolutely free hand"—free, that is, from the gaze of Ethan Allen Hitchcock. Roosevelt claimed his White House desperately needed Pinchot's help on selecting ideal sites for forest preserves and on intelligent habitat management (including selective timber thinning and brush control). As a lure or incentive Roosevelt told Pinchot that his recommendations for forestry would become—in essence—the de facto administration policy. Pinchot, not Hitchcock, would be the ultimate arbiter at Interior. And while ostensibly Pinchot was merely head of a division, in reality he would have more power than Secretary Hitchcock or the GLO commissioner, Binger Hermann (a former Republican congressman who was an attorney in Oregon). Hitchcock would be a useful figurehead—and an ally of sorts—kept only to placate the McKinley's old guard. As for Hermann, he smelled to Roosevelt like an enemy, an Oregonian more interested in pork for river and harbor appropriations than in protecting places like Crater Lake, Three Arch Rocks, or the Cascades. Point-blank reality was that Pinchot (as division head) would be making federal forestry policy.[25] (Later, in 1905 Pinchot would become the first chief of the new forestry service.)

Gladly, Pinchot accepted the president's gracious offer. In the coming years their vigorous friendship continued to blossom. Roosevelt found Pinchot to be a bundle of invaluable insights. Together they would often hike in Rock Creek Park, swim in the Potomac River, play tennis, watch birds, and chop firewood near National Cathedral School. While Pinchot wasn't given to lyrical outpourings like Burroughs or Grinnell, he was a far better conservationist tactician than anybody else orbiting around Roosevelt. In Roosevelt's so-called "tennis cabinet" ("kitchen cabinet" sounded too sissified for Roosevelt), Pinchot was probably his most trusted colleague. Pinchot, in fact, became something of a "faithful bodyguard," always willing to defend Roosevelt from attacks.[26] Seldom did Roosevelt and Pinchot see things through different lenses. (There, however, was one big difference between them: Roosevelt was first and foremost a bird preservationist whereas Pinchot was not.) And they both enjoyed night work and end-of-the-day confidences. "They were appalled by the human destruction of nature everywhere visible in early-twentieth-century America," the historian Char Miller has noted. "The solution, they believed, lay in Federal regulation of the public lands and, where appropriate, scientific management of these lands' natural resources; only this approach, guided by appropriate experts, would ensure the lands' survival. So parallel ran their thoughts that Roosevelt reportedly assured Robert Underwood Johnson, editor of *Century Magazine*, that on questions of conservation the chief forester was in truth the keeper of his conscience."[27]

Not only did Pinchot agree to run the Division of Forestry in Interior, but over Thanksgiving he inserted paragraphs about conservation into the December 3 annual message. To many western senators these insertions were out-and-out heresy. The intensity and boldness of Roosevelt's address, read by a clerk (as was traditional), encouraged conservation enthusiasts on many levels, though the speech was somewhat short on details. And it wasn't just a cranky outburst. It was hard-core Rooseveltian conservationist philosophy, presented on the nation's center stage. For those familiar with Roosevelt's allegiance to the Boone and Crockett Club and various Audubon societies, it wasn't very shocking—but the sheer breadth of the wildlife protection plank *was* unexpected. Even though most New Yorkers had accustomed themselves to the proposition that the sportsman Roosevelt, when it came to wildlife protection or forestry policy, was never content to be a spectator, congressmen on both sides of the aisle were surprised by the piercing vigor of his conservationist agenda.

Nobody has recalled President Roosevelt's First Annual Message with such elegance and insight as historian Edmund Morris in *Theodore Rex*. Combining actual passages of Roosevelt's address with vivid descriptions of individual legislators and the atmosphere, Morris wrote about that frigid December day as if he had been sitting in the visitors' gallery witnessing history.[28] Regardless of its overall eloquence, the annual message consisted of important reports and helpful comments that the White House had received from various departments (in other words, it was cobbled together).[29] For starters Roosevelt, in strong language, condemned filthy anarchists; he was seething because three presidents in his lifetime had been struck down in their prime by lunatics. Thunderous applause arose from Congress as the clerk, reading Roosevelt's bracing prose, exclaimed with pent-up frustration that the American people, usually "slow to wrath," when "kindled" (by an anarchistic abomination like the murder of McKinley) ignited like a "consuming flame."[30] As president he planned on ridding the nation of anarchists, sending them scurrying like mice across the floorboards of national life.

Although Roosevelt offered some uplifting chamber of commerce–like pronouncements about improved business confidence, his address was notable for its stinging language about corporate trust-busting. Industrialists interpreted the address as a sneer from the pulpit. Clearly, Roosevelt planned on restraining the business class, and even openly challenging it over stock market manipulations and monopolist attitudes. Throughout the Gilded Age huge corporations worked overtime to abuse the public welfare, affecting millions of Americans; such abuses were going to be curtailed with Roosevelt in the White House. He promoted immediate federal intervention in regulating corporations. And—like a boot stuck in mud suddenly coming free—he said that workers were no longer going to be treated as industrial wage slaves. Demanding improved labor laws, Roosevelt lambasted, as Morris puts it, politicians that were "fattened at the public trough."[31] Many of those seated—particularly senators from the Deep South still furious over the Booker T. Washington affair—were leaning forward with fingers clasped and heads shaking: no, this traitor to his class and race can't be real. "In the interest of the public, the government should have the right to inspect and examine the workings of great corporations," Roosevelt stated. "The nation should, without interfering with the power of the States in the matter itself, also assume power of supervision and regulation over all corporations doing interstate business."[32] No company was above the law or deserved special treatment from the U.S. government. He outright rejected corporations that wanted rebates and rate fixing. "Great corporations," Roosevelt proclaimed, "exist only because they are created and safeguarded by our institutions; and it is therefore our right and duty to see that they work in harmony with these institutions."[33]

Following these cautionary swipes at corporations President Roosevelt launched into his conservation plank, based on the philosophy of Pinchot and Grinnell. Nothing would palsy his resolution regarding wilderness protection. The West had 6 million inhabitants in 1901; by the time Roosevelt left office there were over 10 million citizens and the population was still growing. Much of Roosevelt's conservationist thinking in the annual message was directed toward this region. With the West so much in the forefront, he was, as Morris noted in *Theodore Rex*, "striking a note altogether new in presidential utterances."[34] Nothing about Roosevelt's conservationist rhetoric could have been misconstrued as give-and-take. He was *telling* Congress the new lay of the land. Disgusted that the United States had cut down almost 50 percent of its timber, and that valuable topsoil had been washed away, Roosevelt was sending a wake-up call.[35] He wanted Congress to save pristine American land while it still existed. Whether they were coniferous forests in the Pacific Northwest or strands of Douglas fir far older than the republic in the Front Range of the Rockies, forests had to be saved. His far-reaching conclusion, after much consideration, was that he wanted the western reserves vastly increased. Decades of study had taught him the symbiotic relationship between timber, soil, and water conservation. "The preservation of our forests is an imperative business necessity," the address stated. "We have come to see clearly that whatever destroys the forests, except to make way for agriculture, threatens our own well being."[36]

And the conservationist creed—albeit carefully modulated—didn't stop there. It would be up to the federal government—not big business—to lease lands for logging or mining, and not just near the famous destinations like Yellowstone and Yosemite. Throughout the West, the prettiest scenery not deforested or contaminated would be on the table for consideration as national parks or forest reserves. Not on his watch would such lovely Pacific Northwest ranges as the Cascades and Olympics be turned into heaping mounds of slag as in Appalachia. No western state would go unaffected. Praising the U.S. Department of Agriculture, Roosevelt promised to sponsor even more science-based studies pertaining to trees, plants, and grasses through its Biological Survey division run by Dr. Merriam. Roosevelt's address was pure radical Americanism—especially the ten paragraphs dealing directly with conservation. That November, just a few weeks before the First Annual Message, John Muir had published the essay collection *Our National Parks*, which included his classic *Atlantic Monthly* articles "The American Forests" and "The Wild Parks and Forest Reservations of the West."[37] Roosevelt had found them highly stimulating and persuasive. Roosevelt did not mention Muir in his annual message but nevertheless sided with Muir's ecologically sensible crusade to save the great forests of the Pacific Slope. Roosevelt, in fact, liked to quote Muir, who wrote in 1897: "The forests of America, however slighted by man, must have been a great delight to God; for they were the best he ever planted."[38]

Roosevelt was the new Delphic oracle of conservation, the political authority of the forestry movement, best-selling author, wilderness trooper, birder, hunter, and moral advocate for nature. For most presidents, give-and-take with Congress was important. Roosevelt, however, believed in only one solution: his own. But, by and large, Congress wasn't persuaded by Roosevelt's far-reaching promotion of forestry in the First Annual Message. Roosevelt, for example, recommended consolidating forest work under the Bureau of Forestry. "This recommendation was repeated in other messages," Roosevelt carped in *An Autobiography*, "but Congress did not give effect to it until three years later. In the meantime, by thorough study of the Western public timberlands, the groundwork was laid for the responsibilities which were to fall upon the Bureau of Forestry when the care of the National Forests came to be transferred to it."[39]

Besides forest reserves Roosevelt spoke out in the First Annual Message on behalf of wildlife protection as formulated by the Boone and Crockett Club, implying that many federal preserves would eventually be created to protect elk, pronghorns, mule deer, and mountain goats. Even though wildlife didn't have the economic importance of timber or water, it was the most endangered resource of twentieth-century America. To Roosevelt the forest reserves, in consequence, should "afford perpetual protection to the native fauna, and flora, safe havens of refuge to our rapidly diminishing wild animals of the larger kind."[40] Eventually Roosevelt would sell suspicious western developers on the need for wildlife protection by offering a quid pro quo. Predator control would serve as an inducement. Once established, federal game reserves would thin out wolves and coyotes in a region, keeping these predators away from domestic livestock in the government lands. This, in turn, would also mean far less predators in communities near forest reserves. Roosevelt had Merriam at the Biological Survey begin printing pamphlets on how best to poison coyotes and wolves.[41]

Diligently, Roosevelt had tweaked drafts of the first annual message, searching for exactly the right phrases. This wake-up speech wasn't a pedestrian tract on the virtues of utilitarian forestry. It was meant for the ages, meant to be bound in gilt-stamped leather. The embryonic wildlife protection movement (best epitomized by the Boone and Crockett Club and the Audubon societies) had now come to fruition on a large scale at the federal level. The U.S. government was headed into the business of saving elk, deer, and buffalo. Even though the address was read by the clerk, listeners could envision the president jabbing his finger at disputants, determined to topple their built-in predispositions. "Roosevelt respected expert opinion and made use of it to a degree which was unmatched among the public men who were his contemporaries," Pinchot explained. "Men of small caliber in public office find scorn of expert knowledge a convenient screen for hiding their own mental barrenness. So true is this that one of the best measures of his own breadth and depth of mind is the degree to which a public man acknowledges the value of expert knowledge and judgment in fields with which he himself, in the nature of things, cannot be familiar. By this standard Roosevelt stood at the very top."[42]

At about the time of the First Annual Message, Roosevelt encouraged Merriam to increase the hiring of so-called "camp men" who could help the Biological Survey's field reporters inventory native plants and animals. For a salary of about twenty-five dollars a month, these camp men (usually hunt guides from the area) would assist the trained scientists working for the Biological Survey. Together the egghead and the rough-and-ready would set traps, prepare skins, and ship species back to Washington, D.C., where they could be carefully studied in laboratories. Roosevelt wanted thorough field notes with biotic summaries accompanying every shipment. One of Roosevelt's favorites among Merriam's "field agents" was J. Alden Loring (who upon his recommendation in 1899 became assistant curator of mammals at the New York Zoological Park). Always encouraging Loring to become a public figure, to stop concealing his genius, Roosevelt tapped him as a talent scout taps a promising athlete. Proud of the way Loring was following in Merriam's estimable footsteps, Roosevelt later had the young naturalist collect for the U.S. National Museum in Europe. Loring also helped reintroduce buffalo back to South Dakota and accompanied former president Roosevelt on his 1909 African safari.[43]

What impressed Merriam about Roosevelt was that even while living the "strenuous life," he never stopped being a faunal naturalist. The microscope had turned a new generation of biologists to studying minute organisms, but Roosevelt stayed focused on what Merriam called the more "obvious forms of life." Starting in November 1901 and continuing until he left office in March 1909, Roosevelt would telephone Merriam quite regularly, particularly during the spring migration, making sure that the warblers in the White House elms were blackpolls or that the flocks of rusty blackbirds along the Potomac basin hadn't decreased in numbers from the previous year. Not long after becoming president, in fact, Roosevelt had asked Merriam to take a twilight bicycle trip with him from the White House to Rock Creek Park to watch a beaver build a lodge. "He was 'delighted' to see the beaver cut a willow and swim with it to a floating log," Dr. Merriam recalled in *Science*, "where he sat up and ate the bark."[44]

Regularly Roosevelt would walk from the White House to Merriam's home to study his world-class collection of mammal bones and skins.[45] He was like a child wandering into F.A.O. Schwartz on Christmas Eve.

Merriam's huge library was *the* "zoological salon" of the District of Columbia. "Few people are aware of Roosevelt's knowledge of mammals and their skulls," Merriam recalled. "One evening at my house (Where I then had in the neighborhood of five thousand skulls of North American mammals) he astonished every one—including several eminent naturalists—by picking up skull after skull and mentioning the scientific name of the genus to which each belonged."[46]

Besides rattling off the genus of skulls, Roosevelt was proud that some of the Biological Survey's best cougars were courtesy of his prodigious hunting efforts. Two weeks after Roosevelt delivered his Annual Message he wrote to Yellowstone's acting superintendent, Major John Pitcher of the Sixth Cavalry, about shooting cougars to control predation. The president wanted to arrive in Yellowstone that June and hunt "varmints" that were "not protected." A backwoodsman, John B. Goff, would serve as a guide, using his pack of hounds to chase the big cats. If Roosevelt collected ten or twelve cougars from Wyoming he could ship them to Merriam to be compared with the ones from Colorado. Being president shouldn't mean relinquishing his reputation as the world's leading expert on cougars. From Major Pitcher's perspective this was an insane request. The president of the United States, busy with international crises, wanted to summer in the wilds of Yellowstone to "thin out" the cougars? But Major Pitcher—particularly as *acting* superintendent—wasn't going to tell his boss no. He started making arrangements.

The rumor that Roosevelt was coming to Yellowstone had C. J. "Buffalo" Jones whooping with excitement. Sidestepping Major Pitcher, Jones started to make plans for a hunt in Yellowstone. He was especially anxious for the president to see some of the buffalo he'd raised like cattle. Most Americans in 1901 could see bison only in Buffalo Bill's Wild West Show, or at Bronx Zoo or the Goodnight Ranch in Texas. Jones was determined to change that sad fact. An old buffalo runner, he had reformed and was among the best bison breeders in the American West. Sickened that his own slaughtering of buffalo had almost brought about their extinction, Jones wanted to show his hero, President Roosevelt, how he had made a 180-degree turn. Roosevelt was greatly interested in Buffalo Jones's claim to have successfully crossbred bison with cattle (producing "catalo") and even reportedly broken a few of the offspring to harness. Unfortunately, one of Jones's captive Yellowstone buffalo—named "Lucky Knight"—had trampled a Wyoming resident to death. Refusing to let Lucky Knight be butchered, Buffalo Jones instead trained it to pull his buckboard. He bragged that he had the only killer buffalo in the West.[47] Unfortunately,

Roosevelt's 1902 trip to Yellowstone was postponed for a year owing to an unexpectedly heavy workload. The anthracite coal strike, heightened Russo-Japanese tension, and other serious presidential concerns forced Roosevelt to postpone seeing Buffalo Jones's bison herd until April 1903.

That Christmas season President Roosevelt grew intrigued by the creation of the new American Scenic and History Preservation Society in New York (it was an outgrowth of Andrew H. Green's Trustees of Scenic and History Places and Objects, which had helped Roosevelt save Palisades Park back in 1900). Just as Roosevelt wanted the Biological Survey to inventory all of America's plants, birds, fish, insects, and wildlife, this new nonprofit organization, modeled after Britain's National Trust, was going to protect both historic sites and scenic places. Roosevelt believed—thanks to Congressman Lacey's inspection tour—that, for example, the Anasazi cliff dwellings in Colorado and the Pueblo Chaco Canyon in New Mexico needed protection. The new trust was going to start doing that and more. Among the places the trust saved were Stony Point Battlefield (thirty-five acres on the shore of the Hudson River near West Point), Lake George Battlefield (thirty-five acres on Lake George), and Fort Brewerton (in Hastings at the foot of Oneida Lake). Wishing he could have founded the trust, Roosevelt began scheming to find new ways for the federal government to interface with it. And Roosevelt started lobbying the trust to save Chalmette Battlefield in New Orleans, the site along the Mississippi River where Andrew Jackson gave the British their comeuppance; he sent Green the appropriate chapter of his *Naval War of 1812*, which documented the historical significance of the battle.* Contained in the trust fund were the seeds of what would become Roosevelt's grand preservationist accomplishment—the Antiquities Act of 1906.

Congressman Lacey had truly gotten President Roosevelt to start thinking in earnest about preserving Chacoan heritage sites in Colorado, Utah, New Mexico, and Arizona. The cultural blossoming of the Chacoan people had begun in the mid-800s (after Christ), well after Saint Augustine was born but long before the supposed "oldest city" in North America was named after him in Florida. The Chacoans built a network of fairly sophisticated villages throughout the Southwest.

*On March 4, 1907, President Roosevelt created the Chalmette Monument and Grounds (site of the Battle of New Orleans), including a cemetery for veterans of the War of 1812. Today the Chalmette National Historic Park—located in Saint Bernard Parish, Louisiana, along the Mississippi River near New Orleans—is a major tourist attraction.

The huge Four Corners high-desert valley was once a hub of Anasazi life. Tribesmen farmed lowlands and constructed elaborate cliff dwellings. The Chaco Anasazi were extraordinary masons, and their towns were monuments to creative architecture. A burning question in archaeological and ethnological circles in 1901 was: what happened to the Anasazi? The answer seemed to be that a great drought had killed them off. The message to Roosevelt was clear: aridity was the death card in the Southwest. To be sustainable, communities had to develop water reservoirs. In 1901 Europeans considered Aztec and Maya ruins in Mexico and South America grand antiquities. The nationalistic Roosevelt scoffed at such boosterism by the European art world on behalf of Mexico. The United States, he said (thumping his chest, as it were), had just as fine ancient rubble in its own Southwest as existed in Peru or Bolivia. The cliff dwellings at Mesa Verde, Colorado, were our Machu Picchu.

Since his days at Harvard, in fact, Roosevelt had been interested in the mysteries pertaining to the vanished Chacoans. The photographer W. H. Jackson of the U.S. Geological Survey had written extensively about how Chacoan stairways were carved into cliffs at Mesa Verde. Unfortunately, the plates of photography he took weren't properly developed, so he brought back to New York only diary notes. But in 1888 the Bureau of American Ethnology spent six weeks in the Four Corners region photographing Chacoan sites for a huge project on Pueblo architecture. It also reported that vandals were looting the antiquities. When Roosevelt was the U.S. civil service commissioner he denounced Chacoan "pot hunters" as being as swinish as the poachers at Yellowstone.

Between 1896 and 1900 the great archaeologist and trail guide Richard Wetherill began excavating the Mesa Verde cliff dwellings. At the same time the American Museum of Natural History in New York began analyzing Pueblo Bonito. When Roosevelt was governor he inspected crates of artifacts from the Southwest when they arrived at the American Museum of Natural History and were eventually put on permanent display. Therefore, by the time Roosevelt became president in September 1901 the fact that the GLO was promoting the idea of a Chacoan national park was old news to Roosevelt. Meanwhile, in Santa Fe, New Mexico, the committed archaeologist Edgar Lee Hewett—a quiet, unassuming intellectual enamored of the Chacoan past—was mapping various sites at Four Corners in preparation for preservation, as head of the School of American Research. Hewett made it a personal crusade to save these amazing archaeological sites. There were dozens of legal hurdles to clear (not the least being Wetherill's claim to the land around Chaco Canyon), but Roosevelt told

Congressman Lacey they'd find a way to preserve Chacoan sites. In 1902 Lacey—hiring Hewett as coconspirator—began working on sensible legislation. It would evolve into the Antiquities Act of 1906.

III

So Roosevelt had started to put the wheels in motion for preservation early in his administration. Polishing up his credentials as a naturalist and one of the four or five popular authorities on North American mammals, Roosevelt also published, on May 7, 1902, a book titled *The Deer Family*. Written while he was vice president, *The Deer Family* was issued as the first volume of the American Sportsman's Library (edited by Casper Whitney of the Boone and Crockett Club).[48] The book was done in collaboration with his fellow naturalists T. S. Van Dyke (on Pacific Coast elk and Columbia black-tailed deer), Daniel G. Elliot (on caribou), and A. J. Stone (on moose), and Roosevelt wrote the first four chapters himself. The publisher of *The Deer Family*—the Macmillan Company—made much of the fact that it constituted, as the *Washington Times* noted, "the first time in the history of the country a book has appeared bearing the name of the President of the United States as that of the author."[49]

Collaboration between authors was fairly commonplace in the academic world of 1902, and scientists usually presented scholarly papers at conferences with three or four names attached to their joint research. But Theodore Roosevelt was president: whom he decided to share his title page with was automatically news. And he didn't mind working with Darwinian eccentrics. The three men Roosevelt chose to be associated with in publishing this historic book were among the very best naturalists in America. All of them held a Darwinian belief in the importance of fossil records and the interconnectedness of environment and life. One of Roosevelt's major motivations for writing his essays in *The Deer Family*, in fact, was to swing a lantern over the names of these naturalists, big game hunters, and explorers, in gratitude for their decades of largely unsung work. And the Boone and Crockett Club, in a congenial way, was circulating petitions in the Great Plains to promote the notion of big game preserves; *The Deer Family* was an important tool in this wildlife repopulation effort.

President Roosevelt had been in awe of Professor Elliot ever since age nineteen, when Theodore Sr. had introduced them in Italy. Honored all over the world for his bravery and his zoological discoveries, and known especially for his astounding papers on birds, Elliot had been vaguely associated with the American Museum of Natural History since its founding. He was also active in the American Ornithological Union. Darwin had detailed the courtship rituals of the Australian bowerbird; Elliot did the same for shorebirds. No fewer than ten nations had decorated Elliot for his first-rate empirical work in the natural sciences. Throughout the 1890s he interacted with Roosevelt socially in New York, swapping stories of bird sightings like a couple of old fuddy-duddies from the British Museum. Playing the eager student, Roosevelt had read Elliot's monographs, all saturated with facts, including *Family of the Pheasants* and *Birds of Paradise*. With his large walrus mustache, which stood out more vibrantly than his pointed beard, Elliot was easily distinguishable in a crowd. When the naturalist Dr. Albert Bickmore heard that Elliot had been chosen by the Field Museum of Chicago to be its zoology curator in late 1894, he lamented that New York had lost "one of America's first scientists." In 1898, while Roosevelt was in Cuba with the Rough Riders, Elliot was the first serious naturalist to systematically study the Olympic Mountains in Washington State (home range to Roosevelt's elk). Elliot had initiated a movement to save the state of Washington's Mount Olympus from the timber conglomerates.

Then there was T. S. Van Dyke, whose book *The Still Hunter* was considered a classic in the sporting genre. Van Dyke was less scientific than Elliot and was blessed with the ability to spin a good yarn about cougars; Roosevelt admired the way he brought wild creatures into the lives of everyday Americans without scientific pretension. He had a tremendous knack for powerful, accurate generalization. Nobody in the United States wrote prose as similar to the president's as Van Dyke. There was a tradition that was passed down from Reid to Van Dyke to Roosevelt. The president was enthralled by a popular article of Van Dyke's, "The Hills of San Bernardino," and recommended it to everyone. "We have left far behind us the mellow flute of the valley quail," Van Dyke wrote, "but his double-plumed and gay cousin of the mountain well supplies his place. From the lowest valley to the loftiest point where vegetation grows, you often see his mottled waistcoat of white and cinnamon, his bluish coat, and long nodding plumes; may hear the gentle patter of his little feet on the pine-needles as he steals softly away, and hear his ordinary *quit-quit-quit-quit queeah* changed into a dismally-anxious *queeeee-awwk*, as he leads the little brood from danger."[50] Just as Elliot was making his mark studying the Olympics, Van Dyke became the preeminent mountain climber–naturalist of California's Palomar mountain range. Fancying himself as the Joaquin Miller of any California landscape south of Big Sur, Van Dyke also wrote *The City and County of San Diego*, published by a small local press.[51]

Unlike that of the other authors in *The Deer Family*, however, some of Van Dyke's work was clearly mediocre. And at times, as when he claimed to have shot four wildcats with one shotgun blast, he defied believability.[52]

Rounding out *The Deer Family*'s quartet was A. J. Stone, the naturalist wunderkind of the moment. As a corresponding member of the Zoological and Ethnological Museum of Natural History and the New York Zoological Society he spent the years 1897 to 1899 living around the Arctic Circle with only his kayak and sled dogs as companions.[53] Taking off from Fort McPherson, the Hudson Bay Company's northernmost outpost, he trudged up through sea ice to forlorn Herschel Island. During one five-month stint Stone hiked 3,000 miles above the Arctic Circle, shattering all previous land travel records. With the bitter wind assaulting him, and temperatures often falling to minus seventy or eighty degrees Fahrenheit, he nevertheless traversed snowdrifts as tall as the White House. Polar bears, northern fur seals, and arctic foxes were all vividly described in his field notes. When the harpooner hero returned to America the *New York Times* saluted his circumnavigation as finishing "one of the most remarkable trips in the history of the North American continent."[54]

The selection of Elliot, Van Dyke, and Stone as coauthors of *The Deer Family* represented three distinct sides of Roosevelt's conservationist persona, though perhaps the president did not realize this. There was the intrepid Elliot, the man of letters, naturalist, and globe-trotter, known for the precision of his scientific work in ornithology and mammalogy (but also heartily equipped to endure leeches and snakebites). The dominant strain of the big game hunter–naturalist was represented more than adequately by the indomitable Van Dyke, who was roaming the West Coast in search of bears, as he had done in Montana. Like Roosevelt's, Van Dyke's prose was action packed, yet careful about wildlife observations. Being a naturalist explorer was an occupation that the president coveted more than any other. Recognizing that the Arctic Circle was one of the last frontiers, Roosevelt, from temperate Washington, D.C., chose Stone, who had exhibited grit, individualism, and adventurousness in the wild taiga and tundra at the top of the world, tethered to a dogsled. They shared a fundamental attitude of no retreat. Clearly Stone, like the president himself, had learned to overcome wind chill, distance, and isolation while still managing to read Tolstoy on an inflated mattress by quiet candlelight in a makeshift outpost shack.

Bookstores throughout the United States set up displays of *The Deer Family*, complete with handsome illustrations of sixty-seven-inch Alaskan moose antlers and Wyoming antelope grazing on the open range. And although Van Dyke, Elliot, and Stone were coauthors, the dark green cloth cover read only: "*The Deer Family* by Theodore Roosevelt and Others." In fact, President Roosevelt had written only one-third of the book, but his lively chapters were far and away the most popular. Upon opening *The Deer Family* the reader immediately encountered a brief "Foreword" by Theodore Roosevelt—written in June 1901, when he was vice president. "This volume is meant for the lover of the wild, free, lonely life of the wilderness," he wrote, "and of the hardy pastimes known to the sojourners therein."[55]

Roosevelt's chapters in *The Deer Family* are beautifully written, combining a nearly childlike rapture for hunting with an adult conservationist philosophy.[56] Not only did the president offer compelling scientific details of the exact bifurcations of a mule deer's main prongs or the pugnacity of elk herds; he made his field observations discernible to the average American. When writing about these mammals the president insisted on biological precision, seamlessly weaving into his narrative a steady succession of scientific facts, big-bored .505 Gibbs flashbacks, and earthy descriptions of Western scenery. In *The Deer Family* were echoes of the naturalist prose that Roosevelt had first showed off in *The Wilderness Hunter*, to the hearty approval of John Burroughs. For example, here is Roosevelt on the North Dakota prairie in *The Deer Family:*

> It was beautiful to see the red dawn quicken from the first glimmering gray in the east, and then to watch the crimson bars glint on the tops of the fantastically shaped barren hills when the sun flamed, burning and splendid, above the horizon. In the early morning the level beams brought out into sharp relief the strangely carved and channeled cliff walls of the buttes. There was rarely a cloud to dim the serene blue of the sky.[57]

Nostalgic passages like these, in which the writer is hungering for open spaces, made *The Deer Family* a minor best-seller for three or four weeks. Literally every review the book received was respectful. From the perspective of time, 100 years after it was written, *The Deer Family*—even more than *The Wilderness Hunter*—may be the most important of all Roosevelt's books for our understanding of his evolved views on conservation. No longer does Roosevelt regale readers with his derring-do across the immensity of the continent. Nor does he champion mountain men in nativistic, white-man's-burden fashion. In *The Deer Family*, Roosevelt—speaking as a U.S. president—became an environmental crusader and scold. Derision toward unsportsmanlike hunters was more amplified than in his previous outdoor books and essays. "The big game hunter should be a field naturalist," Roosevelt wrote. "If possible, he should be an adept with the camera; and hunting with the camera will tax his skill far more than hunting with the rifle, while the results in the long run give much greater satisfaction. Wherever possible he should keep a note-book, and should carefully study and record the habits of the wild creatures, especially when in some remote regions to which trained scientific observers but rarely have access. If we could only produce a hunter who would do for American big game what John Burroughs has done for the smaller wild life of hedgerow and orchard, farm and garden and grove, we should indeed be fortunate."[58]

He urged all sportsmen to immediately become naturalists, keeping detailed notes about wildflowers, prairie grasses, and swamp fronds. Even if the would-be hunter didn't "possess the literary faculty and powers of trained observation necessary for such a task," Roosevelt instructed, he could nevertheless "do his part toward adding to our information by keeping careful notes of all important facts which he comes across." Attempting to create North American field guides from the ground up—something akin to the later regional WPA guides—Roosevelt wanted everyday citizens to partake in inventorying the nation's natural resources. "Such note-books would show the changed habits of game with the changed seasons, their abundance at different times and different places, the melancholy data of their disappearance, the pleasanter facts as to their change of habits which enable them to continue to exist in the land, and, in short, all their traits," he wrote. "A real and lasting service would thereby be rendered, not only to naturalists, but to all who care for nature."[59]

Unbeknownst to Roosevelt's opponents in spring 1902, his desk at 1600 Pennsylvania Avenue had already become a rubber-stamp center for any serious-minded conservationist or natural resources specialist with an honest agenda. Already he was thinking of how best, with a modicum of good sense, to repopulate a federal forest reserve with his Bronx Zoo bison. Regularly, he was staying in touch with William T. Hornaday. Together they also had high hopes of someday creating a national elk reserve near Yellowstone. "Surely all men who care for nature, no less than all men who care for big game hunting, should combine to try to see that not merely the states but the Federal authorities make every effort, and are given every power, to prevent the extermination of this stately and beautiful animal," he wrote of elk. "The lordliest of the deer kind in the entire world."[60]

Although this is a chicken-or-egg situation, the Bronx Zoo had made a special effort to advertise its elk, mule deer, caribou, and moose as all being part of what it called "The Deer Family." Unfortunately, despite the enthusiasm of Hornaday and Roosevelt, the zoo was failing in its efforts to breed moose and caribou in captivity. The sultriness of New York City in summer was, as Hornaday later noted, "decidedly inimical" to the project. "This densely humid and extremely saline atmosphere is about as deadly to the black-tail, caribou and moose as it is to the Eskimos," Hornaday wrote, "and thus far we have found it an absolute impossibility to maintain satisfactory herds of those species in the ranges available for them."[61] It wasn't enough to breed buffalo or elk in captivity. Big game refuges were needed in their western habitats. Worried about a band of dwarf elks in Kern County, California, Roosevelt had the Biological Survey move them to Sequoia National Park, where the Department of the Interior assumed the responsibility for their case.

Furthermore, on January 7, 1902, the executive committee of the Boone and Crockett Club issued its final report on how to create wildlife refuges. Washington insiders called it the Roosevelt Report (knowing full well that the president would adopt the club's recommendations). At the core of the final report was the belief that U.S. game reserves should be established *inside* national forests. For example, President McKinley had established Oklahoma's Wichita Mountains Forest Reserve in 1901 to protect natural resources. The Boone and Crockett Club's report suggested that *part* of this forest could become a buffalo or deer preserve. Because the U.S. government already owned the Wichitas, it had the authority to fence off thousands of acres to save vanishing wildlife. The report also recommended that the economic needs of locals always be factored in when policy recommendation were made.[62]

Prudence, of course, was the tenor of President Roosevelt's directives regarding forest preservation and wildlife protection. As an accidental president, coming into power because of McKinley's death, Roosevelt wanted to avoid unnecessarily kicking over a hornets' nest. Senators from Idaho, Oregon, Montana, Utah, and Colorado already had their long knives out for him. He had to move with caution concerning the regulation of game animals and birds in any new forest reserve he created, to avoid an outcry of states' rights sentiment. His strategy was to start slowly with just one or two forest reserves, where the political opposition would be next to nil. Then, if he was elected in his own right in 1904, depending on the magnitude of his electoral mandate, he would act more boldly and decisively, taking head-on what he called in *An Autobiography* the "great special interests" of the Far West that were destroying nature "at the expense of the public interest."[63] As of 1902 the United States had approximately 43 million acres of forest reserves at its disposal. Roosevelt, with Pinchot at his side, wanted the Forestry Bureau to quickly double or triple that amount. From a long-term planning perspective Pinchot hoped to persuade Roosevelt to transfer the Division of Forestry from the Interior Department to the Department of Agriculture (he accomplished this goal with the Transfer Act of 1905).

IV

Throughout the first six months of 1902 the president fought for the federal government's "conquest" of the arid lands of the West. Increasingly this involved irrigation. He believed instinctively that the huge undertaking of constructing large dams and reservoirs was an obligation of the federal government because its scope was beyond the capacity of private enterprise.[64] Wanting to build on John Wesley Powell's recommendations for the Geographical Survey, Roosevelt fought tooth and nail for land reclamation through hydrological advancements such as dams, reservoirs, and aqueducts as if he himself were the editor of *Irrigation Age*. (Ironically, Powell actually believed in decentralized irrigation, *not* huge, federally run Rooseveltian dams.) Rivers could be redirected, Roosevelt believed, so as to build sustainable western communities in Utah, Arizona, Nevada, California, Texas, and New Mexico. Although he had promoted saving bird habitats, Roosevelt seemed totally ignorant of the potential downside of advanced hydraulic drilling on the desert ecosystem. Filled with good intentions, the shortsighted Roosevelt was, sadly, unable to envision how potentially harmful the dams were to the western environment he so loved.[65] Yet the goal of turning arid land to fields of green was, from a human development perspective, ennobling.

Of course, Roosevelt wasn't working in a vacuum. Many western politicians approved of selling public lands in sixteen western states to fund ambitious irrigation projects. Every politician west of Kansas City or Bismarck, it seemed, was floating a how-to-do-it irrigation bill. The idea was that once settlers prospered on the irrigated western lands, they would help repay the cost of the hydraulic projects by contributing to a revolving fund (something like the later Social Security system). Roosevelt and his followers believed these large-scale irrigation projects would dramatically transform the western economy, landscape, and farming. The days of decaying lumber would be over. As Roosevelt had stated in his First Annual Message, the federal government needed to create "great storage works" for water. Wise irrigation laws should be adopted in the West—laws that issued clear titles for water rights.[66]

Doing all this reclamation legwork for Roosevelt were Pinchot and the young hydraulic specialist Frederick H. Newell, who in June 1902 became chief engineer under Charles D. Walcott, then director of the U.S. Geological Survey. "Pinchot and Newell actually did the job," the president joked, "that I and the others talked about."[67] (Later, in 1907, when Walcott left the Reclamation Service to head the Smithsonian Institution, Roosevelt had Newell serve as director of a new Department of Interior Reclamation bureau.) In his unpublished memoir, written in 1927, Newell explained his commitment to Rooseveltian conservation, inspired, in part, by growing up in the lumber town of Bradford, Pennsylvania. With an aptitude for geology, Newell attended MIT, graduating in 1885 with a degree in mining engineering.

In 1888 he started working for John Wesley Powell and became Powell's right-hand man. A regular at the Cosmos Club, Newell was invited, along with Pinchot, to become a member of the "Great Basin Lunch Mess," where intense discussions were held on western rivers, forestlands, geographical surveying, and soil conservation. As an author Newell was almost as prolific as Roosevelt—only there was no romance of nature in Newell's utilitarian volumes, such as *Oil Well Drillers* (1888), *Agriculture by Irrigation* (1894), *Hydrography of the Arid Regions* (1891), and *The Public Lands of the United States* (1895). When modern-day environmental activists attack Pinchot, they often attack his sidekick Newell as well. Whereas Pinchot enjoyed hiking, Newell found pleasure in dynamiting. Unlike others in Roosevelt's inner circle, Newell never wrote about the inherent beauty of nature. There was the kind of vacancy in Newell's eyes, that a novelist such as Melville might have described as soullessness. As an entrepreneurial engineer he solely wanted to make money off the land. He had an inability to say *no* to western politicians. Newell initiated canals and dam projects, at such a rapid pace, that many failed owing to untested soils and unfeasible transportation. Only on his deathbed did he realize that federal reclamation—to which he had devoted his entire life—was unnecessary and even seriously damaging to much of the arid West.[68]

On June 17, 1902, the Fifty-Seventh Congress created the Reclamation Service (later the Bureau of Reclamation) by approving the Newlands Act (named for Francis Newlands, a Democratic representative from

Nevada). Immediately the act was hailed as a triumph for the Roosevelt administration. "I regard the irrigation business as one of the great features of my administration and take a keen personal pride in having been instrumental in bringing it about," Roosevelt wrote to Hitchcock that very day. "I want it conducted, so far as in our power to conduct it, on the highest plane not only of purpose but efficiency. I desire it to be kept under the control of the Geological Survey of which Mr. [Charles Doolittle] Walcott is the Director and Mr. [Frederick Haynes] Newell the Hydrographer."[69]

The Newlands Act was a revolution for the American West. An overenthusiastic Roosevelt wanted to start with a few large dam projects divided among a few states. Overnight, however, Congressman Newlands was getting great press and Roosevelt grew envious. Why was everybody giving that Democratic fool Newlands all the credit? Roosevelt wanted the western Republicans—for example, William Morris Stewart of Nevada and Francis Emroy Warren of Wyoming—to have the credit for the historic irrigation act. Fuming to Secretary of Agriculture Wilson, Roosevelt threatened to attack Newlands's reputation through back channels.[70] To Roosevelt, Newlands was a shameless grandstander who didn't deserve to have an important act named after him. Truth be told, as the historian Donald Worster points out in *Rivers of Empire*, neither Roosevelt nor Newlands was very instrumental in the federalization of western water issues; they were both, in essence, latecomers. Stubbornly, Roosevelt nevertheless insisted in both writings and public speeches that the landmark western irrigation measures should be called the Reclamation Act, *not* the Newlands Act. (However, at Pinchot's insistence he does toss Newlands a bone in *An Autobiography*.[71])

Why did President Roosevelt throw himself wholeheartedly into the drama of the Newlands Act? Certainly, Roosevelt saw himself as a man of the American West. Even though his views on protecting forests had made him vehement enemies in the region, he was (to his mind) the first western president in American history. (This didn't mean, however, that he abandoned the establishment privileges provided by his aristocratic New York upbringing.) Owing to the dispiriting brouhaha over Booker T. Washington, Roosevelt's name had become a dirty word in the Deep South. With his astute political antennae, Roosevelt knew he needed western support to succeed in national politics. During 1902, with reclamation being debated by the House Committee on Irrigation of the Arid Lands, Roosevelt didn't want to be sidelined.

To Roosevelt, the West was the best hope for America. He rightly foresaw California, Oregon, and Washington as new Edens. Nobody believed more strongly than Roosevelt that the West *had* to be won; it offered landscapes of incalculable value. If the western citizens didn't have water, he worried, they would perish, and their cities would become ghost towns. But dams and reservoirs (built cautiously, without pork-barrel waste) would allow the West to be settled by tens of millions of people. The American cities of tomorrow were Los Angeles, Albuquerque, and Sacramento. The federal reclamation of the West, to Roosevelt, was the next natural step toward conquest. If reservoirs were created, the West Coast from San Diego to Seattle would be humming with jobs. With western populations swelling, Americans, he believed, would turn to the fabled China trade, using Hawaii and the Philippines as stepping-stones. And, finally, as Roosevelt envisioned it, with a proper reservoir system places like the Willamette Valley in Oregon and the San Joaquin Valley in California could become the most productive agricultural lands in the world; of course, he wasn't wrong about this. "The forest and water problems," Roosevelt insisted, "are perhaps the most vital internal questions of the United States."[72]

By 1904 six reclamation projects were up and running. Even critics of the Newlands Act had to admit that Roosevelt had a genius for cutting red tape. Every year exciting projects were launched. For example, the linking of Colorado's wild Gunnison River to the Uncompahgre Valley—a Herculean feat that required constructing a channel ten feet high, ten feet wide, and five miles long by blasting through mountain rock. In Arizona the Salt River was impounded by the 360-foot-high Roosevelt Dam, to create one of the world's largest artificial bodies of water. Such reclamation projects led to agricultural booms in fruits, dates, sugar beets, alfalfa, on and on. More than 3 million acres of the West were cultivated under Roosevelt's reclamation programs. Culverts, bridges, and canals were all engineered, at great expense. "The Roosevelt-Pinchot-Newell vision of millions of desert acres in bloom," the historian Paul Russell Cutright wrote, "was well on its way to reality."[73]

Serious books have been written on the Newlands Act—and this is not the place to do them all justice. It's safe to say, however, that unlike the western agricultural boom, the studies with an eye on the environment don't have a happy ending. The grand irrigation projects—Panama Canals on a reduced scale—destroyed many natural wonders. On the other hand, the engineering done by the Reclamation Service was impressive in both scope and innovation, overcoming mind-boggling obstacles. While Roosevelt had sympathy for western farmers and ranchers worried about drought and rural poverty, one suspects an additional motivation behind his cheerleading for the Newlands Act: it smacked of American triumphalism. To Roosevelt the West—particularly the dry mountain air of the Rockies and the warm climate of California and the Southwest—was a cure for America's industrial ills. Health-seekers by the trainload were moving to Los Angeles, Santa Barbara, San Bernardino, and San Diego, and he knew why.[74]

Roosevelt correctly surmised that someday the population of the West would equal that east of the Mississippi River. But only through efficacious forestry and irrigation, he believed, could the West live up to its limitless potential. Undoubtedly, Roosevelt wanted western greenbelts and scenic wonders saved to enhance the quality of life. This didn't mean, however, that he didn't also want to see large increases in the number of human settlers in the West. And, to repeat, without water, "Go West, young man!" would be foolhardy advice. Therein lay the rub of his advocacy of the Newlands Act. He believed the act would transform the social aspect of the West by substituting "actual homemakers, who have settled on the land with their families, for huge, migratory bands of sheep herded by the hired shepherds of absent owners."[75] (Somehow, the issue always got back to Roosevelt's hatred of sheep.) Writing to future Speaker of the House Joseph Cannon on June 13, 1902, Roosevelt explained his support of western reclamation and irrigation: "This is something of which I have made careful study . . . from my acquaintance with the Far West. . . . I believe in it with all my heart."[76]

V

That August Roosevelt headed to New England for a busy tour, in his private Pullman train compartment, known as the *Mayflower*. Roosevelt had never been very popular in New England, so he considered this trip something of a goodwill tour.[77] Yet he overbooked himself. He always seemed to be saying hellos and good-byes simultaneously. More than fifty reporters and newspapermen followed him, hoping to engage in conversational bouts. It was his first visit to Vermont since McKinley's assassination. For the most part his stump speeches were about the ironclad Monroe Doctrine, trust-busting, and citizenship. For Labor Day weekend in early September, Roosevelt headed to Massachusetts to be with Senator Henry Cabot Lodge and William Moody, son of the famous evangelist Dwight L. Moody. In Springfield more than 70,000 people came to hear the president lecture about not retreating from the Philippines. According to Roosevelt, the United States had a sacred obligation to establish a democracy there. Always trying to sneak beautiful scenery into his itinerary, Roosevelt yielded to an impulse and spent a few days in the Berkshires, traveling in a landau drawn by four gray horses, leaving the *Mayflower* Pullman on the tracks in Stockbridge.

Throughout the New England trip Roosevelt had the Secret Service agent William Craig constantly at his side. Since McKinley's assassination the presidential Secret Service had greatly increased. (In June, though, an armed lunatic had wandered into the White House, waving a pistol about like a drunkard until he was apprehended by the police.) Now, at Pittsfield, Craig ended up giving his life for the president. A runaway trolley, car 29, had run into Roosevelt's carriage at the Howard's Hill intersection, toppling it on its side like a sinking ship. The damage was extensive. Upon impact, Craig, known as "Secret Service Man Extraordinaire, and Plenipotentiary to the President," had risen from his seat and thrown himself directly into the trolley so that Roosevelt wouldn't take the direct hit. Craig was crushed and almost decapitated. Roosevelt was deeply shaken, his face bruised and bleeding. A fist-sized lump swelled on Roosevelt's right cheek, and a coal-black bruise emerged under his right eye. Immediately, Roosevelt, a bit dazed, raced over to Craig, who was dead—the first U.S. Secret Service agent killed in the line of duty. Craig's body was almost unrecognizable.

Once Roosevelt regained full consciousness, he grew angry at the trolley driver, who was arrested but later released on bail. An atmosphere of chaos prevailed, with onlookers screaming in horror and running in all directions. "I am all right," Roosevelt kept saying. "I am unhurt." When people saw that he had survived the crash, they began shouting enthusiastically. "Don't cheer," Roosevelt scolded them. "Don't. One of our party lies dead inside." Sipping brandy to steady his nerves at a physician's office, deeply distraught over the death of his trusted friend, the president nevertheless continued his tour of Massachusetts, but he refused to speak to crowds, opting to instead praise William Craig's courage. The novelist Edith Wharton heard Roosevelt speak in Lenox and noted that what he said was an appropriate response for the grim episode. Roosevelt had developed abscesses on his left leg, turning his ankle a weird purple-green. "This is a dreadful thing," Roosevelt kept saying over and over again, "dreadful."[78] The *New York Times* ran a story with the subhead "Soft Earth Saves President" (he had fallen into a wash from the hill).[79]

Refusing to let the crash at Pittsfield preclude his visit to the Biltmore estate to study its forestry program firsthand, Roosevelt arrived as scheduled on September 9, 1902, following tours of the Civil War battlefields of Chickamauga and Chattanooga. The *Pittsburgh Times* suggested that the president needed to stop traveling so much, that the "strenuous life is sometimes overdone." But onward he went. Local dignitaries in North Carolina poured onto Roosevelt's railway car, eager to shake hands with the president, who, with artificial geniality, kept saying "dee-lighted." His face was still battered and bruised from the accident, so polite people tried not to stare. Heading for Battery Park Hotel, built on the highest point in Asheville, Roosevelt peered out, mesmerized by the Great Smoky Mountains foothills. "Oh, this is magnificent!" he said. "This is indeed a most magnificent country—the grandest east of the Rockies!"[80]

After delivering a patriotic speech Roosevelt headed in his carriage to the Biltmore estate, in a bone-chilling wind. Full of questions, Roosevelt toured the mansion, inspected the lotus ponds, and talked with the level-headed young foresters who had gathered to pay their respects. Ever since Pinchot had promoted the Biltmore at the Columbian Exposition in 1893, the effects of its forestry program (including how best to plant the seedlings of yellow poplar, black cherry, black walnut, and other species) had increased. *Garden and Forest* magazine, for instance, was raving about the experimental station. Under the guidance of Carl A. Schenck, Biltmore's forestry school was setting a standard for scientific professionalism. To Roosevelt the Biltmore was the "cradle of forestry in America" (in 1968 President Lyndon Johnson commemorated it as such by a congressional act).[81] Yet Roosevelt was piqued because Schenck wasn't an American citizen (he kept his German citizenship), so their conversation didn't go well. Although Roosevelt was at the Biltmore for only a few hours, he returned to Washington full of talk about timber physics, dendrology, and wood utilization. And he left all of Asheville abuzz, warmed by his scientific enthusiasm for forestry. "The president came and went yesterday," the Biltmore reported to George W. Vanderbilt, who was in Bar Harbor, Maine. "It had been raining before he came and rained immediately after he left but it was clear while he was here."[82]

Later that month Roosevelt headed to the Midwest. After speaking in Indianapolis he fell ill; his leg looked gangrenous. With the first flash of pain he tried to conceal a cold panic. Listlessness fell over him. He was placed under local anesthesia at Saint Vincent's Hospital, and doctors removed two ounces of serum from a sac in the anterior tibial region. Roosevelt slowly recovered from the makeshift operation, but he was never the same afterward. "I have never gotten over the effects of the trolley car accident six years ago," he wrote to Kermit in September 1908. "The shock permanently damaged the bone."[83] Physicians now believe that the accident in Pittsfield also led to phlebitis and thrombosis, conditions that would eventually become factors in his death.

Refusing to be nursed, Roosevelt threw himself back into the fray. Besides running the White House, he had six children to raise. His eldest, Alice, was sixteen years old; the youngest, Quentin, was four. Promoting the strenuous life for his own brood, the president oversaw pillow fights, wrestling matches, roller-skating, and leapfrog throughout the White House. Furniture and china were regularly broken. All sorts of native plants were ordered, to give certain rooms a more natural feel. Because the White House was under renovation, however, the Roosevelt family had to live at 22 Jackson Place—across from the White House—for several weeks. Whenever T.R. traveled away from Washington, D.C., he wrote his children letters. In the coming years they would receive missives from Yellowstone, the Grand Canyon, the lower Mississippi River, Yosemite, the Painted Desert, and dozens of other extraordinary American outdoor places he was determined to pass on to his progeny as a legacy.[84]

"Will you tell me how things are in the Yellowstone Park as regards game protection?," Roosevelt wrote John Pitcher at Yellowstone on October 24, 1902. "I know the buffalo are almost gone, and I know how difficult it is to protect the beaver. How are the elk being protected? Is there much slaughter of them in the forest preserves outside of the Park, and is there much poaching of them in the Park itself? How are they holding their own? I should be very much obliged if you would give me any information about the game in the Park. What force of rangers have you?"[85]

Undergirding Roosevelt's promotion of the American wilderness from 1901 to 1909 was *Darwinism*—a word uttered reverently by the president. In *Nature's Economy* (1977) the historian Donald Worster writes, convincingly, that ecology after Darwinism became a "dismal science" in America. Roosevelt, it seems, was an exception to this rule. When Darwin, for example, traveled to South America he encountered the "violence of nature," including huge vultures, stalking jaguars, vampire bats, and poisonous snakes. It was a frightful land of volcanoes, earthquakes, and insect swarms. Everywhere Darwin looked in the jungles of South America there were "the universal signs of violence." Ironically, Roosevelt was thrilled by nature's violent side. He wasn't like John Muir studying ferns or John Burroughs praising bluebirds. The blood-and-guts aspect of Darwin's account appealed to Roosevelt. The president, in fact, felt part of the bond of violence. Tumult, cataclysm, horror, and brutality in nature taught Roosevelt to immerse himself in defiance and struggle. On hunting excursions he was engaged in the dark pageant of earth, where death was always looming.[86]

Forget Progressivism or Republicanism. From late 1901 onward President Roosevelt behaved like a Darwinian ideologue, disseminating the great naturalist's ideas as if they were providential. It's impossible to understand anything Roosevelt did or said without taking Darwin, Huxley, and Spencer into consideration. There was never a time when Roosevelt, the politician as inquiring scientist, didn't want to know *everything* about the organic makeup of a forest reserve or a seabird or a moose antler. Besides being a utilitarian conservationist, Roosevelt felt he had a duty to inventory and catalog every type of beetle, lizard, mouse, pine cone, seedling, and wildflower in America. Like a modern-day Thomas Jefferson he wanted all of the American West cataloged as thoroughly as Darwin had cataloged the Galápagos. His handpicked Lewises and Clarks, in this regard, were employees of the Biological Survey, forestry experts like Pinchot, Audubonists of every stripe, and Bullock and Warford outdoors types. It was Roosevelt's obsession with the *truths* of Darwinism and *pragmatism* of Pinchot that made his conservation policies so much more ambitious than those of Cleveland and McKinley. A good equation for understanding our twenty-sixth president is the following: Grinnell (hunting) + Darwin (evolution) + Pinchot (utilitarianism) + Burroughs (tender naturalist) = President Roosevelt.[87]

CHAPTER SIXTEEN

THE GREAT MISSISSIPPI BEAR HUNT AND SAVING THE PUERTO RICAN PARROT

I

Deep in the southern Mississippi Delta near what was then the village of Smedes—on land situated between the Mississippi River to the west and the Little Sunflower River to the east—a historical marker in front of the Onward Store on Highway 61 now commemorates the most celebrated hunt in American history. In mid-November 1902 President Roosevelt, exhausted from mediating between mine owners and the striking members of the United Mine Workers (UMW), was in need of a short vacation. A few weeks earlier, public schools and government offices throughout the Northeast and Midwest had to be closed because there wasn't enough coal to heat them, and the president had threatened to send federal troops to reopen the locked mines of Appalachia. Finally, a settlement was reached with the mine owners through arbitration, and a relieved Roosevelt was ready to go hunting.[1] This particular six-day Mississippi expedition, from November 13 to 18, resulted in a stuffed animal different from the kind produced by taxidermy: the most popular toy ever manufactured—the teddy bear (plus several apocryphal hunting yarns that have masqueraded as fact for more than a century).[2]

After the coal crisis and the carriage crash, President Roosevelt eagerly accepted long-standing invitations from friends to come south for the bear hunting season. No state matters were going to detain him. An open-air vacation was to be the order of the day. The only real outdoors "breaks" he had in 1902 had been the Berkshires and a visit to the Bull Run Historic Battlefield to hunt for Virginia wild turkey on a chilly afternoon. He didn't get one.[3] Naturally, politics also figured into Roosevelt's decision to go south of the Mason-Dixon Line. Mississippi's newspapers and politicians had been attacking the president with a vengeance over the Booker T. Washington affair. The white supremacist James K. Vardaman was a divisive and contentious newspaper publisher, who had lost a race for the governorship to Andrew Longino (a moderate on racial issues). Vardaman had daggers out for the president. In addition to being a bigot, Vardaman was a gaffe machine, unable to achieve even a semblance of political correctness.[4] When Democrats who favored Vardaman heard that Roosevelt was coming to Mississippi to hunt, they denounced the president as that "coon-flavored miscegenist in the White House" and a "nigger lover" hell-bent on destroying the last remnants of Confederate culture. Vardaman—like many white Southern Democrats—was still furious that this *Republican* president had invited Washington to dine at the White House the previous year. It was an unforgivable affront, he said, to Anglo-Saxon culture. Vardaman ran derogatory advertisements in newspapers in Jackson, Vicksburg, and Meridian in hopes of derailing the presidential trip. One read: "WANTED: 16 COONS TO SLEEP WITH ROOSEVELT WHEN HE COMES DOWN TO GO BEAR HUNTING WITH MISSISSIPPI GOVERNOR LONGY." Roosevelt's claws of detractors were not limited to Southern Democrats. An anti-Roosevelt insurgency was brewing among the so-called "lily-white" Republicans, who wanted Mark Hanna to be the party's presidential nominee in 1904.[5]

Roosevelt wasn't intimidated by the vile accumulation of race baiting, but he was acutely aware that this hunt was going to be carefully followed by the press. The Illinois Central Railroad gladly took care of Roosevelt's transportation. He, in turn, cut quite a figure on the 1,000-mile journey from Washington to the Mississippi delta. The towns his train thrummed through—Tunica, Dundee, Lula, Clarksdale, Bobo, Alligator, Hushpuckena, Mound Bayou, Cleveland, Leland, Estill, Panter Burn, Nitta Yuma, Aguilla, and Rolling Fork—are today on or near the American "blues highway," considered by many the birthplace of rock and roll. Throughout the delta that November the fields were covered with bright white bolls—a second cotton harvest. Clad in a fringed buckskin jacket he had acquired in the Dakota Territory, topped off with a brown slouch hat, the president looked like Seth Bullock of Deadwood, and the full cartridge belt around his waist added an air of a Rough Rider ready for action. The Mississippi River valley that loomed in front of him seemed strange, even exotic. The president had already made a request of one of his hosts, Stuyvesant Fish, president of the Illinois Central Railroad: "My experience is that to try to combine a hunt and a picnic, generally means a poor picnic and always means a spoiled hunt," Roosevelt wrote. "Every additional man on a hunt tends to hurt it. Of course I am only going because I want to *hunt*—and do see I get the first bear without fail."[6]

Reporters covering his train ride to Mississippi noted that Roosevelt was reading his friend the French ambassador Jean-Jules Jusserand's *The Nomadic Life*, a history of the Crusades of the Middle Ages, and surmised that the text was meant to inject some intellectual adrenaline and romanticism into the preparations for the "Great Bear Hunt." (Reporters could never account for Roosevelt's eclectic reading tastes.) When the train entered the delta the view from the presidential compartment changed from rolling hills to unhindered flat plains. At each railroad platform were bales of cotton ready for shipping to the textile mills of New England and Europe. The always gregarious Roosevelt waved at the Mississippi field hands who lined the tracks for an unprecedented glimpse of a U.S. president. Blacks recognized that, whatever Roosevelt's shortcomings, cruelty and injustice always moved him to action. Since the Booker T. Washington affair Roosevelt had become a hero to African-Americans and mulattos. Nonsegregationist newspapers in the Mississippi bottom reported the president's trip positively. One headline read: "President Speeds to Bruin Land." A few hamlets along the train route hung patriotic crepe-like paper streamers as a welcoming gesture.

By going to Mississippi, Roosevelt was hoping to accomplish a few things with regard to race. It was the twentieth century, and he felt that the South had to stop seeing the world as a bridge into the burning past. The first step for a new civil rights era, he believed, was to champion antilynching laws throughout the South and Middle West. Anyone lynching a black had to be vigorously prosecuted. Racist vigilantes, the president worried, had gotten out of control. On the economic front what troubled Roosevelt was that African-American cotton pickers were trapped in a dead end: their position as tenant farmers bordered on slavery. The economic situation was unaceptable below the Mason-Dixon Line thirty-seven years after the Civil War. How could he help lift the African-Americans of the Deep South out of their condition of peonage?

But grappling with the "Negro" condition was just part of his agenda in Mississippi. Roosevelt was extremely interested in seeing America's agricultural sector increase under his leadership. Worried about declining farm ownership in the South, particularly in the delta, Roosevelt wanted to educate himself about how the price of the cotton crop could rise up from seven cents a pound to ten cents a pound. (By 1909 he had achieved this objective.) In fact, farm property values, as a result of Roosevelt's agricultural policies, *doubled* throughout the United States between 1900 and 1910. Under the expert management of Secretary of Agriculture Wilson the Roosevelt administration also championed organized food inspection programs and improvements in rural roads. New levees were approved to help control annual overflows. "In many respects," the historian Lewis L. Gould has pointed out, "his administration was an era of unmatched prosperity on the American farm."[7]

In addition to civil rights and his agriculture policy, there was a third factor that influenced Roosevelt to choose the Mississippi Delta for his first high-profile hunt as president: he tacitly acknowledged that he really wanted a black bear. Ever since the 1880s, when he had read two articles in *Scribner's Magazine* by James Gordon—"Bear Hunting in the South" and "A Camp Hunt in Mississippi"—he had itched to explore the Coldwater, Tallahatchie, and Sunflower river floodplains. Such a hunt, of course, included braces of dogs, rough-haired little terriers that could dodge into the canebreak when the bear was enraged. They'd bark and snarl only a few inches from a bear's muzzle. Other ritual activities were likewise followed. Besides Mississippi black bear, Roosevelt hoped to see tall cypresses rising out of the swamps and camp near cottonwoods reported to be ten feet in diameter. His team would cut through bayous with only moss, which grows on the north side of a tree, as a compass. And the southern planters he would be hunting with, he anticipated, were, as the sportsman Frank Forester once wrote, man for man the finest hunters in the western hemisphere.[8] Roosevelt was also hoping to mix with some crack-shot swampers and trampers.

Despite the light rain glazing the rails and the storm-threatening clouds darkening the horizon, when Roosevelt arrived in Smedes on the afternoon of Thursday, November 13, he was ready to hunt. The grayness was somehow eerily appropriate. Buoyantly Roosevelt thanked the engineer, shook hands, signed autographs, and showed off his ivory-handled knife and custom-made Model 1894 Winchester rifle with its deluxe walnut stock. He felt good to be in bear country. For the most part his arrival time had been kept secret, so there weren't many greeters in Smedes other than a large contingent of field workers who had taken the day off to see the president; these were the descendants of slaves.[9] Among those who had joined Roosevelt in Memphis were other members of his hunting party, including John M. Parker, president of the New Orleans Cotton Exchange who later became governor of Louisiana; John McIlhenny, who had been a lieutenant in the Rough Riders and had founded the Tabasco Company in New Iberia, Louisiana; and a local plantation owner, Huger L. Foote, whose grandson Shelby would become one of America's foremost Civil War historians. "My grandfather died before I was born," Shelby Foote has recalled. "But I've got loads of newspaper clippings and photographs from the big hunt. There was no bigger event in our family history."[10]

The main tract of land where Roosevelt would hunt belonged to E. C. Magnum, a shareholder in the Illinois Central and, more important, owner of the sprawling Smedes and Kelso plantations on which the hunt was conducted. Camp was set up on the bank of the Little Sunflower River about twelve miles east of Smedes, reachable after a bushwhacking ride on horseback through a dense tangle of prickly underbrush, stunted pines, sluggish bayous, and canebrake. There were also plenty of fine groves of oak and ash to navigate. Roosevelt had listened to the train chugging for days, and now the delta songbirds immediately provided nourishment to his ears. Supplies were delivered to the camp on mules and by wagon. A-frame sleeping tents had been assembled next to a huge cooking tent that had been erected earlier. That first night, the men swapped bear stories around a roaring bonfire on the bank of the Little Sunflower. Roosevelt's tales of his cowboy adventures in the Wild West usually stole the show, but in this gathering the star raconteur was a fifty-six-year-old African-American, Holt Collier, chosen to lead this hunt because of his reputation as a bear tracker. There was a "glad to be alive" quality about Collier, to which Roosevelt naturally gravitated. "Though the hunt had been planned at high corporate and governmental levels for months," the biographer Minor Ferris Buchanan recalled, "its success was wholly dependent upon the skill and performance of Holt Collier."[11]

Collier had been born a slave in 1846 to the family of General Thomas Hinds, who won fame with Andrew Jackson at the battle of New Orleans. Collier never received a formal education and couldn't even sign his own name. When he was a young boy, his job on the Plum Ridge Plantation had been to provide meat for the Hinds family and their field hands. Accordingly, Collier had killed his first bear with a twelve-gauge Scott shotgun in a wilderness swamp when he was only ten years old. Collier became a runaway slave at the age of fourteen but then, oddly (and intriguingly to Roosevelt), joined the Confederate army. (There was a prohibition against African-Americans serving in uniform in the Confederate army, but an exception was made for Collier.[12]) A brave, gallant soldier with a virile demeanor, he witnessed the death of General Albert Sidney Johnston at Shiloh. He signed up with Company I of the Ninth Texas Cavalry a few weeks later and saw combat in Alabama, Mississippi, and Tennessee.

Like the kind of folk figure Ramblin' Jack Elliott or Woody Guthrie might sing about, Collier became a Texas cowboy during Reconstruction, driving cattle on the open prairie, spitting tobacco on the run. He had gone to Texas after being acquitted of the murder of a Union captain, James A. King of Newton, Iowa, in 1866. Upon hearing that Howell Hinds, his former master, had been murdered in Greenville, Collier came

Holt Collier was the best bear hunter in the Mississippi Delta. William Faulkner later based his story "The Bear," in part, on Collier.

back to Mississippi to avenge his death. Often involved in chasing fugitives, in gunfights, and in horse racing, and having spent decades as an expert guide, Collier had an unsurpassed reputation for being his own man, able to track bears or humans with unfailing instinct. As a marksman he had few peers in the delta. He lived closer to the ground and understood the local geography better than anybody else. Collier epitomized a forest trickster character that the South Carolina Gullahs called "Bur," like "Bur Rabbit" or "Burr Bear."

Mississippi, in the years following the Civil War, was teeming with wildlife; primeval forests and jungle swamps provided ideal habitats for wild game including bear, cougars, and deer. The delta thicket had been a safe haven from slavery and in 1902 was one from Jim Crow. Many of the slaves who escaped north on the Underground Railroad hid in its forestlands on their journey following the "Drinking Gourd" (the Big Dipper). When they returned home as freedmen after Appomattox they considered the forest their friend. (Unfortunately, the Klu Klux Klan also used the forest for secret lynchings and to burn bodies.[13]) The woods were Collier's sanctuary too. Collier could wake up in the woods at dawn and shoot a deer for breakfast within an hour. He often brought forest meat into town for white people. In the late nineteenth century, bears were considered a nuisance in towns such as Greenville and Leland. If you traveled in the delta cane fields a rifle or shotgun was always necessary, because the likelihood of encountering a bear was high. Just as polar bears ruled the Arctic and grizzlies ruled the Rockies, in the Mississippi Delta the black bear was king. Collier, as a boy, had been attacked by a bear, wrestled with it, and eventually stabbed it to death. Whenever he retold this drama, Collier would show off the scar, which had stayed risen on his arm, to admiring sympathizers.

By the time Roosevelt met him at Smedes, Collier had killed more than 3,000 bear; this was considered an American record. Well-dressed—courtesy of a haberdasher in Greenville—and convivial, Collier had piercing brown eyes and very pronounced features. With stolid dignity he wore a Vandyke beard that he had acquired as a Civil War scout, and his cropped salt-and-pepper hair was often covered by a well-worn Confederate cap. His taut muscles seemed to grip his bones. Roosevelt, who was a promoter of Joel Chandler Harris, immediately took to Collier's briar-patch bear tales as if Collier were Uncle Remus come to life. The stories reminded him of the Bear Bob stories his mother had told him when he was a boy. Just hearing the way Collier said "painter" (for panther) brought a smile to Roosevelt's face. Nobody Roosevelt had ever met in the Rockies had as many close personal encounters with bears as Collier. Until the 1890s Collier had earned a very good living as a hunter, evolving into a first-rate guide. Collier hunted through the fall and winter months for thirty-plus years. He sold the meat to railroad workers, timber companies, and levee men. When not hunting he would follow the seasonal spring fairs from town to town, offering his services. In the summer he worked in the stable of his brother Marshall. Mainly, Collier was a survivalist, living off the land with his gun to keep him company. "Money don't buy nothin' in the cane-break, nohow," he used to say, "and a man's dog don't care whether he's rich or po'."[14]

Roosevelt enjoyed such folk wisdom. When John M. Parker, who was choreographing the hunt, commented that its rigors might be too hazardous for a sitting president, a mildly insulted Roosevelt, looking at Collier with a half-embarrassed smile, exclaimed, "This is exactly what I want!" The other hunters, sensing that the president felt insulted, began shifting uneasily, uncomfortable with the ensuing silence.

"Good," Parker shot back, "we will have bear meat for Sunday dinner!"—to which Roosevelt condescendingly replied, "Let us get the bear meat before we arrange for the dinner." Except for this verbal sparring Roosevelt and Parker got on famously, and they formed an important political alliance in coming years.

Roosevelt understood that the secret of Holt Collier's success, as with all good hunters, was that he revered bears and knew all their habits. Black bears, for example, would never sleep in a wet area: they pulled down cane stalks to make a comfortable nesting place. People who thought bears slept in the swamps were wrong. Although constantly maligned in the delta as meat-eating predators, bears actually preferred a diet of acorns, hickory nuts, black walnuts, persimmons, and melons. Sometimes they would wander onto farms to swipe piglets or to raid a hen house, but not often. Though Collier considered it unsportsmanlike, a fairly common bear trap used by others in the region was a pot of honey mixed with whiskey. Lapping up the honey, the bears eventually toppled over drunk; in such an inebriated condition they were easy to shoot or stab. Roosevelt made it clear upon his arrival in the delta that "pothunting," as it was called, wouldn't be tolerated.

Because the press had limited access to the campsite, details of the president's six days in the Mississippi Delta are sketchy. The *New York Times* reported that Roosevelt often simply took to the trails to enjoy nature, preferring gentler episodes to the barking terriers, not particularly interested in rousting a bruin or pulling a ligament. An excellent dinner seemed to always be the main event, with bear paws, opossum, gravy, and sweet potatoes served on tin dishes and accompanied by wine. The clatter of fine cutlery was far more commonplace than gunshot fire during those six days. Bored reporters wrote about dreamy aromas rising from plates of onion fritters, hush puppies, and okra. Everything was served on a rough pine-board picnic table in a clearing; the scene looked like an advertisement for the national parks. The president stuck to his earlier decision not to shoot deer; they weren't predators or nuisances. This was part of Roosevelt's attempt to promote the sportsman's ethic in the South. Despite the bad trails and impenetrable canebreak Roosevelt rode hard, enjoying what the newspapers described as "African jungle" terrain.[15] "The President is enjoying his outing very much," the *Times* reported. "He has not had three days of such complete freedom and rest since he entered the White House."[16]

Collier, his baying hounds, and the terriers first picked up the scent of a bear on the morning of Friday, November 14. Roosevelt had been placed in a stand while Collier, following large misshapen paw prints, tracked the animal through mud gullies and unruly thickets for hours. Eventually, convinced that the hounds had lost the scent, Roosevelt and Huger Foote returned to camp on horseback for a late lunch. Collier continued the pursuit, and around three-thirty PM his dogs caught up with an old 235-pound giant. (Collier said that the bear would have weighed 500 pounds but for a drought that had reduced its food supply.) Immediately Collier bugled for the president to take part in the kill; chasing the exhausted bear into a slough or watering hole, the dogs plunged in after it and refused to let up. Before long the pack had surrounded the doomed bear, lunging at it with their fangs and yelping nonstop in a frenzy. Desperate for its life, with the sweep of a mighty forepaw, the bear seized one of the dogs by the neck and crushed it to death. Collier, irate at the loss—and under strict instructions not to kill the bear but to save the first kill for Roosevelt—leaped from his mount to protect his remaining dogs from the attack. As the bear was at bay with the dogs, Collier lurched close enough to strike the bear's head with a swing of the barrel, stunning the beast. He struck the bear so hard, in fact, that he bent the barrel of his rifle, rendering it useless. With a dog still gnawing at its hind legs, Collier carefully lassoed the bear around the neck and tied it to an oak tree.

Summoned by others with Collier, President Roosevelt and Foote rushed to the slough. The president was dismayed when he took in the gruesome scene: a dog lying dead in the dirt, two others seriously hurt, and a badly stunned, immobile bear tied to a tree, groaning for air. A light rain had become heavier, bringing a chill of evening. The whole scene had a macabre look, or a look of something gone tragically wrong. Seemingly in unison, the hunters cried, "Let the president shoot the bear." For a second, or perhaps a second and a half, a blank-faced Roosevelt thought about what to do. It was an *I-Ching* moment: *I cannot go backward . . . I cannot go forward . . . Nothing serves to further.* Humility fell over Roosevelt. To shoot that bear would be akin to rape, a travesty of the sportsman's ethos. Eventually he shook his head "no" and refused to draw his Winchester. "Put it out of its misery," he ordered, tossing his knife—a gift from the emperor of Japan—to Parker. "I declined to use that knife," Parker recalled in 1924, "but John McIlhenny threw his hunting knife, and I used that, sticking the bear under the ribs while the dogs were in front of him."[17]

After walking away from the scene, Roosevelt later called the afternoon "a most unsatisfactory experience."[18] At best Roosevelt went through the superficial motions of a good guest but, in truth, he was insulted at the roping stunt. According to Minor Ferris Buchanan, a hesitant Parker, following instructions from Collier, plunged the knife into the bear's side; but he failed in his effort to kill it in a single stab, and an obliging Collier had to finish the job, on a very angry animal. The bear's carcass was slung over a horse and brought back to camp by Collier. The whole episode made Roosevelt feel downcast. The bear hunt in the wilderness had turned into an embarrassment.

Hidden for years in the Roosevelt Collection at Harvard University was a photograph of the dead bear strapped to a horse. It is a black eye to Rooseveltian folklore. The president's great-grandson, Tweed Roosevelt, unearthed the photo in 1989 when he was researching the November 1902 hunt in preparation for an address to the Teddy Bear Society of America, a toy collectors' group, in Boston. It's unclear who took the photo—probably Parker. "I told them the truth," Tweed Roosevelt recalls. "I didn't gussy it up. There never was a bear cub, and the bear with T.R. wasn't shot but was knifed to death. They didn't like hearing it."[19]

Nevertheless, by not shooting the bear, Roosevelt stayed true to the "sportsmen's code," the aristocratic European tradition. As the historian Louis S. Warren explained in *The Hunter's Game*, the code frowned on killing young deer or young bears. Some people called such an act "slob shooting." The true conservationists of Roosevelt's era abided by the general premises of the code, which also included *never* shooting any captured animal for recreation. "For many men, hunting became a symbol of masculine strength," Warren writes. "How one hunted and what one killed came to define what kind of man one was." By refusing to shoot Collier's helpless, tied-up bear, Roosevelt was merely abiding by the sportsman's code. But to many average Americans, not killing the bear seemed odd. "Any good hunter realized that to have shot that particular Mississippi bear would have been cowardice," Tweed Roosevelt noted. "As a true hunter-conservationist, T.R. would have never considered engaging in such a sordid act."[20]

Three bears *were* killed on the 1902 hunt, though none by Roosevelt. "There were plenty of bears," Roosevelt later wrote to Philip Stewart, who had hunted cougars with him in Colorado, "and if I had gone alone or with one companion I would have gotten one or two. But my kind hosts, with best of intentions, insisted upon turning the affair into a cross between a hunt and a picnic."[21]

The next morning, Sunday, November 16, the newspapers carried stories about the president's good sportsmanship, as shown in his steadfast refusal to shoot a captive bear. The *Washington Post* ran a front-page article, headlined "One Bear Bagged. But It Did Not Fall a Trophy to President's Winchester." The *Post* reported that the president had been summoned "after the beast had been lassoed" and "refused to make an unsportsmanlike shot." For once compassion overcame single-mindedness in one of Roosevelt's hunts. Then the story took off. The front page of

Perhaps the most famous cartoon of the twentieth century was Clifford Berryman's "The Passing Show," more commonly referred to as "Drawing the Line in Mississippi." It appeared in the Washington Post *and started the "teddy bear" phenomenon.*

the next day's *Washington Post* featured a cartoon by Clifford Berryman, "The Passing Show," that depicted Roosevelt in his hunting regalia, with one hand holding his rifle butt on the ground and the other thrust out in a firm "No!" and a perplexed fellow hunter holding a black bear by a rope around its neck. The caption read "Drawing the Line in Mississippi"—a double entendre that many scholars believe referred to Roosevelt's fierce criticism of the lynchings of African-Americans in the South. The racial inference in Berryman's cartoon must have chilled Roosevelt.[22]

Berryman's cartoon was a hit and was reprinted nationwide, eliciting praise for the president but also chuckles at his inability to bag a bear in Mississippi. Having worked with the *Post* since 1891, Berryman had developed a fine reputation for political satire. He had been raised in the Kentucky bluegrass country and had never heard of a southern bear hunt where the hunter refused to kill the prey—it struck him as funny. There were four or five variants of his cartoon, which is now a classic. One of the disregarded versions portrayed the president as a small boy. Berryman's editor chose the version in which Roosevelt was an adult and the bear was small. But, as Buchanan pointed out in *Holt Collier*, Berryman had two major mistakes in his celebrated cartoon: the bear was not a cub and the man holding the rope wasn't white (it was Holt Collier). "Naturally,"

Roosevelt wrote to a friend, "the comic press jumped at the failure and have done a good job of laughing over it."[23]

Reporters had swarmed around the president all that first day, hoping to send colorful dispatches. But no bear shot by the president was ever hung in the camp. Meanwhile the reporters were stuck in the middle of nowhere, twelve miles from the Smedes telegraph line, with no lively copy to offer from the delta. Some reporters quipped that the president would have been better off fishing for smallmouth bass or chasing after moccasins to exterminate. Any way you sliced it, the story was a non-story: the president didn't bag a bear. Roosevelt took his failure stoically, saying that it was the "nature of the chase." But reporters loath a void. Many of the journalists lampooned Roosevelt for his failed hunt, but several wrote glowingly of Collier, describing his almost superhuman single-handed capture of a large, wild bear at the age of fifty-six. Soon other stories were manufactured. By the time Roosevelt left Mississippi, heading north to Memphis, the buzz was that the president had befriended Holt Collier—a black man. This, combined with the lingering effect of Roosevelt's dinner with Booker T. Washington, perhaps contributed to a boycott by the Tennessee Governor's Guard and Confederate veterans' groups of the president's scheduled parade in downtown Memphis.[24]

Yet the long-term effect was good for Roosevelt's reputation. Berryman's cartoon of the president refusing to shoot a captured bear had captured the public's imagination. A middle-aged Brooklynite, Rose Michtom, impressed by Roosevelt's sportsmanship, made two plush toy bears, stuffed with excelsior and adorned with black shoe-button eyes, as a tribute to the compassionate president who had refused to fire on a captive beast. Her husband, Morris, put the stuffed bears in the window of his stationery and novelty store, and they sold immediately. Then Morris Michtom had a brainstorm: why not seek President Roosevelt's permission to market the toy as "Teddy's Bear"? Michtom sent a letter to the president, apparently in February 1903. The president supposedly wrote back a few lines, essentially saying OK. "I don't think my name will mean much to the bear business," he reportedly said, "but you're welcome to use it."[25] The couple's son, Benjamin Franklin Michtom, remembers that his parents framed Roosevelt's letter and hung it on a wall of their summer home in Florida; after they died and the house was sold, the letter disappeared. No copy has turned up among Roosevelt's voluminous personal papers, housed at Harvard University, or among his presidential papers at the Library of Congress.[26]

Although Roosevelt's letter has been lost and some scholars question whether it was ever written at all, two things are certain: the teddy bear became a rage in the toy business, and the Michtoms made a fortune. Their bears sold for $1.50 apiece, and they couldn't fill the orders fast enough. By 1907 the demand for the cuddly stuffed bears—most with jointed heads, arms, and legs—was so great that the Michtoms formed the Ideal Novelty and Toy Company and moved to a more spacious factory-style building. Coincidentally, in the small medieval town of Giengen, Germany, Margarete Steiff, a seamstress who had been a victim of polio, was also making little plush bears. In the previous few years, she had created a line of appealingly detailed stuffed elephants, donkeys, horses, camels, and pigs. When her nephew, the artist Richard Steiff, began sketching brown bears at the zoos in Stuttgart and Munich and urged her to design a mohair bear toy, she agreed. At first nobody bought the Steiff bears. When one was put on display at the 1903 Leipzig Fair, however, a wealthy American buyer fell in love with it and ordered 3,000 to be shipped to New York. Upon being presented with one of the Steiff bears, Roosevelt supposedly roared his approval and ordered several hundred to be used as table decorations for his daughter Alice's wedding reception. That sealed the deal: the Steiffs, like the Michtoms, officially dubbed their new toy the "Teddy Bear."[27]

The teddy bear craze set off by Steiff and Ideal Novelty and Toy reached its zenith while Roosevelt was president. In 1903 the Steiffs manufactured 12,000 bears; in 1907 the number had soared to 974,000. Dozens of other companies produced their own teddy bears, with various stylistic alterations. Claiming that its version was *the* authentic teddy bear, the Steiff Company began sewing a small metal button into one ear of each of its stuffed toys, to hold the trademark label that still distinguishes the Steiff brand. But Roosevelt always gave credit for the phenomenon to Clifford Berryman, who thereafter included a little bear in all his cartoons of the president. "My dear Mr. Berryman, you have the real artist's ability to combine great cleverness and keen truthfulness with entire freedom from malice," Roosevelt wrote on January 4, 1908; "good citizens are your debtors."[28]

The fad continued after Roosevelt left the White House in 1909. One toy company, eager to cash in by bestowing on the incoming president, William Howard Taft, his own stuffed animal, designed a plush opossum marketed under the slogan, "Good-Bye Teddy Bear. Hello Billy Possum." Unfortunately, with its weird pink eyes, frightening grin, and ratlike tail, Billy Possum was one stuffed critter children refused to hug. The toy was a flop. Other presidents might be remembered as anglers—among them Cleveland, Coolidge, Hoover, Truman, Eisenhower, and George H. W. Bush—but only Theodore Roosevelt was firmly established as an outdoorsman. The teddy bear probably had more to do with this image than his enduring legacy of more than 230 million acres of federal forest and parklands.[29]

Once he was back in Washington, Roosevelt initiated a correspondence with Collier, sending him letters and promising to stay in touch. Presumably a literate friend read these letters to Collier. With so many parvenus in town Roosevelt enjoyed staying in touch with an outdoorsman like Collier. Although it took a few years, Roosevelt would again go hunting with Collier in 1907. Roosevelt pronounced him a better hunter than even John "Grizzly" Adams, Ben Lilly, or Wade Hampton III. At the turn of the twentieth century men used to vie for being considered the best shot; Roosevelt was giving the gold medal to Collier. "He was a man of sixty and could neither read nor write, but he had all the dignity of an African chief," Roosevelt wrote, "and for half a century he had been a bear hunter, having killed or assisted in killing over three thousand bears."[30] A young novelist from Oxford, Mississippi, William Faulkner, drawing on Roosevelt's enthusiasm, later modeled a character in his allegorical short story "The Bear" on Collier, though he made his fictional figure a Chickasaw chief.[31] Capturing the mythical tenor of the bear hunt, Faulkner wrote of his character Sam Fathers that he was an "old man of seventy" and that "the woods" were his "mistress and his wife."[32]

Although the correlation isn't provable, Holt Collier's outdoors acumen may have influenced President Roosevelt in another, unexpected way. Roosevelt had long been an admirer of the buffalo soldiers—the 14,000 African-American men who served in cavalry and infantry units during the Indian Wars on the Great Plains. After his trip to Mississippi Roosevelt suddenly assigned them to patrol the three national parks of California: General Grant, Yosemite, and Sequoia. Additional buffalo soldiers were put in charge of Monterey's Presidio. Roosevelt had developed the idea that African-Americans made fine wilderness police. Captain Charles Young—the third African-American West Point graduate, and a personal friend of Roosevelt—was named acting superintendent of Sequoia National Park. The Buffalo Soldiers became one of Roosevelt's most beloved examples of American democracy at work.

From Collier's perspective Roosevelt's hunt was all upbeat. Fame came to him like a race horse. For Roosevelt, unlike Collier, the Mississippi bear hunt had an unfortunate outcome, which plagued him to the grave and beyond. More than ever, Americans now called him Teddy—the name he loathed. The first sign that individuals knew absolutely nothing about the real Roosevelt was that they dared to say "Teddy Roosevelt charged up San Juan Hill" or boasted that "Teddy Roosevelt was in town." Newspaper columnists were famous for this. Why didn't they call Lincoln "Abie" or Washington "Georgie" while they were at it? Whenever J. P. Morgan, John Hay, and Mark Hanna called Roosevelt "Teddy"—as they often did—he took it as a direct insult. "No man who knows me well calls me by the nickname . . . ," Roosevelt wrote to a friend on December 9, 1902. "No one of my family, for instance, has ever used it, and if it is used by anyone it is a sure sign he does not know me."[33]

Such was the power of a cartoonist and a stuffed toy.

II

For the last two or three months of 1902, President Roosevelt carefully weighed his options for where to create his inaugural New Year forest reserve, taking extra precautions not to set off a firestorm in Colorado and Montana, where timber titans were already on the verge of hanging him in effigy. Right after Christmas Roosevelt had written to Alexander Agassiz, president of the National Academy of Sciences (and son of the great Harvard zoologist), to help him launch a "comprehensive investigation" of the natural history of the Philippines and Puerto Rico. Shrewdly, Roosevelt struck first on January 17, 1903, just after the Christmas holiday season ended, helped by recommendations from Agassiz and a report written by John Gifford of Florida (who was worried about the illegal felling of trees in Puerto Rico).[34] Having risen to fame in the Caribbean as a Rough Rider, and deeply fascinated by the rare tropical wildlife that populated the rain forests, particularly the bright green Puerto Rican parrot (*Amazona vitatta*), Roosevelt created the 28,000-acre Luquillo Forest Reserve (renamed the Luquillo National Forest in 1907).[35]

Nobody in official Washington objected to the Luquillo. Roosevelt would visit Puerto Rico himself in 1906 to see the rainforest firsthand; as an Auduboner he knew it was a famous aviary for parrots and bananaquits. As further evidence of Roosevelt's interest in tropical forests, as ex-president he went to the Amazon and wrote magnificently about them in "A Naturalist's Tropical Laboratory," an article for *Scribner's Magazine.* "In the heat and moisture of the tropics the struggle for life among the forest trees and plants is far more intense than in the North," he wrote. "The trees stand close together, tall and straight, and most of them without branches, until a great height has been reached; for they are striving toward the sun, and to reach it they must devote all their energies to producing a stem which will thrust its crown of leaves out of the gloom below into the riotous sunlight which bathes the billowy green upper plane of the forest. A huge buttressed giant keeps all the neighboring trees dwarfed, until it falls and yields its place in the sunlight to the most instantly vigorous of the trees it formerly suppressed."[36]

From a political perspective creating the Puerto Rican rain forest park was a painless endeavor. In 1900 Puerto Rico had surrendered its sovereignty to the U.S. military authority. President McKinley had issued the Organic Act (known as the Foraker Law) establishing civil government and open commerce between Washington, D.C., and San Juan. Puerto Rico was declared America's first unincorporated territory, and the new Puerto Rican government was assigned a governor appointed by the White House (Charles Herbert Allen), who was helped out by five Puerto Rican cabinet members. Treating Puerto Rico as part of the spoils of the Spanish-American War, the McKinley administration established free trade and a democratic electoral process. During the first full year of Roosevelt's presidency a second round of elections was held (under the Foraker Act), a telephone company was established, and English was made one of the two official languages, along with Spanish. By the authority of a 1902 act of Congress, President Roosevelt was allowed to do as he saw fit with all "crown lands" ceded to America by Spain.

As with the Badlands and the Rockies, Roosevelt had adopted the Luquillo National Forest—the only tropical rain forest in the U.S. National Forest System—as an object of unending fascination and wonder. The Luquillo had a romantic lure that appealed to Roosevelt's image of David Livingstone and to his sense of the lost jungle. With quiet reasoning T.R. studied every biotic aspect of the newly acquired sanctuary located on the east side of Puerto Rico, especially its rain forests. Courtesy of the USDA, Roosevelt had learned that in 1824 Spain had established a forest conservation law, eventually administered by a public forestry commission, to protect the dim, mysterious, green-roofed jungles. Never one to turn down a good idea, Roosevelt felt that America's forestry service could learn a few things about land management from these old, impressive Spanish laws and regulations. In 1876, only a few years after Yellowstone was established, King Alfonso XII of Spain had officially proclaimed the towering Luquillo forests and masses of vines (approximately twenty-five miles from San Juan) a "forest reserve." Lush beyond words, the Luquillo forests received over 200 inches of rainfall annually. This meant that the dark-green ausubo trees and the wide variety of ferns received 100 billion gallons of freshwater a year. Commonly referred to by

Puerto Ricans as "El Yunque"—which roughly translates as "Forest of the Clouds"—Roosevelt's first national forest was a tropical paradise of the first order. Four distinctive forest types were here: the Tabonuco, Sierra Palm, Palo Colorado, and Cloud Forests. Peaks rose over 3,500 feet with trees blanketed with moss, algae, and bromeliads with bright red flowers. San Juanites would picnic and swim at La Mina Falls. But there was trouble in paradise. Throughout the Sabana River valley, like a ring of rust surrounding a jewel, was chronic deforestation due to reckless coco farmers and rubber merchants.

For a conservationist like President Roosevelt the 5,116-acre Luquillo forest was a biotic plum dropped into his lap. There are approximately 100 million species on earth, and half of them exist in tree foliage and trunks. Who knew what undiscovered species lurked in that largely unchartered and unmapped jungle? There were, for example, eleven coqui species (i.e. tiny tree frogs as loud as an opera singer after it rained). Roosevelt immediately assigned a USDA team of botanists, ornithologists, and foresters to write and publish a scientific report on the Luquillo forest. Roosevelt wanted to know everything about it. (A forest ecology report was published in 1905, to Roosevelt's great satisfaction.) Roosevelt marveled at the proliferation of tabonuco trees (which grew at low elevations and could be 100 feet tall), unusual wild palm fruits, and picturesque waterfalls as exotic as something Gauguin might have painted in the South Seas. Sound forest management would be needed to protect this wonderland where more than 240 types of trees coexisted. If the deforestation that had taken place in Haiti was allowed to occur in Puerto Rico, Roosevelt believed, San Juan would lose its fresh water supply. A real disaster. (Likewise, Pinchot was dispatched to the Philippines to write a forest inventory report.)

Roosevelt's preservationist instinct concerning Puerto Rico didn't stop with Luquillo. On July 22, 1902, seemingly arbitrarily, he declared Miraflores Island in the harbor of San Juan off-limits to anything but a forest reserve and a future quarantine hospital for U.S. Marines.[37] In 1906 Roosevelt wrote to Pinchot, asking him to go to Puerto Rico quickly and "oversee what is being done in forestry." Pinchot went and recommended that Culebra Island be declared a wilderness preserve.[38] Following Pinchot's recommendation—and that of the Florida Audubon Society—on February 27, 1909, just before leaving the White House, Roosevelt did something dramatic on behalf of Puerto Rican wildlife. By an executive order he declared the entire island of Culebra a national wildlife refuge. This crab-shaped dollop, about seventeen miles east of the mainland, was (and remains) a pristine reef with a staggering array of Technicolor coral and fish. He was impressed by the large colonies of brown boobies, laughing gulls, and sooty and noddy terns that lived on Culebra; and once he learned that more than 50,000 sea birds used it as a sanctuary he forbade the U.S. Navy to conduct further military exercises there. Even as ex-president, Roosevelt didn't forget Puerto Rico. He worked in tandem with the naturalist Henry Fairfield Osborn to found the New York Zoological Society's Department of Tropical Research. Besides collecting data on endangered species and rare plant life, the new department established Kartabo Station in British Guiana (now Guyana), considered the first on-the-spot rainforest research facility in the western hemisphere.[39]

Owing to President Roosevelt's foresight and action, when the Luquillo National Forest celebrated its centennial in 2003 the Puerto Rican parrot was still surviving—though barely. And the forest had expanded to 28,000 protected acres. In April 2004 Senator Hillary Clinton of New York, urged by environmental groups and Puerto Rican constituents, introduced a bill to add further environmental protection measures to save endangered species in the Luquillo (in 2007 it was renamed El Yunque National Forest). Clinton lamented the decline of the endangered Puerto Rican parrot. "Today," she said, "there are fewer than thirty-five of these parrots."[40] But she added that with the increased financing of two entities essentially created by Roosevelt—the National Forest Service

Theodore Roosevelt Jr. with a favorite parrot.

and the U.S. Fish and Wildlife Service—the parrots might once again thrive in the most spectacular rain forest in the Caribbean. At the USDA-run visitor center in the Yunque National Forest a huge blown-up copy of Theodore Roosevelt's 1903 proclamation declaring Luquillo a national forest has been installed as an exhibit.

Hearing about the beautiful parrots in both Puerto Rico and the Philippines fascinated Roosevelt to no end. Parrots, he believed, were deeply complex creatures with the intelligence of a human three- to five-year-old. Their startling plumage was far more interesting to him than a luminous splash in a painting by Monet or Renoir. Before long, unable to resist, the president acquired parrots as pets. "Loretta, the parrot, has fairly become one of the household," Roosevelt wrote to his son Kermit in January 1904. "I had no idea that parrots could become so social and intelligent. The other day Archie was in bed with a headache. I found Mame sitting beside the bed and Loretta in her cage between them on my bed. She was having a most lovely time, with the feathers on her head and neck ruffled up, chuckling and talking away in low tones, and alternately shaking hands with first one and then the other of her companions. She was evidently as pleased as she could be, and upon my word, of the three I felt as if at the moment she was intellectually taking the lead herself."[41]

Besides Loretta there was also a blue-yellow macaw known as Eli Yale (kept in the greenhouse), which Roosevelt said "looked as if he came out of *Alice in Wonderland*."[42] Roosevelt loved teaching Eli Yale—so named because its colors were those of Yale University—words like "dee-lighted" and his children's names. Sometimes it would scream and make a piercing flock call. Occasionally after White House dinners, Roosevelt would head out to the greenhouse to feed both Eli Yale and Loretta table scraps, particularly dried fruits and vegetables. Both parrots were friends with the well-fed domestic hen Baron Spreckle, who Roosevelt noticed was starting to act like a parrot. Having these birds around the White House and Sagamore Hill helped keep Roosevelt engaged as a Darwinian zoologist—or, as Edith claimed, returned him to his boyhood. "If all the animals and birds which have been sent by admiring friends as gifts to the President and members of his family had been allowed to remain at the White House," a popular magazine surmised, "that historic old structure might easily be turned into a menagerie and the grounds surrounding it into a zoological park."[43]

CHAPTER SEVENTEEN

Crater Lake and Wind Cave National Parks

I

Forest reserves weren't all that President Roosevelt was preserving for prosperity. As a fervent enthusiast of national parks, Roosevelt hoped to establish a few new ones during his tenure as president. Only five national parks existed in the spring of 1902—Yellowstone, Sequoia, General Grant, Yosemite, and Mount Rainier—and he was eager to establish a sixth. The National Park Service would not be created until 1916, when President Woodrow Wilson signed the Organic Act, so all five of these national treasures were managed independently by the Department of the Interior. The Organic Act's high-minded mandate was to "conserve the scenery and the natural and historic objects and the wildlife therein and to provide for the enjoyment of the same in such manner and by such means as will leave them unimpaired for the enjoyment of future generations." But in 1902 the national parks were run by the U.S. Army (Mount Rainier being an exception), with the commanding officers of the troops serving as superintendents, reporting directly to the Secretary of the Interior.[1]

All other things being equal, President Roosevelt's first choice for a new national park was the Grand Canyon plateau—then a national forest in which extraction was allowed. Roosevelt had first learned of the Grand Canyon when he read Major John Wesley Powell's harrowing account of journeying down the Colorado River between 1869 to 1872 as a teenager. There was nothing that President Roosevelt didn't like about the self-taught Powell—a feisty one-armed Civil War veteran and brave explorer who went on to found the U.S. Geological Survey and the U.S. Bureau of American Ethnology. To Roosevelt the Grand Canyon was an immortal landscape. Just as Yellowstone had been ballyhooed in magazines and periodicals during his youth, in 1902–1903 the Grand Canyon was being touted as an unrivaled natural wonder. Gorgeous photographs of the deep gorge with snow around its rim appeared in the popular press, anticipating the heroic work of Ansel Adams (who was born in 1902). One of America's finest landscape painters—Thomas Moran—celebrated the Grand Canyon in canvas after canvas, to great critical acclaim.

Opposition against declaring the Grand Canyon a national park, how-

Roosevelt employed cartoonists like Ding Darling to help promote his grand vision for national forests throughout the American West.

ever, was fierce. Arizona was a mining territory, where rock blasting was pervasive. Mining claims had already been staked (with encouragement from the U.S. government) for the chance to extract from the Grand Canyon zinc, copper, lead, and asbestos. Roosevelt's idea of withdrawing the nearly 300-mile Colorado River gorge from the private sector was anathema to many in Arizona, including the governor of the territory, Nathan Oakes Murphy. Murphy had journeyed to Washington, D.C., in 1902 to lobby against all of Roosevelt's irrigation and federal forestlands projects. An antigovernment zealot, Murphy wanted to oust Arizona's Indians from federal reservations so that the land could be sold to Anglo settlers. Popular in the southern counties of Maricopa and Pima, Murphy fancied himself as the voice of small-time miners and land developers. You might say he was allergic to anything stamped "Interior" or "Agriculture."[2] *The Tucson Daily Citizen*, in fact, deploring his anti-Roosevelt, anti-conservationist bias, fulminated that Murphy "should have retired from the Governorship of Arizona before undertaking to promote the interests of the water stealers and land grabbers. He should have divested himself of his official character before entering the lobby to advocate private monopoly at the expense of public interests."[3]

Realizing that turning the Grand Canyon into a national park was an undertaking strewn with hurdles, Roosevelt looked for a softer, less controversial natural legacy to preserve. Turning to Chief Forester Gifford Pinchot for advice, Roosevelt was told that Crater Lake in southern Oregon was perhaps an ideal choice for a relatively conflict-free national park. In hindsight, Pinchot unquestionably gave the president excellent advice. Pinchot's idea was to save Crater Lake quickly and then have the president journey by train to the Grand Canyon to stir up public sympathy for creating a national park there.

The party of twelve prospectors who discovered Crater Lake in June 1853 saw its loveliness with fresh eyes, and it cast a spell on them.[4] In 1865, the Sprague and Stearns expedition reported on the amazing site to the public at large. Nearly five miles in diameter, situated in the Cascade Mountains, about two hours by horse northeast from Medford (a Klamath County depot juncture for the Oregon and California railroad), Crater Lake was the result of a volcanic eruption. No lake anywhere else was as chameleon-like in changing color as this natural wonder. The lake's edges, for example, owing to the westering light, were sharp turquoise while its center appeared to be a bottomless indigo blue.[5] Even colorists as fine as Marin or O'Keeffe would have been hard pressed to replicate its myriad hues of blinding blue. Everything about the elliptical site suggested geological aberration. In the pre-Columbian era a horrific eruption had capsized the peak leaving an immense cavity. Over the millennia, melting snow and rain filled the 2,000-foot-deep crater.

With a depth of 1,943 feet, Crater Lake was far deeper than any of the Great Lakes—deeper, for that matter, than any other lake in the United States. To Native American tribes—specifically the Klamath and Modoc—this freshwater lake was a sacred site, the opening to an underworld where a giant supposedly ruled with saber and spear. Myths about Crater Lake abounded. The Klamath and Modoc believed that Wizard Island, in the middle of the lake, was the giant's decapitated head. According to another myth the ruling deity of Crater Lake was an oatmeal-colored creature like the Loch Ness monster. In still another myth, the supposedly "unreachable bottom" was where evil spirits or sea devils resided in lodges.[6]

Many a natural site holds a mystery, but Crater Lake was perhaps unique in that people who had looked down at the extinct volcano basin from the twenty-mile circle of cliffs often felt haunted by the visual memory, as if they themselves had witnessed the ancient cataclysm—the lava streams ripping off the mountaintop—which had occurred in the area 7,700 years before. Out of such volcanic disruption in the Cascades was born one of the prettiest lakes in America, the equal of Lake George in New York and Lake Tahoe in California and Nevada. Even though the temperature of Crater Lake was low, it seldom froze in wintertime.[7]

President Roosevelt himself had never visited Crater Lake—and never would. But he had heard from Gifford Pinchot about the extraordinary efforts of an indefatigable Oregonian conservationist determined to save it. Just as Yellowstone had George Bird Grinnell and Yosemite had John Muir, Crater Lake had William Gladstone Steel. Born in Strafford, Ohio, seven years before the Civil War, Steel had first heard about Crater Lake while living in Kansas as a youth. He said he had read a reverential story about the supposedly bottomless freshwater lake in a Topeka or Wichita newspaper, in which a noontime sandwich was wrapped. The story stuck with him. Steel's transient parents, stricken by "Oregon fever," steadily went westward, eventually moving the family to Portland, a regional hub town of about 1,000 people along the alluvial Columbia River. (In the nineteenth century Portland was often called "Stump Town" due to excessive area-wide lumbering.) After growing up in the Midwest region Steel was enthralled by the thought of exploring the green mountain valleys of the Pacific Northwest. Upon graduating from high school, between jobs, Steel took to exploring both the high and the low country of Oregon. No slope or ravine was too mundane for his hiking boots.[8]

Crater Lake Nation Park in Oregon was saved by the Roosevelt administration. Roosevelt called it "an heirloom."

In 1881, at age twenty-seven, Steel started a promotional journal with his brother George. They filled its pages with geographical data and called it *The Resources of Oregon and Washington.* The Pacific Northwest was a densely forested geological wonderland, and the Steel brothers wanted to inventory the far-flung natural resources. Their articles were aimed not at tourists but at mining companies, timber titans, and fish-canning outfits, which were just starting to cast an eye on the region. The brothers' business partner, Chandler B. Watson, had visited Crater Lake in 1873 and was full of its praise. Remembering the story about Crater Lake in the newspaper that had wrapped his sandwich, and always game for a fun week-long trip, Steel traveled 250 miles to the remote site, arriving on August 15, 1885. He wanted to see its reported splendor with his own eyes; it proved to be the turning point of his life. Captain Clarence E. Dutton of the U.S. Army would likewise become bewitched by the lake.

Journeying back to downtown Portland from Crater Lake, Steel developed plans to create a national park out of the "awe-inspiring temple." Gripped by the lake's spellbinding blueness, for the next seventeen years he became a monomaniac on the subject. Intensive cultivation of new conservationist tactics became his focal point. Tall and balding, with the physique of a downhill skier, Steel was extremely well liked in Portland's social circles. Although he wasn't a first-generation Oregon Trail pioneer, he was treated like one, receiving invitations to all the important civic functions and town hall meetings in the Willamette Valley. Officially, Steel was superintendent of postal carriers in Portland, a job which allowed him to rub elbows with everybody of consequence in town. He was respected for his self-control, and his status was enhanced by his ambitious brother George, who had married money and became postmaster. Marshaling data about how the new Yellowstone National Park was attracting tourists from the east coast to Wyoming and Montana, Steel also consulted with lawyers and judges to learn the ropes of the legislative process.[9] "To those living in New York City," he boasted, "I would say, Crater Lake is large enough to have Manhattan, Randall's, Ward's, and Blackwell's Islands dropped into it, side by side without touching the walls, or, Chicago or Washington City might do the same."[10]

After diligently doing his homework, Steel spearheaded an effort to have two bills concerning Crater Lake National Park introduced in Congress. There was a rumor in Portland that homesteaders and developers wanted to acquire Crater Lake so as to log the surrounding tracts of mountain hemlocks, white bark, red firs, and lodgepoles, and even sweeping pockets of ponderosa pine. Steel's first plan of action was to have the

U.S. government reserve the townships around the lake to prevent exploitation or settlement. If that could be accomplished, the next step was to have the U.S. Geological Survey map and scientifically analyze the enthralling terrain. Then, very quickly, perhaps within the year, a national park could be established.

Never one to let a lag develop between his musings and action, Steel boned up on the law and traversed the countryside to find support. He was successful in persuading the U.S. Geological Survey, headed by Powell, to make a complete inventory of Crater Lake. For approximately a month Captain Dutton, accompanied by an able party of geologists and soldiers, lived along the shores of Crater Lake, an area, according to the *New York Times*, rarely seen by white men. The *Times* recounted in vivid detail the hardships endured by Captain Dutton's survey team: donkeys pulling canoes for more than 100 miles up snowbound mountain ridges; pulley ropes dropping the boats down sheer cliffs; measurements of depth taken with the most advanced nautical equipment available west of Denver. After surviving the howling winter winds, Dutton sent Powell a detailed letter about their findings. He declared that as a result of more than 150 soundings aimed at surveying the lake's bottom he believed the depth was 2,005 feet, making Crater Lake, as the *Times* put it, the "Deepest Body of Fresh Water on the American Continent."[11]

Traditionally, Americans like "firsts" and the "biggest," "widest," or "tallest" of anything. So the fact that the *Times* and other newspapers had declared Crater Lake the *deepest* greatly helped Steel's preservation efforts. Oregonians, like everybody else, enjoyed bragging. Dutton helped Steel in another fundamental way. In a long letter to Powell, published in part by the *Times*, Dutton described the geological uniqueness of the lake, mentioning "splendid examples" of glacial striation and rare submerged cinder cones 800 to 1,200 feet high. Many readers probably skipped over Dutton's engineer-like prose about pumice and tufa, but one effect of his findings was to make the entranced Crater Lake the Yellowstone of the Pacific Northwest.[12]

Taking a lesson from the John Muir school of publicity, Steel, with the U.S. Geological Survey report at his side, started introducing Portlanders to the importance of conservation, as simply as possible. Having met with Muir at Mount Rainier, Steel learned how to lobby effectively on behalf of nature. Steel relied on inoffensive efforts (like those of the Sierra Club), aimed at raising consciousness about Crater Lake and other sites in the Cascade Mountains. For starters he organized an Alpine Club (which predated the Sierra Club) and participated in the first nighttime illumination of snowcapped Mount Hood, accomplished with red fire and flares. After having a Portland summit meeting with Muir in August 1888 on strategies of preservation, he released rainbow trout into Crater Lake, hoping to win the support of sportsmen throughout the Pacific Northwest.

Two years later Steel published his first and only book, appropriately titled *The Mountains of Oregon.* Nobody would ever say that Steel wrote with the eloquence of Burroughs or Chapman, but *The Mountains of Oregon* dutifully presented the geological wonders of Crater Lake, reading like a fanciful lawyer's brief for granting it national park status. Yawning chasms, high precipices, weird grandeur, hanging rocks, immense cliffs—the book was filled with enraptured descriptions which leaned toward the style of come-ons for roadside attractions. Mainly, Steel was at pains to explain just how large Crater Lake was; he called it an "immense affair" that would dwarf Chicago and Washington, D.C., combined. Photographs of such Crater Lake sites as Mill Creek Falls and Vidae Cliff were included, complementing the prose. There was even a photograph of himself, sitting with fellow Oregon conservationists, his dark, pointed beard suggesting that he was a Burroughs or Muir in the making. The climax of *The Mountains of Oregon* came when Steel said that even though the Crater Lake area was teeming with game—deer, bears, and cougars—he refused to hunt because the "grandeur and sublimity of the surroundings" filled him with awe.[13]

Upon receiving a copy of *The Mountains of Oregon* Muir wrote to Steel that he was impressed by the "interesting and novel mountain material"; four years later Muir—the bard of Yosemite—published his own riveting work *The Mountains of California.*[14] As propaganda, *The Mountains of Oregon*—a hodgepodge of miscellaneous pieces—worked beautifully. Even timber barons liked seeing the magnificence of their own backyard. In 1893, in large measure as a result of Steel's six years of lobbying, a 4-million-acre Cascade Range Forest Preserve was established in Oregon. The forest reserve encompassed land more than 300 miles long, from the Columbia River to the California state line; it was the largest protected wilderness area in the country. It was a triumph for the intensely focused conservationist against the timber speculators.[15]

When Roosevelt was elected vice president in 1900, Steel was a forty-six-year-old Republican diehard with unimpeachable conservationist credentials. No firm information exists about whether he knew anything of Roosevelt's pro–national park convictions. He was a longtime bachelor known for weekend forays into vagabondage, devoted night and day to the cause of Cascades preservation, and the wheel of fortune was starting to turn his way. That February he had married Lydia Hatch, who was said to be the one and only love of his life. Besides having his wedding reported in the Portland press, Steel started receiving community-wide praise for his program to introduce rainbow trout into Crater Lake. Reports from southern Oregon indicated that his "planting" of rainbow trout had worked; the trout were flourishing, easily surviving the bitter-cold winters (skeptics had feared that this was impossible). Steel, in essence, had provided a welcome twist to wildlife reintroduction efforts, which usually were like those of the Boone and Crockett Club. For in the case of Crater Lake, there was no "re" to be concerned about. Recognizing that the clear waters were ideal for trout, he introduced them to a new habitat in the Cascade Mountains. Congratulatory letters came his way from anglers all over America.

Ironically, Crater Lake's fortunes took a turn for the better when President McKinley was assassinated. With Theodore Roosevelt as president, the chance that it might become a national park was increased tenfold. Over the years Steel had made many friends in the burgeoning conservation movement from coast to coast, mailing copies of *The Mountains of Oregon* to judges and legislators in the hope of teaching them about the pristine Cascades.[16] Of all the valuable contacts Steel had cultivated, however, none outshone Gifford Pinchot. In 1896 Steel had an opportunity to take Muir and Pinchot on a camping excursion to Crater Lake. Although Muir was only moderately impressed, Pinchot was knocked out by the sparkling blue water glistening in the midday sun. As he wrote in his memoir *Breaking New Ground*, "Crater Lake seemed to me like a wonder of the world."[17]

Further bolstering Steel's effort to create a national park was the fact that neighboring Washington state had won a campaign to have 14,411-foot Mount Rainier become a national park in 1899; the park was so large that it created its own weather system and sometimes received 1,000 inches of snow annually.[18] Playing on the rivalry between the two Pacific Northwest states, Steel reminded Oregonians that Crater Lake was every bit as gorgeous as the peaks of the Tatoosh range. Whenever Steel's sales pitch for Crater Lake was falling flat or a setback occurred, he knew Pinchot was available for encouragement. No sooner had Roosevelt delivered his First Annual Message to Congress in December 1901 than Steel solicited letters of endorsement from Muir, Pinchot, and others. Muir begged off—he was preoccupied with his own agenda in Yosemite—but Pinchot rallied to Steel's side like a knight in shining armor. "You ask me why a national park should be established around Crater Lake," Pinchot wrote to Steel on February 18, 1902, in a letter intended for public dissemination. "There are many reasons. In the first place, Crater Lake is one of the great natural wonders of this continent. Secondly, it is a famous resort for the people of Oregon and of other states, which can best be protected and managed in the form of a national park. Thirdly, since its chief value is for recreation and scenery and not for the production of timber, its use is distinctly that of a national park and not a forest reserve. Finally, in the present situation of affairs it could be more carefully guarded and protected as a park than as a reserve."[19]

By procuring this testimonial from Pinchot and using it as exhibit A when presenting the Crater Lake bill to Congress, Steel had the equivalent of a winning lottery ticket. Quite naturally Roosevelt would defer to Pinchot's wisdom about the Oregon backcountry. Steel met a formidable roadblock—Speaker of the House David Henderson, a heavy-fisted politico from Iowa. Ticked off because Steel and Pinchot were foisting a national park on Congress even though T.R. had been president for only a few months, Henderson, representing the antiunion Midwest and western timber-mining interests, offered an adamant "no." By dint of his intimidating seniority he blocked the bill from even being debated. As the first U.S. congressman from west of the Mississippi River to serve as House Speaker, Henderson was mistrustful of federal interference with the free enterprise system. Cranky for a midwesterner, Henderson, whose obstinacy could quickly turn to fury, had worried that Roosevelt had gone too far in pushing natural resource preservation in his First Annual Message to Congress. But at heart Henderson—who had been a valorous colonel in the Civil War and had lost a leg in 1863—was supportive of Roosevelt's militarism.[20] With a little arm-twisting by the new president, Henderson, bristling all the way, reluctantly capitulated. After all, in the context of picking and choosing fights Crater Lake hardly seemed worth crossing swords with the Roosevelt administration.

It didn't hurt that Roosevelt had backup support for the Crater Lake bill from Congressman John Lacey—also from Iowa—who worked mightily on getting Henderson to change his mind. Lacey had visited Oregon in 1887. He was immediately drawn to the Cascades, and he understood that the Pacific Northwest forests were the greatest in the world. Places like Mount Hood and Crater Lake, he knew at once, should be saved for posterity. The slow, sad death of the great trees as a result of wildfires sickened him. "The whole country was covered by a pall of smoke from the burning forests," Lacey recalled to the *Chicago Tribune.* "This was more wicked than the destruction of our forests on the Atlantic only because the great woods of the Pacific are finer, and for the further reason that they are our last."[21]

Double-teamed by Roosevelt and Lacey, Henderson acquiesced and did an about-face, and the Senate passed HR 4393 on May 9. "You give me more thanks than my small share in getting the Crater Lake bill passed deserves, but I am sincerely glad it has got along so far," Pinchot wrote to Steel on May 15. "There is no doubt, in my judgment, that the President will sign it."[22]

Although Pinchot wouldn't lead the U.S. Forest Service for three more years, the impressive power he exerted in creating Crater Lake National Park was a prelude of grand conservation achievements to come. Not only did the utilitarian Pinchot enter the preservationist domain with Crater Lake, but he solidified his alliance with Steel. Mountaineers in arms, both men loathed the reckless way the General Land Office (GLO) of the U.S. Department of the Interior was dealing with the unexplored Cascade Mountains. The saving of Crater Lake was a marker in the conservationists' struggle in the Pacific Northwest, an early indicator that the Roosevelt administration was going to play hardball against the shameful unregulated timbering. Just as important, Pinchot had found a "world wonder" for his boss to establish as America's sixth national park, one which wasn't mired in too much controversy. The word "wonder," in fact, during the first two decades of the twentieth century, was *the* term conservationists used in public discourse to save wilderness sites. A mere forest or lake might seem commonplace, but a "wonder" might bring much needed tourist dollars into local towns. And it worked! Using the U.S. Geological Survey as their base, the tag team of Pinchot and Steel tried to insert the words "deepest" and "wonder" into any public conversation about Crater Lake.

When President Roosevelt signed the Crater Lake bill on May 22—setting aside 240 square miles—he was proud. America had its sixth national park, and its first in Oregon.* Crater Lake was saved for both "great beauty and scientific value."[23] Earlier that day Roosevelt had participated in a ceremony at Arlington National Cemetery honoring the soldiers killed

*There is some debate as to whether Crater Lake was the fifth, sixth, or seventh national park. Because Mackinac Island was decommissioned and General Grant was absorbed into King Canyon National Park in 1940, the ranking can vary. Officially it was the seventh national park established. I'm going with sixth because that was how President Roosevelt saw it in 1902.

in Cuba, Puerto Rico, and the Philippines during the Spanish-American War; he was one of 500 veterans present.[24] The signing of the bill saving Crater Lake took only five or ten minutes of his busy schedule that afternoon; it was incidental. Still, it was a rewarding few minutes. A sentimentalist at heart, Roosevelt now knew what it must have felt like to be Ulysses S. Grant thirty years ago, creating Yellowstone. As a courtesy, Roosevelt had the signing pen shipped to Steel in Portland as a well-deserved souvenir. Applauding Roosevelt and the U.S. Geological Survey for the first-rate report, the *New York Times* described Crater Lake as a natural wonder whose "grandeur" rivaled "anything of its kind in the world."[25]

II

Inspired by the saving of Crater Lake, President Roosevelt looked for another natural "wonder" to designate as a national park. The whole notion of "scenic nationalism" was in vogue.[26] During the 1880s, while living in the Badlands, Roosevelt had perhaps heard about a site in the Black Hills: a "hole that breaths cool air," considered sacred by the Lakota people.[27] It was known as Wind Cave, and the Lakota believed that a beautiful woman—the "buffalo woman"—had once floated out of it to give her people bison. Other tribes believed that a demon or dragon lived in its depths. Some Native Americans believed that the cave had magical powers, that it could predict weather. They weren't completely delusional: the cave opening did serve as a sort of primitive barometer. When the weather was good, air would blow into the cave. However, when a storm approached the low air pressure caused higher pressure to swell inside the underground cavern, causing air to be forced out in a loud, dramatic fashion.[28]

Even though Wind Cave is one of the longest underground mazes in the world (encompassing more than 130 miles of passages), when Jesse and Tom Bingham stumbled upon it while deer hunting in 1881 there was only one entrance: a twelve- by ten-inch "blowhole."[29] As legend has it, the air pouring out of the hole blew the Binghams' cowboy hats off their heads, like a gust off Lake Michigan. Hoping that they had discovered a gold mine, the Binghams returned to Wind Cave the following day with curious friends, only to have their hats now sucked into the maw of the cave; the wind had shifted 180 degrees in less than twenty-four hours. The Binghams were afraid to explore the cave. Perhaps they would be attacked by bats or snakes. But Charlie Crary wasn't hesitant. Boldly he climbed into the blowhole, as if he were Professor Lidenbrock in Jules Verne's *Journey to the Center of the Earth*. He unraveled a reel of twine so that he could find his way out, and he emerged from Wind Cave unscathed and almost stuttering with excitement. His mind was dazzled by the elaborate, delicate, honeycomb-patterned boxwork he had encountered, fragile crystals deposited around like glitter. In his historical study *The National Park*, Freeman Tilden matter-of-factly describes the cavernous rooms as "lacy compartments."[30]

Crary had seen one of the most amazing boxwork formations in the world. More than 300 million years old, Wind Cave is a subterranean wonderland of fractured limestone, crystal fins, and calcium deposits. Words cannot do justice to the diverse geological formations and underground lakes. A whole new geological vocabulary, in fact, was created to describe various types of boxwork: for example, starburst, nail quartz, gypsum flowers, and helictite bushes. With a constant temperature of fifty-three degrees Fahrenheit, the cave network was a perfect retreat from the Black Hills' cold winters and hot summers. Some of the passageways are wet and slippery. An astounding 95 percent of the world's mineral boxwork was found in the cave. Yet much of it was fragile, snapping off when a spelunker's head bumped it or a hand grasped it.

Between 1881 and 1890 South Dakotans hoped that gold could be grubstaked in Wind Cave, as it had been in the Black Hills. Miners tried, but with no luck. Instead, the cave complex started attracting explorers and curiosity seekers. Periodically the local *Custer Chronicle* called Wind Cave a "wonder," and the *Hot Spring Star* added the mystical note that "no bottom" had been found. It was frustrating for locals to sit on a great mineralogical assemblage and not be able to turn a profit. Alvin McDonald of the South Dakota Mining Company considered finding the cave's bottom a challenge akin to Robert E. Peary's going to the North Pole. During 1890–1893 McDonald worked underground, keeping a detailed diary about various caverns and rock formations. He also did some drilling. Cave passages were enlarged for tourists and wooden staircases were erected. McDonald emerged with plenty of artifacts and reams of descriptive prose about stalactites worthy of Verne—he named rooms and mapped trails like an explorer in a work of science fiction—but he was forced to give up the "idea of finding the end of Wind Cave."[31]

When Roosevelt became president, the effort to learn more about Wind Cave was put on a fast track. Homesteading was banned anywhere surrounding the cave's entrance. On April 4, 1902, the GLO commissioned a surveyor in Rapid City to map as many rooms and passageways as he could. Many locals, however, considered this a fool's errand. The U.S. government was also keenly interested in determining the precise amount of minerals available for possible mining. Lawsuits over who owned Wind Cave were put aside as the U.S. government tried to inventory and map the labyrinthine caverns.

As the GLO went about its work, Senator Robert J. Gamble of South Dakota—a Republican—introduced a congressional bill declaring Wind Cave America's "second wonder" (after Yellowstone National Park).[32] Worried that vandals and thieves were stealing crystals and rocks, Gamble claimed that the U.S. Geological Survey and Theodore Roosevelt were on his side in creating the new national park. The *New York Times* aided the cause by publishing remarks by an unidentified U.S. government employee who said that Wind Cave was a "wonderful evolution of nature" exuding "grandeur, grotesqueness and beauty." The *Times* itself said that giving Crater Lake the status of a national park was a wise move and recommended the same for Wind Cave. "There are something like 3,000 chambers and 100 miles of passages," the *Times* enthused, "containing many curious features and formations."[33]

Once again Congressman Lacey came into the act, like a bird going after split grain or thistle seed, backing Senator Gamble's action. Lacey's friends worried that he was fatigued by the legislative struggle; he would arrive at his congressional office at six AM and not leave until nine PM. When he sponsored the bill in Congress in June 1902, fresh from his success with Crater Lake, Lacey argued that Wind Cave *had* to be a national park in order to stop vandals from destroying it. Reportedly, he said, boxwork thieves were out of control. The same grim fate was befalling the Petrified Forest of Arizona and the Anasazi cliff dwellings at Mesa Verde in Colorado. And there was another component to Gamble and Lacey's argument. For meteorological reasons alone, they said, Wind Cave needed to be saved and maybe someday bottled up to furnish electrical power. (More than 1 million cubic feet of air was emitted or taken in every hour at the blowhole entrance.)

When both the House and the Senate approved bills to establish America's seventh national park, Roosevelt was gleeful. On January 9, 1903, without ceremony, he reserved 10,522 acres in western South Dakota to become Wind Cave National Park (eventually it would be enlarged to 28,295 acres). Although it is doubtful that Roosevelt realized this at the time, in addition to the boxwork formations he saved a fine example of mixed-grass prairie, one to which, in coming years, the Bronx Zoo bison would be reintroduced. Wind Cave National Park also became a prime bird-watching location, especially for enthusiasts seeking for the uninhibited song of western tanagers and lazuli buntings. Hunting and fishing were prohibited in the park, and no ponderosa pines could be chopped down for firewood. Rule Number 1 of the General Regulations for Wind Cave was that removal of formations was forbidden and nobody could ever again enter the cave without the approval of the U.S. government. The posted notices read: "All Persons Are Liable to Prosecution."

Wind Cave has proved to be the "wonder" Senator Gamble called it. The mapping and exploring of its passageways continued unabated. When Wind Cave National Park celebrated its centennial in 2003 only an estimated 5 percent of the cave had been explored. Visitors were routinely dumbstruck by its size and complexity. Like the ocean floor it once was, the cave complex remained an unfathomable mystery for future generations to take up. The Carnegie Institution was founded in 1902, and there was some hope that grants could be given to scientists to solve the cave's mysteries. With Roosevelt's new national parks—Wind Cave and Crater Lake—a fait accompli, under the safekeeping of the Interior Department and U.S. Army, two important links in the chain of American "wonders" had been strengthened. And President Roosevelt, as far as western sites were concerned, was just getting started.

III

Contributing to Roosevelt's promotion of the American West was the publication of his friend Owen Wister's novel *The Virginian: A Horseman of the Plains* in 1902. Dedicated to Theodore Roosevelt,* it told the story of a cowboy, a natural aristocrat, who was involved in Wyoming's Johnson County War in the 1890s. Wister's sophisticated use of cowboy imagery and the cowboy culture had echoes of Roosevelt's own days in North Dakota. Wister, a Harvard graduate (summa cum laude) whose short stories in *Harper's* were extremely popular, had hit the mother lode with this action-packed novel, where right (even vigilante justice) prevailed over wrong. *The Virginian* became an overnight sensation; more than 200,000 copies were sold in 1902–1903 alone. While it lacked the humor of Mark Twain and the sophistication of Henry James, *The Virginian* was nevertheless a novel of considerable distinction. (Unfortunately, it is usually remembered only for the Zane Grey pulp western phenomenon it inspired.)

*To be exact Wister's dedication read: "TO THEODORE ROOSEVELT . . . Some of these pages you have seen, some you have praised, one stands new—written because you blamed it; and all, my dear critic, beg leave to remind you of their author's changeless admiration."

President Roosevelt had proofread part of *The Virginian* before publication, wanting Wister to tone down the sadistic violence and gratuitous blood and gore. Roosevelt insisted that the protagonist—the heroic cowboy, who is never given a name—should become *the* arbiter of western culture and morals, a citadel of manly virtue. Amazingly, Wister agreed to make the changes. Like Roosevelt, Wister had traveled all over the West—in Arizona, California, and Washington, always going back to Wyoming—to study cowboys closely. Often depressed, failing to find a producer for his opera about Montezuma, Wister clung to his notebooks, which were filled with reportorial details of his ranch days north of Laramie. In *The Virginian*, Wister, with Roosevelt cheering him on, apotheosized the cowboy as the American hero, a Natty Bumppo writ large, redefining western mythology. Wister was in Wyoming during 1885, which the historian Walter Prescott Webb deemed *the* halcyon year of the "cattle kingdom"—and he masterfully captured the raw essence of that era with a realist's eye for accuracy.

If President Roosevelt could have written one novel in his lifetime, it would have been *The Virginian.* Not only did Wister insist that the West was the mainspring of moral and political regeneration, but his narrative was full of observations about wildlife and "vanishing" big game. "For a while now as we rode," Wister wrote (in the style of the Boone and Crockett Club), "we kept up a cheerful conversation about elk." His writing about the geography of Wyoming also echoed Roosevelt's *Ranch Life and the Hunting-Trail.* For example, Wister wrote, "We had come into a veritable gulf of mountain peaks, sharp at their bare summits like teeth, holding fields of snow lower down, and glittering still in full day up there, while down among our pines and parks the afternoon was growing sombre."[34]

Besides naturalist ideas, Wister hammered home the theme of national unity. The Virginian, for example, came from the south but his love interest, Molly Stark Wood, was from an old Yankee family. Their marriage suggested healing after the Civil War, which had claimed more than 600,000 American lives. What replaced the sectional conflict, for Wister, was the West, which had redemptive power. The chief factor was nature. Unlike James Fenimore Cooper, who saw the forest as a place of danger, Wister promoted national parks, such as Yellowstone, for tourism. *The Virginian* ends with the Virginian and Molly getting married in a beautiful federal forest in Wyoming. No longer were Native Americans a problem in the West—they had been defeated. The new "noble savage" in *The Virginian* was the Virginian himself, who as a cowboy was liberating himself from industrial disease, urban chaos, and boogie-street immorality. Consequentially, for a man like the Virginian (or Roosevelt) to exist, there had to be a wilderness. Owing to Roosevelt's embrace of *The Virginian*, the new American male archetype was off and running, and the concept included the promotion of forestry and humane treatment of horses. "The Virginian is also, despite his skill in violence, a kind man," the literary historian John G. Cawelti noted in a Barnes & Noble edition of the book: "Two of the most striking episodes in the novel—one involving a denuded hen named Emily and the other an abused horse—illustrate his general kindness to animals."[35]

Wister's character was a reserved man who obeyed the law, respected nature, never cursed, was courteous to women, and even befriended Indians. Considerateness and gentlemanliness were part of his charm. And he was a walking, talking exemplification of the Boone and Crockett Club's sportsmen's code, and also a one-man rodeo. Just as the U.S. Forest Service later used Smokey the Bear to teach people about the dangers of forest fires, *The Virginian* could be read as constructive propaganda, i.e. *real men* didn't disobey federal laws by poaching in places like Yellowstone or Yosemite or the Black Hills; *real cowboys* respected the boundaries of national parks and never desecrated these sites. It's indisputable that *The Virginian* was worth more, in terms of public relations for President Roosevelt and his conservationist-preservationist agenda, than any five Madison Avenue firms of later years could have been.

Reveling in *The Virginian* as literature, Roosevelt even approved of the vigilante justice in the novel. Cattle rustlers, plumers, poachers, abusers of wildlife—in Wister's West they deserved to be hanged. Of course, as president Roosevelt couldn't sanction Wister's tracking down and hanging of rustlers, but his sympathies were in that direction. The railroad and barbed wire had ended the open-range era of the West. Gone were the old days of first-generation cowboy lore. President Roosevelt and Wister both scorned the new big business: conglomerates of oil, transportation, and manufacturing which were destroying the wilderness they found so intoxicating. It was up to the new generation of cowboys, Roosevelt believed, to stand against these usurpers. Yesterday's free-range cattlemen had to become the front line in his conservation movement. To Roosevelt the three great values of America—individualism, nationalism, and democracy—needed wilderness to flourish.

As Wister implies in his dedication, he had let Roosevelt line-edit *The Virginian*. Why? If Wister's hero had been prone to wrongheaded violence—eradication of animals, genocide of Indians, brothel morals—

as many first-generation western white men actually were, then wouldn't his book, in effect, be granting mythological status to rogues? To President Roosevelt's thinking there was already too much melodramatic romanticization going on in the American West. There were too many men like Billy the Kid, Jesse James, and John Wesley Hardin. The black hats were dominating the nickel-novel market. The white-hatted scientists who worked for the U.S. Biological Survey naturally seemed boring and tame compared with the Kid. Wasn't it far better to give the lawmakers and biologists a boost in the popular imagination? Wasn't it better to promote Wyatt Earp, Pat Garrett, John Muir, and other men of their type as the heroic face of the American West?* Furthermore, western folklore already had its exploiter gods: Paul Bunyan obliterated forests with his mighty ax and Pecos Bill dug out the Grand Canyon with a huge shovel. All that the Virginian—like a later character, the Lone Ranger—did was try to distinguish right from wrong. Wister's Virginian ushered in a new, better representative western male. Put a pair of field glasses into the Virginian's hands, and he would have been the kind of Auduboner the USDA was promoting in its *Yearbook*.

To Roosevelt, the westerner who most epitomized the Virginian (besides Seth Bullock) was the lawman Pat Garrett, who had shot Billy the Kid at Fort Sumner, New Mexico, on July 14, 1881. Roosevelt had learned from his Rough Riders—half of whom came from New Mexico—that Sheriff Garrett was revered from Las Cruces to the Four Corners. A worshipful Roosevelt loved to meet the renowned sheriffs of the West—they were heroes. Garrett, fairly honest, upright, with a bit of a drinking problem, didn't disappoint him. No sooner had McKinley been assassinated then Roosevelt, in December 1901, hired Garrett to be collector of customs based in El Paso (or one of Roosevelt's "White House gunfighters," as the newspapers began calling such appointments). When Roosevelt and Garrett finally met, around Christmas 1901 in Washington, they got on like long-lost brothers.[36]

The western code that Wister promoted and Garrett lived, President Roosevelt believed, should become a moral code. As an addendum, as Roosevelt said in *The Deer Family*, a backyard naturalist code should be adopted. Wyomingites had to save the Yellowstone forests just as Oregonians had to fight for the environmental integrity of Crater Lake. Concerned about industrialization and urbanization, President Roosevelt

*Roosevelt may have thought Pat Garrett adhered to the western code, but he didn't. Garrett's law enforcement decades had many unethical episodes.

proudly idealized the West as America's last best hope. That was why he pushed for the Newlands Act and for so many irrigation laws. Roosevelt had long understood that Congress could declare places like Wind Cave and Sequoia as parks, but if the locals didn't accept them as such, if county sheriffs continually let poachers go free and laughed as hooligans carved their initials in rocks and stole native American pottery, then the federal reserves would be destroyed. Only if the fighting men of the American West believed in protecting parklands and forest reserves could the West really be the Garden of Eden of his dreams. Although the "Virginian" wasn't what we would now call an eco-warrior, he later became the prototype for Edward Abbey's character Hayduke in *The Monkey Wrench Gang*, which led to the creation of Earth First! in 1980. It's safe to argue that tolerating President Roosevelt's warrior-like romanticization of Wister's cowboy virtue and his mystical belief (like Muir's) in the power of primeval nature is a small price to pay for saving vast acreage of the American West for future generations.

Roosevelt believed that Wister, far from selling out, did America a favor by making his masculine archetype a man of virtue, not vice. (Medicine Bow, Wyoming, continues to boast that it was the model for the setting of *The Virginian*, and that Wister's novel put the "Western-man's ways" on the "straight-and-narrow.") Like *Uncle Tom's Cabin* and *The Jungle*—two other novels with a cause—*The Virginian* helped galvanize Americans to celebrate law enforcement instead of propagating the far more destructive trend of outlaw chic. Big game hunting was going to happen in the West. Therefore, wasn't it far better for settlers to follow the noble sportsman's code of *The Virginian* instead of the butchery of Buffalo Bill? Wister, with President Roosevelt as a guiding light, did the West a favor by making the region's first true literary hero a white hat instead of a black hat. When *The Virginian* wandered the badlands, he was, like Roosevelt himself, a righteous horseman in the "quiet depths of Cattle Land" where every magnificent vista was considered "pure as water and strong as wine." According to Wister, being "a man" meant holding congress with nature—again, like Roosevelt, whom he said was the "greatest benefactor we people have known since Lincoln."[37]

The Virginian continued selling like hotcakes, with a stage production under way. Meanwhile, Roosevelt went on a tear to save western forestlands, catching the pro-development crowd unaware. (It should be noted, though, that Texas had no public lands on which he could declare anything.) If *The Virginian* ended in a shootout, so too would Roosevelt's conservationist battle with western developers; and he planned on winning it. Between the creation of Crater Lake National Park (May 22, 1902) and that of Wind Cave National Park (January 9, 1903), just over seven months, Roosevelt built his conservationist tower strong and tall. His administration established thirteen new national forests. (However, he also reduced the 1.1 million acres of the White River in Colorado by 61,000 acres in 1902; today it's 2.3 million acres.)

Geographically Roosevelt's forests ranged from the great boundary waters of Minnesota and Canada to the foothills of the lush, waterfall-rich Arbuckles in Oklahoma. In New Mexico Roosevelt created Lincoln National Forest in honor of his presidential hero (this 1,103,828-acre reserve straddled four counties and became the birthplace of Smokey the Bear). Heading off the mining industry, Roosevelt declared the Little Belt Mountains of central Montana a national forest, though they were known to contain silver, gold, lead, and zinc. The lush timber forest and grassy meadows of the Little Belts, Roosevelt declared, were far more valuable to America's long-term heritage than a cabal of speculators amassing personal fortunes from the public domain. The Little Belts had formed 70 million years ago; he wanted to keep them intact.

Perhaps the most extraordinary forest reserves President Roosevelt created in 1902 were the huge "experimental" ones in Nebraska. In April he declared Niobrara and Dismal River national forests. (The reserves became jointly known as Nebraska National Forest in 1907.) Within their perimeters were the longest "hand-planted" forests in the world. Starting in the 1890s Charles Edwin Bessey, a botanist at the University of Nebraska, working in conjunction with the U.S. Department of Agriculture, began an innovative tree-planting project in the Nebraska Sand Hills. Because the Sand Hills were semiarid, receiving only about twenty inches of rain annually, the 20,000-square-mile area was terribly deforested. Brushfires were the most menacing threat. In 1901, Pinchot, wanting to assist Bessey's vision of a forested Nebraska Sand Hills, sent a reconnaissance survey team to assess the possibility of planting trees.[38]

Roosevelt had great confidence in the Nebraska Sand Hills rehabilitation project. By creating the two forest reserves on deforested land—208,902 acres, to be exact—Roosevelt was hoping to launch a prototype pilot project that could soon be replicated in Kansas and Iowa. He saw both Dismal River and Niobrara as nurseries for Nebraska farmers who wanted trees for windbreaks and to protect their homes. No longer would sharp summer winds scorch their crops or brutal winter snowdrifts bury homes, as had happened at his North Dakota ranch. Trees would protect farmers from the elements and improve soil humus so that modern agriculture could thrive. Roosevelt also believed that trees in Dismal River and Niobrara would "ameliorate the dryness of the atmosphere," thereby increasing rain.[39]

Roosevelt and Pinchot's Nebraska project was, as the historian John Clark Hunt called it in the journal *American Forests*, "The Forest That Men Made."[40] Building the headquarters in Halsey, Nebraska, President Roosevelt had Forest Service employees plant 70,000 jack pine seedlings (from Minnesota) and 30,000 ponderosa pine seedlings (shipped from the Black Hills Forest Reserve by Seth Bullock). Many trial-and-error experiments commenced. Overcoming prairie fires was extremely difficult. Eventually the workers discovered that native red cedar grew more effectively than pine. From a political perspective the Nebraska project was proof positive that federal forestry efforts were aimed at enhancing western living, not destroying states' rights.[41] Decades later, when America was in the midst of the Great Depression, President Franklin D. Roosevelt would adopt the Nebraska pilot project nationwide as the Civilian Conservation Corps (CCC).

Still, T.R. received a lot of criticism from Nebraskan farmers and stockmen who thought of his forest reserves as socialism—the U.S. government taking lands away from the public domain. Eventually, Roosevelt was forced to reduce his Sand Hills reserves by 2 percent.[42] "Some of the Nebraska Congressmen are uneasy about the growth of the forest reserves in Nebraska," Roosevelt wrote to Pinchot in 1906. "I call your attention to the enclosed account of an interview with [Sylvester] Rush, who was special district attorney at Omaha, who was so active in securing the prosecution of the cattlemen for illegal farming. The article tends to show that some of the cattlemen who have been most prominent in illegal action are pushing this reserve scheme. Of course this does not alter the fact that these reserves are good things where no harm to the homesteader or agricultural settler results; but it also shows that we should be exceedingly careful about going too far with them. Such a course would tend to promote a revulsion."[43]

Besides undertaking the forestry experiment in Nebraska, Roosevelt also preserved millions of acres of Alaskan forestland in 1902. Even though he still hadn't managed to visit Alaska, he had carefully read Merriam's "Harriman Alaskan Expedition Report of 1899." (And he knew that Burroughs and Muir were busying themselves writing memoirs of their Alaskan trip.) Well aware that the Alexander Archipelago—a 300-hundred-mile-long group of pristine islands off the southeastern coast of Alaska, named after a former head of a Russian fur trading company—

was an incubator for thousands of seabirds, seals, walruses, and whales, Roosevelt declared all 1,110 islands a national forest on August 20, 1902. When he was asked how seal and bird rookeries could possibly be a forest reserve, Roosevelt pointed out that the islands were located in a temperate rain forest zone. Roosevelt saw this new Alaska reserve as a preemptive strike against Great Britain, whose sea hunters were constantly killing seals in American waters. "We have taken forward steps in learning that wild beasts and birds are by right not the property merely of the people alive to-day," Roosevelt said, "but the property of the unborn generations, whose belongings we have no right to squander."[44]

Originally conceived by George T. Emmons, a former naval lieutenant who hunted and fished in Alaska regularly, the Alexander Archipelago National Forest Reserve was Rooseveltian conservationism writ large—very large. It was slightly over 4.5 million acres. In 1893 Emmons had overseen the U.S. Navy's Alaska exhibit at the World Columbian Exposition in Chicago.[45] At the time Emmons was considered America's expert on the coastal Alaskan art of the Tlingit and Haida. In 1900, it's safe to say, nobody else, intellectually, knew the islands, inlets, and waterways of southern Alaska with the nautical precision of George Emmons. Although his primary home was Princeton, New Jersey, Emmons spent his summers in Alaska as an "anthropologist without portfolio." All of America's major museums collected Alaskan art from him.[46]

An amateur cetacean biologist also, Emmons wrote a fact-filled report, "The Woodlands of Alaska" (promoting the Alexander Archipelago), and sent it to Roosevelt. Emmons's geological knowledge was based on personal exploration—always a plus with Roosevelt. What impressed Roosevelt so much about Emmons was his scholarly, coolheaded analysis of Alaskan wildlife. His report, for example, admitted that much of Alaska wasn't suited to be a forest reserve. But the forests of southeastern Alaska—the Alexander Archipelago—were a must, if only to protect wildlife. These largely coniferous woodlands were part of a continuous forest which ran through Oregon, Washington, and British Columbia. Because it rained so often, forest fires were rare. These islands were built for the ages. Whetting Roosevelt's conservationist appetite, Emmons boasted that the coast hemlock (*Tsuga heterophylla*) and the Sitka spruce (*Pices sitchernsis*) of Alexander Archipelago were world-class. With scientific prudence Emmons suggested that the islands of the Alexander Archipelago should remain "one immense forest of conifers."[47]

Sounding a lot like Pinchot, Emmons told Roosevelt that he wasn't an idealist with regard to national forests; limited logging would have to take place on the islands of the Alexander Archipelago. Emmons explained how the Roosevelt administration could work around such inconveniences as fishing camps, sawmills, and canneries. Recognizing that Sitka, for example, was inhabited by territorial citizens (white), Emmons didn't rock the boat. He did, however, recommend that the very thinly populated islands of Prince of Wales, Kuiu, Kupreanof, and Chichagof (and hundreds of smaller islets) become federal assets. Although Tlingit and Haida lived on these islands, they weren't considered real citizens in 1902. Unbending about saving the archipelago, Emmons chose forestry over Indian rights as his primary concern.[48]

Another reason President Roosevelt embraced the plan outlined in Emmons's "Woodlands of Alaska" was a boundary dispute between the United States and Canada over a strip of southeastern Alaska. The object of controversy was known as a "Panhandle" along the Alaska-Yukon border. Ever since the discovery of gold along Bonanza Creek in 1896 the United States had denied Canada direct access to the Klondike from the Pacific Ocean. By 1902 the dispute between the nations was enflamed. "[The] claim of the Canadians for access to deep water along any part of the Canadian coast," Roosevelt angrily pronounced, "is just exactly as indefensible as if they should now suddenly claim the island of Nantucket."[49]

Roosevelt had long harbored an expansionist desire to incorporate all of Canada into the United States, though of course he never acted on it. Clearly, in 1902, with Canada a great democratic nation, this wasn't going to happen. Nevertheless, Roosevelt didn't want to have Canadians timbering or fishing in Alaskan waters. The United States, Roosevelt insisted, had legal authority over an unbroken littoral going from the Alaskan Territory to the southernmost part of the Panhandle. In 1901 Roosevelt had ceded about 600 square miles of land to Canada—but now, in 1903, he wasn't going to compromise with regard to the Panhandle. Roosevelt, in fact, belying his reputation as an expansionist, is the only U.S. president ever to shrink the size of American territory. Getting Great Britain to side with the United States in a dispute, an international tribunal voted in favor of the Roosevelt administration in 1902. But the U.S. federal government's seizure of the Alexander Archipelago as a forest reserve was a proactive measure aimed at protecting Alaskan waters from Canadians.

Deeply impressed by Emmons—after all, they shared a love of U.S. naval history and conservation—Roosevelt sent him on an official mission to Alaska in 1902 to iron out a land dispute with Britain over the Alaskan-Canadian border. The sheer professionalism of Emmons on this mission impressed the president mightily. Owing to Emmons's advocacy and diplomacy, on August 20 Roosevelt created the Alexander Archipelago Forest Reserve by a presidential proclamation. "The areas included in the reserve were exactly as George Emmons had proposed them," the historian David E. Conrad noted in an important article in *Pacific Historical Review*, "and his dream of a national forest in Alaska was an accomplished feat."[50]

The Alexander Archipelago Forest Reserve was just one of thirteen reserves President Roosevelt created in 1902—all aimed at also protecting big game. These new federal forestlands totaled 14,276,476 acres. In coming years many of these forest reserves would be increased in acreage, often undergoing many boundary changes before being finalized and renamed. By the end of 1902, however, the reserves broke down as follows:

	Acres
San Isabel, Colorado	77,980
Santa Rita, Arizona	337,300
Niobrara, Nebraska	123,779
Dismal River, Nebraska	85,123
Santa Catalina, Arizona	155,520
Mount Graham, Arizona	118,600
Lincoln, New Mexico	500,000
Chiricahua, Arizona	169,600
Madison, Montana	736,000
Little Belt Mountains, Montana	501,000
Alexander Archipelago, Alaska	4,506,240
Absaroka, Montana	1,311,600*[51]

On December 2, 1902, amid this flurry of national forest legislation, Roosevelt, in his Second Annual Message to Congress, defended the le-

*Not included in this table is the Medicine Bow Forest Reserve (Colorado and Wyoming). Originally created on May 22, 1902, it had about 2 million acres. Medicine Bow Forest Reserve has gone through a number of changes since its creation, including boundary modifications on July 26, 1902, and May 17, 1905. Absaroka is not generally included as one of Roosevelt's 1902 accomplishments, as it was absorbed into Yellowstone Forest Reserve on January 29, 1903. See "100 Years of Conservation and Public Service on the Medicine Bow," U.S. Forest Service Archives, Washington, D.C. See also *Report of the Commissioner of the General Land Office to the Secretary of the Interior for the Year Ended June 30, 1905* (Washington, D.C.: Government Printing Office, 1905), pp. 213–215.

gality of his thirteen new reserves with the passion of John Muir. This generation of Americans, he said, had a duty of handing down natural wonders, not squandering them. Whether it was the temperate rain forests of Alaska; the dazzlingly colorful fall broadleafs of the Great Lakes; the pine barrens of New Jersey; the Joshua tree terrains of the Southwest; or the supreme cottonwoods, tupelos, and bald cypress of America's river bottoms, citizens needed to protect their trees for aesthetic and other reasons. To Roosevelt a single limb of a many-branched oak had more true wisdom than all the congressmen in session combined—except, of course, for Lacey, who was as big as a forest in the president's mind. Furthermore, wildlife survival was, quite simply, completely dependent on trees. In line with the mission of the Boone and Crockett Club and the Audubon societies, Roosevelt also wanted written assurances from Congress that the fish along the White River, the wild turkeys of Wisconsin, and the walruses of Alaska on federal lands would be unmolested in perpetuity. "Legislation should be provided for the protection of the game, and the wild creatures generally, on the forest reserves," he said. "The senseless slaughter of game, which can by judicious protection be permanently preserved on our National reserves for the people as a whole, should be stopped at once."[52]

CHAPTER EIGHTEEN

PAUL KROEGEL AND THE FEATHER WARS OF FLORIDA

I

Starting in March 1903 President Roosevelt engaged in a Herculean effort to save the bird rookeries of Florida. Playing the role of a modern-day Noah, the committed Audubonist insisted that every bird species in Florida was a world unto itself, a masterpiece of Darwinian evolution. And these birds needed habitats to survive. Even though the term "biodiversity" would not be coined until about 1985, Roosevelt had an intuitive grasp of the concept. He worried that for some species in Florida—a hot spot for biodiversity—the death rate was far exceeding the birthrate, threatening them with extinction. The initial showdown over the future of Florida's wildlife took place at Pelican Island, a teardrop-shaped Atlantic Coast islet situated three nautical miles from the hamlet of Sebastian. The island was home to the last breeding colony of brown pelicans on the east coast of Florida. The Indian River Lagoon basin, of which Pelican Island was a part, contained some 4,300 species of plants and animals—more species than any other estuary in the United States—including 685 types of fish and 370 bird species.[1]

Of all the Florida avians, it seemed that the brown pelican was Roosevelt's personal favorite. These pelicans were to him like Keats's nightingale or Wordsworth's cuckoo. The brown pelicans were the finest fishermen Roosevelt knew—a far cry better than any human. That pelicans were such superb natural fishermen, however, may have enthralled the president, but it irritated the dickens out of rural Floridians. Since brown pelicans had an almost insatiable appetite for scooping up finned prey, Florida fishermen saw these birds as unwelcome competition, a hindrance to their livelihood, like woodpeckers devouring corn or mockingbirds incessantly pecking at grapevines. The mere sight of brown pelicans—flying over the low gray river with pouches chock-full of fish to feed their nestlings—made fishermen reach for a gun. The pelicans' fishing grounds were supposed to be humans' fishing grounds. Therefore, the pelicans had to be eradicated. What locusts were to a Nebraskan farmer or gutter rats to a Brooklyn merchant, brown pelicans were to Florida's commercial fishermen. But to Roosevelt, pelicans were a marvelous example of Darwin's evolutionary theory at work, and he wanted them protected. Unlike most

Regularly Roosevelt would row around Long Island Sound on bird-watching adventures. He believed it was a civic responsibility to know the wildlife species that lived in your own backyard.

birds, pelicans weren't masters of evasion; they actually liked people. Unfortunately, that made them even more vulnerable to being slaughtered for their quills, as mounts, for their eggs, and even for target practice. How they had survived so long, given their friendliness, intrigued Roosevelt.

While stationed in Tampa Bay during the Spanish-American War, Roosevelt had studied the brown pelicans' daily routines with an artist's eye for nuance. Often they glided low, within range of a pistol or slingshot, among swarms of gnats (locally known as sand flies) whose buzz may have been inaudible to humans but which nevertheless caused a man to itch. Sometimes the pelicans followed porpoises, which served as their advance agents for detecting schools of fish. Roosevelt had marveled at how good-humored pelicans could be, allowing noisy gulls to use their elongated heads as a resting spot.[2] They built frail twig nests for roosts, and each pelican parent would take a turn sitting on the eggs and then dutifully standing guard, in shifts. They were a highly responsible bird species in this regard, rare and remarkable. And pelicans were just one of Florida's wild creatures Roosevelt admired. Someday he hoped his grandchildren would see loggerhead turtles laying eggs on a Florida beach and manatees patrolling the crystal waters of a spring-fed river. To Roosevelt there was no more nutritive truth than the order of the Abrahamic God four days after Genesis to "let the waters teem with countless living creatures, and let birds fly over the land across the vault of heaven."

Roosevelt first learned about the birds of Pelican Island from the published field notes of Dr. Henry Bryant (of the Boston Society of Natural History), recorded in 1859. Dr. Bryant, in his capacity as a professional ornithologist, scientifically recorded the scores of brown pelicans and other waterbirds he encountered on fantastically misshapen clumps of mangrove in the Indian River Lagoon. "The most extensive breeding place was on a small island called Pelican Island, about twenty miles north of Fort Capron," Bryant wrote in his diary. "The nests here were placed on the tops of mangrove trees, which were about the size and shape of large apple trees. Breeding in company with the pelicans were thousands of herons, Peale's egret, the rufous egret and little white egret, with a few pairs of the great blue heron and roseate spoonbills; and immense numbers of man-o-war birds and white ibises were congregated upon the island."[3] Bryant also reported that a feather hunter had recently killed sixty roseate spoonbills on Pelican Island in a single day.

The word soon spread in ornithologist circles that Pelican Island was an amazing field laboratory; there was little need to venture down to Ecuador or British Honduras. In 1879 Dr. James Henshall, following in Bryant's footsteps, visited the Indian River Lagoon region on a collection trip, expecting the best. Traversing trails that became a trough of swamp water to get to Pelican Island, Henshall was astounded by what he encountered: the mythical rookery was now a dead zone. "The mangroves and water oaks of this island have all been killed by the excrement of the pelicans which breed here," Henshall wrote in his journal. "This guano, which lies several inches deep on the ground, is utilized by the settlers as an efficient fertilizer. At a distance, the dead trees and bushes and ground seemed covered with frost or snow, and thousands of brown pelicans were seen flying and swimming around or perched upon the dead branches. As we passed, we saw a party of northern tourists at the island, shooting down the harmless birds by scores through mere wantonness. As volley after volley came booming over the water, we felt quite disgusted at the useless slaughter, and bore away as soon as possible and entered the Narrows."[4]

Now, more than twenty years after Henshall's disturbing report, Roosevelt was in a position to help save this little piece of Eden from destruction. As an honorary officer of the Florida Audubon Society and a former governor of New York who had helped promote the Lacey Act, he was well acquainted with the so-called Feather Wars going on in the state between bird protectionists and hunters. Now, with the bully pulpit at his disposal, Roosevelt was finally going to put the plumers and eggers out of business once and for all. The main thrust of his policy was to strengthen the hand of the U.S. Biological Survey in Florida, perhaps having it oversee islets like Pelican Island. A new generation of Darwinists could use Florida as an evolutionary laboratory where plants might offer medical cures and breeding habits of birds could be carefully studied. But even if Roosevelt was able to create federal bird reservations on the Atlantic and Gulf coasts of Florida, he would need wardens to protect them. Luckily, the ideal wildlife warden was already living in Sebastian, overlooking Pelican Island: Paul Kroegel, known locally as the "Audubon of the Indian River." Roosevelt, it turned out, was Kroegel's conservationist hero. Essentially, Kroegel became the archetype of the frontline law enforcement wardens Roosevelt wanted hired all over America to help protect birdlife. Therefore, Kroegel's story illustrates a new trend in the wildlife protection movement that Roosevelt initiated in the hope of restoring bird species: federal wildlife protection officers.

II

Paul Kroegel—the man who would become Roosevelt's first wildlife warden—was born in Chemnitz, Germany, on January 9, 1864. When he was three or four years old he became enamored of the gangly white storks (just as T.R. had been enamored of those in Dresden in 1869, when he "drew" Darwin's theory with himself as a stork). The sight of a stork nesting on a chimney was considered a good omen in both Chemnitz and Dresden. Also, because these migratory birds often arrived in Germany around Easter, the coming of the stork was, to northern Europeans of Teutonic ancestry, a symbol of the resurrection of Jesus Christ. Not only did German children like Paul Kroegel refrain from killing white storks; they thought of storks as holy birds imbued with eternal nobility. Germans, of course, weren't the only Europeans to admire storks. According to ancient Greek mythology, a white stork was a fertility symbol. If a woman made direct eye contact with such a stork, she could receive the blessing of a newborn child. In Scandinavia white storks were believed to find human children ("stork children") living in marshes or caves, to clutch them with their red claws, and to deliver them to the doorstep of pregnant women.[5] In northern Germany storks were also thought to bring babies.[6]

This reverential attitude toward the long-legged birds was reinforced in young Kroegel by Hans Christian Andersen's classic children's story "The Storks." Andersen, a Dane, was a veritable storehouse of stork mythology. Danish servants, for example, believed that if people saw a stork fly into town as spring approached they would soon move residences; if they saw a stork standing, they'd retain their current employment.[7] When Andersen wrote "The Storks" in 1838, the rooftops of northern Europe provided nesting areas for them. In Andersen's fairy tale, children who taunted a maimed stork were punished: a stork delivered a dead baby brother or sister in its long beak instead of a live baby. To a child of Kroegel's generation, impressed by Andersen's stories, hurting a stork would be akin to molesting the Easter Bunny, or the Tooth Fairy.[8]

On a cold Christmas Eve in 1870, when Paul Kroegel was six years old, his mother and an infant sister died during childbirth. His distraught father, Gottlob, took Paul and his two-year-old brother, Arthur, and immigrated to the United States the next year. They endured a brutal twelve-day voyage aboard the freighter S.S. *Canada* across the tempestuous Atlantic. Arriving in New York, the Kroegels were deloused and given clean clothes. Paul was immediately homesick for Andersen's storks: all of the tenement roofs in New York were populated by rat nests. After about four years on the Lower East Side of Manhattan, working in the meat business and as a jack-of-all-trades, Gottlob packed up his boys once again and first moved to Chicago and then eventually settled in Florida under the Homestead Act. Gottlob realized that Florida was his best chance of pursuing his idea of the American dream and providing a fine, healthy life for his boys.[9] Lured by stories of eternal sunshine and citrus groves finer than those along the Nile River, Gottlob became a pioneer and homesteaded with Paul in the low-lying central coast of Florida between the Saint Sebastian and Indian rivers (today it is known as Indian River County).* [10] After a brief stay in Fernandina, Florida, Gottlob bought a small skiff to sail the 200 miles of the Saint Johns River. But, encountering a headwind, he and Paul rowed most of the way. They had their boat and supplies hauled six miles on a mule-drawn tram to the shores of the Indian River. From there, they sailed another sixty-five miles south until they came upon a high promontory along the shoreline and decided to stay. They built a palm frond home on top of Barker's Bluff, an old Ais

*Gottlob had dropped off Arthur, then about twelve, in Crestline, Ohio, with his brother Adolf.

Indian mound looking across the lagoon at Pelican Island. (Later that year a summer gale blew the house away, so they constructed a sturdier New England–style cottage, less vulnerable to tropical storms.)

When the Kroegels arrived in 1881, Florida was, as the novelist Wallace Stegner once wrote, a violent dreamland of "six-shooter freedom and orange-grove bliss," a forlorn place where the soil was so luxuriant everything grew wild and the trees didn't drop their leaves in winter.[11] The Indian River region could be bleak, intimidating, and even lethal. But the lure of sunshine meant that every year more and more homesteaders arrived, using machetes to cut away palm sabal and crazy weeds to plant crops. Clearly there was something miraculous about Florida soil if—and only if—you could survive the coral snakes, diamondbacks, mosquito hordes, tropical storms, stark loneliness, and occasional frosts. Basically Florida in 1881 was like the Wild West—a frontier wilderness. Most villages like Sebastian didn't even have a wooden water tank or a one-room schoolhouse to call their own.

By the time Paul Kroegel was a teenager, his nickname around Sebastian was "Pelican Watcher." Just above the tide line of Pelican Island, he enjoyed watching these comical birds cavort with one another.[12] (The brown pelicans reportedly did not share Pelican Island with other species from about 1882 to about 1939.) Protecting pelicans became part of his daily routine in Sebastian. Flocks of brown pelicans, in perfect formation, continually flew over his lookout home, making a steady stream of designs in the sky. Creatures of habit, brown pelicans returned to their favorite rookeries, like Pelican Island, every spring and winter to roost and lived year-round along the central coast, which overlapped the subtropical Caribbean zone and the temperate Carolinian zone. Their presence symbolized the oasis that was the Indian River Lagoon. Birds of all kinds, in fact, including man-of-wars with a wingspan of seven and a half feet, were often in migratory flux around his lagoon. There were white ibis, black-crowned night herons, and great blue herons. Sometimes the sky was so blotted with turbulent streams of wading birds that the flocks appeared like a high dome above the crystal blue lagoon. (The Indian River Lagoon was described by several early explorers and settlers as being crystal-clear and blue.) A celebrated founder of the National Association of Audubon Societies, Thomas Gilbert Pearson, in *Adventures in Bird Protection*, described the mass movement of such flocks as a "great rotating funnel."[13] But because of the plume hunters, eggers, and trigger-happy tourists, every week their numbers were dwindling.

Paul Kroegel, who faced off against plume hunters in the 1900s, was a pioneer in wildlife conservation not only for Florida, but for the entire nation.

Keeping vigil over birds, however, wasn't a livelihood. As with just about all the first-generation settlers in post–Civil War Florida, Kroegel's primary source of revenue came from farming. Besides beans, citrus groves kept the Kroegels in the black. Within eight years of moving to Sebastian—through perseverance and good seed—the Kroegels were among the most prosperous growers in Florida.[14] Sometimes, however, the citrus and bean crops couldn't survive the deadly frost that blew in around Christmas. A single "big freeze" could cover a year's worth of agricultural labor in an icy, deadly crust. Priding himself on beating back the frost, wrapping his trees in burlap, Kroegel defiantly grew grapefruit and oranges on his 143 acres of Indian River property when other, less determined farmers had abandoned their crop.* He also tended more than 100 beehives, selling his homegrown honey at a tiny roadside stand. These Florida bees weren't transported from Europe (like those in New England) but were true "Aborigines."[15] And scarcely a week went by when Kroegel wasn't scything through dense hammocks to cut a trail, sometimes for extra pay and at other times just for the sake of reclamation. "Even though he had a wife and kids," his grandson Douglas Kroegel

*Indian River oranges were and are considered, by many, the best in the world, and are known as the "royalty" of oranges. Three-quarters of all Florida grapefruit comes from the Indian River region.

recalled, "it seemed it was those birds and trails he cared about the most. Even more than his kids. He was on a mission to put the bad guys [plumers] out of business on Pelican Island no matter what it took."[16]

As Kroegel approached his thirty-ninth birthday in 1903, he wondered why the federal government wasn't stopping the massacre of these Pelican Island birds. After all, Washington, D.C., owned the island and half a dozen adjacent smaller keys in Indian River. Didn't the Lacey Act prohibit such reckless slaughter? Couldn't the Roosevelt administration ban it on federal property? Shouldn't the new law against killing pelicans be enforced as the one in Chemnitz was for storks? Wouldn't it be smart to create a bird reserve? Just asking these questions made Kroegel an irregular character in the fishing community of Sebastian. With his steely gaze, huge droopy mustache, and blistered boatbuilder's hands, Kroegel, a father of two, cut an imposing figure for a man only five feet six inches tall. His round face had developed a network of wrinkles, accentuated by a deep year-round tan. Uninterested in clothes, Kroegel often wore heavy wool garments (including sweaters) and Sunday serge in the torrid sun to avoid being "eaten alive" by mosquitoes, which were also considered potential carriers of malaria.[17] He often carried an accordion with him to play at weddings and anniversary parties. To the Indian River fishermen in search of juvenile snappers and snook, the pipe-smoking Kroegel was a misanthrope of sorts. Sebastian back then had something of a Jamaican tang. The air smelled of seaweed, freshwater catch, guano, and horse dung. Local seafarers and dirt farmers sat around with basins of fish in the heat, repairing nets; they ostracized Kroegel for being so indignant about plume hunters bagging egrets and gunning down the fish-stealing pelicans which roosted and nested as thick as bats on the treeless Pelican Island rookery.

Like Audubon—who had tramped around the Gulf's southern tidal swamps—Kroegel would get a careful fix on each bird's unique temperament, silently crawling up close to measure habits such as skittishness, sociability, and cunning. Colonial waterbirds preferred to nest in colonies, grouping closely together for protection from predators. They particularly liked nesting on thick mangrove clumps, where they could easily glance around every inch of the rookery and sound a choir-like alarm if a predator was approaching. So on any given day Kroegel saw herons mingling with pelicans while spoonbills clustered only two or three feet away. As a dreamer—unschooled in ornithology—he imagined this patch of Florida as a northward extension of the tropics, and that idea invigorated him. But owing to guano buildup and "big freezes," particularly the horrific freezes of December 1894 and February 1895, many of the mangroves on Pelican Island had died. With no other options, the birds took to constructing their nests on the ground. Kroegel dutifully recorded in a notebook how this change in nesting affected their hunting habits. But whether the birds were perched high or nesting low, pelican sightings on his beloved islet brought him happiness.

Besides studying birds and farming, Kroegel was a boat builder par excellence. Schooners, skiffs, tall-masted sailboats, yachts, collapsible canoes—he constructed them all. Before long his expertise outclassed that of any other Floridian living within a thirty-mile radius of Sebastian and Wabasso.[18] With no major roads in the sandy-soil Sebastian area, save a few heavily rutted swampy paths, boating in the Indian River was still the main mode of transportation as the nineteenth century wound down.* Whenever a fierce wind blew off the Atlantic coast, the people of Sebastian—even those who thought Kroegel's bird-watching too eccentric—were in full agreement that his boats were sure to survive the coming tempest. His exemplary vessels were—without question—the epitome of nautical craftsmanship. Whenever there was a shipwreck in the Atlantic Ocean—like that of the *Mary Morse* in September 1898—Kroegel would rush out to assist the stranded crew. (In that particular disaster he earned $350 for salvaging the cargo of lumber planks.) Meanwhile, Kroegel had studied hard to earn his captain's papers from the state at age twenty; he was thereafter known as "Captain Paul."[19]

Navigating the shallow Indian River Lagoon—still, according to the Audubon Society, considered the most biologically diverse estuary in America[20]—had become second nature to Kroegel, as much a part of his daily regime as eating, sleeping, and oystering. Kroegel went out patrolling the Indian River Lagoon at midday, when the invigorating ocean breezes brought wind for his sailboat and welcome relief from pesky sand flies and mosquitoes. Salt marshes with the associated mudflats and tidal creeks provided a wealth of food and habitats for fiddler crabs, marsh rabbits, and wading birds. Self-trained in natural history, Kroegel could tell the difference between smooth cordgrass, saltwort, and glasswort at a glance. At nighttime, starlight and the phosphorescence of more than 250 algae species were his illumination as he rowed passed sorrowing cypress.

Starting in 1902, the Florida Audubon Society paid Kroegel a meager

---

*Visitors from New York got to the Indian River by railroad to Jacksonville. Then they took a steamboat up the Saint Johns River, debarking at Salt Lake, near the town of Enterprise. From there it was a mule-drawn wagon tram to Titusville.

wage to patrol the Indian River Lagoon for plumers. He was a conservationist gatekeeper. With his .10-gauge double-barrel shotgun always close at hand, ready to aim, Kroegel was determined to save the approximately 3,000 brown pelicans on Pelican Island from the barbarians. With his sailboat weaving about in a sort of hurried dance, Kroegel began patrolling the lagoon, threatening anybody who dared to disrupt his Indian River rookery. The worst offenders were thugs who sneaked onto the island under cover of darkness. Because the channel was so narrow, all large vessels going up and down the river had to pass within a scant 100 feet of Pelican Island, keeping Kroegel busy. If a boat dared anchor near the island, off Kroegel went, like an arrow shot from a pulled bow.[21] According to family lore, Kroegel slept with one eye open; that's how seriously he took his wildlife protection job.

III

Like Paul Kroegel, and other grassroots activists, Roosevelt had become disgusted that hunters in Florida would shoot a beautiful roseate spoonbill—to name just one species—so unscrupulous shops could sell its gorgeous wings as cooling fans to "snowbird" tourists in Miami and style-conscious debutantes in New York and Paris. Long ago Roosevelt had read about William Bartham's travels through wild Florida in the 1790s, and he wanted patches preserved just as that great eighteenth-century naturalist had encountered them at Alachua Savanna and along the Saint Johns River. Intellectually, Roosevelt was uninterested in compromises and in sidestepping conservation issues. When it came to protecting Florida's endangered wildlife, he was determined to choke the breath from the renegade plumers' business. He wanted the Lacey Act followed to the letter of the law. The Roosevelt administration's threat as of early 1903 was serious: shoot a protected nongame bird; go to jail.

In some ways Roosevelt and Kroegel shared many of the plumers' rough-and-ready Jacksonian attributes. Unhappy away from the outdoors, enamored of even the minutest habits of animal species, they prided themselves on their ability to track down wildlife by hunch and hoofprint. They were both a new breed of outdoorsman who preached wildlife protection.[22] Essentially, Kroegel embodied the new Rooseveltian high-water mark in conservation history: the insistence that wildlife had rights. Emulating the naturalist John Muir—who in 1867 had tramped all over Florida, marveling in his journal that the state's wide rivers did "not appear to be traveling at all"—Kroegel lived to protect "citizen bird."[23] Although Kroegel had never met the president, he knew many of Roosevelt's naturalist friends, and fancied himself as part of their clique.

This, in fact, was another striking characteristic that made Kroegel sui generis in the Indian River Lagoon area: his uncanny ability to cultivate enduring friendships with a wide array of national wildlife protection leaders. Never formally trained as an ornithologist, he nevertheless enjoyed bantering about shorebirds and was up-to-speed on the cutting-edge *Auk* articles of his day. Of the many friendships he formed protecting Pelican Island, one proved life-changing: his encounter in 1900 with Frank M. Chapman, curator of ornithology and mammalogy at the American Museum of Natural History in New York.

By the 1880s Chapman—whose mother spent winters in Gainesville at a family cottage near what is today the University of Florida—was considered America's most prominent popular ornithologist. When Chapman first visited his mother as a young man in his twenties—just after publishing his *Handbook of Birds in Eastern North America*—he made a long "local list" of the hundreds of birds he encountered in north central Florida, as if he were maintaining a sacred scroll. As a complement, Chapman kept a journal of his chance encounters with long-necked ducks, purple gallinules, and a wide range of ibises. Then he had the central, transformative ornithological experience of his life. One Sunday afternoon shortly after Thanksgiving 1886, while hiking in the pinelands around Alachua Lake, he happened upon an idyllic scene that nearly took his breath away. All around him were magnificent birds behaving in an almost magical fashion. Dutifully, he recorded in his diary what he had encountered that November day:

> As I approached the shore numbers of Ducks arose and sought safety in the yellow pond lilies (bonnets) growing some distance from it, and here was a splashing and a calling, a squeaking and squawking, such as I never heard before: odd noises of all sorts and descriptions all unknown to me, and I was without both gun and glass. The place seemed to be alive with birds. Ducks were constantly flying from place to place; coots and Herons were apparently common. On the shore near me birds were just as abundant. A pair of Pileated Woodpeckers, with flaming crests, were pounding away in a tree above my head, and with them were numbers of Flickers, and one Red-bellied Woodpecker. Doves whistled through the woods at my approach, Blue Jays screamed, Mockers chirped, and scores of birds flew from tree to tree. Truly I was in an ornithologist's paradise.[24]

What Chapman had stumbled upon was a patch of pristine Florida. For a budding New York ornithologist it was something akin to a fortune seeker stumbling upon a pile of gold. There, before his eyes, was an embodiment of divine law. At once Chapman started collecting Florida's birds—he shot 581 different ones, to be exact—for the American Museum of Natural History in New York. He always wanted the United States to be a step ahead of the British Museum when it came to ornithological studies. If his conscience bothered him at the sight of torn flesh and flying feathers—and it occasionally did—Chapman just blamed the birds' death on the necessities of modern science. All he could do in good conscience was avoid defacing the dead birds' exotic splendor any more than absolutely necessary when skinning. Meanwhile, Chapman—considered *the* supreme authority on Florida's birds—began exploring new ways for ornithologists to work without guns, using cameras and binoculars.

By the time Roosevelt became president Chapman's advocacy for "citizen bird" found a literary outlet in a revolutionary little book, which tried to turn the "sportsman's ethos" on its head. That year Chapman published *Bird Studies with a Camera*, complete with more than 100 photographs taken by the author himself. This was the kind of evolved birding that Roosevelt approved of—take a photo of the brown pelicans or herons and leave your gun at home. It anticipated eco-tourism and international birding. What Chapman had attempted to do—and by and large succeeded in doing—was to use his camera as "an aid in depicting the life histories of birds." Sometimes, instead of merely photographing a chickadee at a nest hole or young herons on branches seventy feet from the ground, Chapman would also capture images of their habitat—an application of his philosophy regarding museum displays. "A photograph of a marsh or woods showing the favorite haunts of a species," Chapman wrote, "is worth more than pages of description."[25]

Few wildlife enthusiasts could quarrel with the good intentions of Chapman's expansive, habitat-conscious nature photography. Most, however, as the modest sales of *Bird Studies with a Camera* indicated, probably still preferred one of the approximately 430 drawings in Audubon's *Birds of America* to Chapman's photograph of a puffin's burrow (minus the bird). And their reasoning wasn't simply a matter of aesthetic pleasure. Real hunters, not the riffraff who slaughtered nesting birds for money, found something missing from Chapman's philosophy—*the chase.* To Chapman's everlasting credit, he bravely confronted the "chase issue" in his proactive book through a combination of salesmanship (with regard to photography) and ridicule (of plume hunting). The essential message of *Bird Studies with a Camera* was something like "Real men don't kill little helpless uneatable things for sport." But true outdoorsmen, like those responsible for Adirondack Park and Crater Lake National Park, didn't kill shorebirds or songbirds anyway. In some ways Chapman was preaching to the converted. "I can affirm that there is a fascination about the hunting of wild animals with a camera as far ahead of the pleasure to be derived from their pursuit with shotgun or rifle as the sport found in shooting Quail is beyond that of breaking clay 'Pigeons,' " he wrote. "Continuing the comparison, from a sportsman's standpoint, hunting with a camera is the highest development of man's inherent love of the chase."[26]

That argument seemed reasonable enough. But it wasn't convincing to true-blue hunters: they refused to feel ashamed of preferring taxidermy to the darkroom. You could almost hear the dismissive grumble from late-nineteenth-century sportsmen piqued about Chapman's advocacy of the camera. Chapman's argument, they believed, was a ruse, like comparing apples and oranges. About a third of the way into his book—at which point his ideas were already rejected by most sportsmen—the high-minded Chapman lowered the boom regarding wildlife protection. "The killing of a bird with a gun," he wrote, "seems little short of murder after one has attempted to capture its image with a lens." With slow-boiling rage he continued to challenge the very manhood of "so-called

The ornithologist Frank M. Chapman of the American Museum of Natural History did more to save Florida's bird rookeries than anyone else in American history. He successfully lobbied President Roosevelt to create federal bird reservations.

true sportsmen," arguing that rejoicing over a trophy bag of "mutilated flesh and feathers" of nongame birds was utterly obscene.[27]

Reading *Bird Studies* was an inspiration for Kroegel. He quickly recognized that Chapman was a kindred spirit like none other, a self-taught ornithologist of deep learning and forbearance. They were, to his mind, a fraternity of two. All the brown pelicans' habits Kroegel had observed daily from his homestead at Indian River were extolled in gorgeous naturalist prose in Chapman's *Bird Studies*. "No traveler ever entered the gates of a foreign city with greater expectancy than I felt as I stepped from my boat on the muddy edge of this City of the Pelicans," Chapman wrote. "The old birds, without a word of protest, deserted their homes, leaving their eggs and young at my mercy. But the young were as abusive and threatening as their parents were silent and unresisting. Some were on the ground, others in the bushy mangroves, some were coming from the egg, others were learning to fly; but one and all—in a chorus of croaks, barks, and screams, which rings in my ears whenever I think of the experience—united in demanding that I leave town."[28]

There—in precise ornithological prose—was Kroegel's daily reason for skippering a boat around Pelican Island but seldom walking on the sanctuary. In Chapman he had found his own highly personal Thoreau. Chapman had written that brown pelicans had a "dignified way," that the island was their "metropolis." Kroegel liked that kind of descriptive imagery. After spending four full days on the roost, Chapman wrote, "During no hour of the twenty-four did silence reign." The sharp-eyed New Yorker even knew that pelicans liked to flap their broad wings for ten or eleven straight beats. As far as Kroegel knew, nobody, besides himself, had ever done that arithmetic before. And therein is what floored him about *Bird Studies*. Chapman had done the math. The esteemed ornithologist had counted 251 nests during his stay in March 1898 and then broke the total down as follows:

55 nests with 1 egg each
63 " " 2 eggs " "
23 " " 3 eggs " "
63 " " 1 young each;
46 " " 2 young each;

Such field calculations enthralled Kroegel. He thought Chapman was spot-on with his estimate of 2,736 pelicans for the island. The famous ornithologist had also captured the glory of what it *sounded like* to hear a pelican hatch, calling the young chicks' first noise on earth a "choking bark."[29] Besides writing about Pelican Island in *Bird Studies*, Chapman, under Roosevelt's urging, gave public presentations with stereopticon slides in Boston, Washington, D.C., and New York, bringing the plight of Florida's tidewater rookeries to a wider audience. He even kept a "bird count" of species he saw women wearing as apparel on Fifth Avenue. In one hour of one day in 1885, he identified 174 birds of forty different species adorning ladies' hats. In his lectures, he would verbally confront such women for indulging a penchant for precious feathers. Chapman implored them to switch over to domesticated ostrich feathers, which could be plucked or collected without hurting the bird. He had done so himself at an ostrich farm in Florida. The feathers from these flightless birds—celebrated for their fleet-footedness by ancient Egyptian kings—were gorgeous, ideal for hats, decor, and masks. If a woman really wanted to strike a theatrical pose, Chapman would say, then she should don an ostrich feather.

IV

One afternoon Kroegel actually got a chance to meet the mild-mannered, owlish Chapman, who resided at the Oak Lodge in Micco, Florida. The ten-bedroom boardinghouse—situated along the "soothing breeze" belt of the Atlantic Ocean between Melbourne and Vero Beach—was run by Mrs. Frances Latham (known as Ma Latham), a die-hard bird enthusiast who prayed that Pelican Island would someday become a sanctuary. Chapman called Ma Latham a "born naturalist" overflowing with "great enthusiasm and energy" for saving wild Florida.[30] "To me she was a combination of mother and guide," Chapman recalled, "and when . . . my search for *Neofiber* [round-tailed muskrat] was rewarded I believe that her pleasure and excitement equaled my own . . . I never lacked for a sharer of my joys."[31] Salty, no-nonsense, and razor smart, Ma Latham often collected sea turtle eggs to give to herpetologists; once, she collected a full series of loggerhead embryos acquired on daily seashore walks for sixty days.[32] She was among the first U.S. naturalists to truly study the egg-burying habits of loggerheads along the east coast of Florida in what is now the Archie Carr National Wildlife Refuge. With her sun-wrinkled face and no-frills pioneer-style dresses, she epitomized hardscrabble Florida slowly entering the automobile age. Turning her back on the Florida gold rush for feathers and eggs, she challenged all plumers and eggers with a cold glare that caused them to cast their eyes downward in guilt.

By the time Roosevelt was president, Oak Lodge had become a way station for Ivy League–trained scientists, naturalists, and botanists enthralled to find more than 2,000 varieties of plants just a short walk away. After an arduous day of collecting, the wildlife lovers would retreat to the lodge at dusk to watch the sun set over the Indian River Lagoon. Over drinks Ma Latham regaled Chapman and the visiting naturalists with offbeat stories about Florida panthers and black bears, which roamed beaches looking for sea turtle eggs to dig up. Abhorrence, however, came easily to Ma Latham when she thought nature was being violated. For example, she strongly opposed haul-seining, a destructive fishing method in which dories were launched into the Atlantic surf with a long net that was then dragged onto the beach by oxen harnessed to a rope. After such intensive labor it was bounty time for the fishermen. However, Ma Latham believed that such unrestricted harvesting would eventually wipe out the tarpon, red snapper, and other fish species. Over time Chapman had learned to love everything about Ma Latham, as did other distinguished New York conservationists such as William Dutcher, William Beebe, Arthur Cleveland Bent, George Shiras III, Outram Bangs, John Burroughs, Louis Agassiz Fuertes, William T. Hornaday, Herbert K. Job, and Abbott Thayer.[33]

So it was that in 1900, sensing the main chance to save Pelican Island, Kroegel, a friend of Ma Latham, met with Chapman. She had sent word to Kroegel by boat mail that the great New York ornithologist had arrived. It was just over six miles from Sebastian to Micco, and Kroegel made the trek in record time. The meeting apparently went exceedingly well. For all of Chapman's urbane book knowledge of birds, Kroegel actually *lived* amid the cornucopia of rookeries year-round. As a field naturalist Kroegel had studied pelicans longer and harder than Chapman. For obvious reasons Chapman was deeply impressed with the self-taught Kroegel. Certainly the former Wall Street financier and the swamp accordionist weren't cut from the same socioeconomic cloth. But their shared love of birds made them a formidable united front. Together they constituted a sort of two-man Rough Riders cavalry unit—one from the backwoods, the other from the eastern elite—both determined to save the Pelican Island ecosystem. One can imagine them sitting in the amethyst twilight at Oak Lodge among myrtle oaks—a roaring campfire serving as a mosquito repellent—strategizing about how to save the little rookery from the marauders.

Years before Chapman had met Paul he had, in a sense, been a plume hunter himself (albeit for science). In 1898, for example, he took his new wife, Fanny, to honeymoon on Pelican Island. Other New York dandies may have traveled to Niagara Falls or Bermuda on such an occasion, but Chapman (using Oak Lodge as headquarters) went with Fanny to the Indian River Lagoon to shoot and skin pelicans for the American Museum of Natural History. It turned out to be an inspired choice: the Chapmans marveled at the teeming seabird colony they encountered on that beloved lump of mud and mangrove. Late in life Chapman, by then a well-traveled naturalist, wrote that Pelican Island was "by far the most fascinating place it has ever been my fortune to see in the world of birds."[34]

Now, two years after his honeymoon, Chapman had returned to Pelican Island and discovered a 14 percent decline in pelicans. This troubled him greatly. Listening to Kroegel tell about his difficulties protecting the islet, Chapman grew indignant. He recognized that the "Feather Wars" were being fought, and that Kroegel was actually risking his life daily on behalf of the Florida Audubon Society. Then and there Chapman decided that enough was enough. He was now prepared to take the "Feathers War" directly to Roosevelt in the White House. Intuitively Chapman knew that his friend Roosevelt would immediately approve Kroegel's gun, boat, and "badge," sponsored by the Audubon Society. Roosevelt would want to shut the plumers' operation down. Like the cowboys and ranchers Roosevelt admired in the Dakota Territory and the Rough Riders he led into battle—and like what Roosevelt fancied himself to be—Kroegel was a steely, live-off-the-land, never-say-die lover of wildlife. It was as if "the Virginian" had arrived in Vero Beach.[35]

But Chapman was too wise to pester Roosevelt without first having a sensible game plan. He knew he needed to start with his fellow AOU members William Dutcher and Theodore Palmer. Dutcher was chairman of the AOU Bird Protection Committee and Palmer was assistant chief of the Division of Biological Survey. Both men were instrumental in advocating for bird protection throughout the United States. In fact, they were successful in helping persuade twenty-three states, including Florida in 1901, to pass the AOU model law protecting nongame birds. Florida's new law enabled the AOU to begin employing wardens to enforce bird protection. Even before this new law was passed, Chapman understood that at Pelican Island more wildlife protection would be needed. He urged Dutcher and Palmer to investigate the possibility of purchasing Pelican Island. A year later, in April 1902, Dutcher hired Paul Kroegel as one of the new Audubon wardens, on the recommendation of Ma Latham.[36] And the AOU had a cadastral survey of Pelican Island drawn up by J. O. Fries in July 1902.[37]

Here was the problem Chapman, Dutcher, and Palmer faced in trying to save Pelican Island: since the federal Lacey Act passed in 1900 and the AOU model law passed in Florida in 1901, the AOU had failed to purchase Pelican Island with Thayer Fund monies, as a result of a serious technicality. Because Pelican Island was designated "unsurveyed" U.S. government property, the General Land Office couldn't legally authorize its acquisition to create a bird sanctuary. William Dutcher, however, cleverly directed Thayer Fund dollars to commission a survey of Pelican Island acceptable to the General Land Office. He may have been too clever by half. For just as Pelican Island was being officially surveyed—under Dutcher's impetus—the AOU learned that it had opened up Pandora's box: once the General Land Office approved the AOU survey, Pelican Island could then be made available to homesteaders. Free land for all who promised to grow crops or plant grapefruit trees—the worst possible scenario for saving the rookery. Even if the homesteading could somehow be averted, Dutcher feared that the New York millinary industry, with its deep pockets, would purposely outbid him just to spite the bird nuts and stick it to the Audubon Society.[38]

Deeply concerned, Palmer and a fellow AOU member, Frank Bond, asked the Public Surveys Division Chief—Charles L. DuBois—what their options were. It would be suicidal to get into a real estate bidding war against the millinary lobby or try to gain an exemption from homesteading. Wisely, DuBois suggested to them that Pelican Island could legally be made a so-called government reservation by executive order of the president of the United States. The next day Palmer sent that message to Dutcher, in a letter dated February 21, 1903. Palmer urged Dutcher to immediately write to the secretary of agriculture requesting that Pelican Island be set apart as a government reservation.

This would, of course, be unprecedented, but it could nevertheless be done. What a helpful suggestion for DuBois to make! If there was one thing Roosevelt loved, it was setting precedents. You can almost see a cartoon of Dutcher and Chapman thinking "Eureka!" All they needed, to procure Pelican Island for posterity, was (1) to schedule a meeting with Theodore Roosevelt and (2) to convince him on the idea of a bird reservation. Knowing of Roosevelt's insistence that wildlife protection wasn't possible without police protection, Chapman now had Paul Kroegel to present as the ideal warden, a counterpart of Captain Anderson at Yellowstone, Ranger Warford in Arizona, or Seth Bullock roaming the Black Hills.[39]

V

Chapman and Dutcher set up the March 1903 meeting at the White House. And, as noted in the Prologue, Roosevelt handed them their reservation on a silver platter. With little more than a wave of the hand, Pelican Island was established as a federal bird reservation by the president's "I So Declare It." This was a revolutionary moment for biological conservation. Throughout America in 1903 land *was* being set aside for wildlife; but it was for private game preserves. The Biltmore estate in Asheville, North Carolina, for example, had sequestered a pristine 100,000-acre forested preserve, and a resort hotel in Virginia saved 10,000 acres along the Chickahominy River for fishing and hunting. The U.S. Department of Agriculture, in fact, in its 1903 *Yearbook*, indicated that more and more large tracts of wilderness were being sold on the private market to the highest bidder. Pelican Island, the department noted, was an anomaly.[40]

Roosevelt's initial "I So Declare It"—instituted through the Department of Agriculture's U.S. Biological Survey division (or, more simply, "Dr. Merriam's shop")—wasn't difficult to establish. It slipped by essentially unnoticed among reams of government appropriations and bills. Immediately, Roosevelt wanted to know the next steps needed to protect Pelican Island's wildlife. Could Kroegel manage to protect the rookeries in Indian River Lagoon on his own? What other bird sanctuaries needed saving in Florida and elsewhere? These were the kind of probing questions Roosevelt wanted to ask Chapman and Dutcher. Appropriately, the ornithologists, their spirits high, pondered the president's questions and answered them directly. Breeding grounds in Louisiana and North Dakota were high on their list. Both unofficial advisers believed that only lots of game wardens could curtail the relentless slaughter of birds in Florida. Wildlife needed paid guards to protect it from marauders. At Yellowstone National Park in July 1902, Colonel Charles J. ("Buffalo") Jones had been appointed game warden—the first in U.S. governmental history. Now, Kroegel joined him as number two. (And Kroegel was the first on behalf of birds.) Hiring wardens like Jones and Kroegel was ideally suited to Roosevelt's innate "sheriff" temperament. As a law enforcement zealot the president liked to brag that he'd personally track down and shackle bird-killing scoundrels himself, if necessary, to send a broad message throughout Florida that there was a new management in town.

Immediately, Roosevelt appointed Kroegel as his first national wildlife refuge warden for Pelican Island. In a U.S. Department of Agriculture letter dated March 24, 1903, Kroegel was put "in charge" of the rookery effective April 1. He would report directly to Merriam.[41] Roosevelt was going to organize the Biological Survey as his special force on behalf of wildlife protection. "Paul was a convincing person," his granddaughter Janice Kroegel Timinsky recalled. "By the time Theodore Roosevelt was president when he told these hunters to flee he was pure intimidation. It helped him psychologically, I think, to have the Audubon Society on his side. Don't get me wrong—he didn't change. He was still kind of silent. He didn't like shouting or yelling. But after the Lacey Act, and with President Roosevelt in charge, they knew Grandpa wasn't bluffing when he pointed his gun."[42]

There was, however, a hiccup. Because the Department of Agriculture had no money earmarked for wardens, the Audubon Society stepped in and paid for Kroegel's modest salary: one dollar a month. (A couple of years later the Department of Agriculture gave Kroegel a substantial raise, to twelve dollars a month.[43]) But Paul never balked at the low salary. He was now Warden Kroegel. That's what mattered. Although he had voluntarily protected wildlife on Pelican Island for years, Kroegel now held the distinction of being America's first "refuge manager." On April 28, 1903, Dutcher wrote Kroegel with a laundry list of federal instructions, noting that at all costs he was to "prevent the killing of wild birds or taking of their eggs," and adding that any violations of President Roosevelt's executive order should be "reported at once."[44]

Word had been delivered loud and clear: President Roosevelt wanted the plumers and eggers flushed. Following Roosevelt's executive order a ferocious federal crackdown on market hunters rocked Florida. Roosevelt actually relished using the rule of law to incarcerate all the plumers and eggers who could be found. "His sense of right and duty was as inflexible as adamant," John Burroughs wrote in his diary. "Politicians found him a hard customer."[45]

When Kroegel heard that Pelican Island had become a federal bird reservation he matter-of-factly lit his pipe to celebrate. His son Rodney later said he was in a kind of controlled stupor, half-imagining that the Pelican Island decree was a parlor trick from Washington, bound to be revoked. There was, however, no hocus-pocus. The opening salvo in the crusade to save American wildlife had been fired by President Roosevelt. When Warden Kroegel now pointed a rifle muzzle at a pelican poacher, he was doing so with the full authority of the president of the United States. Once again, Roosevelt's genius as a conservationist was that he never listened to other politicians about how to get things done. His instinct was always to turn to the professional biologists, foresters, and field naturalists first. He always consulted with Darwin-minded men like Chapman, Dutcher, or Pinchot and then acted. Once the biological imperative was established he engaged the rough-and-ready outback types like Kroegel. Over and over again, this was the formula Roosevelt used to eventually set aside more than 234 million acres of America for posterity.

With reserved pride, and a sense of genuine responsibility, as of April 1 Warden Kroegel proudly flew a huge American flag—which the USDA had sent him—on a twelve-foot pole at the end of his long dock at Indian River Lagoon. The instructions from the Roosevelt administration had been for Kroegel to put the gigantic flag on the island itself as an unmistakable federal warning to all encroachers. But Kroegel worried that the bright red-white-and-blue flag might scare away the birds. So the flimsy wooden dock it was. The Kroegel homestead now had the look of a U.S. Coast Guard customhouse. "Folks would be more inclined to not mess with the birds if they saw that flag," his granddaughter Janice Kroegel Timinsky recalled. "It was his way of saying, 'Don't tread here.' "[46] The flag also served as a sentinel. When boats sailed past, they would invariably give a patriotic salute by sounding their horns. This alerted Kroegel to head off potential vandals.

Just a couple of months after saving Pelican Island as the first federal bird reservation, Roosevelt decided that he should pay a long-overdue courtesy visit to the author of *Florida Game Birds* at Lotus Lake on Long Island's South Shore. Bringing his oldest boy, Ted, who was fifteen, with him (along with his cousin Emlyn Roosevelt's kids), Roosevelt had much to talk to Uncle Rob about. Together they went bird-watching and drove an automobile around Sayville reminiscing long into the night. Environmentally, two of the states R.B.R. had fought hardest to protect—New York and Florida—were now safeguarded by his nephew. The Roosevelts—Robert and Theodore—were protectors of wild Florida at a time when most rich Americans were searching for development dollars from the coastal state with year-round warm weather.

VI

Kroegel now wore a badge issued by Roosevelt and had framed his diploma-like appointment letter with its raised seal (both courtesy of the Department of Agriculture). But this didn't mean the Feather Wars of Florida were over in the spring of 1903. For starters, Pelican Island had decades before been shot out by plumer gangs. Seeking roseate spoonbill feathers in full spring color, they had sprayed bullets in every direction, leaving shattered bone and guts strewn about the mangrove islet. Not only did the adult roseate spoonbills die, but the young chicks, suddenly parentless, perished too. Essentially, two whole generations of these birds, among other unlucky species, were simultaneously wiped out.

Compounding Kroegel's problem was the fact that plumers and eggers soon owned newfangled motorboats. Sailboats—even fine ones built by Kroegel—simply couldn't keep up with a vessel that could move at forty or fifty miles per hour. Recognizing the disparity in speed, the Florida Audubon Society raised $300 for Kroegel to build a seaworthy twenty-three-foot-long boat fitted with a three-horsepower engine. The power boating era had truly arrived in the Indian River, and the Audubon Society wasn't going to concede the technology edge to the opposition. The vessel commissioned by the Audubon Society was ideal for tropical seas, rain squalls, and storm-swept distances. Fueled by naphtha, an easily flammable oil product, the motorized *Audubon* was operative for warden-guide patrols around Florida's tidal flats and mangrove keys a year before Roosevelt's "I So Declare It" decree. On July 15, 1902, William Dutcher sent Kroegel a telegram expressing an immediately pressing need for a motorboat.[47]

Guy Bradley, a former plume hunter, now a bird protector, was in desperate need of the motorized *Audubon* down in the Everglades. The AOU had just made Bradley a warden, too. He was something of an Everglades yokel, and his primary responsibilities as warden centered on the islands off Cape Sable—a watery prairie ecosystem at the southernmost point on the U.S. mainland. Here, in the shallow turtle grass flats and marshlands, the great white heron—a swan-white relative of the great blue heron—was making an impressive last stand. Nearly 500 nests of these rare birds had been counted, and there were probably many more.[48] The AOU wanted them protected.

Kroegel rendezvoused with Bradley in Miami and turned over the stout little motorboat *Audubon*. The Audubon Society would pay any out-of-pocket costs incurred. Because Bradley kept a diary (even though it was irregular), we know that the handoff of the boat took place, albeit with a lot of hitches. Kroegel was a superior boatbuilder, but his knowledge of naphtha-fueled motors was very limited. As a result, the *Audubon* broke down after a relatively untaxing 230-mile trip. The outboard motor had seized, and the boat was dry-docked to work out the kinks.[49]

By September 1903, the *Audubon*, at long last, was in tip-top shape and Bradley began chasing plume hunters throughout the Ten Thousand Islands around the cape, venturing up the Shark and Rogers rivers, among others. The state authority made him feel empowered and reverent. After dropping a big mushroom anchor and running a stern line to a mangrove, Bradley, like a Wild West sheriff, would post "No Trespassing" and "Do Not Disturb the Birds" signs on every clam shack or fishing camp he encountered. Often his day companions were the peregrine falcons and bald eagles which coursed the savanna searching for small prey. The ospreys and terns hovered around porpoise schools doing the same. The only hamlets in this strange backcountry region—Flamingo, Chokoloskee, and Cape Sable—each had a sleepy population of roughly fifty people. Still, his flyers didn't go unnoticed by locals.

By all accounts Bradley made impressive inroads patrolling the Everglades–Florida Bay–Cape Sable areas: a dizzying complex of freshwater sloughs, sapling thickets, cane fields, sawgrass ridges, and tree islands.[50] The tides, which lashed like a hurricane, made much of Cape Sable appear arid, and the saline soil stunted the growth of hardwood hammocks. The steady sunshine Bradley encountered was often blinding, even unearthly. There were no gentle slopes or saddle ridges in this flat, forbidding landscape. But birdlife abounded. "Citizen bird" enthusiasts, for example, traveled from faraway Boston and New York to see the Cape Sable seaside sparrow, a rare creature that Bradley encountered nearly every day on his beat. Life was good for Bradley, a man who enjoyed being outdoors. Catching silver fish for dinner from the *Audubon* in the quiet evening air—particularly kingfish or mackerel—he lived well off Florida's natural bounty. "Guy's first year as warden had been a busy one," his biographer Stuart B. McIver wrote. "That year the price offered to hunters for egret plumes rose to thirty-two dollars an ounce, more than twice the price for an ounce of gold. Four egrets had to die to yield an ounce of plumes. Bradley's vigilance had helped create a scarcity that was driving the price up—and ultimately making the rookeries all the more tempting to plume hunters."[51]

Excited that the one-two punch of wardens Kroegel and Bradley was producing constructive results in Florida, Dutcher wrote an AOU report in late 1903 claiming that the tide had turned in the good guys' favor. The document could be summed up in two words: imminent victory. Clearly, Dutcher understood that slight disturbances still occurred around Cape Sable, but the systematic avian slaughter (he insisted) had ceased. President Roosevelt was elated. Nothing could please him more than the fact that the U.S. Biological Survey and AOU were starting to seize control of precious rookeries to save them for posterity. Dutcher's enthusiasm, however, was very premature. Bradley—because of the sheer geographical magnitude of his beat—found himself stretched thin and doing the job of ten men.

Chapman—erroneously believing Bradley had taken South Florida from the plumers and eggers around the lower keys, as Dutcher had claimed—journeyed to Florida with his wife to witness the supposedly rejuvenated flocks at the Cuthbert Rookery. Unbeknownst to Bradley, however, either the Smith gang or the "Uncle Steve Boys" (vicious plumer organizations) had been eyeing Cuthbert for a broad-daylight strike. One misbegotten afternoon, when the *Audubon* was nowhere in sight, one of the bandit gangs pillaged the startled rookery, turning the avian nursery into a bloody slaughterhouse. Guns blazed nonstop. Before long the island was strewn with corpses of egrets and great white herons. Cuthbert Rookery has been "shot out," a deeply embarrassed Bradley told the Chapmans upon their arrival to Florida. "You could-a-walked right around the ruke-ry on them birds' bodies, between four and five hundred of 'em."[52]

This news, and some investigative sleuthing of his own, made Chapman fear that the Biological Survey wardens, for no fault of their own, were in a precarious situation. The vast Florida Bay waterways Bradley was being asked to protect were impossible to patrol properly without a motorized fleet. President Roosevelt's warden needed reinforcements and supplies (or at least something more intimidating than a .32-caliber pistol and a single outboard motor). "That man Bradley is going to be killed some time," Chapman wrote in his journal. "He has been shot at more than once, and some day they are going to get him."[53]

Chapman's diary proved prophetic. On the morning of July 8, 1905, Bradley heard rolling gunfire at Oyster Keys rookery. Using binoculars, he could see the familiar blue boat the Smith gang often used for conducting raids, although it was difficult to see through the powder smoke. Hopping into a dinghy, Bradley rowed out toward the tiny island, determined to stop the killings of birds. Stupefied with anger, Bradley quite simply wasn't going to suffer another embarrassing massacre on his watch. The Smith gang, it turned out, ignoring the Roosevelt administration's admonitions, was murdering double-crested cormorants by emptying magazines into their breeding grounds.

Like all colonial birds, double-crested cormorants congregated on islands close to shore. Easily detectable even by a novice bird-watcher because of their shiny black and bronze plumage, cormorants were considered a nuisance by fisherfolk. In the spring and summer many of these long-necked aquatic birds nested along Florida's coast, while others migrated southward to Florida from more northern nesting grounds. What Roosevelt found fascinating about cormorants—the trait which gave him the most delight—was the way they dived and remained submerged for a long time. Scanning underwater for fish or shrimp, they seldom reappeared with an empty beak. As a Darwinian naturalist he was deeply intrigued that the bird had adapted to underwater life so strikingly.

Roosevelt worried that the disreputable plumers of Florida—"sordid bird-butchers" he later called them in his postpresidential *A Book-Lover's Holidays in the Open*—were trying to exterminate the double-crested cormorants just as they did the brown pelicans.[54] Fishermen, he knew, were worried about depleted shellfish harvesting areas and saw cormorants as competition. The future of cormorants, he believed, was imperiled. If federal intervention didn't occur, they were headed toward near-extinction. Proactive measures had to be taken quickly. Consulting with ornithologists like Chapman and Dutcher, Roosevelt made it clear that he wanted cormorants protected, even if the U.S. Biological Survey had to hire more wardens quickly.

Somewhat unsteadily, Bradley approached the Smith gang at Oyster Keys, demanding that they drop their guns. He was the law and had come there to make arrests. From the moment he spoke, he was greeted with resistance. A quarrel ensued over whether an arrest warrant was neces-

Warden Guy Bradley was murdered in Florida for trying to protect bird rookeries from plume hunters.

sary on the waterways; meanwhile, wounded birds, in a frenzy, let out a terrified chant. The initial tension heightened to ferocity. Harsh words were spoken. As the dispute intensified, a sharpshooter in the Smith gang suddenly shot Bradley in the chest, as if he had been wearing a bull's-eye on his work shirt. "He never knew what hit him," Walter Smith, head of the gang, the murderer, later told the police. Bradley slumped forward in the bow, bleeding profusely, motionless. His dinghy drifted westward in the slate-gray water. It journeyed over a reef, away from Oyster Keys and out to sea. The corpse of Bradley disappeared into the distant horizon as the Smith gang stood and watched from the shore.[55] Bradley had died a martyr in the line of duty, murdered trying to stop an outlaw plumer gang.[56]

The National Association of Audubon Societies (NAAS) immediately protested the cold-blooded murder of Guy Bradley, to draw even more attention to the menace of Florida's rookery killers. Outraged, Roosevelt predictably promised not to cower or retreat in the face of the murder. Instead, he appointed more Department of Agriculture wardens in Florida (in a collaborative venture with the Audubon Society) and grew even more determined to create federal bird reservations (U.S. wildlife refuges) to protect cormorants, pelicans, herons, egrets, and other nongame birds. His belief in the Audubon Society's mission, in fact, now increased tenfold. "Permit me on behalf of both Mrs. Roosevelt and myself to say how heartily we sympathize not only with the work of the Audubon Societies generally, but particularly in their efforts to stop the sale and use of the so-called 'Aigrettes'—the plume of white herons," Roosevelt wrote to Dutcher. "If anything, Mrs. Roosevelt feels more strongly than I do in the matter."[57]

Recognizing that the concept of federal bird reservations was the best weapon against pluming, Dutcher staked NAAS's future on creating sanctuaries like Pelican Island across America. Anger over Bradley's death spun the Feather Wars plot. "If the National Association did no other work than to secure Bird Reservations and to guard them during the breeding season," he said, "its existence would be fully warranted."[58]

VII

There is no clear written record of how Paul Kroegel took the murder of Guy Bradley. All we know is that he retrieved the *Audubon* and continued to patrol Pelican Island in the boat he had built for Bradley. Flushed and confident in 1903 he boated out to Pelican Island with his aged father, Gotlobb, and posted two huge signs on the edge of Pelican Island, as instructed: "No Trespassing: U.S. Government Property." They hoped these signs would deter plumers and others who would willfully or unknowingly harm the birds. Unfortunately, the huge signs had a deleterious effect on Pelican Island's wildlife. In November–December 1903, the first winter after President Roosevelt's "I So Declare It" decree, the birds abandoned the rookeries—they just didn't show up. Pelican Island was an avian ghost town, with only three or four ruffled vultures poking around the mudflat. It turned out that the signs had intimidated the pelicans, preventing them from landing. Recognizing the mistake and determined to lure the leery birds back, Kroegel, with help from his father and with the concurrence of Frank Chapman, dismantled the billboards in 1904. And just like that, the pelicans returned.

Meanwhile, Chapman returned to Pelican Island in the spring of 1904, 1905, 1908, and 1914. True to form, he kept detailed records of the rookery and reported his findings directly back to Roosevelt, with a professional air.[59] Chapman's elegant black-and-white pictures from those sojourns, ideal for lantern-slide presentations, constituted a high-water mark of nature photography during the progressive era. Emotionally invested in Pelican Island, Chapman was thrilled to learn that his friend Kroegel was still fearless, issuing citations although less frequently pointing his rifle at would-be encroachers. No longer was Kroegel viewed as a bird kook in Sebastian; after all, he was working for none other than President Roosevelt. In 1905, in fact, Kroegel's local status took another leap upward when he was appointed county commissioner of the new Saint Lucie County by Governor Napoleon Bonaparte Broward. He held the office for the next fifteen years.

And Roosevelt continued pushing his agenda in Florida. One place in particular, Passage Key, seemed to have taken hold of Roosevelt's imagination most firmly. Located offshore from Saint Petersburg, at the mouth of Tampa Bay, reachable only by boat, the Passage Key mangrove rookery had the largest nesting colonies of royal terns and sandwich terns in the entire state. There were so many whitish birds on the island that from above they looked like flocks of sheep corralled for market. Although royal and sandwich terns are difficult to distinguish from each other, royal terns are slightly larger and plumper, with an orange bill instead of a black one (yellow-tipped). Trained ornithologists like Roosevelt could also differentiate between them by the sounds they made. A royal tern made a shrill, rolling "keer-reet" whereas a sandwich tern went "kirr-ick." Both species, however, were known for their wild chirrups when in distress.

When he was based in Tampa Bay in 1898, Roosevelt had grown fond of these terns. In the humid, stifling heat he had watched them fly over the bay with bills pointed downward, plunging into the water for black mullets, gray anchovies, and brown and white shrimp. Now, as president, he had an opportunity to do something permanent to help these pelagic birds survive in the Gulf of Mexico region. Because schools of blackfin and yellowfin tuna were thick around Passage Key, as were blue crabs, Roosevelt feared it might be only a matter of time before the pristine island became a fishing camp; another fear was that it might become a military base. No longer would it look like a deserted tropical orchard—it would be developed. As the gateway island to Tampa Bay, visible with binoculars from both Saint Petersburg (to the north) and Sarasota (to the south), Passage Key was like a natural Statue of Liberty, welcoming seafarers to shore; it was similar in this regard to the Farallon Islands near San Francisco Bay, or to Gibraltar in Spain. If the west coast birds of Florida were to be saved, Passage Key was a fine starting point.

On October 10, 1905, nineteen months after the designation of Pelican Island as a federal reserve, Roosevelt declared Passage Key a federal bird reservation. Signing this executive order whetted his appetite for more preservationist mandates. Not satisfied with having created two biologically intact wonderlands in Florida—Pelican Island (which was enlarged on January 26, 1909)* and Passage Key—Roosevelt asked Chapman, around Thanksgiving 1905, to report back to him on other possible locales in need of preservation.[60] Bit by bit he would save America's finest bird rookeries from molestation. Egrets, herons, pelicans, and dozens of other species could continue being masters of these universes. Before long, the Biological Survey was bombarded with information about ecosystems worthy of federal consideration. Roosevelt was hoping to establish refuges down the entire west coast of Florida. He imagined these sanctuaries as rather like a string of natural pearls dangling downward toward the Caribbean. These new federal bird reservations—which would become "national wildlife refuges" in 1942—were created to demonstrate the Rooseveltian wildlife protection strategy of no surrender, no retreat in Florida.

*The Pelican Island refuge was increased to 616 acres; and in 1968 403 acres more were added to the refuge. As of 2009 it encompassed 4,359 acres of mangrove islands and bottomland in the Indian River (some of which was under lease from the state of Florida).

CHAPTER NINETEEN

PASSPORTS TO THE PARKS: YELLOWSTONE, THE GRAND CANYON, AND YOSEMITE

I

While Pelican Island was being saved as a federal bird reservation in the first months of 1903, President Roosevelt was making last-minute adjustments for a visit to Yellowstone, the Grand Canyon, and Yosemite. The Great Loop tour, as it was called, would be the longest, most elaborate cross-country journey ever taken by a president of the United States. The trek served as an appealing way to present his conservation polices to all regions before the 1904 presidential election. Emphasizing America's natural wonders, the adventure crystallized Roosevelt's already potent belief that the Far West, in all its wildness and rawness, was the least exhausted part of the country. At that time, Yellowstone was interested in promoting popular animals such as elks and bears, while applying a policy of predator control to cougars, wolves, and coyotes. Eager to sneak in some cougar hunting around Yellowstone on the western odyssey, Roosevelt corresponded intensely with the superintendent, Major John Pitcher, about having the proper hunting dogs available for him upon arrival and securing a special U.S. government permit. Wary of repeating the disastrous press coverage of the Mississippi bear hunt, which had been mitigated only by the grace of a cartoonist named Berryman, Roosevelt emphasized that no detail of the itinerary be left to chance. "I am still wholly at sea [as] to whether I can take that trip or not," Roosevelt wrote to Pitcher. "Secretary Root is afraid that a false impression might get out if I killed anything, as of course would be the case, strictly under park regulations and though it was only a mountain lion—that is, an animal of the kind you are endeavoring to thin out."[1]

With unaccustomed suspiciousness, the president surreptitiously asked the secretary of the interior, Hitchcock, to quietly smuggle into Yellowstone three hunting dogs from John Goff's kennel in Colorado. Plotting eight to ten days of clandestine cougar hunting, Roosevelt wrote to Pitcher that if word leaked out, if the reporters discovered his intentions, then he would have to content himself by studying "the game and going about on horseback, or if I get into trim, perhaps snowshoes."[2] If Roosevelt had his way, however, at least a few of the troublesome mountain lions wouldn't get within sniffing distance of an elk or antelope. Meanwhile, the competitive Charles "Buffalo" Jones, who had a bounty hunter mind-set and did not want his role as an exterminator of predators to be co-opted by an-out-of-stater like Goff, imported into Yellowstone two lots of six cougar hounds from Aledo, Texas.[3] As a maxim of the T.R. era went, never pass up a chance to hunt or box or romp with the president, because these activities fostered a lifelong bond.

Meanwhile, in preparation for seeing the Pacific Northwest and California forestlands for the first time, Roosevelt dashed off a note to Frederick Weyerhaeuser, the "lumber king," known for his entrepreneurial acumen and zealous, ruinous de-timbering. Weyerhaeuser, who slurred his Edwardian w's into Bismarckian v's when speaking English, had emigrated from Germany at the age of eighteen to work as a day laborer in Erie, Pennsylvania. To him, America was the land of promises where the bold prevailed. He learned the lumber business from the ground up. He acquired his first sawmill in Rock Island, Illinois, in 1857, and began building a timber empire in the heavily forested Pacific Northwest. Determined to amass a fortune, he clear-cut every tree in sight without the slightest concern over deforestation. Obviously, he had never read *Man and Nature*. In his blinkered outlook, money mattered more than anything else in life. Where others saw a redwood tree or an old-growth hemlock, Weyerhaeuser saw boards and planks and, behind them, dollar signs.

In early 1903, Congressman Lacey had mentioned to Roosevelt that Weyerhaeuser was starting to come around, that he was becoming a forest reserve advocate of sorts, and that he was an untapped potential arborist. Intrigued, Roosevelt wanted to initiate a dialogue with Weyerhaeuser on the ticklish issue of reduced logging, and on the conservationist ethics of Southern Lumber Company: planting a tree for every one chopped down. "Could you come down here sometime next week so I can see you with Mr. Gifford Pinchot?" Roosevelt wrote to Weyerhaeuser on March 5. "I should like to talk over some forestry matters with a practical lumberman. I earnestly desire that the movement for the preservation of the forests shall come from the lumbermen themselves."[4]

When Roosevelt wrote to John Burroughs about his forthcoming Great Loop trip that March, however, the letter was devoid of cougar hunting in Yellowstone, or courting timber barons from Minnesota. Instead, Roosevelt said he wanted to "see," in liberal measure, the elk herds and mountain goats. And he was eager to see the geysers in winter. This was slightly disingenuous of Roosevelt, and further evidence of how skittish the fiasco of the Mississippi bear hunt had made him. Somewhat defensively, however, Roosevelt raised the specter of hunting in his long letter to Burroughs. The novelist Charles Dudley Warner, best remembered for coining the term "gilded age" in a novel of that title cowritten with Mark Twain, had also written an American outdoors classic in the 1870s: *The Hunter of the Deer and Other Essays.* Because Warner—a longtime columnist for *Harper's Magazine* and first President of the National Institute of Arts and Letters—had died in 1900, many of his earlier literary efforts were being reissued in his memory. When Burroughs told Roosevelt how excellent the Warner hunting essays were, the president unctuously doused his enthusiasm. "I think you praise overmuch for its fidelity to life Charles Dudley Warner's admirable little tract on deer hunting," Roosevelt wrote. "[It] was an excellent little tract against summer hunting and the killing of does when the fawns are young. It is not an argument against hunting generally, for as Nature is organized, to remove all checks to the multiplication of a species merely means that every multiplication itself in a few years operates as a most disastrous check by producing an epidemic of disease or starvation."[5]

What triggered Roosevelt's letter was a hard-hitting article that Burroughs had recently written in the *Atlantic Monthly*, "Real and Sham Natural History." Twenty years before, Burroughs had pleaded in *Scribner's Monthly* that poets stop depicting wildlife falsely; they should instead follow the romantic Whitman's fine naturalist example.[6] Now the usually mild-mannered Burroughs lit into Ernest Thompson Seton for completely fabricating bizarre species behaviors in *Wild Animals I Have Known*: Seton had claimed that foxes contemplated suicide, and that deer had sensitive humanlike feelings. What perturbed Burroughs was that Seton advertised his encounters with animals as nonfiction. This purported zoology book wasn't natural history but *fable*. "There are no stories of animal intelligence and cunning on record that match his," Burroughs wrote. "Such dogs, wolves, foxes, rabbits, mustangs, crows, as he has known, it is safe to say, no other person in the world has ever known."[7]

Roosevelt, on reading this article, hurled himself into the fray. On the Executive Mansion's letterhead, Roosevelt wrote Seton a punishing note, haranguing him and challenging him to cough up his facts for the Biological Survey.[8] "Burroughs and the people at large don't know how many facts you have back of your stories," Roosevelt wrote to Seton. "You must publish your facts."[9] Seton, wanting no part of the "bully-boy" Roosevelt (at least in public), wisely stayed mute. Burroughs actually began to feel some pity for Seton, whose ego was shattered and whose literary reputation was taking a pummeling. And Seton, to his credit, took Burroughs and Roosevelt's charges to heart; he later published the scrupulously accurate *Life Histories of Northern Animals* and *Lives of Game Animals*. Both are fine, highly readable zoological contributions to the American naturalist canon.[10] Eventually, after bumping into Seton at an ostentatious literary party hosted by Andrew Carnegie, Burroughs decided he *liked* Seton. "He behaved finely and asked to sit next to me at dinner," Burroughs wrote to his son. "He quite won my heart."[11]

Another of Burroughs's "nature fakers" was Reverend William Long, who had an advanced degree in divinity from Heidelberg University in Germany. Long, who believed in the power of pets to heal the tormented soul, was a popular pastor at First Congregational Church in Stamford, Connecticut, and the author of some celebrated children's books, including *Ways of Wood Folk* and *Fowls in the Air*. A zealous anti-Darwinist, he held that animals' knowledge was based on parental training within each species. Picking apart Long's false descriptions about how robins feed their young, Burroughs called the pastor a deliberate liar. And robins were the least of it. Long also claimed to have seen a woodcock mend its broken leg by making a clay and grass cast for itself. "Why should anyone palm off such stuff on an unsuspecting public as veritable natural history?" Burroughs asked in the *Atlantic Monthly*. "When a man, writing or speaking of his own experience, says without qualification that he has seen a thing, we are expected to take him at his word."[12]

Long came to his own defense that May in the *North American Review*, arguing that wildlife observation was inexact and diverse enough that no one man—not even *le grand* John Burroughs—had the status to say what was true or false with such arrogant definitiveness. Roosevelt was incensed beyond words by this rejoinder. He wanted to go to war, publicly, with both Seton and Long. He wrote to Burroughs a defense of Rudyard Kipling's *The Jungle Books*, which were labeled correctly as fiction. And he launched into an academic discussion with Burroughs that included his views on domestic dogs first introduced to a Pacific Island; differences between the Falkland and the arctic fox; polar bears versus black bears; and Lewis and Clark meeting grizzlies. Roosevelt wanted to convey his support for his friend Oom John in this controversy. He ended the letter by asking Burroughs to accompany him to Yellowstone in April. "I would see," the president wrote, "that you endured neither fatigue or hardship."[13]

Even though the idea of killing cougars to protect elks had a certain merit, it was a biologically naive practice, and a backward view of the predators' role in the ecological order. Furthermore, Roosevelt was right to be concerned about damaging his reputation by hunting *anything* in Yellowstone. He himself viewed hunting as a pastime of great natural and spiritual value, but an increasing number of Americans saw it as a violent vacation. Moreover, the notion of the commander in chief as exterminator in chief at Yellowstone wasn't going to play well with the president's newfound "teddy bear" constituency.[14] Realizing that the cougars weren't a dire threat, Roosevelt decided that "the elk were evidently too numerous for the feed," and the cougars were not "doing any damage."[15] Suddenly, Roosevelt's philosophy became that the cougars in Yellowstone should be left alone. They were, after all, part of the natural balance. Most likely, Roosevelt had known all along that cougars weren't a real problem in Yellowstone; he had just wanted to hunt them for fun. To his credit, when this rationalization broke down, Roosevelt was intellectually honest enough not to hunt the cougars.

Nevertheless, it was becoming painfully obvious to the naturalist community—especially to those like Muir and Underwood, who respected Roosevelt's Harvard training and his knowledge of species traits and coloration—that the president had a blood lust. For all of his promotion of Kodaks, Roosevelt preferred shooting a rifle to clicking a camera. And the president never really disputed this characterization, though he grew tired of continually having to explain himself to animal-rights types such as William Dean Howells and Mark Twain.

Roosevelt, however, had little difficulty in justifying his seemingly paradoxical attitude toward wildlife. Yes, he revered the mountain lion, yet he thrilled to see one treed by his dogs. Roosevelt reconciled his own proclivities by drawing no distinctions between himself and any other forest predator. Nonhunters could indulge in the fantasy that they existed outside the biotic community, as either passive observers or omniscient masters. And yet, most of them ate meat. Hunters shattered this conceit by participating directly in ecological cycles of life and death. The act of hunting forced a re-reckoning of the relationship between the human and natural worlds, and in that sense it was more intellectually honest than all the bleatings of its vociferous critics. Nevertheless, Roosevelt's penchant for the chase was troublesome to his friends in the preservationist community, and has made him a frustrating and somewhat equivocal figure in much of modern environmental history.[16]

II

On March 15 the *New York Times* announced President Roosevelt's Great Loop tour, with a hint that he might hunt in Montana or Wyoming. From Washington, D.C., Roosevelt would travel to Chicago, Milwaukee, La Crosse, Madison, Minneapolis, Saint Paul, Sioux Falls, Yankton, Mitchell, and Aberdeen before reaching the town of Edgeley in his beloved North Dakota. In Fargo, he planned to deliver a major address on American policy toward the Philippines; it was a gift to his favorite state. On April 7 he would greet well-wishers in Jamestown, Bismarck, Mandan, and then Medora. With Burroughs at his side, he would then take the "Roosevelt Special"—six opulently appointed railway cars provided by the Pennsylvania Railroad—to the northern entrance of Yellowstone, where his party would camp from April 8 to 24. From Wyoming, he would backtrack to Saint Louis to dedicate the Louisiana Purchase Exposition Grounds in commemoration of the centennial of Jefferson's acquisition from France. He would then head to the Grand Canyon and Los Angeles and Yosemite to be with Muir. Next he would make campaign-like appearances in a string of major cities along the Pacific Coast before finally returning to the White House. The conductor for the trip would be Roosevelt's friend William H. Johnson, who had taken him to Mississippi in 1902.[17]

The following day the *Times* announced that John Burroughs had officially accepted the president's offer to accompany him to Yellowstone. What intrigued Burroughs most about Roosevelt was that he evinced "radical Americanism" while being a "thoroughgoing naturalist"—nobody else was doing *that*.[18] Because Oom John wouldn't have tolerated a cougar hunt (it was supposed), the president quashed the idea altogether; he would instead act as naturalist in chief. Yellowstone's superintendent, Major Pitcher, issued a stern statement declaring that the president's gun would be sealed by the U.S. Army when he entered the park, just as with every other citizen.[19] In 1903 the U.S. Army, not the Department of the Interior, ran Yellowstone. "I do not know when the trip will be," Roosevelt said, "but I think it will be just as soon as the Senate adjourns. It is doubtful whether there will be any hunting."[20]

Word that President Roosevelt was headed to Yellowstone with the famous naturalist Burroughs drew a sharp backlash from the pro-timber crowd. Governor DeForest Richards of Wyoming denounced Roosevelt as a crazy forest reserve elitist, claiming that his state would work to undermine the president at the coming national Republican convention. (Richards, however, died a few weeks later of acute kidney disease.[21])

THE PRESIDENT AND MR. BURROUGHS NEAR A GEYSER.
[...]yright, 1903, by *The Illustrated Sporting News.*

Roosevelt and Burroughs together at Yellowstone National Park.

Westerners associated with the railroad industry never forgave Roosevelt for his conservation activism in the 1890s when the Boone and Crockett Club campaigned to make the park off-limits to commercial exploitation. But Senator Clarence Clark of Wyoming quickly came to Roosevelt's defense: the colonel who had led a charge up San Juan Hill in the Spanish-American War would be welcome at Yellowstone anytime, night or day. "The people of Wyoming have the most implicit confidence, not only in President Roosevelt personally, but in the wisdom of his Administration," Clark said. "They believe that he knows them, and has a personal interest in the welfare of the State."[22]

Even before Roosevelt's train left Washington, the newspapers were abuzz with gossip about his working vacation. In Washington state a silly tug-of-war developed over whether the president would speak in Seattle or Tacoma.[23] By contrast, the machinists' union in Kansas City asked that Roosevelt not come near their town, as he had done nothing in the White House to help their cause. A group of Texans lamented that Roosevelt was skipping Fort Worth and Amarillo on his way to the Grand Canyon. While Roosevelt was in Arizona, fifty Rough Riders planned to present him with a live black bear, captured in Mexico.[24] The stockmen of the Front Range of the Rockies announced that when Roosevelt's train arrived in Hugo, Colorado, he would be greeted by 200 cowboys in full range regalia shouting "Bully!"[25] Administrators at Yosemite National

Park planned to shoot fireworks into the night sky on the president's arrival.[26] Bored reporters took the time to calculate the impressive figures for Roosevelt's trip: sixty-six days, 14,000 miles, averaging 212 miles a day. He would deliver more than 260 stump speeches and five major addresses. Roosevelt's party would cross every mainland mountain range between the Poconos and the San Gabriels. "With the exception of a fortnight in the Yellowstone region and a few days in the Yosemite," the *New York Times* noted, "he will be pretty steadily in motion."[27]

As departure day neared, Roosevelt acted like a schoolchild anticipating summer vacation. Elated about visiting twenty-five states and showing Oom John the incomparable geysers of Yellowstone, Roosevelt was the happiest he had been in years, in truly high spirits. "I am overjoyed that you can go," Roosevelt wrote to Burroughs. "When I get to Yosemite I shall spend four days with John Muir. Much though I shall enjoy that, I shall enjoy far more spending the two weeks in the Yellowstone with you. I doubt if there is anywhere else in the world such a stretch of wild country in which the native wild animals have become so tame, and I look forward to being with you when we see the elk, antelope, and mountain sheep at close quarters. Bring pretty warm clothes, but that is all. Everything else will be provided in the park."[28]

III

Uninterested in lobbying Roosevelt for anything, Burroughs had accepted the president's offer as a chance to better educate himself about Yellowstone and Rocky Mountain wildlife. "I had known the President several years before he became famous, and we had had some correspondence on subjects of natural history," Burroughs wrote. "His interest in such themes is always very fresh and keen, and the main motive of his visit to the Park at this time was to see and study in its semi-domesticated condition the great game which he had so often hunted during his ranch days; and he was kind enough to think it would be an additional pleasure to see it with a nature-lover like myself. For my own part, I knew nothing about big game, but I knew there was no man in the country with whom I should so like to see it as Roosevelt."[29]

On April 1 the president's train left Union Station with Burroughs on board, and headed for Pennsylvania's famous Horseshoe Curve—near Altoona—in the heart of the Allegheny Mountains. Just before his departure, Roosevelt had written to Dr. Merriam of the Bureau of Biological Survey asking for up-to-date information about the Shoshones, Sioux, Hopi, Apache, and other western tribes. Roosevelt said he would be most grateful if Merriam could put together a box of books, articles, and reports on Native Americans. The esteemed biologist quickly complied with the president's request. Knowing that Merriam was an Indian rights activist, Roosevelt hoped to educate himself about the shabby conditions on the reservations. "In cases where you can do so without interfering with your biological survey work, I should be glad to have you secure for me reliable information concerning the present condition, necessities, and treatment by Government agents of such Reservation and non-Reservation Indians as you may meet," Roosevelt wrote. "Show this to any Government officials as your warrant for inquiry; I shall expect them to give you all possible facilities to find out the facts deemed of interest to me."[30]

In essence Roosevelt seemed to be offering a quid pro quo to Merriam: Roosevelt's chief biologist would provide him with pertinent information on Native Americans while he, in turn, sent status reports back to Washington, D.C., on the Yellowstone cougars, elks, and buffalo. The reports would contain no skins or claws, however: the president would not be exercising his powder finger. As Burroughs confirmed in his 1905 book *Camping and Tramping with Roosevelt*, the president refrained from hunting in or around Yellowstone, taking only field observations and photographs. That didn't protect Burroughs from being teased by his friends. "The other night I met at dinner that fine old John Burroughs," the novelist and editor William Dean Howells wrote to C. E. Norton that April, "whom I congratulated on his going out to Yellowstone to hold bears for the president to kill."[31]

Burroughs consistently defended Roosevelt as a naturalist first and a hunter second. "Some of our newspapers reported that the President intended to hunt in the Park," Burroughs wrote. "A woman in Vermont wrote me, to protest against the hunting, and hoped I would teach the President to love the animals as much as I did—as if he did not love them much more, because his love is founded upon knowledge, and because they had been a part of his life. She did not know that I was then cherishing the secret hope that I might be allowed to shoot a cougar or bobcat; but this fun did not come to me. The President said, 'I will not fire a gun in the Park; then I shall have no explanations to make.' Yet once I did hear him say in the wilderness, 'I feel as if I ought to keep the camp in meat. I always have.' I regretted that he could not do so on this occasion."[32]

Chicago was a highly successful first stop for President Roosevelt on his way to Yellowstone. More than 6,000 people crammed into a downtown Chicago auditorium that had only 5,000 seats. The Halley's comet known affectionately as Teddy Roosevelt had arrived in the flesh and the

Marconi wire was tap-tap-tapping to the world about his visit. Roosevelt passionately defended American hegemony in the Caribbean and an expanded navy, thunderous applause greeting his words.[33] To Roosevelt the Monroe Doctrine wasn't mere diplomatic ornamentation—it was enforceable policy. He received an honorary doctorate of laws from the University of Chicago, which was rapidly becoming one of the best schools in America.[34] Afterward, Roosevelt met with a cheerful group of its students, who sang a specially composed "Dooleyized" song. One verse went:

> There is a sturdy gent who is known on every hand:
> His smile is like a burst of sun upon a rainy land.
> He'll bluff the Kaiser, shoot a bear, or storm a Spanish fort.
> Then sigh for something else to do and write a book on sport.[35]

Roosevelt's "sport" on this Great Loop tour was conservation, not hunting. Among the primary concerns on the first leg were the devastated buffalo herds. Soon, the Bronx Zoo would have a herd to return to the prairies, and the president was shopping for a proper geographical spot. Carefully, Roosevelt inquired about the remnant herd living in Yellowstone. Working closely with Congressman Lacey, Roosevelt was eventually able to appropriate $15,000 to help manage the Yellowstone herd; the improved management included shelter buildings. With Buffalo Jones updating him on bison grazing strategies, the president grew excited about the prospect of reintroducing his Bronx Zoo herd to an experimental range in Oklahoma. If the reintroduction worked, it might be possible to create bison ranges in Kansas, Nebraska, Montana, South Dakota, and Montana.

While Roosevelt appreciated Buffalo Jones's firsthand knowledge of bison, he believed Jones's report on elks "was all wrong." Jones was a self-serving old reprobate whose opinions on Yellowstone's wildlife were usually far off base.[36] Accordingly, a determined Roosevelt tried to hand-count every elk in Yellowstone. He wanted near exact numbers. His "very careful" estimate was that there were more than 15,000 elks (a figure much higher than what Buffalo Jones was claiming). As for cougars, Roosevelt—after careful study—determined that they were actually providing a service in the park, keeping the elk herds thinned down to ideal numbers. "The cougar are their only enemies," Roosevelt noted, "and in many places these big cats, which are quite numerous, are at this season living purely on elk, killing yearlings and an occasional cow; this does no damage; but around the hot springs the cougar are killing deer, antelope and sheep, and in this neighborhood they should certainly be exterminated."[37]

In coming years Roosevelt would modify his harsh views about predator control in Yellowstone and elsewhere. He began seeing cougars and other predators as assets to the park's natural balance. When word got back to the White House in 1906 that Buffalo Jones planned to hunt sixty-five cougars in Yellowstone, Roosevelt surprised Jones by nixing the idea. "I do not think anymore cougar (mountain lions) should be killed in the park," Roosevelt wrote to the superintendent. "Game is abundant. We want to profit by what has happened in the English preserves, where it proved bad for the grouse itself to kill off all the peregrine falcons and all the other birds of prey. It may be advisable, in case the ranks of the deer and antelope right around the Springs should be too heavily killed out, to kill some of the cougar there, but in the rest of the park I certainly would not kill any of them. On the contrary, they ought to be left alone."[38] Before leaving the White House, in 1908, Roosevelt banned the killing of cougars in Yellowstone.[39] (This didn't stop him, however, from encouraging his sons and nephew to hunt cougars around the Grand Canyon when he was an ex-president in 1913.)

During the two spring weeks President Roosevelt was in Yellowstone, and the sheets of ice in the tree-lined rivers were cracking, he wrote a series of long reports to Merriam on how the springtime wildlife was faring. After hiking footpaths, Roosevelt made detailed zoological descriptions of antelope near Gardiner, Montana, and bighorn sheep in Yellowstone Canyon, Wyoming. As if trying to out-naturalist even Burroughs, Roosevelt made Audubonist studies of golden eagles and water ouzels. And while Roosevelt's gun may have been locked up by the U.S. Army, nothing prevented him from collecting a meadow vole for the Biological Survey. These tiny rodents were among the world's most fertile mammals; females were capable of producing three to ten pups every three weeks. Roosevelt, using his hat as a net, scooped one up and skinned it. Unfortunately, his arsenic can was back at Sagamore Hill. "I send you a small tribute, in the shape of a skin with the attached skull, of a *microtus* [*pennsylvania*], a male, taken out of the lower geyser basin, National Park, Wyoming, April 8, 1903," the president wrote to Merriam. "Its length, head and body, was 4.5 inches, tail to tip, 1.3 inches, of which .2 were the final hairs. The hind foot was .7 of an inch. I had nothing to put on the skin but salt."[40]

While Roosevelt was studying birds and animals (and even evidence of insects), Burroughs was analyzing *Homo sapiens Roosevelti*. Since leaving Union Station in the District of Columbia, and all through the Midwest, while Roosevelt was giving stirring speeches in Illinois, Minnesota, Wisconsin, and North Dakota, a watchful Burroughs was keeping copious notes. What amazed Burroughs most was how cordial Roosevelt was to everybody he met, offering good fellowship, firmly shaking people's hands as if he were a next-door neighbor handing out free *Farmer's Almanacs*. His hail-fellow-well-met routine was paying dividends. The trip seemed less like a presidential tour than a triumphant homecoming for a native son. "He gave himself very freely and heartily to the people wherever he went," Burroughs noted. "He could easily match their Western cordiality and good-fellowship."[41]

It seemed that Roosevelt treated the citizens of North Dakota especially warmly. Every old ranch foreman in the state was offered red-carpet hospitality. Roosevelt truly admired these rural folks. North Dakotans never complained about working long hours or giving a neighbor free help. And the unbounded hills and plains hadn't been spoiled by industrialization. Somehow the children of North Dakota seemed purer than children back east, whose heads were filled with false ideas of what constituted success in America. To Burroughs, in fact, it seemed as if Roosevelt were from North Dakota, as if the yeomen planting crops and the village merchants selling wares were somehow his kinfolk.

As his constant companion, with time to while away, Roosevelt regaled Burroughs with stories about western characters he loved, including Hell-Roaring Bill Jones and Hash-Knife Joe. At Saint Paul, Seth Bullock joined the Roosevelt party for a few days of travel. Once they reached Yellowstone, Roosevelt borrowed a sure-footed gray Third Cavalry stallion while Burroughs, hampered by arthritis, rode in a carriage (or ambulance, as he jokingly called it) pulled by two mules. Burroughs had a wild ride because the team got spooked and took off running. Off they went to Mammoth Hot Springs, which Burroughs later described as "the devil's frying pan." Roosevelt sported khaki pants, puttees, a black jacket, and a tan Stetson hat. Burroughs still wore his dark suit—a fashionista from the Whitman catalog of refined dishevelment. Shedding the Secret Service and newspapermen, they explored caves, spied songbirds, inspected pinecones, and studied topographical aberrations. "He craved once more to be alone with nature," Burroughs wrote, "he was evidently hungry for the wild and the aboriginal."[42]

Burroughs understood that America had a lot of weather-vane politicians, but Roosevelt, particularly when it came to conservation, wasn't one of them. He understood that natural resource management was *the* imperative! He understood why species needed to be saved, if only for aesthetic purposes! And nobody Burroughs had ever met knew more about birds. "Surely," Burroughs wrote, "this man is the rarest kind of a sportsman." When it came to conservation, Burroughs understood that Roosevelt was "the most vital man on the continent, if not on the planet, today."[43]

One of the oddest aspects of Roosevelt's trip to Yellowstone was the vigorously enforced ban against journalists. The president wanted sixteen days off from work to study the abundant wildlife without being pestered by reporters. Both Roosevelt and Burroughs wanted to be like old antelopes, straying from the herd of humans. Their headquarters were at Major Pitcher's house, but several U.S. Army camps were also set up deep in the wilderness (but not too far from established roads) so Roosevelt and Burroughs could commune with the outdoors without signs of irritating civilization. Only matters of utmost national importance would be conveyed to Roosevelt through his personal secretary William Loeb, Jr. (whose railway car, Elysian, the "rolling White House," had been unhitched in Cinnabar, Montana). Burroughs was suffering from a head cold, coughing and sneezing, but he gamely trooped onward into the wild with Roosevelt, sleeping in the springtime snow a few miles from Major Pitcher's house.

On April 11, the *Times*, in a mistaken rush to judgment, ran a bogus story under the headline "President Kills Lion in Yellowstone Park" (presumably, he had killed it with his pistol).[44] Because cougars weren't protected in Yellowstone, the *Times* concluded that the president hadn't violated any regulations; he had merely blasted a feline varmint. But the entire story was fabricated: a scoop-hungry reporter had used an unreliable source. When he heard of it, President Roosevelt was livid, because he had worked so hard to avoid giving the impression that he was hunting in Yellowstone. Instead, with Burroughs at his side, Roosevelt had hoped to emphasize his preservationist side. But he could hardly go after the *New York Times*. It was his hometown newspaper, and it had long promoted his political and literary careers. In an article titled "President on the Move," the *Times* clarified the earlier story, explaining that Buffalo Jones—the apparent source—had offered to go cougar hunting with Roosevelt but the president had "declined the offer."[45]

Nearly every day thereafter, the *Times*, covering Roosevelt as best it could from Cinnabar, Montana, would assure readers that the president had "shot no game." By not hunting, Roosevelt was showing that he was a reformed sportsman, that nature could be enjoyed for its own sake. In fact, all Roosevelt and Burroughs hunted for in Yellowstone were voles for the Biological Survey. They also explored the Yellowstone and Lamar rivers; analyzed the shaped balconies and terraces of porcelain-like travertine at Mammoth Hot Springs in the northwest corner; camped near Old Tower Fall Soldiers Station; pondered the great assemblage of petrified wood; rode sleighs to the Upper Geyser Basin; and even tried skiing around the Norris Hotel. What surprised Burroughs was the bizarre erosion to be studied in this patch of Wyoming: aeolian, biological, fluvial, lateral, and sheet were just some of the conditions. You needed a geological textbook to decipher the cycles of erosion, and to differentiate between piping (badlands erosion) and residual boulder (weathered in place). It was a little too much, so he turned to flowers. "I even saw a wild flower," Burroughs wrote, "an early buttercup, not an inch high—in bloom. This seems to be the earliest wild flower in the Rockies. It is the only fragrant buttercup I know."[46]

Spending so much time hiking together, the two naturalists talked about Merriam, who, for all of his God-given talent, had yet to produce a first-rate American zoological book. But Roosevelt believed that gossip was a black art, akin to blasphemy. If one talked badly about friends behind their back, then one had an obligation to tell them to their face. That was the honor code, he believed, of a "real" man. Therefore Roosevelt's letter to Merriam on April 22 can best be classified as tough love, and a long-deferred goad, putting his thirty-year friendship with the biologist on the line:

> Both John Burroughs and I agree that it is very lamentable that you will not produce a really big book. John Burroughs gives me permission to quote him. He says—I entirely agree with him—that you are in danger of taking your place among those men of great natural power and enormous industry, who collect innumerable facts but are somehow never able to do the work of generalization and condensation—that is, to build a structure out of the heap of bricks. It is an awful thing to generalize hastily, and not to pay proper heed to the need of accumulating masses of material. But where one meets a genuine master in his profession—and such I esteem you—it is a loss to the world if he fails to put his discoveries in durable, in abiding, form. This is exactly what I fear will be the case with you. To publish quantities of little pamphlets is merely to take rank with the thousands of small and industrious German specialists. You have it in your power to write the great monumental work on the mammals

of North America, *including their life histories.* If you put it off too long, you will never do it. And if you wait until you are sure you have exhausted the resources of trinomial nomenclature on very obscure shrew or fieldmouse from Florida to Oregon, you will also have to postpone your work indefinitely; for I firmly believe that after you and I are dead there will still be ample opportunity for industrious collectors to secure "new forms" and "probably valid species" from almost any region which it is thought worth while minutely to investigate. But the labors of ten thousand such would not equal one production of a book by you on the lines I have indicated.[47]

IV

Roosevelt's visit to Yellowstone culminated on April 24, when he laid the cornerstone for a basaltic stone railroad archway near the Northern Pacific Railroad depot at Gardiner, Montana. Architecturally similar in style to the Old Faithful and Canyon hotels, the Roosevelt Arch, as it became known, was twenty feet wide and thirty feet high. It looked as if it belonged on the Champs-Elysées in Paris. Carved above the keystone was a phrase Roosevelt fancied: "For the Benefit and Enjoyment of the People." Smaller plaques read "Yellowstone National Park" and "Created by Act of Congress March 1, 1872." Approximately 3,500 people were on hand for the dedication, inducing a group of local Masons, who presented him with a Montana gold nugget mounted on a plaque.[48] In the cornerstone, the Masons also deposited their grand lodge papers, some local newspapers, a handful of rare coins, photos, a King James Bible, and a brief history of Yellowstone.[49]

"The Yellowstone Park," Roosevelt said in his dedication, "is something unique in this world, as far as I know. Nowhere else in any civilized country is there to be found such a tract of veritable wonderland, made accessible to all visitors, where at the same time not only the scenery of the wilderness, but the wild creatures of the Park are scrupulously preserved as they are here, the only change being that these same wild creatures have been so carefully protected as to show literally astounding tameness."[50]

With John Burroughs and Major Pitcher sitting behind him on the platform, Roosevelt offered his own impromptu reflections on the American West, and thanked locals for his tremendous "two-week holiday" in Yellowstone. In what the historian Aubrey L. Haines described as a "rambling speech," Roosevelt talked about buffalo breeding, forest protection, water conservation, and the geological sites that made Yellowstone unique. With a palpable sense of urgency, he warned coming generations to protect Yellowstone from the scars of ore pits and mine tailings—also, forest fires had to be prevented, or fought when they did occur. Flattering the crowd, Roosevelt conveyed his full confidence in their stewardship of a glorious natural setting straight from the hands of God. He praised Montanans, Idahoans, and Wyomingites for their wise protectionist ethics. "I like the country," Roosevelt said in a crowd-pleasing line reported in the *Times*. "But above all I like the men and women."[51]

A few weeks later, *Forest and Stream* magazine published Roosevelt's speech in its entirety. In his talk, Roosevelt had driven home the point that national parks were "essential democracy" at work. America's treasures, like Yellowstone, had to be safeguarded from vandals and exploiters. Here was a place for city dwellers to restore themselves. The president lamented that Europeans were flocking to see Yellowstone more excitedly than Americans. He said that the United States needed to become "awake to its beauties." And he praised the successes of the wildlife protection movement in the park.

"Here all the wild creatures of the old days are being preserved," he said, "and their overflow into the surrounding country, means that the people of the surrounding country, so long as they see that the laws are observed by all, will be able to insure to themselves and to their children and to their children's children, much of the old-time pleasure of the hardy life of the wilderness and of the hunter of the wilderness."[52]

It was a bittersweet occasion for Roosevelt when, bound for Saint Louis, he had to say good-bye to Montana and part company with Oom John (who was going to Spokane for a prearranged lecture) at the Gardiner arch on April 25.[53] Their wonderful times together in the open air were over. As a parting gesture, Roosevelt rallied to Burroughs's defense against a cheap shot at him in the latest issue of *Forest and Stream*. Either in a fit of jealous pique for having been excluded from the Yellowstone trip or, more likely, simply as a result of editorial misjudgment, George Bird Grinnell had run a cruel, devastating personal attack on Burroughs in the magazine. Angrily, Roosevelt responded that Burroughs was a true man, a saint of the woods, a human being of breathtaking sincerity and a naturalist of unparalleled skills. Whitman had once said that Burroughs was "in a sense almost a miracle." For weeks, Roosevelt and Burroughs had observed deer, elk, wild geese, wild mice, chickadees, and red squirrels. Together they had laughed at the jargoning Canadian jays (or camp robbers, as Burroughs called them) in the mornings and watched the sun set over the Yellowstone River gorges at dusk. They had inhaled the fragrance of scattered pines and had been silenced for hours by the beauty of secluded valleys. Now, in Grinnell's magazine, in an article written by someone in New Hampshire who wrote under the name "Hermit," slammed Burroughs—of all people—as a bad naturalist.* [54]

Burroughs, who was averse to conflict, assured Roosevelt that the attack didn't matter; a single day's news of other matters would erase it from people's memory. The president was not so easily mollified, however. He wrote Grinnell a long letter defending Burroughs and lambasting the magazine for allowing the "Sage of Slabsides" to be ridiculed. The letter was postmarked Gardiner, Montana, and immediately fast-mailed to New York. Its reception marked the beginning of a serious rift in Roosevelt's personal relationship with Grinnell. This was a risk that Roosevelt was evidently ready to take. "I have just seen the long letter by 'Hermit' in the *Forest and Stream* attacking John Burroughs, and incidentally furnishing the most ample reason for utter distrust of Hermit's truthfulness in narrating or else his power of accurate observation," Roosevelt wrote. "I will say frankly that I am surprised that a paper of the standing of the *Forest and Stream* should publish such an article, especially unsigned. . . . It is thoroughly discreditable of Hermit not to have attached his real name, and when the *Forest and Stream* permits the article to be published without the name it of course, in the eyes of the public, itself becomes responsible for the attack on Mr. Burroughs." [55]

With this letter Roosevelt reentered the controversy over "nature fakers" yet again (even though Grinnell never leaked the letter to the press). Yellowstone, in general, had energized Roosevelt with regard to his responsibilities as a naturalist. Wandering with Oom John in the park had been good for his soul. On the way to Nebraska, Roosevelt retreated from the miasma of smoke in the dining car to ruminate about possibly breeding Impeyan pheasants to release in the Great Plains states. [56]

Meanwhile, Roosevelt enjoyed seeing the picture-postcard farms as his train rolled through counties that were almost larger than Rhode Island. Somewhere along a handsome open stretch of South Dakota the president once again met up with Seth Bullock, and the two rode around the Black Hills Forest Reserve and attended a cowboy show in Edgemont together. Eating at chuck wagons, talking to farmers about irrigation and tree farming, and petting a tamed buffalo, Roosevelt was certainly in his ele-

*The Hermit was Mason A. Walton, called the "hermit of Gloucester." He spent many solitary years in the woods of Bond's Hill, Massachusetts. Walton was known for his 1903 book *The Hermit's Wild Friends; or, Eighteen Years in the Woods*. His home was called "Ravenswood."

ment. "The President unites in himself powers and qualities that rarely go together," Burroughs wrote of Roosevelt after the Yellowstone trip. "Thus, he has both physical and moral courage in a degree rare in history. He can stand calm and unflinching in the path of a charging grizzly, and he can confront with equal coolness and determination the predaceous corporations and money powers of the country. He unites the qualities of the man of action with those of the scholar and writer—another very rare combination. He unites the instincts and accomplishments of the best breeding and culture with the broadest democratic sympathies and affiliations. He is as happy with a frontiersman like Seth Bullock as with a fellow Harvard man, and Seth Bullock is happy, too."[57]

When President Roosevelt reached Omaha, 50,000 people were waiting to greet him in the swirling dust. Half a dozen horseless carriages were parked nearby, along with thousands of horses at hitching posts. People were scrunched together as if penned up in the Omaha stockyards, and there was an electric current in the air. According to the *Omaha Bee*, merchants in the midtown district and farmers in Douglas County were thrilled by the chance to glimpse a president in the flesh. All they usually got in Omaha was William Jennings Bryan. The overhanging eaves and second-story porches of Queen Anne homes on Wirt Street had been decorated with red-white-and-blue bunting. The Cudahy Packing Company was also gussied up for the president, just in case he made a spontaneous inspection tour. But the ravages of deforestation were evident in this clattering railroad town: there was little greenery growing around the clapboard houses, and birds were in short supply. Construction and cattle seemed to matter most in Omaha.

Roosevelt challenged Nebraskans to start planting more trees and protect the original "scanty forests" of their state. He was proud that his executive orders of 1902, creating the Dismal River Forest Reserve and the Niobrara Forest Reserve, had already proved to be successful. Under Roosevelt's executive orders, 70,000 jack pine seedlings from Minnesota and 30,000 ponderosa seedlings from the Black Hills had been planted that spring in Nebraska. Forest reserves in Nebraska were "good things," as Roosevelt wrote to Pinchot, as long as "the homesteader or agricultural settler" wasn't harmed.[58] Certainly, prairie fires and the semiarid environment were a problem, but Roosevelt insisted that his treeless forest reserves would soon have an abundance of trees. He was right. In 1947, Pinchot, then near death, revisited the Nebraska National Forest (as the two sites were now collectively called) and declared it "one of the great successful tree-planting projects in the world."[59] That April Roosevelt attempted to explain the virtues of modern forestry to farmers. You might say he urged young Nebraskans to start playing Johnny Appleseed instead of Buffalo Bill. Then, it was back to the express train.

President Roosevelt enjoyed stocking his train compartment with only the essentials: toiletries, clean clothes, and a collection of John Burroughs's books. His life was usually so full of clutter that he seemed to relish the enforced sparseness of train travel; his hectic world was reduced to the bare necessities. Known to tip generously, Roosevelt usually had a couple of porters in attendance outside his compartment, ready to fulfill his every wish. He spent many of his travel hours gazing from his open window. Often, the hamlets he saw resembled what we would recognize as Hollywood back-lot sets or exhibits in Dust Bowl museums. Whenever the train stopped at a depot, admirers swarmed the platform. Roosevelt never rejected a handshake, proud that his star was undimmed. A stenographer sometimes came into his compartment so that he could dictate a rambling letter to an ally or a foe. He did this often. The only time Roosevelt became rude was when he learned that a presidential missive of his hadn't been delivered speedily. He reacted *immediately* to communication, and mail delays often triggered a presidential tantrum. One can imagine the joy with which Roosevelt would have used the Internet; his letters about his trips were like high-spirited blogs full of in-the-moment musings.

As a favor to John F. Lacey, his favorite congressman, Roosevelt ordered his Union Pacific train to make a stop in Oskaloosa, Iowa—Lacey's hometown—on April 28. The previous year, Lacey had been reelected to Congress for the seventh time. Nobody in Washington, not even Roosevelt, had a longer, more distinguished career on behalf of wildlife protection than Lacey. Every aspect of Lacey's career held special meaning to Roosevelt. He and Lacey walked around Oskaloosa together, with the congressman pointing out the sights; and Roosevelt christened a new YMCA building by giving a ten-minute speech in front of a hastily assembled crowd of 30,000.[60] Lacey had urged Roosevelt to save the forestlands of Colorado, and he was now similarly animated about saving prehistoric ruins in New Mexico and Alaskan forest reserves. In terms of his personality, Lacey was what's known in the Midwest as a "plain John": *nothing* about his disposition spoke of a will to power or of narcissism. Yet Oskaloosans knew he was fiercely committed to conservation. As a boy, Lacey had played along the meandering streams of southeastern Iowa; his experiences might have come straight from James Whitcomb Riley's idyllic poem "The Old Swimmin' Hole." But owing to deforestation, coal mining, and oil and natural gas drilling these old streams were virtually dead, except for the occasional plump catfish and thick pockets of minnows. This environmental degradation sickened Lacey. "The trees had been felled and the springs had gone dry," he complained. "The streams were gravelly beds, as dry as Sahara, except for a few hours after a big rain had converted them into muddy torrents."[61]

Hugging Lacey good-bye, Roosevelt headed to Keokuk, in the most southeastern part of Iowa. There he was met by the brother of William Hornaday, Calvin, who lived in that Mississippi River town. As a husbandry expert, Hornaday told the president about a new type of steel-link fence being manufactured by the Page Wire Fence Companion of Adrian, Michigan. The new wire had enough holding capability for bison. Immediately Roosevelt had the fencing acquired for both the New York Zoological Society and the National Zoological Park.[62]

Although the main event at Keokuk was Roosevelt's pushing a big button to reopen John C. Hubinger's factory (known for making starch), the president's oratory near the tomb of the Sauk chief Keokuk in Rand Park generated the most lasting memory. Roosevelt liked everything about the town of Keokuk, at the confluence of the Mississippi and Des Moines rivers.

From there it was on to the greatest American confluence city—Saint Louis—where the Mississippi and Missouri rivers met. There, Roosevelt greeted former president Grover Cleveland. Together they participated in the dedication ceremonies at the World's Fair, including a celebration honoring Thomas Jefferson. Roosevelt had little patience for Jeffersonian antifederalism, but he and Jefferson had at least one thing in common: acquiring large tracts of land. By 1909 Roosevelt had preserved more than 234 million acres of America—an area half the size of the Louisiana Purchase. Roosevelt never got the chance he hoped for to visit the American Bottoms around Cahokia, Illinois, where Jim Bridger used to camp. In Saint Louis, instead of preaching the gospel of conservation, Roosevelt concentrated on local road improvement. His worry was that farmers were becoming isolated from the city. "Roads," Roosevelt said, "tell the greatness of a nation."[63]

Compared with the wild graces of Yellowstone, visiting bustling Saint Louis seemed onerous to Roosevelt. He had to endure a Marine Band performance, photos with a cavalry regiment from Oklahoma, a visit to Saint Louis University to discuss Catholic issues, and a private meeting with Secretary of the Interior Hitchcock that took the form of a stroll along the Mississippi River levee. Hitchcock was going after Senator John Hipple Mitchell of Oregon for using political influence to enrich clients with sweetheart land deals. Even though Mitchell was a Republican, T.R. considered him a forest despoiler and wanted him busted for corruption. Hitchcock was happy to oblige.

Too often, environmental historians have given short shrift to Hitchcock's extraordinary work exposing land fraud in the West from 1899 to 1907. Under his watchful eye 1,021 timber depredators were indicted and 126 were convicted. More importantly, Hitchcock, following President Roosevelt's direct order, unearthed collusion, espionage, forgery, bribery, and record falsification in the General Land Office. Hitchcock busted judges, governors, senators, and business tycoons. At issue was the integrity of the Homestead Act, Desert Land Act, and Timber and Stone Act. Hitchcock saw his job as protecting the public domain. To accomplish this, he set up dragnet operations in every state or territory that had public lands, to catch looters. Acting so boldly as an anticorruption reformer, however, had a downside; in August, unsubstantiated charges were leveled at Hitchcock for complicity in a land fraud case in Indian Territory.[64] Roosevelt knew the charges were bogus. He *was* frustrated that Hitchcock had let requests for forest reserves pile up on his desk, unattended to, yet Roosevelt knew that for all his bureaucratic slowness Hitchcock was a man of integrity. "There seemed to be no limit," the reporter Henry S. Brown wrote in a glowing profile of Secretary Hitchcock in *Outlook*, "to the rapacity of the land sharks."[65]

From Saint Louis, the president journeyed west again on the Missouri Central, to Kansas City and Topeka. Joining him all the way to California was Columbia University's president, Nicholas Murray Butler. Roosevelt's train—complete with barbershop, parlor, kitchen, sleeping compartments, and baggage chambers—was an ornate house on wheels.[66] Again huge crowds—often numbering about 20,000—gathered to hear him speak in city after city. As if he were running for president—and, in essence, he was—Roosevelt kissed babies (though he denied doing this), shook thousands of hands, tossed a football, and took photographs with local police departments. Everything about Kansas appealed to Roosevelt: wide-open spaces, clean air, well-maintained farms, sturdy silos, wheat and sunflower fields, an abundance of deer and game birds, a McGuffey *Reader* in every schoolhouse, and God's grace at every supper table. On May 3, Roosevelt arrived on the Pacific Coast Special in the village of Sharon Springs (located on the west edge of Kansas at the border with Colorado). Because it was a Sunday the president followed a parade of kids to a Methodist church (where a Presbyterian minister from Kansas City preached). The stern loneliness of the Book of Job always held Roosevelt's attention in church; but this preacher dwelled on the more philosophical Ecclesiastes, and the president was bored. Roosevelt played peekaboo with two giggling little girls sitting across the aisle. Eventually the girls were summoned into Roosevelt's pew so that the three of them could sing "Amazing Grace" from the same hymnbook.

After the service, Roosevelt went horseback riding along Eagle Tail Creek, accompanied by Kansas's two U.S. senators: Joseph Burton and Chester Long, both Republicans. When he was back in Sharon Springs, ready to board the train for Colorado, a little girl suddenly appeared with a two-week-old badger. Her brother Josiah had trapped it alive, and she wanted President Roosevelt to raise it as a White House pet.[67] To the surprise of the attending dignitaries, Roosevelt roared in delight, saying he would add Josiah (as he named it) to the growing White House menagerie. During the coming weeks, Roosevelt would hand-feed Josiah from a baby bottle.[68]

With its grayish coat, flattened appearance, heavy body, and short tail, Josiah became a favorite of Roosevelt's. At train depots the president would show the cute badger to schoolchildren, pointing out the conspicuous white stripe running down its back. Knowing that badgers were carnivorous, Roosevelt fed Josiah the best meat he could find, although for some reason it was rejecting the meat in favor of starches. "I have collected a variety of treasures, which I shall have to try to divide up equally among you children," Roosevelt wrote home. "One treasure, by the way, is a very small badger, which I named Josiah, and he is called Josh for short. He is very cunning and I hold him in my arms and pet him. I hope he will grow up friendly—that is if the poor little fellow lives to grow up at all. . . . We feed him milk and potatoes."[69]

In Denver, Roosevelt met with cowboys and delivered speeches on irrigation laws and good citizenship.[70] Mayor Robert R. Wright, Jr., proclaimed the day of the presidential visit, May 4, a citywide holiday.[71] A magnificent gold brooch was given to T.R. depicting the Rocky Mountains—he wore it on his winter coat for the rest of the trip. In Denver, he also connected with a former Rough Rider, swapping stories about Cuba. In his Yellowstone journal Burroughs noted how astounding it was that so many Rough Riders wrote to the president about their personal woes. Most were from the territories—Oklahoma, Arizona, and New Mexico. Roosevelt was their spiritual counselor and adviser. One Rough Rider, Burroughs recalled, wrote to the president: "Dear Colonel—

I am in trouble. I shot a lady in the eye, but I did not intent to hit the lady; I was shooting at my wife."[72] What surprised Burroughs was that the president had such time for this nonsense.

After winning the hearts and minds of Denver, Roosevelt was off to the New Mexico Territory, where statehood was still pending. When Roosevelt had attended the Rough Riders' first reunion in 1899 in Las Vegas, New Mexico, he hadn't had time to visit the Old Spanish mission town of Santa Fe, which was off the beaten path. Originally, Santa Fe was meant to be a booming railroad stop on the Atchison, Topeka, and Santa Fe line, but the civil engineers instead chose Lamy, to its south. Now, after finally seeing the fascinating San Miguel chapel (oldest edifice in the United States), Loreto chapel, and La Fonda on the Santa Fe Plaza, Roosevelt better understood why the archaeologist Edgar Lee Hewett (then age thirty-nine) wanted these sixteenth-century structures saved for posterity. Furthermore, to Roosevelt's mind New Mexico was the pocket where many other prehistoric treasures were kept. Hewett was pushing for the timeworn ruins of the Pajarito Plateau, in particular, to become a national park. Congressman Lacey had visited the Pajarito in 1902, with Hewett as guide. Forging a formidable league, Lacey and Hewett reported to Roosevelt that the pot hunters and artifact vandals had to be put out of business. Hewett started working on a special report offering ideas on how to stop the desecration of these New Mexican sites once and for all.

Although he was in Santa Fe for only three or four hours, Roosevelt made it a point to visit the New Mexico Historical Society's museum, probably with the idea of writing another volume of *The Winning of the West*, about Kit Carson, during his postpresidential years. Santa Fe intrigued Roosevelt because it had been permanently settled in 1610, before either the founding of Jamestown or the Pilgrims' landing in Plymouth. Roosevelt knew he would have to move quickly to save New Mexico's earth-toned adobe buildings. That safeguard task would indeed become a priority following the 1904 election. Wearing a white Stetson, he was playing Pat Garrett in the land of Billy the Kid. Speaking to some 10,000 people in front of the territorial capitol, Roosevelt proclaimed the benefits of "forest preservation." Having huge reserves, he said, would be a *prerequisite* to New Mexican statehood.[73]

In the old town of Albuquerque Roosevelt, with Governor Miguel Otero at his side, inquired about the various southwest Indian ruins in the territory that Congressman Lacey had been pestering him to preserve, such as Chaco Canyon and the Gila Cliff Dwellings. The tireless Hewett, in a series of fine articles, had proposed establishing national cultural history parks in the Southwest. Hewett warned that vandals were looting pottery and old cooking utensils from the sites, sometimes using dynamite to blow holes in archaic dwellings. Without federal protection *soon*, there would be nothing left of these antiquities and their sites. In Utah, reports were coming out about petroglyphs 4,000 years old.[74] Hewett, whose nickname in archaeological circles was "El Torro," had legions of friends—and enemies. Roosevelt was squarely in the camp of his friends. Even though Hewett was disliked by governors, ranchers, and landowners in New Mexico, Roosevelt saw him as a territorial treasure in his own right. All of Roosevelt's communication with Hewett was through Lacey. Generally speaking, Roosevelt supported saving all the prehistoric ruins in the Southwest as quickly as possible.

New Mexico's current motto, "Land of Enchantment," is not its first. An earlier motto was "The Land of Sunshine," which Roosevelt found appealing and true.[75] The mild, dry weather of New Mexico served as a balm. Little girls dressed in wedding-dress white made sweet appeals to Roosevelt for New Mexico's statehood, singing patriotic ballads for his pleasure. All the president could do was beam. He gave a spectacular speech in the Old Town followed by luminaries at sunset around the plaza casting everything in a golden glow.

Then, leaving Albuquerque at dusk, the Pacific Coast Special headed for the Grand Canyon. A quick stopover was made in the Painted Desert during the early morning. Roosevelt had time for nothing more than a few inhales and a surveyor-like scan of the flatness. Congressman Lacey had been telling Roosevelt about the Petrified Forest of Arizona and the president now got a feel for the topography under the entrancing moonlight. (A few years later Muir would come to the Petrified Forest to study fossils and draw up a map for upholding the area as a national park.) Arriving in Flagstaff at nine o'clock on the morning of May 6, waking up to the light of the sun, Roosevelt felt well rested. Surrounding him were Merriam's San Francisco Peaks (where Merriam, as head of the Biological Survey, had first discovered this stratification of life zone in 1898). In Flagstaff, the world was full of geological possibilities. Roosevelt had clearly left the hysteria of national politics back in Washington, D.C., 1,900 miles away. He felt isolated and happy. Glory to the West! Glory to John Wesley Powell! Glory to the Arizona Territory! Glory to the Grand Canyon, which at long last he was going to see! Roosevelt was in a glorious frame of mind.

Roosevelt standing on the rim of the Grand Canyon with Governor Brodie of the Arizona Territory.

V

The president's arrival at the Grand Canyon (or the Big Ditch, as locals called it) on the morning of May 6 would, in retrospect, become one of the greatest days in environmental history. The Grand Canyon seemed as if it had been born of a cataclysm, with no eyewitnesses or reliable records. Amaranth in color, with a weird purple-orange glow, it also seemed cosmic, full of yearnings and teachings. In *A Book-Lover's Holidays in the Open* Roosevelt called the Grand Canyon "the most wonderful scenery in the world." He also declared that "to all else that is strange and beautiful in nature the Canyon stands as Karnak and Baalbec, seen by moonlight, stand to all other ruined temples and palaces of the bygone ages."[76]

Many Rough Riders were there that May to stand and gaze at the canyon with him. David Warford was among them.[77] It is not hyperbole to say that Roosevelt's jaw dropped in disbelief. The geologist Clarence Dutton—whose *Tertiary History of the Grand Cañon District* Roosevelt had recently read—called this dynamic chasm of the earth's surface "a great innovation in modern ideas of scenery," adding that "its full appreciation is a special culture, requiring time, patience, and long familiarity for its consummation."[78] Roosevelt now understood what Dutton meant. He had long suspected that the Grand Canyon was the premier natural wonder in America, and now his hunch had been confirmed. He was dying to learn more of its geological secrets. Staring over the ledge of the Grand Canyon made the heart stop at the *immensity* of it all.

What disturbed Roosevelt, however, was that the Arizona territory was *debating* whether to preserve the canyon or mine it for zinc, copper, asbestos, etc. Preservation in this case was so obvious that even engaging in debate seemed almost criminal. To Roosevelt, the Grand Canyon was beyond debate by the locals: it must become the exclusive property of the United States to be saved for future generations. Roosevelt immediately resolved to make it a national park following the 1904 election. Thereafter, only horse trails would be allowed. "In a few places the forest is dense," Roosevelt wrote, "[but] in most places it is sufficiently open to allow a mountain-horse to twist in and out among the tree trunks at a smart canter."[79]

Roosevelt's attitude toward the Grand Canyon was uncompromising: he flat-out refused to let corporate avarice or citizens' ineptitude desecrate the greatest American treasure. He'd go through all the proper motions of getting Congress to designate the Grand Canyon a national park, and if it refused, an executive order would prevail. Roosevelt vowed to make sure that Arizona's developers never drilled an inch of the Grand Canyon. He hoped his presidential visit would start a widespread grassroots movement to preserve it all—every damn acre in the 1,904 square miles—for perpetuity. Public education in Arizona had to be initiated at once. Too many Arizonans simply looked at the Grand Canyon, Roosevelt scolded, instead of *living* within its geological essence. The Rough Riders who greeted Roosevelt—by and large the most popular public figures in Arizona—were his first line of preservationist defense. They would ride over any ridge for their beloved colonel. Now, in a public forum, Roosevelt suddenly found himself asking them to crusade with him again, this time on behalf of preserving the magnificent Grand Canyon, which developers denigrated as useless, like Death Valley.

"I want to ask you to do one thing in connection with it," Roosevelt, at the rim, urged the crowd of Arizonans. "In your own interest and the interest of all the country keep this great wonder of nature as it now is. I hope you won't have a building of any kind to mar the grandeur and sublimity of the cañon. You cannot improve upon it. The ages have been at work on it, and man can only mar it. Keep it for your children and your children's children and all who come after you as one of the great sights for Americans to see."[80]

This speech, for which all the leading Arizonan politicians were in the audience, marked the beginning of Roosevelt's ceaseless determination to save the canyon from destruction. Overawed by its immensity, enjoying even the ground squirrels running across the naked rock, Roosevelt was in rapture. There is something about the Grand Canyon's power that makes one consider immortality. It was grander than all the music Roosevelt had heard; it was finer than all the Transcendental poetry he had read. John Burroughs had written in *Locusts and Wild Honey* about the gospel of the ledge, which was nothing less than "eternity"; Roosevelt now understood what Oom John had meant.[81] To Roosevelt's mind this ledge was a no-growth zone. If Roosevelt had done nothing else as president, his advocacy on behalf of preserving the canyon might well have put him in the top ranks of American presidents. Middle Granite Gorge, the Redwall Cliffs of Havasu Falls, Kaibab Plateau, Marble Canyon, Mount Trumball—all topographically part of what would become the Grand Canyon National Park—became treasured places that miners or loggers would never lay to waste, thanks to Roosevelt's strenuous advocacy. Even industrial activity anywhere near the Grand Canyon wasn't acceptable to Roosevelt. If Carlyle was correct in saying that all history is forged by the deeds of great men, then Roosevelt earned his place in the American pantheon by simply refusing to let commercial interests desecrate the Grand Canyon.

Clearly, to Roosevelt the Grand Canyon was more than a weather-worn chasm or scarred ledge cutting deep into a mountainous region. It was one the world's most spectacular examples of the power of erosion. Time was on display at the canyon as nowhere else. At the comfortable El Tovar Hotel on the South Rim, Roosevelt—seeing the Grand Canyon as a symbol of unifying nationalism after the Civil War—told a reporter for the *New York Sun* that it was one of the "great sights every American should see."[82] Once again, the president sounded like a Baedeker guide to the West. Instead of considering the canyon as just a singular natural wonder (like Crater Lake or Wind Cave), he saw it as an irreplaceable part of the Colorado Plateau landscape, which covered approximately 13,000 square miles in northern Arizona, western Colorado, northwestern New Mexico, and eastern Utah. At meetings of the Boone and Crockett Club he used to insist that Yellowstone was the finest geological site in America, but now he knew better. Taken as a whole, the Colorado Plateau included the Grand Canyon, Zion Canyon, Bryce Canyon, Glen Canyon, Rainbow Bridge, the Canyonlands, Capitol Reef, Arches, Four Corners, Mesa Verde, Petrified Forest, the Painted Desert, San Francisco Peaks, Oak Creek Canyon, and dozens of other sedimentary attractions. What a stunning terrain! What a watershed! Although Roosevelt had extensive experiences in the Dakota Badlands moonscape, the Colorado Plateau was beyond anything he'd ever fathomed.[83]

The plant life around the Grand Canyon also greatly interested Roosevelt, despite his weakness in the realm of southwestern botany. He could identify only 100 or so of the 1,500 species of plants found around the canyon's rim. He could, however, name most of the 300 types of birds that inhabited the area. The sharp-shinned hawk flying above the South Rim he knew would be headed to Central America come fall, and the broad-tailed hummingbirds that whirred about the El Tovar Hotel had arrived because the persimmons had started to bloom. The Townsend's solitaires and Clark's nutcrackers simply did not migrate: they changed elevations as the weather changed.[84] He didn't have to be William Rand or Andrew McNally to know that the abrasive Colorado River didn't start or end in the Grand Canyon. But an impressive 277 twisting miles of the Colorado coursed through the Grand Canyon; it was considered the most exhilarating whitewater run in the American West. Peering down at the immense canyon floor, Roosevelt understood that a working naturalist would try to comprehend the ecosystem from the bottom up. Unfortunately, as a president on the run, he didn't have the time for a careful scientific approach. But the Grand Canyon was the vortex of America's four great deserts—Great Basin, Sonoran, Mojave, and Chihuahuan—and this explained why dozens of cactus species, all amazingly adaptable, many still without names, appeared all around him on the rim. And he vowed to return to the Grand Canyon with his sons.

Insisting on seeing the sun set from the Grand Canyon's north rim, the warm sky ablaze with ragged bands of orange, pink, and purple, Roosevelt leaned over the ledge to soak in the drama. His train didn't leave for Barstow until six o'clock that evening, when the dusk would have thickened. Night comes to northern Arizona fast, as if someone were blowing out a candle. Back at the White House twenty-eight days later, he kicked himself for not having allowed two or three free days to explore the canyon. At the very least, he wished Oom John had been at his side during his unforgettable afternoon there. They could have philosophized about ledges. What was becoming clear from the looping 14,000-mile railroad journey was that the beauty of the American West—the real West—once again had Roosevelt spellbound. From the Grand Canyon onward to Los Angeles, all of Roosevelt's speeches promoted, with intense vitality, the holy trinity of irrigation, forestry, and preservation.

VI

Roosevelt was immediately mesmerized by the high, dry light and yellowish shadows of southern California's deserts, mountains, valleys, forests, and ocean. No wonder so many people were seeking out its charms. Aridity was a problem for California, so the more forest reserves the state had, the better off it would be. After Roosevelt crossed the Mojave, his pulse quickened as his train approached the San Bernardino Valley. A skirting of timber could be seen in all directions. In his speeches, he became blunt and outspoken about keeping paradise intact. Wherever Roosevelt went in southern California that May, his popularity with children was unprecedented for a politician—they adored him. Too young to remember Lincoln or the generals of the Grand Army of the Republic, boys aged five to fifteen had adopted President Roosevelt as their generation's George Washington. They knew nothing of Yorktown or Chancellorsville but everything about the gallantry at Kettle Hill. They were in awe of their top-hatted hero, who shouted "Dee-lighted" and "Bully!" from platforms with a theatrical roar, as if he were giving the Rough Riders the order to "Charrrge!"

Riverside. Pomona. Claremont. Pasadena. Los Angeles. Strangers came up to Roosevelt offering their goodwill. Carrying himself straight as a ramrod, the president enjoyed being a moral exemplar—he was the heir of Lincoln and Emerson with a dash of Boone for good measure. Someday historians, he knew, would tell hundreds of stories about his frenetic time in southern California, so he leaned into his own character even more extravagantly than usual. In city after city he was greeted like an American immortal, never to be laid low. On soapbox after soapbox he whipped up Californian onlookers with fine Bryanesque oratory. They could almost burn calories just watching his body language. No American president had ever expressed America's identity in such a flamboyant, kinetic way. Whether he was lunching at the Westminster Hotel or participating in the annual Fiesta de las Flores, Roosevelt was in love with southern California, particularly the San Gabriel Mountains, which spilled in all directions. Somehow the sky over the 165,000 square miles of California seemed to dance differently from the sky in the east, Roosevelt thought, showcasing shades of blue, purple, red, and gray he'd never known before in his brief acquaintance with Texas, New Mexico, or Arizona.[85]

"I am glad, indeed, to have the chance to visit this wonderful and beautiful State," he told a crowd of 10,000 in front of the Hotel Casa Loma in Redlands. "But for the country itself, though I had been told so much of its beauty and its wonders, I had never realized or could not realize in advance all I have seen. Coming down over the mountain, I was impressed with the thought more and more of what can be done with the wise use of water and forests of this State. The people have grown to realize that it is indispensable to the future of the country to conserve, properly to use, the water and preserve the great mountain forests. . . . I think our citizens are realizing more and more that we want to perpetuate the things both of use and beauty. Beauty surely has its place, and you want to make this State more than it even now is—the garden spot of the continent."[86]

Such rapturous talk about California continued as Roosevelt headed north along the Pacific coast with the powerful Transverse and Santa Lucia ranges looming on the eastward side of the tracks. Ventura. Montecito. Santa Barbara. San Luis Obispo. To promote the Santa Inez and Pile Mountain forest reserves, Roosevelt asked that the U.S. government rangers serve as his special guard. This was a smart, visual way to help locals understand that rangers were akin to the police. According to the *New York Times* Roosevelt was in his "best spirits" ever, enthralled by the Pacific blue and bleached skies, the wildlife-thick Channel Islands, the pock-marked Franciscan missions dating from the late eighteenth and early nineteenth centuries, and the flocks of gulls. And for once, reporters didn't have to hurry up and wait. They could bank on Roosevelt's punctuality. Perhaps because Roosevelt was praising Catholics while he was in California, the Vatican announced that a special goodwill letter was on its way to the White House from Pope Leo XIII.[87]

By the time Roosevelt arrived in Santa Clara County, most Californians were eager to impress him. Already he had become close to the locals. His almost magical personal charm had worked wonders. Californians in effect crowned him king of the sequoias. With U.S. Army forest rangers at his side he toured the gorgeous seaside sites around Santa Cruz and Monterey Bay. There were shifting seascapes, coastal pines, crashing waves, expansive ocean views, hulking rocks, dizzying cliffs, and grass-covered headlands ornamented with purple and yellow wildflowers. Seeing the Pacific Ocean in all its glory buoyed Roosevelt; it allowed him to believe that all his pro-western expansionist views since childhood had been spot-on. Could anybody imagine America without California? With a burst of nationalistic enthusiasm, Roosevelt proclaimed that every seal rock or lone cypress he saw was a U.S. treasure. The groves of sequoias, in particular, filled him with unmitigated joy. No photograph, he said, could possibly do them justice. "You have a wonderful State," Roosevelt proclaimed on May 11. "I am glad to see your big trees and to see that they are being preserved. They should be, as they are the heritage of the ages. They should be left unmarred for our children and our children's children, and so on down the ages."[88]

Accompanying Roosevelt on this leg of the California tour was Columbia University's president Nicholas Murray Butler, whose unobtrusive intelligence the president enjoyed. (Butler didn't seem to mind that Roosevelt, on May 14, received an honorary law doctorate from a rival of Columbia, University of California–Berkeley.[89]) Following an alfresco luncheon in Santa Cruz with naval reservists, Roosevelt, as he would do again in the coming days, threw away his planned remarks and spoke spontaneously about the ethical imperative of saving sequoias and redwoods. They seemed rarer to Roosevelt than a centaur, a hippogriff, or a winged donkey. "This is my first glimpse of the big trees. I desire to pay tribute to the associations, private owners, and State for preserving these trees."[90]

However, Roosevelt believed that the redwoods' protectors—well-meaning conservationist citizens all—had demeaned the lordly trees by hammering placards on them such as "Big Pete," "Old Fremont," or "Uncle John." The sight of these advertisements fretted and annoyed Roosevelt. Even though he posed in front of one bearing such a sign—"Giant"—he thought the practice tasteless. "Let me preach to you a moment," Roosevelt said. "All of us desire to see nature preserved. Above all, the trees should not be marred by placing cards of names on them. People who do that should be sternly discouraged. The cards give an air of ridicule to the solemn and majestic giants. They should be taken down. I ask you to keep all cards off the trees, or any kind of signs that will mar them. See to it that the trees are preserved: that the gift is kept unmarred. You can never replace a tree. Oh, I am pleased to be here among these wonderful redwoods. I thank you for giving me this enjoyment. Preserve and keep what nature has done."[91]

Then Roosevelt requested that he be given private time alone in a redwood grove for reverie. After all, he was in one of God's great cathedrals, and he wanted to wander in solitude, listening only to the song sparrows and orange-crowned warblers in absolute allegiance with enchanted nature. Truly he was in a state of astonishment, looking upward as the elfin light beamed in between the sequoias. The trees had a dwarfing effect; a single redwood weighed about 3,000 tons. To Roosevelt they were priceless when standing tall and irrecoverable if fallen. As the president disappeared deeper into the forest his personal secretary, William

Loeb, led a spontaneous community effort to remove all the commercialized signs that had desecrated the trees. When he returned from his hike, Roosevelt accepted the honor of having a redwood named after him. There was, however, a nonnegotiable condition: no sign reading "Roosevelt Tree" would ever be posted. He couldn't stomach such an insult in his name. Everybody agreed to the terms, and that vaguely comforted the president. Still, he warned that thirsty timber jackals would someday come after the sequoias with industrial saws. Californians, he believed, had a patriotic obligation to defend them.

Following a conservation speech in San Jose, Roosevelt asked that more redwoods be added to the itinerary. He simply couldn't see enough of these ancient sentinels. Each tree—some had a diameter of thirty feet—had its own wondrous personality. The redwoods dwarfed all the trees of the Catskills and the Adirondacks. While he was touring the Santa Cruz mountains, especially those near Boulder Creek and Felton, he started calling the redwoods "giants." During the coming days he would meet fruit growers, ranchers, fishermen, and a woman from Watsonville who had thirty-four children. He spoke at Stanford University and in affluent Burlingame (commonly called "City of the Trees"). But all he could *really* talk about was the utter majesty of the *Sequoia sempervirens*, which John Steinbeck would later call "ambassadors from another time." (Or, as Burroughs was fond of saying, they were "living joys, something to love."[92]) It sickened Roosevelt to think that redwood raiders were clear-cutting these old-growth trees for house decks and unnecessary porches (redwood was a prized luxury wood because it didn't rot). In a speech on May 13 at Stanford University, where the president of the university, David Starr Jordan, introduced him, Roosevelt gave a long address about Congress saving the wilderness heritage. The coastal hills and groves of California, Roosevelt said, needed increased permanent preservation by the federal government.

When Roosevelt arrived in San Francisco for a three-day visit, more than 200,000 people lined the streets to see him. They admired his courage on the battlefield, his reforming spirit, his pluck, his humor, and his ambition, and they were impressed by his national celebrity. San Francisco also held a parade in his honor, but the main event was a dedication of a monument to Admiral Dewey at Union Square. For this dedication, sleeping mats had been unrolled on the sidewalks so that women wouldn't get their dresses dirty. Everyone waited for the president's triumphant wave of the fist—a gesture he had adopted since his days of boxing at Harvard. The entire Bay Area seemed aroused by the event. Civil War veterans offered snappy salutes to Roosevelt as if he were the head of an old soldiers' home—a position far more impressive to them than the presidency. In the most pugnacious speech of his western trip, delivered from a hastily constructed platform at the Mechanics' Pavilion, Roosevelt declared that America's destiny was on the Pacific Ocean. "Before I saw the Pacific slope I was an expansionist," he said, "and after having seen it I fail to understand how—any man confident of his country's greatness and glad that his country should challenge with proud confidence our mighty future—can be anything but an expansionist. In the century that is opening the commerce and the progress of the Pacific will be factors of incalculable moment in the history of the world."[93]

And the Californians cheered: the oyster pirates from Oakland, the grape growers of Napa Valley, the lumbermen from Marin County, the ragtag orphans from North Beach, the horse breeders from the San Joaquin Valley, the naval officers stationed at Treasure Island, the old-time rustics from Point de Reyes, the fishermen from Sausalito, the dandies from Nob Hill, the restaurateurs from Chinatown, the academics from Berkeley, the avocado growers from Fallbrook, the raisin pickers from Fresno, the egg-heads from Menlo Park, the flower merchants from Ventura, the old-time miners from the Sierras, and the buffalo soldiers providing backup Secret Service duty, in addition to every state bureaucrat and politician able to walk. A few Rough Riders had ventured north from Arizona—using their veterans' pensions for train fare—hoping to rekindle remembrance of and pride in the Spanish-American War. And nature helped Roosevelt out. The May inrush of Pacific breeze stimulated the rally like a tonic. Newspapers tried to capture the collective energy of the throng, which hummed with the force of a bass organ pipe from Union Square all the way down to Fisherman's Wharf.

That May 13 in San Francisco marked the apogee of Roosevelt's eventful days as president. All his nationalistic notions, it seemed, were pulled together into a credible narrative for the United States. To Roosevelt the main thrust of American history was western expansionism. The wars with Indians, redcoats, Mexicans, and Spaniards had been worth it. With the building of the Panama Canal the United States would have a two-ocean navy. With Hawaii and the Philippines the nation had stepping-stone ports for the fabled China trade. America wasn't going to be denied its economic empire. And California, he believed, was the gold star of empire. Of course, national politics was full of drum-beating American expansionists, imperialists, and proponents of manifest destiny. What was unique about President Roosevelt was his righteous insistence that

Yellowstone, Yosemite, the Grand Canyon, the redwoods, Mount Olympus, the Painted Desert and so on were the rightful trophies of expansionism. As a conquering conservationist-preservationist he wanted them all saved. At a banquet at Cliff House, overlooking the Pacific Ocean, Roosevelt vowed that the aboriginal American spirit toward the wilderness had to flourish in the twentieth century. Nature was the great replenisher for the American people. His spirit deeply inspired by the beauty of the West, Roosevelt was a rare instance of constructive hyper-Americanism, since his message was that your state has something far more valuable than gold: green forests, sour green glades, box canyons, high plateaus, granitescapes, and lookouts around every bend. When it came to nature preservation, Roosevelt gushed a positively progressive effect onto the collective American psyche.

It's been said by modern environmentalists that Roosevelt had a conflict of loyalties in the West between pro-growth policies like the Reclamation Act and pro-preservation policies like saving the redwoods. This is true. Basically, he wanted to have it both ways. Starting with Roosevelt Dam on the Salt River near Phoenix, Arizona, virtually every major waterway in the West was altered by environmentally destructive engineering projects that T.R. OK'd. Roosevelt saw himself as the master preservationist but also beamed as a master builder. Save the Grand Canyon while building the Roosevelt Dam—that was his conservationist policy. To Roosevelt, conservation could be another form of conquest; development *and* protection working in harmony for an Edenic civilization. When push came to shove, economic growth often took precedence over his preservationism. He made his decisions case by case. Conservation as big business was regularly given precedence over conservation as protection—but there were many exceptions. Over the decades, this has made him something of a bogeyman to the Sierra Club types. Following President Dwight Eisenhower's speech about the industrial-military complex in 1961, for example, Roosevelt's Reclamation Service was denounced by anti-war environmentalists like Dave Brower and Wallace Stegner as scientific capitalism run amok. By the late 1960s and early 1970s, historians lacerated Roosevelt for his imperialistic views, which they equated with the policies that had led to Vietnam. And, unlike John Wesley Powell, a generation of environmental historians led by Donald Worster complained, Roosevelt liked *big* dams—*big* everything, including a *big* intrusive federal government.

But the anti-Roosevelt critics went too far. President Lyndon Johnson's entire "New Conservationism" of the 1960s was purposefully modeled on ideals T.R. had first propounded. And modern environmentalists were aware that Roosevelt had also liked *big* national forests: in 1908, for example, he created the Tongass National Forest, which stretched over 500 miles from north to south in Alaska and included more than 11,000 miles of rugged coastline (a figure equal to nearly 50 percent of the entire coastline of the lower forty-eight).[94] As Roosevelt's attitude toward the redwoods showed during this California trip, he was emotionally a forest preservationist while politically a utilitarian conservationist. It was the right combination for his times. But never for a moment did Roosevelt purposely seek to abuse the American West in any way. But such sincerity had its limits: Roosevelt lacked self-awareness regarding his very real contradiction, his insistence on *bigness* wrought on the western landscape. Yet, always, he wanted to create model cities surrounded by greenbelts of wilderness. He was a promoter of *sustainability* before the concept came into vogue during the Clinton era of the 1990s.

Journalists throughout California commented on how hard Roosevelt was working during his two-week statewide tour to inject conservation into the political bloodstream. The *Los Angeles Times* wrote of his "strenuosity," and the *Oakland Tribune* called him a "drayhorse working every hour."[95] And although she was not involved in policy, the first lady did participate in tree (or at least shrubbery) preservation in the East. Edith Roosevelt was making news by objecting publicly when remodelers at the White House wanted to remove more than seventy bushes from the terrace. She had grown fond of the shrubbery and wanted it to stay put. Even when she was told that the bushes would be carefully removed and replanted in New Jersey, she insisted that they be left unmarred and unmoved.

VII

After shaking so many hands, Roosevelt needed an outdoor adventure in the Sierras. Shortly after midnight on May 15 he left San Francisco for Yosemite Valley with an honorary doctorate from the University of California–Berkeley in hand. Accompanying him was his delegation, which included the Sierra Club's president, John Muir; Governor George C. Pardee of California; and Benjamin Wheeler, president of the University of California–Berkeley. The party enjoyed the scenic mountain ramparts en route, and then Roosevelt's train stopped at Raymond, the railroad depot closest to Yosemite. Three previous U.S. presidents had visited Yosemite—James Garfield in 1875, Ulysses S. Grant in 1879, and Rutherford B. Hayes in 1885—but not while in office, so Roosevelt's visit was a first. And instead of coming to the national park merely as a political gesture, Roosevelt planned to study Yosemite as a naturalist—hence Muir's presence at his side. "Of course of all the people in the world," Roosevelt said, "[Muir] was the one with whom it was best worth while thus to see the Yosemite."[96]

Roosevelt and Muir, taking a buggy, headed straight for the "big tree" section—Mariposa Grove, where some of the oldest redwoods in California grew. A photograph was snapped of them driving through the rather touristy Wawona Tunnel Tree (a towering sequoia that fell in 1969). Mariposa Grove wasn't yet officially part of Yosemite National Park in 1903 but Roosevelt hoped it might soon be. As the *New York Times* reported, Roosevelt and Muir arrived in Mariposa Grove on a bright, perfectly clear day, had lunch, and then wandered off together. Walking around the huge circumferences of the redwoods with Muir, staring upward more than 250 feet to see the top branches, Roosevelt was in his element.[97] While studying the famous "Grizzly Giant," the president blurted out, intensely, that this was "the greatest forest site" he had ever seen.[98] The naturalist Henry Fairfield Osborn had said of Muir that he "wrote about trees as no one else in the whole history of trees, chiefly because he loved them as he loved men and women."[99] Now Roosevelt understood what Muir had been so rhapsodical about over many years. "There are the big trees, Mr. Roosevelt," Muir excitedly said. "Mr. Muir," Roosevelt said with a smile, "it is good to be with you."[100]

Muir had an ethereal quality and his erudition was simultaneously bold and profound. Roosevelt immediately admired him. Muir's eyes were deep blue, his hair was ginger-reddish, and his attitude was life-affirming. While Roosevelt thought in terms of Americanism in nature, Muir thought about the planet in peril. He had even once titled a journal: "John Muir, Earth-planet, Universe." Having read all of Muir's works, and realizing that the great naturalist had spent thirty years studying the trees, rocks, canyons, falls, and glaciers of Yosemite, Roosevelt felt like a student arriving at an academy. Furthermore, Merriam had advised Roosevelt to camp with his friend Muir; he predicted it would be one of the memorable moments of his life.[101]

Because no authoritative account was ever written of Roosevelt and Muir's trip of 1903, it has been pieced together by varied sources over the years. Together, Roosevelt and Muir did explore the park for three days and two nights. Even though Roosevelt was officially booked at the Glacier Point Hotel, he instead camped with Muir in the great outdoors. They would drink in the fresh air, survey the ridgelines, and listen to each other's voice echoing out over the Yosemite Valley. The U.S. Army oversaw the park and was extremely accommodating of Roosevelt's needs. But Roosevelt wanted lots of privacy. Waving a captain and thirty cavalrymen away with a "God bless you," Roosevelt made it clear that he wanted to be alone with Muir among the thickset trees and trailside brush. Only the guides Charlie Leidig and Archie Leanor and the U.S. Army climber Jacker Alder were allowed to untie the saddlehorn rope to be part of the presidential entourage when a hike was in order.[102]

The president treated Muir as his absolute equal throughout the Yosemite adventure. Roosevelt and Muir were both mavericks and shared a strong, rare bond: appreciation of nature. "[Muir] was emphatically a good citizen," Roosevelt noted. "Not only are his books delightful, not only is he the author to whom all men turn when they think of the Sierras and northern glaciers, and the giant trees of the California slope, but he was also—what few nature lovers are—a man able to influence contemporary thought and action on the subjects to which he had devoted his life. He was a great factor in influencing the thought of California and the thought of the entire country so as to secure the preservation of those great natural phenomena—wonderful canyons, giant trees, slopes of flower-spangled hillsides—which make California a veritable Garden of the Lord."[103]

Leaving Mariposa Grove, the Roosevelt party headed to Yosemite's south entrance by carriage, through a handsome glen. As he got out of the carriage, Roosevelt asked for his valise—he didn't like being separated from his personal belongings. When told that the Yosemite Park Commission had brought his baggage to a banquet lunch, he grew enraged. "Get it!" he shouted. According to Muir the two words, barked with an authoritarian air, were like bullets being fired.[104] Although reporters sometimes portrayed Muir as a misanthrope, he made friends quickly. There was never a moment of awkwardness with Roosevelt. Really, the only strange thing about Muir was that he had never once shaved in his life.[105]

On May 15 Roosevelt and Muir mounted horses and trotted off into the vast sequoia lands near the Sunset Tree. The strength and beauty of Yosemite were undeniable. Somehow, there was a summery fragrance in the air even though there was snow. Roosevelt praised the cinnamon-colored sequoias' enduring beauty. Roosevelt recalled in *An Autobiography*, "The majestic trunks, beautiful in color and in symmetry, rose round us like the pillars of a mightier cathedral than ever was conceived even by the fervor of the Middle Ages. Hermit thrushes sang beautifully in the evening, and again, with a burst of wonderful music, at dawn."[106] That evening they built a campfire; continually feeding it wood, they talked until the fire drew down to coals. It was the most famous campfire ever in the annals of the conservation movement. Over the popping and crackling logs Roosevelt and Muir talked about *forest good* and slept soundly without a tent.

At sunrise on May 16 Roosevelt and Muir decided to forgo the day's official itinerary and ride horseback by themselves through the melting snow along an old Indian trail to Glacier Point. There is a marvelous photograph of Roosevelt and Muir standing on a ledgerock overlooking the valley, a respectable 3,200 feet high, with Yosemite Falls thundering at their backs. On close inspection, patches of diminishing snow are noticeable on the thawed ground. Roosevelt looks ready to draw a weapon; Muir is seemingly relaxed, hands behind his back. Over the decades this photograph has become an icon promoting American national parks, for the Sierra Club and the U.S. Fish and Wildlife Service alike. It has been reproduced on book jackets and in magazines. According to historian Donald Worster of the University of Kansas, Roosevelt and Muir had

Theodore Roosevelt and John Muir at Mariposa Grove in California.

good reason to look so satisfied in each other's august company. "They have just agreed that ownership of the much-abused valley below should revert to the federal government and become part of Yosemite Park," he notes in an analysis of the photograph. "Politically, they have forged a formidable alliance on behalf of nature."[107]

Through a blinding snowstorm, Roosevelt and Muir footslogged to Sentinel Dome, a few miles from Glacier Point Hotel. Five feet of snow already lay on the ground. A little base camp was chosen sheltered from the frost heave and glaze ice.[108] Muir built a marvelous bonfire that second evening and made a bed of ferns and cedar boughs. "Watch this," Muir said. Grabbing a flaming branch from the campfire he lit a dead pine tree on a ledge. With a roar, as if a squirt of gasoline had been administered, the flame shot up the dead branches. Suddenly Muir did a Scottish jig around the pine torch. Such ritualistic acts were right up Roosevelt's alley. Leaping to his feet he hopped around the flaming tree, shouting "Hurrah!" over and over again into the night sky. "That's a candle," Roosevelt told Muir, "it took 500 years to make. Hurrah for Yosemite! Mr. Muir."[109]

Muir was born in Dunbar, Scotland, on April 21, 1838. When he was eleven his family emigrated from Glasgow to Marquette County, Wisconsin. Throughout his adolescence he toiled on his father's farm and tinkered with clocks, barometers, hydrometers, and table saws. When he was eighteen he almost died from "choke damp" while digging a well. During the Civil War he enrolled at the University of Wisconsin–Madison, where he invented a study desk that retrieved a book, held it stationary for hours, then automatically replaced it with a different volume. It was a weird contraption, but it indicates how enthusiastic a bibliophile Muir was. With eagerness and diligence Muir read about Henry David Thoreau's rejection of bourgeois society and Robert Burns's revolutionary democracy. The sage Ralph Waldo Emerson, as an old man, encountered Muir and deemed him "one of my men," a true-blue Transcendentalist.[110]

Over time botany became Muir's passion. In 1863 he took his first botanical tramp along the Wisconsin River to the upper Mississippi River. Hunting for plants liberated him from religious orthodoxy and family commitments. He drifted to Ontario, Canada, working for a long spell at a sawmill and a broom and rake factory. In Ontario he discovered the rare orchid *Calypso borealis* (this led to his first published article in the *Boston Recorder*). Odd jobs became Muir's specialty: they were his way to finance his botanical tramps. In 1867, however, a factory accident made Muir temporarily blind. When his vision returned, he made a vow to himself:

he would dedicate his life to nature ("the University of the Wilderness," as he called it). Off he went on a 1,000-mile walk to the Gulf of Mexico and Florida (with South America his eventual destination). When a bout of fever prevented him from tramping south of the Tropic of Cancer, he contemplated the relationship between man and nature in new and profound ways while his temperature soared to over 100 degrees. Like Roosevelt, Muir concluded that all species have an inherent value and a right to exist. Not until 1911, however, would Muir fulfill his dream of exploring the Amazon of Brazil and the mountains of Chile.

Muir's arrival to San Francisco in 1868 forever changed his life. From April to June, he hiked around Yosemite. Walled in by the Sierra range, Muir was captivated by the enduring rocks, slow-moving glaciers, and ancient redwoods, which Yosemite offered up in astonishing numbers. There was a grace to Yosemite which defied language; it was a terrestrial manifestation of the Almighty. There was no denominational snobbishness and no chosen people in nature; there was just one big sky. "His studies in the Sierra, earnestly as they were pursued, were only secondary—his rapt admiration of the dawn and the alpenglow, of majestic trees that wave and pray, of rejoicing waters, and the sacred, history-bearing rocks, of night and the stars on lonely mountain tops," Clara Barrus wrote in an article for *The Craftsman*, "reveal the soul of the mystic."[111]

From 1869 on, Muir's almost wanderlust life was framed by holy Yosemite: making his first ascent of Cathedral Peak; taking Ralph Waldo Emerson to see the great falls; publishing his first article in California on glaciers; and articulating the wilderness protection ethos in *Century* magazine. In between there were all sorts of fine outdoor adventures ranging from climbing Mount Shasta (14,162 feet) to floating 200 miles down the Sacramento River. But somehow he always came back to holy Yosemite. Muir's discoveries in Alaska, his promotion of U.S. national parks like General Grant and Sequoia, and his creation of the Sierra Club in 1892 brought him much celebrity back east. He became wild California personified to the New York literary set. When Muir published his first book in 1894—*The Mountains of California*—he became widely known as the "sage of the Sierras," the West Coast counterpart of John Burroughs. Before long he was writing so much high-quality prose that the term "Muirian" came into academic use.[112]

From the outset, there was much Roosevelt admired about Muir. Although Muir sometimes played the misanthrope, he had a shrewd political instinct. Memories of Yosemite seemed to gush out of Muir once back in the San Francisco Bay area. "Ordinarily, the man who loves the woods and the mountains, the trees, the flowers, and the wild things, has in him some indefinable quality of charm which appeals even to those sons of civilization who care for little outside of paved streets and brick walls," Roosevelt wrote. "John Muir was a fine illustration of this rule. . . . His was a dauntless soul, and also one brimming over with friendliness and kindliness."[113]

Not only did Muir write as naturalist with the authority of someone like Thoreau or Burroughs; he also joined the U.S. Forestry Commission, offering practical advice on land management. He could play the wonk when necessary. Muir's articles in *Harper's Weekly* and *Atlantic Monthly* galvanized popular support for protecting forests. Although history always associates Muir with Yosemite, he was also largely responsible for Mount Rainer's becoming a national park in 1899. So when Roosevelt arrived in Muir's backyard, Yosemite National Park, in 1903 the sixty-five-year-old Muir was celebrated worldwide as a wise man. That year alone, and the next, Muir traveled to London, Paris, Berlin, Russia, Finland, Korea, Japan, China, India, Egypt, Ceylon, Australia, New Zealand, Malaysia, Indonesia, the Philippines, Hong Kong, and Hawaii. Roosevelt had purposefully come to Yosemite before Muir left on his intercontinental tour. The president wanted to pay homage to Muir (and to exploit their high-profile rapport for the history books).

Roosevelt and Muir in Yosemite National Park.

The general goodwill between Roosevelt and Muir that spring was exemplary. Both men had gone after the "malefactors of great wealth" in the West for raping the natural landscape. Muir was thrilled that President Roosevelt—through Secretary of the Interior Hitchcock—was punishing those who abused their power at the GLO (and even forcing the commissioner Binger Hermann to resign in disgrace for covering up the Halsinger Report regarding land fraud in Arizona). Directing Hitchcock to investigate illegal land grabbers such as John A. Benson and Frederick A. Hyde (two lawyers in San Francisco whom Muir deeply distrusted), President Roosevelt waged "historic warfare" against dishonest California copper syndicates, real estate speculators, thieves at the land office, and lumber companies. Under Roosevelt's influence the county indicted Binger Hermann, Senator Mitchell of Oregon, and Benson and Hyde. In other words, besides their insatiable love for the outdoors, Roosevelt and Muir shared enemies lists.[114]

The great three-night Yosemite campout of Roosevelt and Muir almost didn't happen, owing to conflicting schedules. As noted Muir had planned to travel around the world promoting national parks with his conservationist friend Charles S. Sargent that May. But Roosevelt, upon hearing of this, sent Muir a coaxing personal letter. "I do not want anyone with me but you," he wrote, "and I want to drop politics absolutely for four days and just be out in the open with you."[115] Realizing that such private time with the president discussing vulnerable Yosemite would be invaluable to the preservationist movement, Muir wiggled out of his other commitment. "I might be able to do some good in talking freely around the campfire," Muir told Sargent apologetically.[116]

As difficult decisions go, Muir was right to postpone his globe-trotting to spend this time with Roosevelt. Roosevelt and Muir, in the temple of Yosemite, vowed to let their biographies be intertwined for the sake of the conservation movement they were both leading, each in his own way. In effect, the Sierra Club joined forces with the Boone and Crockett Club—hikers and hunters forged an alliance on behalf of California's preservation. Always a biosphere activist, Muir talked nonstop with Roosevelt about the Sierra Club's ambition to get the Yosemite Valley incorporated into the surrounding park. And his stories of reckless timber depredations were ideal for arousing Roosevelt to shout down the "swine"—his new favorite word. Muir proved masterful at riling Roosevelt up. "I stuffed him pretty well regarding the timber thieves," Muir later bragged to a friend, "and the destructive work of the lumbermen, and other spoilers of the forests." As for Roosevelt, he admired Muir's dedication to California's beauty. Muir, he knew, was a hero and a live wire when it came to preserving Yosemite; Muir spoke directly and from the heart at all times. At one point, by the campfire, Roosevelt began telling his yarns about big game hunting. Muir, however, was bored and was singularly unimpressed. "Mr. Roosevelt," he asked, "when are you going to get beyond the boyishness of killing things. . . . Are you not getting far enough along to leave that off?" After a moment's pause Roosevelt, in a softer voice than usual, replied, "Muir, I guess you are right."[117] (But while Roosevelt did start promoting the camera instead of the rifle, he never gave up the sport of shooting big game.)

Because Muir was *the* California mountain man, Roosevelt embraced him as a fellow advocate of the strenuous life. Muir's philosophical concept of God as being found in nature likewise earned Roosevelt's approval. They were joined at the hip in both regards. But Roosevelt was truly at odds with Muir over sport hunting. When Muir, for example, received a solicitation to support a society called the Sons of Daniel Boone (which was like the Boy Scouts), he demurred. Young Americans, Muir wrote, needed to mature away from "natural hunting blood-loving savagery into natural sympathy with all our fellow mortals—plants and animals as well as men." And this wasn't an isolated antihunting statement. Muir's correspondence after 1903 is laden with criticisms of "the murder business of hunting," and with demands that the "rights of animals" be enforced as ethical standards. This was a far cry from Roosevelt's and Burroughs's belief that sportsmanlike hunting and fishing provided "ideal training for manhood" and would in the end "save the nation" from effeminacy.[118]

Hunting wasn't the only intellectual division between Roosevelt and Muir. Roosevelt liked Gifford Pinchot too much for Muir's comfort. Ever since the dispute in Portland Muir saw Pinchot as—for the most part—a deadly enemy. Muir didn't recognize the Pinchot who helped save wonders like Crater Lake or Wind Cave; he saw only a featureless scoundrel who had once said that forests were a factory for trees.[119] And soon to come was Muir's tragic disagreement with Pinchot over Hetch Hetchy—the glacial valley filled by the Tuolumne River in 1923 with the construction of the O'Shaughnessy Dam. Why didn't Roosevelt use an executive order to save Hetch Hetchy? Still, some historians have mistakenly downplayed Roosevelt and Muir's mutual admiration society. There was a very real tenderness between them. Ever since Muir formed the Sierra Club in 1892, Roosevelt had kept a close eye on his courageous actions; Roosevelt was, in fact, a New York cheerleader for Muir. While Roosevelt always saw Ulysses S. Grant as the "father of the national parks," he knew that Muir was California's watchdog. In particular Roosevelt's famous essay "Wilderness Reserves" echoes Muir's 1901 book, *Our National Parks*. However, Roosevelt was disappointed that unlike Burroughs, Muir simply didn't know his birds; he was focused on "the trees and the flowers and the cliffs."[120]

Because Roosevelt considered himself "many-sided" he unhesitatingly and admiringly accepted Muir's self-description as a Californian "poetico-trampo-geologist-bot. and ornith-natural, etc!—etc!—etc!"[121] By 2009 the John Muir National Historic Site had created a Web site featuring dozens of "Muirisms" arranged alphabetically. Whether you looked under "Age" or "Rough It" or "Water Ouzel," all of these pearls of wisdom *could* have been written by Roosevelt; their viewpoints on nature were that closely shared. With great enthusiasm, Roosevelt read Muir, savoring lines like: "Any glimpse into the life of an animal quickens our own and makes it so much the larger and better every way."[122] And Muir wrote to his wife that Roosevelt was "so interesting," overflowing with "hearty & manly" companionship.[123] "Camping with the president was a remarkable experience," Muir told Merriam. "I fairly fell in love with him."[124]

Oh, what a grand time Roosevelt and Muir had together in Yosemite for those three memorable days. They hiked to and camped in many of the most beautiful spots in Yosemite, including Bridal Veil Falls, where they had a fantastic view of El Capitan and Ribbon Falls gushing down from the valley's north rim. Religious metaphors filled Roosevelt's writings about Yosemite, with Muir serving as his Old Testament guide through the wilderness. (Except that Muir's god wasn't the god of ancient Israel.) For starters, there didn't seem to be a sickly face within 100 miles of the park; such human healthiness always appealed to Roosevelt. Even though Yosemite was a national park, bear traps were still laid on the floor of Yosemite Valley; Roosevelt wanted the "setters" arrested. Only hunting bears with rifle or knife was a sport; there should be no steel traps in a national park.[125]

Although Roosevelt changed clothes a few times, he is remembered as wearing jodhpurs with puttees, a thick sweater, a Stetson hat, and around his neck a soiled bandanna. Muir wore an oversize coat and loose-fitting trousers, looking rather like a hobo who had been cleaned up for a photo. Both men later boasted that they were alone in the Sierras, but Leidig and Lenord were constantly with them. There were also two packers and three mules.

Housed in the Yosemite National Park Archive is a detailed report of Roosevelt and Muir's visit of 1903, written by Charlie Leidig, one of the trail guides. It gives a revelatory *insider's* look at the trip. Leidig, for example, claimed that Roosevelt was annoyed when Muir wanted to stick a twig in one of the president's buttonholes. He also noted that "some difficulty was encountered because both men wanted to do all the talking." According to Leidig the president snored loudly, mimicked birds exactly, ate huge amounts of steak-fried chicken, and disdained crowds. Roosevelt's primary order was to "outskirt and keep away from civilization."[126] Highlights, according to Leidig, included seeing the sonorous Bridal Veil Falls, or *Pohono* ("puffing wind"), as the Indians called them.[127]

Roosevelt complained that the botanist and ornithologist Muir was much more interested in the trees than in the deer families they encountered along the primitive trail.[128] Muir explained to Roosevelt on the third day, May 17, that he had an ulterior motive, an agenda item—saving Mount Shasta along the California-Oregon border and enlarging Yosemite National Park to include Mariposa Grove at the Yosemite Valley. Roosevelt was all ears, enjoying himself in the timeless hills and valleys of Yosemite. Always intent on self-mythologizing, Roosevelt had created a "lost in the wild" scenario for himself. It made for good copy. There was something very romantic, indeed, about the president of the United States sleeping outside in a snowstorm, high in the Sierras, with the weather-worn John Muir as a companion. At sunrise Roosevelt and Muir hiked into Yosemite Valley, camping within range of the spray from Bridal Veil Falls. "John Muir talked even better than he wrote," Roosevelt found out in Yosemite. "His greatest influence was always upon those who were brought into personal contact with him."[129]

Back at the Sentinel Hotel, still pumped up with adrenaline, Roosevelt was unbelievably buoyant. He portrayed himself as a surviving backwoodsman, trapped by the harsh winter, eating dusty bread. "We were in a snowstorm last night, and it was just what I wanted," he said. "This is the one day of my life and one that I will always remember with pleasure. Just think of where I was last night. Up there!" President Benjamin Wheeler of the University of California–Berkeley hosted a dinner for Roosevelt at the Sentinel Hotel in the park. Instead of speechifying, Roosevelt recounted his exploits with Muir on Glacier Point "amid the pines and the silver firs in Sierrian solitude, in a snowstorm, too, and without a tent." Again he declared, "I passed one of the most pleasant nights of my life. It was so reviving to be so close to nature in this magnificent forest of yours."

Muir had been a wise, shrewd host. His desired effect had been to galvanize President Roosevelt to save more of wild California from human destruction. The camping in Yosemite clearly worked. Back in Washington, D.C., Roosevelt urged Congress to bring as many California redwoods as possible into the national park system. He wanted both the Yosemite Valley and Mariposa Grove to be part of the Yosemite National Park (at the time, they weren't). Immediately after leaving Yosemite, while he was in Sacramento, Roosevelt fired off a telegram to Secretary of the Interior Hitchcock. "I should like to have an extension of the forest reserves to include the California forests throughout the Mount Shasta region and its extensions. Will you not consult Pinchot about this and have the orders prepared?"[130]

No sooner had Roosevelt sent the order saving the Mount Shasta region than he wrote Muir a thank-you letter; he was already missing Muir's companionship and merry blue eyes. They had achieved a feeling of brotherhood. "I trust I need not tell you, my dear sir, how happy were the days in Yosemite I owed to you, and how greatly I appreciated them," he wrote. "I shall never forget our three camps; the first in the solemn temple of the giant sequoias; the next in the snowstorm among the silver firs near the brink of the cliff; and the third on the floor of the Yosemite, in the open valley, fronting the stupendous rocky mass of El Capitan, with the falls thundering in the distance on either hand."[131] Attached to this letter was his telegram to Hitchcock.

In Sacramento, still full of his Yosemite experience, Roosevelt also spoke publicly on behalf of the Muirian vision of California. Some Californians had demonized Muir as a "fanatic" or "cold-hearted crusader who cared too much for nature and too little for humans"—but Roosevelt was now a defender of the Sierra Club.[132] "Lying out at night under the giant sequoias had been like lying in a temple built by no hand of man, a temple grander than any human architect could by any possibility build, and I hope for the preservation of the groves of giant trees simply because it would be a shame to our civilization to let them disappear," he said. "They are monuments in themselves. . . . In California I am impressed by how great the State is, but I am even more impressed by the immensely greater greatness that lies in the future, and I ask that your marvelous natural resources be handed on unimpaired to your posterity. We are not building this country of ours for a day. It is to last through the ages."[133]

VIII

Sacramento wasn't a very impressive city after Yosemite. It was all dull buildings and mud holes, surrounded by impressive trees. As scheduled, Roosevelt delivered a few speeches in a high clear voice. In Sacramento, at the state capitol, men were wearing wingtips instead of buckskin boots. On leaving Sacramento Roosevelt headed straight to Mount Shasta—known as the "glorious sentinel of the Northern Gateway to California's flowery glades"—which was shrouded in clouds.[134] Rising upward like a mysterious fortress of oneness overlooking the surrounding Klamath Basin terrain, lonely Mount Shasta was seemingly unconnected to any range.[135] The beat poets Jack Kerouac, Lew Welch, and Gary Snyder would later describe Shasta, in cloud and sunshine, as if it were an embodiment of Zen, or California's Fuji. The memory of Shasta stuck in Roosevelt's mind for years to come. The artist Harry Cassie Best, hearing of the president's adulation of Shasta, presented Roosevelt with a realist painting of the snow-clad monarch bathed in eloquent pinks, subdued oranges, and rose-misted purples. The talented Best was able to depict reflected light in the British tradition of J. M. W. Turner. "I appreciate very much your painting, the 'Afterglow on Mount Shasta,' " Roosevelt wrote to Best, "and shall give it a place of honor in my home. I consider the evening twilight on Mt. Shasta one of the grandest sights I have ever witnessed."[136] With President Roosevelt's help many of Best's paintings of California ended up hanging at the Cosmos Club.[137]

In Oregon the president's train headed to the downtown Portland depot, where 20,000 people had come to witness the laying of a cornerstone for a Lewis and Clark Memorial. For hours Roosevelt put his forefinger to the brim of his Stetson instead of shaking all those hands. As always, he was courtly to the women. Roosevelt was able to see the Columbia and Willamette rivers and Mount Hood, but he never made it to Crater Lake, which was too far off the rail line. William Gladstone Steel—called the father of Oregon's first national park—was in the audience for the ceremony at the Lewis and Clark Memorial but was apparently not formally introduced to the president.[138]

In Portland, Roosevelt did meet with the wildlife photographer William L. Finley, the William Dutcher of the Pacific Northwest. It was probably refreshing for Roosevelt to use Linnaean binomials in speaking with Finley; neither Muir nor Burroughs often used these terms, because they seemed pretentious.[139] According to Finley, while Roosevelt had been in the West more than 120 tons of killed wild ducks had been shipped to San Francisco from Oregon; they were a popular dish in the city's booming restaurants. Finley desperately wanted to enact an Oregon model bird law to halt such slaughter. With admirable persistence he was keeping vigilant watch over Oregon's bird population, protecting even old dead stumps because flickers used them as homes.[140] Plume hunting was as horrific in Oregon as in Florida—maybe worse. Millions of Oregon's birds were being slaughtered for this purpose, from Portland to the Klamath Basin.

As a boy growing up in Oregon, Finley had collected bird skins and practiced taxidermy. But in 1899 he became an Auduboner, enamored of the Cascades and intrigued by the mysteries of flight. Then the Pacific Ocean beckoned him. His life mission was now to photograph Oregon's terns, puffins, grebes, and enormous winged pelicans, partly as a form of public relations (to help put the milliners out of business) and partly as an artistic endeavor. Encouraged by the fact that Steel had gotten his Crater Lake National Park from Roosevelt and Pinchot in 1902, Finley, with some advice from Frank M. Chapman, formed a chapter of the National Association of Audubon Societies in Oregon. And along with the photographer Herman Bohlman, Finley began playing the role of "Chapman with Kodak" along the Oregon coast near Tillamook Bay. They had been inspired, in part, by Chapman's revolutionary *Bird Studies with Camera.* Finley's images are now considered pioneering wildlife photography gems; they inspired *National Geographic* to improve its approach to capturing birds up close, even hatching. Finley was part of the first generation to abandon "shoot-skin-record" ornithology in favor of the camera.

William Finley and Herman Bohlman climbing Three Arch Rock. Together they photographed birds all along the Oregon coast and Klamath Basin.

There is no transcription of Roosevelt and Finley's meeting in Portland. Supposedly, Finley showed the president photographic images of Tillamook Bay's bird life (eventually included in his *American Birds*, published in 1907). The genius of Finley (with assistance from Bohlman) was that he'd climb any trees, even wobbly Douglas firs or tilted cedars, to photograph the nests of western tanagers and common bushtits. He was always searching for nature's fair light. Clean-shaven, elegantly slender, with a wild exaltation in his eyes, Finley became perhaps the best ornithologist the Pacific Northwest ever produced. He used ladders, lanterns, ropes, grapnels, rafts, glass plates, tripods, dories, and canoes as the tools of his trade, and no part of nature was off limits to his ingenuity. When camping on the beach, Finley explained, he "reached a sort of amphibian state."[141]

Since the creation of Pelican Island in Florida as a federal bird reservation in March 1903, West Coast ornithologists writing for *The Condor* began telling Roosevelt about Pacific Ocean "bird rocks" that should become refuges. Finley was no different. He wanted Three Arch Rocks—three huge, surf-hammered rocks (plus six smaller ones) half a mile offshore from the town of Oceanside, Oregon—to become the first national wildlife refuge on the Pacific coast.[142] The three principal rocks had arches carved by the wind and waves, making for a dramatic oceanic landmark. Besides their inherent tourist appeal, the rocks were home to Oregon's largest nesting colony of seabirds, and Finley had tried to scientifically document all the varied avian activity there in 1901. As Finley's photographs showed, there were 200,000 nesting common murres on Three Arch Rocks, making this the largest species colony south of Alaska Bay. Pigeon guillemots, rhinoceros auklets, and glaucous-winged and western gulls also came to the rocks. Unfortunately, so too did San Francisco restaurateurs, who raided Three Arch Rocks. In addition, this site was the *only* breeding ground for Steller sea lions on Oregon's coast.[143]

Finley and Bohlman wanted immediate federal bird reservation status for Three Arch Rocks. By documenting the Oregon coast in peril these wildlife photographers rendered a great service to the country. Three Arch Rocks eventually became an iconic site: decades later American Airlines used as its primary travel image a gorgeous color photograph of the arches, reproducing the picture on check-in screens and in-flight magazines.

Pressed for time in Portland, Roosevelt graciously invited Finley to the White House to make a formal presentation of wildlife sites in Oregon and Washington that needed preserving in the near future. Just as Roosevelt had a weak spot for white and brown pelicans, he also had a joyful infatuation with tufted puffins, cute-looking alcids that congregated in large numbers—between 2,000 and 4,000 at a time—on Three Arch Rocks.

By the time Roosevelt left Portland, Finley knew he had a new ally. In the coming years Finley, in collaboration with Bohlman, developed more than 50,000 still nature photographs of the Pacific Coast.[144] And in the summer of 1903, inspired by Roosevelt's policies, Finley and Bohlman literally lived on Three Arch Rocks, determined to capture bird life on film and eventually bring the images to the White House. Fate had done Oregon an immense favor by bringing Finley and Roosevelt together. Meeting Roosevelt transformed Finley from a wildlife photographer to a wilderness warrior. Just a few weeks after meeting T.R. in Portland, Finley went after the operators of the tugboat *Vosberg*, which used to dock in Tillamook Bay, taking passengers on Sunday shooting sprees along the bird rocks. It was slaughter simply for recreation. "The beaches at Oceanside were littered with dead birds," Finley told the Oregon Audubon Society, "following the Sunday carnage."[145]

Armed with the "model bird law," Finley was able to put the *Vosberg* out of business. Furthermore, Finley, using his Rooseveltian alliance with AOU's William Dutcher to Oregon's benefit, arranged for two wardens to be hired with Thayer Fund money in the Klamath Basin. On being elected president of the Oregon Audubon Society in 1906, Finley bought a patrol boat to police the Klamath Basin wetlands against milliners. His commitment to "citizen bird" was total. In a public relations stunt aimed at exposing feather hunting as immoral, Finley tore a plumed hat off a prostitute in Portland, causing bedlam on the street, which local newspapers reported in vivid detail.[146] It was free publicity for the Audubon movement. And Finley also assisted the Roosevelt administration in going after the crooked Senator John Hipple Mitchell's illegal coastal land deals.*[147]

From Oregon Roosevelt headed to Seattle, using the Hotel Washington as his operational base to inventory everything about the town. Shipbuilding and Pacific trade were the themes that dominated Roosevelt's speeches in Seattle. To his chagrin, he never got a chance to see a Roosevelt elk grazing in the Olympic rain forests. He did smell the raw

*On January 1, 1905, Senator Mitchell was indicted, for favoritism regarding land claims, before the U.S. Land Commissioner. On July 5 Mitchell was convicted while the Senate was in recess. He died that December. Mitchell was one of only eleven elected U.S. senators ever indicted. President Roosevelt, cheering on cartoonists who portrayed Mitchell's beard in rich people's pockets, took his death as good riddance.

fir boards for sale along the wharf. Not surprisingly, given his adroitness as a politician, Roosevelt met with the who's who of greater Seattle. No debutante failed to receive a presidential bow. With such a crowded itinerary Roosevelt simply didn't have the time to experience the moss-grown snags of Puget Sound County. In general, he hadn't given himself enough time to explore Washington state. Heading back to Washington, D.C., from Seattle on the "Roosevelt Special," the president was in a sulky mood over this fact. With no appointments to keep, the pace of travel seemed sluglike. Even though Loeb had run a pretty good rolling White House, there was a backlog of bureaucratic work that needed the president's immediate attention. Wall Street financiers were sowing the seed of monopolies, and Roosevelt knew he had to be a regulator of the economy for the sake of the people. He had been spending time with high-caliber men like Burroughs, Pitcher, Muir, Bullock, Lacey, Young, and Finley, and now his heart shriveled when he had to grapple with dry-lipped New York types whose life purpose was making money.

Roosevelt gave hurried conservation-infused speeches in towns like Walla Walla, Washington, and Helena, Montana, as the Roosevelt Special went eastward. Neither community had many buildings from the nineteenth century, and Roosevelt's speeches were stale re-runs. He was starting to feel like an itinerant carny at a medicine show. The flame, spark, and glory of the Great Loop tour were over. The hour for paperwork had returned. No more acting like a child or savage in the wild. On the upside, Roosevelt was eager to get back to Edith and his six children. Nineteen-year-old Alice had just gone to Puerto Rico on a goodwill trip, and her father was as proud as a peacock. "You were of real service down there because you made those people feel like you liked them," he wrote, "and took an interest in them, and your presence was accepted as a great compliment."

Although Roosevelt was "pretty well tired" from the western trip, he wanted to tell stories about all the natural wonders he had explored. Draw your chair up, and Roosevelt would gladly tell an anecdote about Yellowstone, Grand Canyon, or Yosemite. Furthermore, he had taken a real liking to Josiah, keeping the little badger as his constant companion on the train. He knew his children would love the cute little critter. "So far he is very good tempered and waddles around everywhere like a little bear submitting with perfect equanimity to being picked up," he continued to Alice, "and spending much of his time in worrying the ends of anybody's trousers."[148]

When Roosevelt arrived at the White House his wife was surprised to see him looking so healthy and fit. He was also immaculately groomed.

If anything, all that travel seemed to have invigorated him. He was full of incredible stories about the Wyoming Rockies, Nebraska tree nurseries, the whitecapped Colorado River, towering California sequoias, comical Oregon puffins, and the unforgettable gentleness of John Muir in the snow. He was gripped by a hunger for the golden West, which defied description for those who hadn't experienced it. Encounters with bears, fawns, raccoons, and so on had suddenly become well-honed campfire-like yarns at the White House. Heartland citizens had handed animals to Roosevelt as gifts—over a dozen of them—and most were being shipped off to zoos by Loeb. But Roosevelt hadn't come home empty-handed. Josiah—just starting to cut its teeth—was the president's new companion, having been nursed on a bottle all the way from Kansas. With great ceremony, Roosevelt presented the badger to his seven-year-old son Archie as a gift. The family had grown by one.

It seems that every decade some writer, anxious for quick money, publishes a book on White House pets. One star attraction in this lightweight pulp fare is Josiah. (Another is Calvin Coolidge's pet pygmy hippopotamus, Billy.[149]) If anything spoke of eccentricity, keeping a badger at the White House did. As a species, badgers have some undesirable traits: they are unpredictable, frequently carry parasites, have sharp claws, and vomit often in captivity. Sometimes Roosevelt, full of glee, allowed Josiah free rein in both the White House and Sagamore Hill. Over time, Josiah became exhibit A of Roosevelt's over-the-top anthropomorphic enthusiasms. "Josiah, the young badger, is hailed with the wildest enthusiasm by the children, and has passed an affectionate but passionate day with us," the president wrote to a friend that June. "Fortunately his temper seems proof."[150]

Now, with the western trek finished, Roosevelt was ready to make the badger instead of the teddy bear his political symbol. After all, 1904 was an election year and the president wanted to rip the Bryan Democrats and the big-business Republicans to shreds for—among many other things—not understanding the imperatives of preservation, which had become synonymous with the type of progressivism known as Rooseveltian conservationism. The Great Loop journey had convinced Roosevelt not only that his conservationist agenda was on track in the West, but that it needed to be amplified for the ages. And his friendships with Muir, Burroughs, and Finley—key allies—had been enhanced.

CHAPTER TWENTY

Beauty Unmarred: Winning the White House in 1904

I

President Roosevelt started 1904 by writing a spate of letters to family and various friends. The contents were history, and nothing seemed off-limits. Showing a considerable breadth of knowledge, Roosevelt mused about everything from "pagan" Rome to nineteenth-century naval power. Outdoing himself as an intellectual president, perhaps even equaling Jefferson in bookishness, Roosevelt contemplated the lasting achievements of Darwin, Huxley, Spencer, Milton, and a dozen others. It was as if Roosevelt was trying to see where he himself fit into world history. Interestingly, he seemed to feel compelled to insist that Francis Parkman—who had died in 1893—was a more talented historian than the entire American Historical Association membership combined. In Roosevelt's book *Hero Tales from American History*, in fact, Parkman was considered on equal terms with the likes of Washington, Lincoln, and Grant. "He went to the Rocky Mountains, and after great hardships, living in the saddle, as he said, with weakness and pain, he joined a band of Ogallalla Indians," Roosevelt noted after his idol's death. "With them he remained despite his physical suffering, and from them he learned, as he could not have learned in any other way, what Indian life really was."[1]

As president, Roosevelt began using Sir George Otto Trevelyan, author of *Life of Macaulay*, as his chief historian correspondent. Roosevelt, even though he was grappling with international crises in Japan, Russia, Colombia, and Morocco (and in a presidential election year), found time to reflect with Trevelyan on the need for more "faunal naturalists" in the grand tradition of John James Audubon. Perturbed by the germanization (i.e., overspecialization) of American colleges and universities—which Roosevelt continued to believe were strangling the talent out of the new generation of naturalists—the president lamented the unfortunate triumph of the pedantic, petty twentieth-century men who had contaminated history with dullness: there was a "lamentable dearth in America," the president wrote, of new naturalist work of "notable and permanent value" in the tradition of Audubon, Thoreau, and Burroughs.[2]

Forty-five years earlier the world had received Darwin's *On the Origin of Species*, followed by *The Descent of Man.* Roosevelt now wondered what had happened in the naturalist field since then that was truly innovative and exciting. All America had to offer the post-Darwinian world was Dr. C. Hart Merriam of the Biological Survey, who was a "great mammalogist" but apparently had no ability to write triumphant, redefining zoological works. The fact that Merriam wasn't producing a *big book* continued to perturb Roosevelt. "[Merriam] himself suffers a little from this wrong training, and I am afraid he will never be able to produce the work he could, because he cannot see the forest for the trees," Roosevelt confided to Trevelyan. "He cannot make up his mind to write a great lasting book, inasmuch as there continually turns up some species of shrews or meadow mice or gophers concerning which he has not quite got all the facts; and he turns insidiously aside once more to the impossible task of collecting all these relatively unimportant facts. Still, he does understand that we should not leave to storybooks the vital life histories of our birds and mammals."[3]

Roosevelt also seemed somewhat estranged from Grinnell because of the flap between them over the article in *Forest and Stream* that had attacked John Burroughs. By April 1904 Grinnell had finished editing the Boone and Crockett Club's fourth volume, *American Big Game in Its Haunts*—for the first time without Theodore Roosevelt as coeditor. Yet the lead article, "Wilderness Reserves"—later collected in *Outdoor Pastimes*—had been written by Roosevelt; it also appeared in *Forest and Stream.* "Mr. Roosevelt's account of what may be seen [in Yellowstone]," Grinnell wrote, "is so convincing that all who read it and appreciate the importance of preserving our large mammals, must become advocates of the forest reserve game refuge system." As if trying to bridge the gap between Roosevelt and himself, Grinnell went on for several pages about what a "great thing" Roosevelt's "accession to the Presidential chair" had been for forest reserves, national parks, big game, and birds in general.[4] "Aside from his love for nature, and his wish to have certain limited areas remain in their natural condition, absolutely untouched by the ax of the lumberman, and unimproved by the work of the forester," Grinnell wrote, "is that broader sentiment in behalf of humanity in the United States, which has led him to declare that such refuges should be established for the benefit of the man of moderate means and the poor man, whose opportunities to hunt and to see game are few and far between."[5]

Roosevelt's essay "Wilderness Reserves" began with a photograph of tourists at Yellowstone, all dressed in Sunday clothes, watching black bears eat in an open field, and another of Oom John meditating by a clear stream, with traceried wrinkles around his all-seeing, naturalist's eyes. Once again railing against "greedy and shortsighted vandalism," Roosevelt explained why preservation of both forests and wildlife was essential to the long-term health of America:

> The wild creatures of the wilderness add to it by their presence a charm which it can acquire in no other way. On every ground it is well for our nation to preserve, not only for the sake of this generation, but above all for the sake of those who come after us, representatives of the stately and beautiful haunters of the wilds which were once found throughout our great forests, over the vast lonely plains, and on the high mountain ranges, but which are now on the point of vanishing save where they are protected in natural breeding grounds and nurseries. The work of preservation must be carried on in such a way as to make it evident that we are working in the interest of the people as a whole, not in the interest of any particular class; and that the people benefited beyond all others are those who dwell nearest to the regions in which the reserves are placed. The movement for the preservation by the nation of sections of the wilderness as national playgrounds is essentially a democratic movement in the interest of all our people.[6]

"Wilderness Reserves," in which Roosevelt used his 1903 trip to Yellowstone and Yosemite for color and details, was his greatest call yet for preservation. Because there had been no hunting involved in his "western trek," Roosevelt was able to focus his writing on *seeing* such wildlife as golden eagles, magpies, scores of black-tails, and an antelope band numbering about 150. There were many Darwinian food-chain anecdotes, including coyotes feasting on deer carcasses and eagles swooping down for mice. Roosevelt was calling for maintaining the *balance of nature* without unwarranted human intrusions. But the scenic wonders of the West—the groves of giant sequoias and redwoods, the three Tetons glistening in snow, the gulls circling Three Arch Rocks, the Grand Canyon at dusk, the great Mojave Desert with its lonely barren hills—impelled Roosevelt to declare that they should be "preserved for the people forever, with their majestic beauty all unmarred."[7]

Roosevelt was also deeply troubled by the Park Commission's overreach. He rejected the idea of construction along the National Mall from the U.S. Capitol to the Washington Monument in favor of "greenspace."

In a display of outrage that anticipated a later concept, NIMBY, Roosevelt insisted that the Mall should be used for "monumental purposes" only.[8] Too many new buildings, he believed, would ruin the park-like essence of official Washington, turning it into a crass thoroughfare. Critics of Roosevelt's greenspace accused him of hypocrisy, for building a West Wing on the White House (which he wanted architecturally renovated and enlarged) and also a tennis court. "It is true that I have a tennis court in the White House grounds," Roosevelt wrote in his own defense. "The cost of it has been trivial—less than 400 dollars. It has been paid for exactly as the adjacent garden, for instance, is paid for. The cost is much less than the cost of the greenhouses under Presidents Grant, Harrison, Cleveland, etc."[9]

Because there were eight Roosevelts in the White House—Theodore, Edith, Teddy Jr., Kermit, Alice, Quentin, Ethel, and Archibald—not to mention a flood of guests, expansion and remodeling had begun in 1902. Roosevelt wanted the White House divided into living quarters for the first family (East Wing) and office space for the president (West Wing). The architecture firm of McKim, Mead, and White was brought in to do the job. Overseeing it all was Edith, who was determined to transform the old Executive Mansion into "the recognized leader of Washington official Society," and to make it a "moral" factor in the "social life of America." By 1904 Edith had become the most popular first lady since Frances Cleveland. Part of Edith's appeal was that she good-humoredly allowed the six Roosevelt children to collect animals of all kinds. And, as the historian Lewis L. Gould put it, Edith took tremendous care of the president, "the largest child in Mrs. Roosevelt's brood."[10]

One of Edith's wildlife management problems at the White House was Josiah the badger, now full-grown. Because the president allowed Josiah to roam freely, it constantly gnawed into any leg within reach of its teeth, occasionally drawing blood. Furthermore, President Roosevelt built a durable house for Josiah so it could dig tunnels in the White House lawn.[11] "At present he looks more like a small, flat mattress, with a leg under each corner, than anything else," the reporter Jacob Riis recalled after an afternoon at Sagamore Hill. "That is the President's description of him, and it is a very good one. I wish I could have shown you him one morning last summer when, having vainly chased the President and all the children, he laid siege to Archie in his hammock. Archie was barelegged and prudently stayed where he was, but the hammock hung within a few inches of the grass. Josiah promptly made out a strategic advantage there, and went for the lowest point of it with snapping jaws. Archie's efforts to shift continuously his center of gravity while watching his chance to grab the badger by its defenseless back, was one of the funniest performances I ever saw. Josiah lost in the end."[12]

Then there was Jonathan Edwards, an untamable cinnamon bear that had been given to Roosevelt by a group of Republicans from West Virginia and named after the Puritan minister (an ancestor of Edith's). The bear had what Roosevelt called "a temper in which gloom and strength were combined in what the children regarded as Calvinistic proportions."[13] Roosevelt enjoyed taking Jonathan Edwards for walks, feeding it honey and nuts, and playing hard with it, as with an oversize dog. Eventually, when it got too big, the president had to donate it to a zoo.[14] As Riis noted in his *Theodore Roosevelt: The Citizen* (a well-written campaign biography of 1904), the family had so many pets they were hard to count. White rats, opossums, and raccoons were at one time or another White House residents.[15] Peter the rabbit hopped about, sometimes sleeping under sofas or inside closets. In between important meetings, Roosevelt would feed mice to his barn owl, a reddish-chested female. Aquariums were full of horned toads, painted turtles, and salamanders. A medium-size iguana used to wander about the White House corridors as freely as did Maude the pig. Roosevelt's son Theodore Jr. raised a pet lamb named "Teddy" as if it were a family dog. All that was missing was a pushmipullyu, or the White House could have been renamed Puddleby-on-the-Marsh.[16]

When President Roosevelt was awake late, he sometimes recorded habits of his domestic cats, usually for his children to enjoy. "Tom Quartz is certainly the cunningest kitten I have ever seen," he wrote to Kermit from the White House. "He is always playing pranks on Jack and I get very nervous lest Jack should grow too irritated. The other evening they were both in the library—Jack sleeping before the fire—Tom Quartz scampering about, an exceedingly playful wild creature—which is about what he is. He would race across the floor and then jump upon the curtains or play with the tassels. Suddenly he spied Jack and galloped up to him. Jack, looking exceedingly sullen and shamefaced, jumped out of the way and got upon the sofa, where Tom Quartz instantly jumped upon him again. Jack suddenly shifted to the other sofa where Tom Quartz again went after him. Then Jack started for the door, while Tom made a rapid turn under the sofa and around the table and just and away and the two went tandem out of the room—Jack not reappearing at all; and after about five minutes Tom Quartz started solemnly back."[17]

Roosevelt had long been a promoter of guinea pigs as first-rate pets for children; and he kept *five* in the White House: Admiral Dewey, Dr. Johnson, Bishop Doane, Fighting Bob Evans, and Father O'Grady.[18] And then one afternoon a sixth materialized out of thin air. A few weeks after New Year's 1904, the magician Harry Kellar put on a show for the president's family. "I went along and was as much interested as any of the children, though I had to come back to my work in the office before it was half through," Roosevelt reported to his son Kermit. "At one period Ethel gave up her ring for one of the tricks. It was mixed up with the rings of five other little girls, and then all six rings were apparently pounded up and put into a pistol and shot into a collection of boxes, where five of them were subsequently found, each tied with a rose. Ethel's however, had disappeared, and he made believe that it had vanished, but at the end of the next trick a remarkable bottle, out of which many different liquids had been poured, suddenly developed a delightful white guinea pig, squirming and kicking and looking exactly like Admiral Dewey, with around its neck Ethel's ring, tied by a pink ribbon."[19]

Hearing about how President Lincoln had kept a turkey named Jack around the White House after sparing its life one Thanksgiving, the president adopted a one-legged rooster as a favorite pet, prohibiting the cook from even thinking about breaking its neck. Then there was the pet turkey which became friendly with the president's two parrots. The *Washington Evening Star* reported: "There is no home in Washington so full of pets high and low degree as the White House, and those pets not only occupy the

First Lady Edith Roosevelt tolerated her husband's obsession with having live animals around him at all times.

attention of the children, but the President is himself their good friend, and has a personal interest in every one of them."[20] As the White House usher Ike Hoover put it, "A nervous person had no business around the White House those days."[21]

The antics of all these White House pets made colorful newspaper copy. Once Algonquin, a spotted pony, was escorted to the second-floor family quarters to boost the morale of nine-year-old Archie, stuck in bed with measles; the disease had swept through Washington like an epidemic.[22] Ecstatic to see Algonquin, whom he loved, Archie let out an Indian "whoop" and dived to hug his pet. Algonquin was so startled by Archie's abrupt gesture that his legs buckled and all 350 pounds of pony slipped and fell to the floor. The loud thud sounded like a muffled gunshot blast. The whole Roosevelt clan rushed into the bedroom deeply concerned. When the president returned from California he mildly reprimanded Archie, saying that such suddenness was unwise when dealing with ponies; they spooked too easily. The comical episode is now remembered as another Roosevelt first—the first time a horse rode in a White House elevator.[23]

Then there was the famous Kansas jackrabbit affair, which rocked Washington. A Topekan had donated to Roosevelt's White House menagerie two young jackrabbits from his home state. One day while being fed pellets, the rabbits escaped their cage. A wild scramble ensued. The president and his sons chased after them. Escaping from the White House lawn, the rabbits made their way to G Street and Twelfth Avenue, where they parted company, one heading east, the other west. "Newsboys and messenger boys joined in the exciting chase after the rabbits, and for a time business in that vicinity was practically at a standstill," the *Washington Post* wrote in a long feature story. "Both animals were large specimens, and, as they spread out their long limbs, many thought they were young deer." After hours of mayhem one rabbit was captured at Turelane and M Street N.W. The other made its way back to the White House as if wanting to be put back into its hutch. Instead, the president decided to let them both live in the White House shrubbery, "wild and free."[24]

Roosevelt started tossing carrots to the jackrabbits whenever he wandered the White House grounds to feed nuts to the squirrels. Both T.R. and the grounds policeman, named Mr. Curtis, ensconced in a security booth just east of the White House entrance, used to hand-feed the squirrels. Before long the squirrels were as tame as the Angora house cats. Roosevelt would sit on the grass, and the squirrels would scurry up to him to be fed. The squirrels weren't even afraid of the Saint Bernard, named

Rolla, or the retriever Sailor Boy. Sometimes 100 squirrels would line the walk heading into the White House, waiting for the president to come out and apparently knowing that he had pocketfuls of nuts.[25]

Perhaps the most exotic pet President Roosevelt had was a spotted hyena (*Crocuta crocuta*) named Bill, from the plains of Ethiopia. Eventually, Bill weighed about 150 pounds. He had been given to Roosevelt in March 1904 by Menelik II, emperor of Ethiopia, who claimed to be directly descended from King Solomon and the Queen of Sheba. Menelik had brought with him to America such wildlife as elephants, monkeys, tigers, pythons, and rare birds to donate to the Lincoln Park Zoo in Chicago. But he insisted that Roosevelt keep the hyena pup and a lion cub as personal pets.[26] Roosevelt donated the lion cub to the Bronx Zoo, but he kept Bill for a while, teaching it tricks, enjoying its high-pitched cackle, and letting it beg for table scraps.[27]

More than anything else, however, it was dog stories that the press loved. Whether it was Sailor Boy the Chesapeake retriever or Jack the terrier, President Roosevelt always seemed to have a canine friend nearby. He could have made a fortune writing stories about the White House dogs (on the order of later books such as *Old Yeller* or *Marley and Me*) for *Outing*. Even during important cabinet meetings about the Panama Canal or the Ottoman empire, Roosevelt would often pat a dog while he spoke. Sometimes he carried lunch scraps in the pocket of his suit coat to feed them as treats. One of his dogs, Pete, a bull terrier, ended up biting so many White House visitors that the president reluctantly exiled him to Oyster Bay. When Ethel's bull terrier Ace got lost at Sagamore Hill one fall afternoon, a high-profile search was undertaken, as if for a missing person. Eventually the *New York Times* was able to run the headline "Roosevelt Dog Is Found."[28]

Whenever a family pet died, Roosevelt buried it at Sagamore Hill in a special cemetery located north of the house and surrounded by native plants. An inscription on a memorial boulder there read "Faithful Friends." Each buried pet had its name carved in the stone monument, and there was a bench nearby for the mourners. Little American flags were stuck in the ground, as if it were Arlington National Cemetery. In remembrance of Cuba, the famous dog of the Spanish-American War, the president had his name carved into the rock. Eventually, Roosevelt created an arboretum, arching over the burial site of his animal friends.[29] Roosevelt's idea of heaven was a place where all these pets would come and greet him in a grand reunion.

Between meetings, no matter the weather, President Roosevelt would play fetch on the White House lawn with his dogs.

II

That spring a debate raged in Washington regarding what to do with Alaska. More than 100 bills concerning Alaska were presented to the Fifty-Eighth Congress. President Roosevelt wanted virtually all of them—particularly those protecting wildlife—passed. The discovery of gold had caused a rush to Alaska; but Roosevelt hoped to impede development by creating reserves for the big game mammals at Fire Island (which he would turn into a federal game preserve in 1909). He insisted emphatically that harvesting Alaskan wildlife must be regulated, and that a smart plan for managing natural resources must be implemented for the vast territory. He also promoted a court system and infrastructure improvements for Juneau, Skagway, Sitka, and other cities. A variety of books on Alaska—notably *Our New Alaska* by Charlie Hallock (printed by *Forest and Steam*), *A Summer in Alaska* by Frederick Schwatka, *A Trip to Alaska* by George Wardman, and "The Merriam Report" of the Harriman Expedition—had spurred entrepreneurs' interest in the land once derided as "Seward's folly." For the first time Americans were starting to see the acquisition of Alaska for $7.2 million (less than two cents an acre) as a steal. History had vindicated Seward's judgment as, Roosevelt believed, it would someday vindicate his own attempts to save Alaska's caribou herds.

Throughout 1904 Roosevelt also grew interested in the brown bears of Alaska, which could be found in every district. He regularly asked for reports from the Boone and Crockett Club about Alaska's black bears, grizzlies, and glacier (or blue) bears. The polar bear, he learned, was found only along the coast, in ranges of eternal ice, and never below sixty-one degrees north latitude (and it was found at that latitude only when swept down on Bering Sea ice floes). Merriam reported to Roosevelt about the various sizes of brown bears he saw on the Kodiak Islands during the Harriman expedition of 1899. Roosevelt hatched a plan to take a steamer to Alaska and then hunt with an Aleut guide along the salmon streams of Kodiak in search of a bull bear. He even ordered rubber boots and rainproof slickers in anticipation of the journey. Roosevelt envisioned himself not only killing a bear but writing an article about Alaska for *Scribner's Magazine*. Besides the bears, Roosevelt also wanted the newly discovered types of wild sheep and caribou saved. His administration's primary conservation policy initiative in Alaska from 1901 to 1904 was enforcing both the Lacey Bird Act and a wild fowl law (enacted June 6, 1900) for protecting eggs.[30] In 1902 the Boone and Crockett Club, with Roosevelt's support, helped Congress pass an act (32 Stat. L. 327) imposing seasonal hunting and bag limits in Alaska. As the Roosevelt administration structured game laws for Alaska, if you wanted to hunt, you needed a permit—issued by the Biological Survey. Only the secretary of agriculture could permit hides, trophies, carcasses, etc. to be shipped out of Alaska. The federal government under Roosevelt was seizing firm control of the last frontier.

President Roosevelt, in particular, was anxious to bring the districts of Alaska—not officially even a territory until 1912—into the American family. But he simply had no patience for dealing with bureaucrats on Capitol Hill who didn't know rain-fresh ferns from black-green moss. There were individual members of Congress, however, whom he greatly respected. Once again Roosevelt—this time as president—partnered with Congressman John Lacey of Iowa to save Alaskan ecosystems such as Saint Lazaria, the Pribilof Islands, the Yukon delta, and parts along the Bering Sea. Both men wanted the territory's seal and bird rookeries, forests, and fishing streams properly managed. Roosevelt was counting on Lacey (or "the major," as he started calling Lacey after the visit to Oskaloosa in 1903) to figure out how to build a railroad through Alaska's mountain passes while simultaneously protecting the priceless forest reserves. What mattered most to both Roosevelt and Lacey (now being called by the Boone and Crockett Club the "father of federal game protection") was that the incomparable wildlife of Alaska not be molested or its scenic wonders destroyed by reckless industrial capitalism. No American knew more about the dual issues of wildlife protection–forest conservation law and railroad law than Lacey—the author of both *Lacey's Railway*

Digest and the pro-bird Lacey Act of 1900.[31] "Cannot we get the Alaska legislation through?" Roosevelt pleaded with Lacey. "It does seem to me to be very important that this republican Congress show its genuine care for the welfare of Alaska."[32]

An uproar ensued throughout Alaska against the president's tough federal game laws in 1904. A grassroots movement in Juneau sought to repeal 32 Stat. L. 327, and its voice was heard in Congress. Charging that the laws promulgated by the Boone and Crockett Club and the Roosevelt administration favored rich sportsmen from the continental United States who wanted to bag a moose in the Kenai Peninsula, Alaskans flouted the game laws, risking arrest. How dare these elitists flood into Alaska with a perfume-scented permission slip from Secretary Wilson while blue-collar hunters who actually lived in the Brooks Range or the Kenai Peninsula were being rejected by the USDA. The rallying cry of Alaska's pioneers was "home rule." To mitigate, if only slightly, the discord between advocates of conservation and development, the Roosevelt administration was forced to concede to the territory's governor the right to issue permits. But capitulation went only so far. Roosevelt remained vigilant in maintaining federal control of the hunting of fur-bearing animals like seal and fox: USDA managed all fur-bearing land animals while the new Department of Commerce and Labor oversaw seals and walruses.[33] And Roosevelt began developing a strategy to save birds in Alaska just as he had done in Florida, North Dakota, Michigan, and Oregon.

By now George T. Emmons was sending Roosevelt regular reports about wild Alaska. They were all well-crafted, rational, and succinct. Roosevelt seemed to relish learning that the lumber companies of the territory were furious over the Alexander Archipelago Forest Reserve decree of 1902. Letters of protest arrived at the White House from a Protestant missionary in Fort Wrangell and a businessman in Ketchikan. A U.S. congressman took up the crusade to save Alaskan commerce from the conservationism of Emmons and Roosevelt. And, of course, the Indians were opposed to the federal government's engaging in land grabs. Roosevelt's response to all this blowback was predictable. On September 10, 1907, he created the 17-million-acre Tongass National Forest in southeastern Alaska, the largest ever formed. On July 1, 1908, he merged the Alexander Archipelago with the Tongass. The new Tongass National Forest—eventually 17 million acres—was a historical feat. The fjords, glaciers, and Coast Range forest were preserved.

On July 23, 1907, Roosevelt put aside another 5.4 million acres of Alaska as the Chugach National Forest. After the Tongass, it was the second-largest national forest in America. Concerned about the wildlife in the eastern Kenai Peninsula, Prince William Sound, and Copper River delta, Roosevelt was starting to envision Alaska as one vast wilderness refuge. It would become a place for urban dwellers to replenish their spirits. Meanwhile, to protect the Chugach and the Tongass from despoilers Roosevelt approved of ranger boats to patrol the 10,000 miles of gorgeous coast, which would soon become popular with cruise lines. As Roosevelt envisioned them, these patrol boats would be something like traveling ranger stations. In Roosevelt's Alaskan parks the motorboat had replaced the saddle and pack horses. As one early ranger in the Tongass declared, "The Alaskan ranger is just as proud of his boat as the Bedouin horseman is of his steed, and the ranger boats in Alaska are the most distinctive craft sailing the waters."[34] In 1908 the Roosevelt administration had a sixty-four-foot, seventy-five-horsepower yacht designed in Seattle to use for Forest Service duty.

Ever since his 1903 trip to California, where he saw the splendor of fog-bound San Francisco Bay with the Farallon Islands rising out of the blue Pacific, the president had grown even more interested in all things Japanese. In California in 1904 there was a lot of xenophobia and anti-Japanese sentiment, and Roosevelt hoped to curb its ugliest manifestations by talking publicly about Japan's virtues; that is, he hoped to ease the cultural clash. Tension was extremely high between Japan and Russia throughout the election year, over territory in Asia. On February 1 the Russian czar had said, at a dinner in the Winter Palace, "There will be no war"; but a week later Japanese troops landed at Chemulpo in Korea. Japan then torpedoed the Russian fleet at Port Arthur, stunning the Russian authorities. Privately, the president liked seeing Tokyo exhibit its naval power, which was a sign of national greatness. Although, wisely, on February 11 he declared U.S. neutrality with regard to the Russo-Japanese War, he immediately started working the back channels of diplomacy, hoping to arrange for a cease-fire between the two warring countries.

An old friend of Roosevelt's from Harvard, Kentaro Kaneko, was in the United States in early 1904, and the president feted him. To Roosevelt, Kentaro had an irresistible combination of "fine national loyalty" and "Samurai spirit."[35] Roosevelt believed the United States had much to learn from Japan, especially how to properly manage city slums; but that the Japanese, for their part, needed to find "the proper way of treating womanhood." The Japanese island of Hokkaido had also become a hot topic for naturalist discussions at the White House. Photographs of steaming mountain peaks, active volcanoes, and beech forests that captured the Japanese "garden spirit" enthralled the president. On hikes in Rock Creek Park, often with Pinchot at his side, Roosevelt, while rock climbing, would go on and on about samurai literature he had just read. That spring, influenced by Japanese rice-paper drawings on display in Washington, Roosevelt began writing about flowers of the Potomac River basin—the locust trees with white blossoms and honeysuckles moving conspicuously around the south portico. He admired the way the Japanese brought nature into everything they did, filling vases with chrysanthemum flowers and growing miniature bonsai trees in offices. Although he was not a specialist on Asia, he knew something about Japanese folk ways, and he understood that Japanese artists considered humans part of nature; they cultivated the high art of thriving in harmony with animal life. Oddly, Roosevelt encouraged Russian expansion in Asia, hoping that Tokyo wouldn't "lump" Americans together with Russians as "white devils" and as Japan's "natural enemies."[36]

From the White House, the president kept watch on the Japanese fishermen and plumers who were slaughtering wildlife around the Midway atoll. In 1859, Captain Nick Brooks had claimed the guano island of Midway for the United States. There was a marketplace demand for bird fertilizers in California, and at Midway the excrement was readily available for scooping up by the boat load. All was peaceful at Midway until President Roosevelt learned in 1903 that Japanese seafarers were killing the albatross which bred on the island. Immediately, Roosevelt dispatched twenty-one Marines to Midway to protect the albatross from slaughter.[37]

III

In March 1904 Pinchot had presented Roosevelt with a report which claimed that the western states and territories with the most public land were "progressing rapidly in population and wealth." In other words, the larger the forest reserves, the more prosperity for a state or territory. The report recommended that the Timber and Stone Act and the Desert Land Act be repealed, only to prove that land was indeed being irrigated. But as a trade-off the administration called for many new forest reserves. "From 1902 to 1905, over 26 million acres were added to the national forests, and many of the reserves contained good grazing and agricultural land," the historian Donald J. Pisani wrote in *Water, Land, Law in the West: The Limits of Public Policy, 1850–1920*. "No westerner could be sure where the process would end. Because they threatened to limit access to the public domain, both repeal of the land laws and reservation were perceived as threats to economic opportunity."[38]

On April 30, 1904, President Roosevelt officially opened the Saint Louis World's Fair that commemorated the centennial of the Louisiana Purchase (the ribbon-cutting had been delayed for a year owing to construction difficulties at the fairground). A decade earlier at Chicago's World Fair, Roosevelt had been an attraction himself, greeting guests at the Boone and Crockett Club's log cabin, and extolling the virtues of western expansion. Now, in Saint Louis, pushing a golden button to open the Louisiana Purchase Exhibition, he visited some of the nearly 150 miles of exhibits, including a stuffed Roosevelt elk. This fair popularized the hot dog, ice cream cones, iced tea, and sweet rolls. The world's largest pipe organ thundered out songs that Roosevelt heard enthusiastically, including the triumphalist "Hymn of the West," which was sung in his honor.[39]

Although Roosevelt did not overly admire Thomas Jefferson, considering him vastly overrated, he nevertheless sang Jefferson's praises at the fair. If Jefferson had done nothing else, Roosevelt believed, acquiring the Louisiana Territory from Napoleon had been enough to ensure his greatness. The whole continent, from coast to coast, was in Jefferson's debt. This was the same theme he had touched on the previous year when he visited Saint Louis as part of his Great Loop tour. And now, with all eyes on Missouri, that May the Olympic Games opened in Saint Louis. The United States won eighty out of 100 gold medals, though it should be noted that a separate competition was held for "uncivilized tribes" (that is, dark-skinned people).

While all these distractions were going on in Saint Louis, Roosevelt, on June 3, created his third national park: Sully's Hill, on the south shore of Devils Lake in North Dakota, named after the Indian fighter General Alfred Sully—whose fierce battles with the Sioux peoples Roosevelt knew well. The 780-acre parcel along Devils Lake was a densely forested haven for such migratory waterfowl as wood ducks, Canada geese, American white pelicans, mallards, hooded mergansers, and dozens of other species.[40]

Because Sully's Hill National Park was transferred to the National Wildlife Refuge System in 1931 (it is managed today by the U.S. Fish and Wildlife Service), it has been largely ignored by historians of American conservation. Its remote site and the fact that it's named after a general known for Indian massacres have also given this park, in a sense, an orphaned status. Only the WPA guide for North Dakota has ever really done this hilly, serene woodland justice.[41] But Roosevelt felt that by signing a proclamation for Sully's Hill that June, he was accomplishing many goals. For starters, he was beloved throughout North Dakota, like a favorite son. (He won the state in the 1904 presidential election by a landslide.) Roosevelt's act on behalf of Sully's Hill met with virtually no resistance. He was saying that North Dakota had subtle "wonders" equal to those in California, Oregon, and Wyoming. Since first visiting Fargo in 1880, Roosevelt had been enthusiastic about the Red River valley of Minnesota–North Dakota; and the tiny lakes around Sully's Hill offered a chance for him to save a migratory bird area he treasured. Also, the wooded glacial moraine mounds of Sully's Hill were in stark contrast to the surrounding prairies. It was an oasis for wildlife.

But mainly President Roosevelt, from reading so much about ornithology, knew that the Devils Lake area (of which Sully's Hill was a part) provided essential bird breeding grounds in the Central Flyway. Sully's Hill, in particular, was a favorite inland breeding ground of the American white pelican. If his administration was going to save pelicans in Florida and other Gulf states, he likewise needed to preserve the species' northern wetlands and prairie habitats.* Although there is no documentary evidence, Roosevelt may have created Sully's Hill National Park in solidarity with the American Civic Association. That same June, the association, under the conservationist leadership of the newspaperman J. Horace McFarland of Harrisburg, Pennsylvania, initiated a well-publicized effort for America to create more national, state, and municipal parks. A devotee of Roosevelt's philosophy of the strenuous life, McFarland urged that all politicians should adopt parks in their home districts. The Boone and Crockett Club had inventoried all federal and state parks and, embarrassingly, North Dakota had none; making Sully's Hill a national park changed that.[42] "The steep forested hills within Sully's Hill are a unique island of trees in North Dakota's sea of prairies," a local wrote in *Outdoors* magazine, "and many people enjoy the fall colors during September and October."[43]

Even while campaigning Roosevelt found time to stay involved in conservation efforts regarding the Boone and Crockett Club, Sully's Hill National Park, Alaskan lands, California's trout, Virginia's flowers, and the Washington Mall. And his efforts in the movement to protect wild birds increased greatly. Pelican Island had merely whetted his appetite for more federal bird reservations. Edward Howe Forbush, founder of the Massa-

---

*In 1917–1918, elks, bison, and white-tailed deer were reintroduced to Sully's Hill. On March 3, 1931, Congress transferred Roosevelt's North Dakota national park to the National Wildlife Refuge System; today it is managed as a big game preserve.

chusetts Audubon Society and author of *Birds of Massachusetts and Other New England States*, had just released an alarming special report about the diminution of various species along the Atlantic coast. From Oyster Bay that July, the president wrote to Forbush about the wood thrushes, catbirds, meadowlarks, robins, song sparrows, chipping sparrows, and Baltimore orioles he found along the cove near Sagamore Hill. While these species seemed to be thriving on Long Island, Roosevelt worried about New England. "Are the birds," Roosevelt asked Forbush, apparently with fingers crossed, "recovering their ground?"[44]

Roosevelt had encouraged states to form their own bird sanctuaries and forest reserves. Worried about the overharvesting of the Great Lakes pine forests, Roosevelt pleaded with the governors of Michigan, Wisconsin, and Minnesota to create state-run reserves. Following his bear hunt in Mississippi, Roosevelt also pleaded with southern states to develop forestry programs. In 1904 Louisiana became the first southern state to do so; the last one was Arkansas in 1931. Under Pinchot's leadership close cooperation between federal and state forest units was encouraged.[45]

That summer at Oyster Bay, even under the pressure of the 1904 presidential election, Roosevelt found time to write James Rudolph Garfield. It was Garfield who taught Roosevelt the point-to-point walk—how you never let any obstacle get in the way when on a hike. Besides being a hunter and birdwatcher, Garfield loved the outdoors life almost as much as Roosevelt did. And although Roosevelt was only seven years older, he felt very paternalistic toward Garfield, whose father, President Garfield, had been assassinated. "Our imitation of your point-to-point walk went off splendidly," Roosevelt wrote to Garfield on July 13. "I had six boys with me, including all of my own excepting Quentin. We swam the millpond (which proved to be very broad and covered with duckweed), in great shape, with our clothes on. Executed an equally long but easier swim in the bay, with our clothes on; and between times had gone in a straight line through the woods, through the marshes, and up and down the bluffs. The whole thing would have been complete if the Garfield family could only have been along. I did not look exactly presidential when I got back from the walk."[46]

In his letters during the summer of 1904, Roosevelt seemed prouder than ever of his work initiating conservation. Having established three new national parks in fairly short order, President Roosevelt, following Sully's Hill, checked up on Yellowstone regarding wildlife protection management. Buffalo Jones reported back to him that the grizzlies at Yellowstone were thriving, and as a result a new problem had arisen: the bears were rummaging through garbage dumps at an amazing rate and getting tin cans stuck in their teeth and paws. "As many as seventeen bears in an evening appear at my garbage dump," Jones wrote to Roosevelt. "Tonight eight or ten. Campers and people not of my hotel throw things at them to make them run away. I cannot, unless there personally, control this. Do you think you could detail a trooper to be there every evening from say six o'clock until dark and make people remain behind a danger line laid out by Warden Jones? Otherwise I fear some accident. The arrest of one or two of these campers might help."[47]

This bear "mussing," as Roosevelt put it, gave him great joy and only a little concern. All of the Boone and Crockett Club's fights of the 1890s to protect wildlife had paid off at Yellowstone. The bears had rebounded and then some; he looked forward to thinning them out. "Oom John," Roosevelt wrote to Burroughs, laughingly, on August 12: "I think that nothing is more amusing and interesting than the development of the changes made in wild beast character by the wholly unprecedented course of things in the Yellowstone Park. . . . Buffalo Jones was sent with another scout to capture, tie up and cure these bears. He roped two and got the can off of one, but the other tore himself loose, can and all, and escaped, owing, as Jones bitterly insists, to the failure of duty on the part of one of his brother scouts, whom he sneers at as a 'foreigner.' Think of the grizzly bear of the early Rocky Mountain hunters and explorers, and then think of the fact that part of the recognized duties of the scouts in the Yellowstone Park at this moment is to catch this same grizzly bear and remove tin cans from the bear's paws in the bear's interest!"[48]

In that same letter to Burroughs, Roosevelt wrote about soldierly-looking redheaded woodpeckers, in black-red-and-white uniforms, seen flitting about the White House lawn. Honored to be active in a few state Audubon societies, dutifully keeping a "count" of birds seen at the White House, Roosevelt began lunching with ornithologists regularly. Merriam had informed the president that Breton Island was becoming "doable" as a refuge; the Department of Agriculture was ready to declare it a federal bird reservation. All technicalities were cleared up. Even the holes of fiddler crabs and bare mudflats could be protected. After speaking with Frank M. Chapman about some additional specifics, on October 4, 1904, Roosevelt created the Breton Island Federal Bird Reservation of the southeast coast of Louisiana with another "I So Declare It." The reservation was the second unit of what became the U.S. Fish and Wildlife Refuge System (whose stated mission was to "work with others to conserve, protect, and enhance fish, wildlife, plants, and their habitat for the continuing benefit of the American people"). Because nobody lived on the barrier islands—these islands were isolated sixteen miles from Venice, Louisiana, with treacherous Gulf waters in between—most Americans had never heard of the sandy breeding ground where pelicans and herons in the hundreds populated the beach. But plumers in Mississippi and Louisiana had. Regularly gangs made "hits" on nesting wading birds and seabirds.[49]

This changed after Roosevelt's "I So Declare It" of October. Within three or four months "Area Closed" signs were posted all over the new federal refuge. A full-time warden was hired. And in October, the police in Saint Louis had discovered a new investigatory method—fingerprinting. Perhaps, Roosevelt pondered, this new technique could be used against plumer gangs, who were then operating like pirates; three or four months of being locked up and smelling the dungeon stone, the president believed, would quickly turn them into preservationists. "Wreckers are no longer respectable, and plume-hunters and eggers are sinking to the same level," Roosevelt wrote, with regard to Breton Island. "The illegal business of killing breeding birds, of leaving nestlings to starve wholesale, and of general ruthless extermination, more and more tends to attract men of the same moral category as those who sell whiskey to Indians and combine the running of 'blind pigs' with highway robbery and murder for hire."[50]

There was nothing lush or exotic about Breton Island, which had been created from remnants of the Mississippi River's Saint Bernard delta. To some sailors the island was just a long sandbar of broken shells, sargasso weed, and wind-twisted pine boles. Sometimes, though, with the sunset in sharp shades of bright red-orange, the island could look more enticing than a beach at Acapulco. A wide variety of birds crossed and recrossed the island, barely flapping a wing but just gliding in rhythm with the Gulf waters. It was a soothing spot. The Tropic of Cancer vegetation included black mangrove and wax myrtle, both propagated by sprouting up tubers. To President Roosevelt's way of thinking, he had created a bird reservation at the "mouth of the Mississippi" where his beloved pelicans could prosper. It was also a prime place where herons and terns built nests, dived for fish, and hunted for purplish shrimp. All said, thirty-three species of birds—wintering waterfowl, wading birds, secretive marsh birds, and various shorebirds—lived on the island. When the birds were in full plumage Breton Island was quite a sight.

Just as Roosevelt had hired Paul Kroegel to be the warden of Pelican Island, along the Mississippi-Louisiana barrier islands he now employed Captain William Sprinkle, with funds from the USDA and the AOU–

Audubon endowment. Born and bred along the Gulf Coast, Sprinkle was a fine fisherman and shrimper and a professional wildlife protector. Later in life Roosevelt met him and declared that he "knows the sea-fowl" and the island where they "breed and dwell." Sprinkle spent so much time sailing around the Gulf islands that when he returned to the mainland near Biloxi, where he lived, there usually was a quarter inch of dust on the divan. Sprinkle marveled at the iridescence of the Mississippi birds and their musical harmonies. To Roosevelt, "a fearless man" like Sprinkle, who appreciated laughing gulls and skimmers, was worth ten times an Agriculture Department bureaucrat in Washington, D.C., trying to make policy out of paper.

This fisherman-cum-warden with his motorized skiff took to his law enforcement job promptly with a warm feeling for nature in the Gulf. No longer did he eat the green herons' pale-blue eggs; he was tasked with overseeing their hatching. Plumers, in fact, rued the day that Sprinkle had been given the warden's badge. "The Biological Survey does its best with its limited means; the Audubon Society adds something extra; but this very efficient and disinterested laborer [Sprinkle] is worth a good deal more than the hire he receives," Roosevelt wrote. "The government pays many of its servants, usually those with rather easy jobs, too much; but the best men, who do the hardest work, the men in the life-saving and lighthouse service, the forest-rangers, and those who patrol and protect the reserves of wild life, are almost always underpaid."[51]

In Louisiana, unlike Florida, Roosevelt received no immediate criticism for his federal reserve. With the presidential election just a month away, the saving of these barrier islands hardly constituted news outside New Orleans and Mobile. (And even in those communities its news value was scant.) Still deeply disliked in the South for having brought Booker T. Washington into the White House, Roosevelt knew that Louisiana's nine electoral votes would go to the Democrats. Virtually the entire "old Confederacy"—with the exception of West Virginia and Missouri—ended up voting against Roosevelt. Bristling because southern newspapers derided him for not really having fought on San Juan Hill, Roosevelt wrote to a friend that Jefferson Davis was nothing more than an "unhung traitor," worse than Benedict Arnold. Advocates of states' rights were the bane of Roosevelt's political existence. He also got into scrapes with North Carolina's tobacco lobby and Colorado's timber industry; they wanted his hide because of his ceaseless attempts as president to interfere with their profits.

Even though Roosevelt was popular nationally, he wasn't particularly loved by the Republican Party bosses. They'd have preferred Mark Hanna as their nominee. The advantage of Roosevelt as a presidential candidate in 1904 was that he was unhampered by any quid pro quos. Grumbling against Roosevelt usually had to do with his reformist viewpoint. A tough, tireless, bruising campaigner, Roosevelt wrote in early 1904 that to "use the vernacular of our adopted West, you can bet your bedrock dollar that if I go down it will be with colors flying and drums beating and that I would neither truckle nor trade with any of the opposition if to do so guaranteed me the nomination and election." While stumping, he often resembled a boxer more than a politician, punching one fist into the other to make a salient point. Full of anticipation, Roosevelt rolled into Chicago that June, dressed like a rancher and boasting of his successes in Panama and Cuba; regarding conservation, however, only the Newlands Act was cited as a chief accomplishment. There was no mention of new national parks, forests, or bird reservations. Choosing the conservative Charles W. Fairbanks of Indiana as the vice-presidential nominee, Roosevelt was immediately embraced by the old guard in Chicago. By ensuring support from the Republican conservatives, Roosevelt improved his chances for victory in the fall.

Fairbanks was another Ohio Valley frontier type, who was born in

Roosevelt chose Charles W. Fairbanks of Indiana as his running mate in 1904. Although Fairbanks was a conservative, Roosevelt selected him, in part, because he was a fellow conservationist crusader.

a middle-class home and moved west to Indiana for better farming opportunities. At six feet four inches tall, he towered over Roosevelt; and he always looked dignified in his proper Prince Albert coat even while feeding hogs at the Tarrant County Fair. Roosevelt wasn't enthusiastic about Fairbanks, considering him an old guard type in the viselike grip of big business. Also, Fairbanks tended to mumble and was something of a bore. But in the vetting process, Roosevelt had learned that Fairbanks had few if any negatives. He proved to be an excellent choice. For all of their differences, Fairbanks was an ardent conservationist, the founder of the Indiana Forestry Association, and an angler of sorts. And Fairbanks brought both ideological and geographical balance to the ticket. While Roosevelt campaigned with fist pounding a palm, Fairbanks spoke in a clipped, unspontaneous, dullish way. As a team Roosevelt-Fairbanks became known as "the Hot Tamale and the Indiana Icicle."[52]

No matter how cold Fairbanks was, he couldn't compare to the nearly lifeless Judge Alton B. Parker of New York, the Democratic presidential nominee. With the fiery William Jennings Bryan sitting out the 1904 election, Parker had been nominated over William Randolph Hearst. Parker was a decent, fair-minded appeals court judge, but the only real campaign issue he took up was getting the gold standard endorsed in the party's platform. Also, Parker had what modern-day media consultants call an "image problem." He always seemed to be overshadowed by rows of law books—not an uplifting quality in a national politician. And Parker didn't take criticism well: he was thin-skinned. Worse, Parker's choice for vice president was an eighty-one-year-old millionaire, Henry G. Davis of West Virginia, who helped finance the lackluster campaign with his own funds. All things considered, the famous Roosevelt luck was in play. There couldn't have been two less inspiring candidates for Roosevelt and Fairbanks to run against than the humdrum Parker and Davis.

During the campaign, with the media covering little else, Roosevelt appointed his "golden trout watcher," Stewart Edward White, as a special inspector for the California forest reserves. White's job was to stop illegal tree destruction and clear-cutting. Roosevelt also asked him to write a hunter-naturalist's book about big game in California; it would be a sorely needed addition to America's naturalist library. Meanwhile, Theodore's fatherly letters to Kermit, who was attending Groton in Massachusetts, were full of anecdotes about hiking from the White House to Chain Bridge along the Potomac, and descriptions of the autumn foliage—the rusty leaves of the Virginia creepers and the brilliant saffron tones of the beeches, birches, and hickories. Only in the last paragraphs of the letter of October 15, as if embarrassed, did Roosevelt mention the fact that the Democratic Party was besmirching his reputation. "In politics things at the moment seem to look quite right," he told Kermit, "but every form of lie is being circulated by the democrats, and they intend undoubtedly to spring all kinds of sensational untruths at the very end of the campaign."[53]

Roosevelt had a lot to boast about on the campaign trail. For starters, nobody doubted that he was the titular head of the Republican Party. If Roosevelt had jotted down on a three- by five-inch card his list of his historic accomplishments since becoming president, he could have listed the Panama Canal, the forming of a Department of Commerce and Labor (in conjunction with the Bureau of Corporations), settling a boundary dispute with Canada over Alaska, avoiding war with Britain over Venezuela (by going through the Hague Commission), siding with mine workers in the anthracite coal strike, launching numerous antitrust suits against monopolies like the Northern Securities Company (these suits were intended to ensure that rich and poor were equal under the law). With regard to racial matters, he had stood up to bigots in the U.S. Senate such as Edward Carmack of Tennessee and Benjamin Tillman of South Carolina. In the area of conservation, Roosevelt had created three national parks, twenty-nine national forests, and two federal bird reservations. His emphasis on irrigating the West was making human settlement in the arid zones of Arizona, Nevada, and California possible. (Although this was not understood at the time, western reclamation led to overconsumption of water and fertilizer and in that regard proved extremely harmful to the environment.)

So when President Roosevelt wrote to Kermit, a week before Election Day, about a "big sum of substantive achievement" he wasn't boasting falsely. Still, his successes were of the executive kind; his record of working with Congress was mediocre at best. Only his close alliance with Lacey had paid dividends. "Now as to the election chances," Roosevelt wrote to Kermit. "At present it looks as if the odds were in my favor, but I have no idea whether this appearance is deceptive or not. I am a very positive man, and in consequence I both attract supporters and make enemies that he [Parker] does not, in a way that he cannot." Enemies of Roosevelt included *Collier's Weekly* and the *Evening Post*, racist southerners, railroad companies, western developers, timber and mining concerns, Wall Street financiers, and the great capitalists (with the exception of Andrew Carnegie and a few others). The Standard Oil Company had publicly attacked Roosevelt when his administration established the

Bureau of Corporation, seen by the Rockefeller crowd as an insult and as antagonistic to big oil. In the weeks before the election, sensing that Roosevelt was going to win, Standard Oil wrote a $100,000 check for his campaign fund. Boldly, Roosevelt rejected the money, asking that the donation be returned, not wanting to be tainted by oil money. Roosevelt insisted that presidents during the automobile age could not, under any circumstances, afford to take a contribution "from an oil company seeking government influence."*[54]

IV

On November 8, 1904, Roosevelt won a decisive victory over Alton Parker, with 336 electoral votes to 140. The socialist Eugene V. Debs had run as a third-party candidate but didn't earn a single electoral vote. Roosevelt had earned the White House. This was, in fact, the largest plurality for a U.S. president up until that time. As for Congress, the Republicans swept both houses, picking up many new seats. It's been estimated that about thirty of the freshman Republican legislators elected were ardent Rooseveltian conservationists. Because Roosevelt had publicly pledged that he would not run again in 1908 (a decision he came to regret), he was free to push forward his ideas on national forests, wildlife protection, western irrigation, and federal bird reservations. Ironically, Roosevelt's premature pledge not to run again had the beneficial effect of letting him be more aggressive about creating forest reserves. He had learned something from the way President Cleveland had protected 21 million acres before leaving the White House in 1897. Responding to a congratulatory note from Owen Wister, Roosevelt bragged about his successes in irrigation and forestry, claiming he had the "college bred" men of the country on his side.[55] With executive power and no more elections, Roosevelt was off to the races regarding conservation—he was determined to create a new environmental infrastructure for America, one that would become a triumph of twentieth-century policy and planning.

When word of Roosevelt's election went out on the AP and UPI wires, telegrams of congratulations poured into the White House from all over the world: the writers included Kaiser Wilhelm of Germany, Emperor Meiji of Japan, and Prime Minister Balfour of Great Britain. Only one world leader, however, was clever enough to have sent con-

* While Roosevelt may have thought his campaign refused the money a 1912 investigation came to a different conclusion. The check *had* been cashed by the RNC. Although it was determined that Roosevelt hadn't been in the loop.

gratulatory gifts *before* election day, anticipating Roosevelt's victory: Emperor Menelik II of Ethiopia had sent two monkeys, two ostriches, one zebra, and one lioness on the Atlantic Transport liner *Minneapolis*.[56] Menelik wanted Roosevelt to receive the gifts on election night—and he did. Roosevelt was impressed and promptly saw to it that the animals were donated to zoos. A year later three huge elephant tusks arrived from Menelik—one of them was nine feet long. Roosevelt donated two of these to the National Museum and kept one for himself.*[57]

A few days after his election a confident Roosevelt, usually shy about fund-raising, asked Andrew Carnegie directly to fund a forest museum and library that Pinchot had been promoting; it would be a kind of Bronx Zoo for trees. (In 1901, Carnegie had formed the Carnegie Institution of Washington to "encourage in the broadest and most liberal manner investigation, research, and discovery, and the application of knowledge to the improvement of mankind."[58]) As Roosevelt saw it, the "forest life" museum would contain "specimens and models, the material for actual study of the life of the forest firsthand, or as it exists in the woods." The desired effect was to increase "our knowledge of the forest on a new plane and vastly increase the possibility of using it wisely and well." Deforestation was a global curse and Roosevelt wanted to confront it on a global level. "In other words," Roosevelt went on, in a letter to Carnegie, "such a collection, supplemented by a complete library of literature of forestry, and supported by funds for original research, would mark a *wholly* new step in the progress of forestry. Its creation would be a signal service not only to the United States but to every region of the world where trees grow. I'm strongly of the opinion that the plan is a good one."[59]

Never before had Roosevelt written in this way to ask for funds from a rich and powerful man. But Pinchot's ideas of a revolution in forestry were so vital for America that he was willing to approach Carnegie, who was widely celebrated by 1904 for embracing a wide array of educational advancement schemes. Carnegie libraries were springing up on Main Streets all across America. Unfortunately for Roosevelt, however, Carnegie had little or no interest in a tree museum. To his mind, it smacked of a boondoggle. Courteously, the old man rejected the appeal, but he did

*Emperor Menelik was an unusual personality—he lived surrounded by pets—and an unusual leader. He helped Ethiopia create its first modern banks, railroads, postal service, and so on. Anxious to establish modern capital punishment techniques in Ethiopia, he ordered three electric chairs. Unable to produce the electric current necessary for executions, yet not wanting to throw his purchase out, Menelik used one as his throne.

help Roosevelt promote bird rehabilitation projects in Florida. (Roosevelt thanked him for this in *An Autobiography*.) Still, the idea of a tree museum continued to intrigue Roosevelt and Pinchot. Making a return visit to the Saint Louis World's Fair with Edith, the president studied all the buildings with an eye for fine architectural touches, imagining how best to create a forestry museum that would attract visitors by offering modern exhibits. Predictably though, his favorite state-sponsored attraction at the fair was the North Dakota exhibit, which had his "Maltese cross cabin" on display. Roosevelt had succeeded in being the Pike, Carson, Boone, and Crockett of his time in the popular imagination—quite an accomplishment for a Manhattanite of the Knickerbocker aristocracy who had been sickly as a child.

What a thrill it was for President Roosevelt to see his Maltese cabin at the North Dakota display at the fair! Although it was just an ordinary log cabin, it had been carefully dismantled, shipped to Saint Louis, and then reconstructed to look exactly as it did from 1883 to 1886. Two pairs of Sunday trousers, an old straw hat, and high hunting boots once belonging to Roosevelt were put on display. Tourists came to study the ranch brand burned onto one of the logs. Capitalizing on Roosevelt's famous Badlands hunts, expertly hammered onto the side of the cabin were perfectly mounted specimens of deer, elk, eagle, fox, and owl. Besides his own frontier house, a childhood cabin of Lincoln and a dwelling constructed by Grant before the Civil War had also been erected as attractions at the fair; this was exactly the Republican presidential company Roosevelt liked to keep.[60]

During the four-month wait between his reelection in November 1904 and the inaugural ceremony in March 1905, Roosevelt, who reaffirmed his belief in the Monroe Doctrine in tough words, also stayed busy with conservation. President Charles William Eliot of Harvard University, for example, had published *John Gilley, Maine Farmer and Fisherman* for the Christmas season, and Roosevelt promoted it enthusiastically. Gilley was in the rough-hewn American tradition, like Seth Green, Paul Kroegel, and William Sprinkle—so Eliot had produced the very type Roosevelt wanted to hire to defend forests, wildlife, natural wonders, scenic vistas, and waterways.[61]

Roosevelt also meditated, that holiday season, on deer and wolves. Predictably, he had an explicit desire to hunt bear on the outskirts of Yellowstone now that its bear population was increasing. When a discussion turned to national parks Roosevelt, drawing on his 1903 trip with Burroughs, objected fiercely to *ever* allowing "sheepmen, cattlemen, or any other transgressors" into them.[62] Meanwhile, Roosevelt's old ranchhand Bill Merrifield came to see him at the White House one afternoon, shedding his ranchman garb in favor of "a severely correct frock coat, cravat, and top hat." The two men swapped stories for hours about 1880s North Dakota; Roosevelt's chief concern was that this simple, humble plainsman felt comfortable as if in the "people's house."

New animals were continually being added to the Roosevelt family menagerie, and to lull his children to sleep at night the president would read them *The Deerslayer* out loud. An expansive renovation was also under way at Sagamore Hill: Grant La Farge was helping the president redesign the house to look like a Kenyan hunting lodge.[63] And Roosevelt began strategizing about how best to save the Alaskan seal rookeries from British and Japanese fur hunters in the Bering Sea. Every time he read of entire seal colonies being slaughtered, pain struck his heart. The secretary of state, John Hay, was doing his best to get Britain to forgo seal hunting within a sixty-mile radius of the Pribilof Islands and to shorten the hunt seasons. Unfortunately, Hay's negotiations weren't going well.[64] Roosevelt began haranguing the specially formed Bering Sea Tribunal to ban "seal killing" during the spring breeding season. How could Great Britain consider itself a civilized country, Roosevelt fumed, when Britons slaughtered "nursing mother seals on the high sea?"[65]

One of President Roosevelt's first significant postelection acts was to transfer the federal forest reserves from the Interior Department to the Department of Agriculture's Bureau of Forestry on February 1, 1905. This had been Gifford Pinchot's dream since 1898. On March 3, with Inauguration Day approaching, the Bureau of Forestry was renamed the Forest Service. Roosevelt had two major reasons for going along with Pinchot's transfer plan: the GLO was filled with pro-business appointees who knew nothing about scientific forestry, and the centralization of the GLO caused long delays in issuing grazing, mining, and lumbering permits to regional reserve users.[66]

Unlike his boss, Roosevelt, Pinchot was interested in forest administration rather than wildlife protection per se. Wise use of timber resources was his objective. Pinchot, in fact, was extremely hesitant to regulate game animals on the forest reserves (which were yet again renamed National Forests in 1907) for fear of infringing on states' rights and giving western critics such as Senator Mitchell of Oregon reasons to disband the reserves by congressional legislation. This utilitarian attitude regarding forests made Pinchot the bane of Roosevelt's friends who favored wildlife protection, such as Muir, Burroughs, Hornaday, and Finley. According to

Pinchot's "The Use Book"—rules and regulations for rangers to follow—the Forest Service offices would "cooperate with game wardens of the State or Territory in which they serve." A couple of years later Pinchot made this perfectly clear by means of a provision in the Agricultural Appropriations Act of 1907: "hereafter officials of the Forest Service shall, in all ways that are practicable, aid in the enforcement of the laws of the States or Territories with regard to . . . the protection of fish and game."[67] Yet Roosevelt didn't believe only in Pinchot's notion that national forests were to be mainly *conserved*, not preserved. For Roosevelt, always interested in animals, the forests were also "cradles of wildlife."[68]

V

As Inauguration Day, March 4, neared, Roosevelt received the best gift imaginable from sixty-seven-year-old John Hay (still serving as secretary of state), short of saving Alaska's seals or naming yet another elk after him. Hay, who was seriously ill, presented Roosevelt with his precious Lincoln hair-ring. The connotations of this gentlemanly gift brought Roosevelt to tears, especially since Hay had sometimes belittled him. Hay, who was a friend of Roosevelt's father, had also been Lincoln's loyal personal assistant. When Lincoln was shot at Ford's Theatre by John Wilkes Booth and was brought across the street to the Petersen House, the attending doctor had clipped locks of hair from the dying president's head. Hay somehow inherited them and made a special ring with the hairs set like a diamond. "Please wear it to-morrow," Hay had written Roosevelt in his presentation note; "you are one of the men who most thoroughly understand and appreciate Lincoln."

Moved by Hay's gesture, Roosevelt wrote back that he was wearing the ring and would do so when taking the oath of office on March 4. (Later, Roosevelt claimed to have encountered Lincoln's ghost a few times in the White House corridors.[69]) The ring was a farewell gesture. On July 1 Hay died at his summer home in Newbury, New Hampshire, from what doctors believed to be a pulmonary embolism. His death was harder for Roosevelt to absorb than the assassination of McKinley. Without delay, however, Roosevelt appointed the intensely loyal Elihu Root to be Hay's replacement at the State Department. It was an inspired choice, for in the coming years Root would successfully remove the consular service from the "spoils system" by placing it under the direction of the Civil Service, maintain the "open door" policy in the Far East, help create the Central American Court of Justice, and eventually win the Nobel Peace Prize in 1912.

Roosevelt's inauguration parade of March 4, 1905, was quite a spectacle. In the presidential stand 1,200 dignitaries from all over the world sat arm-to-arm. A light blanket of snow was draped over the White House, creating a fine, calming white hush. The half inch of snow wasn't enough to create serious problems.[70] In fact, a mild front had blown into town, making Washington pleasant though still cold. Once again, Roosevelt was lucky. Acting as the impresario for his own inauguration, he had imported the heroic staff figures from the Saint Louis World's Fair representing colorful pioneers, plainsmen, and scouts.[71] At his request, thirty Rough Riders strode next to him as his special guard throughout the day. Representatives from every state and territory participated in Roosevelt's gala parade. Bands blared and boys' glee clubs sang. More than 2,000 American flags were handed out by the War Department so they could be waved by onlookers as the parade went by. A cold northwest wind swept through Washington, but throngs of sunburned cowboys from Texas and Oklahoma arrived to cheer Colonel Roosevelt.

At Roosevelt's request $2,000 from the inaugural planning committee's kitty had been appropriated to bring six renowned Native Americans to participate in the parade. They were the chiefs Geronimo (Apache), Buckskin Charlie (Ute), Little Plume (Blackfoot), America Horse (Brule Sioux), Hollow Horn Bear (Rosebud Sioux), and Quanah Parker (Comanche).[72] Quanah, in a traditional costume, saluted the president flatteringly as the "Great White Chief," initiating a lifelong friendship. Roosevelt promised Quanah—whom he admired greatly—that they would soon hunt wolves together in Oklahoma's Big Pasture–Wichita Mountains. But for the time being, Roosevelt hoped the six chiefs would enjoy themselves in *their* capital city.

Old Seth Bullock—"the Captain"—Roosevelt's conservationist protector of the Black Hills Forest Reserve, arrived in Washington, bringing with him the best broncos from Nebraska and cayuse ponies from South Dakota to march in the grand event. From Bullock's perspective Roosevelt was the only president to understand the American West. For people from the rangelands, riding in Theodore Roosevelt's inaugural parade had replaced Buffalo Bill's Wild West Show as the highest honor. Cleverly, Bullock had asked the cowboy star Tom Mix to ride at his side, causing the girls in the audience to scream with delight. Roosevelt was so naturally close to Bullock that they "spoke" in sidelong looks, understanding everything in a mere glance. Following the parade Roosevelt appointed Bullock the U.S. marshal for South Dakota.[73] "Seth Bullock is really a big fellow," Roosevelt wrote to his son Ted. "He belongs to the Viking age just as much as Harold Hardraade, or Olaf the Glorious, or Gisli Soursop. If ever I went to war I should want him as colonel of a rough rider regiment."[74]

Perhaps the most fitting symbol of the inaugural parade wasn't on the official itinerary. A few Washingtonians, it turned out, had gathered tree limbs and branches from nearby woods in Maryland. Setting up souvenir stands along the Pennsylvania Avenue parade route, these "big sticks" went for ten cents apiece, complete with an inaugural verification card. They sold briskly, revelers waving them in the air to express solidarity with Roosevelt, as they listened to the clip-clopping of the horses' hooves marching down the grand boulevard.[75] Meanwhile, for three and a half hours Roosevelt stood on the reviewing stand, doffing his top hat to the thousands who paraded past him. As the *New York Times* reported, "Old timers agree that in point of picturesqueness, variety, and general interest no inaugural procession in many years" equaled Roosevelt's parade of 1905.[76]

People poured into Washington, D.C., from far and wide to be part of history on that cold March day. More than 35,000 Americans took part in the procession. Most were wearing topcoats and scarves. Not since Andrew Jackson muddied the White House furniture with his "corn liquor" crowd had there been as much fun at an inaugural. People came from the lake regions of the Midwest, and the high plains of Colorado, Wyoming, and Montana. Old-style Rocky Mountain guides from west of Denver suddenly experienced East Coast civilization for the first time. Coal miners from West Virginia arrived by railroad from mountain hollows not even on maps. There were the Scotch-Irish from Kentucky and fishermen from the Georgia Sea Islands. Self-proclaimed cornhuskers ventured into the alien town from the grouse-hunting lands of Illinois and Iowa which the president had first tramped in 1880. Members of the Boone and Crockett Club, proud of their founder, sat together shouting "Roos-e-velt!" "Roos-e-velt!" like children at P. T. Barnum's circus. As one reporter observed, the entire navy base from Norfolk had come to see the author of *Naval War of 1812* in his moment of triumph. A Pennsylvania cavalry regiment played "A Hot Time in the Old Town Tonight," causing Roosevelt to applaud wildly; it was considered a risqué number. A comical note was struck whenever Roosevelt tried to shout out to a parader and the wind blew his glasses-cord into his mouth.

Cultural diversity was a predominant theme in Roosevelt's unusual parade. Eager-faced African-American students from Virginia and Pennsylvania were asked to participate in the parade; so too were Puerto

Ricans and Filipinos, who wore earmuffs to combat the cold. Socialites, carrying canes, arrived by the trainful from Beacon Hill, the Upper East Side, the "gold coast" of Long Island, and Philadelphia's East Falls. It almost seemed as if representatives from every bump, knoll, and stretch of America were present to cheer Roosevelt. The entire event was like the kind of murals Thomas Hart Benton would later paint: no region, epoch, or local type was left out. This was Roosevelt's own Buffalo Bill production, an amazing explosion of Americana choreographed in a way that matched the Saint Louis World's Fair.

This time Roosevelt took his oath of office by pledging on a Bible, the same one he had kissed on January 2, 1899, when he was sworn in as New York's governor.[77] Of all the guests Roosevelt received, it was Robert B. Roosevelt, standing on the same platform, who most strongly plucked the president's heartstrings.* Surely he remembered coughing his way through his Manhattan childhood, with his father trying to open the Museum of Natural History and Uncle Rob lobbying to create New York's fish hatcheries. Until 1905 President Roosevelt had always seen his Uncle Rob as a political liability and was disinclined to be photographed with him—stories about his uncle's flirtations and romances with women served only to hurt his own claim to high personal morality. But Uncle Rob had led the way in the conservation movement—he was, after all, the "Father of Fishes"—and his nephew seemed relieved to welcome the black sheep back into the family. No longer did he have to meet Uncle Rob clandestinely at Lotus Lake to discuss shad or eels. With no future political concerns to worry about, the president now let down his defensive guard. "Dear Uncle Rob," Roosevelt wrote on March 6: "It was peculiarly pleasant having you here. How I wish Father could have lived to see it too! You stood to me for him and for all that generation, and so you may imagine how proud I was to have you here."[78]

At the time of his inauguration, as if presenting himself with a gift, Roosevelt saved yet another North Dakota birding roost near Devils Lake.[79] Stump Lake was ten miles long and two and a half miles wide and teemed with migratory waterfowl. On four islets in the lake, thousands of ducks bred during the nesting season. White pelicans, in particular, bred in the wetlands ecosystem. Unlike brown pelicans, known for diving, the American white pelican, which flocks in a V, feeds by dipping its bill into a lake for fish.[80] Roosevelt knew that Stump Lake was a popular

*Others on the receiving platform with President Roosevelt included Henry Cabot Lodge and Ethan Hitchcock.

spot for oologists, a subgroup of ornithologists who focus on collecting and analyzing bird eggs. Anxious to put the oologists (that is, those not affiliated with the Smithsonian Institution or the American Museum or some similar scientific outfit) out of business, Roosevelt had an executive order prepared for signing. Thousands of eggs were being stripped from Stump Lake, and two generations were killed off in a "hit," as the theft was called in Grand Forks and Fargo. Secretary of Agriculture Wilson had written a letter to Roosevelt suggesting that the North Dakota islets be preserved. "In view of the fact that the season is close at hand when ducks will return to Stump Lake to breed," Wilson wrote on March 5, "I respectfully recommend that the reserve be created at an early date, in order that ample time may be given to make the preparations necessary to afford the birds full protection during the breeding season of 1905."[81]

Four days later, on March 9, Roosevelt declared Stump Lake his third federal bird reservation, placing it on the "same footing" as Pelican Island and Breton Island.[82] Part of his rationale for creating Stump Lake—and Sully's Hill, for that matter—was his sad realization that Canada reared most of North America's wild fowl while the United States did the slaughtering. His thinking would prove to be prescient. Around this same time, encouraged by Congressman Lacey, Roosevelt began envisioning a coordinated system of bird reservations from Lake of the Woods to the Gulf of Mexico and from the Aleutians of Alaska to Nihoa Island in Hawaii.

The bird protection movement was gaining momentum. Pelican Island, Breton Island, and now Stump Lake had been created. Many others were on the agenda. Birds were also being saved in Roosevelt's national forests. And Roosevelt was just getting started with his other preservationist ideas for the territories of Oklahoma, New Mexico, and Arizona.

CHAPTER TWENTY-ONE

The Oklahoma Hills (or, Where the Buffalo President Roams)

I

Following his inauguration in March 1905 Roosevelt became focused on hunting gray wolves (or coyotes) in Texas and Oklahoma. Removed from the rigors of campaigning, he wanted to see the so-called Spirit Trail of the Wichita Mountains and watch the sun drop beyond long reaches of flatlands known as the Big Pasture (a part of the Kiowa-Comanche reservation in southwestern Oklahoma south of the Wichita Mountains and north of the Red River of the South). The time to reinvent himself, during the spring, had again come. As an added incentive, bones of ancient elephants—mammoths—were rumored to have been found around Domebo Canyon in Oklahoma; perhaps he'd be able to inspect a few. Working through Colonel Cecil Andrew Lyon—the chairman of the Texas Republican state executive committee, known for his quick laugh and his lack of artifice—Roosevelt was primarily concerned about not repeating the Mississippi bear hunt. Because in 1905 Texas had no Republican U.S. congressmen or senators—El Paso to Beaumont being hard-core Democratic territory—Lyon was asked by the Roosevelt administration to recruit good Republicans to enter state government posts; the party couldn't completely fold its tent. A shared love of hunting was the brick of Roosevelt and Lyon's relationship; their shared party affiliation was the cement.[1]

"I am delighted with the good news about the wolves, but please do not have any taken alive and turned out," Roosevelt wrote to Lyon on March 16. "I would not care to hunt any that were loosed for that purpose. It would not do, on more accounts than one. If we can put them up and kill genuine live wolves and have a genuine hunt, I shall be very glad. . . . If not we shall take jackrabbits or coyotes; but nothing must be turned loose after having been captured."[2]

Roosevelt had an additional reason to be wary. The day he wrote to Lyon, two white police officers in Mississippi were murdered by a black mob at H. L. Foote's plantation, not far from the birthplace of the "teddy bear."[3] When this news reached the president, he was bound for New

York City to attend the private wedding of Franklin D. Roosevelt and Eleanor Roosevelt (both were his kin). Roosevelt was horrified by the bloodshed: segregationist Mississippi seemed to be a bottomless chasm of violence and injustice. At all costs, Roosevelt instructed his staff, he wanted to avoid galloping into a race war on his trip to Texas and Oklahoma. His advance team was instructed to tread carefully, avoiding political potholes. With regard to other matters, his staffers dutifully found him the right bridles, spurs, and whips, to use in Texas, Oklahoma, and Colorado. A fast, fearless, long-winded hunt horse was also procured. And, of course, what Roosevelt called "outdoors books" were ordered; they dealt with Oklahoma's wildlife species and included range maps and pertinent life histories.[4] What Roosevelt *didn't* want was a hotel with concierge, elevators, and white washbowls with copper fixtures, once he was out of Texas. The White House also announced that Roosevelt intended to spend time with his friend Quanah Parker, the last of the Comanche chiefs to surrender to "the white man's stony road."[5] (A small contingent of Comanche insisted that Parker wasn't their real chief, but that thieving white politicians had anointed him as such.)

According to the plan that was developed, Roosevelt would spend April around Fort Sill, hunting wolves in the surrounding primeval prairie dotted with solitary towers. Then he would venture to the Rockies in search of grizzly bears. (Only T.R. could get away with vanishing for weeks on end and still have the public cheer.) Once again the president wanted to explore the Colorado high country, which was the birthplace of his favorite rivers: the Rio Grande, San Juan, North and South Platte, Yampa, Gunnison, Arkansas, and Dolores—and, most famously, the Colorado. Meanwhile Edith, and three of the Roosevelts' children—Ethel, Kermit, and Archie—would spend a holiday on the Atlantic coast of Florida on the yacht *Sylph*.

Roosevelt, for his part, was advancing the notion of statehood for the Twin Territories (Indian and Oklahoma) as a single entity: Oklahoma. His choosing to hunt wolves in the Wichitas–Great Pasture area rather than in Texas contributed to that message. Until Roosevelt's presidency, the Twin Territories were considered largely no-man's-land to everybody but the mounted warrior cultures of the Kiowa, Apache, Comanche, and Wichitas (considered the ablest bison hunters on the Great Plains). The Spanish explorer Coronado had journeyed through the Oklahoma plains in 1541, looking for a "lost city of gold," but decided there wasn't one. (He carved "Coronado 1541" on Castle Rock where the Santa Fe Trail crossed the Cimarron River.) Thomas Jefferson acquired southwestern Oklahoma in the Louisiana Purchase, but it was considered the least valuable part of the purchase when President Andrew Jackson marched the Five Civilized Tribes from the southeastern United States down the Trail of Tears to the area.

Oklahoma! Everything loose or lost in America seemed to have tumbled into the territories as if through a giant chute. The population included reservation Indians, felons, paupers, drunkards, dirt farmers, fortune seekers, gamblers, and losers of every stripe. The smart pre–Civil War pioneers, those who were not seeking instant wealth, stayed on a network of trails and kept heading west. No pause. No rest in Tulsa, Oklahoma City, or Elk City. Just onward toward Amarillo by twilight and across the great desert lands to the shimmering Pacific Ocean. Others simply followed the Cimarron River, driving longhorns up to Kansas, where Christian settlements had taken hold. But only a rather optionless breed of humanity stayed in Oklahoma for very long.

Because people ventured to the Twin Territories to exploit resources or simply pass through, the basic tenets of conservationism were largely anathema to residents. In 1878, the last buffalo herd was reported to have been obliterated in Oklahoma—all the big game was rapidly disappearing from the Great Plains. Railroad expansion helped end the "golden age" of big game hunting in Oklahoma.[6] There were, however, plenty of smaller animals and birds. And Oklahoma still had vast natural beauty. The Cross Timbers region was a great forestland that stretched from southeast Kansas to northern Texas, dividing the western plains from southeastern forests. In general, the Plains Indians lived to the west and the Five Civilized Tribes to the east. Seldom did the two meet. Wintering waterfowl, nesting wood ducks, and neotropical migratory songbirds all lived in the spurs of the Oklahoma mountains.

There was something still peculiarly western about the Twin Territories, for this hard-living region was unsettled. Oklahoma City, for example, was erected virtually overnight between 1889 and 1895, because of the cheap land. When President McKinley's "great land lottery" was announced in July 1901—Kiowa, Comanche, and Apache reservation land was being given away—homesteaders came pouring into the region by railroad and on horseback. Luckily for posterity, a conservation easement was attached to the lottery. Coinciding with the homesteading act President McKinley—at Vice President Roosevelt's urging—saved 59,019 mountainous acres to create the Wichita Forest Reserve (administered by the Department of the Interior's Forestry Division of the General Land Office). Fortunately, fragments of the Cross Timbers survived at the Wichita Forest Reserve (eighty-nine square miles, considered sacred by Native Americans, just outside Lawton–Fort Sill). The entire Wichita mountain range covered a 1,500-square-mile region that extended into Caddo, Comanche, Kiowa, Jackson, Greer, and Tillman counties.

So, onward to the pure air of Oklahoma! Roosevelt craved a long stay on the prairie before the Twin Territories became one state. His mere presence in any Indian-Oklahoma territorial hamlet would guarantee it a mention in the history books. Roosevelt wanted to inspect the Wichita Mountains landscape, first inhabited by Paleo-Indians more than 10,000 years ago. He would be scouting for an ideal buffalo pasture. The buffalo at the Bronx Zoo, he believed, were ready to return to their home range in Oklahoma, where Coronado first saw these animals in the sixteenth century.[7] According to Charles "Buffalo" Jones, whom Roosevelt appointed as game warden of Yellowstone, the bison then living in Mammoth Valley had become such a popular attraction at the national park that railroad companies were promoting them to tourists. One advertisement exclaimed: "BISON once roamed the country now traversed by the Northern Pacific. The remnant of these Noble Beasts is now found in Yellowstone Park reached directly only by this line."[8]

During the Christmas season of 1904 Roosevelt spent an evening with the naturalist Ernest H. Baynes of New Hampshire, an animal trainer who went everywhere, befriended everyone, and noticed everything. Concerning buffalo, he liked to express original zoological ideas—often sounding brilliant—and then brag about his originality. In September 1904 Baynes had cooked up for Roosevelt a detailed plan to reintroduce buffalo. The initiative called for Roosevelt to remove from private individuals as many of the remaining American buffalo as possible. The U.S. government would offer large-scale ranchers like Charles Goodnight and Howard Eaton a fair price ($1,000) for each buffalo they owned.[9] Baynes even suggested that the government acquire the few remaining Canadian herds. Once all the remaining buffalo were procured, refuges would be established for them throughout the Great Plains. The Wichita National Forest in Oklahoma would be the first. Baynes's plan also offered a compelling reason why numerous reserves (not just the Wichitas) were needed: "If a contagious disease should strike any one of these herds not too large a proportion of the existing animals would be wiped out at the same time."[10]

After the election, Baynes's plan was put into action. Roosevelt even promoted big game preserves in his Fourth Annual Message on December 6, 1904, in a way that would have pleased George Catlin:

> I desire again to urge upon the Congress the importance of authorizing the President to set aside certain portions of the reserves, or other public lands, as game refuges for the preservation of the bison, the wapiti, and other large beasts once so abundant in our woods and mountains, and on our great plains, and now tending toward extinction. We owe it to future generations to keep alive the noble and beautiful creatures which by their presence add such distinctive character to the American Wilderness.[11]

Just after New Year's Day, William Temple Hornaday and Madison Grant of the New York Zoological Society went to work getting the buffalo ready for Oklahoma. The zoo herd had been donated to them by William Whitney of Massachusetts, a wealthy wildlife protection activist. At the time, the federal government owned only two herds: at Yellowstone National Park and National Zoological Park in Washington, D.C. Baynes was authorized by Roosevelt to start identifying private buffalo herds to be acquired by the federal government for a third preserve and a fourth preserve. Meanwhile, Congressman Lacey lobbied every U.S. senator about the new game reserve bill. After a few weeks of congressional seesawing, on January 24, 1905, the bill (33 Stat. L., 614) passed easily. The buffalo would soon be back on the Great Plains.

Essentially, the Wichita Forest Reserve became an experimental laboratory for the Roosevelt administration. A stipulation of the new Lacey Act was that 3,500 to 5,000 livestock grazing permits would be issued annually to local ranchers. Although hunting was strictly forbidden in the Wichitas, a special provision was made to allow wolf and coyote drives as part of a predator control program. At the time, big game managers believed that if buffalo were to survive, the wolves, coyotes, and cougars had to go. Oddly, guns were not allowed on these wolf and coyote hunts—only dogs. Secretary of Agriculture Wilson also made it clear that if these drives scared the livestock or deer, they would be abolished.

But the Wichita rangers had a more menacing threat to the ecosystem than predators. Although homesteading had been banned in the Wichitas, mining was allowed. From 1901 to 1905, prospectors poured into southwestern Oklahoma hoping to strike it rich. Shafts were built in the granite. Assay offices were set up, it seemed, in every one-horse town. Mining camps were established, including in Meers, Oreana, and Craterville. For a fleeting moment the Wichitas were like the Yukon or the Black Hills. But in the end, there was no gold, and luckily, the forest rangers were somehow able to protect the integrity of the Wichita Forest Reserve during the mining boom, with only limited damage.

From January to March 1905, on Roosevelt's instruction, groundwork was done by the New York Zoological Society to establish the Wichita Forest Reserve and National Game Preserve.[12] Secretary of Agriculture Wilson asked the Biological Survey to assist in finding the ideal location in southeastern Oklahoma for the game reserve. The Senate approved the plan. The part of the Wichitas that seemed to be the most highly recommended site was Winter Valley. So on the eve of Roosevelt's inauguration, which was to be followed by his hunting trip to the Wichitas, the buffalo project was on a fast track. Baynes's articles promoting it, first published in the *Boston Evening Transcript*, had been widely syndicated. An Adrian, Michigan fence company had huge spools of product ready to ship off at the president's beck and call.[13] It had been a long, hard fight since the 1880s to establish federal sanctuaries for the vanishing big game of North America. But all the efforts by the Boone and Crockett Club and the New York Zoological Society had finally paid off in the spring of 1905.[14]

On April 1, the president's specific travel plans became known to reporters. Things were happening quickly. Instructions had been received at Fort Sill, Oklahoma, from the White House to have a detachment of soldiers ready at the railroad outpost of Frederick (population 200 at best) for April 8. They would serve as the president's Secret Service detail in the Big Pasture–Wichitas. After touring Texas (Dallas, Austin, San Antonio, etc.), Roosevelt planned to cross the Red River to hunt in the Indian Territory with Chief Quanah. Thoreau had preferred nature in winter, but Roosevelt (like Burroughs) was a springtime man. With "thoroughly congenial company" he would ride on "the flats and great rolling prairies which stretched north from our camp toward the Wichita Mountains and south toward the Red River."[15]

Roosevelt was also eager to visit Fort Sill. On January 8, 1869, Major General Phil Sheridan had chosen this strategic site for a cavalry fort; he named it after a friend. It had been his headquarters during his successful effort to quash the Plains Indians' border raids. Comanche and Apache had been pillaging pioneer communities in Texas and Kansas even after the Chicago World's Fair of 1893 had declared the frontier closed. Fort Sill was a form of revenge. Hiring Buffalo Bill Cody and Wild Bill Hickok as scouts, Sheridan used the fort as U.S. Army headquarters in the Indian wars. Eventually, the army pacified all the Indians on the South Plains. In 1867 the Medicine Lodge Treaty was signed by the chiefs—all under duress—of the Kiowa, Comanche, Apache, Cheyenne, and Arapaho, who thereby agreed to reservations in southwestern Oklahoma. After being recalcitrant for a difficult eight years, in June 1875 Quanah Parker had his Quohada Comanche surrender at Fort Sill. With no buffalo herds present for sustenance, his people couldn't go on. Quanah considered escaping as Chief Joseph had done with Nez Perce, disappearing into the Guadalupe Mountains between Carlsbad, New Mexico, and El Paso, Texas. But he didn't. He couldn't put his people through the necessary deprivations and ordeals. A new, less bloody era had arrived in Comanche country.[16]

With the Comanches pacified and McKinley's homesteading lottery operating, approximately 29,000 Americans moved to Oklahoma. No hill or knoll was unclaimed. Suddenly, Lawton, the town attached to Fort Sill, grew into the third-largest city in Oklahoma. Sensing the closing of the frontier, and aware that raids were a thing of the past, Roosevelt soon transformed the mission of Fort Sill from cavalry to field artillery. On many days in Lawton, china cupboards would shake from cannon blasts. Curiously, after four or five years of mortar rounds, the usually skittish songbirds seemed to decide that the army practice wasn't a menace: they paid little attention to the loud artillery. The birds had adapted; call it Darwin's natural selection at work. As a western historian, Roosevelt was deeply interested in studying the U.S. Army's Seventh Cavalry, Nineteenth Kansas Volunteers, and the Tenth Cavalry buffalo soldiers. And, in keeping with his hobby, he wanted to see with his own eyes where Sheridan had once trained troops. Therefore, Fort Sill was a special American place to Roosevelt—like Valley Forge, the Alamo, the Chalmette Battlefield, and Appomattox Court House—simply because Sheridan had been connected to it.

Surrounding Fort Sill were the rugged Wichita Mountains—a jumbled landscape of rocky outcroppings, rock-crowded peaks, towering sentinels, granite ridges, and huge boulders strewn around 60,000 acres of prairielands. The Wichita Uplift—which occurred during the Pennsylvania Period about 300 million years ago—had created a number of peaks over 2,000 feet above sea level; they were older than the Alps or Himalayas. Four types of habitats could be found in the Wichita Mountains: rocklands, aquatic, mixed grass prairie, and cross-timbers.[17] In this strange range, bunch grass was interspersed with scrub oak forest, cactus, and aloe plants. The Wichitas were a botanical crossroads (*ecocene*) with more than 270 bird species taking advantage of the buffet, fattening up for the winter. It was said by western frontiersmen that the ancestral Wichitas—this series of rock prominences—was where the East met the West at the vertex of the Great Plains.[18] The Red Beds here—formed by prehistoric ocean currents, huge earthquakes, and glacial shifts—resembled the Garden of the Gods outside Colorado Springs. It was a geologist's dream, a rock garden in which boulders hung off cliffs, seemingly ready to tumble in a roaring avalanche with a northerly gust.

The president's lead guide in the Big Pasture and Wichita Mountains was going to be Jack "Catch 'Em Alive" Abernathy, a white, Anglo counterpart of Holt Collier; though his specialty was hunting the prairie wolf (*Canis latrans*) with catch rope instead of hunting the black bear with a knife. Born in 1876 in Bosque County, Texas, colorful Abernathy cut such a swath across the Red River Valley of Texas-Oklahoma that he became a legend to drovers from Amarillo to Tulsa. By the time Abernathy was fifteen years old he had already had successful careers as a cowboy on the A-K-X Ranch and as a bronco buster with the J-A outfit. Nobody, it was said, could saddle a wild horse as skillfully as Abernathy. Some people complained that he overworked and underfed his stallions. But everybody agreed that at dusk, Abernathy was usually the last man to hobble his horse.

Brickish and bullnecked, Abernathy was an excellent pianist and bluegrass fiddler in the fast Smokies style of eastern Tennessee. At his regular gigs in bars in Sweetwater, Texas, "Dixie" was still the most frequently requested song, followed by "Ole Paint" and "The Yellow Rose of Texas." Gregarious to the point of causing irritation, Abernathy had the smile of a dedicated drinker. In reality, he was nearly a teetotaler, taking only an occasional shot or two of bourbon for a toothache. "My six-shooter was my friend," Abernathy wrote of his youthful days, "and constant companion."[19]

Refusing to be an antihero Abernathy, a proud and devout Christian, lived on the straight and narrow. A man who could defend himself in a scrape, he was dutifully married and had five children. Abernathy was always looking for freelance law enforcement work; handcuffs often dangled at his side. Fairly regularly, he served as an under-sheriff or posse leader for day wages. Townspeople looked up to him. He also took on short ranch jobs at regular intervals. Around a campfire, Abernathy was always the guy tossing kindling on the flames, hoping to keep the conversation going. With his white horse, Sam Bass, and his brave dog Catch (a Scottish shepherd), he was as close to the character called "the Virginian" as was possible in north Texas and southwestern Oklahoma. Owen Wister could easily have woven Abernathy's trademark quip—"I am a goner for keeps"—into *The Virginian*. Adding to his western appeal, a kind of Fort Worth drawl colored Abernathy's utterances: "cow" was "kyow" and "woman" was "womb-man." But for all of his decency, Abernathy wasn't a nester. Restlessness was his normal condition.

By 1905, what had really made Abernathy legendary, a one-man show more dazzling than Annie Oakley shooting glass balls out of the sky, was his ability to catch a wolf alive by jumping off a horse, wrestling the wolf bare-fanged to the ground, and jamming his fist into the back recesses of the wolf's jaw. Hence his sobriquet, "Catch 'Em Alive." The wolves (or "loafers," as cowboys called them), unable to bite, would become utterly passive and submissive. With the skill of a wrangler, Abernathy would then wire the captured wolf's muzzle closed and hog-tie its feet.

"Coyote" was the Spanish name to distinguish these wolves from the larger gray wolves that lived in the same Mexico-Texas-Oklahoma range. (In writing about Oklahoma, Roosevelt often used the terms "coyote," "prairie wolf," and "gray wolf" interchangeably.) "Thus was born Catch 'Em Alive Jack," the historian Jon T. Coleman writes in a foreword to Abernathy's autobiography, "the alter ego that would carry Abernathy from Texas to the White House."[20]

Roosevelt first learned of Abernathy's bold, enterprising feats from a doctor friend, Sloan Simpson of Fort Worth, in January 1903.[21] In his customary way, he grunted in disbelief and raised his eyebrows questioningly. Abernathy was brought to Roosevelt's attention again by Colonel Lyon, over Christmas 1904. Apparently, Lyon had witnessed Catch 'Em Alive Jack hand-trap a few coyotes and immediately told the president about the crazy stunt. A vague plan was begun—that he'd go wolf coursing with Abernathy in Texas or Oklahoma soon and make the Wichita Forest Reserve the site of the buffalo preserve inspired by Baynes, Hornaday, and Lacey. Before long Roosevelt would give Abernathy the highest praise he could, saying that the plainsman had "a perfect knowledge of the coyote."[22]

II

As the "Roosevelt Special" pulled out of New York on April 3, 1905, the president was in a fun-loving mood, even though he was grappling with the Moroccan crisis (involving Germany, Spain, and France, whose military and economic interests in Morocco were threatened when its emperor called for Moroccan independence and anticolonial integrity). About foreign affairs in general, Roosevelt joked to the press that he had left the 300-pound William Howard Taft "sitting on the lid keeping [it] down." Reporters covering Roosevelt's trip asked why he had chosen dusty Texas and the Twin Territories, of all places, for a holiday. "I don't exactly say that I need a rest," Roosevelt said, "but I am going to take one in the open, under God's blue heaven."[23]

Excitedly, Roosevelt talked of meeting with Rough Riders at a reunion in San Antonio, riding with Quanah in the Big Pasture, and coursing for wolves with Catch 'Em Alive Jack in the Wichitas near Sheridan's old Fort Sill. The prospect of being caked with cracked mud seemed idyllic to Roosevelt. Other easterners might seek comfort, but he spoke in favor of burrs and smelling the blossom-spray of spring on the prairie. One of his favorite Rough Riders—"Little" (five-foot-two) Billy McGinty of Ripley, Oklahoma—would be around for casual laughs. As a horseman, Roosevelt said, McGinty had no peer. And holidaying in the dry plains had obvious benefits for Roosevelt's heath. Asthmatic wheezing was unheard along the 1,290-mile Red River of the South: this was open-lung country. Unbeknownst to reporters at the time, Roosevelt, as noted, was also planning to see if the Wichita Forest and Game Preserve would be a suitable place to reintroduce bison. In fact, predator control was a major reason why he wanted to hunt wolves. The zoologist William Temple Hornaday at the Bronx Zoo had informed Roosevelt that he was ready—more than ready—to send his hand-fed bison herd by railroad from the Bronx to Oklahoma. All he needed was Roosevelt's go-ahead. Roosevelt, in a sense, was serving as Hornaday's advance scout.

Although this is unprovable, at his inauguration in March 1905 Roosevelt probably told Quanah Parker (the last of the Quahadi Comanche chiefs) about the buffalo repatriation that the New York Zoological Society was overseeing; after all, Congress had (through the Lacey Act) authorized the president to create a big game reserve in the Wichitas, just two months earlier.* There was never a Native American that Roosevelt liked more than Parker. Because he had a white mother—Cynthia Ann Parker, who had been captured by Comanches in a raid near Groesbeck, Texas—Chief Quanah was a so-called half-breed. A tough Plains fighter in his youth, by 1905 he had become an American pragmatist, more of an optimist like William James than a fatalist like Crazy Horse. Yet there was also something mystical about him, an unaccountable reverence in his every utterance. Actually, some of Quanah's guru wisdom may have been drug-induced. The chief used peyote; eating mescaline made him feel equal to every task, able to glide like a bird or crawl like a snake.

* As an additional preservation measure in the Wichita Mountains, the Roosevelt administration levied a $1,000 fine on anybody caught poaching or hunting in the reserve.

The advisability of taking such drugs is always doubtful, but they seemed to work for Quanah. He was bookish and knew that mescaline in the form of peyote buttons had been used along the Rio Grande Valley by Native Americans since 3,700 BC. In the sixteenth century, Spanish priests in Mexico forbade native people to use it; but as the Comanche lost their foothold in the American Southwest in the late nineteenth century, peyote came into vogue among them, like the Ghost Dance.[24] In the face of defeat, why not get high and pray? Around 1890, after a particularly intense vision, Quanah created the Native American church movement. This religious experience led to his denouncing his past raids against white settlers and becoming a committed pacifist. He adopted peyote as a ritual for the Comanche and Kiowa. The bittersweet taste of mescaline could, he believed, free his people's minds from the white man's tyranny, and at the very least it was far better than being locked up at Fort Sill or Fort Leavenworth. The fact that Quanah ate *Lophophora williamsii* was unimportant to Roosevelt, because the drug had brought Parker into the arms of Christ. "The white man goes into his church and talks *about* Jesus," Roosevelt famously said. "But the Indian goes into his tipi and talks *to* Jesus."[25]

No one knew for certain where Quanah was born. However, he claimed it was somewhere around the Wichitas; and just as Roosevelt had learned the North Dakota topography inside-out, Quanah understood every nook and cranny of his beloved Wichita range and the Llano Estacado (Stalked Plains). Even though Quanah wore his hair in long braids and had several wives, he nevertheless adopted many of the white man's ways. An impetuous cordiality informed his dealings with white men. Often, he wore a business suit and derby hat. For example, his home in Cache, Oklahoma—the two-story, twelve-room "Star House," where Roosevelt dined and slept on the porch one evening—was modeled after the U.S. Army general's headquarters in Fort Sill. (Fourteen white stars symbolizing generals in the cavalry were painted on its red roof.[26]) It had been a gift of the Texan rancher Burk Burnett and was equipped with one of the first residential telephones in southwestern Oklahoma.[27]

What Roosevelt most admired about Quanah was his never-failing reverence for buffalo. Unlike Geronimo, the Chiricahua Apache who appeared in William Cody's Wild West Show, Quanah never abandoned his vision of a vast "buffalo common" on the Great Plains. The deeds of the U.S. Cavalry never diminished his fervent belief in the prophecy of a rebirth for the buffalo. A hard worker, not given to idle hours, the soldierly Parker had no concept of quitting. In 1903, Roosevelt had been informed that the 101 Ranch in Bliss, Oklahoma, had hired the capricious Geronimo to lead a hunt for captured buffalo. This was a theatrical pageant: Geronimo, in war paint, would lead on horseback, followed by a convoy of automobiles whose drivers would shoot at terrified buffalo acquired from the Goodnight Ranch in Texas. When Roosevelt heard from the National Editorial Association about this inhumane and unsportsmanlike hunt, he was furious. He wired Governor Thompson B. Ferguson of the Oklahoma Territory to stop the brutality at once. Also, according to the *New York Times*, federal troops from Fort Sill were ordered to arrest the culprits.[28]

Now, heading toward Fort Sill two years later, huge crowds showed up to hear President Roosevelt's speeches in Pittsburgh, Louisville, Saint Louis, and a hodgepodge of depots in between en route to Texas. A high-pitched train whistle always announced his arrival. Children—anonymous and interchangeable—stood along the tracks, waving flags and looking for a glimpse of their hero. Refusing to disappoint them, Roosevelt thrust his head out the train window, shouting hellos and farewells. Courthouse bells usually clanged on his arrival in every village. Water tanks were invariably painted red, white, and blue. Vendors set up wiener and lemonade carts, hoping to capitalize on the whir of anticipation that

Roosevelt formed an important alliance with the great Comanche leader Quanah Parker. Together they worked to bring buffalo back to the Wichita Mountains.

a visit by T.R. brought to town. Usually, Roosevelt's impromptu stages were fairground bleachers, hotel balconies, or railroad sidings. As on all of Roosevelt's whirlwind tours, whistles blew and the atmosphere accompanying each speech was festive. Americans had acquired a new respect for Roosevelt following the 1904 election, and this respect burst forth like the plume of a fountain.

On April 4, in Steubenville, Ohio, however, tragedy struck. The Roosevelt Special accidentally bulldozed into a man who was boarding a freight train and plowed right over him as it went west at thirty-five or forty miles per hour.[29] Roosevelt grew despondent over this; he was also embarrassed. Yet he refused to dwell on the death, and his Roosevelt Special went on. He took in the varied American landscapes from the window of his train compartment: the indolent Ohio hamlets fading, the drainage ditches of Missouri receding, the dark and haunted Oklahoma prairie stretching for hundreds of miles south of Tulsa. On a map, it seemed obvious that Oklahoma had a weird language all its own: Okmulgee, Wetumka, Wapanuka. Staring out into the darkness Roosevelt sought the hills, tall pines, and thicket oaks beyond. The prairielands soothed the neurotic and agonized part of his spirit.

By the time the Roosevelt Special reached Dallas, the president, alert and curious, was in a buoyant mood. Cowboys along the tracks pointed at him in admiration. People craned their necks to see the Rough Rider who hadn't lost his luster. In honor of his visit, Dallas had decorated the large public square near the Oriental Hotel with flags and bunting; an estimated 30,000 people came to hear the president speak about the "American century." Even oil field scouts and get-rich-quick geologists had come to Dallas in wagons. Appealing to the chauvinism of the Lone Star state, he called Texas a "mighty and beautiful state," a "veritable garden of the Lord."[30] Roosevelt promoted improvements for the Trinity River and irrigation projects around Dallas in general. Aridity could be conquered by forest conservation and modern engineering. More water and grass, he said, were needed in Texas. And appealing to the old Confederates' pride, he boasted that he was half southern gray, half northern blue, but *all* Lone Star. This became a standard applause line in his stump speeches in Texas.

At Fort Sam Houston in San Antonio, Roosevelt inspected national troops and spoke of the Monroe Doctrine as a guiding American principle. He also called for restoring integrity to Wall Street. On every street Roosevelt was greeted as if he were a hometown boy who had made good. A major thoroughfare had been renamed Roosevelt Avenue in his honor (only the christening of the polar explorer ship made him prouder). While on horseback in Texas, Roosevelt always made sure a lasso was coiled around the horn of his saddle, just in case a wild horse or runaway cattle entered his domain. What a showman Roosevelt was! Speaking like a cowboy, Roosevelt acted as if dusty ole San Antone were the greatest place on earth. He had learned Texans' lingo, mores, and folkways. "In the old days in Texas I understand that there used to be a proverb that while you would not generally want a gun at all; if you did want it you wanted it quick, and you wanted it very bad," Roosevelt said to a crowd of well-wishers. "That is just the way I feel about the navy. I feel that if we have it the chances are that we will not need it, but that if we do not have it, we might need it very bad."[31]

III

From San Antonio it was on to Fort Worth and Wichita Falls, where many of the largest cattle ranches in the world were situated. Two "old style Texas cattlemen," as T.R. described them—Burk Burnett and Guy Waggoner—were going to lead the presidential wolf hunt to the Big Pasture.[32] Between them, these two pro-Roosevelt ranchers practically owned Fort Worth. Burnett, in particular—who had been a trail boss on the 1,200-mile cattle drive along the Chisholm Trail to Kansas in 1867—was a legend along the Red River of the South. He was known for his financial acumen and hard deals, and his "Four Sixes" brand dominated the North Texas range for decades. He was also a meticulous caretaker of his cattle empire; there was no such thing as a collapsed corral fence on his property. Moving his ranch operations to Wichita Falls—as well as his Longhorns, Durhams, and Herefords—he produced the best cattle strains in Texas. A grin-and-bear-it type, a man of deliberate action who never knew regret, Burnett wasn't enamored of grammar books, dictionaries, or the history of European civilization. Like Roosevelt he was an advocate of direct action.

A close friend of Quanah, Burnett leased 300,000 acres from the Indian Territory to graze cattle. Often, Burnett slept on the wraparound front porch of Quanah's Star House (which Burnett had purchased for Quanah), preferring Indian company to the drifters working in Fort Worth's stockyards district. But for all his Texan down-homeness and honesty, Burnett was extremely rich. His feeder steers were selling at a fair market price, and he had a lot of them. At the time Burnett organized the wolf hunt for Roosevelt he had more than 20,000 head of branded cattle on a spread totaling 206,000 acres.[33]

In anticipation of President Roosevelt's arrival, Catch 'em Alive Jack Abernathy scouted for the most desirable places to camp in the Big Pasture and found an ideal spot along Deep Red Creek in Oklahoma. Some chauvinistic Texans criticized Abernathy for not holding the hunt on Lone Star soil, but the objections passed. Both Burnett and Waggoner did supply Texan "daredevil" riders—their hired ranchhands with the best equestrian skills—to impress the president. (Guy Waggoner, in fact, was an expert rider himself. He became head of the Texas racing commission and eventually moved to New Mexico, where betting on horse races was legal.) Ranchers from some ten or fifteen counties tried to wangle a slot on the hunt party. Every man under fifty wanted to be an extra in the Roosevelt extravaganza.

Roosevelt's longtime physician friend Alexander Lambert was designated as the official photographer for the hunt; no reporters would be allowed to witness it, because there was a strong possibility of an embarrassing moment for the president. Lambert was perhaps Roosevelt's most intimate friend. He specialized in diagnostics, internal medicine, and drug addiction and was also the president's personal physician.[34] His loyalty to the president and to the principle of doctor-patient confidentiality was so great that he left no diaries or reflections about their times together. With his thick goatee, large ears, and an ever-present doctor's bag, he was a fixture on all of Roosevelt's hunts. A professor of clinical

Roosevelt's hunt in the Great Pasture of the Indian Territories (near Oklahoma) attracted the eager participation of the richest ranchers in Texas.

medicine at the Cornell University medical school and director of the fourth division at Bellevue Hospital, he was one of the best doctors of his generation. Besides enjoying his company immensely Roosevelt was beholden to Lambert for his sophisticated knowledge of how mosquitoes carried yellow fever.[35]

On April 5, as arranged, Roosevelt's train arrived in the hamlet of Frederick, Oklahoma. Armed lawmen had the crowd of curiosity seekers under proper control. There was little chance of trouble. A grandstand had been built for the once-in-a-lifetime visit, with patriotic props in place. Wild cheers erupted when the president disembarked from his railroad car, waving his Stetson. About 5,000 to 6,000 people constituted the welcome committee. They came from Fort Sill–Lawton and west from Altus, from north of Hobart and south from Wichita Falls, Texas. Ranchers, farmers, and merchants had dropped whatever they were doing to come hear Roosevelt speak in the middle of nowhere. About twenty deputies patrolled Frederick, making sure that no troublemakers would disrupt the picture-perfect day. In particular, they kept a close eye on Indians. "The next time I come to Oklahoma," Roosevelt said at the outset, to a roar of approval, "I trust I will come to a State."[36]

Abernathy was sitting on his horse, Sam Bass, soaking in Roosevelt's oratory. Accompanying Roosevelt were Colonel Lyon, Burnett, Waggoner, the former Rough Rider Bill Fortesque, Lieutenant General S. B. M. Young, Quanah (with three wives and a baby), Dr. Lambert, and the Rough Rider physician Sloan Simpson.[37] There were also a few Texas Rangers. The hunt was going to be like a caravan in the Sahara. The main worry for local planners was Roosevelt's notorious reckless riding (in October 1904 he had been thrown from his horse and received a serious injury).[38] In the open space of the Twin Territories, T.R. was bound to let loose, galloping his horse's hooves into prairie dog homes with disastrous results.

According to the *Frederick Enterprise*, President Roosevelt spoke boldly about building the Panama Canal and pleaded to be left alone by curiosity seekers while he was on his hunt. "Now I want four days' play," Roosevelt said. "I hear you have plenty of jack rabbits and coyotes here. I like my citizens, but don't like them on a coyote hunt. Give me a fair deal to have as much fun as even a President is entitled to."[39] The *Washington Post* ran a huge front-page headline: "President in Wild." The reporter wrote that the elusive jackrabbits and coy wolves had better watch out: "The distinguished party of hunters will have plenty of elbow room. The whole Territory is theirs for the asking, but the programme is to confine the hunt to the tract of land thirty-six miles square leased by Capt. Burnett from the Kiowa and Comanche tribes."[40]

The scenes of Roosevelt's first acquaintance with Abernathy in Frederick became legendary in the *Washington Post*, the *New York Times*, and elsewhere—the two men shared a boyish predilection for nineteenth-century romance and adventure, and an abiding love of wild things. When Roosevelt spied Abernathy on horseback in Frederick, a wide smile crossed his face. "You look like a man who could catch a wolf," Roosevelt said, shaking Abernathy's hand. "I want to congratulate you, for I know you are going to do what Colonel Lyon says you can do." Given his interest in coyotes (or gray wolves), Roosevelt was utterly fascinated by Abernathy's techniques for catching them. Would the wolves do a somersault when tackled? Did every bite hurt and require stitches? But mostly Roosevelt was baffled by why the wolves became so submissive. "I can't quite understand just all about this yet," Roosevelt said to Abernathy.

"Well, Mr. President," Abernathy responded, "you must remember that a wolf never misses its aim when it snaps. When I strike at a wolf with my right hand, I know it is going into the wolf's mouth. I believe I could shut my eyes and do what you see me do, for I have caught two wolves in my life in inky darkness. However, I prefer not to shut my eyes."[41]

What Roosevelt came to understand was that Abernathy simply out-dominated the wolves. In all coyote or wolf packs a dominant member was genetically predisposed to force the others to submit; usually the conquered animals would roll on their backs as if to say "No more." Abernathy had learned to use this law of the Plains. When he wrestled the wolves into submission, they came to see him as the pack leader and thus became rather docile.[42] Abernathy had negated Roosevelt's preconceived ideas about gray wolves; the heads and nose pads were considerably larger than those of coyotes, and the fur was much longer. Zoology wasn't a fixed science, and Oklahoma had the president rethinking these wolves as a subspecies. Also, because the coyotes around Fort Sill had bred with domestic dogs, the canids in the Big Pasture–Wichitas tended to be larger (with varied coloration differences). There was also a red wolf around southwestern Oklahoma—in between the gray wolf and the coyote in size.[43]

After Roosevelt's brief speech about Oklahoma's statehood, the Monroe Doctrine, and Indian rights, his hunt party went to the Big Pasture. Roosevelt was genuinely enamored of the open range of scrub weeds and tall grasses. About fifteen tents were set up as base camp at Deep Red Creek, a little tributary of the Red River of the South, eighteen miles from Frederick. Roosevelt shared his tent with Lambert. Chuck-wagon food was readily available. A calf with the Four Sixes brand had been slaughtered, and a famished Roosevelt ate from a hindquarter. The first night, the coyotes came up closer to camp to do their howling than Roosevelt had expected. This was the land of broken treaties where cowboys slept in their hat crowns; Roosevelt felt right at home as the moon set on the prairie. As a gesture to the Comanche who didn't speak English, the president used pidgin, and a fumbling sign language. That first evening at camp, however, the president was travel-fatigued and went to bed early. Roosevelt matter-of-factly told Abernathy, Burnett, and Waggoner and the cowboys that he was spent.

IV

On Sunday morning Roosevelt woke at the crack of dawn. It was going to be a fine day for an outing. There was a breathless hush to Oklahoma's light winds of April, which was invigorating. Air wafting up from Old Mexico, warm, sea-scented, almost tropical, filled his nostrils, and he felt at one with the land. Most of his early observations were of the prairieland, shaded by the occasional ash, pine, elm, or black walnut. Intrigued by the Big Pasture and Wichita Mountains ecosystems, Roosevelt inventoried his surroundings: rustling cottonwoods, black-tailed jackrabbits, white-tailed deer, wild turkey, stunted mesquite, tiny swifts, pecan groves, and grasshopper swarms. Around Deep Red Creek alone there were more than fifty mammals and 800 plant species.[44] It didn't take him long to notice that *Didelphis marsupialis*—the ubiquitous Virginia opossums—were everywhere; pretty soon they'd be waddling down Fifth Avenue in New York. Some animals—like opossums—simply knew how to adapt and survive in the face of mankind's intrusion. And Roosevelt's notebooks were quickly filled up with bird sightings. "Cardinals and mockingbirds—the most individual and delightful of all birds in voice and manner—sang in the woods," he wrote; "and the beautiful, many-tinted fork-tailed fly-catchers were to be seen now and then, perched in trees or soaring in curious zigzags, chattering loudly."[45]

Mostly, that first sunshiny day was spent shooting rabbits for supper. As they sat around the fire that Sunday night, sharing wild game, Abernathy was amazed to hear the president talk like a zoologist who had paddled the Red River from the Texas Panhandle to the Mississippi River. It was all firewood and snuff stories, told with the sophistication of a Harvard man. Somehow Roosevelt made even the barest incident interesting. "I was amazed at the President's knowledge of wild animals, snakes, and even the smallest of reptiles and insects," Abernathy recalled in a memoir. "He told of the vinegaroon, the most deadly of the poisonous creatures in Texas and Old Mexico; many of those old time hunters present had never even heard of such a reptile."[46] Clearly, Roosevelt wasn't just a stiff-kneed politician. He came as advertised: a Rough Rider.

Unfortunately, Roosevelt never published his campfire stories. With meat in the pot and log flames jumping high and low, on these outdoor outings Roosevelt would recount moments from the strenuous life with cliff-hanging suspense. There were accounts of sumo-wrestling with a 300-pound Japanese man, tramping toward the Missouri River headwaters, boxing with the heavyweight champ John L. Sullivan, and encountering rattlesnakes in North Dakota. Holding his audience's attention with theatrical gestures, Roosevelt made it seem as if he had strode over the Alleghenies and down the Ohio River valley with Daniel Boone. When Frederick Jackson Turner told stories of western expansion, the farmer was the hero. By contrast, Roosevelt glorified men who had channeled violence into grand deeds such as hunting or boxing. His favorite campfire stories dealt with the ritualistic tradition of hunting grizzly bears: man pitted against beast. What made Roosevelt's stories about grizzlies more interesting than anybody else's was the way he personalized the bears without sweetness: for example, Big Foot Wallace in Wyoming, Old Mose in Colorado, and Old Ephraim in Idaho. And he always spoke of these bears with reverence.

The Wichitas truly were a biotic crazy quilt. Western trees grew in the range that didn't exist even a mile east of Fort Sill. Rounded domes and jagged granite peaks seemed to erupt out of the ground. Here the western meadowlark sang with its counterpart, the eastern meadowlark. As the Oklahoman historian Edward Charles Ellenbrook once put it, this was *the* demarcation line of biological America. Only in the Wichitas could one of John Burroughs's beloved dark-blue eastern bluebirds be seen hopping around with the azure Rocky Mountains bluebird. And Roosevelt, a dedicated Auduboner, hoped to clap his eyes on one of America's most unsociable birds: the Mississippi kite. Oh, what a delicious place the Wichitas were for a naturalist! And they didn't offer just birds and small mammals. Besides collared lizards there was an odd assortment of poisonous snakes, broad-banded copperheads, western diamondback rattlers, prairie rattlers, and western massasaugas among them. The Bronx Zoo's new herpetologist, Raymond L. Ditmars, needed to do field collecting in the Wichitas.

Although Roosevelt wrote an article about his plains hunt, titled "Wolf-

Coursing," for *Scribner's Magazine* that summer, he often used the term "coyotes" while he was in Oklahoma. To Roosevelt the pageantry and novelty of catching a wolf alive were intoxicating. What he admired most about gray wolves and coyotes was their survivalist behavior. Even their cowering had a certain nobility. The reason was fairly simple. During the great slaughter of buffalo by the U.S. Army, wolves and coyotes had prospered. After skinners finished cutting up the buffalo for robes and meat, the canids soon rushed up to the carcass to devour the entrails. With such an abundance of free food, they quickly multiplied throughout the Great Plains. A new breed of hunter—the wolf exterminators—replaced the buffalo hunters in Texas and Oklahoma. These were considered a lowly type in the hunters' pecking order. "A wolfer could manage with just a pack horse, a gun, a bedroll, and a bottle of strychnine crystals, but more often he had two pack horses, or a team pulling a small cart or wagon," the historian Francis Haines noted in *The Buffalo*. "He followed after the buffalo hunters, and at each kill poisoned every carcass within a mile or so of his central point. Then he went away for a few days to let the wolves and coyotes have plenty of time to eat their fill and die."[47] This breed of hunter had two objectives: obtaining wolf furs and being paid by local ranchers.

Roosevelt immediately took a shine to Abernathy (and for that matter to Abernathy's father, a former Confederate soldier, whom he met). There was a softness in Abernathy's eyes that belied his hell-on-wheels reputation. He was like a live coal. Initial conversations between Roosevelt and Abernathy centered on the greyhounds brought along for the hunt. Roosevelt had never seen a breed of dogs so sleek and eager. By contrast, the Mississippi hunt hounds seemed lifeless. "By rights there ought to have been carts in which the greyhounds could be drawn until the coyotes were sighted, but there were none, and the greyhounds simply trotted along beside the horses," Roosevelt wrote. "All of them were fine animals, and almost all of them of recorded pedigree. Coyotes have sharp teeth and bite hard, while greyhounds have thin skins, and many of them were cut in the worries."[48]

The wolf hunt, led by Abernathy, began in earnest on Monday, April 10. The sun was now hot—a fireball low to the ground, causing all the men to change shirts three or four times during the course of the afternoon. The wind was seldom stiff. All the soldiers from Fort Sill and the Texas Rangers assigned to protect the president were surprised that Roosevelt could ride at such breakneck speeds up hills and down gullies. They hadn't expected such equestrian skill from a man in his forties. By high noon the Roosevelt party had caught three wolves. After a lunch of cooked calf, the president finally asked Abernathy to show off his wolf catching. The appointed hour had arrived for Catch 'Em Alive Jack to put up or shut up.

To Roosevelt, watching Abernathy ride a horse was exhilarating: he was an athlete in top form, and there wasn't the faintest hesitation about him. Roosevelt had never before seen such an able rider, even in the Dakotas. Riding beside Abernathy invariably meant falling behind, pulling up the reins breathless. But Roosevelt did pretty well at keeping up. In the East, everybody was obsessed with money and power. Abernathy, although not opposed to either of these, nevertheless seemed organically a part of nature. He had a homegrown quality that Roosevelt associated with authenticity.

Dispatches from Oklahoma about President Roosevelt's grand hunting tour with forty dogs were so colorful that readers might be forgiven for questioning their veracity. Readers were exhilarated. Realizing he had a huge audience for his antics, Roosevelt banged the kettledrum so loudly for Oklahoma's statehood that his congressional enemies said he was pretending to have been born on a carpet of grassland. Roosevelt, it seemed, was acting like an Oklahoman high-plains drifter who understood the bioregional history of the territory from dinosaur fossils to modern dust storms. For its residents—who had an inferiority complex with regard to Kansas and Texas—this presidential gesture was greatly appreciated. When the *New York Times* ran a front-page headline, "President in Foot Races: Also Chases a Wolf Ten Miles and Is In at the Death,"[49] Oklahomans welcomed the national publicity. Roosevelt's animated hunt was far better than grim stories of blasphemy, land steals, and droughts.

What the *Times* didn't tell its readers was that these prairie wolves (unlike the timber wolves of Minnesota) were quite small. Still, they were ferocious when attacking Oklahoman cattle or deer, they had long teeth, their bite was feared (though erroneously) to be poisonous, and a wolf bite did cause throbbing pain, suppuration, and high fever. On the other hand, these wolves were very scared of humans and would slink off and disappear into the brush when humans came into sight. The Kiowa thought these wolves were a trickster spirit because of the way they appeared and then disappeared, like a mirage. Too cunning to be trapped, gray wolves could, however, easily be chased down by fast dogs and fast horses; this is where Abernathy's skill came in.

Secretly, Roosevelt admired coyotes because they were an enemy of the sheep industry. The cowboy in him just couldn't stand sheep. George Bird Grinnell used to say that wolf-coyotes were the smartest animals of all, the only ones who could creep up close to men and not be detected.[50] Whenever camps were abandoned, prairie wolves would soon scavenge around the dying fires and devour the refuse. They wouldn't, however, eat meat if it stank (as a human carcass does) for more than twenty-four hours. The Comanche, who also admired coyotes, often called them "barking wolves" because their evening howls resembled those of dogs. But to Oklahoman cowboys, coyotes were four-legged vultures, varmints to be eradicated. No matter what dire measures ranchers took, however, the adaptable coyotes thickly populated the prairie. Often, ranchers in Oklahoma hung dead coyotes on fences, a grotesque warning to the others to leave the livestock alone. "After nightfall they are noisy," Roosevelt wrote, "and their melancholy wailing and yelling are familiar sounds to all who pass over the plains."[51]

Another reason cowboys in Oklahoma loathed wolf-coyotes was that some were infected with rabies and might therefore lunge at humans. The disease around Fort Sill was more feared than dysentery. The phrase "mad coyote," in fact, was common around Fort Sill. A local told Roosevelt that one night when he slept under the stars a rabid coyote had attacked him. Roosevelt soon learned that this disease was more feared in Oklahoma than in any other place in America. Poisoned bait was set out from Beaver to Greer counties by gentle-faced farmers who felt forced to be exterminators. Strychnine was sprinkled onto rabbit and deer carcasses throughout the prairie; one taste, and the coyotes died instantly.[52]

The Biological Survey's Animal Damage Control Unit assisted with the poisoning, handing out burlap bags of strychnine (a skull and bones was printed on each bag). But it still didn't work on the coyotes, though it did kill border collies and other types of work dogs. So, just as some dirt farmers reject modern chemicals in favor of scarecrows, many Oklahomans continued festooning fences with coyotes.

Roosevelt enjoyed chasing down the wolves with Abernathy and the boys at his side. With the game dogs taking the lead, the Roosevelt party rode behind at breakneck speed over low crested hills and river bottoms, across shallow washouts and the treeless prairie. As a general rule, if the greyhounds didn't corner or tackle a wolf after two or three miles, it would escape. Abernathy had all this down to a science. He never even seemed to perspire. Prairie dog holes were the primary obstacle to his "catch 'em alive" trade. Fortunately, the best horses in the region had acquired a sense for avoiding them. On a beautiful roan cutting horse, Roosevelt raced hard around the Big Pasture–Wichita Forest Reserve, never more than a quarter of a mile behind the greyhounds and some staghounds. For all of their self-assurance, neither Burnett nor Waggoner could keep up with Abernathy on his fast white horse Sam Bass. As a rider, Abernathy had amazing dexterity. Because Roosevelt was so impressed by Abernathy, the ranchmen's envy increased by the hour. They began denigrating Abernathy as a professional attention-hog. Burnett, in particular, wanted to bond with the president; instead, Abernathy had stolen the show. In *Outdoor Pastimes of an American Hunter*, Roosevelt wrote a virtual dissertation on Abernathy's wolf-coursing techniques but gave barely a word to ranchmen, Lyon, or even General Young. Instead of raising his glass of grape juice (his chosen beverage) to the ranchers, he toasted Jack Abernathy, Sam Bass, and a greyhound "blue bitch" with the speed of a cheetah.[53]

"[Abernathy] held the reins of the horse with one hand and thrust the other, with a rapidity and precision even greater than the rapidity of the wolf's snap, into the wolf's mouth, jamming his hand down crosswise between the jaws, seizing the lower jaw and bending it down so that the wolf could not bite him," Roosevelt later recounted, rather breathlessly. "He had a stout glove on his hand, but this would have been of no avail whatever had he not seized the animal just as he did; that is, behind the canines, while his hand pressed the lips against the teeth; with his knees he kept the wolf from using its forepaws to break the hold, until it gave up struggling. When he thus leaped on and captured this coyote it was entirely free, the dog having let go of it; and he was obliged to keep hold of the reins of his horse with one hand. I was not twenty yards distant at the time, and as I leaped off the horse he was sitting placidly on the live wolf, his hand between its jaws, the greyhound standing beside him, and his horse standing by as placid as he was. . . . It was as remarkable a feat of the kind as I have ever seen."[54]

Day after day Abernathy caught 'em alive and Roosevelt glowed with approval. Abernathy always paid dividends as promised. Roosevelt, in fact, came to be strangely in awe of him. In his novel *The Secret Agent*, Joseph Conrad wrote that "shrinking delicacy" exists "side by side with aggressive brutality in masculine nature." That was the case here; Roosevelt the New York bird-watcher had overnight turned into the admirer of an Oklahoman wolf catcher. The conflict between Roosevelt as a preservationist versus Roosevelt as a conservationist and hunter had taken a bizarre turn. John Muir would have been appalled at the grotesque "catch 'em alive" spectacle.

"The fact that I had won the friendship of the president in such a short time," Abernathy later recalled, "naturally raised great popular interest."[55] Unfortunately, Burnett and Waggoner grew more and more envious of Abernathy. They kept muttering half-comical innuendoes at his expense. Ignoring these gibes, the president kept showering Abernathy with attention. Although the press was not present for the wolf-coursing, Lambert was on hand to photograph the catches. When other members of the hunt tried to pose with the president for a photo, Roosevelt waved them off. "I want this picture with just Abernathy and myself in it," Roosevelt said. The loyal Lambert, sensing the value of the publicity, said to Roosevelt excitedly, "You can say that this picture was snapped about a minute from the time Abernathy started the chase and made the catch."[56]

The plains and its sentinels had once again captured Roosevelt's imagination. The more harrowing the encounter, the happier the president was. A six-foot rattlesnake had lunged at Roosevelt four times before he killed it with his eighteen-inch quirt.[57] Even the god-awful sound of a gray wolf being tackled by Abernathy seemingly calmed his nerves. On day three of the hunt, Quanah brought his three wives along for the fun, accompanied by his son and baby daughter. Self-sufficient and uncomplaining, Quanah and his family had their own wagon. With temperatures in the low seventies and bright sunshine turning the prairie grasses different hues of green and blue, Roosevelt was in his element. For lunch he enjoyed eating beef strips by hand, all reeking from wood smoke. He kept blurting out

Fascinated by wolves, Roosevelt here ponders one caught by Jack Abernathy.

bully. In this land where sound preceded sight, Roosevelt was happy that there wasn't a lamppost for miles. "The air was wonderfully clear, and any object on the sky-line, no matter how small, stood out with startling distinctness," he wrote. "There were few flowers on these dry plains; in sharp contrast to the flower prairies of southern Texas, which we had left the week before, where many acres for a stretch would be covered by masses of red or white or blue or yellow blossoms—the most striking of all, perhaps, being the fields of the handsome buffalo clover."[58]

Just in case the taxpayers thought President Roosevelt was on a naturalist adventure at their expense—which he was—he once again invoked the Biological Survey. Roosevelt sent a minutely detailed scientific report to Dr. C. Hart Merriam about the weight and coloration of more than a dozen coyotes. With a uniform collecting technique that would have made Spencer Fullerton Baird proud, Roosevelt recorded facts about Oklahoma's coyotes that are still used for reference. Doing some on-the-spot calculations, Roosevelt wrote to Merriam that the average weight of a coyote in the Wichita Mountains region was thirty pounds. Skulls, skins, and paws were likewise shipped from Fort Sill to Washington, D.C., for the Biological Survey to properly analyze. "All but one are the plains coyote, *Canis nebracensis*," Merriam informed Roosevelt. "They are not perfectly typical, but are near enough for all practical purposes. The exception is a yearling pup of a much larger species. Whether this is *frustor* I dare not say in the present state of knowledge of the group."[59]

What prevented Roosevelt from considering the whole experience perfect, however, was the absence of buffalo. They were gone—and sadly, even the white-tailed deer were vanishing. A wilderness not abounding in game was a contradiction in terms. All Roosevelt could do was ride the buffalo trails, the great highways of Oklahoma, which were the easiest route to water, and imagine the Old Days. Francis Parkman had caught the tail end of them in the summer of 1846 with the Oglala band of Sioux. If you wanted to see a wild buffalo in North America in 1905, Yellowstone and the northern woods of Alberta were your only bets.[60] For now, Roosevelt studied the ancient buffalo herds' well-worn paths. He wore cowboy leggings and felt superior for having left his silk shirt behind in a closet in the East Wing.

Once again Roosevelt was playing the "great natural man." While talking with Quanah one evening at Star House, Roosevelt mentioned Baynes and Hornaday's idea of a bison refuge. "Grandfather wanted to entertain Roosevelt just so-so," Quanah's granddaughter Anona Birdsong Dean recalled of the evening. "He had a table that sat thirty people.

Each woman had a job. Mother went to see if the table was set properly. She found goblets filled with wine setting next to each plate. Grandfather, who never drank, had gotten wine somewhere and told one of the women to fill big glasses with the wine. Mother said, 'Why did you do that?' Grandfather explained that when he went to Washington, Roosevelt served wine in small glasses and he wanted to be more generous than Roosevelt."[61]

But now Roosevelt was truly offering buffalo! Quanah realized that Roosevelt was a generous man at heart, and the very notion of the Wichita Forest Reserve repopulated with buffalo brought tears to his eyes. Although Quanah spoke several dialects and was fluent in English and Spanish, he was nevertheless speechless.[62] Could his peyote vision be becoming true? Could a boyhood dream he had on the lonely Llano Estacado now be reality? Would Oklahoma—or at least a portion of its most scenic terrain on the western side of the Cross-Timbers in the Indian Territories—become a bison refuge?

A few days after that historic dinner, Quanah hung an autographed photo of Roosevelt on his dining room wall. To him, Roosevelt truly was the "Great White Chief." More than any other white man, Roosevelt had his heart in the right place, and Quanah knew that few Americans had ever loved the Wichitas–Big Pasture as Roosevelt did. He and Roosevelt smiled when rattling off the colorful names of the Twin Territories' towns as if they were superior to anyplace Queen Victoria had ever seen: Arapaho, Bowlegs, Etowah, Hydro, Oologah, Talihina. Side by side on horseback, snacking on pecans, the two warriors spoke of black-tailed prairie dogs, armadillos, bald eagles, and the rare black-capped vireo.

Besides bird-watching and wolf hunting Quanah spent time teaching Roosevelt how to properly track horse thieves. Roosevelt, in turn, gave him a thick porcelain cup. Both men enjoyed being conspicuously attired and talking nonstop. In Washington, D.C., the press corps called Roosevelt the "cowboy president." Quanah knew better: Roosevelt was the "buffalo president." As if a spell had been cast over him, Quanah told the Comanche that Roosevelt loved the beauty of southwestern Oklahoma like someone born and raised along the Red River. Roosevelt, in turn, wished that the chief had been a Rough Rider. It had been rumored that William "Buffalo Bill" Cody had offered Quanah $5,000 to perform in Europe with the legendary Wild West Show. Quanah said no. "I'm afraid I would be put in a little pen," Parker informed Roosevelt. "And I no monkey."[63]

In 1901, when President McKinley created the Wichita Mountains as a forest reserve, nobody imagined that it would become the focus of a federal buffalo reintroduction program. But Roosevelt was always thinking and listening. Some of Roosevelt's Rough Riders had come from the Wichitas area and had been full of stories about the landscape's lyrical, magnetic charm. Now, in 1905, as Roosevelt rode amid the strange rock formations with Quanah, he learned why the area was considered sacred by the Comanche, Kiowa, and other tribes. For example, there was a magnificent rock formation 2,464 feet high, designated by the U.S. Army as Mount Scott. Quanah told Roosevelt that a Kiowa medicine woman had prophesied that all the Great Plains buffalo, after being chased by murderous American hunters, had disappeared down Mount Scott's summit in single file. They descended farther and farther, into the earth's core. Inside this safe haven, everything sparkled for the buffalo; the world was "green and fresh" and rivers "ran clear, not red."[64] Someday, the Kiowa medicine woman prophesied, the buffalo would return to the plains from the top of Mount Scott, erupting like a volcano. A gregarious herd would trot out happily, resurrected and with a fierce unity of purpose, for a new day on the Great Plains.[65]

Roosevelt, of course, didn't subscribe to this prophecy, but he nevertheless enjoyed hearing the legend. And he had to admit that the elegant pyramidal sentinel that stood watch at the eastern gate—like an island in the sea—was the energy source of the entire Wichita preserve.[66] For 70 million years, ever since mammals eclipsed reptiles as the dominant vertebrates in Oklahoma, Mount Scott was the home of buffalo or their ancestors. Roosevelt understood the importance of their return. "The extermination of the buffalo," he had lamented, "has been a veritable tragedy of the animal world."[67] Now, at Mount Scott, they were going to make a stand with the help of buffalo men. Roosevelt spoke to the Wichita Forest Reserve rangers—who operated out of a little white clapboard hut with cedar posts—about the imminent prospect that the park would become America's first National Game Preserve for buffalo *and* deer. A wooden archway was soon built in an attempt to keep poachers out; it gave the reserve a legal feel.

Not since his youth had Quanah Parker been so excited about anything as he was about Roosevelt's repopulation scheme. Before long Roosevelt received public support for the idea from the Boone and Crockett Club and the League of American Sportsmen (of which Hornaday was vice president). Hornaday also worked hard to find exactly the right spot in Winter Valley, with grasses and with canyons to provide shelter from the winter weather. Daily, a typical bison herd foraged on vegetation over a two-mile radius, so the acreage of the reserve had to be fairly large. If everything worked according to plan fifteen buffalo—a nucleus herd of breeding cows and young bulls—would be shipped by train to the Wichita reserve in 1907. Hornaday, biologically cautious, selected the buffalo from numerous bloodlines to avoid inbreeding. There was only one chance to get it right.[68]

V

Following the Oklahoma hunt Roosevelt, as a courtesy, asked to meet Mrs. Abernathy and the five children. Approving of their pioneer stock he enjoyed them immensely. The six days at Big Pasture had been *bully*, a brief return to the heroic days before the Indian Wars. Roosevelt extended an open invitation to the entire Abernathy family to visit him in the White House. Clearly, Roosevelt honored Abernathy even though other Texans might consider him a court jester. "The petty rivalry of which I had been the object during the course of the wolf hunt in the Big Pasture was but a small incident in comparison with the experience that was awaiting me when I was now unexpectedly drawn into public life," Abernathy wrote, "as a result of my new friendship with President Roosevelt."[69]

Now that the hunt was over the *Washington Post* cast it in glowing terms. Seldom, if ever, had the *Post* been so enthusiastic about a president. Even Roosevelt's secretary, Loeb, might have blushed at reading the copy. "Mr. Roosevelt acquired in the Indian country a complexion that would do credit to an Apache warrior," one article read. "He is now as brown as a berry and in fine spirits, and the warming up of the past few days has put him in a good trim for the more exciting and hazardous sport which he will experience in Colorado, where for the next four or five weeks he will make life miserable for members of the cat and bear family that happen to come his way."[70]

When Roosevelt arrived in Colorado Springs following a speech in Clayton, New Mexico, with plans to climb Pikes Peak, reporters were still covering his every move. Love him or hate him, Roosevelt made good copy. Once again he used hunting as his hook, arousing a burst of regional western pride. For three weeks he would dwell with the blood of bears, boots in the stirrup, stubble on his face. For Roosevelt the Rockies were always the Alps without handrails, and he promoted them as such. Not since Andrew Jackson had America had a president who was such a celebrity. Wherever Roosevelt went now, he would wave a bandanna to express solidarity with the crowds. But speaking in front of 10,000 people gathered at the Santa Fe station in Colorado Springs, Roosevelt pleaded with both well-wishers and the press to allow him uninterrupted privacy in the wild. "One thing you cannot do on a hunt, and that is to carry a brass band," Roosevelt said. "You cannot combine hunting bears with your Fourth of July celebrations. I am going to beg the people of Colorado to treat me on this hunt just as well as the people of Oklahoma treated me on the wolf hunt."[71]

Roosevelt shelved world affairs and domestic policy while in the Rockies, preferring a wintry saddle blanket to wire reports, and only one major news item seized his attention. The celebrated U.S. senator from Connecticut, Orville H. Platt (not to be confused with Thomas Platt of New York), had died at age seventy-seven. The last time Roosevelt had seen his Republican friend—best known in history for the Platt Amendment of 1901, which offered Cuba self-determination after the Spanish-American War—Platt's face was alarmingly drawn and gaunt, and a rattling cough had somehow caused his complexion to lose its luster: his skin was sepia-tinted. When Platt had tried to laugh, there was only a faint sound, and Roosevelt had known he wasn't long for this world. "It is difficult to say what I think of Senator Platt without seeming to use extravagant expression," Roosevelt had said of Platt at a dinner earlier that year. "I do not know a man in public life who is more loved and honored, or who has done more substantial and disinterested service to the country. It makes me feel really proud, as an American, to have such a man occupying such a place in the councils of the Nation."[72]

Deeply saddened, Roosevelt wanted to create a living memorial for Platt. In the Wichitas, he had heard about a system of freshwater and mineral springs south of Oklahoma City, less than two hours north of Dallas.[73] In 1902 the Chickasaw and Choctaw had ceded the best 640 acres of these freshwater springs to the Roosevelt administration. As at Hot Springs National Park (in Hot Springs, Arkansas), the gateway town of Sulphur, Oklahoma—otherwise a nothingville—had built hotels, hoping to attract tourists to the curative waters. The community, however, had a problem with sinking wells and hoped the federal government would someday come to the rescue.[74]

Roosevelt now decided that the so-called Seven Springs area would make a terrific national park. He refused to wait until Oklahoma became a state, with large acreage carved out for Indian reservations, and there was little political disadvantage to naming the springs after Platt, even though the senator had never visited south-central Oklahoma (he had been a member of a committee on Indian Affairs, however). On June 29,

1906, Roosevelt (through a special act of Congress) declared the Seven Springs area Platt National Park—his latest park after Crater Lake, Wind Cave, Sully's Hill, and Mesa Verde.[75] At a deliberate pace, the U.S. government added infrastructure to the Seven Springs area, including the Lincoln Bridge (completed in 1909). Besides featuring the springs, the park was surrounded by undisturbed grasslands.[76] (In 1976 the National Park Service renamed Platt the Chickasaw National Recreation Area. But as a lasting memorial, a Platt District in Sulphur was designed to recognize Platt National Park's seventy-year history.[77])

On April 24 the *New York Times* reported "The President's Return." Worried about difficulties with the Panama Canal, labor riots in Chicago, anarchy, and high tariffs, Roosevelt had decided to shorten the Colorado leg of his holiday by five or six weeks. The *Times* lamented that Roosevelt, "the hardest working man in the country," couldn't enjoy the Rockies longer. Most presidents are criticized for taking vacations because they may be caught off guard by an unexpected international crisis or a domestic crisis such as a strike. But Roosevelt was immune from such criticism. In fact, the *Times* (and other newspapers) thought he needed *more* time to enjoy himself, catch coyotes, and hunt grizzlies.[78]

Of course, Roosevelt didn't slip out of Colorado in the dead of night. The state threw a huge open-air revivalist meeting in Glenwood Springs to honor him on his departure. With the Old Blue Schoolhouse as a backdrop, ranchers who had ridden in to Glenwood Springs from Newcastle, Rifle, and half a dozen other nearby towns said *adios*. It was quite a pageant. Instead of sprucing up for the farewell, Roosevelt wore filthy blue jeans, a slouch hat, a soiled bandanna, a sheepskin jacket, and a blue cotton workman's shirt. Because it was Sunday, Roosevelt asked that services be held under God's blue sky with the green grass serving as pews. The sunshine in the upper air, the president maintained, was more inspiring than light filtered through stained-glass windows. An organ was rolled out onto the schoolhouse porch, and old-time hymns such as "Rock of Ages" were sung. A Presbyterian minister asked Roosevelt to say a few words to the God-fearing men and women of the Rockies. Thereupon, Roosevelt unleashed a sermon about the strenuous life, peppering his speech with cowboy jargon. At the end of this oration he announced that he would shake every hand offered him in Glenwood Springs. "There are a many of you," he said, "so don't stampede or get to milling."[79]

A week later, on May 7, Roosevelt hosted a good-bye dinner in Glenwood Springs. This was the final good-bye. All Roosevelt could do, however, was talk on and on about Colorado's bears. Lambert had captured some of these bears with a Kodak camera, but other bears weren't so lucky. Roosevelt had been routinely bringing his bear skins to a local taxidermist, Frank Store, who mounted them in record time. Interestingly, Roosevelt requested that his half dozen or so bears be stuffed closed-mouthed. He had these trophies shipped as quickly as possible to Dr. C. Hart Merriam in Washington, D.C., where casts were also made of the bears' footprints, so that the Biological Survey could glean precise scientific data from the samples. As usual, Roosevelt wrote up his outdoor notes to accompany the trophies.

Although Roosevelt did not bring back a grizzly cub to Alice (as promised), he did adopt a small black-and-tan terrier named Skip, whose new home would be the White House. Skip was a gift from Jake Borah to the president during his last few days in Colorado. Roosevelt and the dog, which never barked, had become inseparable friends. To Roosevelt's delight, Skip actually climbed trees to go after chased game.[80] While Roosevelt read books in Colorado for relaxation, sometimes going for three or four hours straight, little Skip would obediently sit in his lap, petted as the pages were turned.

"Archie simply worships Skip, who is developing into a real little boy's dog and accepts with entire philosophy being carried around by Archie in any position," Roosevelt wrote after he returned to Washington. "He has won the hearts of all the family except Mother, who I think resents his presence a little as a slight upon Jack. Yesterday she praised him—you know the kind of praise I mean—'Yes, he is a cunning little fellow and friendly, of course. In fact, he is friendly with *everyone*.' "[81]

VI

While Roosevelt was in Colorado, Edith had decided to purchase a little cabin for her husband in Albemarle County, Virginia, fourteen miles south of Charlottesville. She called it Pine Knot; a favorite phrase of her husband. By a happy coincidence the first lady had journeyed to Keene, Virginia, on May 6 to spend time with Joe and Will Wilmer (family friends). Both Ethel and Archie accompanied her. The Blue Ridge Mountains had long attracted Roosevelt, who particularly liked Jefferson's *Notes on Virginia*. Edith must have known this. Besides enjoying the countryside, she was looking for a wilderness cabin where Theodore could escape Washington's hubbub to "rest and repair." She was tired of seeing him traipse off to Colorado for weeks every time he needed to return to the outdoors. On the Wilmers' horse farm, tucked away among red and white oak, red cedars, dogwoods, red maples, and black cherry trees, was a rustic

Roosevelt loved reading books on nature, hunting, and literature. Here he is with his little dog Skip on his lap in a Colorado cabin.

worker's cabin. Almost on the spot she purchased the cottage, plus fifteen acres, for $280, although the deal didn't go through at the bank until June 15. (She purchased an additional seventy-five acres in 1911.[82])

Returning to the White House with Skip, the president was full of stories about Oklahoma's wolves and Colorado's bears. Wildlife photographs taken by the versatile Dr. Lambert were being developed in a darkroom. Anxiously, Roosevelt waited to see the photos with Catch 'Em Alive Jack. The whole experience at Big Pasture–Wichitas was imprinted on his mind like a brand. Quickly, he ordered the paperwork drawn up to create the Wichita Forest and Game Preserve. He signed the executive order on June 2, 1905, and transferred the newly formed Forest Service headed by Gifford Pinchot to the Department of Agriculture. At Merriam's Biological Survey, last-minute legalities were being completed to create two new federal bird reservations in Michigan: Siskiwit Islands and Huron Islands.[83] Lawyers at the Department of the Interior were also looking into preserving some interesting geological formations and archaeological ruins in the Southwest, such as Chaco Canyon and Mesa Verde. "Arizona and New Mexico hold a wealth of attraction for the archaeologist, the anthropologist, and the lover of what is strange and striking and beautiful in nature," Roosevelt wrote. "More and more they will attract visitors and students and holiday-makers."[84]

Edith was likewise bursting with news of the outdoors. All she could talk about was Pine Knot—the two-story cottage in the middle of a bird paradise. A piazza, she said, offered views of open fields and the Blue Ridge foothills. Everything about the place, she believed, was ideal for solitude. Yet it was also a practical acquisition. While it was isolated from Washington, Pine Knot was not really too far from civilization. Only half a mile away was a general store, and Christ Church could be reached simply by walking across a horse pasture. And Pine Knot had historical significance which Edith knew her husband would cherish: it sat along Scottsville Road, where General Sheridan had marched 5,000 troops in 1865 on a maneuver to destroy the James River canal locks.[85] "Mother," T.R. wrote to Kermit after returning from Colorado, "is a great deal more pleased with it than any child with any toy I ever saw."[86]

It was agreed that after Roosevelt concluded the Russo-Japanese negotiations in early June, he would head south to stay at Pine Knot for a few relaxing days. He could start a Virginia bird count there. Edith went first as an advance scout, setting up furniture and making sure the potbellied stove was in working order. This was not an Adirondacks hunting lodge but a cabin in the rustic tradition of John Burroughs's Slabsides in New York. On June 9, after extracting a promise from Russia and Japan to negotiate face-to-face, Roosevelt took the Southern Railway (train No. 35) to Red Hill, Virginia. A few locals were at the station when he arrived. All T.R. offered them was a boyish declaration that he was glad to become a Virginian homeowner.

After sleeping one night in the ocher-colored cottage—which had a cedar roof, dark green shutters, and a hardwood tree growing *inside*—Roosevelt declared Pine Knot "the nicest little place of the kind you could imagine." Not everybody in Blue Ridge country, however, was happy to have President Roosevelt as a neighbor. Anger over the dinner with Booker T. Washington had made Roosevelt persona non grata in parts of Albemarle County. No reconciliation with these bigots was possible. Also, Roosevelt's admiration for General Sheridan—who was to Virginians what General Sherman was to Georgians—didn't endear him to the locals. Furthermore, Roosevelt's attempts at taking land in the Blue Ridge Mountains for federal forest reserves had angered local timber companies. So, this was hostile country to some degree. Nevertheless, Roosevelt refused Secret Service protection; that was a reluctantly agreed-upon precondition with Edith. He instead chose to sleep with a pistol at his bedside, thumbing it open to check the bullet chambers before blowing out the light. A real man, Roosevelt believed, protected his own family in the woods.

What Roosevelt enjoyed most about Pine Knot was the sound of birds, which reminded him of the Elkhorn Ranch in North Dakota. Ahhh—all of his weariness evaporated in the woods. He was like a refugee who had fled the Washington battlefield, escaping the knife blade to the throat by the grace of God. "It was lovely to sit there in the rocking chairs and hear all the birds by daytime," he wrote to Kermit, "and at night the whip-poorwills and little forest folk."[87] Visiting Pine Knot also gave Roosevelt a chance to cook his favorite food, fried chicken, on a kerosene stove. Just as Roosevelt had played at being an Oklahoma wolf hunter with Catch 'Em Alive Abernathy, he now played at being John Burroughs in the grove. And at Pine Knot he ate like a glutton—a dozen eggs for breakfast, washed down with glasses of milk. No longer was he trim and fit. He was taking on a little of the girth of Grover Cleveland. As Edith's biographer Sylvia Jukes Morris noted, Theodore believed that "a man with a huge brain needed plenty of nourishment."[88]

And President Roosevelt's brain was needed more than ever during the summer of 1905. On the death of Secretary of State Hay, Roosevelt took up the Russo-Japanese peace negotiations himself. The melancholy prospect of endless war had seemed likely, but now diplomats from both sides came to Oyster Bay to search for common ground. Roosevelt then selected the Portsmouth Naval Shipyard in New Hampshire as the site of a further, more intense round of negotiations. What made the Russo-Japanese War so unusual was that the fighting was taking place in neutral nations: China and Korea. Today many military historians call the conflict the first modern war because the telegraph, advanced torpedoes, minefields, and armored battleships were introduced. Roosevelt wanted to prevent a global cataclysm—this was the reason why he threw himself so wholeheartedly into the diplomatic fray. At stake was the balance of power for the entire Pacific.

Using diplomatic muscle, Roosevelt achieved a stalemate. Such large-scale wars, he believed, should be obsolete. His negotiation skills and his aggressive diplomacy, which culminated in the Treaty of Portsmouth (signed on September 5, 1905), won him the 1906 Nobel Peace Prize. Roosevelt had persuaded Russia to stop its expansionism into East Asia, while the Japanese won control of the Korean Peninsula. Neither side was very happy with the negotiated peace. Still, as a warrior-cum-statesman Roosevelt understood the vital importance of stopping a war between such highly armed powers as Japan and Russia; and both parties conceded that he had been fair. Two rare samurai swords were presented to Roosevelt by the Japanese war hero Admiral Togo at Sagamore Hill, as a token of appreciation for his statesmanship.

VII

In September 1905 George Herbert Locke, an editor at Ginn and Company of Boston, sent Roosevelt a complimentary copy of William Joseph Long's new *Northern Trails; Some Studies of Animal Life in the Far North.* It was disingenuous of Locke, who knew that Roosevelt considered Long a nature faker and a prig with epicurean tastes. But controversy sells books. Roosevelt read *Northern Trails* and was infuriated by Long's inaccurate description of wolves. Feeling empowered by his field observations in the Great Pasture, Roosevelt sent a private letter to Locke, lambasting Long as a fraud. "There are statements made in the story which from my own knowledge of animals I am confident are utterly inaccurate," he wrote. "To anyone who knows the relative prowess of a single big wolf and of a lynx, or has seen the ease with which a good fighting dog who knows his business will kill a lynx without himself getting harmed, the whole account of the way two wolves kill a lynx is absurd. Then again, take the account of the killing of the caribou by the white wolf, by a quick snap where the heart lays. The whole account is full of inaccuracies which it is hard to understand in any observer who knows anything about wolves or deer."[89]

Sounding like a fact-checking schoolmaster, Roosevelt listed more than a dozen errors Long had made pertaining to wolves alone. Besides having mastered zoological books on wolves, Roosevelt had listened carefully to Catch 'Em Alive Jack in the Big Pasture. This combination led Roosevelt to believe *he* was the world's foremost authority on wolves. At least, he said, London in *Call of the Wild* and Kipling in the *Jungle Books* made it clear that they were writing fiction. By contrast, Long had *again* written in his preface, audaciously, that "every incident is minutely true to fact." Balderdash! was Roosevelt's response. "Wolves normally kill any large animal by biting at the flanks or haunches," Roosevelt fumed. "Occasionally, but much more rarely, they seize by the throat. I have known them in the hurly-burly of the fight to seize in many different ways, but never under any circumstances have I known them to seize in the way described by Mr. Long."

Roosevelt sent a copy of this letter to Catch 'Em Alive Abernathy. These two outdoorsmen, both astonishingly indifferent to risk, now operated in tandem when it came to wolf studies: the president and the backwoodsman formed a united front. When they were together, they exchanged knowing looks, implying that they shared frontier values which could not be understood by easterners. Roosevelt claimed that Abernathy was an American original. In the fall of 1905 Roosevelt insisted that Catch 'Em Alive, scuffed boots, plain manners, and all, visit the White House as his guest of honor. Roosevelt was eager to show Abernathy off to his Ivy League friends. He treated Abernathy, in fact, like a trophy from the Wild West. Roosevelt held dinners in Abernathy's honor and invited, among others, Mark Twain, Andrew Carnegie, Thomas Edison, and John Jacob Astor. "Although I made no pretensions as a conversationalist—and do not now for that matter—I did not need much initiative in a talk with the aging author of *Innocents Abroad*," Abernathy recalled of meeting Twain. "He had plenty to say and was quite willing to do most of the talking."[90]

On arriving at the White House for the first time, Abernathy was escorted into a cabinet meeting. Only one chair was available, so he took it. The president was not yet there, and the clock ticked on. The cabinet members, including Elihu Root, eyed Abernathy with suspicion. He was wearing a six-shooter and seemed ready to fan the hammer. Suddenly Roosevelt burst into the room, eyes flashing at Abernathy in pure delight. "John you're getting up in the world—occupying the President's chair at a cabinet meeting," Roosevelt laughed. An embarrassed Abernathy, all his swagger evaporated, realizing his mistake, started to get up. Roosevelt forced him down again—they were frontier brothers and shared everything.[91]

What transformed Abernathy into a national celebrity was the publication of Roosevelt's *Outdoor Pastimes of an American Hunter* in November 1905. The book was an omnibus of Roosevelt's best outdoors essays written between 1893 and 1905.[92] His preference had been to call it *Outdoor Pastimes of an American President*, but he worried about commercializing the executive branch. The first half of *Outdoor Pastimes* dealt more with hunting, the second half more with conservation. Chapter 3 was about "Wolf Coursing" with Abernathy and was accompanied by dramatic photographs (taken by Lambert and Simpson). The images included the president and a captured coyote, saddling the Big D cow pony, greyhounds resting in a run, and much more. There were other chapters about hunting in *Outdoor Pastimes*, taking up Colorado's bears and Idaho's mountain sheep, but it was the chapter on wolf-coursing that stole the show.

Five of the chapters in *Outdoor Pastimes* were new; the other six had originally appeared in *The Deer Family*. (The first three chapters were reprints of four articles Roosevelt wrote for *Scribner's* while he was president.) Preserving nature is an overriding theme in the book. In rapturous prose, Roosevelt wrote about the beauty of Yellowstone and Yosemite in his chapter "Wilderness Reserves." Many passages in this essay represent his most glorious writing *ever* about the national park movement. Phrases in the essay were calculated to make the reader want to take a hike or watch birds; basically, the essay was a meditation by America's top wildlife manager. In *Outdoor Pastimes* Roosevelt championed the preservation of buffalo, bears, deer, and wapiti in federal reserves as never before. "The most striking and melancholy feature in connection with American big game," Roosevelt wrote, "is the rapidity with which it has vanished."[93]

The average reader could be forgiven for finding Catch 'Em Alive Jack the most colorful character in *Outdoor Pastimes*, but the book's intellectual muse was Oom John. In a series of long, radiant vignettes Roosevelt recounted his time with Burroughs in Yellowstone. Burroughs, in fact, looms large throughout the book. Not only did Roosevelt dedicate *Outdoor Pastimes* to Burroughs, but he began the book with an open letter to his dear friend, dated October 2, 1905: "Dear Oom John," it started. "Every lover of outdoor life must feel a sense of affectionate obligation to you. Your writings appeal to all who care for the life of the woods and the fields, whether their tastes keep them in the homely, pleasant farm country or lead them into the wilderness. It is a good thing for our people that you should have lived; and surely no man can wish to have more said of him. . . ." From there, Roosevelt went on and on in the same salutary vein.[94]

Burroughs was flattered beyond words, and the *New York Times* said that those, like Burroughs, who "make the study of wild life" would love the book for its utter "accuracy."[95] Furthermore, Oom John wasn't the only outdoorsman whose career the president boosted in *Outdoor Pastimes.* In 1906 Roosevelt, encouraged by the fame *Outdoor Pastimes* had brought Abernathy, appointed him U.S. federal marshal of Oklahoma. Annoyed because easterners didn't believe that Abernathy actually caught wolves alive, Roosevelt also sent a movie crew to Oklahoma in 1907 to make a motion picture of Catch 'Em Alive's feats.[96] Even by the standards of modern filmmaking, which uses sophisticated special effects, footage of Abernathy remains riveting. It's like an early herky-jerky black-and-white western; even *Animal Planet* hasn't yet captured such a crazy episode on film. On this hunt Abernathy came just a hair away from being mauled to death by a wolf. Eventually Abernathy was able to wire the muzzle of the

128-pound wolf and get the animal into a cage. All this was captured on film which Abernathy immediately brought to show the president. "That is the best show," Roosevelt beamed, "that ever has been in the White House."[97]

Roosevelt was so impressed by the footage that he asked Hornaday to allow Abernathy to course for wolves in Oradell, New Jersey. The idea was to get first-class motion picture cameras from New York to film the event. Hornaday would let seven wolves loose on an estate in New Jersey, and Abernathy would recapture them. Sam Bass and a couple of other horses were sent by railroad from Oklahoma, as were Abernathy's best dogs. "The wolves were easily tracked in the new-fallen snow," Abernathy wrote. "Two of them were found within one hundred yards of the cage from which they had escaped. They did not seem wild or disposed to flee at the sight of men with horses and dogs. On the contrary they seemed half tame and inclined to be playful in the snow."[98]

Critics of Roosevelt thought Abernathy was a ridiculous carnival attraction, not worthy of all this attention. Also, members of the U.S. Senate Judiciary Committee were furious at Roosevelt for trying to get Abernathy a marshalship in Oklahoma; they tried to characterize him as a drifter type. According to one charge against Abernathy, he was nothing more than "a bronco rider, and a wolf catcher; was reared in a cow camp; has been a fiddler at country dances, a cotton picker, a patch digger, and a friend of the outlaws—and is not a politician." Telegrams poured into Washington, D.C., from Oklahoma City and Tulsa protesting Abernathy's appointment as marshal. Some senators, adding to these complaints, said that Abernathy played the fiddle at honky-tonks and got into barroom fights. But Senator Freeman Knowles of South Dakota was sympathetic. "Well, Marshal I am going to call you Marshal, since your name already has gone to the Senate," he said, "tell us some wolf stories."

Abernathy gladly obliged Senator Knowles and was easily confirmed. One thing that worked in Abernathy's favor was his knowledge of all the business operations of the Indian tribes in Oklahoma. And Abernathy knew how to let his horse pick the trails along the Cimarron River—a sure sign of a first-class tracker. Many whites considered the Comanche and Kiowa terrorists; Abernathy didn't. "Thus I became Marshal of Oklahoma," he wrote in *Catch 'Em Alive Jack*. "The position was one of great responsibility and trust. The salary was five thousand dollars a year (a big sum to me) and all expenses incurred while in the discharge of duty." Law enforcement, however, wasn't Abernathy's forte. He didn't want to turn down cash offers for catching wolves. "My Dear Marshal,"

Roosevelt wrote to Abernathy June 4, 1906: "I guess you better not catch live wolves as a part of a public exhibition while you are Marshal. If on a private hunt you catch them, that would be alright; but it would look too much as if you were going to show business if you took part in a public celebration."[99]

Roosevelt kept up an intermittent correspondence with Abernathy about all things Oklahoman. He hoped that Abernathy would succeed as a marshal. Before long, however, Abernathy, under a storm of charges of conflict of interest, quit the job. A true Texan-Oklahoman, Abernathy was too lenient toward his outlaw buddies. While this never-turn-on-fourth-cousins attitude was commonplace in Oklahoma City, it didn't play well in the East. Roosevelt remained loyal and found Abernathy employment with the U.S. Secret Service in New York. "My first assignment was in Chinatown and the subways," Abernathy recalled. "I was required to visit the underworld resorts, among them being the opium dens often frequented by rich and fashionably attired society folks. In company with a U.S. Marshal, one night I 'hit the pipe' just to experience the sensation; also to get information from the addicts in the resort."[100]

Catch 'Em Alive wasn't the only colorful westerner appointed to law enforcement jobs by Roosevelt. Bat Masterson of Dodge City ("when Dodge City was the toughest town on this continent," as T.R. put it), famously, was appointed a deputy marshal in New York City. With hopes of saving the Grand Canyon and Petrified Forest, Roosevelt also promoted Ben Daniels—a former Rough Rider in the First U.S. Volunteer Cavalry—as U.S. marshal for the territory of Arizona. All three law enforcement officers—Abernathy, Masterson, and Daniels—epitomized the West as presented in *The Virginian.* Roosevelt saw their core moral values as reflecting his own. What Roosevelt wrote to Senator Clarence Clark of Wyoming about Daniels as being an American "Viking" also applied to Abernathy and Masterson.[101]

There are no letters like this of Roosevelt trying to insinuate himself into eastern establishment traditions except when he was applying to Harvard. Without question, Roosevelt wanted to be accepted as a westerner of the "West that was." He wanted to be associated in history with Quanah Parker, Bat Masterson, Pat Garrett, Wyatt Earp, Ben Daniels, Catch 'Em Alive Abernathy, John Muir, and Joaquin Miller. Even the members of the East Coast elite with whom he interacted—most noticeably George Bird Grinnell and C. Hart Merriam—had made their careers in the West. As for John Burroughs, he was sui generis, Oom John, an earnest American, the Transcendentalist in the modern conservation movement. And, to Roosevelt's mind, his new buffalo pasture in Wichita Forest was a first step in preserving the historical memory of Oklahoma's Wild West spirit. Lawton without buffalo on the plains was just another dull windswept Oklahoma town. Just as buffalo were roaming around Yellowstone National Park (between the Yellowstone and Lamar rivers), Roosevelt envisioned herds grazing along all the tributaries of the Missouri River.

VIII

The return of the buffalo to the Wichita Mountains was officially begun on June 2, 1905, with a "proclamation." A tract of 60,800 acres had been set aside by the Roosevelt administration approximately where the president had coursed for wolves and had ridden with Quanah. Although Roosevelt didn't specifically mention buffalo, he said that wildlife in the Wichita Mountains—including birds and fish—were off-limits to hunters.[102] The USDA's Forest Service would police the game reserve. And the fifteen white-tailed deer in the park needed to become at least 500 strong. A deer rehabilitation project was started. Additionally, the creation of the Wichita Game Preserve (inside the federal forest reserve) was a historic event: America's first national game preserve. The Bronx Zoo was at last fulfilling its mission of saving endangered species from extinction. Baynes's plan was now official: buffalo would again be the lords of the Great Plains. The first wave of buffalo would go to Winter Valley in the west-central portion of the Wichita preserve. With buffalo grass, bluestem, and rock-capped roundabouts for buffalo to hide among in winter, the Wichitas were *the* God-ordained place for the Bronx Zoo's bison.[103]

Credit for the Wichita Forest and Game Preserve must be given to William Temple Hornaday, who never threw in the towel. Ever since his study of 1889 lamenting that only 1,091 bison (wild and captive) remained in North America, he had worked overtime to save the bison as a species.[104] He specialized in understanding herd dynamics and genetic integrity. Besides personally hand-feeding the Bronx Zoo herds, he and John Pitcher had encouraged restoration in Yellowstone National Park with domestic buffalo from the Goodnight Ranch in Texas and the Pablo-Allard Ranch in Montana. President Roosevelt's positive firsthand report about the Wichita Forest Reserve in Comanche Country was important, but Hornaday wanted more specific information about the grasses on the reserve. Roosevelt agreed that this was advisable. On his return from Oklahoma Roosevelt dispatched J. Alden Loring—an energetic naturalist in his mid-thirties who worked for the New York Zoological Society and Biological Survey—to survey the exact part of the range ideal for buffalo.[105] Roosevelt was impressed that Loring held the world record for preparing small mammal specimens in a three-month period: more than 900 collected in Europe for American museums.[106]

On December 8, 1905, at the Bronx Zoo, Hornaday announced the creation of the American Bison Society (ABS), which was to be based in New York. His cofounders included Roosevelt and Charles J. "Buffalo" Jones. With determined earnestness the ABS *demanded* that the American people protect bison herds.[107] Their mission was to numerically increase buffalo herds throughout the Great Plains and Rockies. The ABS simply refused to accept the possibility that buffalo were doomed and could survive only in taxidermy, photographs, or pictographs on cave walls. The Wichita Forest and Game Preserve was the opening salvo of the "buffalo common" movement. Hornaday and Roosevelt were now committed to having large swaths of the bison's historic range restored. Centuries earlier, Father Marquette had drawn a picture of a wild buffalo he saw as far north as Green Bay, Wisconsin; ABS wanted the buffalo to range that far north once again. The ABS would also educate citizens about the bison's endangered status. And large-scale bison conservation was now deemed a national imperative, arousing serious new scientific interest in grassland ecosystems.

On March 4, 1907, all forest reserves were renamed national forests, so the area became the Wichita National Forest and Game Preserve. On October 11, 1907, the fifteen buffalo from the Bronx Zoo were loaded onto a train at Fordham Station in New York, bound for Oklahoma's national forest. Accompanying them on their journey were Frank Rush of Ponca City, Oklahoma, to tend the animals; Elwin R. Sanborn, to write about the event; and H. R. Mitchell, the New York Zoological Park's chief clerk, to manage all the details. A steel woven fence seventy-four inches high was strung up in the Wichita Forest Reserve along oak posts. Sturdy gates were also built, to withstand a charging buffalo that might want to bolt. Congress had appropriated $15,000 to construct the high fence around 8,000 acres of an ideal Oklahoman buffalo habitat. The huge steel fence had been erected in the Winter Valley part of the Oklahoma reserve; what was unclear about this fait accompli was whether it was meant to keep the buffalo *in* the reserve or the poachers *out*. The Biological Survey conducted tests to inoculate the buffalo against the dreaded Texas fever. Large areas of the new buffalo reserve were burned to kill off ticks, which carried the fever. Eventually a dipping vat was constructed to eradicate the ticks. Rush also decided that the best way to protect the buffalo from ticks was to spray them with oil. Eventually the herd would become immune to Texas fever.[108]

Hornaday saw to it that each buffalo had a padded compartment in Arms Palace cars from the Bronx to Ft. Sill, the kind used for the most valuable show horses. No crowded or foul, manure-filled quarters were tolerated; after all, these were *Mr. Roosevelt's buffalo.* While the train was switching lines in Manhattan, Sanborn wrote excitedly about how revolutionary Roosevelt and Hornaday's plan was. "It was a bit awe inspiring," he wrote from Grand Central Terminal, "to realize that in the midst of this vast station with its multitudes of people, its coughing, booming trains, in the center of the greatest city in the new world, were fifteen helpless animals, whose ancestors had been all but exterminated by the very civilization which was now handing back to prairies . . . a tiny remnant born and raised 2,000 miles from their native land. Surely the course of Empire westward takes its way."[109]

Hornaday, who always wrote copiously, left notes about the fifteen bison he selected for Roosevelt's bold reintroduction plan. Because he had hand-fed the herd, these bison had become like pets to him. They had been pampered. And four were named in honor of the great Indian chiefs Lone Wolf, Geronimo, Blackdog, and Quanah.[110]

Seven days after leaving the New York Zoological Society the fifteen bison arrived on rolling boxcars in the hamlet of Cache, Oklahoma. Pacing about and waiting anxiously in full war feathers for the buffalo that October 18 was Quanah. The autumn leaves were beginning to drop from the elm trees. The lakes and other waters were filled with migratory birds. Just seeing the buffalo unloaded at the station choked Quanah up. For a moment he stood in silence. Then he quietly helped load the buffalo onto wagons for their thirteen-mile journey to the fresh green of the Wichita Mountains.[111] When they were unloaded, a great dust cloud arose in the air. And what a homecoming it was to the Plains Indians! Native Americans boulder-hopped, booted it, rim-walked, and hit the trail, to watch these fifteen Bronx-raised buffalo run. To the Plains Indians, they were symbols of how their ancestors had once lived—*free and in harmony with nature.* From the Chickasaw Nation and the Cherokee Strip, from the Kansas border to Palo Duro Canyon in Texas, crowds arrived by the tens of thousands to witness the return of the "Great Spirit's cattle" to the windblown Wichita preserve. Few thought about huge meat piles or dressing the hides. Watching the animals graze in the open air of Winter Valley was their catharsis. It seemed that every Indian lodge within a four-state region was represented—a tribute to the power of word of mouth. Small coals were lit and pipes were smoked. Grass was taken out of saddle packs for feed, but Frank Rush said no—these were buffalo, not pets.

Guardedly, Quanah poked at the buffalo's rib cages like an agricultural inspector at a state fair, examining the Bronx Zoo herd carefully to make sure the "Great White Chief" wasn't playing a trick. Quanah was a shrewd dealer, but he had been hoodwinked before by crooked buffalo hunters, cavalrymen, railroad barons, miners, cattle kings, farmers, and politicians. All of them were culpable in the destruction of the native buffalo. But not Roosevelt—not this time. Yes, these bison had black tongues and cloven hooves. Yes, they had unbranched horns. Knowing that buffalo have four stomachs, Quanah pointed the herd toward the rich grasses of the Wichita Forest and Game Preserve as if saying, "Eat away!" The extinction of the bison was starting to be reversed. Quanah now understood that no war club was strong enough to defeat a man of Roosevelt's honest character. The "buffalo president" hadn't broken his word.

Few presidential gestures meant more to Native Americans than these seven bulls and eight cows. A great thing had happened. This was a true token of peace, generosity, wisdom, and goodwill. It coincided with the departure of the last cavalry regiment from Fort Sill in 1907. At least for this afternoon the Plains Indians, thanks to Roosevelt, felt that their ancestors' spirit had rumbled out of Mount Scott. The highest peak in the preserve—the only one higher than Mount Scott—was Mount Pinchot (2,464 feet), named in honor of the chief forester of the USDA's Wichita Game Preserve. And it wasn't just those lucky enough to be at the Wichitas who celebrated the buffalo's reintroduction. Indians celebrated throughout the Great Plains. Old-timers recalled the days of high-protein bison meat at every meal. Then, every part of the buffalo had been used: the bladder (for food pouches), teeth (ornaments), blood (paints), dung (fuel), tendons (arrow strings), scrotum (containers), tail (switches), brains (hide preparation), and so on.[112] There had been many accomplishments during Quanah's career: fathering twenty-one children; fighting against the onslaught of white civilization; conducting shuttle diplomacy between the Wichitas and Washington, D.C. None, however, equaled the success of helping to bring the buffalo back to the Great Plains.[113] "The buffalo have plenty of good green grass and pure running water," Rush reported to Hornaday. "They did not have a tick on them last summer."[114]

And Hornaday also sought other prairielands for buffalo to return to. In 1906, when Fort Niobrara Military Reservation in Nebraska closed (the U.S. Army was no longer worried about the Indian menace), the ABS stepped into the void. Following Roosevelt's instructions, a private buffalo herd was eventually rounded up and moved to this new prairie range.[115] Meanwhile, Roosevelt issued a second proclamation in 1906, adding 3,680 acres to the Wichita refuge (the Oklahoma City Club hoped that the reserve would therefore soon be elevated to national park status). By 1908, ABS and the Boone and Crockett Club—backed by the Biological Survey—were able to get Congress to appropriate funds for the creation of the National Bison Range on the Flathead Reservation in Montana.[116] (This was, in part, America's answer to Buffalo National Park in Wainwright, Alberta which had opened in 1907).[117] Congress eventually authorized 13,000 acres on the Flathead Indian Reservation for a remnant buffalo herd. The National Bison Range constituted the very first federal appropriation to purchase acreage exclusively for wildlife protection. It eventually grew to be 18,763 acres, protecting not just bison but pronghorns, elk, deer, and bighorn sheep.[118]

By 1911, Hornaday was able to declare that bison were no longer an endangered species. This was a tribute to the Roosevelt administration's proactive resolve. Not only had the Wichita Forest and Game Preserve idea worked; it led to the creation of other bison refuges. By 1912 ABS had been instrumental in the creation of the Wind Cave National Game Reserve, with fourteen buffalo donated by the New York Zoological Society. As Roosevelt envisioned it, Wind Cave National Park, which he had established in 1902, needed a secondary attraction besides underground caves. With the naturalist J. Alden Loring of ABS doing the advance work, buffalo were imported to Wind Cave; visitors to the site are *guaranteed* to see herds.[119] Run by the Biological Survey, the new Wind Cave National Game Preserve consisted of 4,000 acres from Wind Cave National Park, six acres from Harney National Forest, and eighty acres from private ranch lands.

Before long the Sioux (Pine Ridge) and Crow also established herds—following Hornaday and Roosevelt's protocol—in South Dakota and Montana. "The American Bison Society is a splendid organization," Hornaday said when he retired as president in 1911. "It will go from strength to strength, until the time comes that it is no longer necessary to consider movements for saving of the bison. Then I predict your energies will be directed to saving other species of wild life that at present may be as much threatened with extinction as the bison was three or four years ago."[120] He started crusading to save the whooping crane, the trumpeter swan, the great sage grouse, and the prairie sharp-tailed grouse from extinction.[121]

Because the buffalo flourished in the Wichita Forest and Game Preserve, Fort Niobrara, and Wind Cave, Roosevelt—after his presidency—

began promoting the idea that other big game species be reintroduced throughout the West. The initial results were mixed. In 1911 eleven pronghorns were shipped to the national forest from Yellowstone National Park. Unfortunately, they didn't have the high-quality treatment that the buffalo had been given in transport, and two pronghorns died in their railroad car. The others could not survive the harsh Wichita Mountains winter. Wildlife introduction, then, had its limits. However, Roosevelt also wanted to have Rocky Mountain elk repopulate the Wichita outcrops. A herd from the Teton National Forest was shipped in from Wyoming by the Biological Survey in 1911. A bull was also imported from the Wichita (Kansas) Zoo. The elk thrived, a second herd was acquired from Jackson Hole, Wyoming, and a new era of wildlife management had begun.

The Wichita Mountains Wildlife Refuge today is the premier American place to better understand the science of species reintroduction. An odd assortment of wildlife was brought into the Wichitas to have a second chance. A Missouri trapper donated wild turkeys to the refuge, but only one survived for more than a year. The Biological Survey experimented with reintroducing wild turkeys and Rio Grande Bronze hens—they interbred and now thrive on the reserve. Pick your decade, and some animal species was reintroduced (or introduced for the first time) in the Wichitas. Longhorn cattle that had escaped from corrals in Texas were rounded up and shipped to Frank Rush's wildlife park by William C. Barnes and John H. Hatton, two cowboys who didn't want buffalo to be the only symbol of the lost frontier. Rush shared their commitment to longhorns. He personally raised over 300 longhorns year-round on the Wichita Preserve, all safe from slaughter. The 1920s saw bighorn sheep imported from Alberta, but the sheep couldn't survive the still heat of summer.[122]

In 1936 Congress changed the name of the Wichita National Forest and Game Preserve. It would now be called the Wichita Mountains National Wildlife Refuge and placed under the Bureau of Biological Survey. In 1939 all federal refuges moved back under the Department of Interior. From 1940 to today the U.S. Fish and Wildlife Service has controlled the Wichita Mountains National Wildlife Refuge.

By 2009 the Wichitas had so many buffalo that every October U.S. Fish and Wildlife began auctioning off surplus animals, many recently born calves. Tourists came to the Wichitas to see the buffalo herds grazing in a wild environment. But even though these bison appeared tame, they weren't. Their behavior remained unpredictable and was likely to change at any time.[123] As with the bears in Yellowstone, people now visited Oklahoma just to see a herd in the wild. The Wichitas, in this regard, were never disappointing. All over Oklahoma, in fact, in curio shops and general stores, buffalo souvenirs were sold. Roosevelt was right in thinking that Oklahomans needed buffalo in their lives. Just look at all the places named after them in the Sooner State: there is a Buffalo Creek as well as a Buffalo Springs, which isn't far from the hamlet of Buffalo in Garfield County. In Harper County kids fish in another Buffalo Creek. What Lincoln's legacy had become to the Illinois tourist trade, the buffalo has been to Oklahoma.[124]

Even the White House was permanently altered by Roosevelt's embracing of the Oklahoma buffalo. As part of his redesign of the White House in 1902, Roosevelt had the architectural firm of McKim, Mead, and White decorate in a neoclassical style. Construction was also begun on a new executive office building known as the West Wing. He demanded that the East Wing do away with the grand staircase at the west end of the Cross Hall in order to enlarge the state dining room. Roosevelt was quite pleased with the firm's work, but a flare-up occurred with Edith on an issue of interior design. She insisted that stone lions' heads be placed on the mantel in the state dining room. Roosevelt thought these would look un-American. Nevertheless, he capitulated to Edith. But with the success of the Wichita Forest and Game Preserve, with his bison surviving their first winter, Roosevelt grew bolder. Off with the African lion heads! In 1908, in a letter to the architectural firm, Roosevelt demanded that the stone lions be recarved as bison heads because bison "made a much more characteristic and American decoration."[125]

In the summer of 1962, John F. Kennedy had the "bison mantel" buffed and restored. Seventy-eight-year-old Alice Roosevelt Longworth was invited to the White House for the unveiling of the frontier symbol her father had championed.[126] It had become one of the most distinctive features in the White House. And in 2006—to honor Quanah Parker—the recently founded Comanche Buffalo Society, based in Mount Park, Oklahoma, was created to honor the chief Theodore Roosevelt had considered a blood brother for understanding that America without bison wasn't America.

CHAPTER TWENTY-TWO

THE NATIONAL MONUMENTS OF 1906

I

Staring out the White House window one mild spring morning in 1906, President Roosevelt watched his sons Archie and Quentin sculpture little mud monuments in a sandbox. "What a heavenly place a sandbox is for two little boys," Roosevelt wrote to Kermit about his brothers. "Archie and Quentin play industriously in it during most of their spare moments when out in the grounds. I often look out of the office windows when I have a score of Senators and Congressmen with me and see them both hard at work arranging caverns or mountains, with runways for their marbles."[1] With such a crowded indoor schedule, the president lamented that he couldn't join the lads outside to play like a prairie dog. Deadlines and commitments were getting the best of him. Much like his sons, only on a vastly larger scale, Roosevelt was preoccupied with reconfiguring landscapes (by Executive Order, that is) for the United States. Arranging for the designation of wonders, in fact, was an apt description of the Roosevelt administration's conservation policy in 1906. No longer was Roosevelt inventorying possible western landscapes to preserve; he was actually preserving them.

Accordingly, some of Roosevelt's indoor bureaucratic chores of 1906 were of the inspiring outdoors type. For example, paperwork was being drafted at the General Land Office to save Devils Tower in northeastern Wyoming—perhaps the strangest molten rock configuration in North America. This isolated Devils Tower outcropping soared over the Missouri Buttes—five dome-shaped rock formations four miles away from it—in the northwest corridor of the Black Hills. Looming over the Wyoming Valley at 1,267 feet above the river, Devils Tower looked like a huge stone tablet on which the Ten Commandments were said to have been set.

Devils Tower had no cultural features like the prehistoric dwellings in the Four Corners region that Congressman John F. Lacey and Edgar Lee Hewett wanted saved. It was a perpendicular columnar laccolith, a gray horn formed during the Triassic Period about 250 million years ago, when the dinosaurs roamed. Surrounding the main stumplike rock formation of Devils Tower were pine forests, woodlands, and grasslands—in short, an unspoiled sanctuary for teeming wildlife. Devils Tower was a sacred site to more than twenty Plains tribes, including the Arapaho, Crow, Lakota, Cheyenne, Kiowa, and Shoshone. They used it like an altar, a place for prayer offerings, vision quests, marriage ceremonies, and funerals.[2] There was an enduring Indian legend that long ago a huge bear had clawed the tower's side, leaving deep scratches or grooves there. This seemingly otherworldly tower served, appropriately, as a setting in the director Steven Spielberg's 1977 sci-fi movie *Close Encounters of the Third Kind*. To geologists, Devils Tower was an important site: an amazing formation at 5,112 feet above sea level, composed of red sandstone and maroon siltstone, with the oxidation of iron mineral causing the outer surface to look almost rust-colored. It was more difficult to climb than the Tetons (near Yellowstone) or El Capitan (in Yosemite). "There are things in nature that engender an awful quiet in the heart of man," N. Scott Momaday wrote in *The Way to Rainy Mountain*. "Devils Tower is one of them."[3]

In early 1906 Roosevelt considered how Devils Tower could be saved for posterity. Because it was only a mile wide, it didn't seem to be large enough for a national park (although the much smaller Platt and Hot Springs were national parks). And Roosevelt didn't want Devils Tower declared part of a national forest, because it was a "wonder," *not* a utilitarian natural resource for the Forest Service to manage. Roosevelt, who was familiar with the history of the U.S. Geological Survey, knew that Colonel Richard I. Dodge had named Devils Tower in his 1876 book *The Black Hills*. To Dodge, it was "one of the most remarkable peaks in this or any other country."[4] Bully for Dodge! And the geologist Henry

Devils Tower in Wyoming became Roosevelt's first national monument, created in 1906.

Newton had written that it was an "unfailing object of wonder."[5] Bully for Newton, too!

A scholarly debate has continued for decades about whether Roosevelt actually climbed or even visited Devils Tower. Some bogus literature propagated by the tower's boosters claims that he did. At best, that is a half-truth. During his trips to the Black Hills and Bighorns in the 1880s and 1890s he did see it (geologists say it looked about the same then as it had 10,000 years prior).[6] After all, the Little Missouri River flowed within twenty or so miles of it. But Roosevelt probably never stood any nearer to its base. On his Great Loop tour in 1903, his train had made stops in the Wyoming towns of Gillette, Moorcroft, and Sundance, so he probably saw Devils Tower from a distance. And he had visited the South Dakota communities of Edgemont and Ardmore on that trip—both within sight of the tower. But touch it: *no*. Nevertheless, as with Mount Olympus in Washington state, Roosevelt revered Devils Tower as a Wyoming site of great value. And strangely, the "Tree Rock," as the Kiowa called it, was part of Roosevelt's life as a rancher because its lore was persuasive throughout the Dakota Badlands.

Roosevelt wasn't alone in his admiration for Devils Tower. It had its share of cheerleaders during the progressive era. In 1891 Senator Francis E. Warren of Wyoming, a pleasant-looking man in his mid-forties, tried to establish "Devils Tower National Park," introducing Bill S.3364. It was shot down by Congress. Every few years Warren would reintroduce the legislation, only to have it repeatedly rejected.[7] However, Senator Warren was able to have Devils Tower classified as part of a federal timber reserve. This at least bought him some time for negotiation. Roosevelt, who was struggling to get the Grand Canyon officially declared a national park, didn't want to have Devils Tower fail again owing to congressional indifference. But even though he had Warren on his side, politicians in Wyoming were anti–federal government. Since the 1890s, when the Boone and Crockett Club had roared against the segregation of Yellowstone National Park by the railroad company, Roosevelt had been disliked by a substantial number of Wyomingites.

What Roosevelt wanted was an executive order to save places like Devils Tower, the Grand Canyon, Mount Olympus, and the Petrified Forest. And he wanted it without confronting dimwitted legislators whose insatiable craving for profit blinded them to the inspirational value of geological landmarks. Congressman Lacey—as head of the House Committee on Public Lands—ballyhooed Devils Tower's weirdness as an asset for tourism. The nation became intrigued about the tower. By what process had Devils Tower been formed? Was it created by volcanic material, or was it an immense mass of igneous rock? Had it once been part of the bottom of an old sea? Who could solve this geological mystery? Roosevelt, collaborating with Lacey, inquired whether there wasn't a clever way to save Devils Tower in the name of archaeology or paleontology. A concerted legal effort commenced in conservationist circles to devise a new presidential prerogative for preserving extraordinary wonders like Devils Tower from commercial destruction. Congressman Lacey, with his extensive knowledge of land law in the developing West, was a strong advocate of placing this "volcanic plug" in the public domain as a contribution to science.[8] Decode the mysteries of Devils Tower, the thinking went, and geologists would finally understand how the Black Hills came into being.

Part of Roosevelt and Lacey's concern in early 1906 was that "big oil" was an octopus with tentacles that harmed consumers and landscapes alike. That January, in fact, Missouri's attorney general, Herbert Hadley, had initiated court hearings against Standard Oil Company of Indiana, Republic Oil Company, and Waters-Pierce Oil Company for "monopolistic conspiracy." With Roosevelt cheering, John D. Rockefeller, the founder of Standard Oil, was hit with thirty-four subpoenas. For a while, Rockefeller hid from the law. By June 1906 the attorney general of Ohio got into the act, anxious to prosecute Standard Oil of New Jersey for violating Ohio's antitrust laws. Recognizing that oil companies never were good stewards of the land, and worried that natural gas hunters would despoil the Dakotas and Wyoming, Roosevelt encouraged the U.S. attorney general, Charles J. Bonaparte, to prosecute Standard Oil of New Jersey under the Sherman Antitrust Act.

Roosevelt actually wanted Standard Oil of New Jersey (a holding company, which controlled more than sixty other companies) dissolved. He felt virtuously outraged by Rockefeller and others who always had a price and were never concerned about the dignity of land. So while Roosevelt was on the offensive to save places like Devils Tower and the Petrified Forest of Arizona in 1906, he was also on the warpath against the "swine," corporate types like Rockefeller who were interested only in personal enrichment. By contrast, Roosevelt himself was concerned for public enrichment, and his trust-busting was making him wildly popular. Building on the damning evidence amassed by Ida Tarbell in her two-volume *History of the Standard Oil Company* (1904), Roosevelt had his Bureau of Corporations further investigate the oil industries; a report was due in the summer of 1907. This was part of Roosevelt's "campaign against privilege," which was "fundamentally an ethical movement."[9]

II

Saving Devils Tower and the Petrified Forest was on Roosevelt's mind during early January 1906, when he learned that his trusted hunting guide in Colorado, John Goff, had been mauled by a cornered cougar. Roosevelt seemed more grimly interested in how the cougar had attacked Goff than in how many stitches Goff needed. Did the cougar lunge? Was it rabid? Or had it been protecting cubs? How deep were the incisions? "Do let me know about it," President Roosevelt anxiously wrote to Yellowstone Park's superintendent, John Pitcher. "I am interested for Johnny's sake, and besides, I have a zoological interest and am anxious to know how the job was done."[10]

Roosevelt was highly attuned to western affairs that January. Even though he was fighting for appropriations to build the Panama Canal, passing legislation concerning railway rates, and calculating a tariff for the Philippines, he vigorously championed statehood for the three western territories of Oklahoma, New Mexico, and Arizona. In addition, the process for creating Sevier National Forest in south central Utah began that January (eventually, more than 375,000 acres were set aside by the U.S. Forest Service). Utah's Great Basin—known for spectacular canyon scenery—was becoming a federally protected wonderland. In Utah, communities like Ogden, Salt Lake City, or Provo seemed like run-of-the-mill settlements compared with the magnificence of the canyonlands.

There were also hard-fought battles going on in Colorado, pitting railroads against coal companies over land. Roosevelt jumped right into the controversy with shirtsleeves rolled up, like a negotiator for labor *and* management. In Idaho and Montana, angry miners were starting to organize behind radical groups, including the Industrial Workers of the World (IWW, or Wobblies), which had been founded in June 1905. Roosevelt insisted that the so-called Wobbly syndicalists (Big Bill Haywood, Daniel DeLeon, Eugene V. Debs, Mother Jones, and other industrial unionists) must be law-abiding and operate without violence. To Roosevelt the danger of industrial unionism was that its proponents saw it as superior to Americanism. The IWW had been created, in part, because the American Federation of Labor (AFL) had organized only 5 percent of the nation's workers. Roosevelt didn't care for the IWW, for two specific reasons: it was trying to monopolize labor, and it was a threat to free-market capitalism. And Roosevelt, refusing to be intimidated, sent federal troops into Goldfield, Nevada, to crush a miners' strike. He deemed the protest harmful to the nation. "I wish labor people absolutely to understand that I set my face like flint against violence and lawlessness of any kind on their part," Roosevelt wrote a friend, "just as much as against arrogant greed by the rich."[11]

It would take a good psychiatrist to understand why Roosevelt hated anarchy in any guise. If there wasn't *order*, he couldn't function properly. But still he surrounded himself with birds and animals that scurried all around in his homes—macaws squawking, dogs barking, cats jumping on papers, turtles wandering about. His tolerance of animal behavior and his intolerance of human behavior were like night and day. There may have been an inner struggle between his childlike obsession with disappearing into the freedom of the wild—responsibility be damned—and his compulsion to be biologically precise about every songbird, tree, grass blade, and insect antenna. The masculine side of his nature wanted to hunt big mammals while his feminine side wanted to nurture small songbirds. He believed that studying all wildlife had helped sharpen and attune his senses. "Roosevelt loved animals, both wild and domestic," the historian Edward Wagenknecht observed. "Even on the hunting field they were individuals to him. He always hated to shoot a crow, always took care not to frighten a doe away from her babies."[12]

Among the early-twentieth-century conservationists, only John Muir had the temerity to stand up to Roosevelt—at Yosemite in 1903, Muir had challenged Roosevelt to reform his boyish hunting ways. And how was Muir rewarded for his candor? At first, *well*: Mount Shasta, Mariposa Grove, and Yosemite Valley were all saved by the federal government. But over time Roosevelt repaid Muir's casual insult by saying that Muir didn't know birds and then by siding with Pinchot on turning Hetch Hetchy into a man-made reservoir. There were consequences for challenging Roosevelt—and these might involve policy. The sycophant got farther with Roosevelt than the challenger. Nevertheless, Roosevelt continued to admire Muir for rallying to nature's defense. In this regard Muir was embraced by the president as a "radical" in the best sense of the word. Here was the difference, in Roosevelt's mind: the strikers in Goldfield, Nevada, wanted more for *themselves* whereas groups like the Sierra Club of California were fighting for *national betterment.* And there was in Muir's carriage, Roosevelt thought, the radiance of Yosemite itself, which the president truly honored.

That spring, just as Roosevelt was ready to expand the boundaries of Yosemite National Park, a disaster rocked California: at daybreak on April 18, an earthquake destroyed San Francisco. Within a few minutes the streets around Union Square and Chinatown filled with mounds of debris. Three hundred thousand people were left homeless. Gas mains had snapped, and storefronts went up in flames. Boats capsized at Fisherman's Wharf. Broken glass from apartment windows rained down like hail. Beautiful hotels like the Winchester and the St. Francis were wrecked. Merchants shouted in disbelief. People walked about dazed, with fretful eyes, scared that at any minute the entire city would sink into the Pacific Ocean. A human flow out of the Bay Area commenced, under the U.S. Army's leadership. The Chinese had considered 1906 the year of the Fire Horse—a time of mass confusion—and this proved to be prophetic. "The entire event which was to destroy an American City and leave an indelible imprint on the mind of the entire nation," the historian Simon Winchester wrote in *A Crack in the Edge of the World*, "had lasted for just over two and a half minutes."[13]

Reports of the earthquake aroused Roosevelt's martial temperament. This was no middle-size quake like the one in 1868. Unfortunately, he was 2,500 miles away in New York and was unable to order naval action. Shaking an impotent fist at the ground was all he could at first do. Reports of buckling aftershocks came over the telegraph directly into his office in downtown Oyster Bay. Then the telegraph shut off. At best, communication with northern California was hit-and-miss. Telephones weren't working at all in the Bay Area—everything was broken in the stricken city. More than 3,400 people died throughout northern California.[14] Roosevelt issued a national condolence: "I share with all our people the horror felt at the catastrophe that has befallen San Francisco, and the most earnest sympathy with your citizens. If there is anything that the Federal Government can do to aid you it will be done."[15]

Many San Franciscans were in a condition of panic. Social dislocation and even mayhem took over. The San Andreas Fault from northwest of San Juan Bautista to the "triple junction" at Cape Mendocino had ruptured the ground, cracking it open like an eggshell. From above, it looked like a zigzagging chain down the spine of California in the cracked earth. Towns anywhere near the fault line suffered severe damage. Geologists were confounded by the violent power of the vibrating earth. Survivors said that the experience was like walking on a trampoline or falling into a tar pit. Although the event is known to history as the San Francisco earthquake, virtually all towns in northern California suffered extensive damage, and the outlook for a quick recovery was bleak.

Americans had known that California was an earthquake zone but the state's residents had long played ostrich, pretending that their homes weren't actually built along a fault line. Now, terrorized shouts of "Fire! Fire! Fire!" were heard along the same streets where Roosevelt had paraded on his Great Loop tour of 1903. Then, Roosevelt had proclaimed San Francisco the shining white Acropolis of the glorious West Coast, the juggernaut of manifest destiny. Now everything was covered with smoke clouds, and the air was poisonous. People contended for jugs of water, worried about dehydration. Soot-blinded horses frantically neighed and frothed in front of brownstones that had toppled into heaps of rubble. Triage stations were set up in fields. After a while, however, an eerie calm blanketed the city, a collective numbness, as people grew weary of trying to put the fires out. The *New York Times* said that the exact scope of the disaster in terms of terror and damage "will never be known."[16]

Statistics came pouring into the White House about the devastation in California. The quake was felt for about 375,000 square miles from Coos Bay, Oregon, to Los Angeles and well into Nevada's Great Basin. More than 28,000 buildings had been destroyed. Following Roosevelt's direct order, the army and navy quelled public unrest and effectively evacuated residents to safety. The armed forces also provided food and shelter for the homeless. The USS *Preble* was anchored offshore from San Francisco to provide humanitarian relief. At the request of Mayor Eugene Schmitz, martial law was imposed, with orders to shoot looters. The USS *Chicago* evacuated 20,000 people by sea (numerically a world record until Dunkirk during World War II).

On April 22, Roosevelt announced that relief efforts were to be overseen by the Red Cross. Congress had appropriated $2.5 million in aid. Determined to show the world that the United States could handle its own problems, Roosevelt declined foreign aid of any kind (relief money nevertheless trickled in from abroad). When the San Francisco mint was raided by looters, federal troops unloaded their guns, killing more than thirty people. But mostly the recovery efforts went well. More than 1,500 tons of provisions were expertly delivered daily to fifty-two food distribution centers. Exuding optimism, Roosevelt claimed that within the decade San Francisco would be rebuilt. And so it was.

III

Besides grappling with this situation, Roosevelt devoted a lot of energy to Niagara Falls. Since 1904 the State Department had been negotiating with Canada, by means of an International Waterways Commission, to protect the integrity of the falls. With Great Britain brokering the bilateral negotiation, a regulated equitable division of water power between the two countries was obtained. But Roosevelt wanted Canada to agree to protecting the "aesthetic value of the falls."[17] Roosevelt became obsessive about creating an international park between Canada and New York. Not to do so, he said, was sacrilegious. Eventually, in June 1906, a bill passed Congress authorizing the secretary of war to supervise the preservation of the falls. Although there are no documents to prove it, Roosevelt *seemed* to be threatening Canada, by implying that if the Canadians didn't cooperate properly with the War Department, the United States would seize control of Niagara Falls and run it as an American national park.

That spring, Roosevelt also resumed regular contact with the prolific artist Frederic Remington, who had illustrated Roosevelt's *Ranch Life and the Hunting-Trail.* Remington's sketchbooks of heavy-duty paper had themselves become American heirlooms. A few of Remington's illustrations from *Ranch Life*—"An Agency Policeman," "Making a Tenderfoot Dance," and "Cowboy Fun"—had grown in popularity since the 1890s.

And in October 1902 Remington published a novel that the president adored: *John Ermine of Yellowstone* (about a Caucasian boy who is raised by Crow Indians and becomes a scout in the U.S. Army). The *New York Times* said that this novel was reminiscent of Wister's *The Virginian.* Remington's characters included Sitting Bull, Crooked Bear, and the White Weasel. What stood out for Roosevelt was Remington's brilliant description of life among the Crow as they roamed in the western prairies. Remington also did a fine job of illustrating *John Ermine of Yellowstone,* thirty drawings in all.[18] In one drawing, the well-cut John Ermine looks a lot like the blond, blue-eyed George Armstrong Custer before the Battle of Little Bighorn. Conceived as an epic western, *John Ermine of Yellowstone* was an oddly complex tale of an intermixing of European American and Native American strains. The lead character, Ermine, is torn between both cultures, incapable of fully assimilating into either. The novel got solid reviews; one reviewer said that Remington had captured the imperishable quiet of Wyoming's forestlands.

"My dear Remington," Roosevelt wrote to Remington on February 20, 1906. "It may be true that no white man ever understood an Indian, but at any rate you convey the impression of understanding him! I have done what I very rarely do—that is read a serial story—and I have followed every installment of *The Way of an Indian* as it came out."[19] Flattered by Roosevelt's praise of him in the parlors of Washington, D.C., as the Karl Bodner or George Catlin of their generation, Remington made Roosevelt a small wax bronze titled *Paleolithic Man* as an appreciation. It depicted, Remington wrote, a Darwinian representation of a "human figure bordering on an ape, squatting and holding a clam in right hand and a club in left." Remington had created the sculpture at a makeshift studio on Cedar Island in the Chippewa Bay archipelago in the Thousand Islands area, along a scenic stretch of the Saint Lawrence River in New York. Suffering from health problems caused by overeating, Remington, who had become a quirky *odd duck*, was hiding from the world. Jokingly, Remington added in his note that the bronze was modeled after "the original inhabitant of the original Oyster Bay—whenever that was—."[20]

Oh, boy! Did Roosevelt ever fancy that piece of Remingtonia! Hurrah for Darwinian art! Whatever tension and mistrust had developed between Roosevelt and Remington in the 1890s had vanished. Although Roosevelt never purchased a Remington, he had amassed a fine collection of items, which had been bestowed on him as gifts. Even though fanciful artists like Maxfield Parrish were now the rage, Roosevelt stood steadfastly by Remington, whose struggle with obesity had taken on tragic dimensions. (At over 350 pounds he weighed more than Secretary of War William Howard Taft, and his weight was obviously affecting the functioning of his vital organs.) The *Paleolithic Man* statue thrilled Roosevelt. "We hail the coming of the original native oyster," he wrote to Remington. "Mrs. Roosevelt is as much pleased as I am with it. I think it is very appropriate, for undoubtedly Paleolithic man feasted on oysters long before he got to the point of hunting the mammoth and the woolly rhinoceros."[21]

Throughout the spring of 1906 Roosevelt corresponded with John Burroughs, sharing the excitement of spying all the springtime birds around the White House. Oom John was busy raising a few vegetables while writing essays about rural neighbors, salt breezes, and maple syrup. All the features of farm life in the Catskills (and the universe at large) were Burroughs's bailiwick. Of course, the two naturalists chatted about birds. "That warbler I wrote you about yesterday was the Cape May warbler," Roosevelt told Oom John. "As soon as I got hold of an ornithological book I identified it. I do not think I ever saw one before, for it is rather a rare bird—at least on Long Island, where most of my bird knowledge was picked up. It was a male, in the brilliant spring plumage; and the orange-brown cheeks, the brilliant yellow sides of the neck just behind the cheeks, and the brilliant yellow under parts with thick blade streaks on the breast, made the bird unmistakable. It was in a little pine, and I examined it very closely with the glasses but could not see much of its back. Have you found it a common bird?"[22]

There was a warmth and kindness to Roosevelt in spring 1906 that had been missing since the hurly-burly of becoming president. He seemed proud that talented outdoorsmen like Burroughs, Wister, Remington, Chapman, and Merriam were his *real* friends—not those New York money changers deformed by "swinish greed" and by "vulgarity and vice and vacuity and extravagance"[23]; or, for that matter, those Chicago meatpackers whose astonishing workplace uncleanliness Roosevelt called "revolting."[24] Writing to the editor of the *Saturday Evening Post*, Roosevelt, as if taking stock of his friends, boasted that his "intimate" fellows were "men I met in the mountains and backwoods and on ranches and the plains." He meant Bat Masterson, Will Sewall, Joe and Sylvane Ferris, Seth Bullock, John Willis, Jack Abernathy. If you had a biographical history in the West—the old West—Roosevelt was sympathetic.

Echoing Grinnell, Roosevelt insisted that Native American tribes had to be treated fairly under his administration. Sometimes Roosevelt acted as if America had been better off before Wounded Knee, when the Indians still rode freely in the Great Plains. There was something in the president of *John Ermine of Yellowstone*. Roosevelt worked with Inspector James McLaughlin of the U.S. Indian Service to negotiate more than forty tribal agreements on behalf of his administration. Native Americans were opposed to land allotments, but the federal government was allotting land anyway. Roosevelt worked to keep fraud out of the system.* And, almost miraculously, the planned reintroduction of buffalo at Wichita Forest and Game Preserve in southwestern Oklahoma scheduled for 1907 was welcomed by both cowboys and Indians.

Roosevelt also continued to encourage his naturalist friends to write *big* popular zoology. For example, he sent Henry Bryant Bigelow, a junior staff member at the Harvard Museum of Comparative Zoology (who had written about caribou and wolves), a glowing note of endorsement, urging him to try composing gorgeous zoological prose like *On the Origin of Species* or *Birds of America*. "We need that the greatest scientific book shall be one which scientific laymen can read, understand, and appreciate," Roosevelt wrote. "The greatest scientific book will be a part of literature; as Darwin and Lucretius are."[25]

*In 1887 Native Americans owned 138 million acres; by 1934, when the allotments ceased, they had only 48 million acres, much of it not good for farming.

All this natural history and yearning for the high country awakened Roosevelt to new possibilities for preservationism. By June 1906, as Congress adjourned for the summer, Roosevelt grew anxious about the fate of the United States' western sites, particularly the Grand Canyon, the Petrified Forest, Devils Tower, and Mesa Verde. With regard to these, getting commitments out of legislators was like tearing tin. Arrogantly, Congress was also stalling on whether to accept from the state of California two magnificent gifts: Yosemite Valley and Mariposa Big Tree Grove. What in hell were the stingy western senators thinking? That's what Roosevelt wanted to know. Piqued at Congress's hesitation over further preservation of Yosemite, Roosevelt wrote to Senator George Clement Perkins of California, a Republican who was an embarrassment to the Republican Party, a sharp letter demanding that Congress, without delay, seize these old-growth redwoods, spectacular waterfalls, and unparalleled scenic wonders for the enrichment of the public domain. With San Francisco three-quarters destroyed by the earthquake, and people living in tent cities on the outskirts of town, the federal government needed to do something special for California.

"It seems to me that it would be a real misfortune if this Congress adjourned without accepting the magnificent gift of California of the Yosemite Park," Roosevelt wrote. "What is the status of the matter? Is it not possible to have it put through? I earnestly hope you will look it up and let me know. It would be too bad if, either from indifference or because of paying heed to selfish interests, the United States Government fails to act as in my judgement it is morally obligatory upon it to act in view of the generous action of California."[26]

IV

On June 6, 1906, President Roosevelt signed into law "An Act for the Preservation of American Antiquities." It allowed for a president to designate "historical landmarks, historic preservation structures, and other objects of scientific interest" as national monuments. Drafted by the team of Lacey and Hewett, the act was stunning in its exclusion of Congress. It was an unparalleled tool for a president to use.[27] The preservationists involved (who included W. H. Holmes of the Smithsonian Institution and the Reverend Henry Mason-Baum, known for excavations in the Holy Land) had so carefully crafted the language of the legislation that it sounded inoffensive and whisked through the Senate (on May 24) and the House (on June 5) practically unaltered. It gave the president the unencumbered power to unilaterally declare the protection of landscapes of archaeological, scientific, and environmental value federal land.[28] As the historian Robert W. Righter has said in the *Western Historical Quarterly*, now Roosevelt could seek "rapid presidential action" instead of a "dilly-dally" with a "tortoise-paced Congress."[29] At last Roosevelt had a legal way to circumvent Congress in these matters.[30]

More than any other policy Roosevelt adopted as president, the signing of the Antiquities Act has earned him praise from modern environmentalists; it represented the self-proclaimed "wilderness hunter" as a high-minded naturalist statesman. Roosevelt had confounded pro-development interests in the West with a preservationist program for both now and tomorrow. There was no longer a need to negotiate with the timber and mining lobbies over such sites as Devils Tower or the Petrified Forest. The resourceful Roosevelt had given America a way station for these places on the road to national park status. The genius of Lacey and Hewett's effort was that the Antiquities Act didn't limit the acreage a president could designate as national monument lands on behalf of science. Basically, the acreage was entirely up to a president's discretion. In wiggle words, the act stated simply that the monuments were to be "confined to the smallest area compatible with the proper care and management of the objects to be protected." But Roosevelt's idea of "small" was bigger than anybody else's in official Washington.

Until, June 1906, Congress saw Lacey as the bulwark against Roosevelt's overreach. Colleagues knew that Lacey was eager to save abandoned ruins of prehistoric peoples in the Southwest, but he was also a man of compromise. Lacey, imbued with the European social ideal of a strong central state, had pounded on doors on Capitol Hill asking for assistance in saving El Morro (Inscription Rock), Montezuma Castle, and Chaco Canyon. He was always soft-spoken and modest in demeanor. As chair of the House Committee on Public Lands he was, however, a formidable power broker, particularly with the delegations from the Middle West and West. So it was understandable that congressmen and senators from Montana, Wyoming, Washington, Oregon, Idaho, and other western states agreed to Lacey's antiquities project. Why not? Their reaction was as old as politics. They'd scratch Lacey's back and, as a quid pro quo, he would ease up on issues such as timber leasing, mining contracts, and grazing laws in the national forests. These western legislators surely must have worried. But about Roosevelt's overdoing the designation of national monuments, the worst case scenario was no worse than the bird refuges of 1903 to 1905: nothing more than a few hundred acres of prehistoric ruins and natural oddities scattered about the American landscape. That would be a tolerable progressive indulgence compared with the grabs of forest reserves.

Congress, in effect, had been tricked by the otherwise ethical Lacey. The Antiquities Act was a dangerous precedent to set with Roosevelt in the White House. The legislation had placed a new conservationist weapon—the national monument—at T.R.'s disposal. To think that Roosevelt wouldn't stretch his new powers to the extreme was naive. Certainly Roosevelt was honest about the prehistoric ruins in New Mexico and Arizona: these resources *were* preserved for the sake of science. No longer would southwestern pot hunters or tourist vandals have free rein to desecrate these ancient sites. Where Roosevelt grew mischievous, however, was in exploiting the loose language of the Antiquities Act, which stipulated that national monuments were ipso facto of *scientific* value. To Roosevelt a marsh, an arroyo, and a limestone cliff were all of scientific interest. What wasn't a biological or geological birthright to him? And now, as of 1906, the federal government would become the caretaker of historically significant ruins.

At first, the Antiquities Act would permanently protect part of the Four Corners region in the West. Lacey had traveled earlier that spring from Santa Fe to Durango, Colorado, and had been aghast to see thieves taking artifacts from Mesa Verde. He knew that Roosevelt wanted to make life miserable for such heirloom robbers. Lacey began pushing harder for the Anasazi cliff dwellings near Durango, Colorado, to become a national park. Along with Hewett, he also championed preserving the ruins of the Pajarito Plateau in New Mexico near Los Alamos. A grassroots effort was forming to create a "national cultural reservation" on the Pajarito Plateau. When Lacey first visited the region in August 1902, he had been mesmerized by the deserted caves, communal ruins, and adobe villages where Indians still lived. And he knew that the trail guide at Four Corners, John Wetherill, was Roosevelt's idea of a great American. Wetherill was a real-life John Ermine in the Navajo-Apache-Hopi lands.*

Old photographs show Wetherill with deep-set eyes and a pronounced nose, looking like a weathered, desert version of Seth Bullock. He wore a turquoise stone to ornament his favorite belt buckle, and his hair was cut bare on the sides; this midwesterner had clearly adopted the Southwest as his home. The novelist Zane Grey wrote about Wether-

*John Wetherill's brother Richard was the famous archaeologist, based in Santa Fe, who helped promote saving ruins in the Southwest. John and Richard often get confused.

ill, idealistically but simply, in his essay collection *Tales of Lonely Travels*. Once the Antiquities Act was passed, Wetherill made recommendations in the Southwest as requested by Forest Order 19, which asked national forest supervisors to report on prehistoric structures and other artifacts and sites of scientific interest located on the western reserves.

Born in Kansas in 1866, "Hosteen John," as he was called, had moved to Mancos, Colorado in 1880. Although ranching was the family business, Hosteen John became obsessed with the cliff dwellings of Mesa Verde in southwestern Colorado. By 1900 Wetherill, with his wife, Louisa Wade, moved to the Navajo lands of New Mexico. Tired of dealing with droughts and rustlers, he decided to own trading posts at Ojo Alamo, Chavez, and Chaco Canyon. Besides selling trinkets and provisions he became the best-known trail guide for the entire, vast Four Corners region. It was Hosteen John who had taken Hewett and Lacey to see the astounding prehistoric ruins there.[31]

In retrospect, Lacey, Hewett, and Wetherill were together the ideal advocates for southwestern antiquities: a congressman, an archaeologist, and a knowledgeable guide. After gathering information from both Hewett and Wetherill, Lacey felt certain he could get Congress to approve of Mesa Verde. He was more worried about the Petrified Forest of Arizona (soon to become a favorite spot of John Muir and John Burroughs). Thousands of people there were stealing Pliocene fossils, pottery shards, and petrified logs. These thieves would just leave with whatever they wanted. When the Pueblo people had lived in the Painted Desert–Petrified Forest area, they had used fossilized wood for tools; in 1906, travelers en route to California collected chunks for souvenirs, sometimes by the wagonful. "This remarkable deposit has been subject to much vandalism already, and unless permanently reserved and protected, is sure of ultimate destruction," Lacey wrote about the Petrified Forest for *Shield's Magazine*. "The land is useless for agriculture, as it is in the heart of a desert. An attempt was made some years ago to work these trees up into table tops, but the prevalence of small holes in the body of the finest of the logs prevented the success of this commercial enterprise. Otherwise this great national curiosity would have long since become a matter of history only."[32]

Consumed with impatience, Lacey started learning every geological fact about Arizona's Petrified Forest as if he were on assignment from the American Museum of Natural History. To draw attention to the great petrified logs, he wrote reports, delivered speeches, and lobbied the Santa Fe Railway about their value as a stopover attraction for tourists on the way to the Grand Canyon. Lacey also wrote a slogan for the railroad to use: "Come see the Grand Canyon (the greatest scenic wonder in the world) and the Petrified Forest of Arizona (the greatest natural curiosity)." When congressmen from California, Washington, Oregon, and Wyoming told Lacey that their states had petrified forests, too, Lacey grew exasperated. The fools didn't understand. *Of course*, there were other petrified forests. But his was "*The* Petrified Forest of the World," in a class by itself. "Yellow, red, blue, white, black, brown, rose, purple, green, gray, in fact, all the colors of the rainbow, are found in these trees," Lacey said. "Many of them are five feet in diameter and 140 feet in length, and lie just as they were originally deposited, imbedded a few inches in the desert sand."[33]

Since the early 1890s Roosevelt and Lacey had made a lot of conservation deals together. They had become alter egos. But Roosevelt had never seen Lacey so stirred up as he was over the Petrified Forest. Lacey even quoted the poet Samuel Taylor Coleridge, who had said that the great arches of Gothic cathedrals were a "petrified religion." And the Arizona Territory—God bless the United States—had the most holy petrified valley in the western hemisphere. Lacey admitted that he wasn't schooled in the principles of stratigraphy, but he nevertheless knew that the geologic history of the Petrified Forest was worth preserving. Whether they were using petrified wood for tabletops or chopping down old-growth redwoods for decks, Lacey was annoyed by the disrespect that business enterprises and commercial vandals were showing toward the western heritage. "As hard almost as the diamond, as brilliant in colors as the flowers of the field, this ancient forest, which was transformed into stone perhaps before man appeared on the planet, is still to be seen under the sunshine of Arizona," Lacey wrote. "It should by all means be preserved for the admiration and wonder of generations yet to come."[34]

According to Professor Rebecca Conrad, author of *Places of Quiet Beauty*, Congressman Lacey inserted the words "scenic and scientific" into the Antiquities Act as a clever way to preserve places like the Petrified Forest of Arizona. Here wood had been turned to solid silica, rock, and quartz. Lacey also wrote an account of his 1902 trip to Arizona with Wetherill as his guide, and of how his idea for the Antiquities Act came into focus. The archaeological district known as Newspaper Rock Petroglyphs had Pueblo dwellings actually made out of petrified wood. It was all astounding! Borrowing a page from Merriam, Lacey cleverly chose the Roosevelt elk (T.R.'s beloved species) and the Petrified Forest as his original impetus for the Antiquities Act. "It was this trip which led to the introduction and passage of my bill for the preservation of aboriginal ruins and places of scenic and scientific interest upon the public domain," Lacey wrote, "under which the Petrified Forest, the Olympic Range Elk Reserve and about two hundred places of ethnological interest have been designated as 'monuments' and preserved to the public."[35]

What a pity that Congressman Lacey has been left out of most environmental history textbooks covering the Roosevelt era! With the exception of Char Miller, Hal Rothman, Rebecca Conrad, and Patricia Nelson Limerick, few western historians have taken the time to realize all that Lacey did to save prehistoric ruins, desert ecosystems, bird sanctuaries, petrified forests, plug-dome volcanoes, wildlife-rich areas, and national wonders. But Roosevelt, at least, understood that Congressman Lacey was *the man*, the shrewdest pro-conservationist legislator of his time. Lacey's secret had four aspects: he was a committed outdoorsman, amateur ornithologist, and Indian scholar, and he wasn't a credit monger. He championed places like Mesa Verde, the Petrified Forest, Chaco Canyon, and El Morro, even though he earned no votes in Iowa's Sixth District for doing so—Lacey was, therefore, a true American patriot. The Pajarito region of New Mexico, in particular, captivated him with its ancient rock drawings of the sun, snakes, and deer. All the caves and ruins of the Zuni, Taos, and Acoma, he believed, needed to be saved. As for the Petrified Forest, the trees had hardened into a complete and priceless landscape. "Let these trees be protected from vandalism and they will endure forever," Lacey pleaded with the Senate. "It is to be hoped that the public sentiment which has urged and warmly approved of the action of the House of Representatives in thrice passing the bill to set aside this land as a public national park will in the near future bring about favorable action in the Senate. That lover of nature, the President, will be glad to sign such a bill."[36]

Just three days after Roosevelt signed the Antiquities Act of 1906, there was a strange event in Iowa. At the Republican county convention in Fairfield, a prankster let loose an elephant—wearing a banner that read "G.O.P."—to rampage through the crowded hall. Mayhem ensued in the hall as the frightened elephant trumpeted madly about. Republican delegates fled through the windows and doors. According to the *New York Times*, one terrified politician broke an arm in the panic. When the elephant was finally calmed down and the shock of the event had subsided, there was a police inquiry. As it turned out, a group from the pro-Roosevelt and pro-Lacey wing of Iowa's Republican Party had hired the elephant from Robinson's Circus for the prank. "The elephant's name is

'Teddy Roosevelt,' " the *Times* reported, "and the convention was afraid of it."[37]

At Lotus Lake in Long Island that June, Robert B. Roosevelt's health was breaking down: he was seventy-nine and had many ailments.[38] Reports circulated that he wouldn't live long.[39] Nevertheless, R.B.R. led a high-profile campaign on Long Island to replant white pine trees wherever any had previously been chopped down. Even on his deathbed, R.B.R. was engaged in life. Having already planted white pines at his own estate, he implored all his neighbors from Montauk to Brooklyn to do the same. Long before Franklin Roosevelt created the Civilian Conservation Corps in 1933, R.B.R. had started a regional forerunner on Long Island. It was a crusade for him, just as stopping the indiscriminate killing of birds and fish had been following the Civil War.[40]

President Roosevelt was at Oyster Bay on June 14 when Uncle Rob died. Coincidentally, the president had just had a species of trout named after him in California. The obituary in the *New York Times* noted that R.B.R. had been the famous author of *The Game Fish of North America*, *The Game Birds of the North*, *Superior Fishing*, *Fish Hatching and Fish Catching*, and *Florida and Game Water Birds*—all notable conservationist accomplishments. A funeral was scheduled, and the president came to say a proper good-bye. An era had ended, but all of R.B.R.'s conservationist aspirations—everything he had stood for, except his bohemian lifestyle and his philandering—lived on in his nephew in the White House.

Following the funeral, in July 1906, President Roosevelt started planning to save both Devils Tower and Petrified Forest in the fall. The paperwork was now in order. He dashed off a note of gratitude to Congressman Lacey for championing more knolls, buttes, spurs, ruins, and ravines than anybody else in America. It was Lacey who taught Roosevelt to look at petrified logs as gems or precious stones—they were that valuable. Believing that Lacey's methodical approach to saving antiquities was good for the republic, Roosevelt told Lacey that "certain gentlemen" were filled with a "deep sense of obligation" for all his work. This rather dull and stern Iowan, a Civil War veteran of the Mississippi River campaign, who always wore a standard-issue gray suit, had done more for America's environmental and cultural heritage during the progressive era than anybody else. He was a giant like Gifford Pinchot, Jane Addams, or John Muir. Roosevelt suggested that these "gentlemen" wanted to name a park, a monument, or a memorial in his honor for engineering the Antiquities Act of 1906: they wanted to honor him with a mountain, forest, or canyon. The modest Lacey was amused, and he demurred. Nevertheless,

Roosevelt signed "An Act to Protect Birds and Their Eggs in Game and Bird Preserves" into law that June as a tribute to Lacey.[41]

Encouraged now that Mesa Verde had become a national park, Lacey urged Roosevelt to use the Antiquities Act to declare the Petrified Forest a national monument. Time was short. Couldn't the Roosevelt administration somehow circumvent the slow, tedious process of obtaining congressional approval for a national park? The miles of petrified logs, the multihued badlands, the Painted Desert, the historic buildings, and the archaeological ruins would, if *preserved*, be Lacey's legacy. Roosevelt had the Department of the Interior look into it at once. Meanwhile, as the logistics were worked out, Roosevelt wanted trespassers arrested for stealing prehistoric pottery fragments or for setting off a rock slide from a hill in the Petrified Forest. Wetherill was keeping Lacey informed about any syndicates stealing wagons of petrified wood—but the small-time thief was nearly impossible to apprehend.

V

After the success of the Antiquities Act, Roosevelt's intensity in the West increased. On June 19 he signed a joint congressional resolution enlarging Yosemite National Park by 41.67 square miles (nearly 27,000 acres)—no small clump of trees. Suddenly two of California's crown jewels, which Roosevelt had seen on his 1903 western trek with John Muir at his side—Yosemite Valley and Mariposa Big Tree Grove—were acquired by the Department of the Interior. But instead of being elated, Roosevelt grew concerned. Lacey was right. If Congress was so slow to act on behalf of an already established national park like Yosemite, what would it do when he introduced the Grand Canyon, Petrified Forest, and Mount Olympus for consideration as national parks? Would the extractors be able to prevail over the protectors during the congressional process? For a national park designation, Roosevelt needed Congress; but designation as a national monument required only determination.

An ardent believer in statehood for the Territories, Roosevelt now indicated that admittance into the Union entailed a quid pro quo—turning over natural and archaeological wonders like the Grand Canyon, the Canyon de Chelly, and the Petrified Forest to the Department of the Interior to become national monuments). This horse-trading wasn't put in writing—he wasn't that foolish—but the precondition was implied. In territories like New Mexico, Arizona, and Oklahoma the president had the advantage (just as he did in establishing the Luquillo National Forest in Puerto Rico). Consultation wasn't essential for action in de facto colonies. "The Territories are filled with men and women of the stamp of which I grew to feel so hearty a regard and respect during the years that I myself lived and worked on the Great Plains and in the Rocky Mountains," Roosevelt wrote to Mark A. Rodgers, secretary of the Arizona Statehood Association. "It was from these four Territories that I raised the regiment with which I took part in the Cuban campaign. Assuredly I would under no circumstances advise the people of these Territories to do anything that I considered to be against either their moral or their material well-being."[42]

Unquestionably, Roosevelt took a paternalistic attitude toward Arizona. He regarded Arizona's mining, timber, and real estate interests with amused disdain, and with steadily increasing distrust. To most people on the Atlantic seaboard, Arizona seemed far, far away; but Roosevelt considered it his backyard. The Geological Survey had reported, gravely, that Arizona's mineral deposits (except for coal) would be largely extracted by the end of the twentieth century as a result of overmining. This prediction caught Roosevelt's full attention. The insatiable mining outfits would destroy wild Arizona if the federal government didn't intervene.

Congressman Lacey was likewise disgusted by overindustrialization, but he took it as a given in the modern world. As Roosevelt saw it, the true enemies in the West were aridity, adroit political malfeasance, poaching of relics, and thieving of timber. Roosevelt failed to understand that his reclamation projects—especially hydroelectric dams—were aimed at dominating nature on behalf of settlers; they, too, ruined landscapes and made some regions dependent on federal funding. Lacey believed that the solution to western problems was more federal responsibility and preservationist morality, achieved by congressional authorization. But Congress seemed uninterested in the Four Corners region. Action was required. Roosevelt, the "preacher militant," as of the summer of 1906, refused to accept a feather-duster approach to the Southwest. Roosevelt's warrior side wanted to crush his enemies into the dust, not outfox them with legalities. To Roosevelt hate could be a creative impulse for the common good. It's hard to escape the feeling that Roosevelt enjoyed creating national forests and national monuments in part because it was rubbing his opponents' faces in his wilderness philosophy of living.

Still, underlying Roosevelt's hostility toward despoilers was his fear of America without a wilderness. Conservation was a way for Roosevelt to grapple with this anxiety. By saving heritage sites and forests, Roosevelt was providing a way for the body politic to stay healthy. By reclaiming the prehistoric past, Indian relics, volcanic mounds, hidden lakes, fish-filled streams, stands of trees, weird-looking buttes, desertscapes, and petrified wood, Roosevelt believed he could preserve the old pioneer spirit that had made American civilization so special. To Roosevelt, industrialization was a corrosive problem in that it led to urbanization, which in turn stripped citizens of their attachment to the land. A whole generation of youngsters was suffering from what we might now call a nature-deficiency disorder. It was the wilderness, Roosevelt insisted, with reverence, that made American special. The novelist Frank Norris had an octopus to war against—the huge agricultural concerns. Similarly, Roosevelt had the trust titans to rally against, because their concept of laissez-faire economics was unpatriotic. They valued money more than Old Faithful or the Great Smoky Mountains. "If we do not go to church so much as did our fathers," Burroughs commented about the naturalists around Roosevelt in 1905, "we go to woods much more, and are much more inclined to make a temple of them than they were."[43]

VI

It wasn't just the Far West that Roosevelt was worried about. An ugly international incident had occurred in the Alaska Territory, involving Japanese seal hunters wielding clubs, knives, and guns in the Pribilof Islands. On July 16 a small fleet of Japanese vessels attacked the Alaskan seal rookery at Saint Paul Island. Unbeknownst to the Japanese, the Roosevelt administration maintained a small naval–biological research facility on this island, which is in the Bering Sea. The American sailors there were fond of the seals, which had originally been saved by President Grant and were celebrated by Rudyard Kipling in *The Jungle Book*, first published in 1894. A few of the sailors intervened to stop the Japanese poaching raid, and a melee occurred. Sickened because the Japanese had clubbed baby seals and then skinned them alive, the Americans killed five of the raiders, wounded two others, and apprehended another twelve. An international brouhaha erupted over the Japanese butchery and the Americans' heavy-handedness. The *Japanese Times*, for example, said that although seal poaching was a misdemeanor, the U.S. Navy had responded with murder. In contrast, the San Francisco press published gruesome details of the hunt, supported the U.S. Navy, and said that the merciless slashing and beating of American seals in American waters was outrageous.[44]

One side effect of the San Francisco earthquake was a thoughtless increase in anti-Japanese prejudice on the Pacific coast. When people are under duress, they may look for a scapegoat: in San Francisco the recent Japanese immigrants provided one. The Russo-Japanese War had left the United States and Japan as the preeminent powers in the Pacific basin. The negotiated Portsmouth Treaty also bestowed on Japan strategic, political, and economic interests in Manchuria, and these threatened to undermine America's open-door policy as formulated by Hay. Roosevelt greatly respected Japan but feared its rise to power. With nativist emotions running high in San Francisco, an anti-Japanese backlash occurred, manifested in school segregation, riots, and a spate of anti-Japanese legislation in Sacramento. In San Francisco between May 6 and November 5, 1906, for example, there were more than 290 cases of assault, most perpetrated against Japanese immigrants. Two eminent seismologists from Tokyo were stoned for investigating the San Andreas Fault; some San Franciscans didn't want foreigners to tell them not to live on a fault. These racist attacks and stonings angered the Japanese government, particularly because it had given $246,000 to San Francisco for relief after the earthquake. Therefore, a deep distrust already existed between Tokyo and Washington, D.C., when the "Alaskan seal incident" occurred.

The U.S. Department of Commerce and Labor quickly submitted a report confirming that many of the seals had indeed been skinned alive. Aleuts who lived on the island were unbiased eyewitnesses. Even more disturbing were the photographs taken of seals half-skinned, hobbling about maimed and apparently bleating in pain. Many bigots in California used the incident as a pretext for sweeping condemnations of the Japanese character. Roosevelt's own reaction was beyond words. Poaching always set him off like a bomb, and the poaching in this case made him apoplectic. Realizing that the Aleutian Islands were the remotest land in North America, and that policing the 1,200-mile archipelago was an impossible task, Roosevelt nevertheless was proud of the U.S. Navy for attacking these and other raiders. Tokyo wanted the Roosevelt administration to try the sailors for the murder of the five Japanese men. Japanese lawyers, as noted above, argued that according to the Alaskan criminal code, seal poaching was not a felony but a misdemeanor, and that committing murder to stop a misdemeanor was not justifiable in a republic based on democratic principles.

Determined to flummox Roosevelt, the bitter Japanese government developed a legal argument and recommended punitive measures. But Roosevelt was unbending with regard to seal or bird rookeries. Instead of court-martialing the sailors, he congratulated them for being outstanding watchdogs for Alaska's priceless seal herds. However, not wanting to go to war with Japan over this incident, he told Secretary of the Navy Charles J. Bonaparte to remove all U.S. ships from Asian waters. The international incident should be settled by diplomats, not battleships. With tension so high on both sides, Roosevelt privately feared an international incident, even while publicly expressing militaristic bravado.

Realizing that the United States was holding a weak hand, the State Department had Assistant Solicitor William C. Dennis draft a memorandum reflecting Roosevelt's views on the imbroglio; it was submitted on September 10, 1907. "The circumstances of a pelagic seal raid in a wild country like Alaska, carried on by armed raiders and accompanied by a brutal and cruel slaughter of the seal herd, put a severe strain on the common-law doctrine defining the rights of misdemeanants," Dennis wrote. "It has not been so long since Kipling could say 'There is never a law of God or man runs north of Fifty-three,' and it may well be that the methods of those heroic days are still sometimes morally justifiable irrespective of the provisions of the penal code."[45]

Tokyo was furious over Dennis's memorandum. It had about as much legal validity as the hanging of a horse thief in *The Virginian.* Did Roosevelt really think that Japan would accept Kipling as a defense? Dennis was unable to explain convincingly that the president loved seals and that Roosevelt had found the butchering of an Alaskan herd worthy of vigilante action. Worried that the incident might escalate to war, Secretary of State Elihu Root wisely stepped into the fray. The best strategy was to cool down temperatures on both sides. Root, working closely with the Department of the Navy, came up with a different defense of American sailors to present to Ambassador Baron Kogoro Takahira: the poachers were "burglars" and burglary was a felony under Alaskan law. If this premise was accepted by Tokyo, then the killing of the five Japanese during their commission of a crime was justifiable. Wasn't it? Reluctantly, in May 1908, the Japanese government accepted this argument, and the diplomatic crisis ended.

As a conservationist Roosevelt had prided himself on his stewardship of the whole land. This included Alaska's Aleutian Islands. With his literary imagination Roosevelt could hear the waters slapping against ancestral rocks—the sound seemed to travel all the way from the Bering Sea to the corridors of power in Washington, D.C. He could imagine the baby seals Kipling had written about in *The Jungle Book*, the Aleuts killing only what they would eat, and the Japanese poachers believing they could ignore international boundaries. Growing up in Manhattan after the Civil War, T.R. had enriched his naturalist studies by acquiring a seal skull. Now, as president, he had threatened to send battleships to preserve these friendly mammals' Alaskan rookeries.

VII

There was great joy in preservationist circles on June 29, 1906, when President Roosevelt signed Mesa Verde National Park into existence. Credit for the park should probably have gone to the skilled ancestral Pueblo masons who had built the cliff dwellings 700 to 1,600 years earlier. But Roosevelt instead lavished praise on Congressman John F. Lacey, Edgar Lee Hewett, John Wetherill, and others. The motto for saving these enchanting cliff dwellings—rock villages in protected alcoves of the southwestern Colorado canyon walls—was "leaving the past in place." Once the Pueblo tribes of Mesa Verde had migrated south to the Rio Grande region in 1300, the Ute tried to live in the Mesa Verde (Spanish for "green table") cliffs. Much as the Comanche helped bring the buffalo back to Oklahoma, the Ute played a crucial role in protecting the cliff dwellings from pillagers over the decades. From the hundreds of dwellings of Mesa Verde that survived erosion and human defacement, archaeologists in 1906 had saved a hugely important chapter in the saga of prehistoric America.[46]

When Roosevelt created Mesa Verde National Park, it contained 52,073 acres, all rising high above the surrounding mesa. There were more than 4,000 deserted dwellings for archaeologists like Hewett and novices like the Wetherills to analyze now in an appropriate way. Two women—Virginia Donaghe McClurg and Lucy Peabody—had led the successful crusade to preserve these ruins. McClurg was a New Yorker who had moved to Colorado in 1879 to teach. Intrigued by the mysteries of Mesa Verde, she wrote a series of preservationist stories for the *Review of Reviews*, *Cosmopolitan*, and *Century Magazine*. Her partner in championing Mesa Verde—Lucy Peabody—came from Cincinnati. For a while Peabody worked as a secretarial assistant at the Bureau of American Ethnology, where she advocated saving the ruins in the Four Corners area. When she married a major in the U.S. Army, she moved to Denver, where her interest in Mesa Verde's cliff dwellings grew.

Both McClurg and Peabody were devoted Roosevelt Republicans in 1906—i.e., progressives. They saw in Roosevelt the best chance for preserving Mesa Verde from speculators. Influenced by *Uncle Tom's Cabin*, the Chautauqua movement, and Susan B. Anthony, they became a bulwark against silver mining around Durango, Colorado. At that time, Europeans were offering money for ancient relics found around the Mesa Verde excavation site. A recession had developed, and pot hunting became profitable for poor Coloradan farmers and for transients. In 1891 twenty-three-year-old Nils Otto Gustaf Nordenskiöld, of the Academy of Sciences, collected more than 600 items for Sweden from Mesa Verde. (Today they're in a museum in Helsinki and should be returned to the United States at once.) In 1893 Nordenskiöld—whose uncle was a famous Arctic explorer—published a heavily illustrated book, *The Cliff Dwellings of Mesa Verde*. Archaeologists from all over the world now wanted to visit Colorado. "Nordenskiöld's expedition and the loss of a large and valuable collection aroused both admiration and deep resentment among American archeologists," historian Char Miller writes, "and provided strong arguments in Congress for protective legislation."[47] Roosevelt felt that Nordenskiöld had looted American property. Wasn't there any law to stop foreign raiders from stealing U.S. antiquities?

When Roosevelt became president in 1901, the answer to his question was still no. Yet, with Lacey working on the legislative angles (and the activists in Santa Fe who gravitated around Edgar Lee Hewett receiving attention from the press), a federal strategy was incrementally being put in place for the Pajarito Plateau and Mesa Verde. Using the power of the pen, McClurg and Peabody initiated a grassroots progressive movement in Colorado to protect Mesa Verde. Women in Colorado had won the vote in 1893 (they were among the first in America to do so), and these suffragists now made Mesa Verde their cause. They tried to persuade the Ute to cede Mesa Verde to the federal government. They formed a women's association to police the cliff dwellings and protect the site from vandals. With the help of John Wetherill, "No Trespassing" signs were posted—so many, in fact, that they looked frightening. The women also enlisted Hewett to argue the archaeological case in Congress. "These are unquestionably the greatest prehistoric monuments within the limits of the United States," Hewett said. "Aside from their great historic and scientific value they would be of more general interest to the public." A visibly upset Hewett claimed that "irresponsible damage" was being done at Mesa Verde and that the "deterioration progresses very rapidly."[48]

From 1901 to 1903, during the Fifty-Seventh Congress, two bills had been introduced in the House of Representatives to establish a national park at Mesa Verde—both died. Congress did authorize the Department of the Interior to negotiate with the Ute in the hope that they would relinquish the ancient cliff dwellings. But that was a minor issue to Coloradans anxious to save Mesa Verde. With Roosevelt urging Lacey, Hewett, Wetherill, McClurg, and Peabody to stay the course, two bills were introduced in the Fifty-Eighth Congress for a "Colorado Cliff Dwellings National Park." These also failed. It had become clear to Lacey that getting a sweeping act passed to save the southwestern antiquities would be easier than fighting for each ruin separately. So in the late spring of 1906 it was a happy turn of events when the Antiquities Act of June 8 passed and was soon followed by the creation of Mesa Verde National Park.

Roosevelt wasn't passive about his new national park, the first in Colorado. Working closely with the Smithsonian Institution, the Department of the Interior began excavating and repairing the Anasazi sites. Jesse Walter Fewkes of the Smithsonian, for example, had crumbling walls quickly stabilized. Proper roads were soon constructed so that visitors could enjoy the ruins. Cliff dwellings in the park, such as Long House, Mug House, and Step House, were popularized in periodicals including *Harper's* and *National Geographic*. Any vandals who dared touch the ruins would be fined $1,000. The "mothers of Mesa Verde"—McClurg and Peabody—had prevailed.[49]

VIII

That June the Roosevelt administration was on an upswing regarding the preservation of southwestern prehistoric ruins and cliff dwellings: the Antiquities Act and Mesa Verde were steps forward for the progressive movement. However, the Bronx Zoo took a leap backward. It is true that the New York Zoological Society was running the most amazing endangered species program in the world. As Roosevelt had envisioned it in 1895, Charles Darwin had a living memorial in the Bronx. Two Colorado black bears—Teddy B and Teddy G—had been donated to the zoo that spring by an admirer of the president.[50] They were advertised as "teddy bears," and the city's schoolchildren flocked to the zoo as if P. T. Barnum's circus elephants were in town. The rambunctious bear cubs were adorable, with jolly faces and a little white around their muzzles. Every five or ten minutes the cubs, lacking their mother's discipline, tumbled and rolled in a playful wrestling match as crowds gathered around to ooh and aah. Also, British East Africa—particularly the grasslands of Kenya and Uganda—tugged at Roosevelt's mind and he had requested that the New York Zoological Society purchase a baby rhinoceros for $5,000. The board complied. The rhino, only five or six years old and weighing 250 to 300 pounds, was an immediate star attraction at the zoo. And there was another baby star as well: a buffalo calf was born in captivity that June. Hornaday had now successfully raised two generations of calves since the founding of the zoo.[51]

But unfortunately, the New York Zoological Society's success in showcasing small animals led Hornaday to make a fatal error in judgment. At the Saint Louis World's Fair in 1904 Hornaday had been fascinated by a Congolese pygmy, Ota Benga, who was put on public display in an ethnological exhibit called the University of Man. The backstory here is essential. An eccentric missionary and anthropologist, Samuel Phillips Verner, had been hunting for specimens in the Belgian Congo when he stumbled on Ota Benga in a cage. According to Verner, a cannibalistic tribe planned to eat the pygmy. What a find! Immediately, Verner had an idea for a human rights gesture. Why not put the pygmy on display in Saint Louis as an example of *The Descent of Man*? From shrew to spider monkey to chimpanzee to baboon to gorilla to pygmy—the display would be all the rage at the fair. So Verner purchased Benga, thereby, in his own mind, saving him from the boiling pot. Before long Benga found himself in Saint Louis. At the University of Man's display of aboriginals, the "representatives" included Hottentots, Zulus, Eskimos, Filipinos, and Geronimo in the flesh. All the displays included proper species classifications on informative plaques, on the assumption that these would present Darwinian theory in a more visually interesting way and help schoolchildren better understand it. Benga quickly learned to ham it up for coins, dancing and basket weaving like the popular image of a bushman.

Hornaday was taken in by this racist hullabaloo. After seeing Ota Benga in Saint Louis he negotiated to have the pygmy—who had meanwhile been brought back to Africa—delivered to the Bronx Zoo for public display. At first Benga was startled by the diverse animals at the zoo. Being from the Congo had hardly prepared Benga for, say, the huge pythons that hung out of crooked tree limbs and were fed live rats. New Yorkers cheered the new acquisition, which enhanced their civic pride. In the tradition of the Bronx Zoo's educational outreach, Hornaday dutifully wrote an article on Ota Benga for the October 1906 edition of the Zoological Society *Bulletin*; it was called "African Pygmy" and was positioned right before one called "The Collection of Lizards."[52]

Reading that issue of the *Bulletin* is a frightening journey into the perils of Darwinism as applied to human beings. For all of his sophistication in husbandry Hornaday had a deplorable attraction to eugenics. So did Madison Grant, who had approved the Bronx Zoo's abominable display. Hornaday—who was doubtlessly a better man than this ugly episode suggests—called Benga part of the "smallest racial division of the human genus and probably the lowest in cultural development."[53] Kept in a cage next to an orangutan named Dohong, who pedaled around on a tricycle, Benga was provided with straw and rope to weave. No monkey could do that! The pygmy had evolved! "He has much manual skill," the article in the *Bulletin* noted, "and is quite expert in the making of hammocks and nets." Sometimes Benga was encouraged to sleep with the chimpanzees for mutually beneficial socializing. To attract visitors, and hoping to build on the success of Teddy B and Teddy G, Hornaday advertised that something "New Under the Sun" in zookeeping had occurred at the Monkey House. It was as if the Bronx Zoo were trying to explain Mendel's theories of heredity with regard to the trait of smallness versus largeness—using Benga as exhibit A. And the tourists did come in droves.

On September 8, opening day, a huge crowd gathered to see Hornaday's prize exhibit. Expectations were high. And Benga, his teeth filed into arrowheads to add the allure of menace, didn't disappoint the spectators. But the *New York Times* seemed appalled: "Bushman Shares a Cage with Bronx Park Apes." To be fair to Hornaday, Benga had already become a Darwinian specimen at Saint Louis, where headlines such as "Pygmies Demand a Monkey Diet" and "Pygmy Dance Starts Panic in Fair Plaza" appeared. However, in New York the *Times* article of early September 1906 helped raise a public accusation of racism at the Bronx Zoo. The whole spectacle, which included white children laughing and taunting Benga in his cage, turned the "serious minded grave." The *Times* questioned the morality of putting an African on public display in such a pseudoscientific way.

Likewise, African-American ministers in New York protested against putting a human being in a cage with a monkey. The Reverend Dr. R. S. MacArthur of Calvary Baptist Church announced a coordinated "agitation" aimed at freeing Ota Benga. Reports of the whole affair were getting more and more sordid. "It is too bad," MacArthur said, "that there is not some society like [the New York Society] for the Prevention of the Cruelty to Children." MacArthur went so far as to say that Benga was a slave. "We send our missionaries to Africa to Christianize the people," he remarked "and then we bring one here to brutalize him." He also went directly after Hornaday, saying that "the person responsible for this exhibition degrades himself as much as does the African."[54]

In accordance with the Bronx Zoo's educational mission, an informational plaque was placed outside Ota Benga's cage. It read:

> The African Pygmy, "Ota Benga." Age 23 years. Height, 4 feet 11 inches. Weight, 103 pounds. Brought from the Kasai River, Congo Free State, South Central Africa by Dr. Samuel P. Verner. Exhibited each afternoon during September.

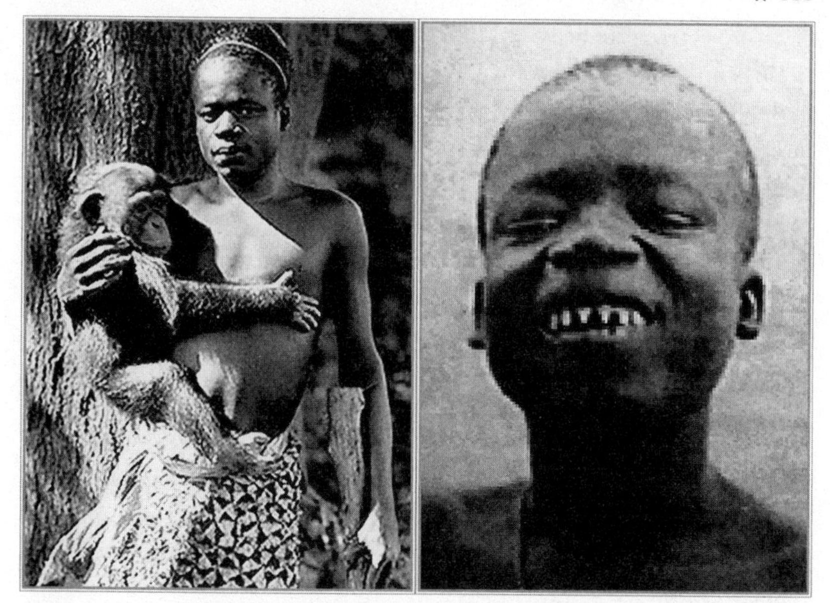

Ota Benga was degraded by being put in a cage as a supposedly Darwinian exhibit at the Bronx Zoo. Benga was often made to pose with monkeys and had his teeth sharpened to look like a cannibal.

That September the tabloids ran stories about Ota Benga—some sympathetic, others mocking. Facts came out: the Bronx Zoo hadn't purchased the pygmy; he was on loan, so the charge of slavery was a guffaw. The *New York Evening News* condescendingly noted that while Benga was black, he wasn't coal-black—he wasn't actually on the bottom rung of the descent of man. That rung was occupied by more dark-skinned blacks. A spirited debate also ensued about whether Benga was a real pygmy or a dwarf or midget. There was great interest also in his sharply filed teeth, which led to speculations about cannibalism. Schoolchildren visiting the zoo goaded Benga to rip at raw meat hurled at him by keepers. Benga's nickname was "Bi," and kids taunted him with it until he waved at them. One afternoon Benga, refusing to wear strange clothes, broke away from his keeper. When he was eventually apprehended he was brandishing a knife; quickly, the zookeepers disarmed him.[55] "We are taking excellent care of the little fellow," Hornaday said in the Bronx Zoo's defense. "He has one of the best rooms in the primate house."[56]

A few courageous Baptist ministers kept coming to the zoo to protest the incarceration of Ota Benga. Although there is no record of President Roosevelt's getting involved in the controversy, Hornaday nevertheless started feeling pressure to reverse course. Charges of zoological quackery were starting to arise. "I do not wish to offend my colored brothers' feelings or the feelings of any one for that matter," Hornaday said. "I am giving the exhibitions purely as an ethnological exhibit. It is my duty to interest visitors to the park, and what I have done in exhibiting Benga is in pursuance of this. I am a believer in the Darwinian theory." However, he insisted that Darwinism wasn't the main reason for displaying the pygmy. Hornaday was a Nebraskan, raised on the frontier, and Ota was his counterpart to Geronimo in the Wild West show. After all, Buffalo Bill had received accolades for parading Apache performers around dusty fairgrounds. Why should Hornaday get pummeled in the press over a Congolese pygmy? Such criticism was selective and hypocritical. Exasperated, and tired of fierce criticism from newspapers and ministers, Hornaday went on to explain that Benga slept in the primate house because it was the most obvious and most "comfortable" place for him to bed down at the zoo.[57] What was Hornaday supposed to do? Have him sleep with the zebras?

On Sunday, September 16, more than 40,000 visitors came to the zoo and went to the monkey house to see Ota Benga. As a special attraction, Bi had been let out of the cage and was free to wander around the zoological park, though with a keeper at his side. "They chased him about the grounds all day, howling, jeering and yelling," the *New York Times* reported of the spectators. "Some of them poked him in the ribs, others tripped him up, all laughed at him." Fearing for the pygmy's life, the keeper put Benga back in his cage. "Me no like America," Benga said. "Me like St. Louis."[58]

Eventually, unable to shake off the criticism, Hornaday cracked. Arriving at work on Monday, with a pack of newsmen firing questions at him, Hornaday threw in the towel. "Enough!" he said. "Enough! I have had enough of Ota Benga, the African pigmy. Ring up the Brooklyn Howard Colored Orphan Asylum. Tell them that they can get busy tinkering with his intellect. I'm through with him here."[59]

Benga was shunted off to the orphan asylum, supposedly as a free man. But he really wasn't free. In 1900, Governor Roosevelt had signed into law an act banning discrimination in public schools; yet, oddly, it wasn't applicable in orphanages.[60] Dressed in a white suit, Benga was kept as a sort of mascot at Howard. He remained something of a celebrity, and he was taught math and a few hundred English words and was introduced (forcibly it seems) to the New Testament. But because there were children at the asylum, Benga was segregated from the mainstream of the institution. The cooks fed him scraps in the kitchen, away from the children's view. Because he became a chain-smoker, he was deemed a bad influence on young people. The relocation to the orphan asylum was becoming a failure for all involved, so another Plan B was tried. In 1910 Benga was shunted off to Lynchburg, Virginia, where he worked on a tobacco farm in the Tidewater region. Deeply disturbed by his experience as a zoo exhibit, and longing for his African homeland, Benga committed suicide in 1916: a pistol shot to the heart. When Hornaday heard about the suicide he was very unsympathetic. "Evidently," Hornaday wrote, "he felt that he would rather die than work for a living."[61]

The small world that is history ridicules Hornaday over the Benga episode. But his views about the University of Man were once taken seriously and given credence throughout America in the early twentieth century. All over the nation, government-run eugenics offices had opened. In 1910, in fact, there was a Eugenics Record Office, created and founded by rich industrialists. An effort was made by the strong to weed out the weak in the "American race." This misguided movement was an outgrowth of a theory called social Darwinism and is often seen as a step toward Nazism. From 1900 to 1935, thirty-two states adopted laws that allowed sterilization of "defective humans." Only half-jokingly, H. L. Mencken said that all the southern sharecroppers needed to be sterilized. As Karl W. Gibson points out in *Saving Darwin*, more than 60,000 Americans were sterilized in the early twentieth century because they had epilepsy or stuttered or were mentally challenged. Ota Benga was, in a sense, a victim of the eugenics movement.[62]

IX

On September 24, 1906, a few days after Ota Benga was transferred from the Bronx Zoo to Brooklyn Howard Colored Orphan Asylum, Roosevelt at last set aside Devils Tower on two square miles of Wyoming wilderness as a national monument. (A clerical error omitted the apostrophe in Devil's Tower, and it has never been reinstated.) Once the Antiquities Act passed in June, Frank W. Mondell, representative at-large from Wyoming, began pushing for Devils Tower to become the first national monument. Although he was vehemently opposed to national forests, Mondell, a resident of nearby Newcastle, Wyoming, correctly surmised that Devils Tower could, if properly promoted, become a first-rate tourist attraction, bringing tourist dollars to Newcastle, Gillette, and Sundance. Although Devils Tower was out of the way, there was a possibility that tourists visiting the Black Hills would make a day's outing to see the bear claw marks.*[63] (Mondell wanted to build an iron stairway from bottom to top: evidently he didn't think it would be obtrusive, possibly because he had no idea what "obtrusive" meant.) As a member of the House Committee on Public Lands, which Lacey chaired, Mondell worked with the GLO all summer to get the size of the site over 1,000 acres so that the tower could be properly cared for and managed.

Depending on where a visitor was standing and what the angle of sunlight was, Devils Tower produced various impressions. At the top reaches the colors were grays; the bottom had soft reds, pastel rust, and yellow-olive combinations. There were almost no roads to the tower in 1906; travelers coming from the east had to ford the swollen Belle Fourche River seven or eight times. From a distance the Tower seemed, deceptively, to be always within grasp.

A cursory look at the Wyoming newspapers of September 25 shows zero interest in the new federal designation. After all, to locals the site was still just forlorn Devils Tower. Unfortunately, the Roosevelt administration had no ranger to assign to the tower. The commissioner of the GLO, Fred Dennett, did provide a "special agent" based in Laramie, Wyoming, whose job included halting commercial vandalism and homesteading on the new national monument property. On the Fourth of July locals were allowed to attempt climbs to the top. What made national monuments so confusing was that, depending on what was expedient on a case-by-case basis, they were under the jurisdiction of one of three departments: Interior, Agriculture, or War. (Eventually, in 1916, they were all brought into the Department of the Interior.)[64]

Because the special agent didn't live at Devils Tower, locals started chipping off hunks of the rock formation for souvenirs. Eventually "No Trespassing" signs were posted all around Devils Tower as a deterrent; these worked, to a limited degree. However, not until the 1930s, when roads were built, did Roosevelt's first national monument finally become a major tourist attraction, with full-time federal protection services provided.

If all Roosevelt had been doing from June to September 1906 was creating Devils Tower National Monument and Mesa Verde National Park, the academic debate over whether he was a *preservationist* or a *conservationist*

*People often mistakenly believe that the Black Hills are only in South Dakota. A third of the ecosystem is in Wyoming. The granite core of the Black Hills crests near South Dakota's Mount Rushmore National Monument.

would not arise. Clearly, his actions on behalf of these western sites, no matter how prosaic the language of the Antiquities Act, made him a thoroughgoing preservationist. Yet throughout the summer of 1906 Roosevelt was also deeply involved with the building of dams, bridges, and reservoirs in the frontier landscape of the West, under the Reclamation Act. Emblematic of the progressive era, his Reclamation Service now had more than 400 engineers and other experts working in Arizona, New Mexico, and Oklahoma. More than 1 million acres of arid land were being irrigated (and this irrigation entailed digging 800 miles of canals, tunnels, and ditches). The Roosevelt administration was giving the American West a concrete and steel reconstruction for the sake of water. This activity by the Reclamation Service caused newspapers in the West to rejoice. Reclaimed land west of the Mississippi River, in fact, could soon sustain 100 million people, owing to wise water policy.[65] "The crowded conditions of the eastern communities will be automatically relieved," the *Ellensburg* (Washington) *Dawn* predicted, "a happy and contented, home-loving and home-owning people will occupy the present arid regime of the west."[66]

Just as the Panama Canal was a triumph of engineering, so too was the hauling of 16 million cubic yards of American earth under the Reclamation Service's guidance. Roosevelt had hired more than 10,000 men and 5,000 horses to reclaim the arid West. The days of general surveys and land examinations were over—the time had come for housing communities to be built in places like Los Angeles, Phoenix, Albuquerque, and Oklahoma City. "We may well congratulate ourselves upon the rapid progress already made, and rejoice that the infancy of the work has been safely passed," Roosevelt wrote to Gifford Pinchot. "But we must not forget that there are dangers and difficulties still ahead, and that only unbroken vigilance, efficiency, integrity, and good sense will suffice to prevent disaster. . . . There remains the critical question of how best to utilize the reclaimed lands by putting them into the hands of actual cultivators and homemakers, who will return the original outlay in annual installments paid back into the reclamation fund; the question of seeing that the lands are used for homes, and not for purposes of speculation or for the building up of large fortunes."[67]

Because 1906 was a midterm election year, Roosevelt tried to push his conservation policy into the slipstream. He boasted about how smart he was to have Pinchot running the Forest Service out of Agriculture (not Interior). To Roosevelt, Pinchot remained a golden boy who could do no wrong. Roosevelt believed that under Pinchot's stewardship the U.S. Forest Service was intensely engaged in making sure all western reserves resources contributed mightily to the "permanent prosperity of the people who depend upon them." Western entrepreneurs lambasted Roosevelt and Pinchot's forest reserve policies as socialism, but the president believed his administration's foresight guaranteed future jobs to stockmen, miners, lumbermen, railroad employees, and small ranches. And it was helping to end rural water shortages. If the western forests were destroyed, there would be no water, and cities would become ghost towns.

Furthermore, Roosevelt believed that Pinchot was doing a superb job of finding innovative ways to put out wildfires, both man-made and natural, which swept over the West in the summertime. He deserved an ovation. Wildfires were most common in zones where the terrain was moist for nine months of the year but then became extremely dry from July through September, creating a tinderbox. When scrub, leaves, twigs, or branches dried out, they became highly flammable. Entire national forests could be transformed almost instantly into smoldering mulch. Sometimes farmers would use fire (circumscribed burning) to clear land. This was often necessary, but it was dangerous. If a light wind picked up the flames and sparks jumped, they could cause an uncontrollable wildfire. Lightning fires were another problem for the dense forests and wind-beaten sagebrush of the West. An entire forest reserve could disappear in a few days. The chaparral in southern California and the lower-elevation deserts in the Southwest were particularly vulnerable to wildfires. Then, of course, there was human carelessness, as well as arson.[68]

What Roosevelt had on his hands in 1906, however, was a feud between forest rangers and stockmen in the western wildlands. To reduce tension Roosevelt formed Forest Service advisory committees aimed at enlightening ranchers on why a steady stream flow and reservoirs were directly correlated with *more* grass for grazing. Forest rangers were their friends, not the enemy. Meanwhile, the USDA's Biological Survey increased its predator control, teaching ranchers that poisoning coyotes was smart but shooting insect-eating songbirds was a mistake.

"There is therefore no longer an excuse for saying that the reserves retard the legitimate settlement and development of the country," Roosevelt wrote to Pinchot in a letter intended for public distribution. "The forest policy of the Government in the West has now become what the West desired it to be. It is a national policy, wider than the boundaries of any State, and larger than the interests of any single industry. Of course it cannot give any set of men exactly what they would choose. Undoubtedly the irrigator would often like to have less stock on his watersheds, while the stockman wants more. The lumberman would like to cut more timber, the settler and the miner would often like him to cut less. The county authorities want to see more money coming in for schools and roads, while the lumberman and stockman object to the rise in the value of timber and grass. But the interests of the people as a whole are, I repeat, safe in the hands of the Forest Service. By keeping the public forests in the public hands our forest policy substitutes the good of the whole people for the profits of the privileged few. With that result none will quarrel except the men who are losing the chance of personal profit at the public expense."[69]

What made Roosevelt so powerful when working to save western ranchlands was that he was speaking from firsthand knowledge. Insofar that landscapes were left to cowboys, places like Texas would be vulcanized into barren zones of burnt grass, as if vast acreage had been shaved by a giant razor. In 1905 Roosevelt had his Public Land Commission issue an alarming report based on data collected by a team of experts. "The general lack of control in the use of public grazing lands has resulted, naturally and inevitably, in overgrazing and the ruin of millions of acres of otherwise valuable grazing territory," the report said. "Lands useful for grazing are losing their only capacity for productiveness, as, of course, they must when no legal control is exercised."[70] But as the historian Deanne Stillman noted in *Mustang*, this dire report was ignored by western congressmen as taste-curdling U.S. federal government babble and "the range got worse."[71]

X

The fall of 1906 was the first occasion when Roosevelt was able to spend much time at Pine Knot that year. The timing of his visit to Virginia was simple: November 1 was the opening of the wild turkey hunting season in Virginia. Arriving on Halloween, Theodore and Edith stayed for the better part of five days at Pine Knot. The woodsy retreat made Roosevelt immediately content. Being in the Blue Ridge Mountains, where the leaves were turning to dazzling fall colors, was sheer joy to the president. He had been consumed with the Cuban revolt, trust-busting, labor disputes, the Panama Canal, Indian matters, and a congressional election, and the thought of bagging a wild turkey for supper was a great relief from these pressures. The hunt was hosted by his friend Dick McDaniel, and the word around Charlottesville-Keene was that the wild turkeys were plentiful. A few well-intentioned local farmers tried to secretly stock plump turkeys on Roosevelt's Pine Knot property to make the president's hunt a guaranteed success. But word leaked out, and the scheme was aborted.

Whenever possible Roosevelt fled the White House to spend time at his rustic cabin, Pine Knot, near Charlottesville-Keene, Virginia.

Traipsing about the woods at Pine Knot, a local physician showed the president to a fine covey of quail in a clearing. Roosevelt waved him off. "I want," he said imperiously, "bigger game than that!"

On November 4 Roosevelt got his bird. The *Washington Post* and the *New York Times* had articles about the wild turkey's weight and colors.[72]

Never before had Roosevelt eaten a turkey that tasted as fine as this one. Free from the White House's tedious schedule, he enjoyed the simple nearby things at Pine Knot. Relieved of encumbrances, he wrote enthusiastic letters about the game bird to both his son Kermit and old Bill Sewall in Maine.[73] In 1907, Roosevelt wrote an essay about the wild turkey hunt, "Small Country Neighbors," for *Scribner's Magazine.* As a literary effort the piece stylistically recalled "Sou'-Sou'-Southerly," written twenty-seven years earlier, or something in a publication of the Boone and Crockett Club. "Small Country Neighbors" was a celebration of the American simple life: a turkey hunt, fresh vegetables, a small cabin or hut in the woods, a tent on the shore. Roosevelt wrote: "Each morning I left the house between three and five o'clock, under a cold, brilliant moon. The frost was heavy; and my horse shuffled over the frozen ruts. . . . It was interesting and attractive in spite of the cold. In the night we heard the quavering screech owls. . . . At dawn we listened to the lusty hammering of the big logcocks, or to the curious coughing or croaking sound of a hawk before it left its roost."[74]

The exaltation of hunting wild turkeys—and killing one—in the crisp air got Roosevelt thinking again about the Appalachians. He very much wanted to create an eastern forest reserve in the Blue Ridge Mountains to match the vast western reserves in the Rockies. But this idea was akin to heresy in West Virginia, Virginia, North Carolina, and Georgia. In 1901, in his First Annual Message, Roosevelt had proposed such an eastern forest reserve to Congress. He wanted it to include vast parts of the Shenandoah Mountains of Virginia, the White Mountains of New Hampshire, and the Great Smoky Mountains of Tennessee–North Carolina. Congress, however, had refused. By late 1902 Roosevelt had grown extremely frustrated that both Democratic and Republican politicians were hindering his plan for an eastern reserve. He wanted to strangle them. The nonauthoritarian part of being president—i.e., working with Congress—annoyed him no end. With regard to natural resource management before the Antiquities Act, Roosevelt's authoritarianism was often foiled in the legislative process. "I should like to see the Government purchase and control the proposed great South Appalachian preserve," he wrote to a friend, "but there are very grave practical difficulties in the way."[75]

From his correspondence during the fall of 1906, it is clear that Roosevelt was deeply worried about the survival of the Blue Ridge Mountains if unregulated timbering persisted. Too many farmers in Virginia were wasting soil, water resources, and forestlands along the James River. Human greed never ceased to amaze Roosevelt. Studying the Round Rock depot outside Charlottesville one afternoon, taking a casual inspection stroll as he'd done as New York's police commissioner, surrounded by green walls of trees, Roosevelt saw that once-pristine forests were girdled, and chopped-up trunks were piled high in lumberyards along the tracks. How utterly unnecessary and regrettable the scene was. To Roosevelt, cutting trees in the Blue Ridge to make room for the plow was a good thing. But massacring mile upon mile of land just for lumber was criminal, as was companies' refusal to replant. "American consumption of lumber was greater than ever before," the historian Roy M. Robbins writes in *Our Landed Heritage*. "It was estimated in 1905, that to supply the Portland [Oregon] mills alone, 80 acres of timber had to be cut every twenty-four hours."[76] The situation was just as bad—if not worse—in Virginia.

XI

While Roosevelt was hunting wild turkey in Virginia, the 1906 congressional elections were held. Generally speaking, it was a good day for Republicans (progressive) in the North and Democrats (progressive) in the South. But in what was interpreted by political pundits, in part, as displeasure with Roosevelt's intense conservationism, twenty-eight Republican seats were lost in the U.S. House of Representatives. This meant that the Republican majority was lessened by fifty-six seats (it was now 222 to 164). One casualty of the election was John F. Lacey of Iowa's Sixth District. Apparently the voters didn't care that Lacey was America's authority on petrified logs in the Arizona desert; they wanted solutions to the economic downturn of 1906 (which would turn into the Panic of 1907). Iowans living in towns like Oskaloosa, Pella, and Eddyville had real problems and spared little thought for Anasazi cliff dwellings, Zuni cave drawings, and the mating habits of little green herons. Congressmen were supposed to bring back pork to the home district not establish federal parks, forests, and bird reservations in other states and territories.

Nothing seized Roosevelt's attention quite like an electoral debacle. Losing Lacey in Congress, for instance, was a blow, because Lacey had so ably aided the conservationist movement in the congressional Committee on the Public Lands. But Roosevelt realized, stoically, that every politician had his day and Lacey's had lasted for thirty-seven years. With his congressional career now terminated, Roosevelt inquired whether Lacey wanted a cabinet appointment or an ambassadorship. Lacey's answer was no. He preferred to practice law in Oskaloosa. What an unsung American hero this Iowan was! Without Lacey, there might have been no model bird laws in Florida, no reintroduction of bison in Oklahoma, and no preservation of cliff dwellings in the Southwest. Mesa Verde might have been destroyed without his intervention. William Hornaday correctly said of Lacey that "he was never elsewhere than on the firing line." A movement was under way in November 1906 to create a "monument as lofty as his own purposes and as imperishable as his fame."[77] The accomplishments of Lacey's governmental career were never forgotten by Roosevelt, who, on returning from Panama, planned to designate as national monuments three southwestern prehistoric sites favored by Lacey.

From Pine Knot the president headed to Norfolk, setting sail for Panama on November 9, 1906, to see "how the ditch is getting along." There had been no cases of yellow fever in Panama City since November 11, 1905; the amazing Dr. William Gorgas had eradicated this scourge from the isthmus. So Roosevelt, with a group of military personnel at his side, was ready for an inspection tour. The first days passed peacefully at sea. Much of his correspondence while he was aboard the USS *Louisiana* dealt with Cuba and America's naval power. Yet he also kept colorful naturalist notes. "All the forenoon we had Cuba on our right and most of the forenoon and part of the afternoon Haiti on our left," Roosevelt wrote to his son Kermit, "and in each case green, jungly shores and bold mountains—two great, beautiful, venomous tropic islands." Among meditations on voodoo, cannibalism, Dutch sea dogs, Vasco Nuñez de Balboa, and the Chagres River flood, Roosevelt wrote about the tropics, proud to think that he had saved wild parts of Puerto Rico, Florida, and Louisiana from destruction. "The deluge of rain meant that many of the villages were knee-deep in water, while the flooded rivers tore through the tropic forests," he wrote of Panama. "It is a real tropic forest, palms and bananas, breadfruit trees, bamboos, lofty ceibas, and gorgeous butterflies and brilliant colored birds fluttering among the orchids. There are beautiful flowers, too. All my old enthusiasm for natural history seemed to revive, and I would have given a good deal to have stayed and tried to collect specimens."[78]

Halfheartedly reporting on the engineering feats associated with the Panama Canal, Roosevelt was proud of his achievement but seemed to prefer being a naturalist. He saw himself as an advance scout for the American Museum of Natural History, which didn't acquire specimens from Panama until 1914.[79] When the *Louisiana* anchored in Puerto Rico, Roosevelt rushed out to inspect the Luquillo National Forest area he had created in 1902. After reading Biological Survey reports about the rain forest and parrots, he now examined them on his own. Returning to his childhood habit of drawing animals, Roosevelt once again doodled parrots and turtles. "The scenery was beautiful," he wrote to Kermit. "It was as thoroly [*sic*] tropical as Panama but much more livable. There were palms, tree-ferns, bananas, mangoes, bamboos, and many other trees and multitudes of brilliant flowers. There was one vine called the dream vine with flowers as big as great white water lilies, which close up tight in the daytime and bloom at night. There were vines with masses of brilliant purple and pink flowers, and others with masses of little white flowers, which at night smell deliciously."[80]

Kermit, now sixteen years old, was full of gratitude that his father sent him such marvelous notes from Panama, from Puerto Rico, and at sea. But he seems to have thought somewhat differently in early December. News came that his father—President Theodore Roosevelt—had won the Nobel Peace Prize for negotiating a settlement in the Russo-Japanese War. Not surprisingly, as the first American ever to win this honor, Roosevelt was pleased. But he was also concerned about the $40,000 check that accompanied the prize. After all, he had spent much of his public career huffing and puffing against bribery and corruption. He had made peace between Japan and Russia because it was his job as president. Ethically, the $40,000 didn't belong to him.

To Kermit, his father was just being unduly foolish. The money could properly be used to build a new wing on Sagamore Hill, to travel around the world, or to earn interest in an inheritance fund for him and his brothers and sisters. Shouldn't the Roosevelt family enjoy this gift? His father, that December, deplored such self-indulgent notions. "Now," the president wrote to Kermit, "I hate to do anything foolish or quixotic and above all I hate to do anything that means the refusal of money which would ultimately come to you children. But mother and I talked it over and came to the conclusion that while I was President at any rate, and perhaps anyhow, I could not accept money given to me for making peace between two nations, especially when I was able to make peace simply because I was President. To receive money for making peace would in any event be a little too much like being given money for rescuing a man from drowning, or for performing a daring feat in war."[81] A prisoner of old ethics, Roosevelt received the check that December. He used it to create a committee in Washington, D.C., for industrial peace.

On December 8, Roosevelt had signed quietistic declarations establishing Montezuma Castle (Arizona), El Morro (New Mexico), and the Petrified Forest (Arizona) as national monuments under the Antiquities Act of 1906. The three monuments were a tribute to Lacey's trademark persistence on their behalf. If Lacey hadn't been a teetotaler, he might have uncorked a bottle of Dom Perignon when told of this order. In Arizona, Wetherill excitedly anticipated more tourists than he could shepherd to see the ancient sites. Roosevelt had created all three monuments as a federal measure to deter artifact thieves and promote scientific study. He understood that these southwestern ruins and petroglyphs were windows to understanding the prehistoric cliff dwellings, pueblo ruins, and early missions discovered by army officers, ethnologists, cowboys, and explorers on the vast public lands in the territories. The story of ancient peoples could be analyzed better in these three monuments than anywhere else in North America; the ruins were that intact. Thanks to the Antiquities Act, nobody was allowed to excavate or appropriate anything from Montezuma Castle, El Morro, or the Petrified Forest without permission from the relevant department (War, Agriculture, or Interior). As Charles F. Lummis wrote in *St. Nicholas*, an illustrated magazine for young adults, these monuments were in a part of the United States which "Americans know [as] little as they do Central Africa."[82] The historian Josh Protas has noted, in *A Past Preserved in Stone*, that the "diversity" of southwestern monuments "set a precedent for the types of monuments that would later be established."[83]

Take, for example, El Morro, on the Colorado Plateau. More than 2,000 inscriptions and petroglyphs were carved into the soft sandstone by explorers, pioneers, and native tribes. Much of western history may have been lost when pioneers headed down the Santa Fe Trail, but at El Morro some clues were left behind. Located ten miles from Ramah, New Mexico, this ancestral Pueblo ruin is at an elevation of 7,219 feet. Probably fewer than 1,000 easterners had seen it by 1906. Apparently, the ancients had used El Morro as a reliable water hole and campsite. At the time T.R. declared El Morro a national monument, the walls carved with signatures looked like a gigantic hotel register. Once again, thanks to Roosevelt, El Morro was now a treasured place, saved for future generations to study and enjoy.

Montezuma Castle National Monument of Arizona featured amazing cliff dwellings that had been molded and lived in by the pre-Columbian Sinagua around AD 1400. The Sinagua had once prospered, developing a sophisticated culture, but then inexplicably vanished. To many people, the Arizona cliff dwellings they left behind were among the wonders of the world. Nobody really knew what to make of them. The Verde Valley area overlooking Beaver Creek had been named by Europeans for the Aztec emperor of Mexico—Montezuma II—in the mistaken belief that he had once lived there, and the misnomer stuck. An early advocate of protecting Montezuma Castle was T.R.'s old Rough Rider friend William "Buckey" O'Neill, who besides being a mayor of Prescott, Arizona, was the editor of *Hoof and Horn*. To O'Neill, the Montezuma Castle cliff dwellings, much like Mesa Verde, raised more questions than they answered. Why did the Sinagua leave? Archaeologists offered conflicting answers to such questions, though warfare and drought seemed the most logical reasons. Now, professional anthropologists and archaeologists could study Montezuma Castle's axes, tools, shells, paints, bone implements, and other artifacts with federal protection.

Because the Roosevelt administration didn't have cash to spare, the Arizona Antiquities Association started repairing Montezuma Castle and put up a protective metal roof. Tourists at Phoenix and Prescott now became enthusiastic about day outings to Montezuma Castle. As a promotional gimmick, it was said that Kit Carson had favored the ruins; he *had* once camped in the area. There were murmurs in the Tucson newspaper that Montezuma Castle should become a national park. All of Arizona was proud of Montezuma Castle. "We were (and perhaps still are) attracted to the ruins, no matter what their size or age," John B. Jackson wrote in *A Sense of Place, A Sense of Time*. "Their shabbiness served to bring something like a time scale to a landscape, which for all its solemn beauty failed to register the passage of time."[84]

But it was Petrified Forest National Monument that created a flash of satori in preservationist circles. If a swath of Arizona's Painted Desert strewn with petrified coniferous trees could be saved, so could Florida's swamps, Louisiana's brackish marshes, and Alaska's tundra. Lacey, who had crisscrossed the country in his effort to save the Petrified Forest—which he considered one of America's five most striking wonders, along with Yellowstone, Crater Lake, Yosemite, and Wind Cave—celebrated at his home in Oskaloosa. Nobody in public life could speak about the Petrified Forest quite like Lacey, even though he came from Iowa. He believed that many of the petrified logs had grown exactly where they now lay. Every log impregnated with silica, stained by iron oxide and other minerals, was a rainbow of colors. "Ages ago, so long that it makes one dizzy to think of it, these trees were alive and growing in the Southwest," Lacey said. "They were coniferous, as shown by microscopic examination of their texture. The species is extinct, and the nearest resembling species now found exists in Asia Minor. The geological history of this forest is easy to read. The trees fell and floated around in some old arm of the sea until the roots and limbs were worn and rounded just as we see like examples on the sandbars of the Mississippi. The trees became heavy and waterlogged and settled to the sea bottom."[85]

Thanks to the guardian spirit of Theodore Roosevelt and John F. Lacey, this ancient sea bottom filled with petrified logs in eastern Arizona was an American treasure for future generations to study and enjoy. And Wetherill was ready to enforce federal protection even in treeless vales where the grass blades had perished due to the pounding sun.

CHAPTER TWENTY-THREE

THE PREHISTORIC SITES OF 1907

I

Even with the administration's designation of Petrified Forest, Montezuma Castle, and El Morro as national monuments in December 1906, President Roosevelt wasn't content. Because John F. Lacey was no longer in Congress, Roosevelt had less clout with the House Committee on Public Land. Furthermore, the relationship between Interior and the USDA's Forest Service was not congenial. By January 1907, Roosevelt had grown increasingly suspicious that his secretary of the interior, Ethan Allen Hitchcock, was too soft on the extraction industries in the West. Hitchcock was like a well-trained bullfighter who made his best passes when there was no bull present (or, as Roosevelt saw it, receded into the shadows when a real goring was possible). Easing Hitchcock, a McKinley man at heart, out of the post, became a priority for Roosevelt in early 1907. Finesse was needed because Roosevelt didn't want to cut the life-line while Hitchcock was still making policy. Another consideration was allowing Hitchcock to reenter the private sector with honor, and this too became a priority for the administration over the holiday season. The situation was especially sensitive because Hitchcock, who was then sixty-one, was feeble (he died in 1909).

As of January 15—the day of the Senate's official confirmation—Roosevelt's new secretary of the interior was James R. Garfield of Ohio. Everybody in Washington knew Garfield as one of Roosevelt's staunchest foot soldiers. As the saying went, he was an old head on young shoulders. Yet he was always something of a messenger boy. And it didn't hurt that Garfield's wife, the former Helen Newell of Cleveland, Ohio, was a prominent Washington hostess (part of the first lady's elegant clique). "Garfield is earnest," the *Saturday Evening Post* wrote. "The President likes earnest persons. Garfield is ambitious. The President likes ambitious persons. Garfield is conscientious, and the President lays much stock by that. In short, Garfield is a clean young man, with a mind that grapples with great problems, no matter what the windup of the encounter may be."[1]

Roosevelt wanted somebody to go after the perpetrators of land fraud in the West, a fellow Republican progressive unafraid of controversy. Garfield was his beau ideal. The handsome forty-two-year-old Garfield

Teddy in Timberland

"Teddy in Timberland" was a popular cartoon that ran in syndication.

had served in the Ohio state senate from 1896 to 1899. He was rara avis because of his old-fashioned sense of bedrock loyalty, always a character trait in short supply. As a silver-star bonus, Garfield was the son of James A. Garfield, the twentieth president of the United States, and had been weaned on national politics. For all their differences, Roosevelt knew that the elder Garfield had been a man of biting intelligence. Young James was a student at St. Paul's School in Concord, New Hampshire, when his father was assassinated and he had witnessed the ghastly shooting, which happened at the Baltimore and Potomac railroad station in Washington, D.C., during his summer break. Somehow he had absorbed the assassination faster than much of the country did, and he refused to let the tragedy derail his ambition. In 1881 he enrolled at Williams College, where he earned straight A grades. Following college Garfield earned a J.D. degree from Columbia University, developing a formidable, Rooseveltian prosecutorial bent. Garfield wasn't afraid to rouse lions from their lairs in the name of good government.

From 1902 to 1903 Garfield was Roosevelt's eyes and ears at the U.S.

Civil Service Commission. His professional attitude was characterized by following orders and saying "Yes sir." Bosses, including Roosevelt, naturally admired that sort of man; and in addition, Garfield was competent. He was promoted to commissioner of corporations at the Department of Commerce and Labor. With the zeal of Lincoln Steffens he lashed out against the corrupt industries of the era: oil, steel, railroads, and meatpacking. Perhaps Garfield was only following orders, but he seemed to relish what we might now describe as being Roosevelt's pit bull. At the very least Garfield understood that the fight for national forestry would be prolonged and intense. Now, in early 1907, Roosevelt wanted to sic Garfield on the timber thieves and land hustlers. As Roosevelt, Pinchot, and Garfield saw it, forestry was a science based on truth—not a political game in which you tried to score points in order to be reelected. "He has a man's job before him now," the *Washington Post* wrote of Garfield that spring. "The Department of the Interior controls the public domain, the forests, the Indians, the patents, the pensions, the Bureau of Education, the Geological Survey, and the Reclamation Service. All the land grafters, all the Indian grafters, all the sharks who are trying to get the timber and the oil and the coal for nothing must come to him and pass under his eye."[2]

The unflappable Garfield immediately was in full stride, walking directly into the line of fire. When the Republican Party lost House seats, pro-business newspapers started attacking Roosevelt for his antagonistic attitude toward Wall Street, big timber, and Standard Oil. For example, Edward Payson Ripley, president of the Atchison, Topeka, and Santa Fe, said that Roosevelt didn't have the public interest at heart in creating forest reserves and national monuments. To Ripley, Roosevelt was simply obsessed with the wilderness. The railroad tycoon E. H. Harriman—who had sponsored the Alaska expedition that included Dr. C. Hart Merriam and John Burroughs—had a similar opinion, deeming Roosevelt a self-promoter and a traitor to his class. Roosevelt, for his part, started instructing Garfield to defend Native Americans from corporate land grabs in Oklahoma, Wyoming, and Montana. George Bird Grinnell was starting to write ethnographic books on Indians—he was doing research for a study that would be published as *By Cheyenne Campfires*. His work was starting to rub off on Roosevelt.[3] In a court order, Roosevelt defined Garfield's job as protecting the "rapidly disappearing timber" for future generations to enjoy. America's forests, Roosevelt believed, belonged to the homeland, but the homeland was under siege by underregulated industrialization (in other words, big business run amok). "Oil and gas,"

Roosevelt wrote to Garfield on February 1, 1907—"I most emphatically believe that we should not permit the lands containing oil and gas to be alienated under conditions which would in effect mean the building up of a great monopoly in oil."[4]

With the success of the Antiquities Act in 1906, Roosevelt had become even more dangerous to western developers, railroad companies, and oil companies. He usually donned a Stetson hat and often wore a bandanna around his neck, and his public rhetoric was full of western toponyms, cowboyisms, and Indian words not often heard in the East. From the White House, he was playing a Rocky Mountain man to help sell his radical conservationism. The oilmen, land developers, and trust titans wanted to see Roosevelt relegated to the sidelines of public life, like John F. Lacey. Instead, they had to confront not only Roosevelt but also Garfield. The timber industry believed that T.R.'s excessive conservationist initiatives were symptoms of his having gone berserk. But such hostility only encouraged Roosevelt to thrust himself forward as the true guardian of America's natural resources. Figuratively, conservationism was simply the wise, righteous preservation of the American way, the prerequisite to Roosevelt's building a republic like none other. "The grazing states, especially Colorado, Wyoming, and Montana, protested vigorously against the new policies," the historian Roy M. Robbins noted in *Our Landed Heritage*. "The stockmen of these states were compelled to use the meadows in the reserves inasmuch as the lower plains gave out during hot weather. The sheepmen were especially anxious, for fear that government regulation would curtail their operations in favor of cattle interests. Both those groups looked with suspicion upon policy which seemed to favor the homesteader."[5]

Working closely with Pinchot, Roosevelt began scheming for innovative ways to create dozens of new national forests before Congress reconvened on March 3. These forests would humanize the soul—if not, the Dark Ages would come to America (or so Roosevelt supposed). On February 23, 1907, in fact, a disgusted senator—Charles Fulton of Oregon, a fellow Republican—introduced the following amendment to an agricultural appropriations bill: "hereafter no forest reserve shall be created, nor shall any addition be made to one heretofore created, within the limits of the State of Oregon, Washington, Idaho, Montana, Colorado, or Wyoming except by act of Congress."[6]

Fulton believed the whole Antiquities Act was nonsense and had to be curtailed. He was sick and tired of arrogant executive orders that gave petrified logs and spotted owls priority over business profits. Also, Fulton said, the lowly settler and the poor farmer were being denied the same rich timberlands by the Roosevelt administration. This was a grim consequence of the government's irresponsible hoarding of resources.

Roosevelt's answer to Fulton was dramatic. On March 2, 1907—four days before the amendment was slated for a vote—Roosevelt released a document to Congress as a presidential fait accompli. Thirty-two new forest reserves had been created, seemingly overnight. Behind each forest listed were snatches of complicated conversations his representatives had conducted with state legislators and land managers about soil erosion, runoff, and deforestation. Numerous papers, passes, exemptions, validations, dues, expansions, and limitations had been issued. Roosevelt's refusal to let Congress inhibit him caused a firestorm against him on Capitol Hill. By contrast, in sleepy Oskaloosa, where he had resumed his law practice on Main Street, Lacey deemed it a great day in the annals of forestry. Roosevelt had delivered a punishing blow to the advocates of states' rights. He had caught Congress flat-footed. And the lumberman's axes had been stilled in certain heavily forested regions, particularly in the Pacific Northwest.

The conservationist pronouncement of March 2 was a fine example of Roosevelt's unappeasable conservationism. Roosevelt issued a long "Memorandum" listing forest reserves either created or enlarged:

Toiyabe Forest Reserve, Nevada
Wenaha Forest Reserve, Oregon and Washington
Las Animas Forest Reserve, Colorado and New Mexico
Colville Forest Reserve, Washington
Siskiyou Forest Reserve, Oregon
Bear Lodge Forest Reserve, Wyoming
Holy Cross Forest Reserve, Colorado
Uncompahgre Forest Reserve, Colorado
Park Range Forest Reserve, Colorado
Imnaha Forest Reserve, Oregon
Big Belt Forest Reserve, Montana
Big Hole Forest Reserve, Idaho and Montana
Otter Forest Reserve, Montana
Lewis and Clark Forest Reserve, Montana
Montezuma Forest Reserve, Colorado
Olympic Forest Reserve, Washington
Little Rockies Forest Reserve, Montana
San Juan Forest Reserve, Colorado

Medicine Bow Forest Reserve, Wyoming, Colorado
Yellowstone Forest Reserve, Idaho, Montana and Wyoming
Port Neuf Forest Reserve, Idaho
Palouse Forest Reserve, Idaho
Weiser Forest Reserve, Idaho
Priest River Forest Reserve, Idaho and Washington
Cabinet Forest Reserve, Montana and Idaho
Rainier Forest Reserve, Washington
Washington Forest Reserve, Washington
Ashland Forest Reserve, Oregon
Coquille Forest Reserve, Oregon
Cascade Forest Reserve, Oregon
Umpqua Forest Reserve, Oregon
Blue Mountain Forest Reserve, Oregon[7]

When Fulton heard that huge tracts of Oregon forestlands had been pickpocketed by the federal government before the agriculture bill could be voted on, he was furious. No serious American, he believed, could have designated so many western forest reserves in such a cavalier fashion. It was a gray, grim day, Fulton lamented, for Willamette Valley's businessmen. According to Fulton's tirade, Roosevelt and Pinchot's team had sneakily withdrawn 16 million acres, ostensibly to prevent overlogging. And eight of the forest reserves were in Oregon: Wenaha, Siskiyou, Imnaha, Ashland, Coquille, Cascade, Umpqua, and Blue Mountain.[8] The whole damn state, Fulton fumed, was becoming a park. The lumber warehouses and industrial storage sheds in his state would be empty if this type of land grab was tolerated. A torrent of accusations followed: Why didn't Roosevelt burn the Constitution while he was at it? Why didn't he just declare Oregon a colony and get it over with? Why didn't he ban sawmills from operating in the West?

Not for a second did Fulton believe that the autocratic Roosevelt was preserving millions of acres for homesteaders or for posterity. The new forest reserves were, to his mind, something the aristocrats of the Boone and Crockett Club and the Audubon Society wanted to have as trophies, at the expense of hardworking, taxpaying citizens. And in general, western interests claimed that this was foul play. Roosevelt, they believed, had acted dishonorably by setting aside the 16 million acres of forest reserves without proper congressional consultation.[9]

As a result of the land withdrawal of March 2, the executive branch was sued. The plaintiffs' lawyers said Roosevelt was acting like a tribal chieftain unaccountable to constitutional law. The defense attorneys said the lawsuits were small-minded. At issue was whether the Roosevelt administration had abused executive powers. Eventually, in 1910, the dispute was brought before the courts, first in *U.S. v. Grimaud* (220 U.S. 506) and then in *Light v. U.S.* (220 U.S. 523). In both cases the U.S. Supreme Court ruled in T.R.'s favor. Big timber had been stymied. The threat of these court cases served only to impel Roosevelt forward with his far-reaching conservationist agenda.[10]

Convinced that the future of America was imperiled, Fulton kept up his dogged pursuit of Roosevelt. That spring Roosevelt had told *Everybody's Magazine* that westerners who didn't understand subspecies of deer and elks weren't good stewards. The president's attitude was plain as day: timber companies were bandits, and westerners incapable of biologically identifying moles were nature fakers. What rubbish! To Fulton, the president was as crazy as a loon—and *dangerous*. But Roosevelt later patted himself on the back for being a political fox. "When the friends of the special interests in the Senate got their amendment through and woke up, they discovered that sixteen million acres of timberland had been saved for the people by putting them in the National Forests before the land grabbers could get at them," Roosevelt bragged in *An Autobiography*. "The opponents of the Forest Service turned handsprings in their wrath; and dire were their threats against the Executive; but the threats could not be carried out, and were really only a tribute to the efficiency of our action."[11]

The combination of the Antiquities Act, the new natural forests, and the debate over nature fakers put Roosevelt in a mood for sparring. He was now forty-nine, and there was about him the cockiness of a gambler who has been winning and is itching for more action. Unleashing Garfield on the corporations was his latest sport. Another favored sport was lambasting faux naturalists untutored in Darwinian biology. "You will be pleased to know that I finally proved unable to contain myself, and gave an interview or statement, to a very good fellow, in which I sailed into Long and Jack London and one or two others of the more preposterous writers of 'unnatural' history," Roosevelt wrote to Burroughs. "I know that as President I ought not to do this; but I was having an awful time toward the end of the session and I felt I simply had to permit myself some diversion."[12]

The seemingly arbitrary forest reserve designations of March 1907 stung the western politicians the most. As the *Congressional Record* noted, even as late as World War I the mention of what Roosevelt had done "still brought forth the wrath from certain quarters."[13] In 1907 the *Walla Walla Weekly* accused Roosevelt of putting the small logging operations out of business in Washington state with his mania for national forests. As this argument went, rich corporations like the Weyerhaeuser Timber Company already owned millions of acres. They weren't adversely affected by T.R.'s conservationism; the little guys were. The *Seattle Post-Intelligencer* sarcastically wondered why T.R. didn't just declare the entire state a national forest. Some people in Seattle and Tacoma argued that Roosevelt's obsession with land fraud made him believe, mistakenly, that national forests were a "panacea." Governor Albert Mead of Washington declared that "Gifford Pinchot, the United States forester, has done more to retard the growth and development of the Northwest than any other man."

Roosevelt scoffed at such criticism as juvenile. There was more to the Pacific Northwest and northern California than fresh-cut boards. The Pacific slope was more wonderful than any place in Europe. The mere thought of Mount Shasta and Mount Olympus made Roosevelt ache, and intensified his love of the United States. America's three West Coast states taken together were larger than any European country. California alone was bigger than Great Britain or Italy. American mammal life also far exceeded that in spent-out Europe. In Oregon, for example, the bears ate both clams and berries and slept in primeval forests and on rock-strewn beaches. Roosevelt called the soil of the San Joaquin Valley the prerequisite for its becoming a God-ordained garden. And the forestlands of these three states, he was convinced, were the finest in the world. "There is nothing quite like the Coast, either in America or anywhere else," Roosevelt would write. "Nature is different from what it is elsewhere. The giant sequoias and redwoods, the wonderfully beautiful isolated mountain peaks and great mountain ranges, the giant chasms like the Yosemite, the forests, the flower meadows, the soft, sunny, luxurious beauty of Southern California, the colder but equable wet climate of the Northwest coast proper, the marvels of Puget Sound, the Valley of the Columbia and of the rivers running into it—all these things, taken separately, may be matched elsewhere, but not when taken together."[14]

One of Roosevelt's strongest conservationist statements was a long letter he wrote on June 7, 1907, to Secretary of Agriculture James Wilson. Obviously composed with posterity in mind, Roosevelt abandoned his usual cheerleading on behalf of "America the beautiful" in favor of a soberminded analysis of the importance of protected forests for national security. "If the people of the states of the Great Plains, of the mountains, and of the Pacific slope wish for their states a great permanent growth in posterity they will stand for the policy of the administration," Roosevelt wrote, disgusted that a Public Lands Convention was being organized in Colorado to overturn his policies. "If they stand for the policy of the makers of this program, they should clearly realize that it is a policy of skinning the land, chiefly in the temporary interest of a few huge corporations of great wealth, and to the utter impairment of its resources so far as the future is concerned. It is absolutely necessary to ascertain in practiced fashion the best methods of reforestation, and only the National Government can do this successfully."[15]

II

Criticism of Roosevelt's national forests of March 2 wasn't confined to California, Washington, and Oregon. Newspapers in the grazing states of Colorado, Wyoming, and Montana also slammed into Roosevelt. Many editors considered the large-scale withdrawal of timber and coal lands completely unacceptable. Inequity was involved because the federal government had grabbed western forestlands while leaving eastern forests in private hands. Demands were made for nullification. An editor of *Denver Field and Farm* fumed that one-fourth of Colorado was now a national forest: soon, decent Coloradans wouldn't even have land left to bury the dead. The *Centennial* (Wyoming) *Post* of March 30 suggested that after March 2, an old cowboy song needed a new verse:

Bury me not on the range
Where the taxed cattle are roaming
And the mangy coyotes yelp and bark
And the wind in the pines is moaning

On the reserve please bury me not
For I never would then be free;
A forest ranger would dig me up
In order to collect his fee![16]

Roosevelt's forest conservationism brought him a lot of hate mail during the spring of 1907. Everybody west of the 100th meridian—which drops from the Manitoba–North Dakota border through Greater Bismarck and straight down to the streets of Laredo—seemed to have a quarrel with him. The White House mailroom grew leery of any letter postmarked Colorado, Wyoming, Montana, Washington, Idaho, Oregon, or California. The *hugeness* of what Roosevelt was doing seemed to be the principal concern. Roosevelt claimed that he was only stopping shifty businessmen, trust titans, and oil hogs from despoiling the national landscape, and that he was surprised when Wall Street called him a "wild-eyed revolutionist." This was disingenuous on his part. Various captains of industry had pleaded with him to ease up on his apparent rancor toward railroads and oil. But Roosevelt refused to capitulate. With muckrakers cheering him on, Roosevelt enjoyed being a wilderness warrior. In his letters, he expressed a somewhat overstated preference for hiking Rock Creek Park to study fauna rather than dealing with unscrupulous robber barons. "The grounds are now putting on their dress of spring," Roosevelt wrote to Kermit about the White House lawns. "The blossom trees are in bloom; perhaps the most beautiful spot at the moment is round the north fountain with the White Magnolia, the pink of the flowering peach, and the yellow of the forsythia."[17]

In the West, cowboys played a game called "chapping," slapping one another with leather chaps to see who would cry uncle first. During the spring of 1907 Roosevelt was engaged in chapping with big timber, in particular. Circumventing Congress, he began appealing directly to the general public in his addresses about conservation. In a sadistic way that no historian, no journalist, and no political commentator can overstate, Roosevelt *enjoyed* making the timber companies suffer. What infuriated his opponents was how he appealed directly to the public, with prosecutorial zeal. It unnerved them. The press always allowed Roosevelt to cloak his conservationism in patriotism and morality—and the newspapers' readers in the nonwestern and nonsouthern states fell for it hook, line, and sinker.

No president ever manipulated the press with the consummate skill of Roosevelt. Part of his cunning was treating even minor journalists as if they mattered. Reporters, as a rule full of self-importance, used the fact that Roosevelt was a man of letters, a member of their tribe, to justify their puffery. A voracious bibliophile, inspired by the Saint Augustine admonition to "Take up, read!", Roosevelt never missed a major article in any contemporary periodical, even an obscure academic journal. As Roosevelt liked to joke, he was at heart a "literary feller." The novelist Ellen Glasgow tried to explain why, against her better instincts, she regularly surrendered to Roosevelt's bravado. She believed that Roosevelt had "dubious literary insight," but she confessed that he also had a strange "human magnetism."[18]

Another factor also aided Roosevelt's career. As the historian Ron Chernow has pointed out indirectly in *Titan*, Roosevelt was a direct beneficiary of "a newly assertive press." Thanks to two technological developments—linotype and photoengraving—the number of glossy magazines proliferated during the Roosevelt era. Too often, Chernow believes, historians have focused on the "strident tabloids" and "yellow journalism" of the period.[19] The circulation battles between Joseph Pulitzer and William Randolph Hearst, for example, were an impetus for sensational copy.[20] But this was also a serious era, when investigative reporting was significant. Periodicals like *McClure's*, *Outlook*, and *Scribner's Magazine* loved Roosevelt for two primary reasons: he wrote for them and he applauded their exposés of corporate corruption. Add into the mix the sheer electricity that Roosevelt produced in his public appearances, the way he sucked the air out of any room, and the trust titans didn't have a chance. Roosevelt didn't lull reporters' sense of right and wrong; he challenged them to write the right thing by flattery and by making good copy.

A case in point occurred on April 14, when Roosevelt delivered a major policy address on Arbor Day, promoting trees. His message was direct: posterity would weave no garland for farmers who overharvested trees and didn't plant new ones. Roosevelt was sure of that. Arbor Day, to Roosevelt, was a holiday to equal the Fourth of July. It had started in 1872, when Nebraska had very few trees: the state board of agriculture had sensibly distributed elms, oaks, and pine seeds for citizens to plant. Arbor Day evolved into a competition in which cash prizes were awarded to whoever planted the most trees. According to the *Omaha World-Herald*, more than 1 million trees were planted on the first Arbor Day. What began as a state holiday in Nebraska soon became a national effort.[21] Many states held annual spring Arbor Day events. Now Roosevelt—with western senators and Rockefeller's supporters clamoring for his head—transformed Arbor Day 1907 into a rallying cry for his visionary conservationist policies. To an audience made up of children from the Washington, D.C., area, Roosevelt preached the wonders of national forests. "It is well that you should celebrate your Arbor Day thoughtfully, for within your lifetime the nation's need of trees will become serious," he said. "We of an older generation can get along with what we have, though with growing hardship: but in your full manhood and womanhood you will want what nature once so bountifully supplied and man so thoughtlessly destroyed, and because of this want you will reproach us not for what we have used, but for what we have wasted."[22]

The eastern press loved this lecture. The *Washington Post* covered it on the front page under the headline "President for Trees."[23] But Senator

Fulton considered it a sickening spectacle of Roosevelt manipulating the press. "A people without children would face a hopeless future: a country without trees is almost as hopeless," Roosevelt had said. "Forests which are so used that they cannot renew themselves would soon vanish, and with them all their benefits. A true forest is not merely a storehouse full of wood, but, as it were, a factory of wood, and at the same time a reservoir of water. When you help to preserve our forests or to plant new ones, you are acting the part of good citizens. The value of forestry deserves, therefore, to be taught in the schools."[24]

Roosevelt kept saying that the "shortsightedness" of deforestation would be solved only by planting trees and reducing lumbering. However, with regard to forest reserves—unlike national monuments—after March 1907 Roosevelt was still forced to work with an irritated Congress on bills aimed at purchasing for the federal government great forest reserves in the White Mountains and Southern Appalachians. Many congressmen felt bruised by Roosevelt's obvious contempt for them. They were hardly in the mood to squander political capital for the sake of his eccentricities. "The only agreement of the bills," Roosevelt lamented to Secretary of Agriculture Wilson, "is that of their great expense." Roosevelt had calculated which states were doing a good job of preserving forests (New York and Pennsylvania) and which states weren't (Michigan and Wisconsin). What brought him great pride was that the western states were far more fortunate than "their eastern sisters" because his administration had shoved "requisite foresight" down their throats.[25] Not on his watch would America become a lumber exporter to the world.

III

That same spring Roosevelt had received the report by the Bureau of Corporations on the unlawful activities of Standard Oil of New Jersey. It infuriated Roosevelt no end: Standard Oil had engaged in price-cutting practices, collusive deals, public misinformation, and so on. How to deal with such abuses? First, Roosevelt increased his calls for much stronger regulation of corporations. This infuriated conservative Republicans, but Roosevelt knew that it was good politics. The banking system and the stock market were going through a severe downturn. Why not make the petroleum industry the scapegoat? The Roosevelt administration issued seven lawsuits against Standard Oil and its subsidiaries (these lawsuits were coupled with numerous antitrust cases that state attorneys general issued). Part of Roosevelt's motivation was trust-busting as nation-building. Criticism was hurled at Roosevelt by Wall Street financiers who

Roosevelt loved Arbor Day because it gave American citizens a chance to do something productive. Every April new trees would be planted across America.

claimed that he was stifling the stock market with his gloomy pronouncements. By dismembering Rockefeller's Standard Oil and Harriman's Santa Fe Railroad, Roosevelt was trying to show that the United States was run by the federal government, not by self-interested capitalists with huge bank accounts and no scruples. To Roosevelt, men like Rockefeller and Harriman were "the most dangerous members of the criminal class—the criminals of great wealth."[26]

Even though gasoline automobiles had infiltrated Washington, D.C., in 1907, Roosevelt insisted on either speed-walking or horseback riding around town. Cars didn't appeal to him—the idea of placing gasoline on top of a hot engine seemed perverse and worrisome, and the rumble of engines scared away the birds. This was the last year before the Model T transformed the American landscape forever—an event which simply bored Roosevelt. Still, he cheered Michigan and Indiana for building better automobiles than anywhere else in the world. As an unapologetic nationalist he liked America to have the automotive edge—or any edge.

Unable to stay idle, Roosevelt began drafting an ornithological report for AOU on how sparrows were different in Long Island and Virginia owing to climate variations. And he was perplexed by some avian mysteries. Why did wood thrushes flourish at Sagamore Hill whereas they were scarce at Pine Knot? He kept detailed bird lists for Virginia about species he encountered: Baltimore and orchard orioles, flickers, redheaded woodpeckers, purple grackles, bluebirds, all nesting "within a stone's throw of the rambling attractive house, with its numerous outbuildings, old garden, orchard, and venerable locusts and catalpas."[27]

For an Audubonist, this was a real feast. But Roosevelt had an eerie experience that May, which he would talk about for years to come. Keenly observant, he saw a flock of passenger pigeons near Charlottesville. Because Darwin, whose name Roosevelt still uttered with reverence, had begun *On the Origin of Species* with a report about experiments conducted on backyard pigeons from around the world, Roosevelt was interested in their evolutionary characteristics. Darwin had successfully bred pigeons, concluding that they were all descendants of *Columba livia* (the rock dove). He had speculated that if the varied pigeon species mated in the wild, the offspring would eventually lose their unique traits and resemble the rock dove: this was due to a process that Darwin called reversion.[28]

What Roosevelt also knew that May 18 about the passenger pigeons (a species on the edge of extinction) was that no flock had been sighted in the wild by an ornithologist for more than twenty-five years. In the era before DNA records containing gene analysis and ancestral chromosome fusions, knowledge about birds' genetic makeup came from detailed field reports in many towns. Roosevelt was thus fulfilling his public duty by reporting on what he saw near Pine Knot. "There were about a dozen, unmistakable with their pointed tails and brown-red breasts, flying in characteristically tight formation to and fro before alighting on a tall, dead pine," the historian Edmund Morris writes in *Theodore Rex*. "He compared them to some mourning doves in the field beyond; and there was no question of the difference between the two species."[29]

Because pigeons were delicious, many species were being driven into extinction by market hunters. For example, in 1904 the Choiseul crested pigeon (*Microgoura meeki*) had vanished; the last one was sighted in the Solomon Islands near New Guinea. In early 1907 two American birds had gone extinct in Hawaii: the Molokai'O'o (*Moho bishopi*) and black mano (*Oreganis funera*). This led Roosevelt to create, in 1909, a huge federal bird reservation in the westernmost Hawaiian islands. But the Molokai'O'o and black mano were rare birds, easily shot by hunters. By contrast, the destruction of the passenger pigeon affected all of North America, where it was known to be the most abundant bird of all time. Before the Europeans arrived in the New World, nearly half of all the birds there were passenger pigeons. To pioneers, they were an unlimited food supply.

Robert B. Roosevelt had tried to stop this slaughter of passenger pigeons in the 1880s, introducing skeet shooting to sportsmen as an alternative, but to little avail. The last recorded passenger pigeon was shot around 1900—nobody had gotten within sight of a flock since then. William T. Hornaday had already shown passenger pigeons in a tombstone cartoon as being extinct (though he added a hopeful question mark). But now Roosevelt had seen a flock at Pine Knot in 1907. Excitedly, Roosevelt hurried back to the cabin and wrote Oom John an effusive letter, insisting that they were "no doubt" passenger pigeons. Burroughs quickly wrote back an encouraging note, saying that the previous year a flock was said to have been seen around Boston, Massachusetts, although the report was unconfirmed. A few weeks later Burroughs said that a flock was seen in Sullivan County, New York.[30] There was no need for Roosevelt to feel diffident: he wasn't the only observer. Perhaps the passenger pigeon could be saved, like the buffalo.

Realizing that passenger pigeons were on Hornaday's endangered species list, and being exceedingly sportsmanlike about this matter, Roosevelt refused to shoot one. But he knew visually that this was the passenger pigeon, as described in Audubon's *Birds of America* (page 25 of Volume 5). All of Albemarle County was abuzz over Roosevelt's sighting which, if true, was the last official report before extinction. That same year W. B. Mershon had published, as a farewell, *The Passenger Pigeon*—a book of memories of the great flocks that constituted an impressive ornithological eulogy.[31] There was also public concern about the impending extinction. As it happened, the last passenger pigeon on earth—named Martha—died in captivity on September 1, 1914, at the Cincinnati Zoo. Thereafter, coffee shops and bars would freakishly boast about having a stuffed passenger pigeon on display.

Roosevelt took pride in the fact that Burroughs's eminence had habituated since their first meeting at the Fellowcraft Club. Likewise he was proud that his illustrator of *Ranch Life*, Frederic Remington, had become famous. Somewhat surprisingly, Remington was perhaps the one western artist willing to denounce the staged rescues in Buffalo Bill's Wild West shows, and Owen Wister's hero in *The Virginian*. Remington called Indians the "aboriginal" Americans, and he was praised by a Crow chief for having white skin but the heart of an Absaroke.[32] In articles, short stories, and two novels Remington had treated Native American warriors as outdoorsmen superior to the "great white hunters" of modernity with their scoped rifles and waterproof sleeping bags. As the *Independent* noted, Remington had been a pioneer in moving away from "mere sentimentality" about Indians to serious ethnography. This intellectual advance in Indian scholarship impressed Roosevelt greatly.[33]

Over the summer of 1907 Roosevelt composed a series of open letters on the White House stationery, honoring Remington's artistic achievements. Roosevelt saluted Remington's painting for the inherent westernness of his broken peaks and purple mountains. Roosevelt was a habitual doodler, but he couldn't draw his beloved West the way Remington could. However, he had the political power to save natural sites in the rutted wagontrail territories. Every Remington rough trapper and graceful Indian radiated a humanity worthy of Rembrandt. "I regard Frederic Remington as one of the Americans who has done real work for this country, and we all owe him a debt of gratitude," Roosevelt declared on July 17. "He has been granted the very unusual gift of excelling in two entirely distinct types of artistic work; for his bronzes are as noteworthy as his pictures. He is, of course, one of the most typical American artists we have ever had, and he has portrayed a most characteristic and yet vanishing type of American life. The soldier, the cowboy and rancher, the Indian, the horses and the cattle of the plains, will live in his pictures and bronzes, I verily believe, for all time."[34]

To Roosevelt, the very talented Remington had made a "permanent record of certain of the most interesting features of our national life." He ranked Remington as high as George Catlin for accurately combining Indian ethnography and western landscapes. By 1904 Remington's bronzes such as *The Bronco Buster*, *The Buffalo Signal*, and *Coming Through the Rough* were themselves virtually national monuments, cycleproof, as recognizable as Leutze's painting *George Washington Crossing the Delaware* or Bingham's *Fur Traders Descending the Missouri*. Remington's bronzes—twenty-one in all, cast at the Henry-Bonnard Company (using the sand-cast method) and the Roman Bronze Work Company (using the latest wax casting method)—were like the Liberty Bell or the Golden Spike of the Transcontinental Railroad: in a word, heirlooms. Lacey had been the legislative genius of the Antiquities Act. Edgar Lee Hewett and other grassroots activists had stirred up preservationist action in the Four Corners region. But it was the spirit of Remington that brought places like Petrified Forest and El Morro out of the bureaucracy at the GLO or the Department of the Interior and animated them with a whiff of the Wild

West for people worldwide. And he did so without succumbing to romanticism—though his work had been treasured as such.

Others were starting to see Roosevelt's saving of wonders as his gift to America. Praising the Antiquities Act of 1906, the *New York Times*, for example, noted that the national monument movement had created America's "conservationist consciousness." The United States stood at the center of a revolution in natural resource management, and Roosevelt was responsible for this positive shift. That was a high compliment to Roosevelt. If the *Times* was correct, then 1907 became the year when Roosevelt's doctrine of conservationism cohered. The four national monuments Roosevelt had founded in 1906—Devils Tower, Petrified Forest, Montezuma Castle, and El Morro—could have just been a flash in the pan to please Hewett and Wetherill. But in 1907, inspired by Remington, Roosevelt kicked up a dust storm, declaring antiquities sites with impressive regularity. Following Hewett's recommendation, Chaco Canyon—which had been a major urban center of the ancestral Pueblo culture—became a national monument through a presidential executive order of March 11. Ruins and hieroglyphics were now folded into the national forestry movement on a permanent basis. As Stanford University's president David Starr Jordan later wrote in the journal *Natural History*, Roosevelt's genius was that "he did not care a straw for precedent."[35]

In 2009 the National Park Service published a detailed time line of Chaco Canyon's history from AD 850 to 1902. In vivid detail it recalled when Hewett first stumbled upon ancient stairways carved into cliffs. But when Roosevelt declared Chaco Canyon a national monument in 1907, very little was known about this prehistoric Four Corners site. Regularly, as Hewett reported, the Hopi and Pueblo of New Mexico made pilgrimages to the ruins as if to a temple. Likewise, Richard Wetherill (the brother of "Hosteen John" Wetherill) had homesteaded in the Chaco Canyon area west of Santa Fe, studying Pueblo Bonito, Pueblo Del Arroyo, and Chet'o Ketl. For the Roosevelt administration to actually acquire the complex of ruins at Chaco Canyon, the GLO had to ask Wetherill to relinquish the valuable land; he enthusiastically did. In an act of high-minded philanthropy, the great southwestern trail guide and Indian trader Richard Wetherill simply handed Chaco Canyon over. Roosevelt later repaid John Wetherill with a personal visit in 1913 to the Betatakin ruins of a "big village of cliff-dwellers" in what is now Navajo National Monument.[36]

Then, on May 6, Roosevelt struck again, in northern California. The

Antiquities Act was starting to take effect. Following the San Francisco earthquake, geologists came to California from all over the world to study the San Andreas Fault and the Lassen Peak volcano (the southernmost one in the Cascade range). What fascinated geologists about Lassen Peak was that it was not a typical mountaintop; it had a cluster of craters on its summit. Situated on the edge of the so-called Pacific plate, Lassen Peak was one of more than 300 active volcanoes that constituted a ring of fire in that part of the world. These volcanoes included Mount Rainier, Mount St. Helens, Alaska's Katmai, Japan's Fuji, and Indonesia's Krakatoa. To Californian poets, Lassen Peak was just a weird and gorgeous mountain in Shasta County. But to concerned geologists, it was a mighty volcano about to blow. Recognizing that this volcano was both a scientific and a natural wonder, Roosevelt granted it national monument status. And it wasn't just the peak that was saved: all the surrounding steaming springs, hissing fumaroles, and gurgling mud pots were saved as well. Roosevelt wanted the entire thermal alley preserved as a monument.

Deeming the Lassen Peak volcano area the Yellowstone of California, Roosevelt also created another national monument on the new park's northeastern border, called Cinder Cone, that same May 6. From above, Cinder Cone looked like a 700-foot-high pottery wheel with a dent on top. According to the U.S. National Park Service, the volcanic cone has been "controversial" since the 1870s "when many people thought it was only a few decades old."[37] They were wrong. Created from volcanic cinders and loose scoria, Cinder Cone was the product of a succession of dramatic eruptions that took place about AD 1700 (or during a 300-year period). "The series of eruptions that produced the volcanic deposits at Cinder Cone were complex," the U.S. Geological Survey reported in 2004, "and are by no means completely understood."[38]

Cinder Cone was particularly striking for its complexity of color. At least five lava flows had occurred at the site, giving the cone a multihued, unweathered surface. Whereas Lassen Peak offered the exquisite beauty of Mount Shasta, Cinder Cone seemed unassuming but was a menacing geological freak. Cross-country skiers were easily fooled: under the silent snow of winter, Cinder Cone was a fiery inferno of red-hot lava—a fact best not forgotten. At night over Cinder Cone, the stars shone with a brightness that pierced through the dark clouds which often hung overhead. But at any given moment, pillars of fire could shoot like a dragon's breath high into the sky from this volcanic hazard, washing away the dwarfish evergreen forests in a cataclysmic sweep of lava—nature at its most brutal. Someday, scientists would have to more fully analyze the paleomagnetic reason for this.

On May 22, 1915, such an event happened at Lassen Peak National Monument, the crossroads of three biological provinces: the Cascades, Sierras, and Great Basin desert.[39] After 27,000 years of dormancy, the volcano erupted, spewing rivers of lava and blowing dark ash all the way to Reno. An avalanche turned trees into debris. The dramatic scene became known as the "Great Explosion." The only other U.S. volcano to erupt in the twentieth century was Mount Saint Helens, on May 18, 1980; this eruption was triggered by a 5.1 earthquake. A thousand years may pass before Lassen Peak or Mount Saint Helens erupts again—or it could happen next year, or tomorrow. That's part of the mysterious appeal of such sites. As for Lassen Peak and Cinder Cone, they were upgraded to national park status in 1916 as a single unit under the Department of the Interior. Lassen Volcanic National Park is considered by many the hidden gem of the California ecosystem.

IV

That June 10, 1907, President Roosevelt, with the Antiquities Act a success, delivered a major address on conservation before the National Editorial Association in Jamestown, Virginia. Roosevelt was appealing to the newspaper world's better nature. The core of the grim problem, the president explained, was that America lacked "foresight" in managing natural resources. Factories polluted the air. Rivers had been turned into cesspools. Lakes were fished out. Crops weren't being rotated. Deforestation without even a slight thought for the future was occurring in county after county. What a dump America could become! In a combination of defiance, humility, schoolmarmish lecturing, and guilt, Roosevelt pleaded with his hundreds of listeners to start a conservation revolution befitting the twentieth century. Journalists had a serious responsibility to the nation to shed light on the problem. By not covering his agenda for national forests they were in effect entering a suicide pact.

There was genuine passion in Roosevelt's remarks. Moreover, he had chosen an ideal venue for this address. At Jamestown, where in 1607 settlers had established the first permanent English colony in the New World, Roosevelt could present himself as an advocate of conservation for the long term. A group known as the Association for the Preservation of Virginia Antiquities was trying to save 22½ acres of the historic sites made famous by Captain John Smith and Pocahontas—some of these places had a real connection to Smith and Pocahontas; others were imagined to have such a connection. The group hoped to generate future tourism. But there was a problem in Jamestown: insidious erosion by the James River was eating away at the historic village. By visiting Jamestown, where red and white mulberry trees had been planted by the first settlers, Roosevelt was sending a strong conservationist message, which included both preservation of antiquities and national forestry. "We have tended to live with an eye single to present, and have permitted the reckless waste and destruction of much of our National wealth," he said. "The conservation of our natural resources and their proper use constitute the fundamental problem which underlies almost every other problem in our National life."[40]

But at this time Roosevelt also continued to push forward his reclamation projects. That same spring of 1907, Roosevelt appointed an Inland Waterways Commission to analyze America's river systems, the development of water power, flood control, and land reclamation. To Roosevelt national monuments were his left punch and reclamation was his right punch. Together they formed the "Roosevelt Doctrine" of conservation. What they had in common was his fervent belief that the federal government, not individuals or corporations, was the best steward of the land. And what both sides of the debate admitted was that water was king. Therefore, everybody thought Roosevelt's Inland Waterways Commission made perfect sense. It was perhaps the one thing Roosevelt did in 1907 that wasn't contentious.

That September, shortly before Roosevelt left to take a journey for the Inland Waterways Commission down the Mississippi River, little Skip died in his sleep at Sagamore Hill. He had been a poem of a dog. The president had owned many pets, but none were as special as Skip. Nights at Sagamore Hill had often found Roosevelt reading history and novels with a snoring Skip in his lap or at his side. They had constituted a harmonious blending of two spirits into one. "We mourn dear little Skip," Roosevelt wrote to Archie, "although perhaps it was as well the little doggie should pass painlessly away, after his happy little life."[41]

At the end of September, President Roosevelt headed to Iowa for a journey on the Mississippi River. His wife had traveled down the Mississippi in the yacht *Mayflower* from Vicksburg to New Orleans earlier that year. It was now his turn. Fulfilling an old dream, getting to play at being a riverboat captain, the president lived on a steamboat for four days (October 1–4). His enthusiasm for the trip was inexhaustible. Boarding the boat in Keokuk, Iowa, he was joined by the former congressman John Lacey,

Gifford Pinchot, and other friends. It was the finest company imaginable, and piquant and witty remarks were the main fare. But no matter how interesting the conversation was, Roosevelt reserved the presidential prerogative of abruptly turning his head (like a lizard following the course of a fly) whenever an unusual bird or a driftwood log appeared. The Mississippi was both the spiritual heart and the economic backbone of America. Roosevelt knew that a tree branch thrown into the Mississippi in Minnesota would float away toward Davenport, Cairo, Greenville, and Natchez; would reach the Gulf of Mexico and go past his bird rookeries and then around the Florida Keys; and might eventually be found by a fisherman in Senegal or Ghana.

Officially, this was an inspection trip on behalf of Roosevelt's Inland Waterways Commission. Incredibly to Roosevelt, who liked infrastructure improvements to be made quickly, the Mississippi River levees, which had ruptured and collapsed in the flood of 1882, still weren't properly fixed. The U.S. Army Corps of Engineers was trying to control the wild waterway, but with only limited success. There was some repair activity: wing dams were being erected to deflect the strong current, and dikes were being built. But, as Twain had prophesied in *Life of the Mississippi*, the riverfront communities would "get left" to ruins the next time the spring rains were heavy.[42] (That is precisely what happened in 1912, 1913, 1927, and beyond.) Nevertheless, sounding like James B. Eads, Roosevelt promoted river engineering over wild, scenic nature for the sake of enhanced commerce on the Mississippi River. Commerce ruled the river. Barges were the gods. The entire Mississippi watershed, Roosevelt believed, needed to be treated as a single unit from sources to stream mouths. Full coordination between the Army Corps of Engineers, Reclamation Service, Forestry Bureau, Division of Soils, Geodetic Survey, and Mississippi River Commission had to commence at once if there was to be even a remote chance of containing the Mississippi.[43]

The river floods were terrible, but Roosevelt liked to brag that the Mississippi Delta had the richest soil in the world. He believed that wherever a Mississippi levee system was built properly, and fears of flooding were removed, the delta would become densely populated, and Memphis and Baton Rouge would become huge transportation hubs. But if the levees weren't secure, if the Mississippi was allowed to rampage, then settlements like Cape Girardeau or Helena would become shells of their former selves. "At present the possibility of such flood is a terrible deterrent to settlement," Roosevelt lamented, "for when the Father of Waters breaks his boundaries he turns the country for a breadth of eighty miles into one broad river, the plantations throughout all this vast extent being from five to twenty feet under water."[44]

Meanwhile, there was plenty of horseplay and suspender snapping aboard the steamer. Every dinner of catfish, hush puppies, and wine was accompanied by bursts of laughter. Churning down the Mississippi, paddle wheel grinding on and on, naturally caused the men to think of Fink, Shreve, Grant, Pike, and all the rest associated with the river called the "Father of Waters." It was fun to watch Pinchot plucking his thick mustache as he told comical anecdotes about his trips to the west coast conifer forests. And each town they passed was of historical interest: Hannibal, Quincy, Saint Louis, Sainte Genevieve, Osceola. All the way to Memphis, Tennessee, the USS *Mississippi* churned, past old Native American mounds, modern locks, and earthen levees built in ancient times.

Formally attired, wearing his top hat on the deck, sitting in a rocking chair and reading Inland Waterways Commission reports until the aperitif hour, Roosevelt prepared for his big address to the Great Lakes–to–Gulf Deep Waterway Association in Memphis. Basically, Roosevelt's speech in Memphis on October 5 was a rehash of his conservationist address at Jamestown earlier that year. That October, in fact, marked Roosevelt's last reclamation project, in Oakland, California. Pinchot had cleverly suggested that in May 1908 Roosevelt hold a White House Governors' Conference to tackle all of America's serious natural resources issues. Without

President Roosevelt and Gifford Pinchot conferring about conservation while traveling down the Mississippi River in October 1907.

hesitation Roosevelt agreed. "It ought to be among the most important gatherings in our history," Roosevelt said, "for none have had a more vital question to consider." Staying in Memphis for only an evening, Roosevelt left the Peabody Hotel, a mid-South institution, for a stroll to the house where Ulysses S. Grant lived before the siege of Vicksburg.

Roosevelt liked the feel of Memphis and how the Chickasaw bluffs rose dramatically several hundred feet along the eastern bank of the Mississippi River. The bluffs afforded protection from floods and access to river commerce. When the Civil War began in 1861 about 1,000 steamboats had plied the river. Now, owing to the advent of railroad traffic, there were far fewer river vessels. But Roosevelt didn't pine for the steamboat era, per se. His romanticism was always tilted more toward horseback riding on the prairie. With his type A personality, he didn't like being confined on a boat. It made him feel antsy, and also helpless—the very thought of boiler explosions, snags, and sandbars made him restless. He was proud that his great-great-uncle Nicholas Roosevelt, a gifted associate of the inventor Robert Fulton, had been the first to steamboat down the Ohio and Mississippi rivers in the *New Orleans*, traversing thousands of miles before the vessel reached its namesake city and anchored across from The Cabildo (where the Louisiana Purchase was signed in 1803, doubling the size of America).[45]

Following his speech in Memphis, Roosevelt headed by train to Stamboul, a hamlet in East Carroll Parish, Louisiana. Stamboul was known for its cypress timber, for its pecan trees, and—henceforth—for President Roosevelt's having set foot there.[46] As usual, Dr. Alexander Lambert was at Roosevelt's side, this time along with two other physicians. Originally Roosevelt was hoping to meet Reverend Herbert K. Job to inspect the Breton Island Federal Bird Reservation, and then to visit Avery Island, which had become a privately managed nursery for the preservation of egrets.[47] But time constraints forced him to settle for a hunt for black bears in northern Louisiana, a short way from the Mississippi River, with John M. Parker and John McIlhenny as his hosts in the canebrakes. Concerned about bad publicity (which he had received in Mississippi five years earlier), Roosevelt banned reporters and gawkers, and even the Secret Service men did not know where his camp was. When reporters followed Roosevelt around on such hunting trips, he saw them as mice looking for sensational copy so as to become rats.[48]

The ever-obliging John Parker took a chance in hiring the uncouth fifty-three-year-old pot hunter Ben Lilly as Roosevelt's guide through the Tensas bayou wilderness in search of black bear. Lilly knew the extensive bottomlands of this part of the alluvial Mississippi Valley better than anyone else. By all accounts Lilly seldom washed; was scraggly, grizzled, and unkempt; and refused to sleep indoors—he preferred hollowed-out logs and switch cane. He had a full beard and intense, wild blue eyes, and was deemed a "goofy old coot" by Roosevelt because of his obsessive muttering. Roosevelt's first impression of Lilly, when the guide arrived in camp dressed like a blacksmith, was a "religious fanatic."[49] As night began to lower, Roosevelt nevertheless strategized with Lilly about their best tracking options come morning. It was as if the president were testing uncertain ice. An early autumn cold had crept into camp, and the men were already getting sniffles and coughs. Roosevelt and Lilly stayed up late and talked about Louisiana black bear and water moccasins, enjoying each other's openness of manner. "I never met any other man," Roosevelt wrote, "so indifferent to fatigue and hardship."[50]

Roosevelt, however, never fully warmed to Lilly. A wise man once said that a person (like Lilly) unable to live in society was either a beast or a god—and Lilly, raised a hunter, was clearly of the first type. But Roosevelt found the whole concept of a wild man anthropologically fascinating. Lilly was called the "most skilled hunter who ever followed a hound," and he was hired throughout the Mississippi Delta to hunt (for a bounty) menacing bears or cougars who had destroyed livestock.[51] Because the canebrakes grew ten to twelve feet high, it was difficult for hunters to see in front of themselves; hence the brakes were a fine cover for bears. Comparing Lilly to James Fenimore Cooper's Deerslayer in "woodcraft, in handihood, in simplicity—and also in loquacity," Roosevelt began sketching Lilly's character traits in rather psychologically complex prose for *Scribner's Magazine*. It was unusual for Roosevelt to write in this way about people, but Lilly was so *like an animal*—the kind of man who knew how to die standing up—that he couldn't resist. "The morning he joined us in camp, he had come on foot through the thick woods, followed by his two dogs, and had neither eaten nor drunk for twenty-four hours; for he did not like to drink the swamp water," Roosevelt wrote. "It had rained hard throughout the night and he had no shelter, no rubber coat, nothing but the clothes he was wearing, and the ground was too wet for him to lie on; so he perched in a crooked tree in the beating rain, much as if he had been a wild turkey."[52]

Rain fell all over the Delta that October. It was pouring on every stretch of the soggy plains, on the cottonfields, on Poverty Point all the way eastward across the Mississippi River to Port Gibson, and on the dark, sinuous Mississippi riptide. It was falling, too, on the federal cemetery on the bluff in Vicksburg, where Union and Confederate soldiers lay under marble slabs. It rained heavily on every interesting wildlife-rich area along the Tensas River bottomland hardwoods, open water pools, and old runs as Roosevelt's party set up a tent camp near Bear Lake, everything turning into a dull mud.

The Roosevelt party was in the thickest patch of this hardwood habitat. Bottomland forests were rapidly being clear-cut for conversion into agricultural areas. So there was a sense that this was the "last" hunt. Garfish were caught. An alligator slid into the water, and a couple of crows pecked for food in a field. Black squirrels made a commotion in the trees, living in easy community with wood rats. The swamp rabbits were amphibious, behaving like muskrats. Bats bawked. Roosevelt described the snapping turtles he encountered as "fearsome brutes of the slime, as heavy as man, and with huge horny beaks that with a single snap could take off a man's hand or foot."[53]

The Louisiana canebrakes were a place where a man could die and not even be noticed. The fleas had disappeared, and even though winter was approaching, the oak trees weren't bare. "Palmettos grow thickly in places," Roosevelt wrote. "The canebrakes stretch along the slight rises of ground, often extending for miles, forming one of the most striking and interesting features of the country. They choke out other growths, the feathery, graceful canes standing in ranks, tall, slender, serried, each but a few inches from his brother, and springing to a height of fifteen or twenty feet. They look like bamboos; they are well-nigh impenetrable to a man on horseback; even on foot they make difficult walking unless free use is made of the heavy bush-knife."[54]

Clearly this fourteen-day hunt wasn't a dignified sport like shooting quail in the sedge. Roosevelt grew perturbed when he was told that there were only four or five bears left in the Louisiana canebrakes (although bears were reported regularly in the more western parishes of Ouachita and Lincoln). As the world's leading advocate of wildlife protection, Roosevelt could have called it quits and headed back to Memphis and the silk bedsheets of the Peabody Hotel. Instead he started insisting that the Mississippi tracker Holt Collier was needed lickety-split. Collier could definitely find a Louisiana black bear (*Ursus americanus luteolus*) in high water. It seemed that Roosevelt had entered the realm of fantasy, barking orders like a colonel while also imagining that he was the type of old-fashioned bayou character Mayne Reid had written about in *The Boy Hunters*. To understand Roosevelt's hunt in the canebrakes, it is necessary to realize that he wasn't after just any bear. He wanted a Louisiana black bear, distinguished by having a longer, narrower snout than most bears, and one of the sixteen recognized subspecies of black bears in America. Long ago this subspecies had been widespread from Mexico to Canada, but now it was dwindling under the pressure of human encroachment. Roosevelt wanted to shoot one to use for a museum display.

Soaked and disgruntled, Roosevelt remained determined. There was no way he was going be denied this trophy, as he had been in 1902. He wasn't going to leave the Tensas River area without killing a bear. Eventually Collier showed up with two planters from Greenville, Mississippi—Clive and Harley Metcalfe—accompanied by a wagon full of bloodhounds. Collier now took charge of the hunt. He instructed Clive and Harvey Metcalfe in a low voice to "take the Cunnel and bum around with him in the woods like you an' me always does, and don't put him on no more stand. He ain't no baby. He kin go anywhere you kin go; jes' keep him as near to the dogs as you kin. Mr. Harley and me'll follow the hounds; when we strike a trail you and the Cunnel come a-runnin."[55]

It's thought that the Roosevelt party pitched their tents at the Bear Lake Hunting Club near Tallulah, Louisiana (the club had been incorporated in 1899 by delta planters). The gentlemen of Louisiana and Mississippi still preferred to hunt from the stand, staying dry from and keeping out of the pneumonia-inducing weather. Roosevelt headed toward the cane thicket and the bogs around Bear Lake. He tracked for hours, but the bear proved elusive. At one juncture a wild boar attacked the hunt dogs, killing two of them.

Spending time with Collier was worth the struggle with briar patches and boars. Much as he had done with Catch 'Em Alive Abernathy, Roosevelt had used Collier as a source of anecdotes to entertain listeners in Georgetown and on Capitol Hill. As Roosevelt told it, Collier was a flawless hunter who wasn't afraid to bloody his knuckles, a black man who had befriended Frank James, killed more bears than Davy Crockett, fought at the battle of Shiloh, traveled with springtime fairs to chase skirts, and gambled in high-stakes poker games on the Mississippi River. Blacks outnumbered whites four to one in the delta, but Collier was the kind of man who transcended racial categorization. Everybody, regardless of color, took a shine to him. "When ten years old Holt had been taken on the horse behind his young master, the Hinds of that day, on a bear hunt, when he killed his first bear," Roosevelt wrote. "In the Civil War he had not only followed his master to battle as his body-servant, but had acted under him as sharpshooter against the Union soldiers. After the war he continued to stay with his master until the latter died, and had then been adopted by the Metcalfs; and he felt that he had brought them up, and treated them with that mixture of affection and grumbling respect which an old nurse shows toward the lad who has ceased being a child."[56]

Frustrated by the thought that he was going to leave the Louisiana canebrakes without a bear trophy, Roosevelt pulled Collier aside, out of earshot from the white planters. "Holt," Roosevelt said. "I haven't got but one or two more days. What am I going to do? I haven't killed a bear." Collier whispered back, "Cunnel, ef you let me manage the hunt you'll sho' kill one to-morrow. One of 'em got away to-day that you ought to have killed." "Whatever you say goes, Holt," was Roosevelt's reply. Collier answered, "All right, Cunnel."

The next day Collier showed the right stuff. The streaming rain had stopped, replaced by mild, clear weather. Collier's dogs got a scent and started following it in the direction of Roosevelt, who was crouched in the hardwood forest waiting for his golden moment. To Roosevelt, this was America's "great forest" of red gums and white oaks, which he called the "Northeast Louisiana Bottoms." "In stature, in towering majesty, they are unsurpassed by any trees of our eastern forests," Roosevelt wrote, "lordlier kings of the green-leaved world are not to be found until we reach the sequoias and redwoods of the Sierras." The greenest mosses of the Tensas River were now surrounding Roosevelt. Worried about encounters with rough thickets, Roosevelt had sensibly worn thornproof gear, which served him well during the hours of pursuit.

Roosevelt would look into hollowed or downed logs for bears. He wasn't worried about other so-called predators. Back in the days when Louisiana was owned by France, there were lots of red wolves and Florida panthers in the primeval Tensas River forest, but they had been wiped out in the effort to control predators. Only the Louisiana black bears—a threatened species in 2009—remained in the thick tangle of creepers and vines. In the adrenaline rush of the hunt, Roosevelt's banged-up knees weren't aching. Eventually the dogs found a she-bear. Leaping up in front of the bear, which was twenty yards away, Roosevelt took aim with his rifle as the animal ran toward him. The shot caught the bear clean in the chest. She moaned as if in surrender or defiance. According to Roosevelt the bear "turned almost broadside" and started walking "forward very stiff-legged, almost as if on tiptoe, now and then looking back at the nearest dogs." She toppled over "stark dead," as Roosevelt put it, "slain in the canebrake in true hunter fashion."[57]

Dancing a jig, Roosevelt dropped his rifle, pulled Harley Metcalfe off his horse, and gave him a hug. What ecstasy Roosevelt felt at killing a five-foot fully mature she-bear with his 45-70 rifle.[58] He had also absorbed the recoil without a bruised shoulder, so the hunt was deemed a complete victory. He talked about it for weeks. Today a lone historical marker commemorating the hunt can be found in the hamlet of Sondheimer, Louisiana.

Now the kennel wagon was loaded up. Holt Collier had earned his pay, and he was toasted by the president as an Olympian of the Tensas River. Dr. Lambert apparently took a series of photographs of Roosevelt and Collier on the hunt, but these have never been found. Others have surfaced, however, courtesy of the Parker estate (one shows the men looking like borderland desperados). The press exaggerated the size of the bear, saying that Roosevelt had shot a 375-pound giant; the president said no, it was only 202 pounds. That evening the hunters ate bear steak, with Roosevelt rattling on about his quarry. The president kept an inventory of what his party had shot: the final count was three bears, six deer, and twelve squirrels, and one each of wild turkey, possum, and duck.

And the birding had proved first-class. Long ago John James Audubon had learned what an avian paradise the Louisiana wetlands could be. Audubon had lived in Saint Francisville for on and off twenty-three months from 1821 until 1830, painting eighty of his exquisite folios there. Using Oakley Plantation as his base, Audubon would regularly live among flocks of egrets and herons. In Louisiana, he drew such fine works as *Carolina Parrot* and *Mocking Birds Attacked by Rattlesnakes*. Roosevelt had expected to see mockingbirds and half a dozen sparrows perking about in the thick woods and sloughs, but nothing had prepared him for the variety of woodpeckers he observed in the groves of giant cypress. Quite famously in the annals of bird-watching, he recorded seeing three great ivory-billed woodpeckers. And dozens of barred owls "hooted at intervals for several minutes at mid-day"—turning their heads sharply when footsteps were heard. To Roosevelt these owls took on a special mystery at night; their cries seemed "strange and unearthly" like the long hoot of the Southern Pacific headed across the flatlands.[59]

Roosevelt spent October 20 at the home of a Delta planter. All in all, despite the deprivations, it had been a fine week of hunting and bird-watching. Fully rested, with a bearskin as a memento for a museum, he headed east to Vicksburg, once called the Gibraltar of the Confederacy. Largely owing to the president's lobbying, the Vicksburg battlefield was being preserved as a 1,800-acre national military park. There was a triumphant procession in the town for the first time since the Civil War, and Roosevelt was given a grand welcome. A new monument was being erected, and the townspeople were excited. Everybody, it seemed, asked Roosevelt about the Louisiana bear hunt—such light conversation helped ease the tension between Democrats and a Republican president. Governor Vardaman of Mississippi, still furious over Roosevelt's dinner with Booker T. Washington, tried to spoil the special event, which was intended to honor the gallantry of both Union and Confederate soldiers during the Civil War. But although Vardaman injected some invective into the gala, Roosevelt was equal to the situation. Because he was in Vicksburg officially, to dedicate a war memorial to Union and Confederate soldiers, he made a concession for the sake of national unity by praising Jefferson Davis for the first and only time in his life. Surrounded by proud soldiers, he basked in every expression of American pride. Vicksburg—the site of a famous Union siege during the Civil War.

Vicksburg—where from 1899 to 1902 the Corps of Engineers diverted the Yazoo River to flow into the old riverbed so that today the Yazoo, not the Mississippi, flows past the town. Vicksburg—the very word had been part of his life since his boyhood. Vicksburg—to Roosevelt it was a sacred site of both the Union's glory and America's eventual healing.

Following an obligatory speech in Leland, Mississippi, Roosevelt made his way back to Memphis and then headed by train to Washington, D.C. The combination of the Mississippi River steamboat trip followed by stalking bears in the Louisiana canebrakes had made him long for the vi-

President Roosevelt in the Louisiana canebrakes with his hunting partners.

tality of his youth. How simple the outdoors life, rich with bird life, was: the search for wood, meat in the pot, the sound of painted finches singing in the dawn. Everything required to feel alive with God was available by walking through a meadow, the woods, or a bayou with an eye out for warblers and vireos. An idea started to percolate in Roosevelt. Perhaps he would lead specimen collecting expeditions to the three A's: the Arctic, Africa, and the Amazon. There were still wild places left where an explorer could make his mark.

As a token of appreciation to Harley and Clive Metcalfe and Holt Collier, Roosevelt had three 45-70-caliber model 1886 Winchester rifles shipped to Mississippi as "treasured keepsakes." On each weapon the hunter's full name was engraved, with "1907" underneath.[60]

V

Reports of Roosevelt's bear hunt sickened Mark Twain. He saw Roosevelt as a hypocrite: a purported conservationist but also an obsessive hunter. To Twain, the spectacle of the president, who was getting rounder by the day, trudging through swamplands in a downpour to kill one of the last black bears in East Carroll Parish was pathetic. Also, why was a busy president disappearing for days to beat the bushes for bears? Twain personally liked Roosevelt, who had once helped him with a tricky customs issue in Europe. He even enjoyed Roosevelt's conversation from time to time. But the hunt in Louisiana was the tipping point. As America's foremost humorist, Twain attacked Roosevelt by writing burlesque versions of the hunt. The gist of the ridicule was that Roosevelt had a rogue hormone, which caused him to light out after animals with deadly intent. A bear or cougar would be better off taking an anesthetic than having to encounter an inglorious death at the hands of a maniac shouting *bully* before pulling the trigger. Yet Twain was dealing with only one side of Roosevelt's multidimensional self. As many others have noted, T.R. had thousands of sides, including bird-watching and forest preservation. But Twain, of course, wasn't looking for balance.

And Twain, as he was apt to do, hit his mark. He had a legitimate ax to grind in this regard. A longtime animal rights advocate who had written *A Horse's Tale* and *A Dog's Tale* and condemned bullfighting, Twain now insisted that killing a bear with a pack of yapping hounds and mastiffs was the equivalent of shooting a cow in a pasture. In *The Adventures of Huckleberry Finn*, he described two loafers in the town of Bricksville who enjoyed pouring kerosene on stray dogs and torching them.[61] Turning to Roosevelt, he now imagined the unfortunate cow looking at the president and saying, "Have pity, sir, and spare me. I am alone, you are many. . . . Have pity, sir—there is no heroism in killing an exhausted cow." But Roosevelt—who Twain said was "still only fourteen years old after living a half century"—coldly refused. Sarcastically, Twain claimed that Roosevelt had shot the Louisiana black bear in an "extremely sportsmanlike manner" and then triumphantly hugged his planter guides as if he just scored the winning touchdown in the Harvard-Yale game. Then Twain essentially severed his friendship with Roosevelt by declaring that Roosevelt was "the most formidable disaster that has befallen the country since the Civil War."[62]

Much as H. L. Mencken or (in our own time) Maureen Dowd cleverly attacked politicians, Twain started eviscerating Roosevelt, deriding the president as one of the most "impulsive men in existence." Twain was baffled as to *why* the president was so widely admired as an embodiment of America malehood. Solve that mystery, and you would know the soul of the American people in all their cruelty and glory. Typically, Twain provided his own answer, saying that Roosevelt would "slap the Devil on his back and shoulder and say 'Why Satan, how do you do? I am so glad to meet you. I've read all your books and enjoyed every one of them.' Who wouldn't be popular with an act like that?"

Twain's feud with Roosevelt dated back to 1898, when they had opposing views of the Spanish-American War. Twain was twenty-three years older than Roosevelt, and the generational gap was a factor in his antagonism. Believing that his own hard-earned wisdom was much richer than Roosevelt's impetuous vitality, Twain differed with Roosevelt on issues such as England against the Boers in South Africa and the revolution in Panama. But something more than foreign policy or disagreements over imperialism erected a barricade between these two colorful figures. Roosevelt's idea of great literature tended to lean toward swashbuckling epics and romantic sagas. He disdained cynicism, irreverence, and irony in books; but all three attributes were Twain's trademarks. Roosevelt did, however, consider Twain a "real genius" and thought that *The Adventures of Huckleberry Finn*, *Tom Sawyer*, and *Life on the Mississippi* were three all-time American "classics." But he disliked *A Connecticut Yankee in King Arthur's Court* because it mocked the noblemen of the Round Table.[63]

What Twain was chastising Roosevelt about in 1907 was the Boone and Crockett Club's "hunting ethos." As Twain saw it, Roosevelt and other elite hunters would always stalk the largest male big game, the animals with the biggest antlers, tusks, heads, or necks. These hunters, then, were obsessed with *bigness*. Some modern scientists now believe that Twain was on to something—that, perversely, such selective hunting causes the decline of the very species that elite outdoorsmen want saved. If Charles Darwin was correct in saying that "a law" was the "ascertained sequence of events," then the shrinking of antler and horn sizes by the twenty-first century was probably an unintended result of hunters' aiming particularly at game with large antlers and horns. In a landmark report in the January 2009 issue of *Proceedings of the National Academy of Sciences*, Professor Chris Darimont of the University of California–Santa Cruz offered startling data about the fate of Canadian bighorn sheep's shrinking curved horns. According to Darimont, modern-day hunters, by aiming for the "mightiest" and "lordliest" big game, had left surviving generations with a noticeably slimmer, less sturdy gene pool. "Human-harvested organisms," Darimont told *National Geographic*, regarding his findings, "are the fastest-changing organisms yet observed in the wild."[64]

What the National Academy of Sciences was claiming in 2009 turned the Rooseveltian sportsman's ethos on its head while proving Darwin's theory of natural selection once again prescient. Natural selection occurred quickly in the hypertechnological twenty-first century. Trophy hunting and pound-fishing for the *biggest* quarry were injuring species' long-term chances for survival.[65] "Hunting, commercial fishing, and some conservation regulations like minimum size limits on fish," Cornelia Dean of the *New York Times* wrote, summarizing the report, "may all work against species health."[66] In a process called "micro-evolution," species were apparently getting smaller, in part because hunters and fishermen were always zeroing in on the biggest game and catches. Not only was natural selection real, but Americans like Roosevelt, culturally obsessed with bigness, as typified by the Boone and Crockett Club, by fishing derbies, and by game management protocols that thinned out the largest species first, were completely wrong. The *Proceedings of the National Academy of Sciences* report didn't mince words: harvesting the largest organisms in the wild (whether these were Colorado bighorn sheep or Oregon salmon) was wrongheaded. Human predators—using guns and nets combined with technology—were causing species to shrink in size. "Our preference for largeness in vertebrates is culturally strong because now we can find the largest fish or the biggest brown bears with new gadgetry," Professor Darimont explained in an interview. "It's the worst thing a hunter or fisherman can do to aim for the biggest game or catch."[67]

Meanwhile, in Alberta, Canada, for example, hunters targeting the largest specimens of Theodore Roosevelt's beloved bighorn sheep had caused horn length and body mass to decrease by about 20 percent from

1979 to 2009. By the time Barack Obama became president, trophy hunting was also pushing polar bears and grizzly bears into the category of endangered species. What a strange twist of fate for hunters inspired by Roosevelt and Grinnell to contemplate! Killing the trophy game, taking out the alpha males, was adversely affecting the species they loved. At least, however, the Boone and Crockett Club had gotten some things right during Roosevelt's presidency: it had fought for huge wildlife refuges, promoted seasonal hunting, insisted on licenses in every state, issued bag limits, and banned the killing of females during the breeding season and of young animals at any time.

Obviously, along the Tensas River in 1907, Roosevelt couldn't have known about "micro-evolution." Given how seriously Roosevelt took evolution, he might have reformed his hunting practices if he had read *Proceedings of the National Academy of Sciences*. But such speculation is moot. All that history recorded of Roosevelt's hunt was his *need* for a Louisiana black bear to donate to America's growing natural history collection (and to satisfy his own desires). And the average American continued to cheer the president onward for his wilderness exploits. Roosevelt appealed to an almost mystical attachment that people had toward bears. The public both adored and feared them. Whether dead or alive, in zoos or as toys, bears were popular. The toy teddy bear remained the rage in 1907. And Roosevelt himself sometimes emitted bearlike grumbles when he was in the outdoors and pretended to be standing on haunches, mainly for comical effect. But now Twain was irritated that Roosevelt could so easily sell his bear act to the American people. It wasn't Roosevelt's charisma that Twain minded, but the way Roosevelt marketed himself as the "great bear hunter." In face-to-face encounters, Twain continued to like Roosevelt. But he was nevertheless nauseated by the carnival atmosphere at the White House—a spotted pony in the elevator, a wolf-catcher at the dinner table, and a pet badger biting the ankles of visitors. Such ridiculous stunts were signs of arrested development—as was shooting a bear for no reason.

"Mr. Roosevelt is the Tom Sawyer of the political world of the twentieth century," Twain wrote, "always showing off; always hunting a chance to show off; in his frenzied imagination the Great Republic is a vast Barnum circus with him for a clown and the whole world for audience; he would go to Halifax for half a chance to show off, and he would go to hell for a whole one."[68]

Twain had a point. But he was blind to all the good work Roosevelt was doing for the wildlife protection movement. Twain simply never mentioned that work in his interviews, articles, or books. No matter what Twain thought of him, Roosevelt didn't have to run to Canada or Louisiana to show off his preservationist side in the fall of 1907. He did that with strokes of the presidential pen—and Twain never even noticed.

On November 16, for example, Roosevelt declared the Gila Cliff Dwellings a national monument. Located in New Mexico's Gila National Forest, the monument had only 553 acres, yet it contained the cave dwellings of the Mogollan people between AD 1257 and 1300. When visitors leaned against the stone there and soaked up the warm sun, they were transported back in time. Reality seemed like an illusion. Pottery pieces . . . knives . . . a cracked crib . . . a doll . . . rattles . . . a slingshot . . . cooking utensils—you could see where babies had been born and elders laid to rest. You didn't need to read Adolph F. Bardelier's *Ancient Society* to be intrigued by the Gila Cliff Dwellings. The Gila wilderness was a place of both concealment and abandonment. And every inch for miles needed to be card-cataloged by the Bureau of American Ethnology. Roosevelt wanted to save not only the cliff houses but also Gila Hot Springs. By 1910 *Harper's Weekly* was promoting Gila Cliff Dwellings National Monument as a tourist site on a par with the Grand Canyon. Roosevelt had applied his promotional magic once again.[69]

Besides the ruins, the national forest along the Gila River was spectacular in its diversity. Wildlife was—and remains—thick around the Gila Cliff Dwellings, which are constructed at an elevation of about 5,700 to 6,000 feet. Even in 1907 the Mexican gray wolf (*Canis lupus bailey*) was associated with this national monument. According to U.S. Fish and Wildlife, it was the most genetically distinct subspecies of the gray wolf. In 1911, disappointed that the Tensas River game had vanished, Ben Lilly moved to the Gila Mountains, where for twenty years he lived in the wild, as indigenous as a mule deer.[70] And it was near Gila Cliff Dwellings National Monument that the conservationist Aldo Leopold had an epiphany about wolves, dramatically rendered in *A Sand County Almanac.* "We reached the old wolf in time to watch a fierce green fire dying in her eyes," Leopold wrote. "I realized then, and have known ever since, that there was something new to me in those eyes. Something known only to her and to the mountain. I was young then, and full of trigger-itch; I thought that because fewer wolves meant more deer that no wolves would mean hunters' paradise. But after seeing the green fire die, I sensed that neither the wolf nor the mountain agreed with such a view."[71]

And on December 19, as if giving America a Christmas gift, Roosevelt created Tonto National Monument to preserve both cliff dwellings and more than 160 species of birds. No more rifling of antiquities would be permitted at Tonto. A land of sandstone formations and rugged wilderness, the Tonto region was carved by wind, sand, and humans. Hidden about the site were caves pocketed into the red rock. Lacey considered the Tonto—situated at the northeastern boundary of the Sonoran desert—the most rugged terrain in Arizona. Unintentionally, saguaro cactus were now saved by the federal government for the first time, for the sake of Hohokam and ancestral Pueblo ruins. (An intermixing of the tribes had probably occurred between the late thirteenth century and the middle part of the fifteenth century when, as the National Park Service put it, the Tonto Basin was *depopulated*.)

Mark Twain—unlike the archaeologists—never appreciated Tonto National Monument. But the archaeologists knew that Roosevelt's monuments in New Mexico and Arizona were important. The late prehistoric pottery found in the upper and lower cliff dwellings of Tonto, in fact, was named "Roosevelt red ware" by archaeologists trained at Harvard and Yale. It was their way of saluting Roosevelt's foresight in the Southwest. Roosevelt red ware was part of a Salado ceramic tradition begun in AD 1280 to 1450 and based on use of red, white, and black paint in geometrically interesting configurations. Most of the gorgeous pottery bowls were found around Roosevelt Lake at Tonto National Monument. These exquisite bowls, now on display at museums in the Southwest, had both interior and exterior decorations. Petrographic analysis of the Salado pottery continues. Each year newfound fragments yield fresh anthropological revelations. Pottery, along with bones, remains our best clue to our ancient past.

Tonto National Monument became part of the U.S. National Park Service in 1933. And the name of Theodore Roosevelt—attached to both the lake and the pottery—remained a major part of the monument's appeal.

CHAPTER TWENTY-FOUR

Mighty Birds: The Federal Reservations of 1907–1908

I

When Roosevelt received an advance copy of Reverend Herbert K. Job's *Wild Wings* from Houghton Mifflin in 1905, illustrated with winsome photographs, it immediately served as an Auduboner-action impetus. Everything always seemed conditional on offshore bird rookeries but Job's black-and-white photos of wild Florida—now housed at Trinity College in Hartford, Connecticut—had the permanence of fine art.[1] To Roosevelt, Job's photograph of man-o-wars wheeling through the air was worthy of the Louvre. With admirable precision, Job captured the biological traits of feisty terns and turbulent skimmers in all sorts of intriguing positions (both on the wing and at rest). Although Job was an amateur photographer, his birds nevertheless seemed alive with joy, and some almost seemed to have human traits. Because the young birds Job spied at Cape Sable and the Florida Keys were vulnerable to predators, he clicked and then scrammed to avoid disturbing their nest incubators.[2]

But photography was not Job's only talent. Accompanying the pictures in *Wild Wings* was painstakingly accurate ornithological prose. (The excellent chapters "Following Audubon among the Florida Keys" and "The Great Cuthbert Rookery" had been previously published in *Outing Magazine*.) All of Job's ornithological observations—whether written or presented as photographs—also offered topographical detail about pathless jungles of red and black mangrove; sea rocks teeming with chattering birds; and sandy white beaches with nesting burrows. This, of course, endeared Job to Roosevelt, who found him the real thing. Job wasn't a fraud like the Reverend John Long, who claimed that egrets built casts for their broken legs and that robins chirped in Morse code. Furthermore, in *Wild Wings*, Job promoted a sensible idea: "Every American Should Be a Game Warden" (a worthwhile precept for the "citizen bird" movement to follow). To Job, birds were windows into the soul of God; killing them for women's hats was akin to blasphemy. Small birds, in particular, Job said, needed protection and love—people should feed them suet and seeds in winter to help survive zero weather.

Job was born in Boston during the Civil War. As a boy he used to take a skiff along Cape Cod, Block Island, and the Connecticut coast to study seabird colonies. The spots where they congregated mesmerized him. Before long he became a dedicated wildlife photographer. After earning an A.B. degree at Harvard in 1888, Job trained at the illustrious Hartford Theological Seminary to become a Congregationalist minister.[3] His first pulpit was in North Middleboro, Massachusetts, before he relocated to Kent, Connecticut. Job was married and had a daughter. He shared with his family an unshakable enthusiasm for birds. Slowly but surely, he started bringing birds into his sermons. To him, birds' nests were sacred incubators ("castles") that needed to be designated "a safe refuge."[4] According to Job, citizens had a "holy obligation" to protect God's little flight machines. Starting in the 1890s, armed with his trusty camera, Job photographed flocks along the Gulf of Saint Lawrence, on the headlands of Nova Scotia, on the wet prairies of North Dakota, and along the banks of the Saskatchewan. Taking a factual approach to ornithology, he promoted the creation of federal bird reservations. He was something of an ornithological writing machine, and no bird was beneath his scrutiny; *Outing Magazine* called him America's "humane sportsman" and a "gentle naturalist" in Burroughs's tradition. "You are one of the Americans I feel particularly proud of as an American," Roosevelt wrote to Job, "because of the excellent work you have done."[5]

By the time Roosevelt was president, Job had established himself as a popular ornithologist of note. This would prove propitious for both men. Job's stock in trade—combining wildlife photographs with outdoorsman-like prose—was indebted to the sportsman style Roosevelt promoted in the Dakota trilogy. Not only had Job written a few poignant articles for *Outing Magazine* but he was an energizing force for the Connecticut Audubon Society. Still, as a writer, Job was only slightly above average: say, seven on a scale of one to ten. So when *Wild Wings* was published, naturalists rubbed their eyes in disbelief. Job had written a minor classic; wherever egrets, herons, and laughing gulls covered every inch of ground, he was in his element. As he wrote in *Wild Wings*, he would set up a pup tent and live with birds on islets for weeks on end. There were, for example, ten amazing shots of Florida pelicans—those Roosevelt favorites—in their black and red mangrove habitat, many taken at close range. For sheer exaltation, Job outdid even Frank M. Chapman. Some of the photos in Chapter 1 ("Cities of the Brown Pelicans") actually came directly from the Pelican Island Federal Bird Reservation. Because Florida was bleached by sunlight, these photos had an impressive clarity.

To Roosevelt, Chapters 2 and 3—"Following Audubon among the Florida Keys" and "In the Cape Sable Wilderness"—were revelatory. Even Job's photo of guano-whitened branches of mangrove interested him.

Cleverly, Job argued that saving the Florida Keys would be the most fitting memorial to the life and legacy of John James Audubon. Here Job found "rare and beautiful water-birds in amazing numbers, tropical islets with their dark mangroves, waving palms, and coral shores, waters prolific in fish and huge sea-turtles, with the soft southern zephyrs playing all over." Job was as slender as a whip. He usually had field glasses at the ready. He wore a cropped mustache. His connection to the great Audubon wasn't accidental. As a hook for *Wild Wings*, he had retraced Audubon's Florida excursion of 1832 and provided a highly accurate bioregional update. The central difference was that Audubon had used a gun (and painted eighteen birds in Key West) whereas Job (with a camera) shot hundreds of photos. Audubon's paintings are enduring classics. Many of Job's photographs were only memorable: spoonbills wary of intruders, snowy egret just hatched, a young ibis nesting, and cormorants leaving a rookery. But Job pulled out all the stops in *Wild Wings*, even excerpting poetry verse from Byron, Bryant, Lanier, and Longfellow to further enhance the reader's emotions. There were also meditations on the secrets of owls, instructions for the new sport of "hawking," and descriptions of shore-patrolling against plumers killing golden plovers. And he championed government protection of rookeries. Like the AOU and the Audubon Society, Job pleaded with fashionable women to stop being induced by vendors to purchase hat and bonnet feathers.

"Here I felt I had reached the high-water mark of spectacular sights in the bird-world," Job wrote after documenting Key West with his camera. "Wherever I may penetrate in future wanderings, I never hope to see anything to surpass, or, in some respects, to equal, that upon which I now gazed. Years ago such sights could be found all over Florida and other Southern States. This is the last pitiful remnant of hosts of innocent, exquisite creatures slaughtered for a brutal, senseless, yes, criminal, millinery folly, decreed by Parisian butterflies, which many supposedly free Americans slavishly follow. Florida has awakened to her loss, and imposes a very heavy fine for every one of these birds killed. Sincerely do I wish that every one who slaughters, or causes to be slaughtered, these animated bits of winged poetry, may feel the full weight of the penalty of the statute and of conscience."[6]

To Roosevelt, *Wild Wings* was filled with fascinating bird lore. It was irresistible to read what an authentic bird-watcher and bird saver—not

Roosevelt was enthralled by the ornithologist Herbert K. Job's Wild Wings. *Job's photographs—like these of baby pelicans and handsome Black Guillemots on sea rocks—were used by Roosevelt to promote the "citizen bird" movement. Roosevelt declared that to kill one of these Florida birds indiscriminately was akin to murder.*

the faker Long—had presented. But Job could never have been a member of the Boone and Crockett Club. The systematic brutality of hunting—all that bubbling blood—never appealed to his Christian sensibility. Leaving his shotgun at home in Connecticut, Job had gone to Florida with a camera which registered in one-thousandth of a second (it had a long-focused four-by-five-inch plate).[7] He was an Auduboner on a mission, and his church, according to the *Congregationalist and Christian World* newsletter, actually paid for his trips to Florida. They were a fine investment.[8] He was going to prove that Floridians were terribly shortsighted for allowing vandals, eggers, developers, and plumers to massacre herons, egrets, pelicans, and other birds. As a leader in Connecticut's Audubon movement, he had an obligation to lead this crusade.

Roosevelt sent Job a heartfelt congratulatory letter on White House stationery. He had admired Job's earlier book *Among the Water-Fowl* (1902) and now thanked his "fellow Harvard Man" for the "exceedingly interesting" book. Not only was Roosevelt inspired to save more of subtropical Florida by creating federal bird reservations, but he also admitted the folly of using guns rather than cameras when engaging in outdoor adventures. "I have been delighted with it," Roosevelt wrote to

Job about *Among the Water-Fowl*, "and I desire to express to you my sense of the good which comes from such books as yours and from the substitution of the camera for the gun. The older I grow the less I care to shoot anything except 'varmints.' I do not think it at all advisable that the gun should be given up, nor does it seem to me that shooting wild game under proper restrictions can be legitimately opposed by any who are willing that domestic animals shall be kept for food; but there is altogether too much shooting, and if we can only get the camera in place of the gun and have the sportsman sunk somewhat in the naturalist and lover of wild things, the next generation will see an immense change for the better in the life of our woods and waters." In a handwritten postscript Roosevelt confessed that he was "still something of a hunter, although a lover of wild nature first!"[9]

An ecstatic Job persuaded his publisher, Houghton Mifflin, to print Roosevelt's letter about *Among the Water-Fowl* as the introduction to *Wild Wings*. The first-edition book cover was an elegant green-turquoise with golden seabirds in flight embossed on the front, back, and spine; it's now a presidential collector's item.

The influence of *Among the Water-Fowl* and *Wild Wings* went far beyond the ornithological community. With the Antiquities Act of 1906 and the National Forest Decree of 1907 as the newest weapons in his preservationist-conservationist arsenal, Roosevelt once again turned hard to wildlife protection by means of federal bird reservations. Spurred on by the Job books, Roosevelt vowed to stop the precipitous decline in avian species such as man-o-wars, albatrosses, and loons. It didn't hurt to have a noted clergyman cheering his efforts on from the sidelines. "The lack of power to take joy in outdoor nature is as real a misfortune," Roosevelt said in *Outlook*, "as the lack of power to take joy in books."[10]

From August 8, 1907, to October 26, 1908, in fact, there were eighteen new "I So Declare It" federal bird sanctuaries. There were no stentorian speeches about these reserves from the president, or from Dr. Merriam or Secretary of Agriculture James Wilson. There was only direct action. In the last five months of 1907 the president—concerned about the Central Flyway from Canada to the Gulf of Mexico—created no fewer than three federal bird reservations in the semitropical waters of Louisiana alone: Tern Islands, mud flats east of the Mississippi River mouth; Shell Keys, seventy-eight unsurveyed acres of low-lying sand-and-shell islets in the Gulf of Mexico south of historic New Iberia; and East Timbalier Island, small marshy islets and sandbars due south of the mainland. All the new reservations were east of the mouths of the Mississippi. If you included windswept Breton Island (established in 1904), Roosevelt had saved much of the biologically diverse Louisiana–Mississippi Gulf Coast island chains from stumpage and slashing.

The remoteness of these Gulf locations greatly intrigued Roosevelt. When there was free time on his calendar, he hoped to inspect the totality of the small Gulf islands and sandy keys thickly covered with seabirds' nests. Roosevelt particularly liked black skimmers (called razorbills in the south Gulf) and wanted to visit these outer islands in June when the great flocks deposited eggs. The heavy surf and heavy onshore wind sounded like a pick-me-up to him. The woolly clouds in the Gulf moved, at dusk, with a galvanic speed that drove up rain. He hoped that Job would soon accompany him on the south Gulf excursion. Roosevelt joked that the only thing better than a clergyman in your corner, covering your action, was a physician. Bird-watching in the Gulf of Mexico with Warden William Sprinkle of Biloxi, Mississippi, and Reverend Herbert K. Job of Kent, Connecticut, as his guides on the outer islands of Louisiana was his latest idea of a splendid open-air holiday on the edge.

Job began doing advance work for this possible trip. Traveling southward, he gathered data for Roosevelt about tarpon and other game fish, the migration of bay birds, and the egg-laying of sea turtles. This scouting also afforded Job the opportunity to befriend Warden Sprinkle, whom he affectionately called "protector of the Gulf birds." As an assignment for *Outing Magazine* in 1907, Job wrote "Curiosities of Louisiana Sea Islands," an article that Roosevelt loved. Job was announcing to the world that his beloved Theodore Roosevelt—the ornithologist president—was heroically protecting America's flyways from unwelcome human encroachment.[11]

According to the Biological Survey's *Annual Report of 1908*, Roosevelt didn't create Tern Island, Shell Keys, and East Timbalier Island at random. He was concerned about effluence from the mouth of the Mississippi River. Besides birds, marine life filled Tern Islands, Shell Keys, and East Timbalier. Single-cell animals, jellyfish, and copepods were omnipresent. Billions of types of plankton—many still not identified by scientists—kept the food chain thriving. As a bonus, a perceptive beachcomber on these Louisiana islands could discover amazing rocks, fossils, shells, coral, and bones. A botanist could marvel at bayberry or wax myrtle. The currents and undertows around these outer islands, however, made them treacherous for the seafaring novice. With just the slightest change in weather a breaker, like a small waterfall, could rush up on the unsuspecting novice's vessel and swamp it. In June 1915, as ex-president, Roosevelt visited these isolated barrier islands with both Job and Sprinkle as companions.

II

From the Louisiana islands Roosevelt, for the first time, moved west of the Mississippi River with his idea of federal bird reservations. Under the instructive guidance of Job, Chapman, and Dutcher, Roosevelt matter-of-factly developed a bold Oregon-Washington strategy for the Biological Survey to implement with environmental interconnectedness in mind. And there was one inflexible rule: once the Roosevelt administration created a federal reservation, it didn't tolerate plumers, seal hunters, or human menaces of any other kind. The refuges were official U.S. government property policed by wardens paid by the Audubon Societies and AOU (with money from the Thayer Fund). Recognizing that America's coastal areas were under siege from the millinery, fishing, and oil lobbies (and anxious to continue to add new rookeries to his conservationist program), Roosevelt established three more federal bird reservations along the Pacific coast in Washington state on October 23, 1907—Flattery Rocks, just off the coast from the town of Ozette; Copalis Rock, an island cluster of bluish sandy clay; and Quillayute Needles, known for its natural sandstone pillars and barking seals. The wildlife was so noisy at these Pacific sites that even the rocks seemed to talk. John Muir had been right in *Steep Trails*. Washington state was, to put it mildly, "strikingly varied in natural features."[12]

When considering these Pacific Northwest refuges, it's important the reader keep in mind that their preservation received little attention from the general public or the press in 1907. It's probably safe to say that 95 percent of Americans had never heard of the Biological Survey, and 99.9 percent had never read a word about Job's bird rookeries. But the Biological Survey and Job were on the front line of the bird protection movement. And from 1885 to 1905 the Biological Survey issued twenty-three separate monographs on North American fauna—though how many were read was another question.[13]

Modernity, in general, didn't work in harmony with the Biological Survey's concern for saving bird flocks and faunal habitats. Forests were being destroyed by logging, depriving birds of essential habitats. An ugly ramification of the telegraph and telephone lines crisscrossing North America was that birds died en masse by flying directly into them. Ernest Harold Baynes—a popular figure in buffalo preservation circles since the successful creation of Oklahoma's Wichita Mountains reserve—wrote in *Wild Bird Guests* that the new Statue of Liberty in New York harbor was a bird trap, killing 1,400 on a single morning.[14] Skyscrapers, in general, were deplored by the Audubon Society. This fear of overindustrialization, long articulated by Burroughs, convinced Roosevelt that he should establish a coordinated system of bird refuges from the Pacific Northwest to the prairie in North Dakota, and from the upper reaches of Lake Huron to the coral reefs of the Florida Keys. Starting in 1903 (with Pelican Island), every year the USDA's Biological Survey issued an annual report, in which bird reservations established for aesthetic reasons were given noticeably more space than animal control. Wildlife protection had taken hold in the Biological Survey in a way that would have been unthinkable before Theodore Roosevelt's presidency.[15] According to Aldo Leopold, in *Game Management*, because of the "Roosevelt Doctrine" of conservation, the "game hog" and the "market hunter" were "duly pilloried in the press and banquet hall, and to some extent in field and wood, but the game supply continued to wane."[16]

Besides poachers, plumers, and eggers the first federal bird reservations along coastal areas—bird cities—had to confront the hazard of bad weather. Even a warden with a shiny USDA badge was no deterrent to winds of 130 or 140 miles per hour. The Biological Survey's annual report for 1907, for example, told of nonhuman difficulties the USDA had at Breton Island in Louisiana. "The islands composing this reservation were visited and somewhat damaged by a severe tropical hurricane which swept the Gulf of Mexico in 1906," the report noted glumly. "Breton Island, six miles in length, was split into three parts, and although normally standing twelve feet above water was flooded throughout its whole extent. Many thousands of pelicans were destroyed by being dashed to the ground by the wind. A beneficial feature of the storm, however, was the extermination of the raccoons and muskrats, which had infested the island and which annually wrought considerable havoc among the nesting birds."[17]

Roosevelt's passionate interest in saving the birds of coastal Washington and Oregon grew out of his trip to Portland, Oregon, and Puget Sound, Washington, in 1903. Not only did plumers want to kill birds, but oil companies wanted to kill seals for heating fuel oil. William Finley began using the model law established by the Lacey Act as a means to prevent the killing of both birds and seals on offshore rocks. Confident, committed, and filled with an enormous sense of purpose, Finley organized the Oregon Audubon Society (today's Portland Audubon Society) using the slogan "Woods, Water, and Wildlife," at around the time of Roosevelt's "Great Loop" tour of 1903. All Finley talked about with T.R. was Three Arch Rocks, off the Oregon shore: how the tides slapped and receded with eternal ferocity (just as Job had described a New Brunswick rookery in *Wild Wings*). Marine life, Finley told Roosevelt, occupied every niche and crevice of these rock mounds. At night, with waves crashing, small ghost crabs wandered about, looking like aliens from outer space. And murres congregated there as nowhere else in America. These unsurveyed Pacific mounds—located only 350 yards from the southern tip of a cape[18]—also teemed with tufted puffins and seals. The mounds ranged in height from 275 to 304 feet. These offshore rookeries were being patrolled, though only haphazardly, by "citizen bird" activists in the state.[19]

The activists in Oregon needed the White House to get involved and establish permanent federal protection. Feeling the burden of responsibility on his shoulders, Finley became a lobbyist. "Finley's contribution to environmental awareness can be equated to that of only one other naturalist who was on the scene in the west during those early years," Roger Tory Peterson wrote—"John Muir."[20] After the 1904 presidential election, Finley began sending Roosevelt private photos of the carnage among wildlife in his state. As the most noted Pacific coast wildlife photographer, Finley took seriously all living creatures that colonized the shore and outer rocks. He was, for the West Coast, like Herbert K. Job and Frank M. Chapman combined—with constant help from his wife, Eileen, and Hermann Bohlman. Roosevelt was enraptured by Finley's images. Whenever possible he talked about the Finleys and Bohlman.

Unfortunately for American environmental history, there are only superficial, potted biographies of Finley available. But we know that Finley took up Roosevelt's invitation to visit the White House. The pictures Finley showed Roosevelt as an inducement for preservation were taken with a 5-by-7 plate camera and are stunning in their high artistic quality. Roosevelt was particularly riveted by the tufted puffins and seals (both favorites with the general public) at Three Arch Rocks. Wisely, Finley had left trunks full of blurrier images back in Oregon. Lobbying Roosevelt, Finley complained furiously that plumers were being allowed to operate along the bird-rich Pacific coast. The Oregon Audubon Society was making headway in policing the area, but Three Arch Rocks needed federal protection—*soon*. Finley spread the photos on a wooden table and explained each one in detail; Roosevelt was nearly jumping with excitement. "Bully, bully," he kept saying, "we'll make a sanctuary out of Three Arch Rocks."[21]

Its reasonable to assume that Finley left Washington, D.C., energized. Roosevelt had told him that the Biological Survey—when all the legal underbrush was cleared—would declare Three Arch Rocks a federal preserve. While it's true that Finley grew impatient as months and then years went by, he nevertheless trusted Roosevelt to do the right thing. Three Arch Rocks was going to be just one piece in the Biological Survey's coordinated preservationist strategy for the West Coast. The reason for the delay regarding Three Arch Rocks was that the Roosevelt administration wanted to first pass anti-trespassing laws in Congress. Finley also had to tolerate a lot of ridicule from congressmen opposed to the idea of a federal bird reservation in Oregon—particularly oceanic bird rocks—a reservation that nobody except a few ornithologists would ever be allowed to visit. But Congressman Lacey, his hair now iron-gray, calmed Finley down. "All in due time," Lacey would say. "All in due time." Disappointment and anxiety occasionally got the best of Finley, but he never forgot his mission. When he had to listen patiently to arguments against birds, Finley, refusing to be baited, remained serene. (For example, Senator Charles Fulton of Oregon once said that birds were like lice. "I really want to know why there should be any sympathy or sentiment about a long-legged, long-necked bird that lives in swamps and eats tadpoles and fish and crawfish and things of that kind," Fulton asked. "Why we should worry ourselves into a frenzy because some lady adorns her hat with one of its feathers, which appears to be the only use it has."[22])

The outcome was well worth the wait. On June 28, 1906, Congress had enacted the Game and Bird Preserves Protection Act (Refuge Trespass Act) to provide "regulatory authority" to managed uses on reservations administered by the Biological Survey. The act made it a misdemeanor to disrupt birds or their eggs on federal wildlife reservations.[23] The syndicates were being shut down.

The naturalist Dallas Lore Sharp, in a marvelous manifesto advocating wildlife refuges—*Sanctuary! Sanctuary!*—wrote about Finley and Bohlman rowing out to the fifteen-acre Three Arch Rocks to study the strange birdlife there. Sharp also brought Roosevelt into the narrative, explaining how the combination of these three bird lovers saved Three Arch Rocks for posterity:

> Swinging their dory (they were practiced now) from her rocky davits, they launched her empty on a topping wave, loaded in their precious freight, and, pulling safely off, headed for shore, making a solemn promise to the old bull sea-lion, and to the flippered herds sprawling along the ledges, and to the flying flocks that filled the air. But none of the multitude heard it above their own raucous screaming, and none of them knew. They did not know how that vow took one of the

> boys across the States to the other ocean shore. They did not see the pictures of their rainy, sea-washed home spread in high excitement over a table in the White House, nor watch an eager man, all teeth and eyes and pounding fists, whanging about and bellowing: "Bully! Bully!" just like an old bull sea-lion. But Finley did. They did not see him study the pictures and vow, "We'll make a sanctuary out of Three-Arch Rocks." But Finley did.[24]

Besides Three Arch Rocks, Finley also fought to save the diversity of wildlife along the Washington state coastline. At Flattery Rocks alone—more than 800 islands, seal rocks, and reefs—special interests were wreaking environmental havoc. The feather and egg mongers were destroying the fourteen species of seabirds that bred along the Flattery Rocks, including fork-tailed storm petrels, double-crested cormorants, black oystercatchers, pigeon guillemots, Cassin's auklets, and tufted puffins. With the Game and Bird Preserves Protection Act on the books, Roosevelt wanted to do something spectacular to save the Washington state rookeries.[25] In 1853 Berthold Seemann—"Naturalist of the Expedition" on board the British explorer Captain Henry Kellett's ship—had written admiringly about Flattery Rocks in *Narrative of the Voyage of H.M.S. Herald*; Roosevelt had made this classic volume a treasured part of his home library.[26] Now, Finley offered visuals to match that exploration text.

Enthralled by Finley's wildlife photographs, Roosevelt worried that before long oil derricks would deface Washington's rugged coastline. And recognizing that the Flattery Rocks system—as well as Copalis Rock and the Quillayute Needles—had been created as a by-product of the Olympic Mountains, he asked Dr. Merriam at the Biological Survey to find a way to preserve these wonders before the population of Tacoma and Seattle and commercial extractors swarmed the coast. Purple starfish, lumpish seals, thrashing whales, crowds of murres, crags, caverns—Roosevelt wanted the biological integrity of these Washington islands preserved unmarred in its entirety. Few oceanic islands in North America, save for those in Alaska, had such steep sides with uncounted colonies of nesting seabirds. If need be, Washington state's cute tufted puffins, which bred on the rock piles (sometimes even underneath boulders), could be used as an appealing symbol. Coastal Washington was their favorite North American congregation point outside Alaska.

After careful consideration the Biological Survey recommended that Flattery Rocks quickly be declared a federal bird reservation; Roosevelt agreed. On October 23, 1907, he signed an Executive Order to that effect. And the Roosevelt administration took its environmental responsibility even farther. On the same day, Roosevelt declared Quillayute Needles (consisting of Hand Rock, Carroll Islets, Bald Island, Jagged Islet, Cake Rock, James Island, Hunting Rock, Quillayute Needles, Rounded Islet, Alexander Island, Perkins Reef, North Rock, Middle Rock, Abbey Island, and South Rock) a federal bird reservation. Then he did the same on behalf of Copalis Rock, unsurveyed rock islands with seabirds, hauling seals, and sea otters. The pelt hunters, fishing organizations, plumers, pot hunters, weekend poachers, oil drillers, and corporate despoilers had been shut down by the federal government along the coast from Seattle to Portland at environmentally sensitive locales. More than any other president, Roosevelt used executive orders without consulting Congress. There was no reason for Washington state to be immune from them. "Roosevelt had the sense to keep wild places in Washington wild," Kevin Ryan of the state's Maritime National Wildlife Refuge Complex has noted. "The shear rocks are too hard for people to climb, so all three Roosevelt reserves have become species indicator zones."[27]

Roosevelt had grabbed land for the Biological Survey and saved the most enchanting part of Washington state coast for future generations. He said that his own great-grandchildren would someday get to smell Washington's salty breeze, see the surf-carved ocean rocks, feel the riptide at the roiled straits, and marvel at God's creation without the stink of gasoline. In Washington state they would be able to see mountains at the ocean's edge with primeval coniferous forests untouched by axes. They would hear the long-drawn call of loons or watch kittiwakes build nests. By boat, they could encounter pristine islands along the international boundary with Canada. They would be able to imagine what Flattery Rocks, Quillayute Needles, and Copalis Rock had been like during the long-gone days when Captain James Cook stumbled on them and Indians built stockade villages at the foot of the towering ocean rocks and searched for lobsters, oysters, crabs, and fish. Over half of America's Pacific coast bird species used these Washington state offshore island rocks, and so the Roosevelt administration shut them off from human encroachment.

Today all three of these national wildlife refuges—Copalis, Flattery Rocks, and Quillayute Needles—work together as the Washington Maritime NWR Complex. Under Roosevelt's executive orders, these islands remain closed to the public—they belong to the seabirds and seals. As Roy Crandall, director of publicity for the Pan American Exposition, wrote in a 1909 edition of *Technical World Magazine*, President Roosevelt (and the Audubon societies)* single-handedly saved "the coasts of Washington, Oregon and California in the Pacific." Owing to Roosevelt's foresight "thousands of protected birds now swarm and multiply in safety and in confidence for they are becoming so tame that bird wardens walk among them and brush them from their paths."[28]

III

Once Finley had persuaded the Roosevelt administration to save bird rookeries in Oregon and Washington, he turned his lobbying efforts to wetlands known for their waterfowl: the marshes of Klamath, Tule, and Malheur (which means "misfortune" in French; the name was given after Peter Skene Ogden of the Hudson Bay Company was robbed of his furs there in 1826).[29] Finley and Bohlman camped for over a month in the marshes of Lower Klamath Lake on the border of Oregon and California during the summer of 1905 and photographed the abundant waterfowl for the National Association of Audubon Societies. Their photographs, as well as their report detailing the tremendous numbers of birds being slaughtered ($30,000 worth of grebes by feather hunters in 1903 and over 120 tons of waterfowl by market hunters in 1904), led to the Klamath becoming the first major waterfowl refuge established and the first refuge associated with the reclamation project.[30]

Writing years later in *Nature Magazine* about the federal bird reservation movement in the Pacific Northwest during the Roosevelt era, Finley offered a succinct rationale for West Coast bird refuges in wetlands (not just on oceanic rocks). "A very large number of lakes and ponds have been drained and many swamps have been dried up under the guise of making agricultural land," Finley wrote. "With the gradual spread of population, each year the migratory flocks return to former nesting sites, only to find them destroyed, and their natural food supply diminished. The vital point today in wild fowl preservation is that a goodly number of the remaining lakes, ponds, and swamps must be preserved. No matter how many game laws we have or how rigidly these are enforced birds, like people, cannot live without homes and many species are sure to be pushed to the point of final disappearance."[31]

*In 1905 the National Committee of Audubon Societies changed its name to the National Association of Audubon Societies for the Protection of Wild Birds and Animals. On October 1, 1940, this would again change to the simpler National Audubon Society (which it remains today).

Because Washington and Oregon weren't overly populated (after all, they got about 140 inches of rain annually) and the extraction operations were just beginning along the Pacific Northwest coast, Roosevelt had been able to make a preemptive strike on behalf of wildlife in 1907 with Three Arch Rocks, followed by the three Washington state bird rock archipelagos and the waterfowl marshes of Oregon. On the same numerologist's dream of a day (08/08/08) that Klamath Lake was established, so too was Key West. Florida, in 1908, proved far more difficult. Nevertheless, Roosevelt was ready to methodically finish the job he had started at Pelican Island, Passage Key, and Indian Key. Outfoxing his opponents every step of the way, confusing them with his embrace of both hunting and preservationist polarities, Roosevelt issued one executive order after another from February to October 1908 to save sites in Florida such as Mosquito Inlet, Dry Tortugas, Key West, Pine Island, Palma Sola, Matlacha Pass, and Island Bay.[32]

Just as Upton Sinclair's book *The Jungle* spurred Roosevelt to issue laws regulating meatpacking, *Wild Wings* saved birdlife in Florida. Understanding that in the tropics residents sometimes resorted to slash-and-burn practices, the president moved quickly to protect "Florida's wildlife heritage." No salt prairie, coral reef, mangrove thicket, or bird rock was precluded from the Biological Survey's consideration. And Roosevelt had the congressional ruling of June 28, 1906, about not disturbing or trespassing on federal bird reservations, to work with as a legal deterrent in Florida. Not that it was foolproof—the congressional order was frequently defied. A deranged Floridian, for example, shot four of Roosevelt's pelicans on Mosquito Inlet, claiming that the refuges were illegal. The courts ruled in favor of the Roosevelt administration.[33] The assailant pleaded guilty and paid a steep fine.

This was a time of profound, positive change in the Florida wildlife protection movement. Numerous islands along Florida's Gulf of Mexico and every bird rookery along the Atlantic Ocean now had aesthetic value to Roosevelt. He would have to save them from the persistent ignorance of the ex-Confederate yokels and from railroad executives like Henry Flagler. In *An Autobiography*, Roosevelt listed his most notable wildlife protection achievements in Chapter II, "The Natural Resources of the Nation." Among them, Florida ranked high. In particular, he saved the West Indian manatees (*Trichechus manatus latirostris*) at Mosquito Inlet in Volusia County, about eighty miles from Orlando. Roosevelt hoped to create "safe havens" throughout Florida where these manatees could live unmolested, as President Ulysses S. Grant had done for the northern fur seals of Alaska. The manatee—whose name is Haitian, meaning "big beaver"—was almost as special to Roosevelt as the buffalo.[34] Although he did not create an American Manatee Association, he did fight for the species' survival. "Wild beasts and birds are by right not the property merely of the people alive to-day," he said with regard to protecting Florida's manatees and seabirds, "but the property of the unborn generations, whose belongings we have no right to squander."[35]

Exactly when President Roosevelt brought manatees into his wildlife protection program remains unclear. (Burroughs, we are told by the ornithologist Charles William Beebe, astonishingly didn't know what a manatee was before visiting Florida in 1903.[36]) Perhaps as a boy Roosevelt encountered manatees or sea cows; they are part of the mermaid myth and had become popular characters in children's books. Maybe he read one of the landmark manatee studies by Outram Bangs or Alfred Henry Garrod.[37] Known for barrel-rolling and playful chases, manatees were hard to dislike. They continually grabbed and kissed each other, and lolled for hours in the warm waters of Florida, evidently with no worries or woes. They were herbivores and socialized with one another by nuzzling playfully. Children's books of the mid-nineteenth century often portrayed the manatee fondly, much as they portrayed the friendly panda. (Fishermen, by contrast, often denounced manatees as homely and shot them on sight.) During the Great Depression they were poached for meat.

Perhaps Robert B. Roosevelt's *Florida and the Game Water-Birds* had affected Roosevelt. In that book, Uncle Rob had briefly diverged from his loosely structured autobiographical narrative to talk about the manatee herds he encountered throughout Florida. He noted how tourists near Mosquito Inlet couldn't believe that "cows feed under water," until they saw a stubby-snouted manatee munching on and sheltering under freshwater plants. "The animals and birds are as queer and unnatural as the herbage," R.B.R. wrote of aquatic Florida. "Or as a climate which furnishes strawberries, green peas, shad, and roses at Christmas."[38]

A close relative of the manatee—Stellar's sea cow—had been hunted into extinction in 1768, and Roosevelt worried that the same fate awaited the West Indian species. Indeed, in 1885, an observer in Florida noted that "ten years ago the meat (of a manatee) could be bought at fifty cents a pound. The animals are becoming far too scarce to admit of it being sold at all. There is no doubt that the manatee is fast becoming an extinct animal." And another factor in the manatees' uncertain prospect was that they reproduced slowly. Male manatees didn't breed until they were nine years old, and females didn't reach sexual maturity until they were five.

A lone calf was then born every three to six years. Mothers insisted on nursing their babies for up to two years.

Given these facts, Roosevelt was concerned about whether manatees had a future in Florida. Even though his Mosquito Inlet Reservation was ostensibly to protect native birds on small mangrove and salt grass islets, shoals, sandbars, and sand spits in Mosquito Inlet and the mouths of the Halifax and Hillsboro Rivers, manatees, he knew, also would receive needed protection in the warden-patrolled waters, especially during calving season. Part of his reason for setting aside Mosquito Inlet (near Daytona Beach) as a federal bird reservation—by means of an executive order issued on February 24, 1908—was to protect the manatees' northernmost range. The purpose of saving Mosquito Inlet, a primordial Darwinian laboratory, was to keep the manatees there free from human molestation; they were the most essential large mammal in the Florida ecosystem. This was the same rationale he used for declaring Three Arch Rocks a federal bird reservation—doing so had the additional benefit of saving seals. If Floridians couldn't rally to protect manatees, then the probable fate of such lesser creatures as Suwannee bass, striped mud turtles, red-cockaded woodpeckers, and southeastern beach mice was dismal indeed. Also, selfishly, Roosevelt wanted manatees saved so that he could enjoy them when he went spearfishing after his tenure at the White House was over.[39]

Sea turtles played a significant role in another of Roosevelt's federal bird reservations. The president liked the fact that sea turtles came from "a different world." Long before the biologist Marston Bates published *The Forest and the Sea* in 1960, Roosevelt understood that the "sea margin"—where marine and terrestrial environments mixed—meant *everywhere* in Florida. Whether it was on tidal flats, sandy beaches, or flyspeck islets, marine species often used land and sea interchangeably.[40] Certainly this was true of sea turtles. The Atlantic coast of Florida, especially from Melbourne Beach to Palm Beach, was the world's most crucial habitat for sea turtles—loggerheads, greens, and leatherbacks.[41] Keeping up on sea turtle biology Roosevelt wanted to protect egg clutches from predators ranging in menace from raccoons to Miami restaurateurs. Roosevelt didn't mind if fishermen caught these turtles to eat (he personally thought turtle fritters were *dee-licious*) but ravaging their breeding grounds indiscriminately needed correction.

On April 6, 1908, Roosevelt declared the Tortugas Keys (now Dry Tortugas National Park) a federal bird reservation. (The Tortugas group, located seventy miles west of Key West, Florida, and 470 miles southeast of New Orleans, consisted of eight little islands: Loggerhead Key,

Bird Key, Garden Key, Long Key, Bush Key, Sand Key, Middle Key, and East Key.) Roosevelt boldly protected both sea turtles' egg-laying and the boobies that congregated by the thousands in the buttonwood trees. When the Spanish explorer Juan Ponce de Leon first visited these coral reefs in 1513, he marveled at the colonies of sea turtles he encountered. His crew recorded catching 160 of them. Ponce de Leon considered them a good omen, and they were also perfect for soup and stew; he was the one who named the remote island group Tortugas ("turtles"). John James Audubon had spent days in Tortugas Key mainly to observe the sooty terns that annually nested here, raised chicks, and then migrated back to the Yucatán peninsula.[42] In Caribbean pirate lore, the key chain had a bad reputation as a site of shipwrecks; Caribbean captains, in fact, called the reefs an "underwater graveyard."[43] Before long, the word "Dry" was added to nautical charts to warn mariners that there was no fresh water on the island chain.

Although one can't be certain, Roosevelt probably first seriously encountered the Dry Tortugas when writing his two-volume history *The Naval War of 1812*, published in 1882. Fort Jefferson—America's largest coastal fort in the mid-1800s, built with more than 16 million bricks—was constructed on Garden Key in that chain (following the battle of New Orleans) to provide a future defense line for Florida, Alabama, and Mississippi. At times the Dry Tortugas were used to quarantine people with yellow fever. After being convicted as a co-conspirator in Abraham Lincoln's assassination, Dr. Samuel Mudd was incarcerated at this remote prison fortress in the Gulf of Mexico, helping to care for these patients. By the time Roosevelt became president in 1901, lighthouses were operating in Tortugas Key warning sloops and schooners of the low-lying staghorn coral, patch reefs, sand flats, and sea grass beds. In *Florida and the Game Water-Birds*, Uncle Rob had declared the Tortugas Keys the center of the greatest fish population in the entire United States. He was indubitably right. Although the Tortugas were difficult to get to, sports fishermen, including the novelist Zane Grey, used to hunt in the warm waters for 300-pound blue marlins and 100-pound wahoos.

The biologist Rachel Carson of the U.S. Fish and Service Wildlife wrote in *The Edge of the Sea* about the fascinating creatures Roosevelt had saved with his federal bird reservations—particularly the Tortugas group. Describing how loggerhead, green, and hawksbill turtles must return to land every year for spawning, Carson noted the majestic seasonal ceremony in the Tortugas group when the turtles "emerge from the ocean and lumber over the sand like prehistoric beasts to dig their nests and bury their eggs."[44] Roosevelt was dead when Carson published *The Sea Around Us* in 1951, but his daughter Alice Roosevelt Longworth wrote a fan letter to Carson, saying that her father would have welcomed this noble literary work with open arms.[45]

The Dry Tortugas were so remote in 1908 that most ornithologists in New York hadn't even visited them. Invariably, visitors to Key West erroneously believed they were at the end of America, not realizing that the Dry Tortugas were seventy miles farther out. The best-informed ornithological studies that had seized Roosevelt's attention were a research paper by Dr. John B. Watson of Johns Hopkins University (it had been funded by the Carnegie Institution, and brevity was its virtue) and, of course, Job's *Wild Wings* (which included photographs of sooty tern swarms estimated at 6,000 or 8,000 strong).[46] Complementing Watson and Job was the Florida Audubon Society, which undertook a bird count on the island chain. The yearly return of the sooty and noddy terns was being touted by some ornithologists as the East Coast's equivalent of the swallows returning to Capistrano mission in southern California. "In other words," the ornithologist Alexander Sprunt, Jr., would write in *The Auk* about the Dry Tortugas, "if not held as a miracle, at least the conviction is extant that the birds arrive and depart on exactly the same day each year and, if it varies at all, it is held to be due to certain phases of the moon!"[47]

Four months after the preservation of the Tortugas group, Roosevelt set his sights on the unpopulated islands of the Key West chain. Developers were eyeing the island chain and hoping to build tourist hotels, so Roosevelt refused to delay. Key West might not seem much different from the other Florida mangrove islands he had established as federal bird reservations, yet it was unique. For one thing, it was a turtle nesting area free of raccoons. This meant that the offshore beaches and sand dunes were an ideal nesting habitat for loggerhead and green turtles (unlike Breton Island in Louisiana, which did have raccoons). Every spring, hard-shelled marine turtles would leave the ocean to bury their round eggs in the coarse-grained sand dunes at Key West. Because there were no raccoons or other egg thieves, the successful hatching rate on Woman Key and Boca Grande Key (both part of T.R.'s Key West Federal Bird Reservation) was extraordinarily high. The turtles' real enemy was fishing nets, and tough laws would have to be enacted to prevent the demise of greens and loggerheads. Later, as a former president, Roosevelt inspected sea turtle eggs in Breton Island; he foolishly dug some up to eat—a sin in the eyes of modern marine biologists. For all his acumen as a naturalist,

Roosevelt—like most of his generation—simply didn't know how endangered they were. It would be another forty years before the plight of sea turtles was discovered and eloquently articulated by Dr. Archie Carr.

Even though the Key West and Dry Tortugas reservations only protected small islands and keys, their protection helped keep the surrounding waters clean and clear. Crater Lake blue, with shades of emerald and dabs of purple, these exquisite waters seemed to roll into infinity. Colonies of soft coral were a pale plum color, and little starfish with tube feet clung to the sides. When Key West islets were saved by Roosevelt by means of an executive order (August 8, 1908), these unparalleled coral reefs were already celebrated among oceanographers all over the world. Most scientists agreed that only the Great Barrier Reef of Australia, which was ten times larger than Key West, was more magnificent. To Roosevelt, Florida's coral reefs—made of living colonies of tube-like animals called polyps—were a world unto themselves. The more than 6,000 shallow coral reefs—from Key Biscayne to Key West to the Dry Tortugas—were breathtaking in their biodiversity. Anybody who has owned tropical fish knows how amazing the Day-Glo colors of the bicolor damselfish, neon goby, clownfish, and foureye butterflyfish can be when viewed in an aquarium tank. Many first-timers in Key West, however, come to experience such wondrous and underwater wildlife in its natural setting. The diver in Key West quickly realizes that an ever-soothing, symbiotic world without human footprints exists in the coral reefs, that the harmonious balance of the ecosystem is awe-inspiring.

If Roosevelt had traveled (as modern visitors do) in a glass-bottom boat a mile off of Key West—over a vast tract of ominous shoals—a coral kingdom would suddenly have appeared before his eyes. As Herbert K. Job understood, this was an important zone for obtaining data on schools of luminous tetra. On closer inspection, Roosevelt would have seen a brown film, or membrane, which blanketed the entire ecosystem together as if a connecting tunic. Polyps with flower-bud mouths, towering barrel sponges, giant octopuses, jellyfish waving their tentacles, porcupine fish ballooning themselves up, angelfish with broad bands of shiny black sailing into and out of coral thickets, all lived in this Key West reef. Their lives were fragile. Roosevelt's nature writings are nearly encyclopedic, but he never wrote about this ecosystem. He surely knew the difference between a lumpfish and a surgeonfish but when it came to differentiating coral species, he was probably clueless. What he did know, however, was that these Florida reefs needed protection, that scientists had still not discovered all the species and plants living on the vegetation-rich bottom of the ocean. The Florida Keys, as *Wild Wings* indicated, was an heirloom as valuable as Yellowstone or the Grand Canyon.

Roosevelt, moreover, considered these reefs national treasures not just because shimmering fish, rays, jellyfish, anemones, big sponges, lobsters, and bull sharks were scientifically fascinating. Roosevelt probably also understood the reefs protected American coasts by reducing "wave energy" from hurricanes and tropical storms. And Key West was the habitat of more than 250 bird species. The national imperative was, therefore, clear. To disregard scientific opinion, aesthetic value, and natural security in favor of fast dollars was, to his mind, reprehensible. When the Key West reservation celebrated its centennial in 2008, only about half of Florida's coral reefs had even been mapped.[48] This didn't, however, mean that the reefs were secure from environmental degradation. A coalition of marine scientists feared that rising carbon emissions might kill off the reefs by 2050 or 2060. "Burning coal, oil and gas adds carbon dioxide—a heat-absorbing greenhouse gas—to the atmosphere," Elizabeth Weise wrote in *USA Today*. "That interferes with the ability of coral, living organisms, to calcify their skeletons, and the coral begins to die."[49]

IV

Each Florida wildlife refuge Roosevelt saved in 1908 had a fascinating story of its own. The last remaining rookeries along the lower Gulf Coast of Florida were documented by the National Association of Audubon Societies Secretary T. Gilbert Pearson in April of 1906, during a trip he made to visit two Tampa Bay bird reservations already established by Roosevelt—Indian Key and Passage Key—and to help the widow of murdered warden Guy Bradley buy a home in Key West. Pearson found a colony of brown pelicans and cormorants at Palma Sola, eight miles south of Tampa Bay, and two large colonies of brown pelicans a few miles north of the Caloosahatchee River, presumably in Matlacha Pass, and Pine Island Sound. He also discovered two large colonies of Louisiana (tricolored) herons in Gasparilla Sound (Island Bay). He learned that the bird laws of Florida were hardly enforced. Only Guy Bradley kept the professional hunters at bay before his murder. Pelican colonies were constantly raided by locals for the eggs. Plume hunting caused the egrets to be so scarce that Pearson only saw a dozen in six weeks of observations. A local bird-skin collector was caught in 1904 selling many different kinds of birds including the now-extinct Carolina parakeet and the extirpated (and possibly extinct) ivory-billed woodpecker. Pearson's reports made their way to Roosevelt's desk and found an attentive audience in the president. The message was clear, these Gulf Coast Florida colonies (Palma Sola, Pine Island, Matlacha Pass, and Island Bay) also needed protection.[50]

First by one bureaucratic trick, then by another, Roosevelt accomplished his overarching goal of protecting birds' habitats. His cleverest tactic was to preserve a group of islets under a single name (for instance, Quillayute Needles in Washington state). Florida's fishermen and its millinery industry accused Roosevelt of grabbing land. The many islands he saved were like a rash, they believed, breaking out wherever an honest land or water bounty might be had. Take Pine Island, for example—the refuge Roosevelt created on the southwest coast of Florida (north of Sanibel Island in Pine Island Sound) in the autumn of 1908. This federal bird reservation was actually three isolated islands inhabited by thousands of brown pelicans.[51]

Roosevelt could have declared a number of refuges around Pine Island, but he didn't. With the agility of a professional politician, and understanding the mental laziness of many uninvolved citizens, he was able to disarm his opponents (whose greed, he felt, was pitted against his own sense of public good). By *not* establishing multiple refuges, Roosevelt made the preservation seem slighter. His detractors now had to object to Pine Island as a single entity. If he had instead issued three separate executive orders, cries against such a "land grab" would surely have been heard throughout Congress instead of merely in the hamlets of Lee County, Florida. A single entity was also convenient. It took only one executive order—Number 939—to save Pine Island as a priceless marine ecosystem where wildlife could flourish. Only a decision by the Supreme Court could have reversed his order, which he issued on September 15. And Roosevelt had powerful allies in the Gulf Coast to make sure this didn't happen.

At Island Bay the Audubon Society's game warden was Columbus B. McLeod—short, bad-tempered, and something of a hermit. His salary of thirty-five dollars a month was paid for by the National Association of Audubon Societies' Thayer Fund. With his motorboat regularly plying the waters of Charlotte Harbor, McLeod also served as the deputy sheriff of Desoto County (present-day Desoto, Charlotte, Hardee, Glades, and Highlands counties), applying warrants and arresting poachers.[52] McLeod—Paul Kroegel's counterpart on the Florida Gulf Coast—was almost sixty, unmarried, and an outdoorsman. He was a friend of the manatees but his primary purpose in life was to protect egrets and roseate spoonbills from extinction. Florida was a bird paradise, he believed, but if the plumer gangs weren't broken up, even the white pelicans would vanish. Warden McLeod would fearlessly motor up to bands of illegal hunters and threaten to have them busted, drowned, run down, hanged, and marched off in shackles—all at once if need be—if they didn't leave the Roosevelt administration's rookeries alone. Some of his rows with these hunters should probably have been covered in the *Tampa Tribune*, but fisticuffs were part of daily life along the Gulf shore. "I protected the Sunset Island Colony for three years in my feeble way without a cent of compensation except the love I had for the wild, free birds and the pleasure it gave me to save the lives of every single bird that I could," McLeod wrote in the fall of 1908, in a report to the Audubon Society. "Since that time you have engaged me as a warden for the Audubon Societies, with a salary and nice little boats, which allowed me to look after their [the birds'] interests more and give them better protection."[53]

McLeod received his mail on the island hamlet of Placida, but he actually lived on Cayo Pelau Island (116 acres of wetland mangrove and ten acres of uplands). With three sandy beaches and rare tropical hardwood hammocks to call his own, he was living the fleabitten life of an outback type. He was poor, but not as poor as the Seminoles. Sometimes he would take a canoe on the lower Peace River, in perfect harmony with his environment. Although biographical information about McLeod's daily activities remains sketchy (he didn't keep a diary as Kroegel sometimes did), he patrolled the Charlotte Harbor rookeries around Gasparilla Sound, chasing away poachers with his boat, badge, and gun. McLeod wrote an emergency report for *The Auk* in 1907, on protecting roseate spoonbills. As a wildlife warden, he worried that enforcement of the Lacey Act was too lax, that the good old boys of Florida simply spat at federal laws. At times, their arrogance made him weep in frustration and dismay. "No Trespassing" signs, he understood, didn't mean much in the matted thickets of southwest Florida. But McLeod had the tough soul of an outdoorsman and lit out after hunters in both boat and prose. "Five years ago there was a fine flock of roseate spoonbills or 'Pink Curlews' that used and did their feeding in the northeast end of Turtle Bay," he said; "only 18 are left now of the flock, and they have for the past two seasons done their feeding on my home island in the fall and winter months. Hunters and tourists killed them, and there are but few left on the Gulf Coast of Florida."[54]

McLeod spoke in a short, curt, unflamboyant way. There was no philosophical complexity about him, and this simplicity added to his credibility. As an environmental activist, McLeod worked with the National Association of Audubon Societies and the Roosevelt administration to go after the illegal hunters in Charlotte Harbor. He was livid because the cold-blooded murderers of Warden Guy Bradley, using expensive lawyers, were never indicted. He complained about this often, calling it an abortion of justice. Herbert K. Job said the same. Regardless of the Lacey Act and the law of June 28, 1906, plumers, he said, were still killing birds for "wings, feathers, and mountings."[55]

Eventually, McLeod himself, while protecting the rookeries in northern Charlotte Harbor and the lower Peace River, became a victim of the "Feather Wars." Shortly after Roosevelt's Executive Order 939 regarding the west coast of Florida, he was murdered by local fishermen and plumers who were furious over federal island grabs. McLeod's patrol boat No. 5 was discovered on November 30, 1908, sunk with sandbags. Blood was splattered all around, as if a can of red paint had exploded. Detectives, believing that McLeod had been hacked to death with an ax, recovered his blood-soaked hat, which had two gashes in the crown. Clumps of hair were also found. A struggle had obviously occurred before McLeod had died. Speculation was that his body had been tossed into the Gulf of Mexico, where sharks and other flesh-eating fish had devoured it. Presumably, this was meant as a message to the Florida Audubon Society that individuals who tried to interfere with the milliners would face dire consequences. The person or persons who killed McLeod were never found.* Once again, as in the murder of Guy Bradley, Job seethed with rage. Notices seeking information about the murder were posted on bulletin boards in fishermen's hotels around the Charlotte Harbor region. The De Soto County Sheriff, however, claimed they couldn't find proof: there were no fingerprints, no witnesses, and no confessions.

McLeod's death strengthened Roosevelt's determination to further safeguard the rookeries and wildlife preserves in Florida. Roosevelt vowed to visit the federal bird reservations that McLeod had protected *soon*; he eventually did go to Pine Island Sound after his presidency, in 1914. And the National Association of Audubon Societies, undaunted, pleaded with citizens of Florida to "awake" and establish a game commission in order to see that the bird laws were enforced.[56] In *Bird-Lore*, an illustrated bimonthly, the editor, Frank M. Chapman, credited McLeod with saving the white pelican colonies in Charlotte Harbor and lamented that the warden "had his head chopped open and his body sunk in the harbor by persons who did not approve of his zeal."[57] McLeod, at the time of his murder, had dutifully kept a bird count the way President Roosevelt had

* According to *Bird-Lore*, in September 1908 L. P. Reeves, an employee of the South Carolina Audubon Society, was also murdered, in an ambush by plumers.

instructed field naturalists to do: he had tallied 1,000 white ibises, 500 pelicans, 250 cormorants, and 150 cranes.[58]

The story of Audubon patrol boat No. 5 sent chills down the spines of those in the "citizen bird" movement. Two Audubon wardens—Bradley and McLeod, both approved by Roosevelt—had been murdered while on duty protecting seabirds. The week McLeod was murdered, conservationists had been lobbying the state legislature in Tallahassee to augment the Lacey Act by hiring state game wardens. Now the Audubon Society in Florida was demoralized. Misdemeanors be damned! Anybody who anchored at a bird rookery should be slapped with a felony charge! If tougher conservationist laws weren't enacted soon, Florida would be deforested, like Palestine or Spain, and bereft of birdlife!

Many Floridians, however, simply didn't want tax revenues used for birds.[59] They'd rather let the landscape become barren than pay a single cent to preserve it. On the other hand, many newspapers proudly promoted avian protection in their editorials. "To proclaim a bird reservation without providing for its protection is a waste of words," the *St. Augustine Record* pointed out. "The Audubon Society asks that citizens of Florida tax themselves by voluntary contributions to maintain wardens on these rookeries during the breeding season. If this be necessary, it is a confession of failure on the part of the State that should be resented. It is an invasion by the citizens of the domain of the State that shall be rebuked. To leave this matter in the hands of a society is to invite murder and will bring the name of Florida into disrepute."[60] And the magazine *Outlook* asked the key question: "Will the people of Florida sleep until it is too late?"[61]

Columbus McLeod didn't die in vain. In 1910, the New York State Assembly outlawed the commercialization of feathers with the passage and signing of the Shea-White Plumage Bill. In 1911, New York State Senator Franklin Delano Roosevelt defeated attempts to weaken the law, thus essentially ending the domestic market for plumes.

V

A zoology book that Roosevelt read one evening in the White House was responsible for his keen interest in the gopher tortoises of Florida, Georgia, and Alabama. In 1907, after poring over the herpetologist Raymond Ditmars's *The Reptile Book*—a beautifully written encyclopedia complete with dozens of vivid and affecting photographs—T.R. grew excited about sea turtles and tortoises of all kinds.[62] "Impatient with the use of scientific language obscure in meaning to all but specialists, [Ditmars] wanted to tell about snakes in an understandable style that would awake the sympathy of the general public," the biographer L. N. Wood remarks in *Raymond L. Ditmars: His Exciting Career with Reptiles, Animals, and Insects*. "It was part of his lifelong campaign to break down the widespread prejudice against reptiles."[63]

Mixing science with popular writing, *The Reptile Book* offered pages of anatomical detail about these slow-moving, solitary tortoises who seemed to be ambassadors from a prehistoric time. To President Roosevelt, this was thrilling Darwinian material, a biological extravaganza between handsome covers. What Pinchot was to trees, Grinnell to biggame, Hornaday to bison, and Chapman, Job, or Finley to birds, Ditmars was to reptiles. Legend has it that Ditmars's office was crammed with more jars full of pickled snakes than anywhere else in the world. When the president fastened on a naturalist or a member of the animal kingdom, he didn't just skim; he read and devoured every biological vagary and nuance. Taking a break from world affairs to ponder reptiles, Roosevelt relished learning about the differences between a tortoise carapace and plastron; about clutches; about terrestrial predilections; and about how "rolling beetles" or tumblebugs lived on their excreta. Roosevelt learned that gopher tortoises had an insatiable appetite for plants, feeding on 400 to 500 different kinds on Florida's islands like Sanibel Island, Captiva Island, and Pine Island. And this figure didn't include mosses or fungi. Clearly, the tortoises needed a lot of habitat to survive.

Characteristically, Roosevelt wrote an exuberant note to Ditmars—a former reporter for the *New York Times* who served as curator of reptiles at the New York Zoological Park from 1899 to 1920 under the direct supervision of William Temple Hornaday. As with Hornaday, Grinnell, Burroughs, Muir, and others, Roosevelt marveled at how Ditmars—a meticulous man of science—"made the present and past life history of this planet accessible in vivid and striking forms to our people generally."[64] Ditmars was clearly part of Roosevelt's tribe. Calling *The Reptile Book* "genuinely refreshing," even though at times a slog to read straight through, Roosevelt invited Ditmars to visit the White House or Sagamore Hill, saying it would be "a great pleasure if I could see you some time."[65] The president wanted to discuss the fate of reptiles in North America with Ditmars, whom he considered the greatest herpetologist alive. "I have a very strong belief," Roosevelt wrote, "in having books which shall be understood by the multitude, and which shall yet be true—in other words, scientific books written for laymen who have some appreciation for science—so that the books will be of value to all men who are interested in the subject. It seems to me that your volume exactly fulfills these requirements. Personally, I have long wanted to have in my library some good books on reptiles."[66]

After his presidency, Roosevelt spent a couple of fine afternoons studying gopher tortoises living on islets around Punta Gorda, Florida. "The burrow was shallow and we speedily dug out the occupant," he reported for the *American Museum Journal.* "It was a fairly large specimen, weighing 11½ pounds, with a shell 13½ inches long, 9 inches wide, and 5¼ inches deep. (Later we secured a small specimen on Captiva Island, which weighed 4¾ pounds, was 8½ inches long, 6 inches wide, and 3½ inches deep.) How this big tortoise got to the island is something of a mystery, as the species is entirely terrestrial; it must have been drifted out by some accident of flood or storm."[67]

A reptile Roosevelt did nothing to protect after reading Ditmars, however, was the alligator (*Alligator mississippiensis*). Alligators populated the swamps of Florida, Georgia, Alabama, Mississippi, and Louisiana,[68] and Roosevelt encountered alligators when he journeyed to the South in 1907, but he never took one for a trophy at Sagamore Hill. An article by John Mortimer Murphy in *Outing Magazine* titled "Alligator Shooting in Florida" didn't endear the species to Roosevelt.[69] Murphy compared baby alligators eating prey to a terrier shaking a rat, and Roosevelt's southern notebooks recounted incidents of humans being bitten by alligators along sandbanks and in mudflats. "In the lakes and larger bayous we saw alligators," Roosevelt wrote in *Outdoor Pastimes of an American Hunter.* "One of the planters with us had lost part of his hand by the bite of an alligator."[70] When he traveled to Panama aboard the USS *Louisiana* in November 1906—becoming the first U.S. president to visit a foreign country during his term of office—he wrote to Kermit about what he *thought* was a cunning gator with a carnivorous mouth. "There are alligators in the rivers," he reported. "One of the trained nurses from a hospital went to bathe in a pool last August and an alligator grabbed him by the legs and was making off with him, but was fortunately scared away, leaving the man badly injured."[71]

Here is a very rare example of Roosevelt misidentifying a species. Alligators weren't found in Panama. Instead, the nurse was probably attacked by the common caiman (*Caiman crocodilus*). There is no good explanation why Roosevelt was so sloppy with this field observation. Later, when he was traveling in both Africa and Brazil, Roosevelt's disdain for crocodilians of any kind and his utter revulsion regarding their biological traits became obvious. Traveling down dark rivers he would contemptuously shoot at these creatures as if they were rats on a garbage heap. What disgusted Roosevelt most about crocodilians was their insatiable appetite and bizarre digestive system. Abandoning any pretense of objective Darwinian analysis, Roosevelt deemed them evil monsters that destroyed ecosystems (crocodilians, especially alligators, actually protected egrets by feeding on their predators, such as raccoons and snakes). Knifing open a shot Nile crocodile in Africa, for example, Roosevelt was nauseated by the varied contents of its stomach. "The ugly, formidable brute had in its belly sticks, stones, the claws of a cheetah, the hoofs of an impala, and the big bones of an eland, together with the shell plates of one of the large river-turtles."[72]

That Roosevelt had little use for swamps in general becomes clear from his response to a ridiculous scheme for draining the Everglades as part of a program by the U.S. Reclamation Service. Despite all his good work for "citizen bird" in the southern latitudes, Roosevelt almost made a serious blunder in the Everglades: he directed the Reclamation Service to investigate the possibility of draining them. "Turn-of-the-century conservationists stopped the annihilation of the birds of the Everglades," the reporter Michael Grunwald of *Time* wrote in *The Swamp: The Everglades, Florida, and the Politics of Paradise*. "But they had no problem whatsoever with the drainage of the Everglades."[73] Just as Roosevelt allowed Hetch Hetchy in Yosemite to be flooded for the sake of San Francisco's water supply, he wanted the Everglades drained so Miami could grow southward into affordable housing villages—a forerunner of suburbia.

The initiator of the Everglades scheme was Dr. John Gifford of Coconut Grove, the first American to earn a doctorate in forestry. As editor of the magazine *Conservation*, Gifford vigorously promoted the wise use of natural resources. Gifford insisted that conservationism meant "reclamation of swamplands and the irrigation of deserts." In 1901 he wrote the important *Practical Forestry*, which made it clear that swamps tangled with palmetto didn't impress Gifford like birch, oak, and elm. Yet, ironically, Florida—as Gifford's narratives *The Everglades of Florida* (1911) and *Billy Bowlegs and the Seminole War* (1925) attest—was his lifelong passion as a conservationist.[74] As the chief promoter of saving Puerto Rico's Luquillo National Forest, Gifford was all for preserving vast patches of jungles for eco-tourists to use. He was also an advocate for "citizen bird." Eventually, he became a professor of tropical forestry at the University of Miami, where he promoted various uses for the cypress, maple, and pine. Gifford's articles regularly appeared in the magazine *Tropics*. Unlike Henry Flagler, who was involved in railroads, hotels, and real estate in

Florida, Gifford considered that *some* of wild Florida had to remain. But the Everglades? A 200-square-mile alligator swamp? To Gifford, the Everglades were a putrid wasteland. So he concocted plan after plan to drain the great swamp. After all, Washington, D.C., had been a straggling village until its swampland was drained; now it had theaters and museums and was the finest capital in the western hemisphere.[75] Gifford's most improbable scheme entailed importing sacks full of cajeput (melaleuca) seeds from Australia. He hoped that these water-absorbing trees would thrive and dry up the Everglades.[76]

Roosevelt himself never became engaged in these drainage schemes, but his Forest Service under Gifford Pinchot did. Pinchot considered draining the Everglades by planting Australian melaleuca trees a noble idea. When John Gifford quit *Conservation*, his replacement, Thomas Will, was even more of an advocate for drainage. As a former president of Kansas State Agricultural College, Will envisioned citrus groves and housing developments instead of little blue herons and alligators. With the apparent approval of Roosevelt's Forest Service, Will offered his visions to the Florida Everglades Homebuilders Association, the Everglades Farming Association, and the South Florida Development League. Aiding Will's boosterism was a legendary promoter of development in Florida: Napoleon Broward. A huge fan of Roosevelt's, Broward was opposed to railroads, corporations, and Flagler's populism. He claimed to be a Rooseveltian conservationist, but he favored only reclamation. And although he was a dedicated outdoorsman, he nevertheless led the campaign to drain the Everglades with considerable audacity. His conservationist rationale was that the wildlife could live *around* the newly drained Everglades communities. Broward thought of himself as standing up for the "little folks." He believed that Flagler and other rich railroad men were "draining the people" from Florida, "instead of the swamp." Roosevelt's fingerprints aren't found on any of the documents in this episode, but he clearly sided with Broward in the hope that south Florida could be made "fit for cultivation."

A generation of Florida environmentalists never forgave Roosevelt for embracing Gifford, Will, and Broward's drainage scheme. They pointed out that Florida had plenty of land available for settlement without destroying the Everglades ecosystem. In *The Swamp*, Grunwald inventoried anti-drainage comments that became widespread in Florida: "a wildcat scheme," "a sinful waste," "nonsensical," and so on. Roosevelt ignored such complaints, unconcerned that a big project (on the scale of the Panama Canal) in the Everglades might bankrupt the state, and that its effect on nature would be devastating. Because Broward was an ambitious reformer—opposed to child labor, opposed to the millinery industry, in favor of road expansion, and a committed educational activist—the Roosevelt administration approved of Florida's building four canals in the Everglades. In 1908, the same year Roosevelt saved Mosquito Inlet, Tortugas Keys, Key West, Pine Island, Matlacha Pass, Palma Sola, and Island Bay Federal Bird Reservations in Florida, he named Napoleon Broward president of the National Drainage Congress.

Perhaps President Roosevelt supported the idea of draining the Everglades simply for reasons of political expediency. After all, Broward was a Rooseveltian reformer. The president had enough problems in the South without squaring off against Broward. And keep this in mind: even though Roosevelt approved twenty-four federal irrigation-drainage projects as president, not one was in Florida. After the Newlands Act of 1902, all his projects were in the West.[77]

Roosevelt, however, never approved a major project for draining the Everglades, and the plan never got off the ground. What's curious about his implicit support of the idea, however, is that he simultaneously embraced the opposite logic with regard to the Panhandle and central Florida. There, Roosevelt created large national forests instead of approving drainage projects. Federal bird reservations were usually not more than five to 100 acres in area (there were exceptions, though) because they were confined to isolated islands and swamps. But the national forests sprawled over whole counties. On November 24, 1908, Roosevelt created the Ocala National Forest in central Florida. Pelican Island was only 55 acres; Ocala covered 607 square miles. It ran nearly from the Atlantic to the Gulf. According to Frank M. Chapman, who knew the area inside-out, the name "Ocala" came from the Timucuan Indians and meant "big hammock." This region was where Chapman had cut his teeth as an ornithologist in the 1870s. The Ocala ecosystem was like a recharge battery for the entire Floridan aquifer. The new national forest contained more than 600 natural lakes and ponds. The management of Ocala National Forest was similar to that of the forests of the West,[78] but Roosevelt wanted the waterways to remain pristine, with no urban pollution.

Today Florida treasures the Ocala National Forest. Situated between Marineland on the Atlantic and Silver Springs in the interior, it attracts tourists because of its renowned mirrorlike waters and their cypress trees, bending ferns, and water lilies. It's the premier home of the Florida black bear (*Ursus americanus floridanus*). For Americans seeking "wild Florida," camping at Ocala in places such as Doe Lake and Big Bass Lake has become a must. Ocala is about as close to the wild Florida of the days of Ponce de Leon as one can comfortably get without heading into the deepest recesses of the Everglades.

Three days after the Ocala National Forest was created, Roosevelt created another national forest, in the Panhandle. Established on November 27, 1908, the Choctawhatchee National Forest was a flatland full of longleaf and pine trees. It looked almost like a tree nursery or an arboretum. Turpentine workers and small-time fisherfolk lived in the Choctawhatchee—the Roosevelt administration didn't mind them. Cattlemen were another story. They were burning down the pinelands to create grazing areas—a practice T.R. deemed reprehensible. American and European tourists were heading to Miami and Palm Beach in droves, and many people thought the Panhandle worthless. But Florida was a big state with over 1,200 miles of tidal shoreline.[79] Just going from Key West to Pensacola—near where the Choctawhatchee National Forest was formed—was a 1,000-mile journey. If Ditmars had spent a day there, he'd have given up counting reptiles; the number of lizards and tree frogs was in the millions. While south Florida was becoming a center for land-promotion schemes and nightclubs, the Panhandle, still fairly rural in 1908, was considered Florida's best-kept secret. An outdoorsman could collect specimens in the Panhandle waterways and pinelands without many distractions.

Because the South was anti–federal government, the mere fact that the Roosevelt administration had created the Choctawhatchee National Forest there was notable. The forest reserve had about 467,000 acres and was situated on the extreme western arm of Florida and Choctawhatchee Bay, Santa Rosa Sound, and East Bay. It went from the Gulf of Mexico to about twenty miles into the Panhandle's interior. The longleaf and pines and the dense undergrowth of blackjack and turkey oak made the forest tract commercially valuable. The Choctawhatchee National Forest was President Roosevelt's last great conservation initiative in Florida. It wasn't far from where John Quincy Adams had preserved his tree farm for the U.S. Navy in 1828.

VI

Just eleven days after Pine Island, Roosevelt issued Executive Order 943 (September 26, 1908), creating the Matlacha Pass Federal Bird Reservation. Originally, he protected three teardrop-shaped Matlacha Pass islands overflowing with mangrove vegetation—red, black, and white.

Later, when he visited the area, he commented on the Florida figs and pawpaws.[80] There was also a lot of buttonwood. Sea grape grew wild on the Matlacha Pass islands, as did strangler fig and gumbo-limbo. So many unusual plants grew at Matlacha Pass, in fact, that a pharmaceutical company dispatched botanists to Florida looking for possible cures. On any given day a visitor to Matlacha Pass circa 1908 could see birds nesting, eastern indigo snakes hunting for mice, and American crocodiles sunning in mudflats. There were numerous hummingbirds, whose metabolism and oxygen consumption always fascinated Roosevelt. (Hummers have been clocked at 200 wing-beats a second.[81]) But it was the West Indian manatee of Matlacha Pass that impelled Roosevelt toward bold preservation measures.[82] On the same day Matlacha Pass was saved, he created yet another federal bird reservation for seabirds near Sarasota. This was Palma Sola, an island on an isolated lake where plants sprang up and grew at an astonishing rate.

Not far from T.R.'s federal bird reservations at Pine Island, Matlacha Pass, and Palma Sola was Island Bay, an intricate complex of mangrove keys including some that were unnamed. When T.R. signed Executive Order 958 on October 23, 1908, creating Island Bay (including the refuge islands Gallagher Key and Bull Key) as a "preserve and breeding ground for native birds," he considered it his last one in Florida, the pièce de résistance there.[83] Surrounded by brackish waters, the bottom around Island Bay offered fine examples of such rich Florida marine vegetation as widgeon grass and shoal grass. Besides the more commonplace shorebirds, gulls, and terns, the rare little blue herons often congregated on Island Bay. For centuries the Calusa Indians lived on these islets, and for good reason: an abundance of shellfish was available. West Indian manatees also congregated in the warm waters around Island Bay where vegetation was plentiful.[84] Fittingly, Roosevelt's last bird reservation in Florida had the additional value of protecting manatees from human encroachment.

During this period Roosevelt, hoping to guide public opinion, supported William Dutcher's request to the AOU and—following the lead of Pennsylvania and Delaware—appointed state ornithologists in Florida. A concerted effort was also under way for state Audubon societies to affiliate with the SPCA and other humane organizations. Some states, as a result of Dutcher's lobbying, outlawed the shooting of birds on Sundays. The AOU chastised Arizona, Hawaii, Oregon, Michigan, and other states with a high concentration of birds for not adopting this prohibition, which gave "absolute rest to bird life for the one day per week."[85]

With Roosevelt's support, Dutcher also wrote to the U.S. Navy, asking it to protect the rookeries of the Philippines, and of the Midway atolls, which were owned by the United States and were a station of the Pacific Cable Company. Roosevelt had already sent U.S. Marines to Midway to protect the albatross, and he was ready to do the same for terns in the Dry Tortugas. "I am informed that the Japanese people have been in the habit of visiting these islands for the purpose of killing birds for their plumage," Dutcher wrote to T.R.'s secretary of the navy, William Moody, about Midway. "It is known that during the past few years enormous numbers of seabirds have been killed by the Japanese and have been shipped to the Paris, London, and New York markets for millinery ornaments; among these birds were great numbers of a very beautiful form of the tern family known as *Gygis alba*. Our Society is under many obligations to your Department for your hearty cooperation in our work for the preservation of sea-birds, the latest and one of the most notable instances being your order of April 24 [1903] *in re* the birds on the Dry Tortugas, Florida."[86]

Once the Dry Tortugas became a federal bird reservation in 1908, Roosevelt personally asked the secretary of the navy to make sure that the Tortugas group, including every key and shoal, would never be disturbed. No traps, torpedoes, maneuvers, or mock invasions would be allowed to turn this paradise into an ash heap. Roosevelt wanted the Tortugas group astir with birds flying along the ocean's edge. A special warden, W. R. Burton, was assigned to Bird Key by AOU. Burton's job was to report to Dutcher anybody encroaching on the Tortugas bird sanctuaries. Dutcher in turn would report the matter to Roosevelt, who would inform the secretary of the navy. If any U.S. sailors dared touch a sooty tern's egg or nest, they would be severely punished. Herbert Job went to the naval station at Key West and spoke personally with the coolheaded Commander George Bicknell. Bicknell understood what the president wanted and expressed to AOU the Navy Department's deep regret that some shortsighted Florida residents seemed "determined to make of their beautiful state a lifelong, treeless desert as fast as they possibly can."[87]

Roosevelt scoffed at the notion, expressed by people in Florida's chambers of commerce, that the White House's approval of AOU-Audubon wardens in Florida smacked of socialism. Collective action on behalf of "citizen bird" was a good thing, he said. "Every civilized government which contains the least possibility of progress, or in which life would be supportable, is administered on a system of mixed individualism and collectivism and whether we increase or decrease the power of the state, and limit or enlarge the scope of individual activity, is a matter not for theory at all, but for decision upon grounds of mere practical expediency," Roosevelt argued. "A paid police department or paid fire department is in itself a manifestation of state socialism. The fact that such departments are absolutely necessary is sufficient to show that we need not be frightened from further experiments by any fear of the danger of collectivism in the abstract."[88]

VII

Creating seven federal bird reservations in Florida from February to October 1908 brought Roosevelt unexpected accolades from an up-and-coming political cartoonist: Jay Norwood ("Ding") Darling. A short résumé of Darling's life will help us to better understand how Roosevelt influenced a new generation of bird protectionists.

Although Darling was born in Michigan, he grew up in Sioux City, Iowa. As a teenager he often explored the flatlands of Nebraska and South Dakota like a cowlicked Tom Sawyer. He was employed as a cattle herder but had ambitions to earn a college degree. When he went to Yankton College in South Dakota, however, his smart-aleck side got the best of him, and he was kicked out in 1894 after taking the school president Henry Kimball Warren's horse and buggy on an unauthorized ride. Rebounding, however, was part of Darling's nature. He developed a fascination for Darwinian biology—an outgrowth of his infatuation with the ecosystems of the Missouri River and the Big Sioux River—and the prowling habits of cougars intrigued him (he and Roosevelt were kindred spirits in this regard). He learned the Latin taxonomy of the Great Plains creatures, and he enrolled in Beloit College (Wisconsin), where he became the art editor of the yearbook. But he remained mischievous and an incurable class clown, and it didn't take him long to get suspended for ridiculing faculty members in a series of cartoon strips.[89]

Darling eventually graduated and was hired as a political cartoonist at the *Sioux City Journal*. He deeply admired Roosevelt's Dakota trilogy on hunting, and he rallied to the president's side in 1900 owing to their shared affinity for preservationism-conservationism. To some people, Darling's allegiance to Roosevelt was nearly insufferable. His inaugural political cartoon in the *Journal*, in fact, showed McKinley and T.R. on an elephant, lording it over a broken-down donkey carrying an imbecilic-looking William Jennings Bryan.[90] But Darling's satire was popular, and he was soon snagged by the *Des Moines Register and Leader* and given creative license. With doglike devotion to all things Rooseveltian, Darling used his carte blanche to promote all aspects of the conservation movement in his cartoon strips.[91]

When Darling spoofed the developers of his day, he always knew that Roosevelt was cheering him on. In particular, Darling took a deep, personal interest in Roosevelt's attempts to stop the wanton destruction of bird habitats around America, appreciating that the president seldom hesitated to exercise executive power on behalf of wildlife. To Darling the federal bird reservations were a masterstroke against ignorance and greed. Few Americans even noticed their establishment, but Pine Island, Dry Tortugas, Stump Lake, and Three Arch Rocks became magical places to Darling. A great friendship developed between T.R. and Darling, with conservation as the link.[92] Darling relished the fact that it took Congress many decades to fully understand the permanency of Roosevelt's "I So Declare It" executive orders: and by then it was too late to reverse. Darling also claimed the avian photographs of William L. Finley—particularly his iconic Californian-shot images of golden eagles, condors, and great blue heron—as galvanizing influences on his wildlife protection crusade.[93]

Darling believed that the federal bird reservations, even more than the national parks or national forests, were the enlightened, sensible way to save aviaries. Claiming that President Roosevelt was his mentor, Darling, a Republican, became an important warrior in Florida's land issues. By 1917 Darling's pro-conservation cartoons were syndicated in some 150 dailies by the *New York Herald Tribune*. When T.R. died two years later, Darling was shattered—losing Roosevelt was like losing a father. But he didn't mope for too long. He quickly moved to fill the void left by the great man—especially in Florida. Preservationist leadership always meant a lifetime of knife fights with rich companies and their soulless lawyers.

Wisely, President Franklin D. Roosevelt recruited Darling in 1934 to serve on the President's Committee for Wildlife along with Aldo Leopold and Thomas Beck. A year later he hired Darling to lead the Bureau of Biological Survey into a productive period after Merriam's era. In this capacity, Darling struck on an interesting conservation awareness scheme. He designed (and Franklin Roosevelt approved) a "duck stamp" that generated extra income for the U.S. wildlife refuges.[94] Growing ever more fervent about protecting T.R.'s achievements in wild Florida, Darling founded the National Wildlife Federation, largely keeping the "wildlife protection legacy" of T.R. alive for decades to come. Just as Theodore

Roosevelt saved bison and elks on preserves, Darling led efforts to rescue Nevada's dwindling antelope herds. And during World War II, while other cartoonists focused on the war, Darling worked around the clock to save various islands near Fort Myers from overzealous developers.[95]

Owing to Ding Darling's intense lobbying, in December 1945 President Harry Truman approved a lease with the State of Florida creating the Sanibel National Wildlife Refuge, adjacent to the sites T.R. had saved in 1908. Besides birds, Darling had sought protection from the Truman administration for all estuarine habitats including sea grass beds and mudflats.[96] Keeping field notes and making sketches, as Theodore Roosevelt would have wanted, Darling recorded dark shadowy terns whose colors were indiscernible, black skimmers clouding the sky like dimly seen bats, and reddish egrets wading in flats looking for small fish. There were no flamingos to record, however, for these magnificent Phoenicopteridans had been killed off in Florida by plumers, who later bragged about it at night in taprooms. Darling's cartoons also highlighted his efforts to create a refuge for the "toy" deer of the Keys from marauding hunters. His efforts helped to create the National Key Deer Refuge in 1957. His life was testimony against the destructive lunacy of Floridians.

When Darling died in 1962, at eighty-six, his foundation proposed renaming Sanibel National Wildlife Refuge after him. In 1967, with the approval of President Lyndon Johnson, the J. N. "Ding" Darling National Wildlife Refuge was officially created as part of a larger, more easily administered complex that encompassed three of Roosevelt's refuges of 1908—Pine Island, Matlacha Pass, and Island Bay. Even though this new refuge wasn't declared until more than sixty years after T.R.'s death, its existence can be seen as his crowning achievement for wildlife protection in Florida. Nowadays it is populated by great egrets, snowy egrets, wood storks, roseate spoonbills, great and little blue herons, white and brown pelicans, tri-color herons, yellow-crowned night herons, short- and long-billed dowitchers, lesser and greater yellowlegs, anhingas, cormorants, blue-winged teal, ospreys, and bald eagles. Today, also, Ding Darling National Wildlife Refuge is one of the top ten birding spots in America.[97] "Ding idolized Roosevelt," Darling's grandson, Christopher Koss, recalled. "They both shared an interest in ecology and refused to waver when it came to protecting birds. Roosevelt inspired not just Darling but an entire generation to fight for conservation. With Roosevelt as leader there became a meeting of the young minds."[98]

Theodore Roosevelt's "wild Florida" strategy of 1908 might have failed if it hadn't been for the support of people like McLeod, Bradley,

Dutcher, Chapman, and Darling. Recognizing that for victory in Florida, pro-wildlife troops were necessary, Roosevelt had recruited them. When he approved the appointment of wardens in the tradition of Kroegel and Bradley, he felt the same unswerving conviction he once exhibited as a Rough Rider in Cuba. Prototypes of future on-site employees of the U.S. Fish and Wildlife Service, these biologically informed wardens enlisted eagerly in Roosevelt's cause. Roosevelt encouraged his disciples on how to win through a combination of public education, grassroots work, alliance-building, and scathing ridicule. The key factor was bringing poachers and plumers to account. Some of these Rooseveltians later became Bull Moosers, paying T.R. homage well into the 1940s, when World War II caused the conservationist movement to temporarily taper off.

What's most impressive about Roosevelt's bird reservations is how coordinated the system became. Signing executive orders on behalf of birds became a habit for Roosevelt during his last eighteen months in office. The Biological Survey's sanctuaries were like latticework, linked by regional offices. What Roosevelt asked Floridians to do between 1901 and 1909 was think about the future. The industrial growth of Jacksonville, Tampa, and Miami was a good thing. The Reclamation Service might consider draining the Everglades and building canals throughout south Florida. People had to live and improve. Yet, Floridians also needed to develop "the right kind of a civilization."[99]

Typically, Roosevelt envisioned Florida's big cities surrounded by big greenbelts. He knew that Florida was a fragile, hurricane-lashed ecosystem and that it needed perennial care. Florida couldn't be stripped of its greenness. Manatees, roseate spoonbills, greens and leatherbacks, marlins, sooties, and mangrove forests—all were a heritage to be passed down to future generations. Did Floridians not want their children to see colonies of interesting waterbirds? And coral beds? Outsiders would always try to swoop into Florida and extract natural resources for profit, leaving behind environmental degradation. Shouldn't real Floridians protect their state's biological bounty of tropical forests, pristine beaches, and coral reefs from corporate molestation?

VIII

The Kentucky poet Wendell Berry once wrote, "When despair for the world grows in me and I wake in the night at the least sound, in fear of what my life and my children's lives may be, I go and lie down where the wood drake rests in his beauty on the water, and the great heron feeds."[100] Berry might almost have been communicating with William L.

Nobody did more to save the birds of Oregon and Washington than William L. Finley, pictured here patting a Common Murre. Finley, a brilliant wildlife photographer, led the West Coast Audubon Society movement.

Finley's spirit. For Finley, as Oregon's pioneering wildlife photographer, waded in scum ponds and slept soaking wet in ocean-rock crevices in order to document the habits of migratory birds. His stolid ornithological concentration had already become legendary in rural Oregon. The establishment of Three Arch Rocks on October 14, 1907, inspired Finley to preserve yet another waterfowl concentration site in Oregon: the Klamath basin, a series of lakes and marshes that were a stopover for approximately three-quarters of the Pacific Flyway waterfowl.[101] These extensive wetlands attracted more than 6 million waterfowl, including the American white pelican, the double-crested cormorant, and numerous heron species (including the kind that brought Wendell Berry such comfort).[102] Roosevelt didn't write an introduction to Finley's first book, *American Birds*, as he had done for Herbert Job's *Wild Wings*, but he nevertheless marveled at Finley's ornithological accuracy.[103] The two simultaneous events—Three Arch Rocks becoming a federal reserve and *American Birds* being published—were not accidental.

In 1905 Finley and Bohlman—fresh from studying golden eagles in California—started spending a lot of time in the lower Klamath basin, on instructions from William Dutcher, now head of the National Association of Audubon Societies. Nowhere in North America were there so many jungles of floating tules (pronounced *too-lees*) as in the Klamath basin. Tules are a huge species of sedge, part of the family Cyperaceae. Little clusters of sprouting tules—a plump, rounded green stem with grass-like leaves often clustered around beige flowers—were once so common around Tulelane Lake that in the California-Oregon wetlands there was a popular expression, "out in the tules," meaning "beyond way-away." Along the shorelines of the California-Oregon lakes the tule marshes served as a buffer against water surges and high winds. Even the haze that beset the state borderline area, particularly along the Pacific coast, was known as tule fog. Walking on a tule was akin to hiking in snowdrifts—you never knew when you'd plunge downward. Treading carefully in the tules, Finley and Bohlman once again delivered photographic gems. Their straight-on black-and-white portrait of a canvasback in tule was as artful as an Audubon print. Some of the photographs, such as four Caspian terns screaming at each other, were comical. Others, like some tiny spotted sandpipers dancing around a spring flower, were aimed at winning children over to the Audubon Society's cause.[104]

Worried that President Roosevelt—his chief ally—was consumed by the ceaseless distractions of Washington, D.C., Finley focused his activism on counterbalancing the San Francisco plume hunters who were killing off birdlife in the Klamath basin. Keeping detailed notes about the plumers he encountered, Finley recorded that the going rates per dozen birds were: teal ($3), mallard ($5), pintail ($7), and canvasback ($9). Bales of skins were being shipped out, like hay. Camping on marsh tules, which made a fine mattress, Finley started writing up his field notes as articles for *The Condor* magazine in 1907; these articles included "Among the Pelicans," "The Grebes of Southern Oregon," and "Among the Gulls on Klamath Lake."[105] As Roosevelt had intuited, Finley was a naturalist writer almost as good as Muir and Burroughs. And what made Finley sui generis among wildlife photographers of the early twentieth century was that he also took motion pictures of the lower Klamath. He had footage of great migratory waterfowl swarms and close-ups of pelicans nesting. (Roosevelt liked that, too: after all, he had filmed Jack "Catch 'Em Alive" Abernathy wolf-coursing in Oklahoma.) Today, Finley's grainy black-and-white films are prized possessions in the archives of the U.S. Fish and Wildlife Service in Shepherdstown, West Virginia.

Much like Paul Kroegel at Pelican Island, Finley at the lower Klamath and Lake Tule was a master of walking softly among the birds. The western grebe rookeries he studied were like none other in the world. He would sit quietly for hours in the low reeds, camouflaged, spying as grebes dived and swam underwater with chicks on their backs. Even when the wind ruffled the surface of the tule and sheets of rain blew eastward horizontally, Finley didn't abandon his blind. Once he got lucky and photographed western grebes hatching from eggs amid dry tules. "We watched as one of the little Western Grebes cut his way out of the shell and liberated himself," Finley wrote in *The Condor*. "The wall of his prison is quite thick for a chick to penetrate, but after he gets his bill through in one place, he goes at the task like clock work and it only takes him about half an hour after he has smelled the fresh air to liberate himself. After the first hole, he turns himself a little and begins hammering in a new place and he keeps this up till he has made a complete revolution in his shell, and the end or cap of the egg, cut clear around, drops off, and the youngster soon kicks himself into the sunshine."[106]

On behalf of the Oregon Audubon Society, Finley, armed with field notes, photography, and a home movie, journeyed back to Washington, D.C., in 1906 to personally show off the Oregon lake region to Roosevelt at the White House. Roosevelt congratulated Finley on Oregon's model law for birds. Finley inquired about Three Arch Rocks. The conversation went back and forth like that for an entire evening. The two naturalists discussed how to preserve lower Klamath Lake as a breeding ground for native birds. The Klamath basin was a large area of land which dwarfed Crater Lake. In Florida, Roosevelt was creating federal bird reservations of forty to 200 acres. If Klamath was declared a reservation, it would be something like 80,000 acres. Further complicating the situation was the fact that the Reclamation Service was draining the wetlands for large-scale agricultural farming. Finley told Roosevelt that his Klamath irrigation project—while obviously a well-intended offshoot of the Newlands Act—was turning wetlands into mudflats. "We move to conserve or develop one resource," Finley complained, "while at the same time, we are destroying another."[107]

Roosevelt thought Finley had the right stuff. He was a well-spoken, accomplished young man, rather overly serious, but otherwise with no discernible flaw. He had a wonderful scientifically inclined mind, which appreciated that even maggots had an important role in nature, feeding grackles and nighthawks. Also, Finley could make even a magpie nest sound as interesting as the Taj Mahal. Like Chapman, he was doing a fine job of popularizing birding. *Sunset* magazine—aimed at middle-class families—had published a few of his well-written ornithological pieces. Perhaps knowing that Roosevelt had a soft spot for the American white pelican (Roosevelt had, in fact, saved both Stump Lake and Chase Lake in North Dakota largely for their benefit), and much like Pinchot and La Farge in 1900 regaling Roosevelt with stories about Mount Marcy, Finley told of rowing in puffs of wind, making treacherous landings, putting up rough campsites, setting up a blind near Rattlesnake Island, and seeing half-grown pelicans and hearing their cries. Combined with the home movies, all this was quite a pitch. What image could have appealed to Roosevelt more than Finley's colorful remark that these pelicans looked "like a squadron of white war-ships"?[108]

Bravo! Bully! Wow! Roosevelt loved the whole presentation. Perhaps he had discovered a new American original like the young Audubon. Clearly, Finley wasn't a fringe ornithologist but a main voice. Once again T.R.'s famous exuberance was called forth by the incredible photographs of young burrowing owls perched on Bohlman's lap and Finley hand-feeding double-crested cormorants at Tule Lake. There was nothing bland about Finley's photos. Having already saved, for John Muir, the 14,162-foot Mount Shasta, which was prominent on the horizon in much of the Klamath basin area, Roosevelt wanted to solve the political problems associated with creating a huge federal bird reservation in the middle of his Reclamation Service project. This was seen by Auduboners as Pinchot's public revenge. Finley, however, took the compromise with a certain grace, as though, having toiled so long to bring attention to the Klamath basin, he felt a distinct relief in having gotten Roosevelt to establish something big for birds.

As if in a great wave of protectionism, Roosevelt rescued more than 37 million waterbirds in Oregon. On August 8, 1908, by means of an "I So Declare It" executive order he created the Klamath Lake Reservation, consisting of 81,619 acres of lakes and marshes. (This would be only the first of six national wildlife refuges set aside in the Klamath basin.*) He wanted the habitat preserved—particularly the ten- to fifteen-foot-high tule—for the pelicans and grebes. But a serious error had also been made. The entire Klamath basin should have been declared a national park, like Crater Lake or Mesa Verde, and the Reclamation Service should have been booted out of southern Oregon once and for all. However, this didn't happen. Spurred on by the Klamath Waters Association, the raping, dredging, and draining of the wetlands ecosystem continued. At best, the reclamation project was a product of its time. Although Auduboners were (and still are) grateful that Roosevelt had helped rescue the white peli-

*Like many of Roosevelt's federal bird reservations the one at Klamath Lake grew enormously. By 2009, under the designation Malheur National Wildlife Reservation, it was 186,500 acres. In the spring the Malheur Refuge offers stunning concentrations of birds. By April, for example, there are often as many as 300,000 snow and Ross's geese. Come May the Malheur is overrun with neotropical songbirds.

cans, cormorants, grebes, and great blue herons, his conservation policy mainly failed in the Klamath basin. It was the worst example of Roosevelt's trying to reconcile agricultural utilitarianism and waterfowl preservation. But since then, rehabilitation efforts have succeeded.[109] "I hope the marsh," Finley wrote, "will defy civilization to the end."[110]

Also, Finley succeeded in getting 81,619 acres of the Klamath basin preserved inside the Bureau of Reclamation's reservoir. Eventually Roosevelt created many of these "overlay" refuges in the West—seventeen were established on February 25, 1909, alone, by Executive Order 1032. Some environmental historians think that these "overlay" bird reservations within reservoirs were largely a waste of time. Finley, they claim, was hoodwinked by Pinchot and those around Pinchot.

When in 1911 Governor Oswald West of Oregon, a Democrat, selected Finley to become the state's first Fish and Game Commissioner, Roosevelt celebrated. It was a victory for "citizen bird." But besides women's suffrage and prison reform, wildlife protection was one of the issues dearest to West's heart. With unflinching determination, he insisted on public access to Oregon's beaches.[111] "In Governor West of Oregon, I found a man more intelligently alive to the beauty of nature and of harmless wild life, more eagerly desirous to avoid the wanton and brutal defacement and destruction of wild nature, and more keenly appreciative of how much this natural beauty should mean to civilized mankind, than almost any other man I have ever met holding high political position," Roosevelt wrote in *Outlook*. "He had put at the head of the commission created to express these feelings in action, a naturalist of note, Mr. Finley."[112]

Praise from a former president is always a good thing for a writer. But Finley felt especially blessed when Roosevelt commended West's environmentalist ethos in rhapsodic terms: "He desired to preserve for all time our natural resources, the woods, the water, the soil, which a selfish and shortsighted greed seeks to exploit in such fashion as to ruin them and thereby to leave our children and our children's children heirs only to an exhausted and impoverished inheritance; he desired also to preserve, for sheer love of their beauty and interest, the wild creatures of woodland and mountain, of marsh and lake and seacoast."[113]

In the years after Roosevelt and Finley's initial collaboration of 1903 to 1909, the U.S. Fish and Wildlife Service decided to pay tribute to these conservationists. In 1964, 5,325 acres of the Willamette Valley were saved as the William L. Finley National Wildlife Refuge. Here was one of the best spots to see Canada geese, mallards, northern pintails, and great blue herons. A botanist could study broad-leafed pondweed, water plantain,

American slough grass, and Engelmann's spike rush in the refuge. Anybody who wants to experience the Willamette Valley ecosystem before Portland's sprawl reaches it can today wander around the white oak savanna in all its primordial splendor. And in the refuge (which is near Corvallis, Oregon), among bottomland ash forest and native prairie, is a herd of Roosevelt elk—a truly fitting tribute to the two naturalists' work together.

And who had the idea for this refuge? Ding Darling. He was intent on having the Willamette Valley refuge, deep in the foothills of the coast range, named for Finley and populated with Roosevelt elk. The story of Finley and Roosevelt's collaboration now lives on in the Oregon landscape, thanks to Darling's foresight. As for Governor Oswald West, there is a spectacular Oregon state park—situated on the Pacific Ocean between Hug Point and Nehalem Bay—named to honor his crusade for public access to beaches and for bird sanctuaries.

The Malheur National Refuge—spanning a forty-mile area in the southeastern corner of Oregon—celebrated its centennial on April 4, 2008, by brewing a micro beer manufactured by Rogue and called Great Egret Pale Ale.[114] U.S. Fish and Wildlife had good reason to celebrate. After decades of setbacks in the Klamath, rehabilitation was succeeding. Now tens of thousands of visitors come to the northern Great Basin to watch migratory birds feed in the high desert.

Three Arch Rocks became—even more than Crater Lake—a symbol of Oregon's pristine beauty. The largest of the surf-pounded mounds was officially named Finley Rock and is home to the largest colony of tufted puffins in the state. Tourists come to view the puffins with binoculars from Oceanside Beach and Cape Meares on the mainland. But no trespassing is allowed at the sanctuary. And from May 1 to September 15 no boats are allowed within 500 feet of the mounds.[115]

Early on, some critics of Roosevelt's reservations at Klamath basin said that only a lunatic would have the audacity to declare Oregon's tule land a breeding grounds for birds. As their argument went, this wasn't Mount Rainier or Crater Lake but a swamp splotched with bird excrement! Other critics said that Roosevelt's penchant for bird-watching was warped, *occultism*. Scoffing at such thinking, Roosevelt said that preserving the Pacific coast's wildlife was democratic in spirit. There was nothing warped about protecting canvasback and white pelicans. In fact, he wanted to protect endangered birds all over American territory, from the eskimo curlews in Alaska to the whooping cranes in Michigan to parrots in Puerto Rico. That some fellow Americans couldn't understand the inherent morality of species survival was troubling to Roosevelt. But so what? Before leaving the White House he planned to create even more than the twenty-five bird sanctuaries of 1903 to 1908. It was as if he had found in wildlife protection his autumnal passion. And he had Grover Cleveland's famous "midnight reserves" to use as a presidential precedent.

CHAPTER TWENTY-FIVE

THE PRESERVATIONIST REVOLUTION OF 1908

I

Old John Muir could barely believe the stunning news. On January 9, 1908, President Theodore Roosevelt issued an unexpected proclamation designating a wondrous 295-acre strand of coast redwoods (*Sequoia sempervirens*) as the Muir Woods National Monument in northern California.[1] It was a magnificent tribute to the self-described "poetico-trampo-geologist."[2] Situated across the Golden Gate Bridge near Mill Valley in Marin County, Muir Woods was an ideal forest to honor the "sage of the Sierras." The giant trees along Redwood Creek, each casting a hulking shadow on its way skyward, seemed a trenchant retort to the unchecked capitalism of the gilded age. Sunlight filtered down to create a living silence in the forest. This fine uncut coastal redwood strand had been donated to the federal government by William Kent, a wealthy disciple of Muir's, originally from Chicago, who had purchased it in 1905 for $45,000.[3] A businessman, philanthropist, and amateur naturalist (who would later be a congressman from California), Kent couldn't stomach lumbermen clear-cutting the shoulder of Mount Tamalpais for a "few dirty dollars" and criminally depriving "millions of their birthright."[4]

Muir Woods was really something special. Eternity was somehow present in the delicate greenery of the redwood foliage, and the thick fog of the Pacific dripped into the forest offering moisture year-round. In fall, ladybugs swarmed Redwood Creek, and the big-leaf maples turned yellow. During the winter months, steelhead and silver salmon migrated up the creek to spawn. Sweet berries ripened in the springtime meadows, interspersed with a dazzling display of purple and pink wildflowers. In the Bohemian and Cathedral groves, visitors could see sequoias more than 250 feet high and fourteen feet in diameter.* Many were 2,000 years old or more. Because the rather slender redwoods had little resin, they were insect-resistant. They had survived windthrows, wildfires, and the march of progress. And although the coastal redwood trees were always

*The tallest tree ever recorded was a 414-foot Douglas fir discovered in British Columbia during the late nineteenth century. The largest in the United States was an 800-year-old tree called Hyperion in Redwood National Park, 379 feet tall.

the main year-round attraction at Muir Woods, the site wouldn't have been complete without its stands of Douglas fir, tanbark oak, and bay laurel. On the forest floor, mushrooms often emerged after the frequent rain showers and helped regenerate the soil. For bird-watchers, there were squawking Steller's jays, melodic thrashers, cawing ravens, and the "zree-zee-zee" of the golden-crowned kinglet. Around every bend there were spontaneous bird sounds piercing the divine silence. It was poetically appropriate that this wonder of biological diversity in California bore Muir's imprint—a perfect confluence of place and name.

Kent had stipulated that the forest be dedicated to Muir. Already in California, individual sequoias were named after great men: for example, General Grant in General Grant National Park and General Sherman in Sequoia National Park.[5] Even burned, hollowed "goose pens" where early pioneers used to hide poultry often had individual names. Although Muir himself was deeply averse to the materialism and filthy lucre of his age, he nevertheless recognized that the national monument bearing his name was a gift only wealth could have bestowed. In California the words "John Muir" had become part of the land. Glaciers advanced and retreated, but Muir—and his woods—would last for the ages. "Saving these woods from the axe & saw, from money-changers & water-changers, & giving them to our country & the world is in many ways the most notable service to God & man I've heard of since my forest wanderings began," Muir wrote to Kent with profound gratitude, "a much needed lesson and blessing to saint & sinner alike. That so fine divine a thing should have come out of money-mad Chicago! Wha wad a'thocht it! Immortal sequoia life to you."[6]

Except for naming a large swath of the Catskills the John Burroughs National Monument, little could have made Roosevelt happier than the gift of these California redwoods—one of the greatest accumulations of biomass in the world. Of the original 2 million acres of coastal redwoods in America, more than 80 percent had been logged; the sawmills' blades had to slow down. But Roosevelt's satisfaction with Muir Woods was also tinged by envy. Thus far in his illustrious career his only namesake in nature was a rare species of Olympic elk. "I have just received from Secretary Garfield your very generous letter enclosing the gift of Redwood Canyon to the National Government to be kept as a perpetual park for the preservation of the giant redwoods therein and to be named the Muir National Monument," Roosevelt wrote to Kent. "You have doubtless seen my proclamation of January 9th, instantly creating this monument. I thank you most heartily for this singularly generous and public-spirited action on your part. All Americans who prize the undamaged and especially those who realize the literally unique value of the groves of giant trees, must feel that you have conferred a great and lasting benefit upon the whole country. I have a very great admiration for John Muir; but after all, my dear sir, this is your gift. No other land than that which you give is included in this tract of nearly 300 acres and I should greatly like to name the monument the Kent Monument if you will permit it."[7]

But Kent modestly declined the honor. He considered himself a mere instrument in the whole affair, and insisted that Muir's name was a much worthier one for these woods. "I thank you from the bottom of my heart for your message of appreciation, and hope and believe it will strengthen me to go on in an attempt to save more of the precious and vanishing glories of nature for a people too slow of perception," Kent replied to Roosevelt. "Your kind suggestion of a change of name is not one that I can accept. So many millions of better people have died forgotten, that to stencil one's own name on a benefaction, seems to carry with it an implication of mandate immortality, as being something purchasable." As for the Kent family name, he had "five good husky boys" to carry it forward. Should they fail to make something of themselves, Kent concluded, "I am willing it should be forgotten."[8]

Attached to Kent's letter were exquisite photographs of the gargantuan trees at Muir Woods, which led Roosevelt to reflect on his own legacy. His oldest son, Ted, was a few months away from graduating from Harvard University. Like many fathers, the president realized that his brood had suddenly grown up on him. Kermit—little Kermit—was now almost six feet tall. Roosevelt wondered if he'd been as good a father as his own, Theodore Sr., had been. Why hadn't he taken his boys to see Crater Lake or Key West or Mesa Verde? "Apparently, I have saved up more last year than I ever have before," Roosevelt wrote to Douglas Robinson the day after establishing Muir Woods, "and I am mighty glad of it, for next year Ted will go out into the big world, and from that time right along the little birds will hop off one after another out of the nest."[9] It saddened him to recall that he had left both Ted and Kermit at home during his sojourn of 1903 to Yellowstone, Yosemite, and the Grand Canyon. Roosevelt now resolved to bring his teenage boys along when he bivouacked in remote parts of Africa, Arizona, or the Amazon.

Except for the pursuit of wealth, one's legacy is often the strongest motivator for powerful figures. Everybody supposedly had at least one memorable sermon in him, and Roosevelt's was American conservationism. He had far exceeded any other individual in U.S. history in his efforts to preserve the natural wonders of the West. But he was also uncomfortably aware that his preservationist accomplishments were geographically scattered and isolated, and that his objectives were still hobbled by a slow-moving legislative process. Roosevelt needed a better weapon against congressional lethargy in matters of conservation, and he found it in federal bird reservations, big game commons, and national monuments (as set forth by the Antiquities Act of 1906).

Accustomed to pushing against limits, Roosevelt was determined to put the theoretical power of the "national monument" as a legislative maneuver to an audacious practical test. He wasn't going to allow the size of a national monument to be a stumbling block with regard to the Grand Canyon. On January 11, 1908, just two days after the Muir Woods initiative, the president snatched the Grand Canyon from the preservation-versus-development debate by declaring it a national monument. His goal was straightforward: to save the Grand Canyon, unmarred. It was, by any measure, a bold step. Until then Roosevelt, with persistence, had put aside only monuments of limited acreage such as Devils Tower, El Morro, Montezuma Castle, the Petrified Forest, Chaco Canyon, Lassen Peak, Cinder Cone, and the Gila Cliff Dwellings. And Muir Woods had been a generous gift by Kent to the federal government; who could argue with such beneficence? But the Grand Canyon wasn't a small site preserved for scientific interest, as stipulated by the Antiquities Act. This was 808,120 acres of mineral-rich land in Arizona, larger than some New England states. The poet Carl Sandburg wrote that "each man sees himself in the Grand Canyon."[10] This may be true, but only Roosevelt could claim, on a return visit to the canyon in 1913, that he was its presidential protector. "The importance of the canyon will likely outlive the parochial American idea of wilderness designation as world heritage site and mass tourism," the historian Stephen J. Pyne surmised in *How the Canyon Became Grand*. "A place that can hold a score of Yosemite Valleys and in which Niagara Falls would vanish behind a butte, that could absorb the shock of American expansionism and democratic politics, that could transcend a century of intellectual inquiry from Charles Darwin to Jacques Derrida, has not exhausted its capacity to refract whatever light nature or humanity casts toward it provided a suitable overlook exists from which to view it."[11]

By 1908, Roosevelt's characteristic impatience had made it difficult for him to abide by the old rules and wait for legislators to see the light. Most politicians of his day made backroom deals, but the supremely self-confident Roosevelt did most of his political maneuvering in full view of the public. He always welcomed scrutiny. He would just declare something and let the chips fall where they may. And if Congress and Arizonan lawyers were confused about the Grand Canyon's irreplaceable aesthetic value, he would *make* them see it. If Congress was drowsy, Roosevelt was going to wake it up; if it was operating in the gutter, he was going to teach it to look at the stars. His job as president was to procure the most happiness for the most people. The Grand Canyon was a truth for all time, not to be denied to future generations—a holy spot.

So Roosevelt abandoned noisy ideas and went for a premeditated fait accompli in northern Arizona. Purposely oblivious of the legal obstacles, Roosevelt aimed to evict the Kaibab Cattle Company, for example, from the scenic and scientifically invaluable area. In the nineteenth century, explorers such as Robert Stanton, John Hance, and William W. Bass had promoted the Grand Canyon as a magnet for tourists. As tourism increased, the calls for its becoming a national park did too. But Congress was too scared to move, worried about the reaction from miners and ranchers who were against any alteration that would limit their access to public land.[12] As the historian Hal Rothman noted in *Preserving Different Pasts*, designating the Grand Canyon as a national monument allowed Roosevelt to "circumvent the fundamentally languid nature of congressional deliberation and instantaneously achieve results he believed were in the public interest."[13] When the dust cleared, the Grand Canyon was a national monument. More important, Roosevelt had conclusively demonstrated the elasticity—and thus the power—of the Antiquities Act, the new favorite instrument of the conservation movement. As Rothman put it, "no piece of legislation" had ever before (or since) "invested more power in the presidency," than the Antiquities Act. The elastic clause "objects of historic or scientific interest" made it "an unparalleled tool" for Rooseveltian conservationism.[14]

The Antiquities Act was to Roosevelt a contraption with which he could dictate land policy in the West, circumventing Congress. Nature may not have proceeded by leaps, but Theodore Roosevelt now did. Roosevelt claimed that those 800,000 acres of Arizona contained prehistoric ruins and hence had scientific value. True, there were ruins in the Grand Canyon, but only very meager ones. There was nothing to match Mesa Verde, Chaco Canyon, or Montezuma Castle. As for ethnographical research concerning the residents, the Havasupai were a dwindling tribe content with foraging around the Colorado River, as they had done since the arrival of Spanish conquistadors. Yet Roosevelt wasn't wrong when he claimed this vast part of Arizona had ruins and had Indians. If the ruins were "diminutive by regional standards," they nevertheless provided a political pretext for Roosevelt to invoke the Antiquities Act and remove the area for special protection within the public domain.[15] As ex-president, in *A Book-Lover's Holidays in the Open*, Roosevelt declared all of Arizona and New Mexico an anthropologist's dreamland.[16]

The apparent ease with which Roosevelt designated the Grand Canyon a national monument belied what a long slog it had actually been to save the site. Local opposition to such a huge national park at the Grand Canyon—particularly among corporate cattle, sheep, lumber, and mining outfits—was fierce. Communities surrounding the Grand Canyon, such as Flagstaff, Williams, and Peach Spring, saw the resource-rich proposed parklands as their own. The concept of a "monument," they believed, was a shenanigan, which the courts would rule unconstitutional. Truth be told, their proprietary claims to parts of the canyon were not groundless. The federal government, after all, had encouraged these Arizona pioneers to displace the Hualapai and Havasupai, and to construct their own wagon roads and stagecoach lines. Over time, they had legally purchased grazing lands, mineral deposits, and spring holes. They had developed an interior network of trails at considerable cost. They had already constructed gateway villages along the approach to the Grand Canyon to welcome tourists. Now Roosevelt, disregarding the wishes of an adjourned Congress, was telling the pioneers to step aside for the Department of the Interior.

For those Arizonans displaced by the Antiquities Act, Roosevelt could muster little sympathy. He saw only their greed or simplicity, and he firmly believed that when archaic land claims from the mid-nineteenth century clashed with the moral imperatives of the progressive era, morality must win out. Beyond that, however, this was a struggle between personal profit and the public good. The country's future depended on the outcome. So he took a stand. The resources of the Arizona Territory belonged not to local residents only, but to all Americans. Arizonans would have to shake off ignorance and accept that the Grand Canyon was a national treasure. His former Rough Riders who lived in the territory had already done so.

When Benjamin Harrison was a U.S. senator, after Reconstruction, he had introduced legislation in 1882, 1883, and 1886 to preserve the Grand Canyon as a "public park." The bills had all perished in committee. When he became president, he used his new executive powers to create the Grand Canyon Forest Reserve. But Harrison's legislation had allowed entrepreneurs to construct commercial monstrosities on the rim, and these had cheapened the canyon's appeal. With hotels now appearing along the rims, and tourists arriving in unprecedented numbers by both road and rail, Roosevelt maintained no illusions that the Grand Canyon would remain completely "unmarred," as he had called for in 1903. However, Roosevelt insisted that the north rim stay exactly as it had been when Spanish explorers first encountered it in the early sixteenth century. He believed that by repeatedly refusing to declare the Grand Canyon a national park, Congress had thrust the responsibility for protecting it on his shoulders. This time, he wouldn't give Congress a chance to stop him. *Somebody* had to change the old land-use ways for the new progressive ethos. As Mark Twain once quipped, Roosevelt may have been a "spotless" character, but he was always ready to "kick the Constitution into the back yard."[17]

Weighing the options presented to him by the Department of the Interior, Roosevelt decided to again implement the Antiquities Act of 1906, using it to confer national monument status on the Grand Canyon on January 11, 1908. In the following decades, Arizonan libertarians would claim (accurately) that the Antiquities Act had been intended only to preserve scientifically valuable sites of 5,000 acres or less, and that Roosevelt's edict was an abuse of executive power. But regardless of the legality of Roosevelt's maneuver, the Grand Canyon—the great American temple of nature—would never again return to private hands. From 1908 to 1918, even as the land-use status of the new national monument was hotly debated in both Arizona and Washington, D.C., the increasing tourist appeal of the Grand Canyon made its reprivatization politically unfeasible. The Grand Canyon finally became a national park by an act of Congress on February 26, 1919. The now legendary announcement came just a month after Roosevelt's death. Furthermore, in 1975 President Gerald Ford signed the Grand Canyon Enlargement Act, placing the entire Colorado River corridor under the management of the National Park Service.[18] In 1990, the area between Bright Angel Point and Cape Royal was renamed Roosevelt Point to honor the twenty-sixth president.[19] And tourists come by the millions to experience what is now a World Heritage Site, just as Roosevelt predicted during his presidency. "To none of the sons of men," Roosevelt wrote of his new national monument, "is it given to tell of the wonder and splendor of sunrise and sunset in the Grand Canyon of the Colorado."[20]

Not that Roosevelt was anxious about how his preservation of the Grand Canyon would be evaluated in history books. Certitude was his greatest political strength. In fact, on the same afternoon that he declared the Grand Canyon a national monument, he began threatening to do the same with large parts of the Appalachian and White Mountains, an action certain to cause tremendous resistance by congressmen from Maine to Georgia. One notable exception was Governor Robert Glenn of North Carolina, who committed himself politically to Roosevelt's conservationist crusade, hoping that the Great Smoky Mountains would emerge as a national monument.[21] For his part, Roosevelt intended to take the Antiquities Act to its limit not just in the West, but everywhere in America. He envisioned the act as a federal hand with numberless fingers. It was obvious that his last fifteen months in office were going to be filled with conservationist action.

And as of January 11, 1908 ("Grand Canyon Day"), it was also clear that no parcel of wilderness—private, public, or other—was immune from potential seizure by the federal government. Roosevelt's preservationist initiatives would be as simple as they were decisive. Orthodoxy was being shattered, and many more projects were in the works. With no worries about reelection in 1908, Roosevelt was ready to take on "the American goliath" (which he later defined as a vicious plutocracy, with morals of "glorified pawnbrokers," that owned both political parties along with "ninety-nine percent at the very least of the corporate wealth of the country, and therefore the great majority of the newspapers").[22]

II

Roosevelt, in fact, wasted little time in pressing the Antiquities Act into service yet again in California in early 1908. Stanford University's president, David Starr Jordan, had previously written to Roosevelt about the Pinnacles, a fantastic landscape of jutting rocks and volcanic formations near Soledad, California. The spires and crags of the Pinnacles were awe-inspiring. Jordan was a leading eugenicist and ichthyologist, whose *Guide to the Study of Fishes* sat prominently on one of Roosevelt's bookshelves at Sagamore Hill. The inspiration for this book had been Robert B. Roosevelt. No fewer than 1,000 genera and 2,500 species of American fish were named after Jordan. A founding member of the Sierra Club, he was a devotee of Darwin and Huxley's biology, and he developed his own laws of biogeography. Fascinated by the adaptability of species, in 1907 Jordan had cowritten *Evolution and Animal Life*, which President Roosevelt found illuminating.[23] If Jordan—an expert on organic evolution—believed that the Pinnacles were worth saving, then Roosevelt needed no other authority. He could now wave his wand—the Antiquities Act—and immediately guarantee federal protection to whatever Jordan wanted.

On January 16, 1908, Roosevelt turned these 2,500 acres of grandeur into Pinnacles National Monument, on Jordan's recommendation. (By 2009 the site had been expanded to more than 26,000 acres.) Only superlatives can describe the Pinnacles ecosystem. Oddly, the Pinnacles had more types of bees—400 species—than anywhere else known to entomologists. There were also 149 species of birds, forty-nine of mammals, twenty-two of reptiles, eight of amphibians, sixty-nine of butterflies, and forty of dragonflies and damselflies. And there were also precipitous bluffs, talus caves, crags of volcanic rock, and little canyon hideaways where the lizards seemed almost as ancient as the brontosaurus. The clincher, to Roosevelt, was that *The Condor* (a periodical) reported California condors using the Pinnacles as a primary roosting site. And the red rock formations, courtesy of an extinct volcano, were more than 23 million years old. If that didn't constitute antiquities, what did? As Jordan boasted, the Coast Range chaparral (the finest examples in the national park system) and the riparian, xeric, and foothill woodlands were ideal for getting away from the cityscapes of San Diego, Los Angeles, and San Francisco. One of the prettiest sites in the natural world was in the Pinnacles: the acmon blue butterflies, in spring, congregating on coyote brush flowers. And saving all this was as easy as signing a declaration!

Holed up in the White House because of snowstorms and freezing weather, Roosevelt began methodically marking on a map of the United States sites he wanted preserved by the federal government before he left office on March 3, 1909. His desks at the White House and at Sagamore Hill were crossroads for plans to rehabilitate species. From a political standpoint, the Antiquities Act was a revelation, freeing the Department of the Interior from having to squabble with Congress. Much like federal bird reservations for the USDA, national monuments soon became an idée fixe at Interior. Even more significantly, bureaucrats and politicians alike were beginning to see national monuments as a way station to national park status. And even Pinchot, chief of the Forest Service, wanted his reserves studied for Indian artifacts. "The importance of taking steps to preserve such objects has become very apparent," Pinchot wrote to his on-site employees, "and as soon as possible I wish you to report specifically upon each ruin or natural object of curiosity in your reserve, recommending for permanent reservation all that will continue to contribute to popular, historic, or scientific interest."[24]

Roosevelt was temperamentally well suited to conflict and acrimony, and his presidency had already had more than its share of both. By February 1908, he had made an impressive number of political and corporate enemies, including Standard Oil Company, the Louisville and Nashville Railroad, the E. H. Harriman conglomerate, and J. P. Morgan, among many others. But as the biographer Ron Chernow wrote in *Titan: The Life of John D. Rockefeller, Sr.*, Roosevelt had "no more potent ally than the press," in his corner.[25] And the clashes of Roosevelt versus the titans made good copy. The corporations of the gilded age spent millions of dollars on advertising, trying to smear Roosevelt's reputation, cripple him politically, and exhaust him personally. They had failed on every count. Each swipe had the reverse effect—bolstering Roosevelt's obstinacy. He licked them at their own game. Although he had never been solicitous of the opinions of his political antagonists, by February 1908 Roosevelt had lost all patience for anti-forestry. He now saw his enemies as Dickensian villains, full of "bosh and twaddle and vulgarity and untruth."[26]

America's corporate leaders (and their army of bosom chums) were not the only unfortunates on the receiving end of Roosevelt's wrath. When he had established Wind Cave National Park in 1902, a clamor had arisen for him to do the same with Jewel Cave, 143 miles of unmapped underground passageways in the Black Hills just west of Custer. Even though it was the second largest cave in the world, this labyrinth hadn't been discovered until 1900, when a couple of small-time prospectors, the brothers Frank and Albert Michaud, had felt a blast of chilly air emanating from a rock crevice in Hell Canyon one warm summer afternoon. It was a good place to protect the carcasses of slaughtered cattle from rotting in the heat. But just maybe there was gold beneath where they stood!

Excited by their find, the Michauds purchased dynamite in Rapid City, South Dakota, blasted a big opening, and then crawled into a cavernous room aglow with calcite crystal. Much to their chagrin, there was no gold to be found. But there were numerous caverns filled with stunning, gemlike calcite crystals, which caught the light from their lanterns and returned it in varied patterns and colors. The brothers rushed to procure a mining claim for Jewel Tunnel Lode, as they named the site, and began advertising it as a tourist attraction. They even used the caves to hold a number of dances for local couples—the crystalline walls were a natural forerunner of the disco ball. A few geologists who studied the site determined that the passageways were part of an extinct geyser channel.*[27]

* While not dismissing the Michauds' story, Gail Evans-Hatch and Michael Evans-Hatch in *Place of Passages: Jewel Cave National Monument Historic Resource Study* raise the possibility that Burdett Parks, a cowboy at the nearby X-4 ranch, actually made the discovery first.

The Roosevelt administration soon got involved, using the almost unlimited power of the executive branch to establish national monuments for the permanent preservation of places deemed to have historic and scientific interest. Certainly Jewel Cave met this criterion. In 1907, he authorized Harry Neel and C. W. Fitzgerald to survey Jewel Cave. Roosevelt imagined the underground site to be part of a larger preservation initiative in the Deadwood and Rapid City area, which would include the Badlands, Devils Tower, Wind Cave, and the Black Hills National Forest. Neel and Fitzgerald's report recommended that Jewel Cave be declared a national monument—despite the mining claim of the Michaud brothers.[28] Like Wind Cave, this new find was situated in the Pahasapa limestone rock layer of the Black Hills, and the federal government wanted to study its geological history.[29]

When Roosevelt established Jewel Cave National Monument on February 7, 1908—his thirteenth designation thus far under the Antiquities Act—he knew there could potentially be decades of legal problems.[30] But Roosevelt was apparently untroubled by the possible consequences. There was no restraining him. As far as he was concerned, the 1,280-acre Jewel Cave Monument was a national treasure. If only the Michauds weren't such self-interested money-grubbers, they would have understood that such a unique natural wonder belonged to *all* Americans. Furthermore, the Michauds had to stop making additional openings to Jewel Cave, because they were desecrating the site. This was tough medicine for a couple of presumably lucky prospectors, who thought they had stumbled on a fortune. But as Roosevelt saw it, if soldiers gave their lives for the democracy, surely land could be deeded to the federal government for the sake of science. With federal lawyers breathing down their necks, the Michaud brothers sold their claim to the U.S. government for $500.[31]

To Roosevelt's way of thinking, the notion that a little knot of men in the Black Hills saw Jewel Cave as a source of profit was depressing to contemplate. The labyrinth of chambers and tunnels belonged to the U.S. government: end of story. Conservation was, above all else, a moral issue to Roosevelt—a cause he believed he shared most intimately with men like John Muir and Seth Bullock, who were uninfected by the greed of New York and Chicago. By contrast, the South Dakota miners seemed to Roosevelt rather like lowly English sparrows—deemed a pestilential invasive species by the Biological Survey and thus deserving neither understanding nor accommodation.[32] Well, if sparrows they were, then the president would deal with the Michauds, and all those like them, in his typically expedient fashion. "Is there any kind of air gun which you would recommend which I could use for killing English sparrows around my Long Island place?" Roosevelt wrote to Dr. C. Hart Merriam. "I would like to do as little damage as possible to our [other] birds, and so I suppose the less noise I make the better."[33]

Ridding Sagamore Hill of English sparrows wasn't the only hunting Roosevelt had in mind that spring. While last-minute legal maneuvering was taking place in the Department of the Interior to establish Natural Bridges in Utah and Wheeler in Colorado as national monuments, Roosevelt began planning a post-presidential safari to British East Africa. His romantic notion was that he would leave civilization in favor of the roar of the lion and the pleasant odor of buffalo. Roosevelt was going to invite two celebrated trophy hunters whom he highly admired—R. J. Cunninghame and Frederick Courtney Selous (both of whom collected big game specimens for the British Museum)—to join him. Selous would later dedicate his *African Nature: Notes and Reminiscences* (1908) to Roosevelt. Roosevelt also struck up a lively correspondence with the British hunter-explorer Lieutenant Colonel John Henry Patterson, about possibly seeing the great African fauna and flora, rhinoceros, gnu, water buffalo, and giant eland. Roosevelt actually looked forward, he wrote, to being served up as "food for ticks, horseflies, and jiggers."[34]

A veteran of the Boer War, the well-bred Patterson had written an adventure saga, *The Man-Eaters of Tsavo*, in 1907. (Decades later, it was made into two Hollywood movies.) With narrative verve and vividness, Patterson's book told of his quest to track down and kill two male lions that had eaten approximately 140 railroad workers over the course of nine months in the Tsavo province of Kenya. During the 1890s, rinderpest (a bovine disease) had killed millions of buffalo, zebras, and gazelles, the primary food source for lions. Their food supply thus limited, the starving maneless lions turned to humans as prey. Inspired by Patterson's action-filled prose, particularly his references to weapons like the Martin-Enfield double-barreled rifle and the .303 Lee Enfield, Roosevelt now wanted a lion head for his library wall at Oyster Bay. (Roosevelt, in fact, worked behind the scenes to eventually help the Field Museum of Chicago acquire the stuffed Tsavo lions and put them on permanent display.[35]) "A year hence I shall leave the Presidency, and, while I cannot now decide what I shall do, it is possible that I might be able to make a trip to Africa," Roosevelt wrote to Patterson on March 20. "Would you be willing to give me some advice about it? I shall be fifty years old, and for ten years I have led a busy, sedentary life, and so it is unnecessary to say that I shall be in no trim for the hardest kind of explorer's work. But I am fairly healthy, and willing to work in order to get into a game country where I could do some shooting. I should suppose I could be absent a year on the trip."[36]

While Roosevelt hastened to assure Patterson that he wasn't a "butcher," he nevertheless hoped to acquire a multitude of specimens for American museums, entering Africa in the Mozambican port of Mombasa and boating down the Nile with shotgun in hand. Roosevelt arrived in Mombasa in April 1909, and departed for home from Khartoum in March 1910. This was no bear hunt in Mississippi and no holiday in Louisiana's canebrakes. He would be away from the United States for over a year. That letter of March 20, in fact, started an epistolary exchange between the two men that would continue throughout 1908. The president interrogated Patterson repeatedly about the prospects of bagging specimens of zebras, giraffes, and cheetahs. As the tenor of his letters made clear, he could hardly wait to be armed to the teeth and free from the shackles of the White House. The Smithsonian Institution had been eager to officially sponsor Roosevelt and his party in obtaining all types of specimens, from the diminutive Kenia dormouse to the colossal white rhinoceros, for its collection. With a retinue of hundreds, Roosevelt would fulfill a lifelong dream of visiting Kenya and Uganda in the heart of British East African safari land.*

There was nothing odd about Roosevelt's desire to hunt big game in Africa. Only the Boone and Crockett Club's taxidermist, Carl Akeley, who had constructed the world's first habitat diorama in 1890 at the Milwaukee Public Museum, had begun to biologically inventory the wildlife along the Nile River with seriousness of purpose. Working out of the Field Museum in Chicago, Akeley had developed a new technique for taxidermy, which did a better and truer job of preserving texture and musculature. Roosevelt truly admired the way Akeley displayed wildlife in a group setting. The gorilla and the elephant were his specialties. Roosevelt thought Akeley could make a very distinctive contribution to the annals of scientific exploration by collecting with him in Africa. Roosevelt and Akeley would risk their lives for the right gorilla. Poets called nature a mother, but Roosevelt knew it was also a grave. Regardless of his worries about exertion by a middle-aged man in the heat, and about sleeping sickness (the primary concern), with proper funding Roosevelt believed he could revolutionize the African exhibits at all of America's top museums.

*In 1908 British East Africa combined Uganda and Kenya. Tanganyika (Tanzania), where T.R. also visited, was a German colony.

It was strange, though, that he was planning to disappear from the American political scene for virtually a full year. In considering Roosevelt's maverick conservationist agenda of 1908–1909, it is important to remember that the president didn't foresee a future political career. Having already won a Nobel Peace Prize for diplomacy, Roosevelt increasingly envisioned his postpresidential role in part as being a global spokesperson for big game animals, wildlife protection, and natural resources conservation. Didn't South Africa need antelope preserves? Shouldn't India find ways to protect its tigers? Couldn't China develop its own fish hatcheries? This had always been Roosevelt's hope for the Boone and Crockett Club—that it would go global with its preservationist and conservationist message and ideas. Akeley, a new club member, actively promoted this global approach. And Roosevelt was about to write his first non-American centric book, even though the content was about an American naturalist expedition traveling in the so-called dark continent. "I speak of Africa and golden joys," Roosevelt later wrote in *African Game Trails*. "The joy of wandering through lonely lands; the joy of hunting the mighty and terrible lords of the wilderness, the cunning, the wary, and the grim."[37]

III

The success of Mesa Verde National Park in 1906 had gotten Roosevelt extremely interested in the Four Corners states and territories: Colorado, Utah, Arizona, and New Mexico. Guided by Congressman Lacey, Roosevelt had read about Montezuma Castle, El Morro, and other prehistoric ruins of the Southwest. And at some point Roosevelt also read an article in *National Geographic* magazine titled "The Colossal Natural Bridges of Utah," which had piqued his curiosity.[38] Pinchot's U.S. Forest Service—at the time, a branch of the USDA—was in charge of Utah's two national forests, Sevier and Manti. And Utah also had many more spectacular wonders. Word had reached the White House that southeastern Utah had the largest number of natural bridges in the world. The novelist Edward Abbey later wrote rapturously about the three bridges, which retained their original Hopi names. The first, Kachina, meant "spirits that had lightning snake symbols on their bodies"; the second, Uwachomo, signified "flat rock mound"; the third, Sipapu, meant "place of emergencies." The bridges, the largest of which was 222 feet high, had been formed by streambed erosion (unlike many other arch formations in Utah, which were created by wind, rain, and ice). Surrounded by piñon forest, Anasazi ruins, and a gorgeous slickrock canyon, Natural Bridges met every criterion for national monument status.[39]

Priding himself on knowing *everything* about the West, Roosevelt started reading every book he could find on Hopi and Navajo culture, and later contributed an essay about the southwestern tribes of the Four Corners region in his *A Book-Lover's Holidays in the Open* (1916).[40] Ultimately Roosevelt was taking a tricky line: preserving Hopi-Navajo culture in the southwest while also encouraging soldiers, agents, missionaries, and traders to Americanize these peoples. Roosevelt never hesitated to promote the education of Native Americans, and he seemed unaware of the humiliation being inflicted on tribes by the imposition of the Bible and reservation life. "The Indian should be encouraged to build a better house," Roosevelt wrote of this policy, "but the house must not be too different from his present dwelling, or he will, *as a rule*, neither build it nor live in it."[41]

On April 16, 1908, Roosevelt signed into existence Natural Bridges National Monument, his first such designation in Utah. These three natural sandstone bridges, spanning a desert canyon and isolated from any town or hamlet, met Roosevelt's scientific qualifications for monument status because they were "extraordinary examples of stream erosion."[42] Over the decades, Natural Bridges National Monument became a popular hiking and camping destination, especially for families. Together with *National Geographic*, Roosevelt helped launch the movement to save Utah's splendid canyons from private development.

Certainly, Roosevelt's heavy-handed federalism angered people besides corporate bigwigs and disgruntled ranchers. Many westerners had long felt that businessmen on the east coast had treated the West like a colony, extracting wealth and resources without giving anything back. They were tired of the east coast stuffed shirts telling them what to do. As the episodes involving the Grand Canyon National Monument showed, they could also turn their blistering ire on the federal government. But as the historian G. Michael McCarthy argued in *Hour of Trial: The Conservative Conflict in Colorado and the West, 1891–1907* (1977), there were plenty of westerners who approved of federal regulation of resources. At a public lands convention in Denver in 1907, for example, many of the attendees favored the Roosevelt administration's deployment of rangers to protect national forests from wildfire throughout Colorado. They understood that Colorado was special because of its mountains, forests, and water resources.[43]

The president of the American Scenic and Historic Preservation Society, Dr. George F. Kunz, also backed Roosevelt's national monument initiatives every step of the way. Kunz and Roosevelt had been allies since the battle of 1899–1901 to preserve the Palisades cliffs in New York, and Roosevelt thought that Kunz was among the most cogent voices for protection of American scenery. One might almost name an American site at random—from Yellowstone to the Catskills, from Watkins Glen to Bunker Hill—and find that Kunz had fought hard for its scenic integrity. Scholarly, clear-minded, and committed to the cause, Kunz considered all significant ruins as part of the cultural heritage. He had been offended by Oscar Wilde's offhand comment that the United States "had no ruins." Whether Wilde had spoken out of ignorance, arrogance, or flippancy, Kunz knew better. The United States had Mesa Verde, Montezuma Castle, Chaco Canyon, the Gila Cliff Dwellings, and numerous other "ruins." But Wilde was right in one respect: most Americans did not value their ruins as Europeans did. Rome, for example, celebrated its ancient ruins, such as the Colosseum, whereas by contrast, it had taken a small group of Colorado women, fighting tooth and nail, to protect the stunning cliff dwellings of Mesa Verde—and their battle had taken place largely unnoticed by America's newspapers.

According to Kunz, there were hundreds of antiquities sites, both natural and architectural, that needed preservation—places filled with arrowheads and pottery shards whose designation as monuments wouldn't disrupt commerce in the slightest. Grateful to Roosevelt for raising the nation's consciousness on this issue, Kunz became his strongest public defender, countering criticisms of the president's conservationist and preservationist agenda. "Niagara Falls, Letchworth Park, the Hudson River, the Yellowstone Park, the Grand Canyon and the Colorado, the Agatized Trees, the Giant Redwoods, the Columbia River, and the prehistoric remains of the Southwest, are the poetry of our possessions," Kunz wrote. "What nation is rich without a poet, and what country has such grand natural objects to inspire the poet as ours?"[44]

Roosevelt and Kunz were deeply suspicious of anyone foolish enough to register any kind of concern about the creation of national monuments like Pinnacles or Natural Bridges. Legal and constitutional objections, Roosevelt was convinced, gave false legitimacy to a much baser desire: protecting entrenched interests. "The very luxurious, grossly material life of the average multimillionaire whom I know, does *not* appeal to me in the least, and nothing would hire me to lead it," Roosevelt wrote to Cecil Arthur Spring-Rice on April 11. "It is an exceedingly nice thing, if you are young, to have one or two good jumping horses and to be able to occasionally hunt—although Heaven forfend that anyone for whom I care should treat riding to hounds as the serious business of life! It is an exceedingly nice thing to have a good house and to be able to purchase good books and good pictures, and especially to have that house isolated from others. But I wholly fail to see where any real enjoyment comes from a dozen automobiles, a couple of hundred horses, and a good many different houses luxuriously upholstered. From the standpoint of real pleasure I should selfishly prefer my old-time ranch on the Little Missouri to anything in Newport."[45]

Medora, North Dakota, wasn't a village to Roosevelt anymore, but an entire world unto itself, a time and place of magic. Memory is selective, and Roosevelt's seemed to fasten on the pleasant shade, tranquil skies, and meadowlarks of summer, and to excise the long, cold Dakota winters. In fact, Roosevelt was arguably as proud of his exploits at Elkhorn Ranch as he was of any other chapter in his biography. After killing a buffalo in Montana, he had rushed home to New York City to cofound the Boone and Crockett Club along with his friend George Bird Grinnell. According to *Forest and Stream*, that was the opening salvo of the conservation movement. Now, in mid-May of 1908, President Roosevelt was scheduled to hold a summit at the White House on conservationism, with most of America's governors present (a few states, including Texas, sent their lieutenant governors instead). To clear his mind of clutter, and prepare himself for the challenging conference ahead, Roosevelt, together with Edith, escaped to Pine Knot. Usually, at this retreat, Roosevelt liked to live strictly by himself with Edith. But now the Roosevelts' houseguest—the only nonrelative ever invited to sleep at the cabin in Albemarle County—was Oom John. It had been a last-minute invitation, and Burroughs had accepted.

At Pine Knot both Roosevelt and Burroughs enjoyed the first appearance of new organisms springing up from the old Virginian earth. Soon the cicadas would descend on the trees and the lightning bugs would take to the air. The land was clothed in new plants and wildflowers. Earthworms were plowing the soil, and the naturalists admired these lowly creatures with new eyes. But it was bird-watching that most pleasantly consumed their energy at Pine Knot. Owing to the spring migration, the two companions were able to identify seventy-five species.[46] Always wearing a scrunchy suit, Burroughs was quicker to spot the birds; the more casually attired Roosevelt was better at identifying them by their twitters.[47] The president was hoping to show Oom John a passenger pigeon, but that sighting never occurred. They did see a swamp sparrow, and rare warblers only four and a half inches long. Some of the warblers had migrated more than 3,000 miles from Canada to wintering spots in South

America, and they were resting in Virginia for a few days on their return. Likewise, some of the blackpolls Roosevelt and Burroughs saw had just made a nonstop flight of 2,300 miles over water, perhaps from Venezuela or Cuba. "It was really remarkable," Burroughs later recalled, "how well [Roosevelt] knows the birds and their notes."[48]

What a fine time the two naturalists had, poking around the countryside, marveling at migratory birds whose flight was directed, often, by the magnetic field of the earth. Oom John came to appreciate Roosevelt's fair-mindedness and skill as a raconteur even more than he did during their Yellowstone trip of 1903 or their wanderings in the Catskills or their jaunts on Long Island. "The President is a born nature-lover, and he has what does not always go with that passion—remarkable powers of observation," Burroughs wrote in *Camping and Tramping with Roosevelt.* "He sees quickly and surely, not less so with the corporeal eye than with the mental. His exceptional vitality, his awareness all around, gives the clue to his power of seeing. The chief qualification of a born observer is an alert, sensitive, objective type of mind, and this Roosevelt has to the preeminent degree."[49]

But Roosevelt's penchant for turning everything into a competition grated on Burroughs, who was proud of his humility—proud that he didn't boast. Roosevelt's ego, by contrast, would have made Napoleon flinch. As if engaged in a footrace, Roosevelt challenged Oom John over who would see the most species of sparrow or woodpecker, or who would first spot an eastern bluebird. Talking incessantly, Roosevelt would quote poetry by Longfellow and Tennyson, and evince surprise after forcing Burroughs to admit that he hadn't memorized the precise verses. Too much time spent in the company of Roosevelt, Burroughs decided, could try the patience of the Old Testament's Job. "I rather shrank from him," Burroughs admitted later, ". . . his dominating qualities, his strenuousness—his mood always antipathetic to my own."[50]

Shockingly, to Burroughs, Pine Knot was genuinely rustic. It had no amenities aside from an old woodstove. His own Slabsides was regal by comparison. The "barn-like structure" of Pine Knot made it too primitive for a gentleman to sleep in.[51] Oom John looked around and let out a great sigh. If Burroughs had had his former energy and confidence, a local inn in Charlottesville would have been hurriedly found. But as it was, bizarrely, Roosevelt had two northern Virginia flying squirrels (*Glaucomys sabrinus fuscus*) living indoors, and Burroughs remained unable to sleep because of the racket these nocturnal critters made (they were active at night because their large eyes allowed them to see in extremely low light). Roosevelt, however, seemed to genuinely enjoy their midnight madness, tossing them nuts the way he threw nuts to the scampering grays on the White House lawn. Delightedly Roosevelt watched the squirrels glide from the rafters to the bed and from the bed to the floor, like trapeze artists. As an evolutionist, he was intrigued by their membrane, called a patagium, which allowed them to glide as far as 240 feet through the air. Roosevelt erupted in disapproval when Burroughs gingerly moved the flying squirrels' nest outside the cabin, and a quarrel ensued—those were Roosevelt's indoor flying squirrels! How dare he! Eventually, Roosevelt compromised by placing the nest in his own bedroom, minimizing Burroughs's interaction with the squirrels.[52] Still, the high-pitched chirps continued to annoy Burroughs; even holding a pillow over his ears didn't work. At one juncture Roosevelt himself was bitten by one of the squirrels; blood trickled down his hand from a real puncture wound. What shocked Burroughs was that Roosevelt seemed to admire the squirrels even more for this hostile act.[53]

"Mr. Burroughs, whom I call Oom John, was with us and we greatly enjoyed having him," Roosevelt wrote to his son Archie. "But one night he fell into great disgrace! The flying squirrels that were there last Christmas had raised a brood, having built a large nest inside of the room in which you used to sleep and in which John Burroughs slept. Of course they held high carnival at night-time. Mother and I do not mind them at all, and indeed rather like to hear them scrambling about, and then as a sequel to a sudden frantic fight between two of them, hearing or seeing one little fellow come plump down to the floor and scuttle off again to the wall. But one night they waked up John Burroughs and he spent a misguided hour hunting for the nest, and when he found it took it down and caught two of the young squirrels and put them in a basket. The next day under Mother's direction I took them out, getting my fingers somewhat bitten in the process, and loosed them in our room, where he had previously put back the nest. I do not think John Burroughs profited by his misconduct, because the squirrels were more active than ever that night both in his room and ours, the disturbance in their family affairs having evidently made them restless!"[54]

IV

On May 11, 1908, two days before the participants were to arrive in Washington for the governors' conference, Roosevelt arrived there from Pine Knot, with Burroughs at his side. Deciding to do something to honor the founders of western naturalist studies in America—Meriwether Lewis and William Clark—Roosevelt signed into being the first national monument in Montana. If John Muir could have a monument named after him in Marin County, then surely one should be dedicated to the great explorers of the Jeffersonian era in Jefferson County, Montana. Roosevelt had a deep-seated memory of reading Lewis and Clark's diaries as a boy, and he recalled thinking that their 4,134-mile journey up the Missouri River and onward to Oregon was a more stirring epic than Gilgamesh. They had hunted in the prairies, wintered with the Mandan in North Dakota, traversed the Rockies, and navigated the length of the Columbia River to marvel at the Pacific Ocean. To Roosevelt, it was important to memorialize them with a natural site. He settled on a 600-foot-long and 400-foot-deep series of caverns in Montana.

The Lewis and Clark National Monument, which overlooked the Lewis and Clark Trail along the Jefferson River, was Roosevelt's salute to an earlier age of American exploration. Aside from setting aside a unique cavern, Roosevelt took distinct pleasure in reintroducing the two great American explorers to popular consciousness.

On May 13, Roosevelt brought the Conference of Governors at the White House to order. As he had promised in Memphis during his inspection tour of the Mississippi River, nearly every governor summoned (from both states and territories) attended the conference.[55] Roosevelt had asked each governor to bring along with him three competent advisers on natural resource management.[56] The nine justices of the U.S. Supreme Court were also in attendance—something unprecedented for a blue-ribbon commission. In addition to the politicians and judges, Roosevelt had invited scores of biologists, geologists, ornithologists, and advocates of forestry science to offer their informed input. "Any right thinking father earnestly desires and strives to leave his son both an untarnished name and a reasonable equipment for the struggle of life," Roosevelt had said, as a rationale for the conference. "So this Nation as a whole should earnestly desire and strive to leave to the next generation the National honor unstained and the National resources unexhausted."[57]

When Roosevelt took to the podium in the East Wing on the morning of May 13 for the opening session of the conference of governors, there was a palpable sense of unity among the attendees. The White House had given a multiple-course dinner party the previous evening, so the governors and the other dignitaries were in a fine mood. According to Plutarch's *Lives*, Alexander was once asked if he craved any service and snapped in reply, "Stand a little out of my sun." That's how it was for Roosevelt that afternoon. Starting with Genesis, Roosevelt went on to relate how civilization began on the banks of the Nile and Euphrates. Organic evolution was always on the march. Leaping over millennia, he swung his oration quickly to the founders of America at Independence Hall in Philadelphia in 1776. Then, the natural resources of the young country, ranging from anthracite coal to vast timberlands to streams full of fish, had seemed limitless. But owing to unwise land management, America was losing its gifts. Utilization of mineral fuels and metals had transformed America into a steel empire, but overmining was turning the nation's wilderness into an eyesore. "The mere increase in our consumption of coal during 1907 over 1906," Roosevelt warned, "exceeded the total consumption in 1876, the Centennial Year."

Roosevelt's speech was a kaleidoscope of times and places: here were Kentuckians felling forests and Mississippians watching as their riverine modifications washed the delta away. It was a quasi-public lecture on causes and effects of land misuse. He included many doomsday predictions about the depletion of natural resources (Jimmy Carter's famous "malaise" speech of 1978 seems cheery by comparison). Here were new and frightening concepts: timber famine, choked rivers, denuded fields, obstructed navigation, exhausted oil fields. But Roosevelt also wanted to introduce phrases like "sustainable growth," "renewable resources," and "a future undiminished for our children" into the vocabulary of the twentieth century. "We have become great in a material sense because of the lavish use of our resources, and we have just reason to be proud of our growth," Roosevelt said. "But the time has come to inquire seriously what will happen when our forests are gone, when the coal, the iron, the oil, and the gas are exhausted, when the soils shall have been still further impoverished and washed into the streams, polluting the rivers." To Roosevelt, these questions didn't relate only to the next century or his grandchildren's generation. "One distinguishing characteristic of really civilized men is foresight," Roosevelt said. "We have to, as a nation, exercise foresight for this nation in the future; and if we do not exercise that foresight, dark will be the future."[58]

Reading the *Proceedings of the Conference of Governors* is quite telling, in a number of ways. Andrew Carnegie spun out dozens of statistics about iron ore and copper production, and he ended his address with praise for President Roosevelt's Reclamation Service but not a peep about national monuments or forests. Geologists offered charts of riches still in the ground. The new governor of California, George C. Pardee, spoke about clear-cutting redwood forests so as to export timber, not of saving stands like those preserved in Muir Woods and Mariposa Grove. Governor

Charles E. Hughes of New York praised Roosevelt for his work in Albany in 1900 to preserve the Adirondacks, and Governor James O. Davidson of Wisconsin bragged about the pine, hemlock, oak, and maple of the Dells, even though the Dells had been brutalized by overcommercialization. But these were oratorical exceptions to the norm. Most of the presentations at the governors' conference came from technocrats who were terrified by the prospect of America's vanishing natural resources.

There were some eruptions of anti-Roosevelt sentiment during the conference. Examples, in fact, were plentiful. Although Governor Edwin C. Norris of Montana effectively used humor to attenuate the sharpness of his speech, he strongly objected to federal land grabs for national monuments and forests. Montana needed Reclamation Service projects, but not the Lewis and Clark National Monument—absolutely not! Grumbling that he was tired of misinformed easterners claiming that Montana was nothing but "chill icebergs, cold weather, and blizzards," Norris launched into a measured denunciation of the Forest Service. To begin with, he knew that some reserves were necessary for the watershed. But why was Montana, more than any other state, bearing the brunt of federal land seizures? Why were New Yorkers and Yalies so willing to declare acreage in Montana federal property—21 million acres, in fact—while holding back their own tree parks in places like the Berkshires, Adirondacks, and White Mountains? Norris supplied his own answer: President Roosevelt was treating Montana as little more than a protectorate because Montanans had little recourse on the federal level. "I would suggest, Mr. Secretary of the Interior, that there be no more [forest reserves]", Governor Norris concluded with quiet indignation. "We have sufficient."[59]

Everybody who attended the conference, including Norris, ended up with wonderful stories to tell. Although Muir was worried that Pinchot would try to own the resource management movement, he had no quarrel with Roosevelt's developing a national dialogue on conservation. Roosevelt himself was clearly in his element at the conference, using his characteristic political shrewdness to mentally assess the men surrounding him. The president cut a stylish figure in a tailor-made suit and hat, looking as elegant as the best man at a wedding. In his memoir, *Fighting the Insects*, the entomologist L. O. Howard, known as Roosevelt's "exterminator" at the USDA, recalled with a mixture of fondness and amazement the personal welcome he received from the president at the conference. "I found myself in line immediately behind Dr. C. Hart Merriam, the animal and bird man," Howard recalled. "Immediately in front of him was William J. Bryan. As we reached the President, Mr. Bryan, in a pompous and somewhat condescending way (at least it seemed so to me), said 'Mr. President, I congratulate you, sir, on having started this conservation movement, which, in my opinion, has tremendous possibilities of good for the future of the country. I assure you, sir, that it meets my entire approval and will receive my hearty support.'"

According to Howard, the president, with a fiercely playful gleam in his eyes, barely acknowledged Bryan's smug compliment. Instead, he looked over Bryan's shoulder and waved to Merriam. "I am pleased," Roosevelt muttered to Bryan perfunctorily, and then quickly pivoted toward Merriam, as if dismissing the Democratic Party's contender to succeed him in 1908. "How are you, Hart?" Roosevelt greeted Merriam with pointed warmth. "What do you suppose John [Burroughs] and I saw on the twenty-fifth of March at Pine Knot? A yellow warbler, by George!" Roosevelt then turned to Howard. "Hello, Doctor!" the president said, "How are the bugs?"[60] As Roosevelt moved off into the crowd, away from Bryan, Howard looked back to see Bryan, still rooted in place, realizing that the president had just gotten the better of him. Howard's anecdote revealed two things about Roosevelt: his allegiance to the naturalist community and his contempt of Bryan.

To the president, increased interstate cooperation on behalf of conservationism was a stark necessity. If, for example, Missourians dumped sewage in the Mississippi River, then it would travel downstream, adversely affecting citizens in Tennessee, Arkansas, and Louisiana. Garbage didn't respect state lines. Likewise, it would do no good to save some white pelicans at Stump Lake, North Dakota, only to let them be slaughtered while they were roosting in Charlotte Harbor, Florida. Migration made state game laws pointless. "One of the most useful among the many useful recommendations in the admirable Declaration of the Governors relates to the creation of state commissions on the conservation of resources to cooperate with a federal commission," Roosevelt wrote to one friend. "This action of the Governors cannot be disregarded. It is obviously the duty of the Federal Government to accept this invitation to cooperate with the states in order to conserve the natural resources of our whole country."[61]

A cloth-bound three-volume report of the National Conservation Commission would be made available to the public in January 1909. As an accommodating gesture, Congress would receive an advance copy in December. Roosevelt believed that the publication of the findings would be a historical landmark of his administration, and a turning point in conservation history (though in truth it never acquired any true cogency). He glowed in the aftermath of the conference. "The grounds are too lovely for anything, and spring is here, or rather early summer, in full force," Roosevelt wrote to his son Archie on May 17. "Mother's flower-gardens are now as beautiful as possible, and the iron railings of the fences south of them are covered with clematis and roses in bloom. The trees are in full foliage and the grass brilliant green, and my friends, the warblers, are trooping to the north in full force."[62]

When Roosevelt was at Pine Knot with Burroughs, he had reported to Chapman that their trip was a feast of bird-watching worthy of Audubon. His cabin sat smack in the middle of a mecca of Virginia bird-watching, where every streak and eye mark could potentially signify a new species. Unbelievably rare birds could be seen—his passenger pigeons were a case in point. Eagerly, Roosevelt listed for Chapman the various gnatcatchers and summer redbirds they'd seen. Burroughs and Roosevelt had used Chapman's accessible guide, *Birds of the Eastern United States*, and had spent hours analyzing the noted ornithologist's various disquisitions. "When I see you again I am going to point out one or two minor matters in connection with the song of the Bewick's wren and the looks of the blue grosbeak, where we were a little puzzled by your accounts," Roosevelt wrote to Chapman on May 10. "I suppose that there is a good deal of individual variation among the birds themselves as well as among the observers."[63]

Chapman was gratified to receive a letter from the White House detailing Roosevelt and Burroughs's adventures at Pine Knot. But he couldn't help feeling that Roosevelt was questioning the veracity of his ornithology with respect to the grosbeaks. He was already unusually defensive, partly because in early 1908 a Biological Survey report on grosbeaks had deviated from some of his published ornithological observations of the species.[64] Chapman had taken a series of photographs of blue grosbeaks, showing their resplendent blue plumage and the pale gray bill and black and chestnut wing bars. He had also taken photos of female blue grosbeaks, with their brown body and pale beige breast. He volunteered to come to Oyster Bay and put on a slide show for Roosevelt. Unfortunately, Sagamore Hill still had no electricity and no suitable screening room—Roosevelt wanted it to feel rustic—but he invited Chapman to visit the White House in August. Because Roosevelt was planning an African trip for March 1909, he wanted advice on photography from the author of *Birds with a Camera* for Kermit. Perhaps sensing Chapman's distress, Roosevelt took pains in his reply to put the ornithologist at ease.

"As regards the blue grosbeak, your description of the habits was exactly borne out by the conduct of the individuals we saw," Roosevelt wrote on June 7. "They did not behave at all like indigo buntings or rose-breasted grosbeaks, but stayed by preference along the bushy sides of a ditch in the middle of an open pasture, frequently going out into the open grass. Both males and females would sit solemnly on the tops of some thin stalk or small twig a couple of feet high beside the ditch. The Bewick's wrens were very tame and confiding. To our ears not only their song but their subdued conversational chirping had a marked ventriloqual effect; seeming to be much further away than it was. It had no resemblance to the song of the house wren, and none whatever to the Carolina wren. I do not understand the principles upon which the sparrows are generically divided. The swamp sparrow seems to me in color scheme and even in voice to be more like a spizella than a zonotrichia."[65]

V

That Roosevelt continued to make birds his hobby in 1908 is indisputable. They were the reason why he had no patience with symphonies or operas. For Roosevelt, warblers were harmonicas, the doves flutes, the jays clarinets, and certain combinations string quartets. There was a feeling among many of the U.S. governors, in fact, that the president was more enthusiastic about birds than about the Constitution, limited government, private property, or corporations. Many business interests interpreted Roosevelt's national forests, federal bird reservations, buffalo parks, and national monuments as yet another unneeded expansion of his already huge federal regulatory blanket, smothering land development and entrepreneurship. Roosevelt's real object with these monuments, these critics claimed, was to broaden executive power by sabotaging checks and balances. In the Arizona Territory, for instance, not only had he turned some 800,000 acres of private land into the Grand Canyon National Monument by presidential fiat, but he had also dispatched armed former Rough Riders to oversee it. Also, Alexander Brodie, Arizona's governor from 1902 to 1905, was one of Roosevelt's former lieutenant colonels and had been appointed by Roosevelt. (Three territorial governors appointed by Roosevelt, in fact, had served with him in the Cuban campaign.) After stepping down from the governorship in 1905, Brodie had joined the War Department at Roosevelt's request.

Over the decades confusion has reigned over a group of nine former Rough Riders who joined the Arizona Rangers. They weren't employed by either the Department of the Interior or the U.S. Army; they were troubleshooters for Roosevelt in Arizona. These men drifted around like a Secret Service outfit within the Texas Rangers, ferreting out information about illegal mining in the Kaibab National Forest and about vandals at Montezuma Castle near Sedona. The Arizona Rangers were known to have direct access to Roosevelt through Governor Brodie, and they were determined to quash illegal activities in the territory. Operating with little publicity, tough as nails, they brought a conservationist ethos to Arizona because Roosevelt wanted them to. Some of these men were the ones who had given Roosevelt Remington's *Bronco Buster* bronze as a departing gift in Montauk, Long Island, in 1898. When T.R. declared the Grand Canyon a national monument, it was great for him to have the Arizona Rangers (not to be confused with park rangers) on his side.[66]

On June 23, 1908, as if rubbing salt into the wounds of westerners opposed to the Antiquities Act, Roosevelt designated thousands of acres of the Grand Canyon as a game preserve. Not a single oil well, ore mine, or asbestos vein was permitted. The Executive Order was clear. As the critics put it, Roosevelt's idea of commercial activity in the territories involved tens of thousands of deer and elk frolicking about, so that future members of the Boone and Crockett Club could shoot them. This criticism was patently unfair. Without the Roosevelt Dam along the Salt River, settlements never could have prospered in Arizona's central valleys. It was an engineering wonder, rising 220 feet out of a canyon gorge. Phoenix grew into a metropolis because of Roosevelt's large-scale irrigation projects. And Roosevelt had encouraged copper-mining towns like Jerome, Globe, and Bisbee to prosper. As for deer and elk, Arizona needed to think of them as game resources, which enriched human civilization in numerous ways.

The National Governors' Conference had bolstered Roosevelt's resolve to create new forest and game reserves and enlarge old ones. Roosevelt had taken the tone of the governors and, in general, liked what he had heard. Taken as a whole they seemed to understand that conservation of natural resources was *the* issue of the era. For every grumbler like Governor Norris, there were three other governors who supported *increased* federal forestlands. The philosopher Edmund Burke once wrote about the "wisdom of ancestors." The activists at the governors' conference deserved this accolade. They articulated a visionary conservationist agenda for America in the twentieth century. At least the issues of forestry were no longer hidden from public view. At Pine Knot, a worried Burroughs had told Roosevelt that Taft, whom Roosevelt had chosen as his successor, was too "weak" to sustain such Roosevelt innovations as the wildlife refuge system. According to Burroughs, Roosevelt dismissed his warnings at the time. He had made the choice because he believed that Taft, then his secretary of war, would faithfully uphold Rooseveltian values, and he was confident that he had chosen wisely.[67]

But Roosevelt wasn't leaving anything to chance. While America was consumed with the election of 1908—Taft versus Bryan—that June the president prepared a massive preservationist initiative, scheduled for announcement on July 1. The forestry movement would be forced down his opponents' throats. Even more than his own birthday, Roosevelt loved the first of July. On that day in 1898, he had won glory in the Battle of San Juan Hill in Santiago, Cuba, leading the two famous charges against the Spanish army. Roosevelt had led the first charge, up Kettle Hill (on horseback). The second and more famous of the two, up San Juan Hill proper, he had led on foot. It was his "crowded hour," as he famously phrased it. In addition, other important historical events that resonated with Roosevelt had happened on July 1. In 1858, the year he was born, Charles Darwin and Alfred Russel Wallace had offered joint papers on evolution to the Linnaean Society on July 1; these had been a thunderclap of scientific progress whose reverberations can still be felt. As every American schoolchild of Roosevelt's generation knew, July 1, 1863, was when the Battle of Gettysburg began, with the Union and Confederate forces colliding as Robert E. Lee desperately consolidated his forces.

So on July 1, 1908, in the last full year of his presidency, Roosevelt began a grand expansion of federal forestlands that was stunning in both its scale and its breadth of vision. It was Roosevelt's presidential "crowded hour." Forty-five new national forests—scattered throughout eleven western states—were declared that day. New boundaries were also created for existing forests. All day long, Roosevelt signed documents creating or recognizing national forests. A total of ninety-three federal forest sites were effected within a twenty-four-hour period. Innovative protocols for range management, wildfire control, land planning, recreation, hydrology, and soil science were introduced throughout the American West. Because of Roosevelt's "crowded hour," much of the Rocky Mountains region, as well as the Pacific Northwest, would no longer be vulnerable to the lumberman's ax (although, of course, forest fires would remain a serious threat).

Undoubtedly, the "Crowded Hour Reserves" had been under consideration by the Forest Service since 1905. Pinchot had worked intensely with the GLO, state agencies, business interests, and governors to iron out the legalities. Many of the new national forests bore well-known Indian names: Nez Perce National Forest in Idaho, Cheyenne National Forest in Wyoming, Uncompahgre National Forest in Colorado, and Sioux National

Forest in Montana and South Dakota, among others. A few were named to honor great Anglo-Americans who had contributed to the opening of the West, including Lewis and Clark, Madison, Jefferson, and Custer in Montana; Powell in Utah; and Pike in Colorado. Other "Crowded Hour Forests" took the names of nearby cities: for example, Santa Barbara in California and Boise in Idaho. The new Columbia National Forest in Washington state near the Mount Saint Helens volcano—home to bald eagles, bull trout, chinook salmon, grizzly bears, northern spotted owls, gray wolves, and marbled murrelets, among other species—should have been granted national park status; it was certainly spectacular enough. But at least in 1949 it did become the Gifford Pinchot National Forest.

Roosevelt also announced on July 1 the creation of Kaibab National Forest (in what he called the Buckskin Mountains), home to high-altitude flora, mountain lions, and scores of mouse subspecies.[68] This was another of Merriam's special patches of wild Arizona. It was a geographical area beloved by both the Biological Survey and the Boone and Crockett Club. And now, after years of contention, it would be preserved as a national forest. A Texas character, Uncle Jim Owens, who had grown up on the Goodnight Ranch, was tapped to oversee the wildlife protection efforts in the national forestlands around the Grand Canyon. Uncle Jim Owens was a passionate supporter of the American Bison Society, working tirelessly to bring bison back to the West. "He was keenly interested not only in the preservation of the forests," Roosevelt noted, "but in the preservation of the game."[69]

Many of the "Crowded Hour Reserves" were in places Roosevelt knew personally, particularly the Bighorn Forest in Montana, the Bitterroot in Montana-Idaho, and the Teton in Wyoming. In the 1890s, Roosevelt had written a series of essays about all three for his celebrated Dakota trilogy. New Ranger stations were ordered built in 1908, sometimes three or four per forest. Some of the Crowded Hour Reserves forests had begun as suggestions by his friends. For example, the ornithologist William Finley of Oregon successfully sponsored Malheur, Umatilla, Suislaw, Cascade, Umpqua, Siskiyou, Crater, and Wallowa national forests in his state. In the exhilarating rush of creation on July 1, two prime national forest sites in Oregon had been left off the list: Deschutes and Fremont. Pinchot worked quickly to correct the situation, and added them to the list of forest reserves on July 14. Judged purely on the basis of their size and scope, Roosevelt's Crowded Hour Reserves were among the most important designations in the annals of American forest and wilderness preservation.

The jam of Crowded Hour Reserves animated the U.S. Forest Service as never before. Waves of new applications for positions as rangers and superintendents came pouring onto Pinchot's desk. Working for Roosevelt to protect national forests was suddenly considered a great honor. Take, for example, Clinton G. Smith of North Cornwall, Connecticut, a first-generation student of Yale Forestry School. On August 1, 1908, he was made supervisor of Pocatello National Forest in Utah (which had been established in March 1907 but was expanded on July 1, 1908, to include Port Neuf National Forest and part of Bear River National Forest). Smith's prescribed duties were to protect vast acres of the Crowded Hour Reserves; he also started teaching people in Idaho and Utah about silviculture. Over time, Smith became a member of the District Investigative Committee of the Forest Service, prosecuting abusers of the public domain in towns like Malad City, Montpelier, and Soda Springs.[70]

What should historians make of the Crowded Hour Reserves of July 1, 1908? In March 1903, Roosevelt had lectured before the Society of American Foresters (of which he was an associate member) about his intentions regarding resource management. He had forcefully advanced the argument that forestry was vital to all Americans, particularly those who lived in the arid West. To Roosevelt, the dryness of the West was the worst impediment to the region's long-term prospects. But by a wise combination of forest reserves and man-made reservoirs, westerners could transform their arid homesteads into lush, Edenic communities. Because Roosevelt was creating national parks, national monuments, and federal bird reservations, all of which were preservationist measures, he didn't want his Crowded Hour Reserves to be misrepresented as just another sop to wilderness activists. These dozens of new and enlarged national forests would, in the long run, economically sustain a thriving American West. "The object is not to preserve the forests because they are beautiful, though that is good in itself, nor because they are refuges for the wild creatures of the wilderness, though that, too, is good in itself," Roosevelt explained. "[B]ut the primary object of our forest policy, as of the land policy in the United States, is the making of a prosperous home."[71]

VI

His correspondence during the summer of 1908 shows that Roosevelt grew excited by the prospect of once again being a naturalist, after the end of his presidency. From Sagamore Hill, Roosevelt made arrangements with Charles Doolittle Walcott, secretary of the Smithsonian Institution, to sponsor his field collecting in Africa just a few weeks after he stepped down as president. Roosevelt planned to write articles for *Outlook* and a book for Scribner's about his experiences in British East Africa, the Congo, and Egypt. He wanted Walcott to suggest professional taxidermists and naturalists to accompany him on his safari. All specimens collected by the Roosevelt expedition would be donated to the Smithsonian Institution and other important American museums. Roosevelt's mission was simple: for America to have the best natural history collections in the world. Because 1908 was an election year, Roosevelt was looking forward to handing the reins of his party to the presumptive Republican nominee, William Howard Taft. "I would a great deal rather have this a scientific trip, which would give it a purpose of character, than simply a prolonged holiday of mine!" Roosevelt wrote to Henry Cabot Lodge over the summer. "I am no longer fit to do arduous exploring work, and this will probably be about the last time that I shall be fit even for the moderate kind of trip I have planned. But it seems to me that there is something worth doing to be done along the lines I have laid out—something that is still the work of a man of action; and I should like to remain a man of action as long as possible."[72]

Roosevelt felt confident that the American people would elect Taft over William Jennings Bryan, the Democratic nominee. Perhaps with this in mind, the president spent almost as much time, it seemed, hiring his African expedition team as he did stumping for Taft. He gave no outward signs of guilt, hesitation, or concern over misplaced priorities. After reading the zoologist Edgar Alexander Mearns's *Mammals of the Mexican Boundary of the United States* (1907), Roosevelt appointed Mearns as chief naturalist and physician for the African expedition. Mearns had acquired his impressive reputation while on duty as a major in the Philippines. Wielding a machete with the same prowess as Bill Sewall using an ax, he was the ideal lead scout for jungle exploration. Roosevelt also extended an invitation to two younger American Darwinian naturalists: J. Alden Loring (an authority on small mammals), and Edmund Heller (the Chicago Field Museum's specialist on big game). Heller in particular interested Roosevelt, as he had been trained on an extended expedition to the Galápagos.[73] Mearns, Loring, and Heller were all fine taxidermists, and taxidermy was a skill crucial to the expedition's serious work. As this was taking place, Roosevelt wrote more than thirty letters about the African trip to Henry Fairfield Osborn at the American Museum of Natural History, oftentimes asking him for such hard-to-find items as books on Ugandan gorillas and snakes of the Nile River.[74] And Roosevelt gladly accepted an invitation, offered annually to naturalists of high esteem,[75] to deliver the prestigious Romanes Lecture at Oxford University for 1910. The world-renowned biologist Thomas Huxley had previously delivered the lectures, so Roosevelt was immensely pleased.

When he returned to the White House from Sagamore Hill in September 1908, Roosevelt met with ex-Congressman John Lacey to discuss possible southwestern Indian ruins in need of rescue by the federal government. Lacey also promoted the admirable idea of preserving the crumbling Spanish colonial missions of Arizona in Tumacacori, along the border with Mexico: Santos Angeles de Guevavi and San Cayetano de Calabazas. To save these adobe-style architectural gems, the federal government would need to seize them. Until this time, no historic Spanish buildings or structures had been declared national monuments, but the Anasazi cliff dwellings of Mesa Verde had been designated part of a national park, and Roosevelt was now smitten with the idea of Tumacacori. The memory of Spanish rule in Arizona shouldn't be effaced by misplaced national pride or triumphalism. Americans needed to better understand the decades when imperial Spain colonized parts of their Southwest. On September 15, Roosevelt signed the Tumacacori National Monument into law. One signature designated three landmarks.

The first weeks of September were exceptionally busy for Roosevelt. *Collier's* was publishing an essay by Jack London, "The Other Animals," in which London attempted to cut Roosevelt down to size for attacking the realism of his fictional dogs—a point of great pride for him—and for calling him a "nature faker." London skewered Roosevelt for supposedly stating that animals did not reason, were below mankind in the biological pecking order, and could perform only mechanical and reflexive actions. London believed that accident counted for much in nature and that Roosevelt's certainty was arrogant.

London insisted that his two novels about dogs—*Call of the Wild* and *White Fang*—were consistent with evolution. He had been in Hawaii when he heard that Roosevelt considered him a nature-faker. Embarrassed, London said he had "climbed into my tree and stayed there." But by the time he sailed to Tahiti on the *Snark*, London was ready to exchange blows with both Roosevelt and Burroughs. For starters, he insinuated that they had old-school European tendencies: "They believe that man is the only animal capable of reasoning and that ever does reason," he wrote. "This is a view that makes the twentieth-century scientist smile. It is not modern at all. It is distinctly mediaeval. President Roosevelt and John Burroughs, in advancing such a view, are homocentric. . . . Had not the world not been discovered to be round until after the births of President Roosevelt and John Burroughs, they would have been geocentric as well in their theories of the Cosmos. They could not have believed otherwise."[76]

"When London says this," a furious Roosevelt wrote to the editor of *Collier's*, Mark Sullivan, "he deliberately invents statements which I have never made and in which I do not believe."[77] Roosevelt considered *White Fang* a decent work as fiction. But zoologically, its behavioral descriptions of wolves and lynx irked him enough that he publicly called it "mischievous nonsense." Its purported facts were all wrong. In his letter to *Collier's*, Roosevelt inventoried all of London's offenses against biological accuracy. Sounding less like a president than like a fact-checker, he offered exact page references in *White Fang* where London's veracity failed to pass muster.

Had Roosevelt really enough free time to engage in such a hypothetical, picayune debate with a writer of fiction? Nobody came out ahead in these "nature-faker" controversies, but London did get in a few truly memorable lines at Roosevelt's expense:

> Now, President Roosevelt is an amateur. He may know something of statecraft and of big-game shooting; he may be able to kill a deer when he sees it and to measure it and weigh it after he has shot it; he may be able to observe carefully and accurately the actions and antics of tom-tits and snipe, and, after he has observed it, definitely and coherently to convey the information of when the first chipmunk, in a certain year and a certain latitude and longitude, came out in the spring and chattered and gambolled—but that he should be able, as an individual observer, to analyze all animal life and to synthetize and develop all that is known of the method and significance of evolution, would require a vaster credulity for you or me to believe than is required for us to believe the biggest whopper ever told by an unmitigated nature-faker. No, President Roosevelt does not understand evolution, and he does not seem to have made much of an attempt to understand evolution.[78]

Words are slippery, but London was a master of them. Nothing, short of insulting his family, cut into Roosevelt's sense of self more than a claim that he didn't understand Darwinism. But it was 1908, a presidential election year, so Roosevelt let the matter go. When pressed, Roosevelt could often behave like a cuttlefish which, when unable to extricate itself from a dangerous situation, blackens the surrounding waters, eventually becoming invisible in the murk. That's how he dealt with London in the end, writing him off as a black-hearted fraud.* Now, a century later, the "nature-faker" debate smacks of childish egoism, particularly on the part of a sitting American president. But to dismiss Roosevelt's obsession with proper descriptions of wildlife as mere conceit is to misread a central facet of his complex personality—his all-encompassing belief that he understood evolution. Roosevelt was a faunal naturalist, steeped in Darwinian biology, and he knew more about wolves than perhaps anyone else in the country. If London was going to count himself as a literary realist, then lynxes twice the size of the Biological Survey's heaviest specimen had no business in his art. In this instance, as often before, Roosevelt was more than happy to risk being exposed as an intellectual bully if doing so meant that he could set a record straight.

That September, Roosevelt was also championing farmers: he was deeply involved in getting his new Country Life Commission up and running. Although the reference work *Roosevelt Cyclopedia* doesn't mention it, the Country Life Commission was a direct outgrowth of his reading of John Burroughs's and Ralph Waldo Emerson's works. Roosevelt's motto came from Jonathan Swift's *Gulliver's Travels:* "Whoever can make two ears of corn, or two blades of grass, to grow upon a spot of ground where only one grew before, would deserve better of mankind, and do more essential service to his country, than the whole race of politicians put together."[79]

Over the summer, Roosevelt had persuaded Liberty Hyde Bailey, director of the College of Agriculture at Cornell University, to serve as chairman of the commission tasked with analyzing the status of rural life in America. At first, Bailey turned Roosevelt down—he was too busy—but Roosevelt wouldn't take no for an answer. "Your letter is not only a great disappointment but a great surprise to me," Roosevelt wrote to Bailey on August 4, from Sagamore Hill. "I would not have gone into this thing at this time if I had not been assured as a matter of course from our conversation that you would accept. I believe you are entirely right when you say in your letter that this is the greatest opportunity that has yet presented itself for the influencing of country life conditions and the setting in motion of movements that shall organize and vivify the affairs of the open country. Yet my dear Mr. Bailey, by your action you are doing all you can to hurt this great opportunity. You have no right to do it, my dear sir. It is imperative from the standpoint of the work that you and I

*Roosevelt and London eventually patched up their differences. London, in fact, endorsed T.R. for president in 1912 believing the Bull Moose Party was a variant of democratic socialism.

have so much at heart that you should accept the chairmanship of this Commission, no matter how little work you do with it."[80]

The Country Life Commission was Roosevelt's attempt to broaden the definition of the American town so that it would be more than just blocks with buildings on them. Small towns held the key to the perpetuation of American democracy; lose them and you would have, say, Brussels or Berlin. When Roosevelt was living in North Dakota, he had disapproved of people's saying that the city boundaries were two miles long and four miles wide. Just as with migratory birds or roaming animals within the United States, boundaries weren't the be-all and end-all. The American town had to include clean rivers and lakes outside the business district. Wilderness myths were part of a town. What was Helena, Montana, without stories of grizzlies or North Platte, Nebraska, without buffalo tales or Douglas, Arizona, without reports of a jaguar sneaking over the Mexican border? The town had to know the name of the old man dwelling in the shack down by the creek along the verdant meadow. The town didn't need the city as much as it needed the country.

While not quite a boondoggle, Roosevelt's Country Life Commission never really got off the ground. Although the commission had outstanding directors, including Henry Wallace (editor of *Wallaces' Farmer* in Des Moines, Iowa) and Kenyon L. Butterfield (of Massachusetts Agricultural College in Amherst), Roosevelt's folksy rural vision seemed antiquated in the brave new world of the Model T. People were thinking in terms of the cylinder, the engine, and the conveyor belt, not of ponies' oats. Congress, in fact, scoffed at Roosevelt's commission as romantic nonsense reminiscent of *Currier & Ives*. Both Democrats and Republicans had grown weary of listening to Roosevelt pontificating about the strenuous life, wilderness parks, national heritage sites, and the eternal virtues of the family farm. City nights were in vogue, as was owning a new car—not five o'clock suppers by candlelight in cabins like Pine Knot. But after his presidency, Roosevelt insisted in *Century Magazine* that his commission, though dismissed in 1908–1909, would someday be recognized and acknowledged for its agrarian wisdom. "The first step ever taken toward the solution of these problems [of rural life] was taken by the Country Life Commission, appointed by me, opposed with venomous hostility by the foolish reactionaries in Congress, and abandoned by my successor," Roosevelt wrote. "Congress would not even print the report of this Commission, and it was the public-spirited, far-sighted action of the Spokane Chamber of Commerce which alone secured the publication of the report."[81]

Roosevelt's own sentiments, like those of the people whom he most admired (John Burroughs chief among them) were decidedly antiurban and anti-industrial. Now, as his tenure at the White House wound down, Roosevelt wanted to do something for the "uplifting of farm life."[82] In essence, the commission was to teach farmers, with their close connection to the soil, to become America's front line of conservationists—to protect forests from clear-cutting and streams from pollution. No matter how holy so-called *rural simplicity* might be, it could go only so far in the twentieth century. Recognizing that farms in America were worth $30 billion (with annual produce worth about $8 billion), Roosevelt boasted that the heartland was the greatest breadbasket the world had ever known.[83] "Important tho the city is," Roosevelt wrote to Herbert Mynick, editor of *Good Housekeeping*, "and fortunate tho it is that our cities have grown as they have done, it is still more important that the family farm, where the homemaking and the outdoor business are combined into a unit, should continue to grow. In every great crisis of our Government, and in all the slow, steady work between the crises which alone enables us to meet them when they do arise, it is the farming folk, the people of the country districts, who have shown themselves to be the backbone of the nation."[84]

What disturbed Roosevelt about American farmers, however, was their lack of science education. Farmers mismanaged forests by overburning and overlumbering. That was an economic waste. The twentieth century was unmistakably going to be a century of science, and for farmers it would entail learning the newest techniques for increasing yields, as well as protecting their crops against disease and pests. A scientific examination of water had to take place in every town. Community water supply systems needed to be developed for the sake of public health. Farmers didn't have to read Freud's *The Interpretation of Dreams* to be modern; but they did have to read the Department of Agriculture's annual *Yearbook*. Roosevelt wanted Congress to appropriate funds to open new agricultural colleges, establish conservation training camps, and increase the emphasis on tree planting in public schools.

In essence, Roosevelt was positioning himself as pro-farmers, on behalf of Taft and in opposition to the Democratic presidential nominee, William Jennings Bryan. While Bryan appealed to the populist Deep South by attacking "government by privilege," Roosevelt wooed Iowa, Kansas, Minnesota, and Wisconsin—farm states that greatly valued education—by promoting agricultural science. Taft's platform also included reforming the federal bureaucracy, increasing tariffs, and developing a sound currency—all positions that Roosevelt adhered to. But Roosevelt thought that Taft could score points against Bryan (the first three-time presidential candidate for a major party) by promising family farmers federal benefits and better methods of growing, based on the "new conservationism."

When Bryan dared to publicly chastise Roosevelt as soft on corporations at the expense of everyday folks, the president roared back. In a pamphlet-length letter to Bryan postmarked September 27, 1908, Roosevelt presented evidence of his warring against beef packers, Federal Salt Company, General Paper Company, Otis Elevator Company, American Tobacco Company, Powder Trust, Virginia Carolina Chemical Company's conglomerates, and Standard Oil Company, among numerous others. Roosevelt boasted that he was the real progressive, whereas Bryan was a poseur. "I believe in radical reform," Roosevelt said, "and the movement for such reform can be successful only if it frowns on the demagogue as it does on the corrupt; if it shows itself as far removed from government by a mob as from government by a plutocracy. Of all corruption, the most far-reaching for evil is that which hides itself behind the mask of furious demagoguery, seeking to arouse and to pander to the basest passions of mankind."[85]

With a gift that fine writers often possess, Roosevelt was sometimes able to size men up—and he didn't like what he saw in William Jennings Bryan. Ever since William Kent had donated Muir Woods to the federal government, Roosevelt corresponded with him about politics. Now, when many reporters were saying that Roosevelt was being too brutal in his public attacks on Bryan as a left-wing demagogue, claiming that the Democrats were sympathetic to the IWW and Russian radicals, the president explained his sledgehammer tactics to Kent. "I felt it was imperative to put aggressive life into the campaign," Roosevelt wrote. "It seems to me incredible that people should fail to understand Taft's inherent worth." By contrast, Roosevelt said, Bryan was the "cheapest faker we have ever had proposed for President."[86]

From Roosevelt's perspective, the difference boiled down to this: he himself was a champion of Darwinism, agricultural science, conservationism, and irrigation. By contrast, Bryan was promoting silver, the Russian Revolution, and a half-baked Christianity. But not everybody in the Taft campaign was happy with Roosevelt's presenting himself as a hunter-conservationist. During the fall of 1908, for example, at the Irrigation Congress held in Albuquerque, Roosevelt and Pinchot's forestry ethos was itself taken as demagoguery. Western Republicans at the Irrigation Congress accused extremist forestry policies of possibly costing them the Rocky Mountain states of Colorado, Montana, and Wyoming in the November presidential election. At issue were the Crowded Hour Reserves. "No policy of recent years has done so much to alienate the friends of Government as the mistaken policy of the forest service, and if any of the mountain states shall go Democratic this fall, it will be chiefly for that reason," D. C. Beaman of Denver told the gathering in New Mexico. "If a State Government had treated its people in such a manner it would have been ousted at the next election. Shall we stop mining coal, shut down our steel works, gas and electric plants and go back to the blacksmith shop and the tallow candle?"[87]

Election day was a relief to Roosevelt. He had been occupied by politics throughout the fall, and even as he planned his African safari he had been coaching Taft on how to win Catholic votes, state by state, away from the "small Protestant bigots." He had also tried promoting conservationism to American farmers. Few major politicians had ever advocated keeping religion out of politics with quite the fervor of Roosevelt. In his role as coach during the campaign, Roosevelt had always wanted Taft to attack Bryan with more fury, to skin him like a skunk. That was eventually Roosevelt's general attitude toward Bryan. Taft, disregarding Roosevelt, played his cards just right. November 3 was Taft's day, not Bryan's. The electoral count was 321 to 162: a landslide. The election couldn't have turned out better for the Republicans, all around. According to the French ambassador, Jean-Jules Jusserand, Roosevelt's "joy" over Taft's victory was "overflowing." Roosevelt deemed it a vindication of his own policies. "We beat them," Roosevelt gloated, "to a frazzle."[88]

Even though Taft had won, the general public seemed more enamored of Roosevelt—a rare American political leader who willingly turned his back on power. Excerpts from the Country Life Commission report now ran in rural newspapers all over the country. And (with the exception of Colorado) virtually all the western states where Roosevelt had created national forests, national monuments, and federal bird reservations voted for Taft. Roosevelt's far-reaching conservation policies were winning over what he called "the base and sordid materialism" of Bryan's evangelical buffoons.[89]

VII

Nobody that holiday season in America could really envision T.R. as a private citizen. Life after the White House has always been entirely a matter of personal preference. John Quincy Adams, for example, became a U.S. congressman for eighteen years. Ulysses S. Grant barnstormed around the country for some time. For Roosevelt, stepping down meant returning to his life as a big game hunter, wilderness wanderer, and faunal naturalist. But he hoped that his conservationist acolytes—including his fourth cousin Franklin D. Roosevelt, who was a die-hard advocate of forestry—would continue fighting the crusade in Washington.

"Every now and then solemn jacks . . . tell me that our country must face the problem of what it will do with its ex-Presidents," Roosevelt wrote to Ted Jr. after Thanksgiving, "and I always answer them that there will be one ex-President about whom they need not give themselves the slightest concern, for he will do for himself without any outside assistance; and I add that they need waste no sympathy on me—that I have had the best time of any man my age in all the world, that I have enjoyed myself in the White House more than I have ever known any other President to enjoy himself, and that I am going to enjoy myself thoroly [*sic*] when I leave the White House and what is more, continue just as long as I possibly can to do some kind of work that will count."[90]

As 1908 wound down, Roosevelt noted that despite his activist progressive agenda he had left America's finances in better shape than they were in 1901. He had cut taxes slightly and reduced the interest-bearing debt. He had also reduced the amount of interest paid on America's debt. His administration had a net surplus of $90 million, and receipts over expenditures for all seven and a half years. "I am especially pleased, because the average reformer is apt to embark on all kinds of expenditures for all kinds of things," he wrote to the historian George Otto Trevelyan, "good in themselves, but which the nation simply cannot afford to pay for."[91]

Thinking about his own presidential legacy had a tendency to set Roosevelt spinning. Insisting that he was sure to be a winner in the contest of history, he churned out letter after letter extolling his successes in Cuba, Panama, Santo Domingo, and Venezuela. Internationally, his primary worry was Kaiser Wilhelm II of Germany, whose militarism bothered Roosevelt's sense of decency. "The German attitude toward war," Roosevelt lamented to the British editor John St. Loe Strachey, "is one that in the progress of civilization England and America have outgrown."[92] As for Russia, it was guilty of "appalling . . . well-nigh incredible mendacity."[93]

Roosevelt, of course, was full of praise for his foreign policy advisers. Besides Root as secretary of state, Roosevelt had relied on Taft as secretary of war; Senator Lodge of Massachusetts; the ambassador to Russia, George Von Lengerke Meyer; the ambassador to Britain, Whitelaw Reid; and the troubleshooting diplomat Henry White. There were three foreigners living in Washington, D.C., whom Roosevelt treated like cabinet members: Cecil Arthur Spring-Rice and Arthur Lee of Great Britain, and the French ambassador to the United States Jean-Jules Jusserand. Working with these men, Roosevelt had developed his "big stick" diplomacy, based on what we would now call American realpolitik and a huge modern navy. Other Rooseveltian principles included always acting justly toward other countries; never bluffing; striking only if prepared to strike *hard*; and, finally, always allowing an adversary to save face in defeat.[94]

But Roosevelt's conservation activism didn't die out merely because of an election. On December 7, 1908, he created a national monument in Colorado. It was called Wheeler, after Captain George Wheeler, who had explored the area in 1874 for the U.S. Army, and its eroded outcropping of volcanic ash geologically resembled the badlands of North Dakota. Roosevelt was intrigued by the jagged spires, reminiscent of organ pipes, and he also deemed Wheeler a scientific site of great importance. But as it turned out, Coloradans weren't good custodians of this unique 30-million-year-old area, and in 1950 Wheeler lost its monument status. It was transferred from the National Park Service to the U.S. Forest Service. (Currently it's called the Wheeler Geologic Area and is part of the La Garita Wilderness Area.[95])

Next, inspired by Muir Woods, the Pinnacles, Cinder Cone, Lassen Peak, and other California sites that he had recently saved, Roosevelt began strategizing to make the six Farallon Islands off the coast of California a federal bird reservation, safe from human predation. The Farallons were a favorite roosting area for the common murre, black oystercatcher, and Leach's storm petrel, but the Farallones Egg Company had plundered these islands by raiding nests and had taken an estimated 14 million eggs. Studying photographs of the islands, and enjoying the charming intimacy of seals and otters at play, Roosevelt decided to have the Biological Survey draw up documents for him to sign in early 1909. Even though San Franciscans were used to ransacking the islands, Roosevelt was going to cut them off as an act of charity and wisdom.

Roosevelt, however, let California's preservationists down in one profound way. For all of his thoroughness, he didn't back Muir's wise opposition to flooding the Hetch Hetchy Valley in Yosemite. Instead, he favored the short-term need for water rather than the long-term aesthetic value of Hetch Hetchy. In contrast to his self-congratulatory boasts during December, Roosevelt's letter to the *Century*'s editor, the prominent conservationist Robert Underwood Johnson, exuded self-doubt. Was it smart to have approved a petition by San Francisco to convert Hetch Hetchy into a reservoir, following the earthquake of 1906? Had he let the Sierra Club and John Muir down with regard to this conservation issue? "As for the Hetch Hetchy matter," Roosevelt explained to Johnson, "it was just one of those cases where I was extremely doubtful; but finally I came to the conclusion that I ought to stand by Garfield and Pinchot's judgment in the matter."[96] Neither Johnson nor Muir held a noticeable grudge against Roosevelt for his disastrous decision concerning Hetch Hetchy (although they did hold a grudge against both Garfield and Pinchot).

And Roosevelt took his conservationism global that December. At the Joint Conservation Congress he rallied against the disease of global deforestation. Roosevelt invited Canada and Mexico to participate in the North America Conservation Congress on February 18, 1909, in Washington, D.C., for a simple reason: nature didn't recognize artificial boundaries. If Mexico polluted the Rio Grande River, that would hurt the citizens of Texas. Similarly, if Canada overfished Lake Superior, the effect on Minnesota would be horrific. Roosevelt wanted Arbor Day to be international, because deforestation was a global curse. Also, every country needed tough antipollution laws to regulate the handling of sewage and industrial waste. Migratory birds needed protection like the Lacey Act everywhere from the Arctic to Antarctica, from Lassen Peak to the Himalayas. Roosevelt's hope was that his conference of February between the United States, Canada, and Mexico might be the precursor to a global conference. "It is evident that natural resources are not limited by the boundary lines which separate nations," Roosevelt said, "and that the need for conserving them upon this continent is as wide as the area upon which they exist."[97]

A final 1908 brouhaha occurred when Roosevelt declared Loch-Katrine in Wyoming a federal bird reservation, to protect a wide range of aquatic fowl as well as nesting bald eagles and peregrine falcons. Congressman Frank W. Mondell protested that Loch-Katrine Federal Bird Reservation was undemocratic. Wyoming had so sparse a population that it had only one congressman—Mondell, who with his long face, pointed eyebrows, and handlebar mustache, served in that capacity for twenty-six years. Mondell was the champion of Wyoming's oil fields and coal mines. Imposing one of T.R.'s federal bird reservations on his state—not far from Teapot Dome—was, to him, nothing short of an act of war. On February 11, 1909, Mondell, a member of the Committee on Public Lands (he would later be its chairman), wrote a scathing letter to Dr. Merriam, which was published in the *Congressional Record*. "I desire to dissent most emphatically," he wrote about Loch-Katrine, "and to register my protest against the order in question."[98]

By January 1909 Roosevelt felt a great thirst rising in him. The idea of being a lame-duck president, sitting in a chair and letting his paunch expand, when he had nine weeks to achieve *American things*, was ludicrous. Roosevelt believed that instead of pardoning people, outgoing presidents should compile a list of social ills in need of solving. There would be no idling in the White House corridors. When the editor of *Saturday Review* asked Roosevelt how to sum up his executive modus operandi, the president's answer was revealing. "My business was to take hold of the Conservative [Republican] Party and turn it into what it had been under Lincoln," Roosevelt wrote, "that is, a party of progressive conservatism, or conservative radicalism; for of course wise radicalism and wise conservatism go hand in hand."[99]

As ex-president, Roosevelt hiked the canyonlands of the American Southwest, many of which he had saved by designating them as national monuments.

CHAPTER TWENTY-SIX

DANGEROUS ANTAGONIST: THE LAST BOLD STEPS OF 1909

I

To some, President Roosevelt looked tired during his last days in the White House; his exhausted face seemed to consist of only the two eyes, with dark shadows like a raccoon's mask. Unapologetic for riding roughshod over Congress, Roosevelt seemed to enjoy being an all-around nuisance on the Hill. In fact, he had developed a sense of mischievousness in early January 1909. While reporters were gossiping about William Howard Taft's cabinet appointments, the fifty-one-year-old Roosevelt started to put into motion his last bold conservationist acts as president. Owing to his intense connection with the outliers of America, he seemed to feel immune from the disapproval of Washington's opinion-makers. The *people* were with him. Conjuring up the ghost of Grover Cleveland—who had set aside 7 million acres of forest reserves as a parting gesture in 1897—Roosevelt was ready to trump his predecessor. "Ha ha!" Roosevelt had written to Taft on New Year's eve: "While you are making up your Cabinet, I, in a lighthearted way, have spent the morning testing the rifles for my African trip."[1]

Playing kingmaker, Roosevelt had selected the fifty-one-year-old Taft over the more senior Charles Evan Hughes, Elihu Root, and Joe Cannon to be his successor as the Republican candidate for president. The voters agreed that Taft was the logical choice; his experience as the chief civil administrator of the Philippines and as secretary of war qualified him to be the commander in chief. Keep in mind, though, that whoever was chosen by Roosevelt in 1908 would have been likely to defeat William Jennings Bryan, so Taft clearly owed his good fortune to Roosevelt. One evening during his second term, following his surprise announcement that he wouldn't be a presidential candidate in 1908, Roosevelt invited Taft (and Taft's wife, Nellie) to a private dinner at the White House. After dessert Roosevelt and the Tafts repaired to the library for serious conversation. Leaning back in his leather armchair Roosevelt began to pretend that he had prophetic powers. Squeezing his eyes shut, gazing

Roosevelt enjoyed chopping his own firewood and clearing his own paths in the wild.

upward, he jokingly chanted: "I am the seventh son of a seventh daughter. I have clairvoyant power. I see a man before me weighing three hundred and fifty pounds. There is something hanging over his head. I cannot make out what it is; it is hanging by a slender thread. At one time it looks like the presidency—then again it looks like the chief justiceship."

"Make it the presidency!" Nellie burst out excitedly.

"Make it the chief justiceship!" Taft quickly interjected, not wanting history to record that he had *asked* to be president.[2]

Of course, Taft welcomed Roosevelt's nod. But already in January 1909, Roosevelt was becoming lukewarm about his handpicked successor. Taft wasn't offering administrative jobs to young, progressive Republicans as T.R. had hoped. Many holdovers from Roosevelt's administration, in fact, were being dismissed. And Taft didn't even seem to know what the Biological Survey and the Forest Service were. On a number of occasions Taft treated Pinchot like someone to be brushed aside. Roosevelt, still only middle-aged, perhaps realized that he was going to miss wielding power. "I should like to have stayed on in the Presidency, and I make no pretense that I am glad to be relieved of my official duties," he wrote to Cecil Spring-Rice. "The only reason I did not stay on was because I felt that I ought not to."[3]

Adding to Roosevelt's discomfort with president-elect Taft was that Taft dismissed James R. Garfield as secretary of the interior in January 1909. Roosevelt kept his composure. But in his eyes Taft now had two strikes against him. "No one knows exactly why Taft chose to drop Garfield," historian M. Nelson McGreary writes in *Gifford Pinchot*, "but many of the various explanations boil down to a general feeling on the part of Taft that Roosevelt's Secretary of the Interior was a bit overzealous in his support of conservation and was inclined to stretch the exact letter of the law on occasion in order to protect what Garfield felt was the public interest."[4] Moreover, Roosevelt worried that Taft would be too frightened to initiate antitrust suits.

Some of Roosevelt's friends, sensing an impending rift between the outgoing and incoming presidents, scrambled to find a meaningful position for Roosevelt; the possibilities included becoming a U.S. congressman and becoming president of Harvard University. Civic organizations also sought ways to memorialize him in plaques, bas-reliefs, and newsletters. Only half-jokingly, Senator Philander Knox declared that Roosevelt "should be made a Bishop." And magazines offered Roosevelt substantial sums of money to write for them. He declined the offers from major magazines, on the principle of not selling out the presidency, but he did sign on with *Outlook*, a minor but important periodical for which he would write approximately a dozen long articles a year (like those that now appear on op-ed pages), for $12,000.[5]

Roosevelt's big money came from an arrangement with *Scribner's Magazine* to write a series of articles that would eventually become the book *African Game Trails*. Determined that his specimen collecting not be written up as a "game butchering trip," and wanting to frame his adventure as Wildlife Conservation, Roosevelt had Andrew Carnegie donate $75,000 for the presumably scientific expedition. With reporters, in fact, Roosevelt incessantly stressed that he was working for the Smithsonian Institution; only a few duplicate trophies would be kept for his wall at Sagamore Hill. And with due diligence Roosevelt started reading everything on such African explorer-hunters as Gordon Cummins and Cornwallis Harris.[6] Instead of mocking the thirty-four-year-old Winston Churchill as he had done in the fall, Roosevelt now carefully read Churchill's *My African Journey* (1908) and was impressed that the author had bagged a white rhinoceros.[7] Roosevelt also wanted to hunt a rhinoceros, because

America couldn't let the indefatigable Britons have the edge in natural history anymore.

To get into physical shape for Africa, Roosevelt would ride horseback every day to the point of exhaustion. "The last fifteen miles were done in pitch darkness and with a blizzard of sleet blowing in our faces," the president wrote his son Kermit. "But we got thru safely, altho we are a little stiff and tired nobody is laid up."[8] And he raced his favorite horse over fifty miles a day that January, as if preparing to charge up San Juan Hill. He would need to make the American scientific societies and explorers' clubs proud when he was abroad. Friends from the Cosmos Club were concerned that by trying to get into trim, Roosevelt would suffer a heart attack or stroke. Jokes circulated on Capitol Hill that "crazy Teddy" was going to die from the strenuous life, now that he was a fat ex-president loaded down with guns. Some people scoffed that with his White House tenure winding down, he was little more than a rusted bolt, hard to shake loose.

With no congressional legislation pending, Roosevelt, as interregnum presidents are likely to be, was written off as a lame duck that winter. Big business, in particular, anticipating revenge, couldn't wait for Roosevelt's passport to be stamped in some godforsaken African port. "Congress of course feels that I will never again have to be reckoned with," Roosevelt noted, "and that it's safe to be ugly with me."[9] In focusing so much on his forthcoming African safari, congressional Democrats, in particular, hoped Roosevelt's radical conservationist crusade would peter out. However, regardless of what dark thoughts Taft harbored about Gifford Pinchot personally, he nevertheless had to embrace Roosevelt's conservation on the campaign trail in 1908. To challenge Roosevelt on forestry would have been a death knell to Taft's candidacy. "If I am elected President," Taft said in Sandusky, Ohio, "I propose to devote all the ability that is in me to the constructive work of suggesting to Congress the means by which Roosevelt policies shall be clinched."[10]

After enduring T.R. in the White House for seven and a half years, the legislators should have known that he would be setting traps in early 1909. "I have a very strong feeling that it is a President's duty to get on with Congress if he possibly can, and that it is a reflection upon him if he and Congress come to a complete break," the president wrote to Ted Jr. "This session, however, they felt it was safe utterly to disregard me because I was going out and my successor had been elected; and I made up my mind that it was just a case, where the exception to the rule applied and that if I did not fight, and fight hard, I should be put in a contemptible position. While inasmuch as I was going out on the 4th of March I did not have to pay heed to our ability to cooperate in the future. The result has, I think, justified my wisdom. I have come out ahead so far, and I have been full President right up to the end—which hardly any other President ever has been."[11]

So, while wags laughed about Roosevelt's African trip, the president acted. Congressmen had been asleep in 1903 when Roosevelt had created his first federal bird reservation at Pelican Island, Florida. Now, with Dr. Merriam of the Biological Survey still his steadfast ally, Roosevelt began rapidly declaring federal bird reservations during the last six weeks of his administration. He began with the Hawaiian territory: on February 3, he signed Executive Order 1019, declaring an entire archipelago of the Hawaiian Islands a federal bird reservation.

In 1903 Roosevelt had sent U.S. Marines to safeguard the seabirds of the Midway atoll, securing it as an American possession. Now he was saving the relatively nearby islands around Nihoa Island to Kure atoll from plumers and Japanese poachers. If you looked in an atlas for these flyspeck islands—west of the world's largest lighthouse, on Kauai—the phrase "far-flung" might come to mind. But although human civilization might not have been thriving in this island group, the birds were there in magnificent numbers.[12] "To many these remote, shimmering, uninhabited islands are devoid of interest; to the naturalist, however, every square foot of the surface, and all the life that inhabits them, has an interesting story to tell," a professor of zoology, William Alanson Bryan of the College of Hawaii, remarked in 1915. "The geologist finds in them subjects of the greatest interest and importance."[13]

When Mark Twain wrote *Roughing It* in 1871, he presented Hawaii to the American reading public as an exotic place. Geologically, Twain had focused on a volcano (a "muffled torch"), claiming that in Hawaii lava flowed like a "pillar of fire."[14] To Roosevelt this was a sure sign of a lazy journalist resorting to clichés: Twain made no mention of humpback whales, spinner dolphins, koa trees, or yellow hibiscuses. And then there was Jack London, who went to Hawaii in the yacht *Snark* but knew nothing of sea urchins or the numerous types of dolphins and sadly believed that he grew funnier as the bottle emptied. Robert Louis Stevenson had spent time in Hawaii during 1889, but he had chosen to focus on leprosy.

From Roosevelt's perspective any indigenous Hawaiian chant—for example, the Kauai prayer "Hanohano Pihanakalani"—had more connection to the natural environment of Hawaii than a page by Twain,

London, or Stevenson. Natives sang of the exquisite uplands, mountain shells, mokihana trees, and rare terns.[15] But the literary "nature fakers" gave false pre-Darwinian descriptions of tropical vegetation that didn't even exist on the islands. Didn't Twain have eyes for the Hawaiian stilt, with its long pink legs? Couldn't London comment on the green sea turtles? Wasn't Stevenson competent enough to write about one of the forty species of sharks in the waters of Hawaii? The point was obvious: these literary men couldn't have distinguished a hammerhead from a whitetip reef shark. Why didn't important American writers study Hawaiian coral reefs, home to more than 7,000 marine species? Roosevelt believed that an accurate inventory of Hawaii's fauna and flora in the form of a popular book was sadly needed.

Roosevelt's intense interest in Hawaii went back to his expansionist fever of the 1890s. In 1893, in fact, Roosevelt had cheered as American colonists, mainly sugarcane growers, toppled a kingdom and replaced it with the republic of Hawaii. As assistant secretary of the navy he had lobbied Congress fervently to annex the island for strategic reasons. That effort had become a standard part of his public business in 1898. On April 30, 1900, Hawaii finally became an official U.S. territory; and a jubilant Roosevelt couldn't wait to loll on the sand beaches, watching the surf roll in, perhaps using the Hawaiian Islands as stepping-stones on his way to Japan or Australia.

Doing his homework, Roosevelt now homed in on saving the far northwestern Hawaiian Island chain in the mid-Pacific, including Nihoa, Necker, French Frigate Shoals, Gardner Pinnacles, Maro Reef, Laysan Island, Lisianski Island, and Pearl and Hermes atolls, before leaving the White House. There were twenty-one islands plus smaller ones contiguous to the big nine. Executive Order 1019 designated the entire chain as a single federal bird reservation.* While Pearl Harbor was the supreme port, the Hawaiian Islands Federal Bird Reservation had such stunningly diverse wildlife—including blue gray noddies, wolf spiders, and monk seals—that as an ecosystem it defied classification.[16] Geologists supposed that these islands were the summits of submerged mountains, but that was only a guess. An intruder's foot would step on a bird burrow with practically every stride.[17] Everywhere there are birds," ornithologist William Palmer wrote on the islands of Executive Order 1019, "thousands upon thousands of albatross, white and brown, in great, distinct colo-

*The exception was Midway, which in 1909 was the relay station for the commercial Pacific Cable Company's trans-Pacific wire.

nies; great rookeries of terns and petrels and frigate birds; countless rail run everywhere in the long grass; bright red tropical honey birds, bright yellow finches flutter in the shrubs; curlews scream; ducks quack; crake chirp all the day."[18]

Today, 100 years after Roosevelt created the Hawaiian Islands Federal Bird Reservation, many of the small lagoons have still not been mapped. Biologists called the archipelago a wonder of ornithological activity. On French Frigate Shoals alone, in 1909 there were eighteen species of seabirds including the rare black-footed albatross, Laysan albatross, Bonnien petrel, Bulwer's petrel, and wedge-tailed shearwater, among other unusual species. On French Frigates Shoals a 120-foot volcanic rock rose over the lagoon, which was alive with growing reefs.[19] At Gardner Pinnacles—two barren-looking rock outcroppings—new spider species would soon be discovered. The first aerial photograph of Maro Reef showed a porkchop-shaped atoll in which the coral reefs shot out like spokes from a huge wheel. Discovered by an American whaling ship in 1920, Maro Reef was a "rough quadrangular wreath of white breakers."[20]

A real Robinson Crusoe adventure could be had at Laysan Island, where few if any human footprints could be found. The island was discovered in 1828 by an American captain sailing to the Orient, and the ornithologist Walter K. Fisher had spent a week there during Roosevelt's presidency, declaring in *National Geographic* that it was one of the most remarkable places on the planet; bold young albatrosses came up to him to be patted.[21] The island—three miles in length and two and a half miles in breadth—was a wondrous kingdom for ornithologists because it had a highly saline lake (one of only a few natural lakes in all of Hawaii around which birds congregated).[22] Just as there is now regular or premium gasoline, guano from Laysan Island was once considered the best fertilizer. The specific mix of bird excrement and coral sand formed a rich calcium phosphate coveted for fuel in California.[23]

What angered Roosevelt in February 1909 was that shiploads of rabbits had been released on Laysan Island by a Honolulu slaughterhouse firm in the hope that these hares would get fat on the thick vegetation—then, when they were at their maximum weight, they would be killed for a shish kebab eaten at Honolulu luaus. The problem was that the rabbits were devastating the tropical ecosystem of the island. Rare plants were being eaten down to nubs.[24] By issuing Executive Order 1019, Roosevelt officially banned such releases of rabbits. Preservationist discipline had arrived in the Hawaiian Territory. Because Laysan Island was so remote, however, policing it to enforce the law against plumers and rabbit raisers would be difficult. Since the entire Hawaiian Islands Federal Bird Reservation was administered by USDA, which didn't own boats in Hawaii, Roosevelt employed vessels of the U.S. Revenue Cutter Service to patrol against poachers and rabbit breeders. The U.S. fish commissioner likewise was ordered to protect the new federal bird reservation, including the ocean whales, from commercial fishing and hunting companies.[25]

There was also an archaeological component to the Hawaiian Islands Federal Bird Reservation. On the small basalt island of Necker, only forty-six acres in area, numerous religious relics had been discovered. Fifty-five "cultural places" were unearthed there. Many of the discoveries were stone enclosures designated as *wahi pana* (religious shrines) and filled with *makamae* (cultural artifacts). Supposedly, Necker Island had been the last refuge for a Pygmy-like race, the Menchune, who had been chased there by Polynesians. Over the centuries, native Hawaiians made Necker a sacred ceremonial site. In 1989, the George H. W. Bush administration would list all of Necker Island on the National Register of Historic Places.[26] Every square inch of the island—700 miles northwest of Honolulu—was an antiquities site.

Marine biologists today consider Roosevelt's Executive Order 1019 a stupendous moment in oceanographic history because it preserved the great bird and seal rookeries of Hawaii from human exploitation. Now called the Hawaiian Islands National Wildlife Refuge, the chain continues to serve as nesting areas for more than 14 million breeding seabirds, waterfowl, wintering shorebirds, endangered turtles and seals, and legions of whales. It was Roosevelt's counterpart of the Galápagos, a gift to marine biology, the world's largest oceanic conservation area, where evolution was happening incrementally in a discernible way. Here, in the westernmost islands of Hawaii, the food chain remained intact.

Yet if you pick up a Hawaiian guidebook at, say, Barnes & Noble and thumb through the index, you probably won't find Roosevelt's name. This omission is based on Roosevelt's never having visited Hawaii in a fishing craft, yacht, or naval vessel. But you will learn in Hawaiian guidebooks that Twain said that Oahu "beseechingly" haunted him, that Stevenson found the leprosy colony "a land of disfigurement and disease," and that London had called himself a *kama'aina* ("child of the island"). Only Roosevelt himself, in his *A Book-Lover's Holidays in the Open* (1916), noted the importance of Executive Order 1019 to the marine biology of what he called "the western extension of the Hawaiian archipelagos."[27]

Because Hawaii was a territory in 1909—it did not become a state until 1959—there was no dissent from Capitol Hill over Executive Order

1019. Only the Hawaiian rabbit breeders and eggers were up in arms. The territorial governor, Walter Frear of California, likewise had a few grumbles. None of this amounted to more than a few angry letters for public consumption. In 1911 the Field Museum of Chicago sent Charles A. Corwin to the Hawaiian Islands Federal Bird Reservation to inventory species. His report on Laysan Island alone was so impressive that it made national news. "It has been established that the island is inhabited by at least 8,000,000 birds, most of which consist of two species of Albatross," Corwin wrote in the *Christian Science Monitor*, a publication that took ornithology seriously. "There were so many birds on the ground, nesting, that we had to crowd our way through to avoid stepping on them. . . . We can fully verify the stories that these strange birds have a peculiar dance which resembles the cake walk. They clap their bills together and waddle about with high stepping antics, ducking their heads first under one wing, then under the other. All through the dance they whistle and utter weird sounds."[28]

With the Hawaiian Islands Federal Bird Reserve created, the Biological Survey sent Roosevelt an additional twenty-five bird reservations to sign before March 4. While the reporters were abuzz over Taft's choices for his cabinet and about the inaugural festivities, the Roosevelt administration would stealthily save birds' breeding grounds. Unnoticed by the *Washington Post* and *New York Times*, the final "I So Declare It" hour had arrived for American ornithologists. Suggestions from Job, Finley, Dutcher, Chapman, Beebe, and other enthusiasts came pouring into the Biological Survey office at the Department of Agriculture. There was a real sense of urgency. On February 25, it virtually rained federal bird reservations in America, and nobody in interregnum Washington had the power to object. For ornithologists of the twenty-first century, these areas are hallowed not only because of their wildlife but also for their green underbrush and marshlands: Salt River (Arizona), Deer Flat and Minidoka (Idaho), Willow Creek (Montana), Carlsbad and Rio Grande (New Mexico), Cold Springs (Oregon), Belle Fourche (South Dakota), Strawberry Valley (Utah), Keechelus, Kachess, Clealum, Bumping Lake and Conconully (Washington), Pathfinder and Shoshone (Wyoming).

What fantastic American names! Stephen Vincent Benét should have written a poem about them. Just reading them aloud is poetry. The two federal bird reservations that the Roosevelt administration likewise created in California were very special, unique in their biodiversity. First, on February 25, came East Park, located in Colusa County an hour north of Sacramento. Here thousands of swans were struggling to coexist with humans around the cool water of a medium-size lake. A very rare tricolor blackbird also used the watering-hole as a home base for much of the year.[29] In springtime the rolling hills of East Park bloomed with the rare adobe lily (*Fritillaria pluriflora*) and residents of Sacramento came to camp along the lake free of charge. In 1921 the Bureau of Reclamation of the U.S. Department of the Interior took over control of East Lake. Nowadays, even with a hydroelectric dam, the thriving birdlife and flower fields remain federally protected.

But the true gemstone of February was the Farallon Islands, a spectacular cluster of rocky islands in the Pacific Ocean twenty-five miles from Golden Gate Bridge. An inventory of wildlife in the Farallons runs for pages. The islands remain the largest nesting colony south of Alaska and the world's biggest colony of western gulls. Half of the ashy storm petrels continue to reside on these offshore bird rocks. Marine whales congregate around the Farallons, as do southern sea otters, Dall's porpoises, minke whales, orcas, and northern elephant seals. Every square foot is full of special designs. Although vegetation is limited, there are clusters of pigweed, tree mallow, false clover, plantain, curly dock, and Monterey cypress. Little skiffs from the San Francisco Bay area often circled around the islands (the largest island is seventy acres in area) to see the colorful tufted puffin or watch two-ton elephant seals at Saddle Rock dispute during the breeding season over harems. A biological site of the rarest kind, the Farallons were influenced by the California current, westward winds, upwelling mixing, deep cold water, and finally, in February 1909, by the Roosevelt administration's "citizen bird" policy.[30]

From Hawaii to Florida and Michigan to Alaska, birding areas and habitats were put off-limits and set aside for the future. Their status would be enforced by the Biological Survey, National Audubon Society, and AOU. In Hawaii, Oregon, Washington, Louisiana, and Florida the plumers were isolated and suppressed, forced to operate clandestinely, like criminals. And when Roosevelt wrote that birds should be saved for reasons "unconnected with dollars and cents," he wasn't speaking just for himself.[31] Groups like the National Audubon Society and AOU had swept across the land. The unstated presumption was that weird beaks, webbed claws, and wingspreads had aesthetic value for a generation of Americans who championed "citizen bird" and were fed up with market butchers. When Senator George C. Perkins, (among other things) a forestry advocate, questioned T.R. over wildlife protection policy the president snapped. "I would like to break the neck of the feebly malicious angleworm who occupies the other seat as California's Senator," Roosevelt wrote to his son Ted; "he is a milk-faced grub named Perkins."[32]

With the exception of Henry Ford, Andrew Carnegie, and a few others, industrialists of the time were perplexed by Roosevelt's passion for protecting loons, cormorants, and herons. What would lead a hero of the Spanish-American War to insist that shooting a dark-eyed junco was akin to grand larceny? Didn't Roosevelt's infatuation for birds constitute fanaticism? Roosevelt, for example, had shut down prime Florida real estate to make life comfortable for egrets. "I have no command of the English language that enables me to express my feelings regarding Mr. Roosevelt," the railroad tycoon Henry Flagler said. "He is shit."[33]

Flagler was correct in considering Roosevelt a vicious adversary. As Dwight D. Eisenhower once noted, T.R., despite his disarming toothy smile, "feared nobody" and was a "dangerous antagonist." And there was a cost to interfering with Roosevelt's love of pelicans and petrels: T.R. would blast his anticonservationist critics as hopeless "spoilers," "fools," "demagogues," "exploiters," and "idiots." The novelist Booth Tarkington, long after Roosevelt died, reflected on how T.R. had subjected people he perceived as anticonservationist to verbal tirades. "The Colonel had his own way of saying 'swine,' and it gave that simple Gothic word a peculiarly damning power," Tarkington wrote. "The swine he had in mind seemed to be incomparably more swinish than the ordinary swine that other people sometimes mention."[34]

Likewise John Burroughs explained in his *Journals* how Roosevelt could be a cutthroat political adversary when it came to the defense of songbirds. "He was a live wire, if there ever was one, in human form," Burroughs wrote in his journal. "His sense of right and duty was as inflexible as adamant. His reproof and refusal came quick and sharp. His manner was authoritative and stern. He was as bold as a lion and, at times, as playful as a lamb."[35] Every word Roosevelt uttered on behalf of wildlife protection seemed to have an exclamation point. When it came to squaring-off with politicians he was a specialist at haranguing, never crying uncle, giving his opponents the back of his hand. When challenged about the legality of his national monuments, federal bird reservations, or national forests, he took on that lean, ravenous look of a famished wolf. America had a new conservationist code to promote, thanks to Roosevelt: pummel the exploiters until they were licked.[36] He believed that his successor, William Howard Taft, would take just such a line.

Taft had even promised to keep Pinchot as the head of the Forest Service—a real concession following the dismissal of Garfield. Roosevelt had created his own circle or set of outdoorsmen in government, including Pinchot and Garfield, who agreed with his every sales pitch, plea from the bully pulpit, executive order, sermon about outdoorsmen, and writ concerning wildlife. And the first eight weeks of 1909 were their golden time. These acolytes weren't Luddites ready to sabotage the cutting blades in defense of nature. Many were evolutionists—such as David Starr Jordan and Henry Fairfield Osborn—who understood the arcane biology of how conjunction took place in colonial protozoa and cared deeply about species recognition marks. Roosevelt had nevertheless kicked down the door of the wildlife protection movement, knocking it off its hinges, by giving evolutionists places of honor at his table in the White House. As an ex-president, Roosevelt would travel to Florida and Louisiana to inspect his revolutionary rookeries, pleased at his accomplishments on behalf of "citizen bird." "The number and tameness of the big birds showed what protection has done for the bird life of Florida of recent years," he wrote after inspecting his federal bird reservations in the Gulf of Mexico. "The plumed lesser blue herons, and more rarely the great blue heron and the lovely plumed white egret, perched in the trees or flapped across ahead of the boat."[37]

II

While Merriam remained Roosevelt's go-to biologist in early 1909, Pinchot was first among equals in T.R.'s conservationist group. Pinchot, one of the most effective administrators to ever operate within the federal bureaucracy, expanded the U.S. forest reserves from approximately 43 million acres to about 194 million acres from 1901 to 1909. Besides serving on such presidential commissions as the Organization of Government Scientific Work and Public Lands (1903), Department Methods (1905), Inland Waterways (1907), and Country Life and the National Conservation Commission (1908), Pinchot had also led the highly successful governors' conference of 1908. A tireless worker, he crisscrossed America promoting forestry, and he established the first press bureau ever within a federal agency. Much like Roosevelt—his boss and hero—Pinchot manipulated the press like a puppeteer, knowing exactly which strings to pull. Pinchot's home on Rhode Island Avenue in Washington, D.C., became a veritable salon where smart journalists could have a drink and talk about the western reserves. Roosevelt noted of Pinchot that "among the many, many public officials who under my administration rendered literally invaluable service to the people of the United States, he, on the whole, stood first."[38]

On February 18, 1909, Roosevelt convened the North American Conservation Conference at the White House, following Pinchot's directive that conservation had to go global.[39] "The keynote of the conference was that international streams are affected by cutting forests on either side of the boundary line," the *Washington Post* wrote, "and that conservation plans, to be the most practical, must be international."[40] The U.S. delegates at the North American Conservation Congress included Chief Forester Pinchot, Secretary of State Bacon, and Secretary of the Interior Garfield (all members of the "tennis cabinet"—the term that reporters had given to Roosevelt's inner circle at the White House). Ottawa and Mexico City sent their appropriate counterparts. Meetings were held at the White House, at the State Department, and at Pinchot's home. After first agreeing to some "conservation measures" of "continental good" the statesmen announced a trilateral "Declaration of Principle" aimed at the creation of a World Conservation Congress to meet at The Hague in September 1909 to push Rooseveltian conservationism forward on a global level. Pinchot's dream, in fact, was for President Taft to go to The Hague and champion issues like international wildlife reserves, parks, and tree planting; the banning of poaching and overfishing of waterways; and forest managing techniques. As Pinchot envisioned it that February, the global conference, to be held while T.R. was in Africa in the fall, would start a conservation revolution around the world.[41]

Before Roosevelt left office, fifty-eight nations had, in fact, received invitations from the White House to meet at The Hague. Immediately, Great Britain, France, and Germany accepted. Other nations soon followed suit.[42] But after only a few weeks in office, President Taft balked and called off the conference, feeling that Pinchot had overreached himself. The World Conservation Congress never had a chance to succeed. As the historian Paul Russell Cutright put it in *Theodore Roosevelt: The Naturalist*, "the project died aborning."[43] The rift between Taft and Pinchot was becoming unbridgeable. "Pinchot, bitterly disappointed, never gave up hope of holding such a conference," McGeary wrote in *Gifford Pinchot*, "especially since he later became firmly convinced that conservation of natural resources was a primary means of insuring a permanent peace."[44]

That Taft was uninspired by Pinchot's idea of a World Conservation

Congress disappointed Roosevelt greatly. A breach in policy was starting to manifest itself between the incoming (conservative) and outgoing (progressive) GOP administrations. In December, for example, during a meeting on conservation at the Belasco Theatre in Washington, Pinchot had quite innocently introduced Taft as the "president elect." "I'm not the President-elect," Taft had snapped when he stood on the podium. "That is merely the imagination of Mr. Pinchot. I'm just a private citizen."[45]

Determined to keep Roosevelt's torch lit in public policy, Pinchot and Garfield co-wrote *The Fight for Conservation* (1910). When the first draft of the book was finished that February, Garfield asked the departing president to write a "foreword." Basically they were looking for a letter of recommendation, a strong show of support, an embrace (in writing) from their boss. Somehow Roosevelt found time to compose a fine summation of the sterling qualities of his chief forestry advocates, giving his colleagues his indelible stamp of approval in glowing language. "Both have stood for absolute honesty, for absolute devotion to the needs of the public," Roosevelt wrote. "Both have stood no less for entire sanity and for farsighted understanding of the many diverse needs of the Nation. They have been fearless in opposing wrong, whether by a great corporation or by a mob; by a wealthy financier, or by a demagogue."[46]

January to March was a very busy period for the U.S. Forest Service. An eager Pinchot had decided to challenge the anti-Roosevelt forces one last time with many new forest reserves in the West. Pinchot and Garfield had been gathering data on what forestlands were doable. Roosevelt was well past worrying about the political repercussions of any last-minute reserves. "Keep it legal" was his only direction to Pinchot. Echoing Thoreau, Roosevelt planned on leaving the White House promoting forests and meadows and even corn as a sort of ecumenicism.[47] Pinchot and Garfield, working with a few like-minded congressmen, made their mentor proud. Starting on January 20 with Humboldt, Nevada, and not stopping until the last hours of the last days of his administration on March 2 with Sequoia, California, Roosevelt declared thirty-seven new national forests: Humboldt, Moapa, Nevada, and Toiyabe (Nevada); Cleveland, Calaveras Bigtree, Modoc, California, Shasta, Trinity, Lassen, Plumas, Tahoe, and Sequoia (California); Klamath (California and Oregon); Mono (California and Nevada); Pecos, Gila, Datil, Lincoln, Alamo, and Carson (New Mexico); Prescott, Tonto, Sitgreaves, and Apache (Arizona); Zuni (Arizona and New Mexico); Dixie (Arizona and Utah); Marquette and Michigan (Michigan); Superior (Minnesota); Black Hills (South Dakota and Wyoming); Sioux (Montana and South Dakota). And for the first time Arkansas, without a battle, was brought into the fold with two new forest reserves: Ozark and Arkansas.[48]

III

By far the biggest new national forests that season, in terms of acreage, were in the Alaska Territory. Since his creation, on August 20, 1902, of the Alexander Archipelago Forest Reserve, Roosevelt had steadily added Alaskan land to the federal government's holdings. On September 10, 1907, he had created the Tongass National Forest, an immense area in southeastern Alaska with huge strands of western red cedar, Sitka spruce, and western hemlock. The reserve also included thousands of islands and salmon-rich streams. On July 1, 1908, Roosevelt merged the Alexander Archipelago with Tongass to create what is today Tongass National Forest, a monstrous eco-zone of snowcapped peaks, blue glaciers, clear streams, muskeg, forests, and fjords. More than 17 million acres of pristine Alaska were preserved between the two forests, making Tongass National Forest the largest reserve in the United States. The diversity of salmon alone made the Tongass the crown jewel of the USDA's forest reserves. King (chinook); coho (silver); pink (humpback); sockeye (red); and chum (dog) salmon flourished in these icy waters, as nowhere else on earth except Canada. And where there were salmon, grizzly bears and black bears came in huge numbers.

The creation of Tongass National Forest was a tremendous presidential accomplishment. But as Roosevelt's days in the White House waned during early 1909, he yearned to do even more to both protect and develop the "last frontier." To Roosevelt the "spell of Alaska" wasn't gold mining but the rain forests and wildlife breeding areas saved for posterity. Anybody who shot a northern spotted owl, for example, should be arrested. But Roosevelt always wanted *smart* community development. After all, it was the Roosevelt administration that in 1905 had pushed an act through Congress calling for a first-class road system to be laid out in Alaska.[49] The act was an example of how the U.S. government—not railroad titans or oil companies—could help Alaska develop. The enemy in Alaska, as Roosevelt saw it, was huge companies that bought up land tracts to mine, log, or drill. Roosevelt thought the U.S. government instead of the private sector should build and operate a short-line railway to connect Controller Bay with Alaska's great coalfields.[50] And Roosevelt wanted federal protection of sea otters and stringent government regulations on fur seal hunters and fisheries.

The promotion of the Elkins Act of 1903, the creation of the Bureau of Corporations, and the bills regulating railroad rates all showed Roosevelt's desire to enhance executive powers. Roosevelt believed that because Seward had purchased Alaska in 1867 with federal funds, the U.S. government should dictate how the territory was developed. "Roosevelt's belief in the benevolent power of executive action found particularly vivid expressions in his mature conservation program," Joshua David Hawley wrote in *Theodore Roosevelt: Preacher of Righteousness*; this statement is especially true regarding U.S. territories like Alaska. "The ruinous overharvesting of the country's timberlands, severe water shortages in the arid West, the rapid deterioration of the famed open ranges, and the more gradual but steady disappearance of suitable wildlife habitats nationwide—all these developments convinced Roosevelt as president that the United States needed a rational, unified policy to manage the nation's natural resources. As in the field of economic regulation, Roosevelt concluded here as well that, when it came to conservation policy, Congress was not up to the task."[51]

Roosevelt had received an advance copy of Major-General A. W. Greely's *Handbook of Alaska: Its Resources, Products, and Attractions*, published later in the spring of 1909. It had a moose as its frontispiece and was filled with photographs shot by members of Roosevelt's administration in the Coast and Geologic Survey, the Geological Survey, the Bureau of Fisheries, and the Biological Survey. At the end there was a handsome pull-out map of Alaska. With his stumpy forefinger Roosevelt would trace down the Porcupine River's course from Yukon Province to Tanana, Alaska, proud that the landmass was part of the United States. Had anybody ever hiked the Kokrines Hills? Or floated down the Kuparuk River from the Chukchi Sea? How exciting—America owned fjords to rival Norway's, and glaciers larger than England. To Roosevelt, Alaska was the hardiest explorers' paradise. According to measurements by the Boone and Crockett Club, the vast territory was one-third greater than all the Atlantic states from Maine to Florida. Roosevelt had read a great deal about the southern two-fifths of Alaska, which encompassed the densely wooded Tongass National Forest, but he was somewhat ignorant about the watersheds of the Yukon and the lesser Kuskokwin rivers, and the large upper-fifth (essentially treeless) shores of the arctic coast. The *Handbook* filled in the blanks for the departing president.[52]

IV

While Roosevelt opened the White House to ornithologists like Job, Chapman, Finley, and Dutcher in February 1909, he also never turned down an opportunity to consult with renowned hunters from around the world. As he prepared for Africa, he found these hunters marvelous company and was almost greedy for their companionship. "He had a great respect for genuine hunters—the kind who endure hardship, exhibit prowess and tell the truth—such men as Selous, Warburton Pike, George Bird Grinnell and Charles Sheldon," Merriam recalled. "Early in his career in the White House he asked to be notified when out-of-town hunters came to the city. So when the Canadian hunter and sub-arctic explorer Warburton Pike had arrived, Sheldon and I were requested to bring him to dine at the White House. The time happened to be a particularly busy one politically, and we were warned that the President must excuse himself directly after dinner. But instead, he took us upstairs and kept us in his den till midnight. He was several times interrupted by messengers, but declined to see them. Finally his son-in-law (Nicholas Longworth) came with an important telegram. Roosevelt waved him away with the remark that he was not to be interrupted—that for this one night he felt entitled to enjoy himself."[53]

During his last weeks at the White House Roosevelt wrote his old editor at the Boone and Crockett Club, Caspar Whitney, a long letter about hunting and the strenuous life. Roosevelt recounted how over the decades he had effectively compensated for being a poor shot by having great will in stalking prey. "In short, I am not an athlete; I am simply a good, ordinary, out-of-doors man," Roosevelt wrote; his statement was passed around to fellow club members. "The other day I rode one hundred and four miles. Now this was no feat for any young man in condition to regard as worth speaking about; twice out in the cattle country, on the round-up, when I was young I myself spent thirty-six hours in the saddle, merely dismounting to eat, or change horses; the hundred-mile ride represented what any elderly man in fair trim can do if he chooses. In the summer I often take the smaller boys for what they call a night picnic on the Sound; we row off eight or ten miles, camp out, and row back in the morning. Each of us had a light blanket to sleep in, and the boys are sufficiently deluded to believe that the chicken or beefsteak I fry in bacon fat on these expeditions has a flavor impossible elsewhere to be obtained. Now these expeditions represent just about the kind of things I do. Instead of rowing it may be riding, or chopping, or walking, or playing tennis, or shooting at a target. But it is always a pastime which any healthy middle-aged man fond of outdoors life, but not in the least an athlete, can indulge in if he chooses."[54]

Eighteen months after leaving the White House, Roosevelt wrote "The Pioneer Spirit and American Problems," a piece for *Outlook* about the American West, praising Cheyenne, Denver, and Omaha. Sounding rather like Walt Whitman, Roosevelt also celebrated ranchmen, miners, cowpunchers, mule skinners, and bull whackers. He boasted about Indians, pioneers, and buffalo herds. Symbolically, Roosevelt was turning his back on the east coast yet again. He called Abraham Lincoln a westerner because Lincoln had the western trait of "undaunted and unwavering resolution." And again, Roosevelt claimed that the federal government knew what was best for the West with regard to water power, forest reserves, land leasing, grazing rights, and so on. According to Roosevelt, America always had to have a national system of government, and anybody promoting states'-rights or territories'-rights schemes were charlatans.

"The man of the West throughout the successive stages of Western growth has always been one of the two or three most typical figures," Roosevelt wrote, "indeed I am tempted to say the most typical figure in American life; and no man can really understand our country, and appreciate what it really is and what it promises, unless he has the fullest and closest sympathy with the ideals and aspirations of the West."[55]

How earnestly Roosevelt still believed that the American character was connected to the restorative power of outdoors life was evident in recommendations he gave the U.S. Army War College before stepping down as president. Having learned that the army's regulations for field service were being modernized, he inserted himself into the internal debate over the revisions. Even though the forerunners of jeeps were being introduced to the army, Roosevelt stood by the symbolic power of horses. Fixing cars was fine, and mechanical skills were always helpful, but all army servicemen, Roosevelt believed, needed to be skilled climbers and equestrians. "We must not allow packing to become a lost art in the army," Roosevelt wrote in a report. "The old timers in the Rocky Mountains are passing away and we must look to the army to keep up the knowledge of this essential and invaluable service art."[56]

One American range that Roosevelt was itching to explore in February 1909 was Washington's fog-shrouded Olympic Mountains. Before he headed to Africa, Roosevelt saved Mount Olympus (7,980 feet in elevation) as a national monument. Ever since Merriam had named the huge elk in the Olympic range *Cervus roosevelti*, the president had hungrily read whatever he could about the region. As *Time* would later note, geologists were unsure exactly how snow-mantled Mount Olympus had been formed. Folklore in Washington state claimed that Paul Bunyan had journeyed to Puget Sound to milk an orca to cure Babe (his devoted blue ox) of a life-threatening illness. When Bunyan thought Babe was going to perish, he dug a huge grave, and the dirt pile became Mount Olympus.[57] The very fact that such a folktale survived in the Pacific Northwest was a sure sign that wilderness still existed in the Cascades circa 1909—and the same would be true of the later legend of Bigfoot. Somehow it was fitting that on March 3, Paul Bunyan, Babe, Bigfoot, the giant elk, and Theodore Roosevelt became linked in the Cascades folklore with the creation of Mount Olympus National Monument. Furthermore, Professor Daniel G. Elliot, a co-author of *The Deer Family*, had been promoting national monument status for Mount Olympus for over two years.

Since the successful reintroduction of bison on the Wichitas and Flathead reserves, Roosevelt had tried to create the National Elk Reserve in the Olympics. However, Congress had prevented him from saving remnant bands as part of Olympic Elk National Park. Roosevelt was furious at being stymied. His Grand Canyon National Game Reserve was a model of proper wildlife reserve management. So, Roosevelt circumvented Congress on March 3, declaring Mount Olympus and 615,000 acres of lush valley, spectacular mountains, fish-filled streams, and old-growth forests off-limits to any kind of hunting, timbering, or extraction.[58] Quite literally, Mount Olympus was North America's most amazing temperate rain forest. In Proclamation No. 869 creating the national monument, Roosevelt described the ecosystem as follows: "the slopes of Mount Olympus and the adjacent summits of the Olympic Mountains . . . embrace . . . numerous glaciers, and . . . the summer range and breeding grounds of the Olympic Elk . . . a species peculiar to these mountains and rapidly decreasing in numbers."[59]

Corporations in Seattle and Tacoma erupted in protest over Mount Olympus National Monument. As with the Grand Canyon, the dissenters said, T.R. was abusing the intent of the Antiquities Act. The onetime mayor of Seattle, Richard A. Ballinger, a reformer and low-key conservationist, thought Mount Olympus should be timbered. Comical remarks circulated in newsrooms that Roosevelt was leaving the White House to dwell on Mount Olympus with his buddies Jupiter and Apollo and his supersize species of Irish elk. The *New York Times* eventually defended Roosevelt's new national monument, addressing the humorists by concluding,

"Mr. Roosevelt is always right here on earth. If he were on Olympus there would be no room there for Jupiter or Apollo."[60]

What a wondrous, moss-draped world the Mount Olympus rain forest was! For a lover of big game, seeing a 700-pound Roosevelt elk browsing in the lush primeval forest constituted an experience unequaled in North America. Mount Olympus, with glaciers descending its rugged slope, was to this national monument what Old Faithful was to Yellowstone. The northern spotted owl, a threatened species, could be found easily on Mount Olympus by just hiking the trails. Three sui generis Olympic species—the Olympic marmot, Olympic snow mole, and Olympic torrent salamander—weren't found anywhere else in the world. For sheer picturesque beauty, Soleduck Falls, swollen from frequent rains and melting snows, was also in a class of its own. Sometimes naturalists described Olympic as three parks in one—dramatic glacier-capped mountains, miles of wild Pacific coast, and a temperate rain forest. Mount Olympus also had the largest Alaskan cedar and Douglas firs in America.

When Woodrow Wilson became president in 1913, he deemed Roosevelt's Proclamation No. 869 excessive and downsized Mount Olympus National Monument from 615,000 to 300,000 acres. Roosevelt was furious. When he died in 1919, a movement was under way to return the ecosystem to its original 615,000 acres. But presidents Wilson, Harding, Coolidge, and Hoover were unwilling to do it, wanting to secure the "timber vote" of the Olympic Peninsula. At last, in 1938 Franklin D. Roosevelt, by presidential proclamation, returned his cousin's acreage to the monument. And he changed Mount Olympus's designation to Olympic National Park. In the early 1950s about seventy miles of rugged Washington coastline was added to the park by President Harry Truman. Joining the Roosevelt elk in federal protection were whales, dolphins, sea lions, and sea otters. The entire Olympic Peninsula, with the surf slapping hard on the offshore rocks, was Bar Harbor and then some. Four valleys facing the ocean—the Queets, the Quinault, the Hoh, and the Bogachiel—produced endless botanical surprises for visitors. Juncos flit about the grasses and woodpeckers excavate old logs. Deer are everywhere. The tide pools full of invertebrates of countless shapes, sizes, colors, and textures at Olympic National Park could have kept Charles Darwin busy for 100 years.[61]

V

During his last days in the White House, Roosevelt thought a lot about Charles Darwin. His knowledge of the great naturalist had matured since he drew men as storks when he was a teenager in Dresden. In preparation for his African expedition Roosevelt had *On the Origin of Species* wrapped in a waterproof cover to avoid damaging it on safari. It was part of what he called his "pigskin library" of favorite titles. On February 25, just two days before he created his federal bird reservations, Roosevelt had reflected on Darwin. "I think the trouble about Darwinism is that people confound it with evolution," Roosevelt wrote to James Joseph Walsh, a professor of physiological psychology who had just dedicated his book *Catholic Churchmen in Science* to President Roosevelt. "I suspect that all scientific students now accept evolution, just as they accept the theory of gravitation, or the general astronomical scheme of solar system and the stellar system as a whole; but natural selection, in the Darwinism sense, as a theory, evidently does not stand on the same basis. It must be tested, as the atomic system is tested, for instance."[62]

Other biological books added to the "pigskin library" included Darwin's *Voyage of the Beagle* and Huxley's *Essays*.[63] Roosevelt had also started reading a book on neo-Darwinism by Hans Driesch, *The Science and Philosophy of the Organism* (lectures delivered in 1907–1908 at the University of Aberdeen). This work, which was both factual and meditative, proved to be a helpful scaffolding for Roosevelt's Oxford University lectures, "Biological Imperatives in History," in 1910. A German biologist known as a pioneer in embryology, Driesch spent years operating out of the Marine Biological Station in Naples, Italy, experimenting with the division of embryo cells of sea urchins. These were complicated experiments encompassing biology, physics, mathematics, and philosophy. Roosevelt strove to understand Driesch's findings in depth. He was enthralled to find that Driesch had included the issue of coloration of bears in *The Science and Philosophy of the Organism*. "Do we understand in the least why there are white bears in the Polar Regions if we are told that bears of other colours could not survive?" Driesch inquired. "In denying any real explanatory value to the concept of natural selection I am far from denying the action of natural selection. On the contrary, natural selection, to some degree, is self evident."[64]

Roosevelt also scrounged around for grant money to help Professor John A. Lomax of the University of Texas continue his fieldwork collecting cowboy ballads. Roosevelt went hat in hand on Lomax's behalf to potential benefactors. Writing to the president of the Carnegie Institution, Roosevelt asked that Lomax be given $1,000, quickly, for his "very original and instructive study into a phase of native American literary and intellectual growth which of course has been totally neglected."[65] However, his request was rejected.

Frustrated that some intellectuals rolled their eyes at the idea of cowboy culture as serious academic fare, Roosevelt turned to the French ambassador, Jean-Jules Jusserand, to find financial assistance for Lomax. "A Texas professor is doing some really good work in collecting frontier ballads in the cow country of Texas," he wrote to Jusserand. "They are of course for the most part doggerel (as I believe to be true with the majority of ballads as they were originally written); but these are interesting because they are genuine. The deification of Jesse James is precisely like the deification of Robin Hood and the cowboy is a hero exactly as the hunter of the greenwood was a hero. Also, the view taken of women seems to be much the same as that taken in many of the medieval ballads."[66] This time Roosevelt found Lomax a grant.

Filled with growing enthusiasm for Texan culture, Roosevelt wrote an introduction for Lomax's *Cowboy Songs and Other Frontier Ballads* (1910).[67] At the same time he drafted speeches that he planned to deliver at Oxford, the Sorbonne, and Berlin University in 1910, once he emerged from the African wilds. Darwinian themes were apparent in all three addresses, accentuated by his own philosophy of the strenuous life. Then, on March 2, with just forty-eight hours left of his presidency, he wrote heartfelt thanks to Merriam, Garfield, and Pinchot. Together, these warriors had challenged the titans of the Gilded Age and beyond—railroads, timber companies, and mine owners.[68] Only in writing to Edith or to his children had Roosevelt ever been so sentimental and emotional. "As long as I live I shall feel for you a mixture of respect and admiration and a general affectionate regard," he wrote to Pinchot, saluting his work as first chief of the Forest Service. "I am a better man for having known you. I feel that to have been with you will make my children better men and women in after life; and I cannot think of a man in the country whose loss would be a more real misfortune to the Nation than yours would be. For seven and a half years we have worked together, and now and then played together—and have been altogether better able to work because we have played; and I owe to you a peculiar debt of obligation for a very large part of the achievement of the administration."[69]

Predictably, Roosevelt didn't forget about his obligations to American ornithology on his way out of power. In 1908, Lucy Maynard, a member of the Audubon Society, had met with him at the White House for a nature lecture. She was updating a new edition of her work on birds in Washington, D.C., and wondered whether the president had any interesting recent sightings to report. "Why, yes," Roosevelt answered gleefully. "But I'll do better than that. I'll make you a list of all the birds I can remember having seen since I have been here."[70]

A few days later, Roosevelt sent Maynard a long list, which she published in 1909 as the introduction to her *Birds of Washington and Vicinity.* In all, ninety-three species were listed, some with brief commentary. For example:

> Night Heron. Five spent winter of 1907 in swampy country about one-half mile west of Washington Monument.
>
> Sparrow Hawk. A pair spent the last two winters on and around the White House grounds, feeding on the Sparrows—largely, thank Heaven, on the English Sparrows.
>
> Screech Owl. Steady resident on White House grounds.[71]

Inauguration Day, March 4, brought a swirl of snow. Ten inches had fallen, and everything was shrouded in white that concealed rock-hard ice. Horse-drawn plows were working overtime to at least keep Pennsylvania Avenue passable for Washingtonians and other guests. Carpenters had worked until the last minute erecting a reviewing stand for the parade outside the North Gate of the White House. But the snow-removal efforts and the carpenters couldn't keep up with the snowfall. For the first time in U.S. history the inauguration ceremonies were moved indoors, to the Senate chamber.[72] The essence of Roosevelt as a phenomenon was evident in the way his departure was receiving more attention in the newspapers than Taft's arrival. Even with the bad weather, Roosevelt was exuberant: "I knew there'd be a blizzard when I went out," he had said on inauguration eve. Clapping his hands, he declared the storm the "Roosevelt Blizzard," and said that's what history would call it. Trying to hold his own, determined from the outset not to be Roosevelt's "creature," Taft remarked that he thought not.[73] "You're wrong," Taft said. "It's my storm. I always said it would be a cold day when I got to be President of the United States."[74]

At ten o'clock on Inauguration morning, Roosevelt and Taft headed to the Capitol in a twelve-team carriage. Snow was swirling about, and many of the bleachers lining Pennsylvania Avenue were empty owing to the inclement weather. Both men usually had a hearty sense of humor, but it wasn't on display that day, although T.R. waved to the shivering spectators. Nellie Taft broke all precedent by riding in a carriage with her husband.[75] At the Capitol, Roosevelt signed some last-minute bills, hugged some close friends, and prepared to relinquish power. Vice President James S. Sherman of New York had already been sworn in. Just after noon, Roosevelt and Taft walked into the Senate chamber, receiving enormous foot-stomping cheers. For a few minutes they looked like a united front. Then a century-old Bible was held out and Taft took the oath.[76] "Observers were struck by Roosevelt's immobile concentration as his successor was sworn in," the historian Edmund Morris wrote in *Theodore Rex*. "Those who did not know him thought that the stony expression and balled-up fists signaled trouble ahead for Taft."[77]

Not since Lincoln had America had such a folk figure as Roosevelt for its president. He was beloved. Groups from all over America wanted to memorialize Roosevelt, chisel his face in granite, or cast a bronze of his likeness. But such gestures were hardly commensurate with his accomplishments, such as saving the Tongass and Mount Olympus. "For millions of contemporary Americans, he was already memorialized in the eighteen national monuments and five national parks he had created by executive order, or cajoled out of Congress," Morris maintained. "The 'inventory,' as Gifford Pinchot would say, included protected pinnacles, a crater lake, a rain forest and a petrified forest, a wind cave and a jewel cave, cliff dwellings, a cinder cone and skyscraper of hardened magma, sequoia stands, glacier meadows, and the grandest of all canyons."[78] In seven years and sixty-nine days, Roosevelt had saved more than 234 million acres of American wilderness. History still hasn't caught up with the long-term magnitude of his achievement.

All of Roosevelt's cabinet dutifully came to see him off at Union Station, but he lingered longest with Pinchot.[79] In coming years Pinchot would become governor of Pennsylvania, forestry advisor to F.D.R., and the co-author of a Darwinian travel odyssey from New York to Key West and on to the Galapagos. Pinchot would have the huge burden of keeping the conservation movement kinetic while Roosevelt was in British East Africa. Soft-spoken, almost tearful, Roosevelt was attentive and considerate to everybody at Union Station: children carrying teddy bears; army troops; porters; bystanders; and congressmen with whom he no longer had to negotiate. Already, scholars were trying to determine precisely where Roosevelt would fit in the spectrum of American presidential history. Roosevelt himself believed that he was a smart hybrid of both Jeffersonian and Hamiltonian impulses with modern Darwinism added for good measure. "I have no use for the Hamiltonian who is aristocratic or for the Jeffersonian who is a demagogue," Roosevelt wrote to William

Allen White shortly before leaving office. "Let us trust the people as Jefferson did, but not flatter them; and let us try to have our administration as effective as Hamilton taught us to have it. Lincoln, and Washington, struck the right average."[80]

At three-twenty that afternoon, Roosevelt left for Oyster Bay as the youngest ex-president in American history. There was about T.R. an air of moral satisfaction. Like Washington and Lincoln, he had accomplished much. He was still walking singular among America's political class. Regarding conservation alone he had left two watchdogs strategically behind to mind the store. The first was Gifford Pinchot, who would be a gadfly every time Taft failed to protect a Roosevelt natural wonder or forest reserve. And, devilishly, Roosevelt had left a big game trophy at the White House: the head of a huge bull moose, shot in Maine, still adorned a wall in the executive dining room. For weeks that bull moose would loom over every presidential meal or conference, until eventually it was taken down. Both the bull moose and Gifford Pinchot were harbingers of difficult days ahead for William Howard Taft. The reign of Theodore Roosevelt hadn't really ended on that snowy March afternoon. The conservation movement had spread all over America, and his acolytes had just begun to fight for the inheritance of unmarred public lands.

There was no going gently into retirement for Roosevelt. He remained America's hubristic flywheel and nationalistic sage. British East Africa. Egypt. Rome and Berlin. Paris and London and Oxford. Brazil. Chile. Uruguay. Argentina. The Grand Canyon and the Federal Bird Reservations of the Gulf of Mexico. He visited them all. He dined with European princes and prayed in the Hopi kivas of northern Arizona. And every single day, like an unbroken stream, he crusaded for conservation to prevail over the global disease of hyper-industrialization. "We regard Attic temples and Roman triumphal arches and Gothic cathedrals as of priceless value," Roosevelt decreed, full of wilderness warrior fury. "But we are, as a whole, still in that low state of civilization where we do not understand that it is also vandalism wantonly to destroy or to permit the destruction of what is beautiful in nature, whether it be a cliff, a forest, or a species of mammal or bird. Here in the United States we turn our rivers and streams into sewers and dumping-grounds, we pollute the air, we destroy forests, and exterminate fishes, birds, and mammals—not to speak of vulgarizing charming landscapes with hideous advertisements."[81]

Freed of the restraints of public office, Roosevelt amped up his recriminations against despoilers, finding solace in the world's deepest dark forests. Swollen with courage, he created the Bull Moose Party in 1912, in part to defend his Alaskan forest reserves from exploitation. Even as his sunlight dimmed, he held firm to his visionary stances on wildlife protection and sustainable land management. He saw the planet as one single biological organism pulsing with life and championed the interconnectedness of nature as his own Sermon on the Mount. As forces of globalization run amok, Roosevelt's stout resoluteness to protect our environment is a strong reminder of our national wilderness heritage, as well as an increasingly urgent call to arms.

Roosevelt's greatest White House accomplishment was encouraging young people to join the wildlife and forestry protection movements. Here, the cowboy conservationist reaches out to a Colorado girl.

Flattery Rocks
Quillayute Needles
Copalis Rock
Bumping Lake
Three Arch Rocks
Conconully
Keechelus
Kachess
Clealum
WASHINGTON
Cold Springs
Willow Creek
National Bison Range
Stump Lake
Chase Lake
MONTANA
NORTH DAKO
OREGON
Deer Flat
Lake Malheur
Shoshone
Loch-Katrine
Belle Fourche
IDAHO
Klamath Lake
Minidoka
WYOMING
SOUTH DAKOT
Pathfinder
East Park
Strawberry Valley
NEBRASKA
NEVADA
Farallon
UTAH
CALIFORNIA
COLORADO
KANS
Grand Canyon
OKLAHOM
ARIZONA
NEW MEXICO
Wichita Forest
Salt River
Rio Grande
Carlsbad
TEXA
ALASKA
Behring
Yukon Delta
Pribilof
Tuxedni
Bogoslof
Fire Island
Saint Lazaria
NOT TO SCALE OF MAIN MAP
Hawaiian Islands
HAWAII
NOT TO SCALE OF MAIN MAP

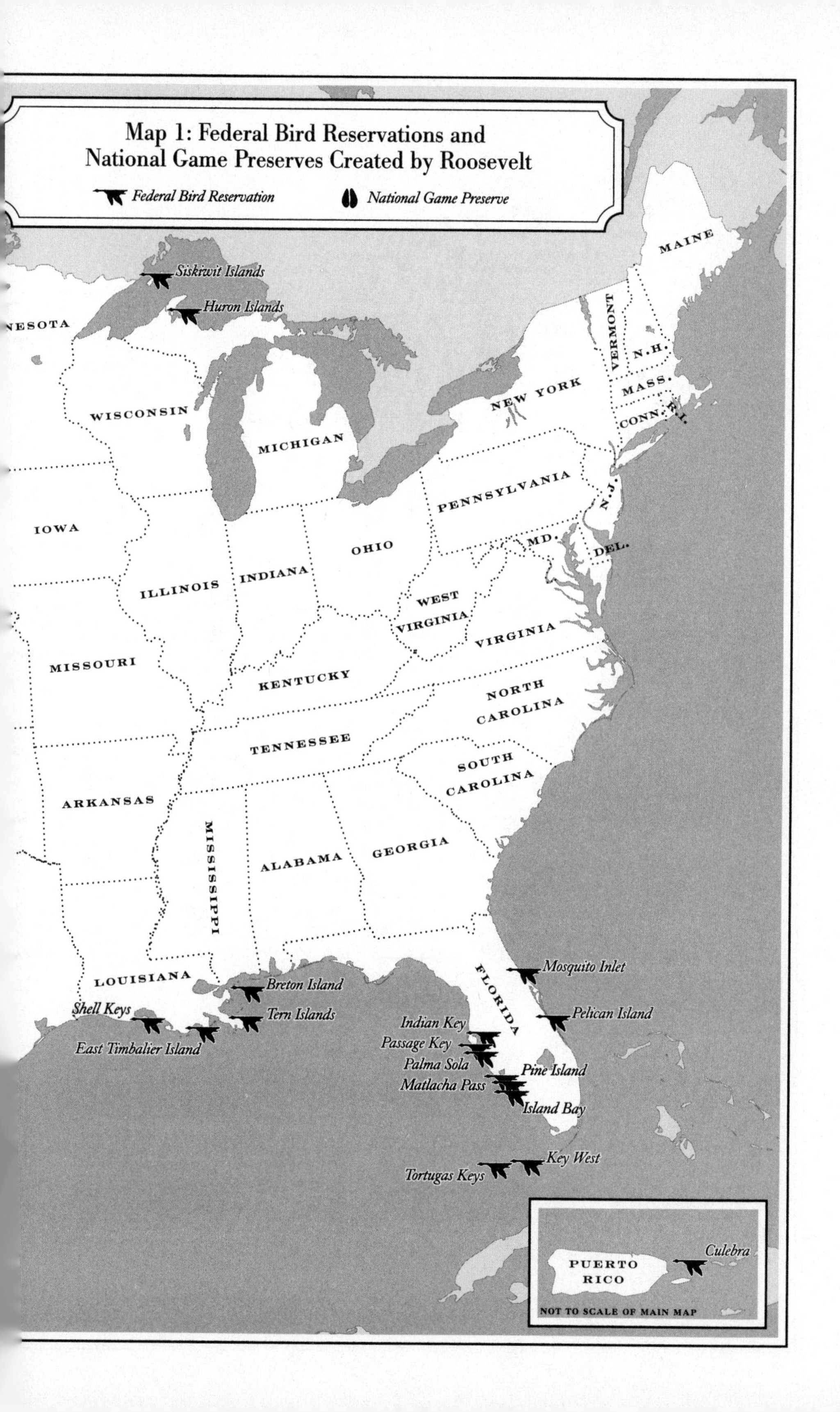
Map 1: Federal Bird Reservations and National Game Preserves Created by Roosevelt
Federal Bird Reservation
National Game Preserve
Siskiwit Islands
Huron Islands
NESOTA
WISCONSIN
MICHIGAN
MAINE
VERMONT
N.H.
MASS.
NEW YORK
CONN.
R.I.
N.J.
PENNSYLVANIA
IOWA
OHIO
MD.
DEL.
ILLINOIS
INDIANA
WEST VIRGINIA
VIRGINIA
MISSOURI
KENTUCKY
NORTH CAROLINA
TENNESSEE
SOUTH CAROLINA
ARKANSAS
MISSISSIPPI
ALABAMA
GEORGIA
LOUISIANA
FLORIDA
Mosquito Inlet
Breton Island
Shell Keys
Tern Islands
Pelican Island
East Timbalier Island
Indian Key
Passage Key
Palma Sola
Pine Island
Matlacha Pass
Island Bay
Key West
Tortugas Keys
PUERTO RICO
Culebra
NOT TO SCALE OF MAIN MAP

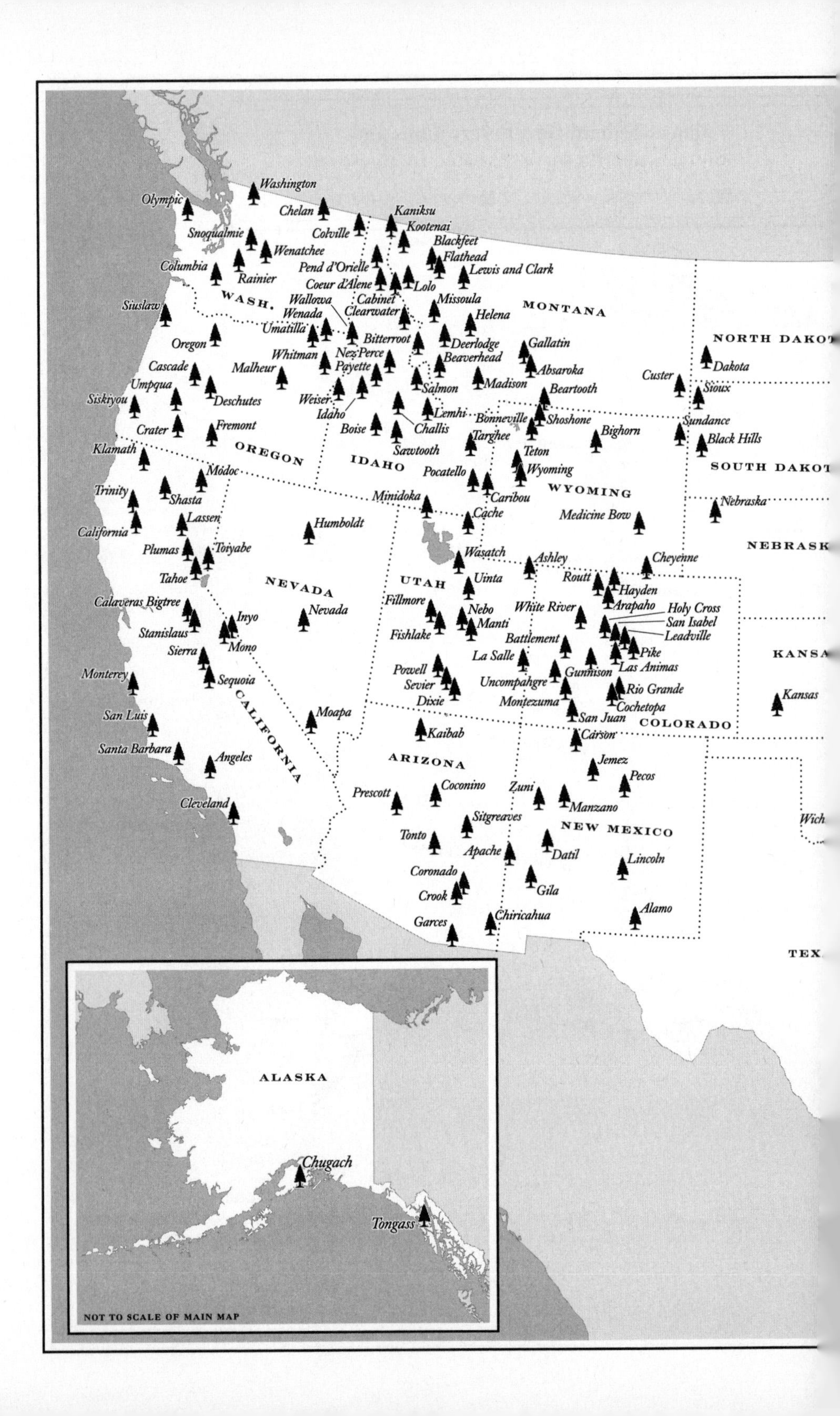

Washington
Olympic
Chelan
Kaniksu
Colville
Kootenai
Snoqualmie
Blackfeet
Wenatchee
Flathead
Columbia
Pend d'Orielle
Lewis and Clark
Rainier
Coeur d'Alene
Lolo
WASH.
Wallowa
Cabinet
Missoula
Siuslaw
Wenada
Clearwater
Helena
MONTANA
Umatilla
Oregon
Bitterroot
Deerlodge
Gallatin
Whitman
Nez Perce
Beaverhead
NORTH DAKOT
Malheur
Payette
Dakota
Cascade
Absaroka
Umpqua
Salmon
Madison
Custer
Siskiyou
Weiser
Beartooth
Sioux
Deschutes
Idaho
Lemhi
Bonneville
Shoshone
Sundance
Fremont
Boise
Challis
Crater
Bighorn
Targhee
Black Hills
Sawtooth
Klamath
OREGON
Teton
IDAHO
SOUTH DAKOT
Modoc
Pocatello
Wyoming
Trinity
Shasta
WYOMING
Caribou
Minidoka
Nebraska
Lassen
Cache
Medicine Bow
Humboldt
California
Plumas
Toiyabe
Wasatch
Ashley
Cheyenne
NEBRASK
Tahoe
Routt
NEVADA
UTAH
Uinta
Hayden
Calaveras Bigtree
Fillmore
Nebo
Arapaho
White River
Holy Cross
Inyo
Nevada
Manti
San Isabel
Stanislaus
Fishlake
Battlement
Leadville
Mono
Sierra
Pike
La Salle
KANSA
Las Animas
Powell
Monterey
Gunnison
Sequoia
Sevier
Uncompahgre
Rio Grande
Dixie
Montezuma
Kansas
Cochetopa
San Luis
Moapa
San Juan
COLORADO
Kaibab
Carson
Santa Barbara
Angeles
ARIZONA
Jemez
CALIFORNIA
Pecos
Coconino
Zuni
Prescott
Manzano
Cleveland
Sitgreaves
Wich
NEW MEXICO
Tonto
Datil
Apache
Lincoln
Coronado
Gila
Crook
Alamo
Chiricahua
Garces
TEX
ALASKA
Chugach
Tongass
NOT TO SCALE OF MAIN MAP

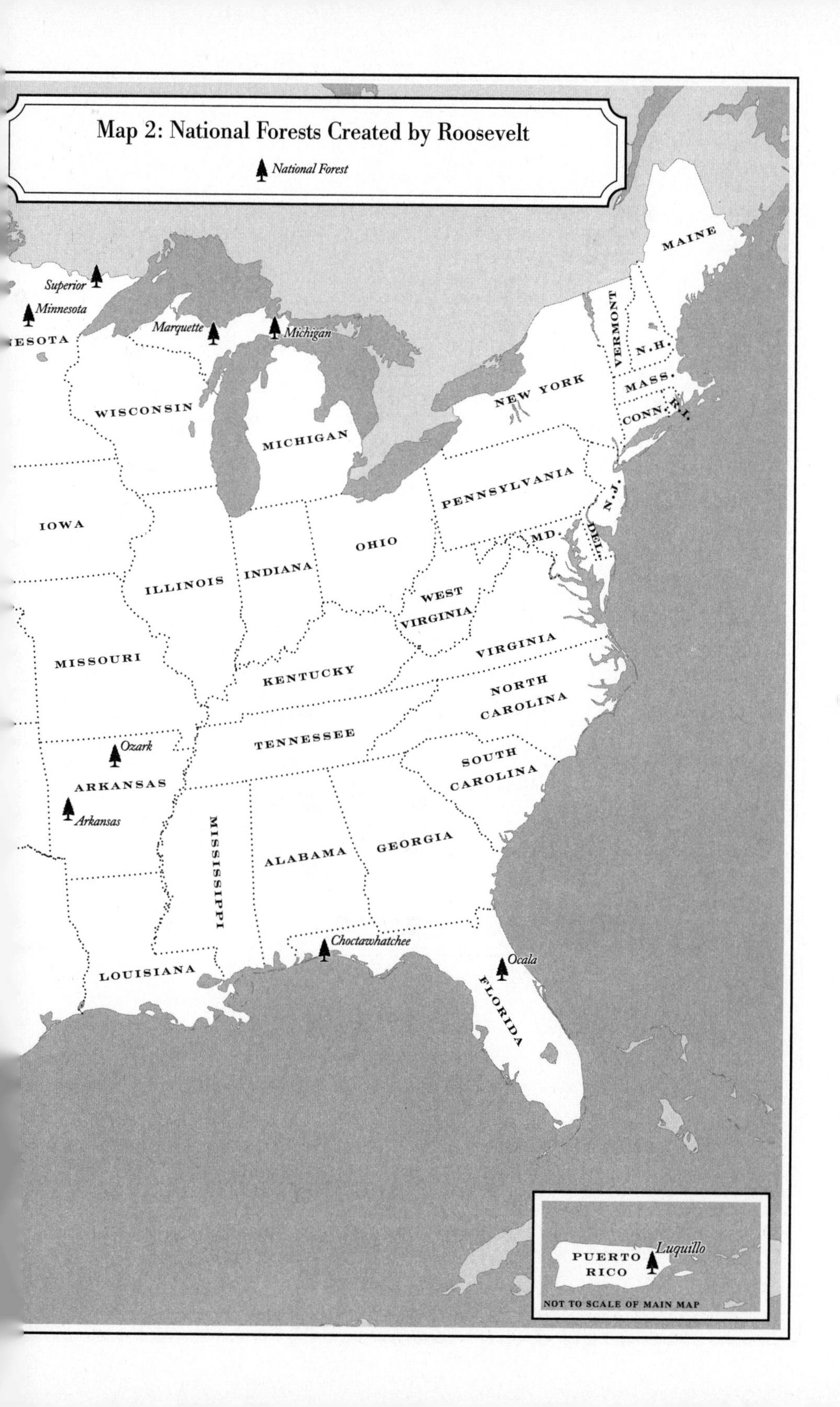
Map 2: National Forests Created by Roosevelt
National Forest
Superior
Minnesota
ESOTA
Marquette
Michigan
MAINE
VERMONT
N.H.
MASS.
NEW YORK
CONN.
R.I.
WISCONSIN
MICHIGAN
PENNSYLVANIA
N.J.
IOWA
OHIO
MD.
DEL.
ILLINOIS
INDIANA
WEST VIRGINIA
MISSOURI
VIRGINIA
KENTUCKY
NORTH CAROLINA
TENNESSEE
Ozark
SOUTH CAROLINA
ARKANSAS
Arkansas
MISSISSIPPI
ALABAMA
GEORGIA
Choctawhatchee
Ocala
LOUISIANA
FLORIDA
PUERTO RICO
Luquillo
NOT TO SCALE OF MAIN MAP

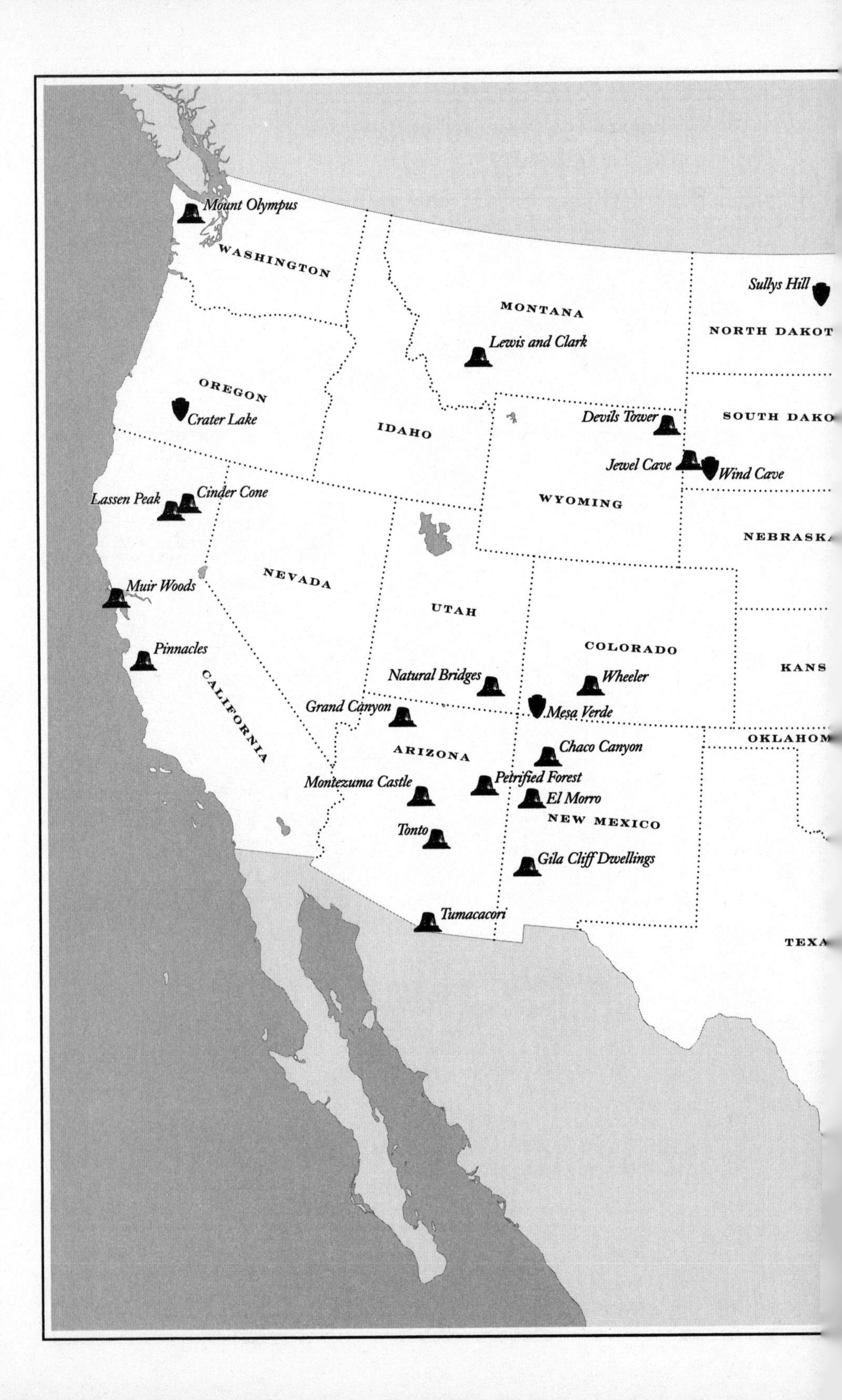

Mount Olympus
WASHINGTON
MONTANA
Sullys Hill
NORTH DAKOT
Lewis and Clark
OREGON
Crater Lake
IDAHO
Devils Tower
SOUTH DAKO
Jewel Cave
Wind Cave
WYOMING
Lassen Peak
Cinder Cone
NEBRASKA
NEVADA
Muir Woods
UTAH
COLORADO
Pinnacles
CALIFORNIA
KANS
Natural Bridges
Wheeler
Grand Canyon
Mesa Verde
OKLAHOM
ARIZONA
Chaco Canyon
Montezuma Castle
Petrified Forest
El Morro
NEW MEXICO
Tonto
Gila Cliff Dwellings
Tumacacori
TEXA

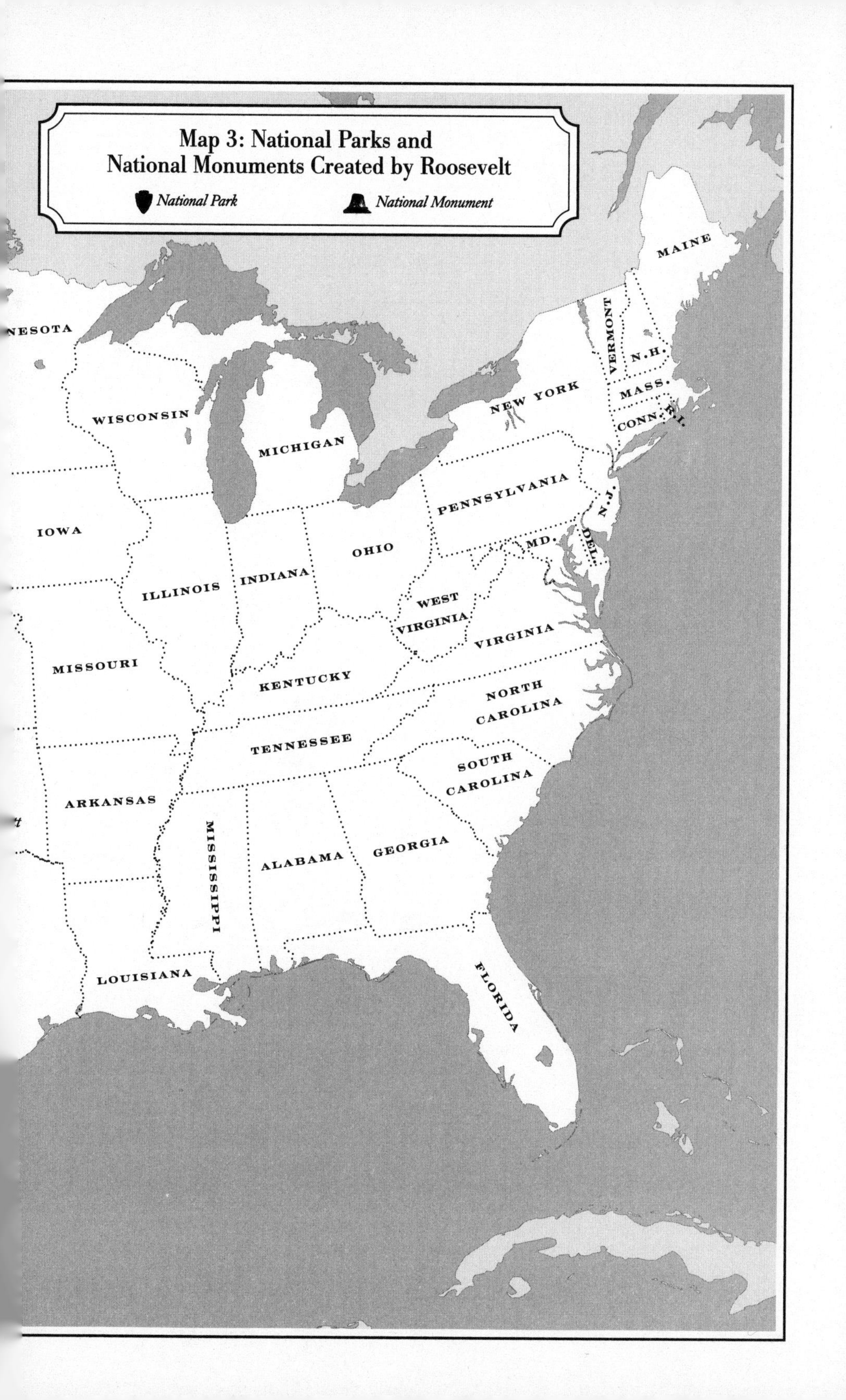
Map 3: National Parks and
National Monuments Created by Roosevelt
National Park
National Monument
NESOTA
WISCONSIN
MICHIGAN
IOWA
ILLINOIS
INDIANA
OHIO
MISSOURI
KENTUCKY
TENNESSEE
ARKANSAS
MISSISSIPPI
ALABAMA
GEORGIA
LOUISIANA
FLORIDA
SOUTH
CAROLINA
NORTH
CAROLINA
VIRGINIA
WEST
VIRGINIA
MD.
DEL.
N.J.
PENNSYLVANIA
NEW YORK
VERMONT
N.H.
MAINE
MASS.
CONN.
R.I.

APPENDIX

This list was compiled from the *Establishment and Modification of National Forest Boundaries: A Chronological Record* (1891–1973); the annual reports of the Division of Forestry (1886–1901); Bureau of Forestry (1902–1903); U.S. Geological Survey's *Annual Reports* (1897–1900); and my own additions.

National Forests Created or Enlarged by Theodore Roosevelt, 1901–1909

	Forest	Date
1.	Luquillo (Puerto Rico), renamed El Yunque National Forest in 2006	January 17, 1903
2.	White River (Colorado)	May 21, 1904
3.	Sevier (Utah)	January 17, 1906
4.	Wichita (Oklahoma)	May 29, 1906
5.	Lolo (Montana)	November 6, 1906
6.	Caribou (Idaho and Wyoming)	January 15, 1907
7.	Colville (Washington)	March 1, 1907
8.	Las Animas (Colorado and New Mexico)	March 1, 1907
9.	Wenada (Oregon and Washington)	March 1, 1907
10.	Olympic (Washington)	March 2, 1907
11.	Manti (Utah)	April 25, 1907
12.	Manzano (New Mexico)	April 16, 1908
13.	Kansas (Kansas)	May 15, 1908
14.	Minnesota (Minnesota)	May 23, 1908
15.	Pocatello (Idaho and Utah)	July 1, 1908
16.	Cache (Idaho and Utah)	July 1, 1908
17.	Whitman (Oregon)	July 1, 1908
18.	Malheur (Oregon)	July 1, 1908
19.	Umatilla (Oregon)	July 1, 1908
20.	Columbia (Washington)	July 1, 1908
21.	Rainier (Washington)	July 1, 1908
22.	Washington (Washington)	July 1, 1908
23.	Chelan (Washington)	July 1, 1908
24.	Snoqualmie (Washington)	July 1, 1908
25.	Wenatchee (Washington)	July 1, 1908
26.	Fillmore (Utah)	July 1, 1908
27.	Nebo (Utah)	July 1, 1908
28.	Lewis and Clark (Montana)	July 1, 1908
29.	Blackfeet (Montana)	July 1, 1908
30.	Flathead (Montana)	July 1, 1908
31.	Kootenai (Montana)	July 1, 1908
32.	Routt (Colorado)	July 1, 1908
33.	Cabinet (Montana)	July 1, 1908
34.	Hayden (Colorado and Wyoming)	July 1, 1908
35.	Challis (Idaho)	July 1, 1908
36.	Salmon (Idaho)	July 1, 1908
37.	Clearwater (Idaho)	July 1, 1908
38.	Coeur d'Alene (Idaho)	July 1, 1908

39. Pend d'Orielle (Idaho)	July 1, 1908
40. Kaniksu (Idaho and Washington)	July 1, 1908
41. Angeles (California)	July 1, 1908
42. San Luis (California)	July 1, 1908
43. Jemez (New Mexico)	July 1, 1908
44. Sundance (Wyoming)	July 1, 1908
45. Santa Barbara (California)	July 1, 1908
46. Weiser (Idaho)	July 1, 1908
47. Nez Perce (Idaho)	July 1, 1908
48. Idaho (Idaho)	July 1, 1908
49. Payette (Idaho)	July 1, 1908
50. Boise (Idaho)	July 1, 1908
51. Sawtooth (Idaho)	July 1, 1908
52. Lemhi (Idaho)	July 1, 1908
53. Siuslaw (Oregon)	July 1, 1908
54. Cheyenne (Wyoming)	July 1, 1908
55. Medicine Bow (Colorado), enlarged and renamed Roosevelt National Forest in 1932 as an honor to T.R.	July 1, 1908
56. Cascade (Oregon)	July 1, 1908
57. Oregon (Oregon)	July 1, 1908
58. Umpqua (Oregon)	July 1, 1908
59. Siskiyou (Oregon)	July 1, 1908
60. Crater (California and Oregon)	July 1, 1908
61. Beartooth (Montana)	July 1, 1908
62. Holy Cross, Colorado	July 1, 1908
63. Targhee (Idaho and Wyoming)	July 1, 1908
64. Teton (Wyoming)	July 1, 1908
65. Wyoming (Wyoming)	July 1, 1908
66. Bonneville (Wyoming)	July 1, 1908
67. Absaroka (Montana)	July 1, 1908
68. Beaverhead (Montana)	July 1, 1908
69. Madison (Montana)	July 1, 1908
70. Gallatin (Montana)	July 1, 1908
71. Deerlodge (Montana)	July 1, 1908
72. Helena (Montana)	July 1, 1908
73. Missoula (Montana)	July 1, 1908
74. Bitterroot (Idaho and Wyoming)	July 1, 1908
75. Ashley (Utah and Wyoming)	July 1, 1908
76. Uncompahgre (Colorado)	July 1, 1908
77. San Juan (Colorado)	July 1, 1908
78. Rio Grande (Colorado)	July 1, 1908
79. Pike (Colorado)	July 1, 1908
80. Montezuma (Colorado)	July 1, 1908
81. Leadville (Colorado)	July 1, 1908
82. Gunnison (Colorado)	July 1, 1908
83. Cochetopa (Colorado)	July 1, 1908
84. Arapaho (Colorado)	July 1, 1908
85. Battlement (Colorado)	July 1, 1908
86. Shoshone (Wyoming)	July 1, 1908
87. Uinta (Utah)	July 1, 1908
88. Crook (Arizona)	July 1, 1908

89. Coconino (Arizona)	July 1, 1908
90. Inyo (California)	July 1, 1908
91. Stanislaus (California)	July 1, 1908
92. Sierra (California)	July 1, 1908
93. Chiricahua (Arizona and New Mexico)	July 1, 1908
94. Coronado (Arizona)	July 1, 1908
95. Garces (Arizona)	July 1, 1908
96. Monterey (California)	July 1, 1908
97. San Isabel (Colorado)	July 1, 1908
98. Minidoka (Idaho and Utah)	July 1, 1908
99. Jefferson (Montana)*	July 1, 1908
100. Custer (Montana)	July 1, 1908
101. Nebraska (Nebraska)	July 1, 1908
102. Wallowa (Oregon)	July 1, 1908
103. Fishlake (Utah)	July 1, 1908
104. La Salle (Utah)	July 1, 1908
105. Wasatch (Utah)	July 1, 1908
106. Powell (Utah)	July 1, 1908
107. Bighorn (Wyoming)	July 1, 1908
108. Kaibab (Arizona)	July 1, 1908
109. Deschutes (Oregon)	July 14, 1908
110. Fremont (Oregon)	July 14, 1908
111. Ocala (Florida)	November 24, 1908
112. Dakota (North Dakota)	November 24, 1908
113. Choctawhatchee (Florida)	November 27, 1908
114. Humboldt (Nevada)	January 20, 1909
115. Moapa (Nevada)	January 21, 1909
116. Cleveland (California)	January 26, 1909
117. Pecos (New Mexico)	January 28, 1909
118. Prescott (Arizona)	February 1, 1909
119. Calaveras Bigtree (California)	February 8, 1909
120. Tonto (Arizona)	February 10, 1909
121. Marquette (Michigan)	February 10, 1909
122. Nevada (Nevada)	February 10, 1909
123. Dixie (Arizona and Utah)	February 10, 1909
124. Michigan (Michigan)	February 11, 1909
125. Klamath (California and Oregon)	February 13, 1909
126. Superior (Minnesota)	February 13, 1909
127. Gila (New Mexico)	February 15, 1909
128. Black Hills (South Dakota and Wyoming)	February 15, 1909
129. Sioux (Montana and South Dakota)	February 15, 1909
130. Tongass (Alaska)	February 16, 1909
131. Toiyabe (Nevada)	February 20, 1909
132. Datil (New Mexico)	February 23, 1909
133. Chugach (Alaska)	February 23, 1909
134. Modoc (California)	February 25, 1909
135. Ozark (Arkansas)	February 25, 1909
136. California (California)	February 25, 1909
137. Arkansas (Arkansas)	February 27, 1909

*Merged with Lewis & Clark National Forest in 1932.

138. Mono (California and Nevada)	March 2, 1909
139. Sitgreaves (Arizona)	March 2, 1909
140. Lincoln (New Mexico)	March 2, 1909
141. Shasta (California)	March 2, 1909
142. Alamo (New Mexico)	March 2, 1909
143. Carson (New Mexico)	March 2, 1909
144. Zuni (Arizona and New Mexico)	March 2, 1909
145. Trinity (California)	March 2, 1909
146. Apache (Arizona)	March 2, 1909
147. Lassen (California)	March 2, 1909
148. Plumas (California)	March 2, 1909
149. Tahoe (California)	March 2, 1909
150. Sequoia (California)	March 2, 1909

Federal Bird Reservations Created by Theodore Roosevelt, and Admitted by the Bureau of Biological Survey, USDA

Most of Roosevelt's bird reserves are now part of the U.S. Fish and Wildlife's National Wildlife Refuge System (NWR) 1901–1909. Special thanks to William Reffalt, a U.S. Fish and Wildlife volunteer, for helping compile this list.

Name of Bird Reservation	Date	Status
1. Pelican Island (Florida)	March 14, 1903	NWR
enlarged	January 26, 1909	
2. Breton Island (Louisiana)	October 4, 1904	NWR
3. Stump Lake (North Dakota)	March 9, 1905	NWR
4. Siskiwit Islands (Michigan)	October 10, 1905	Natl. Park
5. Huron Islands (Michigan)	October 10, 1905	NWR
6. Passage Key (Florida)	October 10, 1905	NWR
7. Indian Key (Florida)	February 10, 1906	No. Fed. Land
8. Tern Islands (Louisiana)	August 8, 1907	No. Fed. Land
9. Shell Keys (Louisiana)	August 17, 1907	NWR
10. Three Arch Rocks (Oregon)	October 14, 1907	NWR
11. Flattery Rocks (Washington)	October 23, 1907	NWR
12. Copalis Rock (Washington)	October 23, 1907	NWR
13. Quillayute Needles (Washington)	October 23, 1907	NWR
14. East Timbalier Island (Louisiana)	December 7, 1907	No. Fed. Land
15. Mosquito Inlet (Florida)	February 24, 1908	No. Fed. Land
16. Tortugas Keys (Florida)	April 6, 1908	Nat'l. Park
17. Key West (Florida)	August 8, 1908	NWR
18. Klamath Lake (Oregon and California)	August 8, 1908	NWR
19. Lake Malheur (Oregon)	August 18, 1908	NWR
20. Chase Lake (North Dakota)	August 28, 1908	NWR
21. Pine Island (Florida)	September 15, 1908	NWR
22. Matlacha Pass (Florida)	September 26, 1908	NWR
23. Palma Sole (Florida)	September 26, 1908	No. Fed. Land
24. Island Bay (Florida)	October 23, 1908	NWR
25. Loch-Katrine (Wyoming)	October 26, 1908	No Fed. Land
26. Hawaiian Islands	February 3, 1909	NWR
27. Salt River (Arizona)	February 25, 1909	Bur. Recl.
28. East Park (California)	February 25, 1909	Impt. Recl.

29. Deer Flat (Idaho)	February 25, 1909	NWR
30. Willow Creek (Montana)	February 25, 1909	Other NWR
31. Carlsbad (New Mexico)	February 25, 1909	Bur. Recl.
32. Rio Grande (New Mexico)	February 25, 1909	Bur. Recl.
33. Cold Springs (Oregon)	February 25, 1909	NWR
34. Belle Fourche (South Dakota)	February 25, 1909	Impt. Recl.
35. Strawberry Valley (Utah)	February 25, 1909	No. Fed. Land
36. Keechelus (Washington)	February 25, 1909	Bur. Recl.
37. Kachess (Washington)	February 25, 1909	Bur. Recl.
38. Clealum (Washington)	February 25, 1909	Bur. Recl.
39. Bumping Lake (Washington)	February 25, 1909	Bur. Recl.
40. Conconully (Washington)	February 25, 1909	Impt. Recl.
41. Pathfinder (Wyoming)	February 25, 1909	NWR
42. Shoshone (Wyoming)	February 25, 1909	No Fed. Land
43. Minidoka (Idaho)	February 25, 1909	NWR
44. Tuxedni (Alaska)	February 27, 1909	Other NWR
45. Saint Lazaria (Alaska)	February 27, 1909	Other NWR
46. Yukon Delta (Alaska)	February 27, 1909	Other NWR
47. Culebra (Puerto Rico)	February 27, 1909	NWR
48. Farallon (California)	February 27, 1909	NWR
49. Bering Sea (Alaska)	February 27, 1909	Other NWR
50. Pribilof (Alaska)	February 27, 1909	Other NWR
51. Bogoslof (Alaska)	March 2, 1909	Other NWR

Status Note: *Other NWR* means the original reservation has been absorbed into a larger NWR. *No. Fed. Land* indicates valid state/private claims area eroded away completely, or reservoirs transferred to water users. *Bur. Recl.* indicates the Reclamation Project became more important than the secondary withdrawal for wildlife. *Impt. Recl.* indicates recreation became a dominant use and the refuge was revoked. (Code by Reffalt.)

National Game Preserves Created by Theodore Roosevelt, 1901–1909

1. Wichita Forest, Oklahoma—June 2, 1905. Land added May 29, 1906. This was the first federal game preserve.
2. Grand Canyon, Arizona—June 23, 1908. Note that the Grand Canyon also became a national monument in 1908.
3. Fire Island Moose Reservation, Alaska—February 27, 1909.
4. National Bison Range, Montana—March 4, 1909.

National Parks Created by Theodore Roosevelt, 1901–1909

1. Crater Lake National Park, Oregon—May 22, 1902.
2. Wind Cave National Park, South Dakota—January 9, 1903.
3. Sullys Hill, North Dakota—June 2, 1904; became a national game preserve in 1914.
4. Platt National Park, Oklahoma—June 29, 1906; now part of Chickasaw National Recreation Area.
5. Mesa Verde National Park, Colorado—June 29, 1906.
6. Dry Tortugas National Park—saved as a federal bird reservation it became a national monument in 1935 and then a national park on October 26, 1992.

Theodore Roosevelt National Park, near Medora, North Dakota, was established in 1947 as a memorial to the great "conservationist president." Located in the Badlands of western North Dakota, where T.R. was a cattle rancher in the 1880s, Theodore Roosevelt National Park consists of three units with a total of about 110 square miles.

National Monuments Created by Theodore Roosevelt, 1901–1909

1. Devils Tower, Wyoming, September 24, 1906
2. El Morro, New Mexico, December 8, 1906
3. Montezuma Castle, Arizona, December 8, 1906
4. Petrified Forest, Arizona, December 8, 1906 (became a national park in 1962)
5. Chaco Canyon, New Mexico, March 11, 1907
6. Lassen Peak, California, May 6, 1907 (became part of Lassen Volcanic National Park in 1916)
7. Cinder Cone, California, May 6, 1907 (became part of Lassen Volcanic National Park in 1916)
8. Gila Cliff Dwellings, New Mexico, November 16, 1907
9. Tonto, Arizona, December 19, 1907
10. Muir Woods, California, January 9, 1908
11. Grand Canyon, Arizona, January 11, 1908 (became an enlarged national park in 1919)
12. Pinnacles, California, January 16, 1908
13. Jewel Cave, South Dakota, February 7, 1908
14. Natural Bridges, Utah, April 16, 1908
15. Lewis and Clark, Montana, May 11, 1908 (later given to the Forest Service, in 1950)
16. Tumacacori, Arizona, September 15, 1908
17. Wheeler, Colorado, December 7, 1908 (transferred to the Forest Service in 1950)
18. Mount Olympus, Washington, March 2, 1909 (now part of Olympic National Park)*

---

*Today Roosevelt's Hawaiian Islands Federal Bird Reservation is a National Monument of over 88 million acres, encompassing both terrestrial and marine ecosystems. Within its confines remain two NWR sites. If the acreage of this national monument was fully included in T.R.'s legacy portfolio, one could say he saved over 300 million acres of America.

NOTES

* Unless otherwise noted all Theodore Roosevelt letters cited are at the Library of Congress and Harvard University. All will soon be available online courtesy of Dickinson State University's Theodore Roosevelt Center in North Dakota.

Prologue: "I So Declare It"

1. *Theodore Roosevelt's America: Selections from the Writings of the Oyster Bay Naturalist* (Greenwich, Conn.: Devin-Adair, 1955), p. xviii.

2. T.R. to Frank M. Chapman (March 22, 1899), quoted in Frank M. Chapman, *Autobiography of a Bird Lover* (New York: Appleton-Century, 1933).

3. Frank M. Chapman, "Birds and Bonnets," *Forest and Stream*, Vol. 26, No. 5 (February 25, 1886), p. 84. (Letter to the editor.)

4. John T. Zimmer, "Frank Michler Chapman," *American Naturalist*, Vol. 80, No. 793 (1945), p. 476.

5. For an explanation of early ornithologists' museum strategies see Nancy Pick and Mark Sloan, *The Rarest of the Rare: Stories Behind the Treasures at the Harvard Museum of Natural History* (New York: HarperCollins, 2004).

6. Oliver H. Orr, Jr., *Saving American Birds: T. Gilbert Pearson and the Founding of the Audubon Movement* (Gainesville: University Press of Florida, 1992), p. 74.

7. Williams R. Adams, "Florida Live Oak Farm of John Quincy Adams," *Florida Historical Quarterly*, Vol. 51 (1972), pp. 129–147. Adams's preserve is now called the Naval Live Oaks–Gulf Islands National Seashore.

8. John F. Reiger, *American Sportsmen and the Origins of Conservation*, 3rd ed. (Corvallis: Oregon State University Press, 2001), pp. 5–7.

9. George Catlin, *North American Indians: Being Letters and Notes on their Manners, Customs, and Conditions, Written during Eight Years' Travel Amongst the Wildest Tribes of Indians in North America, 1832–1839* (Philadelphia, Pa.: Leary, Stuart, 1913), pp. 294–295. Orig. published 1844.

10. Kathryn Hall Proby, *Audubon in Florida* (Coral Gables: University of Miami Press, 1974), p. 51.

11. Frank Graham, Jr., *The Audubon Ark: A History of the National Audubon Society* (New York: Knopf, 1990), p. 10.

12. Henry David Thoreau, "Chesuncook," *Atlantic Monthly*, Vol. 2 (August 1858). Also see Thoreau, *The Maine Woods* (Boston: Ticknor and Fields, 1864), p. 160.

13. Doug Stewart, "How Conservation Grew from a Whisper to a Roar," *National Wildlife* (December–January 1999).

14. "The Bisby Club's Resort." *New York Times* (June 8, 1890), p. 12. Also see, Ken Sprague, "History and Heritage Remembering 19th and 20th Century Life," *Adirondack Express* (July 25, 2006), p. 4.

15. William T. Hornaday, *Our Vanishing Wild Life* (New York: New York Zoological Society, 1913), p. 248.

16. Gifford Pinchot, *The Fight for Conservation* (New York: Doubleday, 1910), p. 48.

17. T.R. to William Adolph Baillie-Grohman (June 12, 1900).

18. John Burroughs, *Signs and Seasons* (Boston, Mass.: Houghton Mifflin, 1886), p. 213.

19. George Laycock, *Wild Refuge* (Garden City, N.Y.: American Museum of Natural History Press, 1969), pp. 12–20.

20. Jonathan Weiner, "Darwin's Delay," *Slate* (May 3, 2007).

21. John M. Blum, "Theodore Roosevelt: The Years of Decision," in Elting E. Morison (ed.), *The Letters of Theodore Roosevelt*, Vol. 2 (Cambridge, Mass.: Harvard University Press, 1951), Vol. 4, p. 1486.

22. Orr, *Saving American Birds*, p. 1.

23. Hornaday, *Our Vanishing Wild Life*, p. 15.

24. Erick Gill, "Pelican Island: 10 Years in the Making," *Vero Beach Magazine* (February 2003), pp. 7–14.

25. Thomas Gilbert Pearson, *Adventures in Bird Protection*, p. 52. Also see, Robin W. Doughty, *Feather Fashions and Bird Preservation: A Study in Nature Protection* (Berkeley: University of California Press, 1975), p. 11.

26. Doughty, *Feather Fashions and Bird Preservation*, pp. 10-11, 23. Also see Mary Van Kleeck, *A Seasoned Industry: A Study on the Millinery Trade in New York* (Philadelphia, Pa.: Russell Sage Foundation, 1917), pp. 10–23.

27. Doughty, *Feather Fashions and Bird Preservation*, p. 3. Also see Stuart B. McIver, *Death in the Everglades: The Murder of Guy Bradley: America's First Martyr of the Environment* (Gainesville: University of Florida Press, 2007), p. 4.

28. Doughty, *Feather Fashions and Bird Preservations*, p. 15.

29. "Maxim's New Machine Gun," *New York Times* (December 5, 1884), p. 3. New weapons such as the LeFever semiautomatic hammerless shotgun (1883) and the Maxim machine gun (1884) began replacing the outdated Civil War Gatling gun. Relatively easy to carry, these guns occasionally made their way into the hunting scene. Most of the time, however, plume hunters used shotguns.

30. Doughty, *Feather Fashions and Bird Preservation*, p. 12.

31. Frank Graham, Jr., *The Audubon Ark: A History of the National Audubon Society* (New York: Knopf, 1990), p. 18.

32. Florida Audubon Society, "Who We Are: History of Audubon of Florida" (1999). Pamphlet.

33. Orr, *Saving American Birds*, p. 74.

34. "Audubon of Florida Timeline," National Audubon Society (2009).

35. U.S. Department of Agriculture, *Yearbook 1902* (Washington D.C.: Government Printing Office, 1903).

36. "Society at Home and Abroad," *New York Times* (October 20, 1901), p. 7. T.R. renamed the Executive Mansion the White House on October 12, 1901.

37. William Allen White quoted in Hermann Hagedorn and Sidney Wallach, *A Theodore Roosevelt Round-Up* (New York: Theodore Roosevelt Association, 1958), pp. 159–160.

38. "William Alford Richards: Cadastral Survey" (Cheyenne: Bureau of Land Management, Wyoming Archive).

39. William Reffalt, "Prologue to Pelican Island" (February 2003). Unpublished, Pelican Island National Wildlife Refuge Archive, Vero Beach, Fla.

40. There are numerous versions of the "I So Declare It" story all with slight variations, including Patricia O'Toole, *When Trumpets Call* (New York: Simon & Schuster, 2005), pp. 32–33, Lewis L. Gould, *The Presidency of Theodore Roosevelt* (Lawrence: The University of Kansas Press, 1991), p. 111, and William H. Harbaugh, *The Life and Times of Theodore Roosevelt* (New York: Collier Books, 1967), p. 315. I combined these with information gathered by Paul Tritaik of U.S. Fish and Wildlife from his Florida files.

41. T.R., "My Life as a Naturalist," *American Museum Journal*, Vol. 18 (May 1918), p. 321. See also Hermann Hagedorn (ed.), *The Works of Theodore Roosevelt, Memorial Edition* (New York: Scribner, 1923), p. 443.

42. Frank Chapman to T.R. ([n.d.] 1908). Chapman Papers, American Museum of Natural History, New York.

43. Frank M. Chapman, *Camps and Cruises of an Ornithologist* (New York: Appleton, 1908), pp. 85–95.

44. Paul Russell Cutright, *Theodore Roosevelt: The Naturalist* (New York: Harper, 1956), p. 144.

45. Kathleen Dalton, *Theodore Roosevelt: The Strenuous Life* (New York: Knopf, 2002), p. 16.

46. Reffalt, "Prologue to Pelican Island."

47. Pelican Island Federal Bird Reservation Declaration (March 14, 1903). U.S. Fish and Wildlife Archive, Vero Beach, Fla.

48. Charles Alexander, "A Life with Birds," *Birder's World* (April 2003), p. 42. The figure changes annually.

49. Sidney P. Johnston, *A History of Indian River County* (Vero Beach, Fla.: Indian River County Historical Society, 2000), p. 39.

50. T.R., *An Autobiography* (New York: Macmillan, 1913), p. 436.

51. "Passage Key and the American Wildlife Conservation Movement" [n.d.] (Crystal River, Fla.: U.S. Fish and Wildlife Services Archives).

52. Bill McKibben (ed.), *American Earth:*

Environmental Writing Since Thoreau (New York: Library of America, 2008).

53. T.R., *A Book-Lover's Holidays in the Open* (New York: Scribner, 1916), pp. 316–317.

54. John L. Eliot, "Roosevelt Country: T.R.'s Wilderness Legacy," *National Geographic*, Vol. 162, No. 3 (September 1982), pp. 340–362.

55. Aida D. Donald, *Lion in the White House: A Life of Theodore Roosevelt* (New York: Basic Books, 2007), p. 193.

56. "The President Helps Lay a Cornerstone," *New York Times*, April 25, 1903, p. 1.

57. "Conservation as National Duty," President Theodore Roosevelt's Opening Address in *Proceedings of a Conference of National Governors* (May 13, 1909).

58. T.R., *Outdoor Pastimes of an American Hunter* (New York: Scribner, 1905), p. 317.

1: The Education of a Darwinian Naturalist

1. Oliver H. Orr, Jr., *Saving American Birds: T. Gilbert Pearson and the Founding of the Audubon Movement* (Gainesville: University of Florida Press, 1992), p. 18.

2. David McCullough, *Mornings on Horseback* (New York: Simon and Schuster, 1981), p. 114.

3. Paul Russell Cutright, *Theodore Roosevelt: The Naturalist* (New York: Harper, 1956), p. xiii.

4. T.R. to Edward Sanford Martin (November 26, 1900).

5. T.R., *African Game Trails* (New York: Scribner, 1910), p. xi.

6. *The Works of Ralph Waldo Emerson*, Vol. 1 (London: Macmillan, 1884), p. 6.

7. Janet Browne, *Darwin's Origin of Species* (New York: Grove, 2006), pp. 1–5.

8. T.R., "The Pigskin Library" in *Literary Essays*, National Edition, Vol. 2, pp. 337–346. This originally appeared in *Outlook*, Vol. 94, Issue 18 (April 30, 1910).

9. Edmund Morris, *The Rise of Theodore Roosevelt* (New York: Coward, McCann, 1979), p. 23.

10. T.R. to James Joseph Walsh (February 23, 1909).

11. Christian Fichthorne Reisner, *Roosevelt's Religion* (New York: Abingdon, 1922), p. 32.

12. Jacob A. Riis, *Theodore Roosevelt: The Citizen* (New York: Grosset and Dunlap, 1904), pp. 7–8.

13. Darwin, *The Descent of Man* (Akron, Ohio: Werner, 1874).

14. T.R., "My Life as a Naturalist," *American Museum Journal*, Vol. 18 (May 1918), p. 321.

15. Richard W. Etulain, *Telling Western Stories: From Buffalo Bill to Larry McMurtry* (Albuquerque: The University of New Mexico Press, 1999), p. 5–30.

16. Mayne Reid, *The Scalp Hunters; Or, Romantic Adventures in Northern Mexico*, Vol. I (London: Charles J. Skeet, 1851), p. 2.

17. Mayne Reid, *The Land Pirates; or The League of Devil's Island* (New York: Beadle's Half-Dime Library, 1879), pp. 4–14.

18. Mayne Reid, *The Boy Hunters, Or Adventures in Search of a White Buffalo* (London: David Bogue, Fleet Street, 1852), p. 17.

19. Ibid., pp. 27–28.

20. Ibid.

21. Joan Steele, *Captain Mayne Reid* (Boston, Mass.: Twayne, 1978), pp. 104–106.

22. T.R., *The Rough Riders* (New York: Scribner, 1899), pp. 104–112.

23. Reid, *The Boy Hunters*, p. 424.

24. Mayne Reid, *The White Chief: A Legend of Northern Mexico*, Vol. 2 (London: David Bogue, Fleet Street, 1855), p. 145.

25. Steele, *Captain Mayne Reid*, pp. 66–67.

26. Edgar Allan Poe quoted in *The Handbook of Texas* (Austin: Texas State Historical Association, 1997–2002).

27. Kathleen Dalton, *Theodore Roosevelt: A Strenuous Life* (New York: Knopf, 2002), p. 43.

28. T.R., *An Autobiography* (New York: Macmillan, 1913), p. 16. (In 1899 the *New York Times* ranked the fifteen greatest "Books for Boys" of the nineteenth century; in first place was Reid's *Ran Away to Sea* of 1858—which dealt with the slave trade—followed by *The Swiss Family Robinson.*)

29. J. G. Wood, *The Common Objects of the Country* (London: G. Rutledge, 1858), p. 33.

30. Reverend J. G. Wood, *Home without Hands: Being a Description of the Habitations of Animals, Classed According to Their Principle of Construction* (New York: Appleton, 1866), pp. 362–369.

31. Corrine Roosevelt Robinson, *My Brother, Theodore Roosevelt* (New York: Scribner, 1921), p. 2.

32. T.R., *An Autobiography*, p. 6.

33. Carl Safina, *Song for the Blue Ocean: Encounters along the World's Coasts and Beneath the Seas* (New York: Holt, 1997), pp. 399–400.

34. Paul Russell Cutright, *Theodore Roosevelt: The Making of a Conservationist* (Urbana: University of Illinois Press, 1985), p. 2.

35. T.R., *An Autobiography*, pp. 14–16.

36. Theodore Roosevelt Museum Inventory List for 1867. Also published in Cutright, *Theodore Roosevelt: The Naturalist*, p. 3.

37. T.R., "My Life as a Naturalist," pp. 321–325.

38. David McCullough, *Mornings on Horseback* (New York: Simon and Schuster, 1981), p. 119.

39. Jonathan Rosen, *The Life of the Skies: Birding at the End of Nature* (New York: Farrar, Straus, and Giroux, 2008), p. 128.

40. T.R., "My Life as a Naturalist," pp. 321–324.

41. T.R. to Martha Bulloch Roosevelt (April 28, 1868).

42. T.R. to Theodore Roosevelt, Sr. (April 30, 1868).

43. T.R. Childhood Drawings. Houghton Library, Harvard Universiy.

44. Deborah Solomon, "Inspiration on the Hudson," *New York Times* (August 21, 1988).

45. T.R. Boyhood Diaries, "My Life: Three Weeks of My Life, Age Nine Years, August 1868" (August 10 to September 5, 1868).

46. T.R., *An Autobiography*, pp. 4–5.

47. Robert E. Bieder, *Bear* (London: Reaktion, 2005), pp. 74–101.

48. Paul Schullery, *American Bears: Selections from the Writings of Theodore Roosevelt* (Boulder, Col.: Robert Rinehart, 1997), p. 59.

49. T.R. Boyhood Diaries, "Journal of Theodore Roosevelt of U.S.A." (May 12 to September 9, 1869).

50. Ibid., "My Journal in Switzerland" (August 15, 1869).

51. Ibid., diary entry (August 6, 1869).

52. Ibid., "My Journal of Northern Italy" (September 9 to October 20, 1869), diary entry (September 13, 1869).

53. Ibid., "My Journal of Northern Italy" (September 9 to October 20, 1869).

54. "Leopold and Rudolf Blaschka, Design Museums Touring Exhibition," Dresden National History Museum Catalog. Dresden Historical Center, Germany. The museum was torn down in 1944 (online: no author).

55. T.R. Boyhood Diaries, "My Journal in Prussia" (October 21 to October 28, 1869).

56. Ibid. (October 26, 1869).

57. Ibid. (December 3 to December 31, 1869).

58. Ibid.

59. Ibid., "My Journal in Italy" (December 14 to March 9, 1870), entry (December 21, 1869).

60. Ibid., "My Journal in Italy," entries (January 17, 18, 19, 1870).

61. Ibid. (December 14 to March 9, 1870), entry (March 1, 1870).

62. Cutright, *Theodore Roosevelt: The Making of a Conservationist*, p. 20.

63. T.R. Boyhood Diaries, "My Journal in England" (May 25 to September 10, 1870).

64. Cutright, *Theodore Roosevelt: The Naturalist*, p. 4.

65. T.R. Boyhood Diaries, "Now My Journal in the United States" (May 25 to September 10, 1870), entry (June 6, 1871, Spuyten Duyvil, New York).

66. W. H. H. Murray, *Adventures in the Wilderness; Or Camp-Life in the Adirondacks* (Boston, Mass.: Fields, Osgood, 1869).

67. Philip G. Terrie, *Forever Wild: A Cultural History of the Adirondacks* (Syracuse, N.Y.: Syracuse University Press, 1994), pp. 68–71.

68. Thomas Jefferson quoted ibid., p. 22.

69. T.R. Boyhood Diaries, "In the Adirondacks and the White Mountains" (August 1 to August 31, 1871), entry (August 4, 1871, Plattsburgh, New York).

70. T.R. to Josephine Dodge Daskam (May 7, 1901).

71. James Fenimore Cooper, *The Pioneers* (Riverside, Cambridge: D. Appleton & Company, 1876), p. 247. Also see Hugh C. MacDougall, "James Fenimore Cooper: Pioneer of the Environmental Movement," James Fenimore Cooper Society Archives (online). This talk was first written in April 1990 for a program on Earth Day; since then it has been given, with minor changes, before a number of audiences in the Cooperstown area. The version cited here was given in 1999 to the Adirondack Club in Oneonta.

72. T.R. Boyhood Diaries, "In the Adirondacks and White Mountains" (August 1 to August 31, 1871), entry (August 18, 1871).

73. David W. Blight, *Race and Reunion: The Civil War in American Memory* (Cambridge, Mass.: Harvard University Press, 2001), p. 2.

74. T.R., *An Autobiography*, p. 9.

75. Emlen Roosevelt quoted in Edmund Morris, *The Rise of Theodore Roosevelt* (New York: Coward, McCann, and Geoghegan, 1979), p. 35.

76. T.R., *An Autobiography*, pp. 7–8.

77. David McCullough, *Mornings on Horseback*, pp. 29–30.

78. History of American Museum of Natural History, New York, Founding Documents. (File.)

79. "New York's New Museum," *New York Times* (December 23, 1877), p. 1.

80. Carter B. Horsley, "The Museum of Natural History," in *The Upper West Side Book* (City Review, 2007).

81. "Natural History Museum: Costly Building in Central Park," *New York Times* (December 20, 1877), p. 2.

82. "New York's New Museum."

83. Joseph Wallace, *A Gathering of Wonders: Behind the Scenes at the American Museum of Natural History* (New York: St. Martin's, 2000), p. 142.

84. McCullough, *Mornings on Horseback*, p. 118.

85. American Museum of Natural History, Founding Documents.

86. For an interpretation of racism, imperialism, and sexism in the American Museum of Natural History's Roosevelt memorial, see Donna Haraway, "Teddy Bear Patriarchy: Taxidermy in the Garden of Eden, New York City, 1908–1936," *Social Text*, No. 11 (Winter 1984–1985), pp. 20–64.

2: Animal Rights and Evolution

1. Stephen Zawistowski, "Companion Animal Population—Historical Context and Future Directions," SPAY USA Conference (July 7, 2000). (Transcript.)

2. Edmund Morris, *The Rise of Theodore Roosevelt* (New York: Coward, McCann, 1979), p. 98.

3. Steve Zawistowski, *Companion Animals in Society* (Clifton Park, N.Y.: Thomas Delmar Learning, 2008), pp. 53–55.

4. T.R. to Mark Sullivan (September 9, 1908).

5. T.R., *An Autobiography* (New York: Macmillan, 1913), pp. 434–35. The quotation first appeared in *Outlook* (January 25, 1913).

6. Ibid.

7. Gary Francione, *Rain without Thunder: The Ideology of the Animal Rights Movement* (Philadelphia, Pa.: Temple University Press, 1996), p. 6.

8. T.R. to Philip Bathell Stewart (July 16, 1901).

9. Donald G. McNeil, Jr., "When Human Rights Extend to Non Humans," *New York Times* (July 13, 2008), p. 3.

10. Henry Bergh Clipping File, American Society for the Prevention of Cruelty to Animals Archives, New York.

11. Mildred Mastin Pace, *Friend of Animals: The Story of Henry Bergh* (New York: Scribner, 1942), pp. 25–27.

12. Thomas Paine, *The Age of Reason* (New York: Eckler Edition, 1915), pp. 67–68.

13. Stephen Zawistowski to Douglas Brinkley, May 7, 2008.

14. Henry Bergh Clipping File, ASPCA Archives, New York.

15. Bergh quoted in Pace, *Friend of Animals*, p. 31.

16. William C. Spragens (ed.), *Popular Images of American Presidents* (Westport, Conn.: Greenwood, 1988), p. 187.

17. ASPCA Chapter Archive, New York.

See also Letters to the Editor, *New York Times* (July 23, 1868), p. 2.

18. Murat Halstead, *The Life of Theodore Roosevelt: The Twenty-Fifth President of the United States* (Akron, Ohio: Saalfield, 1902), pp. 28–29.

19. Roswell Cheney McCrea, *The Humane Movement* (New York: Columbia University Press, 1910), p. 150.

20. A. H. Saxon, *P. T. Barnum: The Legend and the Man* (New York: Columbia University Press, 1989), pp. 233–238. Also see Lane and Zawistowski, *Heritage of Care*, pp. 24–25.

21. Pace, *Friend of Animals*, pp. 34–118. Also see Henry Bergh Clipping Files, ASPCA Archive.

22. Ibid., pp. 41–48.

23. "The Real Story of Mary Ellen Wilson" (October 2008), American Humane Archive, Englewood, Col. (Online pamphlet.)

24. "Protection for Children," *New York Times* (December 17, 1874), p. 3.

25. Halstead, *The Life of Theodore Roosevelt*, p. 30.

26. Morris, *The Rise of Theodore Roosevelt*, pp. 39–40.

27. T.R. Boyhood Diaries, "Journal of Travels to Europe, Including Egypt and Holy Land" (October 16, 1872 to May 12, 1873), diary entries (October 15 to October 25, 1872), Ship.

28. Joel Ellis Holloway, *Dictionary of Birds of the United States: Scientific and Common Names* (Portland, Ore.: Timber, 2003), p. 25.

29. T.R., "My Life as a Naturalist," *American Museum Journal*, Vol. 18 (May 1918), pp. 321–329.

30. Paul Russell Cutright, *Theodore Roosevelt: The Making of a Conservationist* (Urbana: University of Illinois Press, 1985), p. 32.

31. T.R. Boyhood Diaries, "Journal of Travels to Europe, Including Egypt and Holy Land" (October 16, 1872, to May 12, 1873) entry (October 25, 1872), Liverpool.

32. Paul Russell Cutright, *Theodore Roosevelt: The Naturalist* (New York: Harper, 1956), p. 7.

33. "On the Return of the Arab Courier," Carnegie Museum of Natural History (Clipping File), Pittsburgh, Pa.

34. Ibid.

35. T.R. Boyhood Diaries, "Journal of Travels to Europe, Including Egypt and Holy Land" (October 16, 1872, to May 12, 1873), entry (November 1, 1872), Liverpool.

36. Ibid. (November 28, 1872), Alexandria.

37. Ibid. (November 30, 1872), Cairo.

38. Ibid. (December 1, 1872), Cairo.

39. T.R., *An Autobiography*, pp. 19–20.

40. T.R., *My Life as a Naturalist*, pp. 321–333.

41. Ibid., pp. 321–325.

42. T.R. Boyhood Diaries, "Journal of Travels to Europe, Including Egypt and Holy Land" (October 16, 1872, to May 12, 1873), entry (December 29, 1872).

43. Ibid. (February 24, 1873), Jerusalem.

44. Ibid. (March 4, 1873), Mart Saba.

45. Ibid. (March 17, 1873), Damascus.

46. Ibid. (March 6, 1873), several miles from Hebron.

47. Morris, *The Rise of Theodore Roosevelt*, p. 69.

48. Mayne Reid, *The Boy Hunters* (London: David Bogue, Fleet Street, 1852), pp. 63–67.

49. T.R. Boyhood Diaries, "Journal of Travels to Europe, Including Egypt and Holy Land" (October 16, 1872 to May 12, 1873), April.

50. Ibid. (April 28, 1873), Vienna.

51. Corinne Roosevelt Robinson, *My Brother Theodore Roosevelt* (New York: Charles Scribner's Sons, 1921). pp. 77–78.

52. Henry James, *Madonna of the Future* (London: Macmillan, 1883), p. 38.

53. Janet Browne, *Darwin's Origin of Species* (New York: Grove, 2008), pp. 99–100.

54. Ernst Mayr, "Darwin's Influence on Modern Thought," *Scientific Review* (July 2000), pp. 79–83.

55. T.R. to Oliver Wendell Holmes (October 21, 1904).

56. T.R. to George Otto Trevelyan (January 23, 1904).

57. Charles Darwin, *On the Origin of Spe-*

cies (1859 edition), pp. 665–666. Original first edition published in November 1859.

58. T.R., "My Life as a Naturalist."

59. Reverend Alfred Charles Smith, *The Attractions of the Nile and Its Banks, a Journal of Travel in Egypt and Nubia* (London: John Murray, Albemarle Sheet, 1900).

60. Charles Darwin, *On the Origins of Species*, 1st ed. (Cambridge, Mass.: Harvard University Press, 1964), pp. 62–81. (Facsimile reprint.)

61. Browne, *Darwin's Origin of Species*, p. 14.

62. T.R. to Philip Bathell Stewart (July 16, 1901).

63. T.R. to Martha Bulloch Roosevelt (October 5, 1873), and *Theodore Roosevelt's Diaries of Boyhood and Youth* (New York: Scribner's Sons, 1928).

64. Edward J. Larson, *Evolution: The Remarkable History of a Scientific Theory* (New York: Modern Library, 2004), p. 88.

65. Ibid.

3: Of Science, Fish, and Robert B. Roosevelt

1. T.R. to Martha Bulloch Roosevelt (July 13, 1873).

2. Paul Russell Cutright, *Theodore Roosevelt: The Making of a Conservationist* (Urbana: University of Illinois Press, 1985), p. 70; and Edmund Morris, *The Rise of Theodore Roosevelt* (New York: Coward, McCann, 1979), p. 75.

3. T.R., *An Autobiography* (New York: Macmillan, 1913), pp. 19–20.

4. Steve Zawistowski, *Companion Animals in Society* (Clifton Park, N.Y.: Thomas Delmar Learning, 2008), p. 54.

5. Nancy Pick, *The Rarest of the Rare: Stories behind the Treasures at the Harvard Museum of Natural History* (New York: HarperCollins, 2004), p. 17.

6. Roosevelt Museum Minutes (December 26, 1873). Also quoted in Cutright, *Theodore Roosevelt: The Making of a Conservationist*, pp. 6–81.

7. Roosevelt Museum Minutes (April 6, 1874).

8. Morris, *The Rise of Theodore Roosevelt*, p. 75.

9. T.R., *An Autobiography*, pp. 23–25.

10. Aaron Sachs, *The Humboldt Current: Nineteenth-Century Exploration and the Roots of American Environmentalism* (New York: Viking Adult, 2006).

11. T.R., "My Life as a Naturalist," *American Museum Journal*, Vol. 18 (May 1918).

12. Morris, *The Rise of Theodore Roosevelt*, p. 65.

13. Pick, *The Rarest of the Rare*, p. 8.

14. Morris, *The Rise of Theodore Roosevelt*, pp. 75–76.

15. Jesse Merritt, "A Brief History of the Town of Oyster Bay," *Oyster Bay Historical Society* (July 2003).

16. John Hammond, "The Early Settlement of Oyster Bay," *Freeholder: The Oyster Bay Historical Society* (September 2003).

17. John Rather, "Notable 'Firsts,' " *New York Times* (September 28, 1997).

18. T.R. Boyhood Diaries (1874–1876), Houghton Library, Harvard University.

19. T.R., "My Life as a Naturalist."

20. Roderick Frazier Nash, *Wilderness and the American Mind*, 4th ed. (New Haven, Conn.: Yale University Press, 2001), pp. 1–3.

21. Aldo Leopold, "The Wilderness and Its Place in Forest Recreational Policy," *Journal of Forestry*, Vol. 19 (1921), p. 719.

22. Patricia Nelson Limerick, *Something in the Soil* (New York: Norton, 2000), p. 277.

23. Wilderness Act of 1964 (16 U.S.C. 1131–1136, 78 Stat. 890)—Public Law 88–577 (approved September 3, 1964), U.S. Fish and Wildlife Service Archives.

24. "Roosevelt's Boyhood Life in Adirondacks Is Recalled by Guide," *Adirondack Enterprise* (January 28, 1930). Special thanks to Michele Tucker, Curator of the Adirondack Research Room in the Saranac Lake Free Library, Saranac Lake, N.Y.

25. T.R., "Journal of a Trip to the Adirondacks."

26. T.R., "My Life as a Naturalist."

27. T.R., "Notes on the Fauna of the Adirondack Mountains."

28. Paul W. B. Joslin, "Movements and Home Sites of Timber Wolves in Algonquin Park," *American Zoologist* (1969), pp. 279–288.

29. Charles Darwin, *On the Origin of Species*, 1st ed. (Cambridge, Mass.: Harvard University Press, 1964), p. 490. (Facsimile.)

30. Anthony DePalma, "A Rising Number of Birds at Risk," *New York Times* (December 1, 2007).

31. Fred J. Alsop, *Birds of North America—Eastern Region* (New York: Dorling Kindersley, 2001), pp. 545–550.

32. Mayne Reid, *The Boy Hunters, Or Adventures in Search of a White Buffalo* (London: David Bogue, Fleet Street, 1852), pp. 8–9.

33. Janet E. Buerger, "Ultima Thule: American Myth, Frontier, and the Artist-Priest in Early American Photography," *American Art*, Vol. 6, No. 1 (Winter 1992), pp. 82–103.

34. Robert M. Utley, *A Life Wild and Perilous: Mountain Men and the Paths to the Pacific* (New York: Holt, 1997), pp. 156–285.

35. Institute for Government Research, *The U.S. Geological Survey: Its History, Activities, and Organization* (Baltimore, Md.: Johns Hopkins Press, 1919), p. 9.

36. Clarence King, U.S. Geological Survey 1st Annual Report (1880), p. 4. Also quoted in "The Four Great Surveys of the West," *United States Geological Survey*, C1050 (Washington, D.C.: U.S. Geological Survey, Department of Interior, 2000).

37. William H. Goetzmann, *Exploration and Empire: The Explorer and the Scientist in the Winning of the American West* (New York: Knopf, 1966), pp. xiii–xiv.

38. Ibid. Also "History of Yosemite" Files, Yosemite National Park Archive, California.

39. Lary M. Dilsaver, *America's National Park System: The Critical Documents* (Lanham, Maryland: Rowman and Littlefield, 1994), p. 28.

40. Stacey Bredhoff, *American Originals* (Seattle: University of Washington Press, 2001), p. 58.

41. Ira N. Gabrielson, *Wildlife Refuges* (New York: Macmillan, 1943), pp. 74–81.

42. Text of Benjamin Harrison's Official Chugach Proclamation, Chugach National Forest. U.S. Forest Service Archives, Washington, D.C.

43. "The National Wildlife Refuge System: Promises for a New Century" (Shepherdstown, W. Va.: U.S. Fish and Wildlife Service Archive, 2003).

44. Scott Weidensaul, *Return to Wild America* (New York: North Point, 2005), pp. 321–345.

45. David McCullough, *Mornings on Horseback* (New York: Simon and Schuster, 1981), pp. 20–21.

46. Bill Bleyer, "The Forgotten Roosevelt," *Newsday* (October 6, 1985).

47. Ibid., p. 11.

48. McCullough, *Mornings on Horseback*, p. 22.

49. Nathan Miller, *The Roosevelt Chronicles: A Story of a Great American Family* (Garden City, N.Y.: Doubleday, 1979), p. 142.

50. William S. Spragens, *Popular Images of American Presidents* (Westport, Conn.: Greenwood, 1988).

51. John A. Gable, "Robert B. Roosevelt," *Theodore Roosevelt Association Journal*, Vol. 9, No. 4 (Fall 1983), p. 13. Dr. Gable, former executive director of the Theodore Roosevelt Association (TRA) in Oyster Bay, N.Y., gave me access to two blue-bound "House of Roosevelt" volumes filled with R.B.R.'s letters and diaries. When I cite R.B.R. Papers, TRA, this is what I am referring to.

52. Robert B. Roosevelt, *Five Acres Too Much* (New York: Harper, 1869), pp. 19–20.

53. Horace Greeley to R.B.R. (April 5, 1871), R.B.R. Papers, TRA.

54. "President's Uncle, R.B. Roosevelt, Dead," *New York Times* (June 15, 1906), p. 9.

55. Bill Bleyer, "The Forgotten Roosevelt," *Newsday* (October 6, 1985), p. 27.

56. Miller, *The Roosevelt Chronicles*, p. 147.

57. Bleyer, "The Forgotten Roosevelt," p. 11. Also see Robert B. Roosevelt to Rutherford B. Hayes (April 4, 1880), R.B.R. Papers, TRA.

58. M. Fortescue Pickard, "The House of Roosevelt" (unpublished, October 12, 1936). (Given to author by John A. Gable.) [n.d.] pp. 120–124.

59. Robert B. Roosevelt, "Is the Turtle

a Fish?" Diary Entry (or Notes), R.B.R. [n.d.] Papers, TRA.

60. Survey of Historic Sites and Buildings," Theodore Roosevelt Birthplace National Historic Site, New York, N.Y.

61. Author interview, P. J. Roosevelt, New York, N.Y. (June 1995).

62. McCullough, *Mornings on Horseback*, p. 21.

63. T.R., *An Autobiography*, p. 12.

64. John F. Reiger, *American Sportsmen and the Origins of Conservation*, 3rd ed. (Corvallis: Oregon State University Press, 2001).

65. Frank Forester, *The Warwick Woodlands* (Philadelphia, Pa.: G. B. Zieber, 1845).

66. Henry William Herbert, *Frank Forester's Field Sports of the United States*, Vol. 1 (New York: Stringer and Townsend, 1848), p. 12.

67. James A. Tober, *Who Owns the Wildlife? The Political Economy of Conservation in Nineteenth Century America* (Westport, Conn.: Greenwood, 1981), pp. 48–54. Also see George Bird Grinnell, "American Game Protection: A Sketch" in George Bird Grinnell and Charles Sheldon (eds.), *Hunting and Conservation: The Book of the Boone and Crockett Club* (New Haven, Conn.: Yale University Press, 1925), pp. 221–224.

68. Bigelow quoted in Pickard, "The House of Roosevelt."

69. Paul Schullery, "Hope for the Hook and Bullet Press," *New York Times* (September 22, 1985), sec. 7, page 1.

70. Charles Hallock, *Vacation Rambles in Michigan* (Grand Rapids and Indiana Railroad, 1877).

71. George Bird Grinnell and T.R., *Trail and Camp-Fire: The Book of the Boone and Crockett Club* (New York: Forest and Stream, 1897), p. 332.

72. Robert Barnwell Roosevelt, *Game Fish of the Northern States of America, and British Provinces* (New York: Carleton, 1862).

73. Pickard, "The House of Roosevelt," p. 101.

74. Parker Gillmore to R.B.R. (1862), R.B.R. Papers, TRA.

75. Pickard, "The House of Roosevelt," p. 102.

76. Robert Barnwell Roosevelt, *Game Fish of the Northern States of America and British Provinces*, p. 36.

77. John F. Reiger, *American Sportsmen and The Origins of Conservation*, pp. 150–151.

78. Richard P. Harmond, "Robert Barnwell Roosevelt and the Early Conservation Movement," *Theodore Roosevelt Association Journal*, Vol. 14 (Summer 1988), pp. 2–11.

79. "The Reed Draper Collection of Angling Books," Central Michigan University Archives, Mount Pleasant.

80. Sylvia R. Black, "Seth Green, Father of Fish Culture," *Rochester History* Vol. 6, No. 3 (July 1944); and Pickard, "The House of Roosevelt," pp. 102–103.

81. "The Legacy of Seth Green" (Arlington, Va.: Trout Unlimited Archive, 2007).

82. Ibid.

83. Clinton E. Atkinson, "Feeding Habits of Adult Shad (Alosa sapidissima) in Fresh Water" *Ecology*, Vol. 32, No. 3 (July 1951), pp. 556–557.

84. Seth Green, *Trout Culture* (Caledonia, N.Y.: Green and Collins, 1870).

85. Arthur D. Welander, "Notes on the Dissemination of Shad, Alosa sapidissima, along the Pacific Coast of North America" (Wilson), *Copeia*, Vol. 1940, No. 4 (December 27, 1940), pp. 221–223.

86. R.B.R. quoted in Pickard, "The House of Roosevelt," pp. 104–106.

87. Robert Barnwell Roosevelt, *Florida and the Game Water Birds of the Atlantic Coast and the Lakes of the United States* (New York: Orange Judd, 1884), p. 24.

88. Ibid., p. 12.

89. Ibid., p. 10.

90. Seth Green to Robert B. Roosevelt (November 10, 1884), R.B.R. Papers, TRA.

91. *Suffolk County News* (June 15, 1906).

92. Richard Hammond and Donald H. Weinhardt, "Robert Barnwell Roosevelt on the Great South Bay," *Long Island Forum* (August–September 1987), p. 167.

93. Spencer F. Baird to Robert B. Roosevelt (May 16, 1874), R.B.R. Papers, TRA.

94. Pickard, "The House of Roosevelt."

95. Robert B. Roosevelt, "The Roosevelt Coat-of-Arms," R.B.R. Papers, TRA.

96. Robert B. Roosevelt, "Frog Notes" (n.d.), R.B.R. Papers, TRA.

97. Hammond and Weinhardt, "Robert Barnwell Roosevelt on the Great South Bay," p. 161.

98. Robert B. Roosevelt quotes this slogan in "Notes on the Old New York of His Day" (unpublished), R.B.R. Papers, TRA.

99. Bleyer, "The Forgotten Roosevelt," p. 11.

100. Ernest Schwiebert, Introduction, in *Superior Fishing: Or, the Striped Bass, Trout, and Black Bass of the Northern States* (St. Paul: Minnesota Historical Society, 1985), pp. xii–xiii. (Reprint edition.)

4: Harvard and the North Woods of Maine

1. T.R., *An Autobiography* (New York, Macmillan, 1913), pp. 29–30.

2. Everett Parker, "A Historical Perspective of the Moosehead Lake Region," *Moosehead Historical Society* (2007).

3. T.R., *An Autobiography*, pp. 29–32.

4. T.R., "My Life As a Naturalist," *The American Museum Journal*, Vol. XVII (May, 1918), No. 5.

5. Jonathan Rosen, *The Life of the Skies: Birding at the End of Nature* (New York: Farrar, Straus, and Giroux, 2008), p. 126.

6. T.R., "My Life As a Naturalist," *The American Museum Journal.*

7. T.R. to Anna Bulloch Grace (July 7, 1872).

8. Paul Russell Cutright and Michael. J. Brodhead, *Elliott Coues: Naturalist and Frontier Historian* (Urbana: University of Illinois Press, 1981).

9. "Dr. Arthur H. Cutler, School Founder, Dies," *New York Times* (June 22, 1918), p. 9. T.R. became the first graduate of the Cutler School of New York.

10. Arthur Cutler, "Reminiscences," in Stefan Lorant, *The Life and Times of Theodore Roosevelt* (Garden City, N.Y.: Doubleday, 1959), p. 136.

11. T.R., "Remarks on the Zoology of Oyster Bay (1874–76)," T.R. Collection, Harvard University.

12. Robert B. Roosevelt, "Notes on the Old New York" (unpublished), R.B.R. Scrapbook, TRA.

13. Kathleen Dalton, *Theodore Roosevelt: A Strenuous Life* (New York: Knopf, 2002), pp. 59–60.

14. Edward J. Renehan, Jr., *The Lion's Pride: Theodore Roosevelt and His Family in Peace and War* (New York: Oxford University Press, 1998), p. 42.

15. T.R. quoted in Lorant, *The Life and Times of Theodore Roosevelt* (New York: Doubleday, 1959), pp. 135–136.

16. T.R. Boyhood Diaries, "Notes on Natural History" (July 24–30, 1874).

17. T.R., *An Autobiography*, p. 25.

18. Public Broadcasting Service, "People and Events: The Centennial Exposition of 1876," *The American Experience*. Transcription on website from 2002; and "Centennial Exposition of 1876," Pennsylvania Historical and Museum Commission Files, State Museum Building, Harrisburg.

19. David Starr Jordan, *Manual of the Vertebrates of the Northern United States* (Chicago: Jansen, McClurg, 1876).

20. Paul Russell Cutright, *Theodore Roosevelt: The Making of a Conservationist* (Urbana: University of Illinois Press, 1985), p. 97.

21. Anna Roosevelt Cowles (ed.), *Letters from Theodore to Anna Roosevelt Cowles, 1870–1918* (New York: Scribner, 1924), p. 12.

22. R. W. G. Vail, "Your Loving Friend, T.R.," *Collier's* (December 20, 1924).

23. Ralph Waldo Emerson, "The Superlative," in *Lectures and Biographical Sketches* (Cambridge, Mass.: Edward W. Emerson, 1883), p. 139.

24. David McCullough, *Mornings on Horseback* (New York: Simon and Schuster, 1981), pp. 200–201.

25. Nancy Pick and Mark Sloan, *The Rarest of Rare* (New York: HarperCollins, 2004), p. 16.

26. Carleton Putnam, *Theodore Roosevelt: The Formative Years* (New York: Scribner, 1958), pp. 137–138.

27. Robert Barnwell Roosevelt, *Five Acres Too Much* (New York: Harper, 1869), p. xi.

28. Jim Reis, "Pieces of the Past," Northern Kentucky University Archives, Vol. 2, pp. 56–59. Also see Nathaniel Shaler, *The First Book of Geology* (Boston, Mass.: Ginn, Heath, 1884).

29. "Obituary: Nathaniel S. Shaler,"

Bulletin of the American Geographical Society, Vol. 38, No. 5 (1906), p. 336.

30. Donald Wilhelm, *Theodore Roosevelt as an Undergraduate* (Boston, Mass.: J. W. Luce, 1910), p. 35.

31. "Theodore Roosevelt, Student," *New York Times* (June 12, 1907), p. 8.

32. T.R. quoted in Joshua David Hawley, *Theodore Roosevelt: Preacher of Righteousness* (New Haven, Conn.: Yale University Press, 2008), pp. 35–36.

33. McCullough, *Mornings on Horseback*, pp. 213–214.

34. T.R. to Gifford Pinchot (March 14, 1907). Library of Congress (microfilm), Series 2, Vol. 71, Real 345, p. 335.

35. Richard Welling, "My Classmate Theodore Roosevelt," *American Legion Monthly* (January 1929), pp. 9–11.

36. T.R. letter quoted in Cutright, *Theodore Roosevelt: The Making of a Conservationist*, p. 100. T.R.'s other ornithologist friend at Harvard was Frederic Gardiner, a graduate of the class of 1880. He later became a minister, and the love of birds became part of his sermons.

37. Edmund Morris, *The Rise of Theodore Roosevelt* (New York: Coward, McCann, 1979), p. 90.

38. T.R., *Letters from T.R. to Anna Roosevelt Cowles, 1870 to 1918* (New York: Charles Scribner's Sons, 1924), pp. 22–23. Letter to Father and Mother, April, 1879.

39. "Review of Minot's *The Land and Game Birds of New England*," *Harper's New Monthly Magazine* (April 1877), p. 772.

40. T.R. journal (June 23, 1877). Also see Morris, *The Rise of Theodore Roosevelt*, pp. 90–91.

41. T.R., *Outdoor Pastimes of an American Hunter* (New York: Scribner, 1905), p. 339.

42. C. Hart Merriam writing in *Bulletin of the Nuttall Ornithological Society* (April 1878). Quoted in Paul Russell Cutright, *Theodore Roosevelt: The Naturalist* (New York: Harper, 1956), p. 18.

43. For biographical information on C. Hart Merriam see Keir B. Sterling, *Last of the Naturalists: The Career of C. Hart Merriam* (New York: Arno Press, 1977), and "Dr. Merriam, Famed Natural Scientist, D.C.S," *Washington Post*, March 21, 1942, p. 9.

44. Cutright, *Theodore Roosevelt: The Making of a Conservationist*, pp. 102–103.

45. T.R., "Small Country Neighbors," *Scribner's Magazine* (October 1907) Vol. XLII, No. 4.

46. "Senator Hill's Condition," *New York Times* (July 28, 1882), p. 1.

47. Putnam, *Theodore Roosevelt*, pp. 147–148.

48. T.R. Boyhood Diaries (December 25, 1877).

49. Nathan Miller, *Theodore Roosevelt: A Life* (New York: Morrow, 1992), p. 81.

50. T.R., *An Autobiography*, p. 26.

51. Roderick Nash, *Wilderness and the American Mind* (New Haven, Conn.: Yale University Press, 1967), pp. 88–90.

52. T.R., *Outdoor Pastimes of an American Hunter*, p. 322.

53. McCullough, *Mornings on Horseback*, p. 205.

54. Morris, *The Rise of Theodore Roosevelt*, p. 109.

55. "Deer and Caribou in Maine: From the *Bangor Commercial*, Jan. 20," *New York Times* (January 29, 1888). The *Bangor Commercial* reported that Mr. H. O. Stanley, one of the fish commissioners, said that deer and caribou were so plentiful in Maine that hunting permits should be allowed.

56. T.R., "My Debt to Maine," in *Maine, My State* (Lewiston, Maine: Journal Printshop, 1919), p. 17. Also see "Deer and Caribou in Maine: From the *Bangor Commercial*, Jan. 20," *New York Times* (January 29, 1888). The *Bangor Commercial* reported that Mr. H. O. Stanley, one of the fish commissioners, said that deer and caribou were so plentiful in Maine that hunting permits should be allowed.

57. Ibid., p. 19.

58. Ibid., p. 21.

59. William Wingate Sewall, *Bill Sewall's Story of Theodore Roosevelt (T.R.)* (New York: Harper, 1919), p. 5.

60. T.R., "My Debt to Maine," p. 19.

61. Sewall, *Bill Sewall's Story of Theodore Roosevelt*, p. 5.

62. Ibid., p. 4.

63. Charles G. Washburn, *Theodore Roosevelt: The Logic of His Career* (Boston, Mass.: Houghton Mifflin, 1916), p. 5.

64. Cutright, *Theodore Roosevelt: The*

Making of a Conservationist, pp. 79–83. Coues had signed *Birds of the Colorado Valley* to: "Theodore Roosevelt (from the author), Jan. 1879."

65. Sewall, *Bill Sewall's Story of Theodore Roosevelt*, p. 6.

66. T.R., "My Debt to Maine," p. 20.

67. T.R., *Outlook* (July 27, 1912); and address at Saint Louis, Mo. (May 31, 1916), *Mem. Ed.* 24, p. 483.

68. T.R. to Martha Bulloch Roosevelt (September 14, 1879).

69. Ibid.

70. Thoreau, *The Maine Woods*, p. 120.

71. Steven M. Cox and Kris Fulsaas, *Mountaineering: The Freedom of the Hills* (Seattle, Wash.: Mountaineers, 2003), pp. 16–17. First printed 1960.

72. Yagyu Munenori, "Martial Arts: The Book of Family Traditions" in Thomas Cleary (ed.) *Soul of the Samurai* (North Clarendon, VT: Tuttle Publishing, 2005), pp. 78–79.

73. Lewis Carroll, *The Hunting of the Snark: An Agony in Eight Fits* (New York: Pantheon, 1966), p. 26. (Originally published 1876; *Alice through the Looking-Glass* was earlier, 1872.)

74. Carleton Putnam, *Theodore Roosevelt*, p. 163, Also see John Watterson, *The Games Presidents Play* (Baltimore: Johns Hopkins University Press, 2006), p. 68.

75. T.R., "My Debt to Maine," p. 17.

5: Midwest Tramping and the Conquering of the Matterhorn

1. Edmund Morris, *The Rise of Theodore Roosevelt* (New York: Coward, McCann, 1979), p. 112.

2. Castle Freeman, Jr., "Owen Wister: Brief Life of a Western Mythmaker, 1860–1938," *Harvard Magazine* (July–August 2002), p. 42.

3. Owen Wister, *Roosevelt: The Story of a Friendship 1880–1919* (New York: The Macmillan Company, 1930), pp. 4–8.

4. David McCullough, *Mornings on Horseback* (New York: Simon and Schuster, 1981), pp. 210–211. McCullough believes very strongly that Wister was playing Parson Weems when writing up the boxing story in his memoir.

5. Wister, *Roosevelt: The Story of a Friendship 1880–1919*, pp. 4–7.

6. Carleton Putnam, *Theodore Roosevelt: The Formative Years* (New York: Scribner, 1958), p. 178.

7. Ibid., p. 179.

8. Morris, *The Rise of Theodore Roosevelt*, p. 122.

9. Kay Redfield Jamison, *Exuberance: The Passion for Life* (New York: Random House, 2004), pp. 8–21.

10. Ibid., pp. 131–132.

11. Winthrop Chandler, *Roman Springs: Memoirs* (Boston, Mass.: Little, Brown, 1934), p. 195.

12. T.R., *An Autobiography* (New York, Macmillan, 1913), p. 7.

13. Putnam, *Theodore Roosevelt: The Formative Years*, p. 134.

14. "Theodore Roosevelt, Student," *New York Times* (June 12, 1907), p. 8.

15. T.R. College Diary (May 5, 1880).

16. Louis Hawes, "A Sketchbook by Thomas Cole," *Record of the Art Museum, Princeton University*, Vol. 15, No. 1 (1956), pp. 2–23. The diaries of Thomas Cole have been underappreciated by environmental historians. Take, for example, his eloquent entry about the significance of trees in his life: "Treading the mosses of the forest, my attention has often been attracted by the appearance of action and expression in trees. I have been led to reflect upon the fine effects they produce, and to look into the causes. They spring from some resemblance to man. . . . Exposed to adversity and agitations, they battle for existence or supremacy. On the mountain, exposed to the blasts, trees grasp the crags with their gnarled roots, and struggle with the elements with wild contortions." In Rev. Louis L. Noble, *The Life and Works of Thomas Cole* (New York: Cornish, Lamport, 1853), pp. 125–126.

17. T.R. to Corinne Roosevelt (July 24, 1880).

18. Anna Eleanor Roosevelt (ed.), *Hunting Big Game in the Eighties: The Letters of Elliott Roosevelt* (New York: Scribner, 1933), pp. ix–x.

19. Eleanor Roosevelt, *The Autobiography of Eleanor Roosevelt* (New York: Da Capo, 1992), p. 5.

20. Elliott Roosevelt to Theodore Roosevelt Senior (February 20, 1876) in Anna Eleanor Roosevelt (ed.), *Hunting Big Game in the Eighties*, p. 20.

21. Kathleen Dalton, *Theodore Roosevelt: A Strenuous Life* (New York: Knopf, 2002), pp. 58–61.

22. Francis Parkman, *Oregon Trail* (New York: Charles E. Merrill, 1910). Originally published in 1849 as *The California and Oregon Trail*, though Parkman had never visited California. Later he denounced that title as a "publisher's trick" designed to increase sales.

23. T.R. quoted in *Independent* (November 24, 1892), *Mem. Ed.* 14, p. 286.

24. T.R., "Midwest Tramp Diary" (August 17, 1880).

25. William Cronon, *Nature's Metropolis: Chicago and the Great West* (New York: Norton, 1991), p. 19.

26. Washington Irving, *A Tour on the Prairies* (Norman: University of Oklahoma, 1956). Originally published as part of *The Crayon Miscellany*, 3 vols. (Philadelphia, Pa.: Carey, Lea, and Blanchard, 1835).

27. T.R. to Anna Roosevelt (August 22, 1880).

28. T.R., "Midwest Tramp Diary" (August 19, 1880).

29. T.R., *The Wilderness Hunter* (New York and London: Putnam, 1893), p. 450.

30. T.R. to Anna Roosevelt (August 22, 1880).

31. T.R. to Martha Bulloch Roosevelt (August 25, 1880).

32. Ibid.

33. T.R., "Midwest Tramp Diary" (August 24, 1880).

34. T.R. to Corinne Roosevelt (September 12, 1880).

35. T.R., "Midwest Tramp Diary" (August 27, 1880).

36. Vachel Lindsay, "Abraham Lincoln Walks at Midnight," in *The Congo and Other Poems* (New York: Macmillan, 1914).

37. *History of Western Iowa: Its Settlement and Growth* (Sioux City, Iowa: Western, 1882), p. 505.

38. T.R., "Midwest Tramp Diary" (September 5, 1880).

39. *History of Western Iowa*, p. 534.

40. T.R., "Midwest Tramp Diary" (September 8, 1880).

41. Emily Dickinson, "A Narrow Fellow in the Grass" (number 986, "The Snake"), *Springfield Republican* (February 14, 1866).

42. T.R. to Liberty Hyde Bailey (August 10, 1908).

43. T.R. to Corinne Roosevelt (September 12, 1880) in *Letters*, Vol. 1, p. 46. Also in Corinne Roosevelt, *My Brother, Theodore Roosevelt* (New York: Charles Scribner's Sons, 1921), p. 114.

44. "Residential Notes" (Visitors' Center, Fargo-Morehead Convention and Visitors' Bureau, 2008).

45. Bruce Watson, "World's Unlikeliest Bestseller," *Smithsonian* (August 2005).

46. "Red River State Park" (Minnesota Department of Natural Resources, 2008). (Pamphlet.)

47. T.R., "Midwest Tramp Diary" (September 14, 1880).

48. Ibid. (September 24, 1880).

49. Ibid. (September 21, 1880).

50. Ibid. (September 22 and 23, 1880).

51. Thomas L. Altherr and John F. Reiger, "Academic Historians and Hunting: A Call for More and Better Scholarship," *Environmental History Review*, Vol. 19, No. 3 (Autumn 1995), pp. 39–56.

52. "Hunting Trips of a Ranchman: Part II," in *The Works of Theodore Roosevelt: Hunting Trips on the Prairie and in the Mountains* (New York and London: Putnam, 1902), p. 121. (Originally printed by Putnam, 1885.)

53. "History of Cattle Ranching: Cattle Industry," Bill Lane Center for the Study of the North American West, Stanford University.

54. T.R., "Midwest Tramping Diary" (September 30, 1880).

55. H. W. Brands, *T.R.: The Romantic* (New York: Basic Books, 1997), pp. 108–09.

56. T.R. Private Diaries, 1878–1885 (October 27, 1880).

57. Ibid. (March 24, 1881).

58. T.R., "Sou'-Sou'-Southerly," Introduction by John Rousmaniere, pp. 70–75. Also see D. J. Philippon, "Theodore Roosevelt's 'Sou-Sou'-Southerly': An Unappreciated Nature Essay," *North Dakota Quarterly*, Vol. 64, No. 1 (Winter 1997), pp. 83–92.

59. T.R., "Sou'-Sou'-Southerly," *Gray's Sporting Journal*, Vol. 13, Issue 3 (Fall 1988), p. 75. (Originally written in March 1881, the article in *Gray's* includes, as noted above, a brief introduction by John Rousmaniere.)

60. Ibid.

61. Putnam, *Theodore Roosevelt*, p. 224.

62. T.R. Honeymoon Diary (July 5, 1881).

63. T.R. to Bill Sewall (September 5, 1881).

64. T.R. to Anna Roosevelt (September 5, 1881).

65. Edward Whymper, *The Ascent of the Matterhorn* (London: John Murray, Albemarle Street, 1880).

66. T.R. to Anna Roosevelt (August 5, 1881).

67. Louis S. Warren, *The Hunter's Game: Poachers and Conservationists in the Twentieth Century* (New Haven, Conn.: Yale University Press, 1997), p. 180.

68. T.R. to Bill Sewall (September, 1881).

69. Isaac Hunt, Oral History, Theodore Roosevelt Birthplace, N.Y. Also see Lisa Slaski, "Hon. Isaac L. Newton: From Salisbury, New York, to Jefferson County, New York" (Herkimer County Historical Society). (June 2008, online.)

70. Morris, *The Rise of Theodore Roosevelt*, p. 162.

71. Nathan Miller, *Theodore Roosevelt: A Life* (New York: HarperCollins, 1994), p. 124.

72. Elting E. Morison and John Blum (eds.), *The Letters of Theodore Roosevelt* (Cambridge, Mass.: Harvard University Press, 1951–1954), Vol. I, p. 1450.

73. T.R., *An Autobiography*, pp. 67–68.

74. William Healey Dall, *Spencer Fullerton Baird: A Biography* (Philadelphia, Pa.: Lippincott, 1915), pp. 396–419.

75. Elmer Charles Herber, *Correspondence between Spencer Fullerton Baird and Louis Agassiz: Two Pioneer American Naturalists* (Washington, D.C.: Smithsonian Institution, 1963), pp. 6–9.

76. Dall, *Spencer Fullerton Baird*, pp. 416–432.

77. E. F. Rivinus and E. M. Youssef, *Spencer Baird of the Smithsonian* (Washington, D.C.: Smithsonian Institution Press, 1992), p. 1.

78. Spencer F. Baird to T.R. (April 25, 1882), quoted in Paul Russell Cutright, *Theodore Roosevelt: The Making of a Conservationist* (Urbana: University of Illinois Press, 1985), p. 136. Cutright claims that these letters were housed with the Baird Collection at the Smithsonian Institution, but the Smithsonian simply doesn't have them in its archive.

79. T.R. to Spencer F. Baird (April 27, 1882), ibid., p. 136.

80. Spencer F. Baird to T.R. (April 28, 1882), ibid.

81. Brands, *T.R.: The Last Romantic*, pp. 119–120.

82. Spencer F. Baird to T.R. (May 26, 1882), ibid., p. 137.

83. Ibid., pp. 138–139.

6: Chasing Buffalo in the Badlands and Grizzlies in the Bighorns

1. T.R., *Hunting Trips of a Ranchman: Sketches of Sport in the Northern Cattle Plains* (New York and London: Putnam, 1885), pp. 240–269.

2. Frances Theodora Parsons, *Perchance Some Day* (New York: Privately published, 1951). Also Nathan Miller, *Theodore Roosevelt* (New York: William Morrow, 1992), p. 146.

3. T.R. Private Diaries (January 3, 1883).

4. Hermann Hagedorn, *The Roosevelt Family of Sagamore Hill* (New York: Macmillan, 1954), pp. 5–10.

5. Sagamore Hill National Historic Site, House History (Archives), Oyster Bay, New York. Also see "Theodore Roosevelt at Home," *American Monthly Review of Reviews*, Vol. 18 (July–December 1898), pp. 594–595.

6. Henry F. Pringle, *Theodore Roosevelt: A Biography* (New York: Harcourt, Brace, 1931), pp. 54–55.

7. "The Northern Pacific," *New York Times* (September 9, 1883), p. 2.

8. Eugene V. Smalley, *History of the Northern Pacific Railroad* (New York: Putnam, 1883), p. v.

9. Hiram Rogers, *Exploring the Black Hills and the Badlands* (Boulder, Col.: Johnson, 1999), p. 179.

10. John Roach, "Dinosaur Mummy Found; Has Intact Skin, Tissue," *National Geographic News* (December 3, 2007).

11. John P. Bluemle, "North Dakota's Petrified Forest," *North Dakota Notes Number 3* (North Dakota Geological Survey, 2002). (Online.)

12. "Henry H. Gorringe Dead," *New York Times* (July 7, 1885), p. 5.

13. T.R. quoted in Edmund Morris, *The Rise of Theodore Roosevelt* (New York: Coward, McCann, 1979), p. 198.

14. Chester L. Brooks and Ray H. Mattison, *Theodore Roosevelt and the Dakota Badlands* (Washington, D.C.: National Park Service, 1958), p. 3.

15. Tom McHugh, *The Time of the Buffalo* (Lincoln: University of Nebraska Press, 1972), p. 278.

16. T.R. to Martha Bulloch Roosevelt (February 20, 1883).

17. T.R., *Ranch Life and the Hunting-Trail* (New York: Century, 1888).

18. Carleton Putnam, *Theodore Roosevelt: The Formative Years* (New York: Scribner, 1958), pp. 309–310.

19. T.R., *Hunting Trips of a Ranchman*, pp. 32–33.

20. T.R. to Martha Bulloch Roosevelt (September 4, 1883).

21. Robert M. Utley, *The Lance and the Shield: The Life and Times of Sitting Bull* (New York: Holt, 1993).

22. T.R., *An Autobiography* (New York: Macmillan, 1913), p. 54.

23. James S. Brisbin, *The Beef Bonanza; or, How to Get Rich on the Plains* (Philadelphia, Pa.: James Lippencott, 1881), p. 90. Also see David Dary, *Cowboy Culture: A Saga of Five Centuries* (Lawrence: University Press of Kansas, 1989), pp. 308–331.

24. Hermann Hagedorn, *Roosevelt in the Bad Lands* (Boston, Mass.: Houghton Mifflin, 1921), p. 40.

25. Dary, *Cowboy Culture*, pp. xi–xiii, and p. 83.

26. Harold E. Briggs, "The Development and Decline of Open Range Ranching in the Northwest," *Mississippi Valley Historical Review*, Vol. 20, No. 4 (March 1934), pp. 521–536.

27. Peter Applebome, "Wrangling over Where Rodeo Began," *New York Times* (June 18, 1989).

28. Morris, *The Rise of Theodore Roosevelt*, p. 206.

29. David A. Dary, *The Buffalo Book: The Full Saga of the American Animal*, rev. ed. (Athens: Swallow/Ohio University Press, 1989), p. 42.

30. Henry Remsen Tilton, "After the Nez Perces," *Forest and Stream*, Vol. 9, No. 21 (December 27, 1877), pp. 403–404. Cited in Dary, *The Buffalo Book.*

31. Hagedorn, *Roosevelt in the Bad Lands*, p. 10.

32. Champ Clark, *The Badlands* (New York: Time-Life Books, 1974), pp. 112–113.

33. Lincoln Lang, *Ranching with Roosevelt* (Philadelphia and London: Lippincott, 1926), p. 31.

34. Joel Berger and Carol Cunningham, *Bison: Mating and Conservation in Small Populations* (New York: Columbia University Press, 1994), p. 45.

35. J. A. Allen, "The Little Missouri 'Bad Lands,' " *American Naturalist*, Vol. 10, No. 4 (April 1876), pp. 207–216.

36. Ibid., p. 135.

37. Lewis F. Crawford, *Badlands and Bronco Trails* (Bismarck, N.D.: Capital, 1922), pp. 11–12.

38. T.R., *A Book-Lover's Holidays in the Open* (New York: Scribner, 1916), pp. 31–32.

39. T.R., *Hunting Trips of a Ranchman*, pp. 10–12.

40. T.R., *An Autobiography*, p. 100.

41. Hagedorn, *Roosevelt in the Bad Lands*, p. 14.

42. T.R., *Hunting Trips of a Ranchman*, p. 230.

43. Thomas Berger, *Little Big Man* (New York: Delacorte/Seymour Lawrence, 1964), p. 47.

44. T.R., *Hunting Trips of a Ranchman*, pp. 249–250

45. T.R. Diary (August 24, 1884).

46. T.R., *Hunting Trips of a Ranchman*, pp. 13–16.

47. Brooks and Mattison, *Theodore Roosevelt and the Dakota Badlands*, p. 18.

48. Lang, *Ranching with Roosevelt*, pp. 101–102.

49. Putnam, *Theodore Roosevelt*, p. 330.

50. Hagedorn, *Roosevelt in the Bad Lands*, p. 24.

51. Lang, *Ranching with Roosevelt*, pp. 366–367.

52. Ibid., p. 105.

53. T.R. to Casper Whitney (January 31, 1908, a form statement about hunting).

54. T.R., *Hunting Trips of a Ranchman*, p. 263.

55. Hagedorn, *Roosevelt in the Bad Lands*, p. 36.

56. Ibid., p. 37.

57. T.R., *Hunting Trips of a Ranchman*, p. 268.

58. Hagedorn, *Roosevelt in the Bad Lands*, p. 45.

59. Putnam, *Theodore Roosevelt*, p. 345.

60. Gail Bederman, *Manliness and Civilization: A Cultural History of Gender and Race in the United States 1880–1917* (Chicago, Ill.: University of Chicago Press, 1995), pp. 170–184.

61. Ibid., p. 61.

62. Lang, *Ranching with Roosevelt*, p. 364.

63. Larry Barsness and Ron Tyler, *Heads, Hides, and Horns: The Compleat Buffalo Book* (Fort Worth: Texas Christian University Press, 1985), p. 132.

64. Dary, *The Buffalo Book*, pp. 196–197.

65. T.R. to Jonas S. Van Duzer (November 20, 1883).

66. T.R., *An Autobiography*, p. 87.

67. T.R. to Alice Lee Roosevelt (February 6, 1884).

68. Quoted in Stefan Lorant, *The Life and Times of Theodore Roosevelt* (Garden City, N.Y.: Doubleday, 1959), p. 196.

69. Steven J. Peitzman, "From Dropsy to Bright's Disease to End-Stage Renal Disorder" *Milbank Quarterly*, Vol. 67, Suppl. 1, *Framing Disease: The Creation Negotiation of Explanatory Schemes* (1989), pp. 16–32.

70. Morris, *Rise of Theodore Roosevelt*, p. 241.

71. Putnam, *Theodore Roosevelt*, pp. 386–388.

72. *New York Sun* and *New York Times* (February 17, 1884).

73. T.R. Private Diaries (February 14 and 16, 1884).

74. William Sewall, *Bill Sewall's Story of Theodore Roosevelt* (New York: Harper, 1919), p. 11.

75. Quoted in Putnam, *Theodore Roosevelt*, pp. 390–393.

76. T.R. Private Diaries (June 9, 1884).

77. Putnam, *Theodore Roosevelt*, p. 452.

78. T.R., *Ranch Life and the Hunting-Trial*, p. 81.

79. H. W. Brands, *T.R.: The Last Romantic* (New York: Basic Books, 1997), p. 182.

80. T.R. to Bill Sewall (July 6, 1884), in Sewall, *Bill Sewall's Story of Theodore Roosevelt*, p. 14.

81. *New York Tribune* (July 28, 1884).

82. T.R. to Anna Roosevelt Cowles, Chimney Butte Ranch (August 12, 1884).

83. Bill Sewall quoted in David McCullough, *Brave Companions: Portraits in History* (New York: Simon and Schuster, 1992), p. 62.

84. Elers Koch, "Big Game in Montana from Early Historical Records," *Journal of Wildlife Management*, Vol. 5, No. 4 (October 1941), pp. 357–369.

85. Don G. Despain, "Vegetation of the Big Horn Mountains, Wyoming, in Relation to Substrate and Climate," *Ecological Monographs*, Vol. 43, No. 3 (Summer 1973), pp. 329–355.

86. T. R. Bighorns Diary (August 21, 1884).

87. Merrifield quoted in Putnam, *Theodore Roosevelt*, p. 67.

88. T. R., *Hunting Trips of a Ranchman*, pp. 119–120.

89. T. R., *The Wilderness Hunter* (New York: Putnam, 1893), p. 146.

90. T. R., *Hunting Trips of a Ranchman*, p. 294.

91. U.S. Fish and Wildlife Archive, "Grizzly Bears" Files, Shepherdstown, W. Va. Special thanks to the Grizzly Bear Outreach Project for biological and behavioral information on grizzlies.

92. T. R., *Hunting Trips of a Ranchman*, p. 313.

93. Bessie Doak Haynes and Edgar Haynes, *The Grizzly Bear: Portraits of Life*

(Norman: University of Oklahoma Press, 1966).

94. T. R., *Hunting Trips of a Ranchman*, pp. 158–160.

95. "The Occidental Hotel History," Archive, Buffalo, Wyo. (June 2009).

96. T. R., Bighorns Diary (September 18–19, 1884). Fort McKinney had been established in 1878, after the Indian wars had ceased.

97. Putnam, *Theodore Roosevelt*, p. 487.

98. Clark, *The Badlands*, p. 39.

99. T. R. Bighorns Diary (October 1, 1884).

100. Elizabeth Royte, "Night Moves," *New York Times Book Review* (June 22, 2008), p. 9.

101. T. R., *Hunting Trips of a Ranchman*, p. 125.

7: Cradle of Conservation: The Elkhorn Ranch of North Dakota

1. Roderick Frazier Nash, *Wilderness and the American Mind*, 4th ed. (New Haven, Conn.: Yale University Press, 2003), pp. 152–153.

2. *New York Sun* (October 11, 1884). T.R. said that if it became "necessary" to make comments on Cleveland's lapsed morals, he would "not hesitate to express them."

3. "Cleveland's Electoral Vote," *New York Times* (November 7, 1884), p. 1.

4. T.R. to Henry Cabot Lodge (November 7, 1884).

5. Jack London, *The Call of the Wild* (New York: Regent, 1903), p. 67.

6. T.R., *Hunting Trips of a Ranchman*, p. 84.

7. T.R., *The Wilderness Hunter*, p. 386; T.R. Private Diaries (November 1884); and Carleton Putnam, *Theodore Roosevelt: The Formative Years* (New York: Scribner, 1958), p. 508.

8. T.R., *Hunting Trips of a Ranchman*, pp. 75–76.

9. H. W. Brands, *T.R.: The Last Romantic* (New York: Basic Books, 1997), p. 181. Also see T.R., *An Autobiography* (New York: Macmillan, 1913), p. 98. Edmund Morris, *The Rise of Theodore Roosevelt* (New York: Coward, McCann, 1979), p. 287; Hermann Hagedorn, *Roosevelt in the Bad Lands* (Boston, Mass.: Houghton Mifflin, 1921).

10. John Burroughs, *Locusts and Wild Honey* (Boston, Mass.: Houghton, Osgood, 1879), p. 80.

11. T.R., *Ranch Life and the Hunting-Trail* (New York: Century, 1888), p. 73.

12. T.R., *Hunting Trips of a Ranchman*, pp. 223–224.

13. Becky Lomax, "Tracking the Bighorns," *Smithsonian* (March 2008), pp. 21–22.

14. T.R., *Hunting Trips of a Ranchman*, pp. 226–227.

15. Ibid., p. 113.

16. T.R. to Anna Roosevelt (December 14, 1884).

17. Paul Grondahl, *I Rose Like a Rocket: The Political Education of Theodore Roosevelt* (Lincoln: University of Nebraska Press, 2007), pp. 158–159.

18. T.R. to Bamie Roosevelt (April 29, 1885). Also see Morris, *The Rise of Theodore Roosevelt*, p. 300.

19. T.R., *Hunting Trips of a Ranchman*, p. iii.

20. Unlike most of T.R.'s books *Hunting Trips of a Ranchman* wasn't serialized; it was published first in book form. Over the years, however, it was frequently reprinted in various fashions. It appeared in *Big-Game Hunting* (1898) and as the first part of *Hunting Tales of the West*, a four-volume set (1907). Individual chapters have been reprinted in more than a dozen periodicals.

21. T.R., *Hunting Trips of a Ranchman*, p. 19.

22. Ibid., p. 140.

23. Ibid., p. 147.

24. Ibid., p. 140.

25. T.R., *The Wilderness Hunter* (New York: Putnam, 1893, 1909), pp. 381–382.

26. "The Game of the West," *New York Times* (July 13, 1885), p. 3. (Review.) Also see *London Spectator* (January 16, 1886) and *New York Tribune* (September 7, 1885).

27. T.R., *The Wilderness Hunter*, p. xiii.

28. Morris, *The Rise of Theodore Roosevelt*, p. 299.

29. Quoted in John F. Reiger, *American Sportsmen and the Origins of Conservation*, 3rd ed. (Corvallis: Oregon State University Press, 2001), p. 115.

30. Ibid., p. 116.

31. George Bird Grinnell, "Introduction," in *The Works of Theodore Roosevelt*, Memorial ed., Vol. 1 (New York: Scribner, 1923), p. xv.

32. Michael Punke, *Last Stand: George Bird Grinnell and the Battle to Save the Buffalo, and the Birth of the New West* (New York: HarperCollins, 2007), pp. 164–167.

33. Grinnell, "Introduction," p. xvi.

34. "Dr. G. B. Grinnell, Naturalist, Dead," *New York Times* (April 12, 1938), p. 23.

35. John F. Reiger, *The Passing of the Great West: Selected Papers of George Bird Grinnell*, updated ed. (Norman: University of Oklahoma Press, 1985), pp. 6–7.

36. George Bird Grinnell, "Recollections of Audubon Park," *Auk*, Vol. 37 (July 1920). See also Witmer Stone (ed.), *The Auk: A Quarterly Journal of Ornithology*, Vol. 37 (Lancaster, Pa.: American Ornithologists Union, 1920), p. 373.

37. Richard Rhodes, *John James Audubon: The Making of An American* (New York: Knopf, 2006), p. 416.

38. George Bird Grinnell, "Memoirs" (unpublished). Written between November 26, 1915, and December 4, 1915, these memoirs recount his life to 1883. They are housed at Birdcraft Museum of the Connecticut Audubon Society, Fairfield. (Thanks to John F. Reiger for bringing this to my attention.)

39. Maria R. Audubon (ed.), *Audubon and His Journals* (New York: Scribner, 1897), p. 107.

40. Ibid., p. 131.

41. Grinnell, "Memoirs," p. 37.

42. George B. Ward and Richard E. McCabe, "Trail Blazers in Conservation: The Boone and Crockett Club's First Century," in *Records of North American Big Game*, 4th ed. (New York: Scribner, 1980), p. 9.

43. See George Bird Grinnell, *American Duck Shooting* (New York: Forest and Stream, 1901); George Bird Grinnell, *American Game-Bird Shooting* (New York: Forest and Stream, 1910).

44. "True Indians Stories," *New York Times* (March 27, 1893), p. 3.

45. Reiger, *The Passing of the Great West*, p. 2.

46. Margaret Mead and Ruth Bunzel, *The Golden Age of American Anthropology* (New York: George Braziller, 1960), pp. 113–114.

47. George Bird Grinnell, *By Cheyenne Campfires* (Hartford, Conn.: Yale University Press, 1926).

48. Mari Sandoz (ed.), *The Cheyenne Indians: Their History and Ways of Life* (New York: Buffalo-Head Press, 1962), p. v.

49. Grinnell, "Introduction," p. xvii.

50. Reiger, *The Passing of the Great West*, p. 3.

51. Mike Thompson, *The Travels and Tribulations of Theodore Roosevelt's Cabin* (San Angelo, Tex.: Laughing Horse Enterprises, 2004), pp. 22–25.

52. T.R., *Ranch Life and the Hunting-Trail*, p. 25.

53. Donald Dresden, *The Marquis de Mores: Emperor of the Badlands* (Norman: University of Oklahoma Press, 1970), pp. 111–112.

54. Sylvia Jukes Morris, *Edith Kermit Roosevelt: Portrait of a First Lady* (New York: Coward, McCann, and Geoghegan, 1980), p. 79.

55. Morris, *The Rise of Theodore Roosevelt*, pp. 319–320.

56. "Literary Notes," *New York Times* (September 20, 1886), p. 10.

57. T.R. to Corinne Roosevelt (March 20, 1885).

58. Morris, *The Rise of Theodore Roosevelt*, p. 329.

59. Ibid., p. 330.

60. T.R. to Bamie (June 19, 1886).

61. T.R. to Anna Roosevelt (May 15, 1886).

62. "Review of *The Life of Thomas Hart Benton*," *Zion's Herald* (February 16, 1887) Vol. 64, Issue 7. Also see James Freeman Clarke, "Benton and his Times" *The Independent*, December 15, 1887, Vol. 39, Issue 2037.

63. T.R., *Thomas Hart Benton* (Boston, Mass., and New York: Houghton, Mifflin, 1886), p. 225.

64. Ibid., p. 268.

65. Lowell E. Baier, "The Cradle of Conservation: Theodore Roosevelt's Elkhorn Ranch, an Icon of America's National Identity," *Theodore Roosevelt Association Journal*, Vol. 28, No. 1 (2007), pp. 15–22.

66. Hagedorn, *Roosevelt in the Bad Lands*, pp. 407–411.

67. All six articles for *Outing* were collected under the title *Ranch Life and Game-Shooting in the West* in *The Works of Theodore Roosevelt* (Memorial ed., Vol. 1).

68. A. R. Crook, "Misrepresentation of Nature in Popular Magazines," *Science*, New Series, Vol. 23, No. 593 (May 11, 1906), p. 748.

69. Ben Merchant Vorpahl, *Frederic Remington and the West: With the Eye of the Mind* (Austin: University of Texas Press, 1978), p. 51.

70. T.R., "Water-Fowl and Prairie Fowl," *Outing* (August 1886).

71. T.R., "The Ranch," *Outing* (March 1886).

72. T.R., "A Tame White Goat," *Harper's Round Table* (July 27, 1897). This essay was later reprinted in an omnibus of T.R.'s *Harper's Round Table* articles: *Good Hunting in Pursuit of Big Game in the West* (New York: Harper, 1907). The book was published while T.R. was president.

73. Hagedorn, *Roosevelt in the Bad Lands*, pp. 419–420.

74. Grinnell, "Introduction," p. xxi.

75. Putnam, *Theodore Roosevelt*, pp. 590–591.

76. Lincoln Lang, *Ranching with Roosevelt* (Philadelphia and London: Lippincott, 1926), pp. 241–243.

77. Brands, *T.R.*, p. 208.

78. Morris, *The Rise of Theodore Roosevelt*, pp. 363–366.

79. T.R., *Ranch Life and the Hunting-Trail*, p. 79.

80. T.R. to Anna Roosevelt, Medora, Dakota (April 16, 1887), in Elting Morison (ed.), *The Letters of Theodore Roosevelt*, Vol. 1 (Cambridge, Mass.: Harvard University Press, 1951). For his net loss see Hagedorn, *Roosevelt in the Bad Lands*, p. 482.

81. T.R., *Hunting Trips of a Ranchman*, pp. 211, 47.

82. Quoted in Putnam, *Theodore Roosevelt*, p. 596.

83. Frederick Wood, *Roosevelt as We Knew Him* (Philadelphia, Pa.: John C. Winston, 1927), p. 12.

84. T.R., *The Winning of the West*, Vol. 3 (New York: Putnam, 1894), pp. 45–46.

85. Brands, *T.R.*, p. 215.

86. T.R., *The Winning of the West*, Vol. 1 (New York: Putnam, 1889), p. xxii. (Presidential Edition.)

87. Clay S. Jenkinson, *Theodore Roosevelt in the Badlands: An Historical Guide* (Dickinson, N.D.: Dickinson State University, 2006), pp. 104–105.

8: Wildlife Protection Business

1. Paul Russell Cutright, *Theodore Roosevelt: The Naturalist* (New York: Harper & Brothers, 1956), p. 69.

2. Nelson Bryant, "Unveiling a Whitetail Buck, in the Spirit of Boone and Crockett," *New York Times* (February 25, 1996).

3. George Bird Grinnell, *American Big Game in Its Haunts* (New York: Forest and Stream, 1904), p. 495.

4. George B. Ward and Richard E. McCabe, *Records of North American Big Game* (New York: Scribners, 1952), pp. 62–63.

5. "Snap Shots," *Forest and Stream* (February 16, 1888), Vol, 30, Issue 4.

6. Founding documents, in Charles Sheldon, "A History of the Boone and Crockett Club: Milestone in Wildlife Conservation" (unpublished), Boone and Crockett Club Archive, Missoula, Mont.

7. Thomas L. Altherr and John F. Reiger, "Academic Historians and Hunting: A Call for More and Better Scholarship," *Environmental History Review*, Vol. 19, No. 3 (Autumn 1995), pp. 39–56.

8. Lowell Baier, "Note to Reader," in "Boone and Crockett Club: Past and Present Roles 1887–1992" (unpublished), Archive, Missoula, Mont.

9. George Bird Grinnell (ed.), *Hunting at High Altitudes* (New York: Harper, 1913), pp. 435–439. Also see Michael Punke, *Last Stand* (New York: HarperCollins, 2007), p. 166.

10. H. Duane Hampton, *How the U.S. Cavalry Saved Our National Parks* (Bloomington: Indiana University Press, 1971).

11. Paul Russell Cutright, *Theodore Roosevelt: The Making of a Conservationist* (Urbana: University of Illinois Press, 1985), pp. 73–74.

12. James B. Trefethen, *An American*

Crusade for Wildlife (New York: Winchester Press, 1975), pp. 81–82.

13. Ward and McCabe, "Trail Blazers in Conservation: The Boone and Crockett Club's First Century," in *Records of North American Big Game*, p. 49.

14. Ibid.

15. T.R. to the Editor, *Forest and Stream* (December 3, 1892), reproduced in "A Standing Menace: Cooke City vs. the National Park." (Pamphlet. There is a copy in Yellowstone National Park Library.) Also Rocky Barker, *Scorched Earth: How the Fires of Yellowstone Changed America* (Washington, D.C.: Island, 2005), pp. 77–78.

16. George Bird Grinnell, "Editor's Note," *Forest and Stream* (January 17, 1889).

17. "Snap Shots," *Forest and Stream* (February 16, 1888), p. 8.

18. Estelle Jussim, *Frederic Remington, the Camera, and the Old West* (Fort Worth, Tex.: Amon Carter Museum, 1987), pp. 19–21.

19. Joseph G. Rosa and Robin May, *Buffalo Bill and His Wild West: A Pictorial Biography* (Lawrence: University Press of Kansas, 1989), pp. 102–137.

20. Roscoe L. Buckland, *Frederic Remington: The Writer* (New York: Twayne, 2000), p. 5.

21. Peggy and Harold Samuels, *Frederic Remington: A Biography* (Garden City, N.Y.: Doubleday, 1982), pp. 72–75.

22. T.R., *The Wilderness Hunter* (New York and London: Putnam, 1893), p. 131.

23. Olin D. Wheeler, *6,000 Miles through Wonderland: Being a Description of the Marvelous Region Traversed by the Northern Pacific Railroad* (Saint Paul, Minn.: Chas S. Fee, General Passenger and Ticket Agent, Northern Pacific Railroad, 1893), pp. 34–40.

24. T.R., *The Wilderness Hunter*, pp. 135–136.

25. T.R., *The Winning of the West*, Vol. 2 (New York: Putnam, 1894), p. 71. T.R. was intrigued at being called "Boston Man" because, he claimed, Indians around the upper Ohio used to call frontiersmen "Virginians."

26. T.R., *The Wilderness Hunter*, p. 136.

27. Ibid., p. 145.

28. Ibid., pp. 120–142.

29. John Allen Gable (ed.), "President Theodore Roosevelt's Record on Conservation," *Theodore Roosevelt Association Journal*, Vol. 10 (Fall 1984), pp. 2–11.

30. T.R. to J. P. Morgan (September 18, 1899).

31. H. W. Brands, *T.R.: The Last Romantic* (New York: HarperCollins, 1997), p. 448.

32. T.R. to Henry Cabot Lodge (October 19, 1888).

33. T.R. to Cecil Arthur Spring Rice (November 18, 1888).

34. Joseph Bucklin Bishop, *Theodore Roosevelt and His Time: Shown in His Own Letters*, Vol. 1 (New York: Scribner, 1920), pp. 43–52.

35. T.R., *Ranch Life and the Hunting-Trail* (New York: Century, 1888), p. 6.

36. Remington quoted in John Gabriel Hunt, "Foreword," *Ranch Life and the Hunting-Trail* (New York: Gramercy, 1995), p. vi.

37. T.R., *Ranch Life and the Hunting-Trail*, pp. 147, 131.

38. Ibid., p. 134.

39. Ibid., p. 186.

40. "History and Organization of the Biological Survey Unit," United States Geological Survey Archives, Washington, D.C. Also see Jenks Cameron, *The Bureau of Biological Survey: Its History, Activities, and Organizations* (Baltimore, Maryland: The Johns Hopkins University Press, 1929), pp. 21–27.

41. W. W. Cooke, "Bird Migration in the Mississippi Valley," Bulletin 2, Biological Survey (1889). Walter B. Barrows, "The English Sparrow in America," Bulletin 1, U.S. Department of Agriculture, Division of Economic Ornithology and Mammalogy (1889).

42. Wilfred H. Osgood, "Clinton Hart Merriam," *Journal of Mammalogy*, Vol. 24, No. 4 (November 17, 1943), pp. 421–436. Also C. Hart Merriam, "Two New Shrews," *Proceedings of the Biology Society of Washington*, Vol. 15 (March 22, 1902), pp. 75–76.

43. "Officers of the Biological Society," *Washington Post*, (January 11, 1991), p. 6.

44. *Yearbook of the Department of Agriculture 1909* (Washington, D.C.: Government Printing Office, 1910), pp. 115–119.

9: Laying the Groundwork with John Burroughs and Benjamin Harrison

1. Clifford Johnson (ed.) *John Burroughs Talks: His Reminiscences and Comments* (Boston: Houghton Mifflin, 1922).

2. Ed Renehan, Jr., *John Burroughs: An American Naturalist* (Hensonville, N.Y.: Chelsea Green, 1992), p. 178

3. Elizabeth Custer, *"Boots and Saddles" or Life in Dakota with General Custer* (New York: Harper, 1885); *Tenting on the Plains or General Custer in Kansas and Texas* (New York: Charles L. Webster, 1887).

4. William Hard quoted in William Davison Johnston, *TR: Champion of the Strenuous Life* (New York: Farrar, Straus, and Cudahy, 1958).

5. John Burroughs to Julian Burroughs (October 12, 1920), Vassar Library Collection, Poughkeepsie, N.Y.

6. Charles Dickens, *Hard Times for These Times* (New York: Hurd and Houghton, 1870), p. 24.

7. John Burroughs to Louis Untermeyer (June 4, 1919), John Burroughs Collection, Vassar College Library Collection, Vassar University, Poughkeepsie, N.Y.

8. Renehan, *John Burroughs*, pp. 7–10.

9. John Burroughs, *My Boyhood*, 2nd ed. (Garden City, N.Y.: Doubleday, Page, 1922), pp. 5–6.

10. John Burroughs, *John Burroughs's America: Selections from the Writings of the Naturalist* (New York: Devin-Adi, 1951), pp. 3–20.

11. Clara Barrus, *Life and Letters of John Burroughs* (Boston, Mass.: Houghton Mifflin, 1925), pp. 1–42.

12. Paul Brooks, *Speaking for Nature* (San Francisco, Calif.: Sierra Club Books, 1980), p. 6.

13. John Burroughs, *John JamesAudubon* (New York: Small, Maynard and Company, 1902).

14. Renehan, *John Burroughs*, pp. 77–78.

15. John Burroughs, *Notes on Walt Whitman as Poet and Person* (New York: American News Company, 1867).

16. Daniel Mark Epstein, *Lincoln and Whitman: Parallel Lives in Civil War Washington* (New York: Ballantine Books, 2004), pp. 279–298.

17. Walt Whitman, *Leaves of Grass* (Garden City, N.Y.: Doubleday, Page, 1902), pp. 94–95. (Originally published in 1855.)

18. John Burroughs, *Notes on Walt Whitman as Poet and Person* (New York: American News Company, 1867).

19. John Burroughs, *Wake-Robin* (Boston: Houghton Mifflin Company, 1871), pp. 188–196.

20. Burroughs quoted in Brooks, *Speaking for Nature*, p. 8.

21. Renehan, *John Burroughs*, pp. 178–179.

22. Burroughs quoted in Clifton Johnson, "Introduction," in John Burroughs, *In the Catskills* (Boston, Mass: Houghton Mifflin, 1910), p. xii.

23. John Burroughs journal entry (March 7, 1889), Berg Collection, New York Public Library.

24. H. W. Brands, *T.R.: The Last Romantic* (New York: Basic Books, 1997), p. 228.

25. Clara Barrus (ed.), *The Heart of Burroughs's Journal* (Boston, Mass.: Houghton Mifflin, 1928), pp. 32–33.

26. Burroughs, quoted in Foreword, in *American Bears: Selections from the Writings of Theodore Roosevelt* (Boulder, Col.: Robert Rinehart, 1997), p. ix.

27. T.R., *Ranch Life and the Hunting-Trail* (New York: Century, 1888), p. 59.

28. Brands, *T.R.*, pp. 258–259.

29. Ibid., pp. 24–48.

30. T.R. to Anna Roosevelt (June 17, 1891).

31. John Reiger, *American Sportsmen and the Origins of Conservation*, 3rd ed. (Corvallis: Oregon State University Press, 2001), p. 168.

32. "Gen. John W. Noble Is Dead; Secretary of the Interior in Harrison's Cabinet Dies at 80," *New York Times* (March 23, 1912), p. 13. (Special to the *Times*.)

33. George Bird Grinnell, "Brief History of the Boone and Crockett Club" (unpublished), Boone and Crockett Club Archive, Missoula, Mont. It had been partially published in *Forest and Stream*.

34. Elliot Coues, *Birds of the Northwest: A Hand-Book of the Ornithology of the Region Drained by the Missouri River and its Tribu-*

taries (Washington, D.C.: Government Printing Office, 1874); *Birds of the Colorado Valley: A Repository of Scientific and Popular Information Concerning North American Ornithology* (Washington, D.C.: Government Printing Office, 1874); *Fur-Bearing Animals: A Monograph of North American Mustelidae* (Washington, D.C.: Government Printing Office, 1877).

35. William T. Hagan, *Theodore Roosevelt and Six Friends of the Indian* (Norman: University of Oklahoma Press, 1997), p. 36.

36. *Letters of Theodore Roosevelt, Civil Service Commissioner 1890–1895* (U.S. Civil Service Commission, 1958), p. 44.

37. T.R. to Alice Roosevelt (July 2, 1891).

38. George B. Utley, "Theodore Roosevelt's Winning of the West: Some Unpublished Letters," *Mississippi Valley Historical Review*, Vol. 30, No. 4 (March 1944), pp. 495–506.

39. John Milton Cooper, Jr., "Introduction," in T.R., *The Winning of the West: From the Alleghenies to the Mississippi 1769–1776*, Vol. 1 (Lincoln: University of Nebraska Press, 1995), p. x.

40. Kathleen Dalton, *A Strenuous Life* (New York: Knopf, 2002), pp. 131–132.

41. Ibid.

42. T.R., *The Winning of the West*, Vol. 1, p. 133.

43. Clara Barrus (ed.), *The Heart of Burroughs's Journals* (Boston, Mass.: Houghton Mifflin, 1928), p. 32

44. Frederick Jackson Turner, "The Winning of the West," *Dial* (August 1889), p. 73.

45. William Frederick Poole, "Roosevelt's The Winning of the West," *Atlantic Monthly*, Vol. 44 (November 1889), pp. 693–700. Also see Utley, "Theodore Roosevelt's Winning of the West: Some Unpublished Letters," pp. 495–496.

46. "Pushing Their Way," *New York Times* (July 7, 1889), p. 11.

47. George B. Utley, "Theodore Roosevelt's Winning of the West: Some Unpublished Letters," pp. 495–506.

48. Cooper, "Introduction" in T.R., *The Winning of the West*, p. xii.

49. T.R., *Biological Analogies in History* (London: Oxford University Press, 1910), p. 6.

50. James R. Gilmore in *New York Sun* (September 29 and October 10, 1889). (Included in T.R. Scrapbooks at Harvard.) T.R. was accused of plagiarism but he was considered innocent by most fair-minded scholars.

51. Parkman quoted in W. R. Jacobs (ed.), *Letters of Francis Parkman*, Vol. 2 (Norman: University of Oklahoma Press, 1960), pp. 209–232.

52. Francis Parkman, "The Forests of the White Mountains," *Garden and Forest*, Vol. 1 (February 29, 1888).

53. Wilbur R. Jacobs, "Francis Parkman: Naturalist-Environmental Savant," *Pacific Historical Review*, Vol. 61, No. 3 (May, 1992), p. 341.

54. T.R. to Francis Parkman (July 13, 1889).

55. Dalton, *A Strenuous Life*, pp. 132–133.

56. Ibid., p. 134.

57. T.R. to Gertrude Elizabeth Tyler Carow (October 18, 1890).

58. Arnold Hague, "The Yellowstone Park," in Theodore Roosevelt and George Bird Grinnell (eds.), *American Big Game Hunting: The Book of the Boone and Crockett Club* (New York: Forest and Stream, 1893), p. 259.

59. Michael L. Collins, *That Damned Cowboy: Theodore Roosevelt and the American West 1888–1898* (New York: Peter Lang, 1989), p. 117.

60. William Frederick Poole to T.R. (November 1889) in George B. Utley, "Theodore Roosevelt and the Winning of the West: Some Unpublished Letters," pp. 495–497.

61. William T. Hornaday, *Our Vanishing Wild Life* (New York: New York Zoological Society, 1913), p. x.

62. United States Statutes at Large, xvii.32, quoted in *Publications of the Colonial Society of Massachusetts: Transactions 1902–1904* (Boston: Published by the Society, 1906), p. 377.

63. Thomas Wolfe, *Of Time and the River: A Legend of Man's Hunger in His Youth* (Garden City, N.Y.: Sun Dial, 1944), p. 155.

64. George Bird Grinnell to Archibald Rogers (December 24, 1890), quoted in Reiger, *American Sportsmen and the Origins of Conservation*, p. 157.

65. T.R., "Hunting in Cattle Country" in T.R. and Grinnell (eds.), *Hunting in Many Lands*, pp. 292–293.

66. Corinne Roosevelt Robinson, *My Brother Theodore Roosevelt* (New York: Scribner, 1921), p. 144.

67. T.R., *The Winning of the West*, pp. xli–xlii.

68. Diary quoted in Corinne Roosevelt Robinson, *My Brother Theodore Roosevelt*, p. 149.

69. Ibid., pp. 146–147.

70. Sylvia Jukes Morris, *Edith Kermit Roosevelt: Portrait of a First Lady* (New York: Coward, McCann, and Geoghegan, 1980), p. 130.

71. Collins, *That Damned Cowboy*, p. 122.

72. T.R. to Gertrude Elizabeth Tyler Carow (October 18, 1890)

73. Blaine Harden, "In the New West, Do They Want Buffalo to Roam?" *Washington Post* (July 30, 2006), pp. A 8–9.

74. "The Treasures of Yosemite," *Century*, Vol. 40, No. 4 (August, 1890); "Features of the Proposed Yosemite National Park," *Century*, Vol. 40, No. 5 (September 1890).

75. Jeremy Johnston, "Preserving the Beasts of Waste and Desolation: Theodore Roosevelt and the Predator Control in Yellowstone," *Yellowstone Science* (Spring 2002), p. 15.

76. Christine Macy and Sarah Bonnemaison, *Architecture and Nature: Creating the American Landscape* (New York: Routledge, 2003), p. 51. Also see Donald J. Pisani, "Forest and Conservation in 1865–1890," in Char Miller (ed.), *American Forests: Nature, Culture, and Politics* (Lawrence: University Press of Kansas, 1997), pp. 16–17.

77. Reiger, *American Sportsmen and the Origins of Conservation*, pp. 157–158.

78. "The Rapid Destruction of Our Forests," *Scientific Monthly* (December 1887), pp. 225–226.

79. "The Week in the Club World," *New York Times* (January 2, 1898), p. 15.

80. Reiger, *American Sportsmen and the Origins of Conservation*, pp. 168–170. Also *Compilation of Public Timber Laws and Regulations and Decisions Thereunder* (Washington, D.C.: Government Printing Office, January 21, 1897), p. 131. For further data on the history and development of forest reserves in the northwestern United States see E. H. MacDaniels, "Twenty-Five National Forests of North Pacific Region," *Oregon Historical Quarterly*, Vol. 42 (September 1941), pp. 247–255.

81. George Bird Grinnell, "Secretary Noble's Monument," *Forest and Stream* (March 9, 1893).

82. Gifford Pinchot, *Breaking New Ground* (New York: Harcourt, Brace, 1947), p. 85. Edward A. Bowers of the General Land Office also deserves credit for his fierce lobbying efforts on behalf of Section 24.

83. Roger A. Sedjo, "Does the Forest Service Have a Future?" *Regulation*, Vol. 23, No. 1 (2000), pp. 51–55.

84. Harold K. Steen, "The Beginning of the National Forest System" in Miller (ed.), *American Forests*, pp. 49–50.

85. Udall quoted in Reiger, *American Sportsmen and the Origins of Conservation*, p. 153.

86. John W. Noble to T.R. (April 16, 1891), Yellowstone National Park Archives (Doc. No. 254), Wyoming. Also see Sarah E. Broadbent, "Sportsmen and the Evolution of the Conservation Idea in Yellowstone: 1882–1894," MA thesis, Montana State University, 1997.

87. Francis G. Newlands, "Irrigation Congress," *Irrigation Age*, Vol. 1 (October 1891), pp. 195–196.

88. T.R. and George Bird Grinnell, *Hunting in Many Lands: The Book of the Boone and Crockett Club* (New York: Forest and Stream, 1895), p. 44.

89. T.R., "The Northwest in the Nation: Biennial Address before the State Historical Society of Wisconsin" (January 24, 1893), T.R. Collection, Harvard University. (Reprint.)

90. Collins, *That Damned Cowboy*, pp. 131–137.

91. Frederick Jackson Turner, *The Frontier in American History* (New York: Holt, Rinehart, and Winston, 1920), p. 92, 178.

Also Patricia Limerick, "The Forest Reserves and the Argument for a Closing Frontier," in Harold K. Steen (ed.), *The Origins of the National Forests: A Centennial Symposium* (Durham, N.C.: The Forest History Society, 1992), pp. 10–18.

92. Walter La Faber, *The New Empire: An Interpretation of American Expansion 1860–1898* reissue (Ithaca, N.Y.: Cornell University Press, 1998), p. 64.

93. T.R. to Frederic Remington (December 28, 1897).

94. T.R. and George Bird Grinnell, "Our Forest Reservations," in T.R. and Grinnell (eds.), *American Big-Game Hunting* (New York: Forest and Stream, 1893), pp. 326–330.

95. Ibid., pp. 326–330.

96. Limerick, "The Forest Reserves and the Argument for a Closing Frontier," p. 13. Limerick notes that this argument about "white people" being "scared" originated with Professor Richard White.

97. T.R., *The Winning of the West*, Vol. 1, p. 139.

98. Limerick, "The Forest Reserves and the Argument for a Closing Frontier," pp. 13–18.

99. T.R. to George Bird Grinnell (August 30, 1897).

100. George Cotkin, *Reluctant Modernism* (Lanham, Md.: Rowman and Littlefield, 2004), p. 4.

101. T.R., "Biological Analogies in History," *Outlook*, June 11, 1910, Vol. 95, Is. 6.

102. H. Paul Jeffers, *An Honest President: The Life and Presidencies of Grover Cleveland* (New York: Morrow, 2000), p. 6.

103. T.R., letter to the editor of *Forest and Stream*, in "A Standing Menace: Cooke City vs. the National Park" (pamphlet) quoted in Robert Underwood Johnson, *Remembered Yesterdays* (Kessinger, 1923), p. 309.

104. H. W. Brands, *The Money Men: Capitalism, Democracy, and the Hundred Years' War over the American Dollar* (New York: Norton, 2006), pp. 160–161.

105. Collins, *That Damned Cowboy*, p. 127.

106. "Two Ocean Pass," National Park Service, National Natural Landmark (October 1965).

107. T.R., *The Wilderness Hunter*, pp. 182–184.

10: The Wilderness Hunter in the Electric Age

1. T.R. to Madison Grant (March 3, 1894).

2. T.R., *The Wilderness Hunter* (New York and London: Putnam, 1893), p. 351.

3. Laura Tangley, "Birding in the Texas Tropics," *National Wildlife* (February/March 2007), pp. 38–45.

4. T.R., *The Wilderness Hunter*, p. 354.

5. Ibid., pp. 354–359.

6. T.R. to Cecil Arthur Spring-Rice (May 3, 1892). By 2008 there were still 2 million feral pigs in Texas (half the U.S. total). According to the *Dallas Morning News* they were mangling the state's pastures, crops, and waterways.

7. T.R. to Anna Roosevelt (August 26, 1892).

8. "The Last of Sitting Bull: The Old Chief Killed While Resisting Arrest," *New York Times* (December 16, 1890), p. 1.

9. Alvin M. Josephy, Jr. (ed. in charge), *The American Heritage Book of Indians* (New York: Simon and Schuster, 1961), p. 348.

10. Dee Alexander Brown, *Bury My Heart at Wounded Knee*, Thirtieth Anniversary ed. (New York: Macmillan, 2001), pp. 440–445.

11. Sherman quoted in Brandon (ed.), *The American Heritage Book of Indians*, p. 366.

12. T.R. to Charles Collins (January 21, 1891), Indian Rights Association Papers (Microfilm), Reel 6.

13. William T. Hagan, *The Indian Rights Association: The Herbert Welsh Years* (Tucson: University of Arizona Press, 1985).

14. Laurence M. Hauptman, "Theodore Roosevelt and the Indians of New York State," *Proceedings of the American Philosophical Society*, Vol. 119, No. 1 (February 21, 1975), pp. 1–7.

15. William T. Hagan, *Theodore Roosevelt and Six Friends of the Indian* (Norman: University of Oklahoma Press, 1997), p. 32.

16. George Bird Grinnell, "In Buffalo Days," in T.R. and Grinnell (eds.), *American Big-Game Hunting* (New York: Forest and Stream, 1893), p. 159.

17. Hugh Chisholm (ed.), *The Encyclopaedia Britannica: A Dictionary of Arts, Sci-*

ence, Literature, and General Information 11th ed., Vol. 6 (New York: Encyclopaedia Britannica Company, 1910), p. 502.

18. *The Historical World's Columbian Exposition and Chicago Guide* (St. Louis: James H. Mason, 1892), p. 270.

19. Norman Bolotin and Christine Lang, *The World's Columbian Exposition: The Chicago's World Fair of 1893* (Champaign: University of Illinois Press, 2002), p. 106.

20. Marjorie Warvelle Bear, *A Mile Square of Chicago* (Oakbrook, Ill.: TIPRAC, 2007), p. 205. See also Lincoln Ellsworth, *Beyond Horizons* (New York: Doubleday, Doran, 1935), pp. 3–4.

21. Letter published in T.R., *The Wilderness Hunter*, p. 425.

22. Peter Hassrick, *Wildlife and Western Heroes* (Fort Worth, Tex.: Amon Carter Museum, 2003), pp. 136–137. Proctor had met Pinchot in New York in the 1880s and they became friends.

23. Jesse Donahue and Erik Trump, *Political Animals: Public Art in American Zoos and Aquariums* (Lanham, Md.: Lexington, 2007), pp. 20–23.

24. Trumball White and W. M. Igleheart, *The World's Columbian Exposition* (J. W. Ziegler, 1893), p. 514.

25. Kenneth Frampton, *Modern Architecture: A Critical History* (Oxford: Oxford University Press, 1980), p. 62.

26. Paul Andrew Hutton, "Col. Cody, the Rough-Riders, and the Spanish American War," *Points West* (1998 Fall Issue), pp. 8–11.

27. Bobby Bridger, *Buffalo Bill and Sitting Bull: Inventing the Wild West* (Austin: University of Texas Press, 2002), p. 442.

28. John Patrick Barrett, *Electricity at the Columbian Exposition* (Chicago: R. R. Donnelly and Sons, 1894).

29. T.R. to James Brander Matthews (June 8, 1893).

30. Theodore Whaley Cart, "The Lacey Act: America's First Nationwide Wildlife Statute," *Forest History*, Vol. 17, No. 3 (Oct. 1973), p. 413.

31. J. Anthony Lukas, *Big Trouble: A Murder in a Small Western Town Sets Off a Struggle for the Soul of America* (New York: Simon and Schuster, 1997), p. 241.

32. John F. Reiger, *American Sportsmen and the Origins of Conservation*, 3rd ed. (Corvallis: Oregon State University Press, 2001), p. 106.

33. Alfred Runte, *Trains of Discovery: Western Railroads and the National Parks* (Lanham, Md.: Roberts Rinehart, 1998), p. 49.

34. "History of the Boone and Crockett Club Books as Recalled by G. B. Grinnell" (1925), Boone and Crockett Club Archives, Missoula, Mont.

35. T.R. to George Bird Grinnell (August 24, 1897), Boone and Crockett Club Archives, Missoula, Mont.

36. Matthew Baigell, *Albert Bierstadt* (New York: Watson-Guptill, 1981), pp. 8–14.

37. T.R. to Albert Bierstadt (February 7, 1893), Joseph M. Roebling Collection of the American Heritage Center of the University of Wyoming, Laramie.

38. Albert Bierstadt, "A Moose Hunt" (February–April 1893), Roebling Collection of the American Heritage Center, University of Wyoming. Attached to the original essay is T.R.'s Sagamore Hill calling card.

39. Eric Nye and Sheri Hoem (eds.), "Big Game on the Editor's Desk: Roosevelt and Bierstadt's Tale of the Hunt," *New England Quarterly*, Vol. 60, No. 3 (September 1987).

40. T.R. to Albert Bierstadt (June 8, 1893), Roebling Collection of the American Heritage Center of the University of Wyoming. But by not blowing the whistle on Bierstadt, by allowing his fib to stand, T.R. had protected a friend. More than a decade later, when T.R. was in the White House, Grinnell wrote up the moose story in his *American Big Game in Its Haunts*, in a way the president would have approved, claiming that the sixty-four-and-a-half-inch antlers were "in the possession" of the late painter.

41. Grinnell, "In Buffalo Days," p. 169.

42. G. Edward White, *The Eastern Establishment and the Western Experience: The West of Frederic Remington, Theodore Roosevelt and Owen Wister* (New Haven, Conn.: Yale University Press, 1968).

43. "Gen. Anderson Dead at University

Club," *New York Times* (March 8, 1915), p. 9.

44. T.R., "Coursing the Prongbuck," in *American Big-Game Hunting*, p. 129.

45. T.R., "Literature of American Big-Game Hunting" in *American Big-Game Hunting*, p. 325. (Unsigned.)

46. Reiger, *American Sportsmen and the Origins of Conservation*, pp. 150–151.

47. Dick Baldwin, "Trapshooting with D. Lee Braum and the Remington Pros," (*Remington* Vandalia, Ohio: Trapshooting Hall of Fame and Museum, 1967).

48. "Trap Shooting in Saratoga," *New York Times* (May 10, 1893), p. 3. The *New York Times* used to promote trapshooting in the 1880s as a way to downplay the mass killing of birds. While the sports page would mention all-day shoots with live birds in places like the League Island Gun Club of Philadelphia, it gave more ink to trapshooting events.

49. George Bird Grinnell, "Editorial," *Forest and Stream* (July 14, 1881). Also see William B. Mershon, *The Passenger Pigeon* (New York: Outing, 1907), pp. 223–225.

50. Reiger, *American Sportsmen and the Origins of Conservation*, pp. 150–151.

51. Corinne Roosevelt Robinson, *My Brother Theodore Roosevelt* (New York: Scribner, 1921), p. 127.

52. T.R., "Preface," in *The Wilderness Hunter*, p. xiii.

53. T.R., "Preface," in *The Wilderness Hunter*, p. xiv. (The preface was written in June 1893 at Sagamore Hill.)

54. T.R., *The Wilderness Hunter*, p. 174.

55. "Mr. Roosevelt's Americanism," *New York Times* (August 6, 1893), p. 19.

56. "New Publications: The Wilderness Hunter," *Forest and Stream*, Vol. 41, No. 4 (July 29, 1893).

57. "Dangers of Moose Hunting," *Youth's Companion* (November 23, 1893).

58. T.R. to Hoke Smith (April 7, 1894).

59. Denis Tilden Lynch, *Grover Cleveland: A Man Four-Square* (New York: Van Rees, 1932), p. 191.

60. "Hoke Smith's Appointment," *New York Times* (February 16, 1893), p. 5.

61. G. Michael McCarthy, "The Forest Reserve: Colorado under Cleveland and McKinley," *Journal of Forest History* (April 1976), p. 80.

62. U.S. Department of the Interior, *Annual Report, 1893* (Washington, D.C.: Government Printing Office, 1894), p. 555.

63. Captain George S. Anderson to Secretary of Interior Hoke Smith (March 17, 1894), Vol. V (Letters Sent) National Archives, pp. 1–9, Yellowstone National Park. Quoted in H. Duane Hampton, "U.S. Army and the National Parks," *Forest History* (October 1966), p. 14.

64. "Save the Buffalo," *Forest and Stream*, Vol. 42, No. 15 (1894). (Editorial.)

65. "The Lacey Act of 1894," *U.S., Statutes at Large*, Vol. 28, p. 73.

66. Mary Annette Gallager, "John F. Lacey: A Study in Organizational Politics," PhD dissertation, University of Arizona, 1970.

67. Samuel Johnson Crawford, *Kansas in the Sixties* (Chicago, Ill.: A. C. McClurg, 1911), p. 146.

68. "John F. Lacey: Champion of Birds and Wildlife," Iowa National History Foundation, Des Moines.

69. Michael L. Tate, *The Frontier Army in the Settlement of the West* (Norman: University of Oklahoma Press, 1999), p. 233.

70. George S. Anderson, "Protection of the Yellowstone National Park" in T.R. and George Bird Grinnell (eds.), *Hunting in Many Lands* (New York: Forest and Stream, 1895), p. 388.

71. Alice Wondrak Biel, *Do (Not) Feed the Bears: The Fitful History of Wildlife and Tourists in Yellowstone* (Lawrence: University Press of Kansas, 2006), p. 7.

72. T.R. and George Bird Grinnell, "Preface," in *Hunting in Many Lands*, pp. 11–12. Although the preface had a shared byline it almost certainly was written by T.R.

73. Runte, *Trains of Discovery*, pp. 1–12.

74. Muir quoted in Alfred Runte, "Foreword," in John Muir, *Our National Parks* (San Francisco, Calif.: Sierra Club Books, 1991), p. x. (Muir originally published the book in 1901.)

75. John Burroughs, *The Last Harvest* (Boston, Mass.: Houghton Mifflin, 1922), p. 220.

76. W. Hallett Phillips to George A. Anderson (March 31, 1894), 6, LR, No. 1217-A, Yellowstone National Park Archives, Wyo.

77. T.R. to George S. Anderson (January 21, 1895), Letter Box 6, Doc. No. 1282, Yellowstone National Park Archives.

78. Aubrey L. Haines, *The Yellowstone Story: A History of Our First National Park*, Vol. 2 (Yellowstone: Yellowstone Library and Museums Association, 1977), pp. 68–69.

79. T.R., "Wilderness Reserves," *Forest and Stream*, September 3, 1904, Vol. LXIII, Issue No. 10, p. 1. This is one of the chapters in the Boone and Crockett Club book *American Big-Game in its Haunts*.

80. Quoted in H. W. Brands' *T.R.: The Last Romantic* (New York: Basic Books, 1997), p. 259.

81. T.R., "Hunting in the Cattle Country," in T.R. and Grinnell (eds.), *Hunting in Many Lands*, p. 297.

82. T.R. to Henry Cabot Lodge (September 30, 1894).

83. T.R., *The Winning of the West*, Vol. 3 (New York: Putnam, 1905), pp. 44–45.

84. William T. Hagan, *Theodore Roosevelt and Six Friends of the Indian*, pp. 22–23.

11: The Bronx Zoo Founder

1. "Madison Grant, 71, Zoologist, Is Dead," *New York Times* (May 31, 1937), p. 15. In 1993 the New York Zoological Society changed its name to the Wildlife Conservation Society. Its mission is to "advance the study of zoology, protect wildlife, and educate the public." As of 2009 it operated the following public attractions in the New York area: the Bronx Zoo, New York Aquarium, Central Park Zoo, Queens Zoo, and Prospect Park Zoo.

2. Madison Grant, *The Vanishing Moose and Their Extermination in the Adirondacks* (New York: Century, 1894).

3. "Killing of the Buffalo," *New York Times* (July 26, 1896), p. 20.

4. George Bird Grinnell and T.R. (ed.), *Trail and Camp-Fire* (New York: Forest and Stream, 1897), p. 313.

5. George Bird Grinnell, "In Buffalo Days," in T.R. and Grinnell (eds.), *American Big-Game Hunting* (New York: Forest and Stream, 1893), p. 171.

6. Richard Manning, *Grassland: The History, Biology, Politics, and Promise of the American Prairie* (New York: Viking, 1995); Tom McHugh, *The Time of the Buffalo* (Lincoln: University of Nebraska Press, 1972), p. 39; and Christopher Ketcham, "They Shoot Buffalo, Don't They," *Harper's Magazine* (June 2008), p. 74.

7. T.R. to Madison Grant (March 3, 1894).

8. "Zoo Plans Are Approved," *New York Times* (November 23, 1897), p. 12.

9. Boone and Crockett Club Report, "Past and Present Roles of the Boone and Crockett Club 1887–1992," Missoula, Mont. (Unpublished.)

10. "Zoo Plans Are Approved," *New York Times* (November 23, 1897), p. 12. Also see William T. Hornaday, *Popular Official Guide to the New York Zoological Society*, 11th ed. (New York: New York Zoological Society, June 1, 1911), p. 136.

11. Lincoln Lang, *Ranching with Roosevelt* (Philadelphia, Pa.: Lippincott, 1926), pp. 360–365.

12. Lowell E. Baier, "The Boone and Crockett Club: A 106-Year Retrospective," (Boone and Crockett Club Archives, Missoula, Mont.), p. 9.

13. "Black Mesa Reserve," Boone and Crockett Club Archives, Missoula, Mont.

14. T.R., *A Book-Lover's Holidays in the Open* (New York: Scribner, 1916), p. 32.

15. "William Temple Hornaday" (1996), University of Iowa Museum of Natural History Archive, Iowa City.

16. William T. Hornaday, *Two Years in the Jungle: The Experiences of a Hunter Naturalist* (London: Kegan Paul, Trench and Co., 1885), p. 1.

17. William T. Hornaday, *The Extermination of the American Bison* (Washington, D.C.: National Museum Report, 1889).

18. "History of the Wildlife Conservation Society," Wilderness Conservation Fund Archives, New York.

19. William T. Hornaday, *Our Vanishing Wild Life* (New York: New York Zoological Society, 1913), p. 92.

20. Paul Verner Bradford and Harvey

Blume, *Ota Benga: The Pygmy in the Zoo* (New York: Dell, 1992), p. 173.

21. "History of the Wildlife Conservation Society."

22. "Zoo Plans Are Approved: Park Board, after Long Study of the Proposed Zoological Park, Commends It," *New York Times* (November 23, 1897), p. 12.

23. T.R. quoted in Paul Russell Cutright, *Theodore Roosevelt: The Naturalist* (New York: Harper, 1956), p. 73.

24. T.R. to James Brander Matthews (December 21, 1893).

25. T.R. to Madison Grant (March 3, 1894).

26. Casper W. Whitney, "The Cougar," in T.R. and George Bird Grinnell (eds.), *Hunting in Many Lands* (New York: Forest and Stream, 1895), p. 253.

27. T.R. and Grinnell, "Preface," in *Hunting in Many Lands*, p. 12.

28. Charles E. Whitehead, "Game Laws," in *Hunting in Many Lands*, pp. 370–372.

29. T.R. and Grinnell (eds.), *Hunting in Many Lands*, pp. 424–432.

30. Edmund Morris, *The Rise of Theodore Roosevelt* (New York: Coward, McCann, 1979), p. 25.

31. T.R. quoted in Van Wyck Brooks, *John Sloan: A Painter's Life* (New York: Dutton, 1955), p. 55.

32. H. Paul Jeffers, *Roosevelt the Explorer* (Lanham, Md.: Taylor Trade, 2003), p. 8.

33. William Harbaugh, *Power and Responsibility: The Life and Times of Theodore Roosevelt* (New York: Farrar, Straus, and Cudahy, 1961), pp. 82–83.

34. Jeffers, *Roosevelt the Explorer*, p. 84.

35. Owen Wister, *Roosevelt: The Story of a Friendship, 1880–1919* (New York: Macmillan, 1930), p. 51.

36. Jeffers, *Roosevelt the Explorer*, pp. 82–84.

37. "The Summer Plans of Authors," *New York Times* (June 21, 1896), p. 27.

38. T.R. quoted in H. Paul Jeffers, *Colonel Roosevelt: Theodore Roosevelt Goes to War* (New York: Wiley, 1996), p. 20.

39. "Electoral Vote Counted; McKinley and Hobart Formally Declared to Have Been Chosen as President and Vice President," *New York Times* (February 11, 1897), p. 4.

40. Roderick Frazier Nash, *Wilderness and the American Mind*, 4th ed. (New Haven, Conn.: Yale University Press, 2001), p. 136.

41. George R. Leighton, *Five Cities: The Story of Their Youth and Old Age* (New York: Harper, 1939), p. 269.

42. George Bird Grinnell and T.R., "Preface," in *Trail and Camp-Fire* (New York: Forest and Stream, 1897), pp. 7–8.

43. Gerald W. Williams, *The Forest Service: Fighting for Public Lands* (Westport, Conn.: Greenwood, 2006), pp. 404–405. Williams documented the total acreage saved by the Washington Birthday Reserves: San Jacinto, California, 737,280 acres; Stanislaus, California, 691,200 acres; Washington, Washington, 3,594,240 acres; Mount Rainier, Washington, 1,267,200 acres; Olympic, Washington, 2,188,800 acres; Priest River, Idaho and Washington, 645,120 acres; Bitterroot, Idaho and Montana, 4,147,200 acres; Lewis and Clark, Montana, 2,926,080 acres; Flathead, Montana, 1,382,400 acres; Big Horn, Wyoming, 1,198,080 acres; Teton, Wyoming, 829,440 acres; Uinta, Utah, 705,120 acres; Black Hills, South Dakota, 967,680 acres.

44. T.R. to George Bird Grinnell (August 24, 1897).

45. Char Miller, *Gifford Pinchot and the Making of Modern Environmentalism* (Washington, D.C.: Island, 2001), p. 120.

46. Leighton, *Five Cities*, p. 269.

47. "After Us, the Deluge," *New York Times* (May 8, 1897), p. 6.

48. John Muir, "The American Forests," *Atlantic Monthly*, Vol. 80 (August 1897).

49. Grover Cleveland, *Fishing and Shooting Sketches* (Philadelphia, Pa.: Curtis, 1901), pp. 3–6.

50. John Muir, *Our National Parks* (Boston, Mass. and New York: Houghton Mifflin, 1901), p. 1.

51. *Sixth Annual Report of the Bureau of Agriculture and Industry of the State of Montana for the Year Ending November 30, 1898* (Helena, Mont.: Independent, 1898), p. 51.

52. T.R. to George Bird Grinnell (August 24, 1897), Boone and Crockett Club Archives, Missoula, Mont.

53. "The Cabinet Confirmed," *New York Times* (March 6, 1897), p. 3.

54. Gifford Pinchot, *Breaking New Ground* (New York: Harcourt, Brace, 1947), p. 123.

55. "The Big Federal Domain," *New York Times* (November 19, 1897), p. 3.

56. Gretel Ehrlich, *John Muir: Nature's Visionary* (Washington, D.C.: National Geographic, 2000).

57. *Sixth Annual Report of the Bureau of Agriculture and Industry of the State of Montana for the Year Ending November 30, 1898*, p. 52.

58. Rexroth quoted in David Taylor, *A Soul of a People* (Hoboken, New Jersey: John Wiley & Sons, 2009), p. 133.

59. *Twentieth Annual Report of the United States Geological Survey to the Secretary of the Interior, 1898–1899; Part V—Forest Reserves* (Washington, D.C.: Government Printing Office, 1900), p. 143.

60. Morris, *The Rise of Theodore Roosevelt*, p. 554.

61. T.R. to Anna Roosevelt Cowles (November 13, 1896).

62. Nathan Miller, *Theodore Roosevelt* (New York: William Morrow, 1992), p. 246.

63. Morris, *The Rise of Theodore Roosevelt*, pp. 548–563.

64. Keir B. Sterling, *Last of the Naturalists: The Career of C. Hart Merriam*, rev. ed. (New York: Arno, 1977), p. ix.

65. T.R. to Henry Cabot Lodge in Cutright, *Theodore Roosevelt: The Naturalist*, p. 80.

66. Sterling, *Last of the Naturalists.*

67. "North American Bears," *New York Times* (April 22, 1896), p. 14.

68. T.R. to Henry Fairfield Osborn (May 18, 1897).

69. Ibid.

70. T.R., "Social Evolution," in *The Works of Theodore Roosevelt*, Memorial ed. (New York, 1923–1926), Vol. 14, pp. 109–128.

71. T.R., "A Layman's Views on Specific Nomenclature," *Science* (April 30, 1897), pp. 685–688.

72. Sterling, *Last of the Naturalists*, p. 242.

73. Ibid. Also see Cutright, *Theodore Roosevelt: The Making of a Conservationist* (Urbana: University of Illinois, 1985), pp. 192–196.

74. Sterling, *Last of the Naturalists*, p. 176.

75. T.R. to Charles Addison Boutelle (June 22, 1897).

76. T.R. to George Bird Grinnell (August 2, 1897), Boone and Crockett Club Archive, Missoula, Mont.

77. T.R. to George Bird Grinnell (August 24, 1897).

78. Ibid.

79. T.R. to Henry Fairfield Osborn (September 14, 1897).

80. Merriam quoted in *Science* (May 14, 1897).

81. Cutright, *Theodore Roosevelt: The Naturalist*, pp. 80–88. Also see C. Hart Merriam, "Natural History: Roosevelt's Wapiti," *Forest and Stream*, January 1, 1898, Vol. L, Issue No. 9, p. 5.

82. Dr. C. Hart Merriam, "Cervus roosevelti," *Proceedings of the Biological Society of Washington* (December 17, 1897).

83. Cutright, *Theodore Roosevelt: The Naturalist*, p. 85.

84. T.R., "Wapiti," in Hedley Peek and Frederick George Aflalo, *The Encyclopaedia of Sport Vol. II* (London: Lawrence and Bullen, 1898), p. 530.

85. Cutright, *Theodore Roosevelt: The Naturalist*, p. 85.

86. T.R. to C. Hart Merriam (February 22, 1899).

87. T.R., "List of Books," in *Trail and Camp-Fire*, p. 339.

88. Burnham quoted in Frank Graham, Jr., *The Adirondacks: A Political History* (New York: Knopf, 1978), p. 148.

89. Grinnell and T.R., *Trail and Camp-Fire* (New York: *Forest and Stream*, 1897) p. 153.

90. T.R., "On the Little Missouri," in *Trail and Camp-Fire*, pp. 219–220.

91. George Bird Grinnell, "Introduction," *Works*, Memorial Edition, Vol. 1, p. xix.

92. T.R. to John A. Merritt (December 23, 1897).

12: The Rough Rider

1. Richard H. Collin, *Theodore Roosevelt, Culture, Diplomacy, and Expansion: A New View of American Imperialism* (Baton Rouge:

Louisiana State University Press, 1985), p. 123.

2. Ibid, p. 120.

3. T.R. to Henry Cabot Lodge (August 10, 1886). See also Henry Pringle, *Theodore Roosevelt: A Biography* (New York: Harcourt, Brace, 1931), pp. 166–167.

4. T.R., *American Naval Policy as Outlined in the Messages of the Presidents of the United States, from 1790 to Present Day* (Washington, D.C.: Government Printing Office, 1897).

5. Edmund Morris, *The Rise of Theodore Roosevelt* (New York: Coward, McCann, 1979), p. 598.

6. "The Maine at Havana," *New York Times* (January 25, 1898), p. 6.

7. T.R. to William Sheffield Cowles (March 29, 1898).

8. Henry Pringle, *Theodore Roosevelt* (New York: Harcourt, 2003), p. 124.

9. T.R. to William Sturgis Bigelow (March 29, 1898).

10. T.R. to Robert Bacon (April 8, 1898).

11. Daniel Henderson, *"Great-Heart": The Life Story of Theodore Roosevelt*, 3rd ed. (New York: Knopf, 1919), p. 62.

12. Akiko Murakata, "Theodore Roosevelt and William Sturgis Bigelow: The Story of a Friendship," *Harvard Literary Bulletin*, Vol. 23, No. 1 (January 1975), p. 93.

13. Morris, *The Rise of Theodore Roosevelt*, p. 612.

14. Mrs. Winthrop Chanler, *Roman Spring* (Boston, Mass.: Little, Brown, 1934), p. 285.

15. Robert Lee, *Fort Meade and the Black Hills* (Lincoln: University of Nebraska Press, 1991), pp. 160–161.

16. Morris, *The Rise of Theodore Roosevelt*, pp. 613–620. Also see Leonard Wood, "Roosevelt: Soldier, Statesman, and Friend" in *The Rough Riders and Men of Action* (New York: Scribner's, 1926), pp. xv–xvi.

17. Marilyn Bennett, *It Happened in San Antonio* (Guilford, Conn.: Twodot, 2006), pp. 53–56.

18. Buckhorn Saloon Museum Archive, San Antonio, Tex.

19. T.R. to Henry Cabot Lodge (May 25, 1898).

20. Sarah Lyons Watts, *Rough Rider in the White House: Theodore Roosevelt and the Politics of Desire* (Chicago, Ill.: University of Chicago Press, 2003), p. 163.

21. G. Edward White, *The Eastern Establishment and the Western Experience: The West of Frederic Remington, Theodore Roosevelt, and Owen Wister* (New Haven, Conn.: Yale University Press, 1968), pp. 149–153.

22. Harbaugh, *Power and Responsibility*, p. 104.

23. Michael L. Collins, *That Damned Cowboy: Theodore Roosevelt and the American West, 1883–1898* (New York: Peter Lang, 1989), p. 146.

24. "The Rough Riders Land at Montauk," *New York Times*, (August 16, 1898), p. 1.

25. H. W. Brands, *T.R.: The Last Romantic* (New York: Basic Books, 1997), p. 344.

26. Owen Wister, "Balaam and Pedro," *Harper's Monthly* (January 1894).

27. Peggy Samuels and Harold Samuels, *Teddy Roosevelt at San Juan* (College Station: Texas A&M Press, 1997), p. 58.

28. Lydia Kingsmill Commander, *The American Idea* (New York: A. S. Barnes, 1907), p. 75.

29. Henry Castor, *Theodore Roosevelt and the Rough Riders* (New York: Random House, 1954), p. 45.

30. Samuels and Samuels, *Teddy Roosevelt at San Juan*, pp. 58–59.

31. Jack [John] Willis, *Roosevelt in the Rough* (New York: Ives Washburn, 1931), pp. 36–37. Reprint.

32. David H. Burton, "Theodore Roosevelt's Social Darwinism and Views on Imperialism," *Journal of the History of Ideas*, Vol. 26, No. 1 (January–March 1965), pp. 103–118.

33. T.R., "Social Evolution," *North American Review* (July 1895). Republished in *American Ideals, and Other Essays* (New York: Putnam, 1897), pp. 293–317.

34. Ibid., p. 296. Also see Patrick Sharp, *Savage Perils: Racial Frontiers and Nuclear Apocalypse* (Norman: University of Oklahoma Press, 2007); and John Morton Blum, "Theodore Roosevelt: The Years of Decision," *The Letters of Theodore Roosevelt* (Cambridge, Mass.: Harvard University Press, 1954), Vol. 2, p. 1486.

35. John Burroughs, "The Biological Origin of the Ruling Class," cited in Renehan Jr., *John Burroughs* (Post Mills, Vt.: Chelsea Green, 1992), p. 199.

36. Edward J. Renehan Jr., *John Burroughs* (Post Mills, Vt.: Chelsea Green, 1992), pp. 198–200.

37. Morris, *The Rise of Theodore Roosevelt*, p. 630. Also see "The War: Expected Naval Battle, Firing at Cabanas," *The Observer*, May 1, 1898, p. A5.

38. T.R. to Henry Cabot Lodge (June 12, 1898).

39. "Colt Machineguns in the Spanish American War," (2008), Fort Sam Houston Museum, San Antonio, Tex.

40. William McKinley Executive Order (March 28, 1898) from the Executive Mansion; William McKinley Proclamation (May 27, 1898); William McKinley Proclamation (June 29, 1898); McKinley's third State of the Union address, in James D. Richardson, *A Compilation of the Messages and Papers of the Presidents 1789–1907*, Vol. 10 (Washington, D.C.: Bureau of National Literature and Art, 1908), pp. 343, 253, 121.

41. T.R., *Letters*, Vol. II, p. 843. Also see Brands, *T.R.: The Last Romantic* (New York: Basic Books, 1997), p. 346.

42. T.R. to Corinne Roosevelt Robinson (June 15, 1898).

43. Owen Wister, *The Virginian: A Horseman of the Plains* (New York: Macmillan, 1902), p. 334.

44. Kathleen Dalton, *Theodore Roosevelt: A Strenuous Life* (New York: Knopf, 2002), p. 173.

45. T.R., *The Rough Riders* (New York: Scribner, 1899), 1905 reprint, p. 73.

46. Charles Darwin, *The Descent of Man*, 2nd revision (London: John Murray, Albemarle Street, 1890), p. 54.

47. Joseph Bucklin Bishop, *Theodore Roosevelt's Letters to His Children*, p. 16.

48. Nathaniel Lande, *Dispatches From the Front: A History of the American War Correspondent* (New York: Henry Holt, 1995), p. 151.

49. T.R., *The Rough Riders*, pp. 15–16.

50. Ibid., "Appendix A: Muster-Out Roll."

51. Edward Marshall, *The Story of the Rough Riders* (New York: G. W. Dillingham, 1899), p. 127.

52. John Hay letter to T.R. (July 27, 1898), quoted in "Credit 'Splendid Little War' to John Hay," *New York Times* (July 9, 1991), p. A18.

53. Davis, *Badge of Courage*, pp. 259–261.

54. T.R. to William Rufus Shafter (August 3, 1898).

55. Jeff Heatley (ed.), *Bully! Colonel Roosevelt, the Rough Riders, and Camp Wikoff* (Montauk, N.Y.: Montauk Historical Society, 1998), pp. 55–94.

56. Marshall, *The Story of the Rough Riders*, p. 23.

57. Cara Blessley Lowe, "Introducing Cougar," in Marc Bekoff and Cara Blessley Lowe (eds.), *Listening to Cougar* (Boulder: University of Colorado Press, 2007), p. 5.

58. "The Rough Riders Land at Montauk," *New York Times* (August 16, 1898), p. 1.

59. Marshall, *The Story of the Rough Riders*, p. 24.

60. T.R., *The Rough Riders*, p. 222.

61. T.R. to his children (June 6, 1898).

62. N. Scott Momaday, *House Made of Dawn* (New York: HarperCollins, 1968), pp. 14–20.

63. T.R., *The Rough Riders*, p. 222.

64. Robert C. V. Meyers, *Theodore Roosevelt: Patriot and Statesman* (Philadelphia, Pa.: Ziegler, 1902), p. 284.

65. Joseph Bucklin Bishop, *Theodore Roosevelt's Letters to His Children* (New York: Scribner, 1919), pp. 15–16.

66. Author interview with James Stringer (July 15, 2008), Santa Fe, N.M. (Stringer is the great-grandson of Cuba's later owner, Samuel Black.)

67. Heatley (ed.), *Bully!* p. 485.

68. T.R., *The Rough Riders*, pp. 221–223.

69. Albert Smith, *Two Reels and a Crank* (Garden City, N.Y.: Doubleday, 1952), p. 57.

70. T.R. to Corinne Roosevelt Robinson (June 27, 1898).

71. Morris, *The Rise of Theodore Roosevelt*, p. 643.

72. Stephen Crane, "Roosevelt's Rough Riders, Loss Due to a Gallant Blunder," *New York World* (June 26, 1898).

73. Virgil Carrington Jones, *Roosevelt's*

Rough Riders, (Garden City, N.Y.: Doubleday, 1971), p. 6.

74. T.R. to Corinne Roosevelt Robinson (June 15, 1898).

75. T.R., *The Rough Riders*, p. 92.

76. Ibid., pp. 104–105.

77. Ibid.

78. T.R., "Kidd's Social Evolution," *The North American Review* (July 1895). Also included in *The Works of Theodore Roosevelt* (Nation Edition), Vol. XIII (New York: Charles Scribner's Sons, 1926), pp. 223–241.

79. Samuels and Samuels, *Teddy Roosevelt at San Juan*, p. 296. Also see John M. Blum, *The National Experience* (New York: Harcourt, Brace, World, 1963), p. 495.

80. T.R., *The Winning of the West*, Presidential Edition (New York: Putnam, 1889), p. vii.

81. Patrick Sharp, "The Darwinist Frontier," in *Savage Perils*.

82. T.R., *The Rough Riders*, p. 15.

83. Robert Gearty, "Park Is Teddy Terrain; Renaming in Montauk for Roosevelt," *New York Daily News* (January 4, 1998).

84. Michael Pollak, "Screen Grab; Remembering Rough Rider Who Was a President," *New York Times* (February 1, 2001). I was one of fourteen historians who had written President Bill Clinton a letter on March 31, 1999 urging the president to award T.R. the medal he so richly deserved. Others included Stephen E. Ambrose, John A. Gable, Nathan Miller, Edmund and Sylvia Morris, William N. Tischin, and Geoffrey C. Ward. Also see "Medal of Honor Awarded to Theodore Roosevelt," *Theodore Roosevelt Administration Journal*, Vol. XXIV, No. 2 (2001), pp. 3–9.

85. "An Exciting Night in Camp," *New York Times* (September 15, 1898), p. 2.

86. James H. McClintock, *Arizona: Prehistoric, Aboriginal, Pioneer, Modern*, Vol. 2 (Chicago, Ill.: S. J. Clarke, 1916), p. 522.

87. "Rough Riders' Mascot Dead," *Chicago Times Herald* (June 13, 1899). Rough Riders Museum Archive, Las Vegas, N.M. Special thanks to Pat Romero for bringing this to my attention.

88. Author interview with James Stringer (July 15, 2008), Santa Fe, N.M. Mr. Stringer kindly read to me the *Arizona Daily Sun*'s obituary of Cuba the dog (n.d.).

89. T.R. to Francis Ellington Leupp (September 3, 1898).

90. Morris, *The Rise of Theodore Roosevelt*, p. 670.

91. Bishop, *Theodore Roosevelt's Letters to His Children*, p. 17.

92. White, *The Eastern Establishment and the Western Experience*, p. 58.

93. Robert Hendrickson, *Happy Trails: A Dictionary of Western Expressions* (New York: Facts on File, 1994), p. 34.

94. T.R., *The Rough Riders* (Appendix D, Revised Edition), p. 320. Also see "Mens Gift to Roosevelt," *New York Times* (September 14, 1898), p. 3.

95. White, *The Eastern Establishment and the Western Experience*, pp. 168–169.

96. Virgil Carrington Jones, *Roosevelt's Rough Riders*, p. 277.

97. Leonard Wood, "Roosevelt: Soldier, Statesman, and Friend," *The Works of Theodore Roosevelt*, Memorial Edition, Vol. 13 (New York: Scribner, 1924), p. xiii.

98. T.R. to John Ellis Roosevelt (March 31, 1898).

99. John A. Correy, *A Rough Ride to Albany: Teddy Runs for Governor* (New York: Fordham University Press, 2006).

13: Higher Political Perches

1. Paul Russell Cutright, *Theodore Roosevelt: The Making of a Conservationist* (Urbana: University of Illinois Press, 1985), p. 199.

2. "History of Executive Mansion," New York State Historical Society, New York City.

3. T.R., *New York* (New York: Longmans, Green, 1891). Also see "Historic New York, New York, by Theodore Roosevelt," *New York Times* (March 29, 1891), p. 19.

4. Donald M. Roper, "The Governorship in History," *Proceedings of the Academy of Political Science* Vol. 31, No. 3 (May 1973), pp. 16–30.

5. "Gov. Roosevelt Shut Out," *New York Times* (January 3, 1899), p. 2.

6. *Public Papers of Theodore Roosevelt, Governor, 1899* (Albany, N.Y.: Brandow Printing Company, 1899), p. 25.

7. G. Wallace Chessman, *Governor Theodore Roosevelt: The Albany Apprenticeship, 1898–1900* (Cambridge, Mass.: Harvard University Press, 1965), p. 5.

8. "Fish, Forests, and Politics," *Forest and Stream*, Vol. 53 (December 9, 1899). Also see "Gov. Roosevelt Is Inaugurated," *New York Times* (January 3, 1899), p. 1.

9. T.R., "The New York Fish Commission" *Field and Stream* (December 9, 1899).

10. Charles Earle Funk, *What's the Name, Please?* (New York: Funk and Wagnalls, 1936), p. 129.

11. "James Wallace Pinchot," Grey Towers National Historic Site, Archive, Milford, Pa. (Biography profile.) Special thanks to Richard Paterson.

12. Char Miller, *Gifford Pinchot and the Making of Modern Environmentalism* (Washington, D.C.: Island, 2001), p. 70.

13. Ibid.

14. "Gifford Pinchot Dies Here at 81," *New York Times* (October 6, 1946), p. 56.

15. George Perkins Marsh, *Man and Nature* (New York: Scribner, 1864), p. 44.

16. Owen Wister, *Roosevelt: The Story of a Friendship* (New York: Macmillan, 1930), p. 174.

17. T.R. to Gifford Pinchot (May 22, 1894), Gifford Pinchot Papers, Library of Congress, Washington, D.C.

18. M. Nelson McGeary, *Gifford Pinchot: Forester-Politician* (Princeton, N.J.: Princeton University Press, 1960), p. 53.

19. Gifford Pinchot, *Just Fishing Talk* (New York and Harrisburg, Pa.: Telegraph, 1936), pp. 72–74.

20. "A Clan Hangs," *Time* (March 23, 1931).

21. Frank W. Carpenter, "Heins & La Farge," *New York Architecture* (April 26, 1988).

22. Gifford Pinchot, *Breaking New Ground* (New York: Harcourt, Brace and Co., 1947), pp. 144–146.

23. Archie Butt, *The Letters of Archie Butt* (New York: Doubleday Page & Company, 1924), p. 147.

24. McGeary, *Gifford Pinchot*, p. 47.

25. Cutright, *Theodore Roosevelt: The Making of a Conservationist*, p. 203.

26. T.R., *An Autobiography* (New York: Macmillan, 1913), p. 409.

27. "Roosevelt's Annual Message," *New York Times* (January 4, 1900), p. 6.

28. Sandra Weber, *Mount Marcy: The High Peak of New York* (Fleischmanns, N.Y.: Purple Mountain, 2001), p. 9.

29. Ibid., pp. 9–12.

30. C. Grant La Farge, "A Winter Ascent of Tahawus," *Outing*, Vol. 36, No. 1 (April 1900).

31. Pinchot, *Breaking New Ground*, pp. 144–146.

32. Gifford Pinchot, Diary (1899). Library of Congress, Washington, D.C. Also see Sandra Weber, "Gifford Pinchot: Walrus of the Forest," *Highlights* (August 2005), pp. 34–35.

33. Miller, *Gifford Pinchot and the Making of Modern Environmentalism*, p. 148.

34. Ibid., p. 149.

35. T.R. to John Hay (February 7, 1899).

36. Richard O. Weber, "How T.R. Handled Being Governor," *Theodore Roosevelt Association Journal*, Vol. XXIV, No. 2 (2001), p. 17.

37. Clara Barrus (ed.), *The Heart of Burroughs's Journal* (Boston, Mass.: Houghton Mifflin, 1928), pp. 320–322.

38. Gail Bederman, *Manliness and Civilization* (Chicago, Ill.: University of Chicago Press, 1995), p. 93.

39. "The Strenuous Life," speech given by T.R. at the Hamilton Club (April 10, 1899). See "Gov. Roosevelt in Chicago," *New York Times* (April 11, 1899), p. 3. Also T.R., "Speech before the Hamilton Club, Chicago, Illinois, April 10, 1899," in *The Works of Theodore Roosevelt*, Memorial Edition (New York: Scribner's Sons, 1925), Vol. 15, p. 281.

40. "Gov. Roosevelt in Chicago," *New York Times*, April 11, 1899, p. 3.

41. Bederman, *Manliness and Civilization*, p. 174.

42. Richard O. Weber, "How T.R. Handled Being Governor," *Theodore Roosevelt Association Journal*, Vol. XXIV, No. 2 (2001), pp. 19–20.

43. T.R.'s letter of November 28 was published in T.R., "The New York Game Commission" and "The New York Game Protectors," *Forest and Stream* (December 9, 1899).

44. T.R. to Tiffany and Company (February 2, 1899).

45. T.R. to Frank M. Chapman (February 16, 1899).

46. Ibid.

47. Chessman, *Governor Theodore Roosevelt*, pp. 242–243.

48. T.R. Chronology as New York Governor, T.R. Collection, Library of Congress, Washington, D.C. Also see H. K. Bush-Brown, Letter to the Editor, "The Palisades Park Movement," *New York Times* (January 30, 1900), p. 6.

49. Fuller quoted in Thomas R. Slicer, "Famous Visitors at Niagara Falls," *The Niagara Book*, new rev. ed. (New York: Doubleday, Page, 1901), p. 290.

50. Ibid.

51. Russell D. Butcher, *America's National Wildlife Refuges* (Lanham, Md.: Roberts Rinehart Publishers, 2003), p. 428.

52. Jack Demattos, *Garrett and Roosevelt* (College Station, Tex.: Early West, 1988), pp. 1–62.

53. T.R. to William Allen White (July 1, 1899).

54. *Public Papers of Theodore Roosevelt, Governor, 1899* (Albany, N.Y.: Brandow, Department Printers, 1899), p. 323. Also see T.R., *Campaigns and Controversies* (New York: Scribner, 1926), Vol. 14, p. 319.

55. Edmund Morris, *The Rise of Theodore Roosevelt* (New York: Coward, McCann, 1979), p. 712.

56. David Magie, *Life of Garret Augustus Hobart, Twenty-Fourth Vice President of the United States* (New York: Putnam, 1910), p. 231.

57. Ibid.

58. T.R. to John Davis Long (December 2, 1899).

59. T.R. to Bradley Tyler Johnson (November 21, 1899).

60. Pinchot, *Breaking New Ground*, p. 158.

61. Philip C. Jessup, *Elihu Root*, Vol. 1 (New York: Dodd, Mead, 1938), p. 210.

62. "Governor Roosevelt's Annual Message," *New York Times* (January 4, 1900), p. 6.

63. Curt Meine, "Roosevelt, Conservation, and the Revival of Democracy," *Conservation Biology*, Vol. 15, No. 4 (August 2001), pp. 829–831.

64. T.R., *Annual Message of the Governor*, Albany, N.Y. (January 3, 1900), Memorial Edition, Vol. 17, p. 63; and National Edition, Vol. 15, pp. 54–55. Also see "Roosevelt's Annual Message," *New York Times* (January 4, 1900), p. 6. (The newspaper printed the address in its entirety.)

65. "Gov. Roosevelt's Annual Message," *New York Times* (January 4, 1900), p. 6.

66. Cutright, *Theodore Roosevelt: The Making of a Conservationist*, p. 2.

67. T.R. to Grant La Farge (February 9, 1900).

68. Audubon's picture can be found in Richard Rhodes, *John James Audubon: The Making of an American* (New York: Knopf, 2004).

69. Jeff Wells, "What We Buy Hurts Birds We Watch," *Philadelphia Inquirer* (October 29, 2007).

70. "Bird Protection Bill Signed," *New York Times* (May 3, 1900), p. 6.

71. Ibid. Also see "Forest, Field and Streams," *Time* (June 16, 1930).

72. "Plumage of Birds on Hats," *New York Times* (May 4, 1900), p. 8.

73. Barrus (ed.), *The Heart of Burroughs's Journals*, p. 321.

74. T.R. to Frank M. Chapman (May 8, 1900).

75. Fred J. Alsop III, *Birds of North America: Eastern Region* (New York: DK, 2001), pp. 6–21.

76. John Burroughs, *Signs and Seasons* (Boston, Mass.: Houghton Mifflin, 1886), p. 246.

77. Alexander Wilson, *American Ornithology; or, The Natural History of The Birds of the United States*, Vol. 3 (London: Whittaker, Treacher, and Arnot, London; Edinburgh: Stirling and Kenney, 1832), p. 200.

78. Florida Audubon Society Files, Miami.

79. Leslie Poole, "The Women of the Early Florida Audubon Society," Tampa Bay History Center (2007). (Pamphlet.)

80. "Origin of the U.S. Fish and Wildlife Service," U.S. Fish and Wildlife Archive, Shepherdstown, W. Va. (2008 update.)

81. *Summary of Federal Wildlife Laws*

Handbook with Related Laws (Government Institutions, 1998), New Mexico Center for Wildlife Law, University of New Mexico School of Law Archives.

82. Doug Stewart, "How Conservation Grew from a Whisper to a Roar," *National Wildlife* (December–January 1909).

83. *Report of the Secretary of Agriculture: Report of the Chiefs* (Washington, D.C.: Government Printing Office, 1919), p. 418.

84. Stewart, "How Conservation Grew from a Whisper to a Roar."

85. Oliver H. Orr Jr., *Saving American Birds: T. Gilbert Pearson and the Founding of the Audubon Movement* (Gainesville: University of Florida Press, 1992), pp. 74–75.

14: The Advocate of the Strenuous Life

1. T.R. to John Burroughs (May 1, 1900).

2. Bryce quoted in Charles G. Washburn, *Theodore Roosevelt: The Logic of His Career* (Boston, Mass.: Houghton Mifflin, 1916), p. 33.

3. Edward J. Renehan, Jr., *John Burroughs: An American Naturalist* (Post Mills, N.Y.: Chelsea Green, 1992), p. 182.

4. T.R. to John Burroughs (May 5, 1900).

5. John Burroughs, *Far and Near* (Boston, Mass.: Houghton Mifflin, 1904).

6. Ibid., pp. 215–222.

7. T.R. to John Burroughs (May 21, 1900).

8. Edward Evans, "Ethical Relations between Man and Beast," in Donald Worster (ed.), *American Environmentalism: The Formative Period, 1860–1915* (New York: Wiley, 1973).

9. Clara Barrus, (ed.), *The Heart of Burroughs's Journals* (Boston, Mass.: Houghton Mifflin, 1928), pp. 97–98.

10. Renehan, *John Burroughs*, p. 201.

11. Clara Barrus, *The Life and Letters of John Burroughs*, Vol. 1 (Boston, Mass.: Houghton Mifflin, 1925), p. 256.

12. John Burroughs, *Under the Apple-Trees* (Boston, Mass., and New York: Houghton Mifflin, 1916), p. 265.

13. Perry D. Westbrook, *John Burroughs* (New York: Twayne, 1974), p. 106.

14. John Morton Blum, *The Republican Roosevelt* (New York: Atheneum, 1962), p. 25.

15. James Perrin Warren, *John Burroughs and the Place of Nature* (Athens: University of Georgia Press, 2006), pp. 150–193.

16. T.R. to Henry L. Sprague (January 26, 1900).

17. Barrus, *The Heart of Burroughs's Journals*, pp. 297–298.

18. G. Wallace Chessman, *Governor Theodore Roosevelt: The Albany Apprenticeship: 1898–1900* (Cambridge, Mass.: Harvard University Press, 1965), p. 253.

19. T.R. to George McAneny (June 5, 1900).

20. Ibid.

21. Kathleen Dalton, *Theodore Roosevelt: A Strenuous Life* (New York: Knopf, 2002), p. 190.

22. T.R. to William Henry Lewis (July 26, 1900).

23. Edmund Morris, *The Rise of Theodore Roosevelt* (New York: Coward, McCann, and Geoghegan, 1979), p. 722.

24. David Henry Burton, *Theodore Roosevelt, American Politician: An Assessment* (Teaneck, N.J.: Fairleigh Dickinson University Press, 1997), p. 88.

25. T.R. to William Adolph Baillie-Grohman (June 12, 1900).

26. T.R. to Senator Hanna (June 27, 1900), quoted in Joseph Bucklin Bishop, *Theodore Roosevelt and His Time Shown in His Own Letters* Vol. 1 (New York: Scribner, 1919), p. 139.

27. T.R. to Theodore Roosevelt, Jr. (February 19, 1904).

28. Dalton, *Theodore Roosevelt: A Strenuous Life*, p. 101.

29. Don Russell, *The Lives and Legends of Buffalo Bill* (Norman: Oklahoma University Press, 1960), p. 419.

30. Nathan Miller, *Theodore Roosevelt: A Life* (New York: William Morrow, 1992), p. 344.

31. Stefan Lorant, *The Life and Times of Theodore Roosevelt* (New York: Doubleday, 1959), p. 335.

32. Robert B. Roosevelt to Charles Hallock (July 1900), R.B.R. Papers, TRA—Oyster Bay. (Thanks to the late John A.

Gable for providing me a copy of this fascinating note.)

33. Lorant, *The Life and Times of Theodore Roosevelt*, p. 335.

34. David S. Barry, *Forty Years in Washington* (Boston, Mass.: Little, Brown, 1924), p. 246.

35. T.R. to Senator Marcus A. Hanna, quoted in Bishop, *Theodore Roosevelt and His Time Shown in His Own Letters*, pp. 139–140.

36. T.R. to the National Irrigation Congress (November 16, 1900).

37. T.R. to Percy S. Lansdowne (December 7, 1900).

38. T.R. to Frederick Courteney Selous (November 23, 1900).

39. T.R. to Edward Sanford Martin (November 26, 1900).

40. T.R. to Elihu Root (December 5, 1900).

41. T.R. to Philip Bathell Stewart (December 6, 1900).

42. Frank Donaldson Biography, Medical and Chirurgical Faculty, University of Maryland, College Park.

43. Clara Barton, *The Red Cross: A History of This Remarkable International Movement in the Interest of Humanity* (Albany, N.Y.: J. B. Lyon, 1898), p. 617.

44. "Roosevelt at Home," *New York Times* (October 17, 1898), p. 2; and "History of Red Crags" (courtesy of Red Crags Bed and Breakfast).

45. T.R., *The Wilderness Hunter*, (New York: Putnam, 1893), p. 344.

46. T.R., *Outdoor Pastimes of an American Hunter*, p. 2.

47. Because the White River National Forest was more than 1 million acres, as president Roosevelt, capitulating to developers in Meeker, reduced the size by 61,000 acres in 1902 and 159,000 acres in 1904. U.S. Department of Agriculture History File on White River National Forest (October 29, 2007).

48. T.R., *Outdoor Pastimes of an American Hunter*, pp. 2–3.

49. C. S. Forbes, "President Roosevelt," *Vermonter*, Vol. 7, No. 4 (November 1901).

50. T.R., *Outdoor Pastimes of an American Hunter*, p. 3.

51. Ibid., pp. 3–30.

52. Ibid.

53. T.R., *The Wilderness Hunter* (New York: G. P. Putnam's Sons, 1893), p. 344.

54. Jeremy Johnston, "Preserving the Beasts of Waste and Desolation: Theodore Roosevelt and Predator Control in Yellowstone," *Yellowstone Science* (Spring, 2002), pp. 15–16.

55. T.R. to Frederick Courteney Selous (March 8, 1901).

56. Blum, *The Republican Roosevelt*, p. 29.

57. Henry F. Pringle, *Theodore Roosevelt: A Biography* (New York: Harcourt, Brace, 1931), p. 241.

58. H. W. Brands, *T.R.: The Last Romantic* (New York: Basic Books, 1997), p. 407.

59. T.R. to W. H. Taft (April 26, 1901).

60. T.R. to Charles Emory Smith (April 3, 1901).

61. T.R. to Winthrop Chanler (March 8, 1901).

62. T.R. to Caspar Whitney (March 16, 1901).

63. T.R. to Florence Bayard Lockwood La Farge (March 29, 1901).

64. T.R. to Hamlin Garland (April 4, 1901).

65. T.R. to C. G. Gunther's Sons (April 23, 1901).

66. C. Hart Merriam to T.R. (May 3, 1901).

67. T.R. to Gifford Pinchot (April 16, 1901).

68. T.R. to Eugene Hale (May 13, 1901).

69. T.R. to Caspar Whitney (June 7, 1901).

70. T.R. to William Wells (June 17, 1901).

71. James B. Trefethen, *Crusade for Wildlife: Highlights in Conservation Progress* (Harrisburg, Pa.: Stackpole: and New York: Boone and Crockett, 1961), pp. 67–69.

72. Alden Sampson, "The Creating of Game Refuges," in George Bird Grinnell (ed.), *American Big Game in Its Haunts* (New York: Forest and Stream, 1904).

73. T.R. to Erwin Brown (June 13, 1901).

74. T.R. to W. H. Taft (April 26, 1901).

75. Lorant, *The Life and Times of Theodore Roosevelt*, p. 357.

76. "Our History," Vermont Federation of Sportsmen's Clubs (November 7, 2005). (Pamphlet.) Between 1878 and 1920 the league helped create the Vermont Fish and Wildlife Protection.

77. C. S. Forbes, "President Roosevelt," *The Vermonter* (Essex Junction, Vermont), Vol. 8, No. 4 (November 1901).

78. Charlotte Mehrtens, "Chazy Reef at Isle LaMotte," *Geology of Vermont* (1998). (Pamphlet.)

79. Christina and Diane E. Foulds, *Vermont* (Woodstock, Vt.: Countryman, 2006), pp. 481–487.

80. Forbes, "President Roosevelt."

81. Ibid.

82. Ibid.

83. "Mr. Roosevelt en Route," *New York Times* (September 7, 1901), p. 1.

84. Edith Roosevelt is quoted in Arthur H. Masten, *Tahawus Club 1898–1933* (Burlington, Vt.: Free Press Interstate, 1935), p. 54.

85. Jon Krakauer, *Into Thin Air* (New York: Villard-Random House, 1997), p. 270.

86. "Hunt over Mountains for Mr. Roosevelt," *New York Times* (September 14, 1901), p. 1.

87. Morris, *The Rise of Theodore Roosevelt*, p. 741.

88. Ibid.

89. Margaret Leech, *In the Days of McKinley* (New York: Harper, 1959), p. 601.

90. Special to the *New York Times*, "Mr. Roosevelt Is Now the President," *New York Times* (September 15, 1901), p. 1.

15: The Conservationist President and the Bully Pulpit for Forestry

1. Arthur M. Schlesinger Jr., *Journals 1952–2000* (New York: Penguin, 2007), pp. 760–761.

2. George H. Lyman to Henry Cabot Lodge (November 13, 1901), Henry Cabot Lodge Papers, Massachusetts Historical Society, Boston.

3. Lewis L. Gould, *The Presidency of Theodore Roosevelt* (Lawrence: University Press of Kansas, 1991), pp. 8–9.

4. William Allen White to Cyrus Leland (December 19, 1901), William Allen White Papers, Library of Congress.

5. T.R., *Hunting Trips of a Ranchman* (New York: G. P. Putnam's Sons, 1885), p. 121.

6. Donald Worster, *Nature's Economy: A History of Ecological Ideas*, 2nd ed. (Cambridge: Cambridge University Press, 1994), p. 270.

7. La Follette quoted in Farida A. Wiley, "Introduction," in *Theodore Roosevelt's America* (New York: Natural History Library Edition, 1962), p. xxiii.

8. T.R., *A Book-Lover's Holidays in the Open* (New York: Scribner, 1916), pp. 96–97.

9. Charles R. Farabee Jr., *National Park Ranger: An American Icon* (Lanham, Md.: Roberts Rinehart, 2003), pp. 17–18.

10. Ibid.

11. T.R. to David E. Warford (August 20, 1901).

12. Kenneth C. Kellar, *Seth Bullock: Frontier Marshall* (Aberdeen, S.D.: North Plains Press, 1972), p. 120.

13. "A New Cabinet Member," *New York Times* (December 22, 1898), p. 1.

14. Kellar, *Seth Bullock*, p. 120.

15. T.R. to Seth Bullock (September 24, 1901).

16. Charles G. Washburn, *Theodore Roosevelt: The Logic of His Career* (Boston and New York: Houghton Mifflin Company, 1916), p. 120.

17. T.R. to Ethan Allen Hitchcock (January 25, 1902).

18. T.R. to Booker T. Washington, September 14, 1901 in Emmett Jay Scott and Lyman Beecher Stowe, *Booker T. Washington: Builder of Civilization* (New York: Doubleday, 1916), p. 49.

19. *New Orleans Statesman* quoted in Washburn, *Theodore Roosevelt: The Logic of His Career*, p. 73.

20. Pearl Kluger, "Progressive Presidents and Black Americans" (PhD dissertation, Columbia University, 1972), pp. 311–312.

21. *Richmond Times* quoted in H. W. Brands, *T.R.: The Last Romantic* (New York: Basic Books, 1997), p. 423. See also Louis R. Harlan, *Booker T. Washington: The Making of a Black Leader, 1856–1901* (New York: Oxford University Press, 1978), p. 314.

22. "The Night President Teddy Roosevelt Invited Booker T. Washington to Dinner," *Journal of Blacks in Higher Education*, No. 35 (Spring 2002), pp. 24–25.

23. John Ise, *The United States Forest Policy* (New Haven, Conn.: Yale University Press, 1920), p. 161; Samuel T. Dana, *Forest and Range Policy* (New York: McGraw-Hill, 1956), pp. 102–104.

24. John Allen Gable, ed., "President Theodore Roosevelt's Record on Conservation," Vol. 10. *Theodore Roosevelt Association Journal* (Fall 1984), pp. 2–11. Also see Theodore Roosevelt Association Online Archives, "Conservationist, Establishment and Modification of National Forest Boundaries: A Chronological Record, 1891–1973" (compiled and edited from research done by the National Geographic Society and the Theodore Roosevelt Association staff, November 2005).

25. T.R. to James Wilson, (October 18, 1901).

26. M. Nelson McGeary, *Gifford Pinchot: Forester-Politician* (Princeton, N.J.: Princeton University Press, 1960), pp. 65–67.

27. Char Miller, *Gifford Pinchot and the Making of Modern Environmentalism* (Washington, D.C.: Island, 2001), pp. 147–150.

28. Edmund Morris, *Theodore Rex* (New York: Random House, 2001), pp. 70–80. (All of chap. 4 of this biography deals with December 3, 1901.)

29. Lewis L. Gould, *The Presidency of Theodore Roosevelt* (Lawrence: University Press of Kansas, 1991), p. 29.

30. "President Roosevelt's First Message," *New York Times* (December 4, 1901), p. 6.

31. Morris, *Theodore Rex*, p. 75.

32. "President Roosevelt's First Message," *New York Times*, p. 6.

33. Ibid.

34. Morris, *Theodore Rex*, p. 76.

35. Paul Russell Cutright, *Theodore Roosevelt: The Naturalist* (New York: Harper, 1956), pp. 164–165.

36. "President Roosevelt's First Message," p. 6.

37. John Muir, *Our National Parks* (Boston, Mass.: Houghton Mifflin, 1901). For Muir's original essays see "The American Forests," *Atlantic*, Vol. 80 (August 1897); and "The Wild Parks and Forest Reservations of the West," *Atlantic*, Vol. 81 (January 1898).

38. Muir quoted in Austin Considine, "Fall Colors without the Crowds," *New York Times* (October 19, 2007), p. D1.

39. T.R., *An Autobiography* (New York: Macmillan, 1913), p. 415.

40. *Outing Magazine*, Vol. 39 (1902).

41. Worster, *Nature's Economy*, p. 262.

42. Cutright, *Theodore Roosevelt: The Naturalist*, p. 93.

43. "John A. Loring, 76, Noted Naturalist," *New York Times* (May 9, 1947), p. 21.

44. C. Hart Merriam, "Roosevelt the Naturalist," *Science*, New Series, Vol. 75, No. 1937 (February 12, 1932), pp. 181–183.

45. Ibid.

46. Ibid.

47. Robert B. Pickering, "Return of the Buffalo: An American Success Story," *Points West* (Fall 2000).

48. "Notes and News," *New York Times* (May 3, 1902), BR 14.

49. *Washington Times* (June 22, 1902).

50. T. S. Van Dyke, "The Hills of San Bernardino," *Californian*, Vol. 4, No. 21 (September 1881), p. 220.

51. T. S. Van Dyke, *County of San Diego: The Italy of Southern California* (National City, Calif.: National City Record Steam Print, 1887).

52. T. S. Van Dyke, "Those Four Wild-Cats with One Bullet," *Forest and Stream* (November 17, 1881), p. 309.

53. "Arctic Travel Record Broken," *New York Times* (September 18, 1899), p. 2.

54. Ibid.

55. T.R., T. S. Van Dyke, D. G. Elliot, and A. J. Stone, *The Deer Family* (New York: Macmillan, 1902), "Foreword."

56. John Spears, "All about Deer by President and Others," *New York Times* (May 31, 1902), p. BR9.

57. T.R. et al., *The Deer Family*, p. 117.

58. T.R., *Outdoor Pastimes of An American Hunter*, p. 188.

59. T.R. et al., *The Deer Family*, p. 27.

60. Ibid., pp. 134–135.

61. William T. Hornaday, *Popular Official Guide to the New York Zoological Park*, 17th ed. (New York: New York Zoological Society, 1899), p. 57.

62. Alden Sampson, "The Creating of Game Preserves," in George Bird Grinnell (ed.), *American Big Game in Its Haunts* (New York: Forest and Stream, 1904), p. 41.

63. T.R., *An Autobiography*, p. 419.

64. Cutright, *Theodore Roosevelt: The Naturalist*, p. 167.

65. Donald Worster, *Rivers of Empire: Water, Aridity, and the Growth of the American West* (New York: Pantheon, 1985), pp. 169–171.

66. Gould, *The Presidency of Theodore Roosevelt*, p. 41.

67. *Works*, Mem. Ed., Vol. 15, p. 558.

68. Donald J. Pisani, "A Tale of Two Commissioners: Frederick H. Newell and Floyd Dominy," presented at *History of the Bureau of Reclamation: A Symposium*, Las Vegas, Nev. (June 18, 2002).

69. T.R. to Ethan Allen Hitchcock (June 17, 1902).

70. T.R. to James Wilson (July 2, 1902).

71. T.R., *An Autobiography*, p. 408.

72. "President Roosevelt's First Message."

73. Cutright, *Theodore Roosevelt: The Naturalist*, p. 168.

74. David Dary, *Frontier Medicine: From the Atlantic to the Pacific, 1492–1941* (New York: Knopf, 2008), p. 222.

75. T.R., *Works*, Mem. Ed. Vol. 22, pp. 450–452.

76. T.R. quoted in Lawrence H. Budner, "Hunting, Ranching, and Writing" in Natalie A. Naylor, Douglas Brinkley, and John Allen Gable (eds.), *Theodore Roosevelt: Many Sided American* (Interlaken, N.Y.: Heart of the Lakes, 1992), pp. 161–169.

77. Brands, *T.R.: The Last Romantic*, p. 448.

78. Steven E. Siry, "President Theodore Roosevelt's Brush with Death in 1902," *Theodore Roosevelt Association Journal*, Vol. 25, No. 1 (2002), p. 5.

79. "President's Landau Struck by a Car," *New York Times* (September 4, 1902), p. 1.

80. Bob Terrell, "Roosevelt's Visit a 'Red-Letter Day' in Asheville's History," *Asheville Citizen-Times* (April 9, 2000). Also see "Last Day in Dixie," *The Washington Post*, September 11, 1902), p. 1.

81. Ovid Butler (ed.), *Carl Alwin Schenck, The Birth of Forestry in America: Biltmore Forestry School, 1898–1913* (Santa Cruz, Calif.: Forestry History Society, 1974).

82. George W. Vanderbilt letter, Biltmore Company Archives, Presidential Visit File, Asheville, N.C.

83. T.R. to Kermit Roosevelt (September 1908).

84. Cutright, *Theodore Roosevelt: The Naturalist*, pp. 89–99.

85. T.R. to John Pitcher (October 24, 1902).

86. *Nature's Economy*, pp. 125–129.

87. Ibid., pp. 167–171.

16: The Great Mississippi Bear Hunt and Saving the Puerto Rican Parrot

1. "Coal Miners Declare the Big Strike Off; Arbitration Plan Accepted by a Unanimous Vote," *New York Times* (October 22, 1902), p. 1.

2. Paul Schullery, *American Bears: Selections from the Writings of Theodore Roosevelt* (Boulder, Colo.: Roberts Rinehart, 1997), p. 10.

3. "President on Hunting Trip Near Bull Run, Virginia," *New York Times* (November 2, 1902), p. 5.

4. William F. Holmes, *The White Chief: James Kimble Vardaman* (Baton Rouge: Louisiana State University Press, 1975), pp. 105–111.

5. " 'Lily White' Plan to Boom Mr. Hanna," *New York Times* (November 17, 1902), p. 1.

6. T.R. to Stuyvesant Fish (November 6, 1902).

7. Lewis L. Gould, *The Presidency of Theodore Roosevelt* (Lawrence: University Press of Kansas, 1991), p. 42.

8. Clarence Gohdes, *Hunting in the Old South: Original Narratives of the Hunters* (Baton Rouge: Louisiana University Press, 1967), p xii.

9. Minor Ferris Buchanan, *Holt Collier: His Life, His Roosevelt Hunts, and the Origin of the Teddy Bear* (Jackson, Miss.: Centennial, 2002), p. 157. This is a fine biography. Buchanan, a litigation attorney in Jackson, Mississippi, helped me understand the great Mississippi bear hunt of 1902 in many ways. All my writing on Holt Collier has been influenced by his research.

10. Author interview with Shelby Foote (April 6, 1997), New Orleans.

11. Buchanan, *Holt Collier*, p. xiii.

12. Ibid., pp. 3–150.

13. "A Brief History of African Americans and Forests," *Celebrating a Century of Service, A Glance at the Agency's History U.S. Forestry Service*, Issue 25, Bi-Weekly Postings, U.S. Forest Service, International Programs Archives, Washington, D.C.

14. Buchanan, *Holt Collier*, p. 140.

15. "Bears in Combine," *Washington Post* (November 18, 1902), p. 1.

16. "The President's Sunday," *New York Times* (November 17, 1902), p. 1.

17. John Parker to Judge J. M. Dickerson (February 26, 1924).

18. T.R. to Philip Bathell Stewart (November 24, 1902).

19. Author interview, Tweed Roosevelt (February 11, 1998).

20. Author interview with Tweed Roosevelt (May 17, 1999).

21. T.R. to Philip Bathell Stewart (November 24, 1902).

22. Edmund Morris, *Theodore Rex* (New York: Random House, 2001), p. 172.

23. Buchanan, *Holt Collier*, pp. 179–180.

24. "Snub the President," *New York Times* (November 18, 1902), p. 1.

25. Gregory Wilson, "How the Teddy Bear Got His Name," *Washington Post Potomac* (November 30, 1969), pp. 33–35.

26. Douglas Brinkley, "The Myth of the Great Bear Hunt," *Oxford American*, Issue 36 (November/December 2000), pp. 116–121.

27. Peter Bull, *The Teddy Bear Book* (New York: Random House, 1970). The information I give here is a synthesis from this work.

28. T.R. to Clifford Berryman (January 14, 1908).

29. H. Paul Jeffers, *Roosevelt the Explorer: Teddy Roosevelt's Amazing Adventures as a Naturalist, Conservationist, and Explorer* (Lanham, Md.: Taylor Trade, 2003), p. 125.

30. T.R., *Outdoor Pastimes of an American Hunter* (New York: Macmillan, 1908), p. 366. Second Edition.

31. Buchanan, *Holt Collier*, p. xi.

32. William Faulkner, "The Bear," in *Go Down, Moses, and Other Stories* (New York: Random House, 1942).

33. T.R. to John Moulder Wilson (December 9, 1902).

34. *El Yanque National Forest* (Greendale, Indiana: The Creative Company, 1996), p. 8.

35. Kathyrn Robinson, *Where Dwarfs Reign: A Tropical Rainforest in Puerto Rico* (San Juan: Editorial de la Puerto Rico, 1977), p. 186.

36. T.R., "Naturalist's Tropical Laboratory," *Scribner's Magazine* (January 1917), Vol. 1, LXI, No. 1, p. 53 and see Gerald D. Lindsey, Wayne J. Arendt, Jan Kolina, and Gray W. Pendleton, "Home Range and Movements of Juvenile Puerto Rican Parrots," *The Journal of Wildlife Management*, Vol. 55, No. 2 (April 1991), pp. 318–322.

37. Theodore Roosevelt Executive Order—Reserving Miraflores Island in Puerto Rico (July 22, 1902). (Transcript.)

38. T.R. to Gifford Pinchot (November 28, 1906), LC, Series 2, Vol. 618, Reel 343, p. 398.

39. Joseph Wallace, *A Gathering of Wonders: Behind the Scenes at the American Museum of Natural History* (New York: St. Martin's, 2000), p. 39.

40. *Congressional Record, Senate*, S 4302 (April 22, 2004), Library of Congress.

41. T.R. to Kermit Roosevelt (January 23, 1904).

42. T.R. to Theodore Roosevelt, Jr. (November 1, 1901).

43. "Pets in the White House," *Zion's Herald* (February 24, 1909), p. 240.

17: Crater Lake and Wind Cave National Parks

1. Quotation from the Organic Act, 16 U.S.C. §1. See also Duane Hampton, *How the U.S. Cavalry Saved Our National Parks* (Bloomington: Indiana University Press, 1971).

2. Jay J. Wagoner, *Arizona Territory 1863–1912: A Political History* (Tucson: University of Arizona Press, 1970), pp. 362–364.

3. *Tucson Daily Citizen* (January 23, 1902).

4. *Reports of the Department of the Interior* (Washington, D.C.: Department of the Interior, 1917), p. 806.

5. Edison Pettit, "On the Color of Crater Lake Water," *Physics: E. Pettit*, Vol. 22 (1936), pp. 139–146.

6. Winthrop Associates Cultural Research (comp.), "Crater Lake: The Klamath Indians of Southern Oregon Cascades" (1993). (Housed in the archive at Crater Lake National Park, Oregon.)

7. Pettit, "On the Color of Crater Lake Water." Also J. S. Piller, *The Geology of Crater Lake National Park*, U.S. Geological Survey Professional Paper No. 3 (Washington D.C.: Government Printing Office, 1902) pp. 47–48.

8. Stephen R. Mark, "William Gladstone—Mazamas Founder (Chronology)," Crater Lake National Park, Oregon (Archive). Mark's historical writings greatly informed this entire chapter.

9. W. G. Steel, *The Mountains of Oregon* (Portland, Ore.: David Steel, Successor to Himes the Printer, 1890), pp. 17–18.

10. Ibid., p. 32. Also *Administrative History Crater Lake National Park, Oregon, USDI—National Park Service* (Denver, Colo.: National Park Service, 1988), pp. 27–28. Also see William Gladstone Steel, "Crater Lake and How to See It," *West Shore*, Vol. 12, No. 3 (March 1886), pp. 104–106; and Alfred Runte, *National Parks: The American Experience* (Lincoln: University of Nebraska Press, 1997), p. 67.

11. "Crater Lake Explored," *New York Times* (August 30, 1886), p. 1.

12. Ibid.

13. Steel, *The Mountains of Oregon*, pp. 20–32.

14. John Muir to William Gladstone Steel (October 2, 1892), Steel Letters, Box 1, Crater Lake National Park, Oregon.

15. Stephen R. Mark, "Crater Lake: Seventeen Years to Success: John Muir, William Gladstone Steel, and the Creation of Yosemite and Crater Lake National Parks," *Mazama*, Vol. 72, No. 13 (1990), p. 5.

16. Ibid., p. 6.

17. Gifford Pinchot, *Breaking New Ground* (New York: Harcourt Brace, 1947), p. 101.

18. Timothy Egan, "Respecting Mount Rainier," *New York Times* (August 22, 1999), p. 17.

19. Gifford Pinchot to William Steel (February 18, 1902), in Stephen R. Mark, "Seventeen Years to Success: John Muir, William Gladstone Steel, and the Creation of Yosemite and Crater Lake National Parks," U.S. Department of Interior, National Park Service, Crater Lake Archives (May 2001).

20. "D. B. Henderson Dies; Was Ill Nine Months," *New York Times* (February 26, 1906), p. 9.

21. John Lacey, quoted in *Chicago Tribune* (June 18, 1905).

22. Gifford Pinchot to William Steel, May 15, 1902 in Stephen R. Mark, "Crater Lake: Seventeen Years to Success."

23. U.S. Congress, House, *Congressional Record*, 57th Cong. 1st Sess. (April 19, 1902), pp. 4450–4453. Also Steve Marks, "A National Park in the State of Oregon," *Southern Oregon Today* (January 2001), Vol. 1, No. 1.

24. "President Roosevelt on Citizens' Duties," *New York Times* (May 22, 1902).

25. "Crater Lake National Park," *New York Times* (November 16, 1902), p. 28.

26. Runte, *National Parks*, p. 71. Also Conversation with Stephen R. Mark.

27. "The Buffalo Woman," Wind Cave National Park Archives, National Park Service, Wind Cave, S. Dak.

28. "Birth of a National Park—The Winds of Wind Cave," Wind Cave National Park Archives, National Park Service, Wind Cave, South Dakota. This online history was invaluable in writing the Wind Cave sections of this book.

29. "Wind Cave Exploration," Wind Cave National Park Archives, National Park Service, Wind Cave South Dakota.

30. Freeman Tilden, *The National Parks* (New York: Knopf, 1979), p. 250.

31. Alvin McDonald Diary (1891–1893), Wind Cave National Park Archive, Wind Cave, S. Dak. (Unpublished.)

32. Gamble was a staunch supporter of T.R. See "Want Roosevelt Again," *New York Times* (March 24, 1907), p. 1.

33. "South Dakota Cave: Senator Gamble Wants to Preserve the Wonder in a Park" *New York Times*, (June 22, 1902), p. 23.

34. Owen Wister, *The Virginian* (New York: Macmillan, 1902), p. 340.

35. John G. Cawelti, "Introduction," in Owen Wister, *The Virginian* (New York: Barnes and Noble, 2005), p. xxvii.

36. Jack DeMattos, *Garrett and Roosevelt* (College Station, Texas: Creative Publishing Company, 1988).

37. Owen Wister, "Rededication and Preface," in *The Virginian* (New York: Macmillan, 1911), pp. 285, 50, vii.

38. R. Douglas Hurt, "Forestry on the Great Plains, 1902–1904," Lecture for Kansas State University's People, Prairies, and Plains, N.E.H. Summer Teachers' Institute on Environmental History (July–August 1996).

39. Carlos G. Bates and Roy G. Pierce, "Forestation of the Sand Hills of Nebraska and Kansas," *USDA Forest Service Bulletin* Vol. 121 (1913), pp. 8–11; and Raymond J. Poole, "Fifty Years of the Nebraska National Forest," *Nebraska History*, Vol. 34 (September 1953), p. 145.

40. John Clark Hunt, "The Forest That Men Made" *American Forests* 71 (December 1965), p. 32.

41. "Wildlife Management in the Forest Service," in *Celebrating a Century of Service: A Glance at the Agency's History*, Bi-Weekly Postings, Issue 22, Washington, D.C.: U.S. Forest Service, International Programs Archives.

42. R. Douglas Hurt, "Forestry on the Great Plains, 1902–1904."

43. T.R. to Gifford Pinchot (April 9, 1906), Library of Congress, Pinchot Papers (microfiche), Series 2, Vol. 62, Reel 341, p. 444.

44. Quoted in *Outlook*, Vol. 109 (January 20, 1915).

45. Polly Miller and Leon Miller, *Lost Heritage of Alaska: The Adventure and Art of the Alaskan Coastal Indians* (New York: Bonanza, 1967), pp. 243–252.

46. David E. Conrad, "Creating the Nation's Largest Forest Reserve: Roosevelt, Emmons, and the Tongass National Forest," *Pacific Historical Review*, Vol. 46, No. 1 (February 1977), pp. 65–83.

47. George T. Emmons, "The Woodlands of Alaska," Tongass National Forest Archive, Ketchikan, Alas.

48. Conrad, "Creating the Nation's Largest Forest Reserve."

49. William N. Tilchin, *Theodore Roosevelt and the British Empire: A Study in Presidential Statecraft* (New York: St. Martin's, 1977), pp. ix–xi.

50. Conrad, "Creating the Nation's Largest Forest Reserve," *Pacific Historical Review*, pp. 65–82.

51. Frederick Converse Beach and George Edwin Rines (eds.), *The Encyclopedia Americana*, Vol. F–H (New York: Scientific American Compiling Department, Frederick Converse Beach, 1904–1905), table listed under "Game Preserves."

52. "Message of the President," *New York Times* (December 3, 1902), p. 2.

18: Paul Kroegel and the Feather Wars of Florida

1. George Keyes, "Pelican Island," *More Tales of Sebastian* (Vero Beach, Fla.: Sebastian River Area Historical Society, 1992). (Originally published in *Vero Beach Press Journal*, August 15, 1990). For number of species, see Frank J. Thomas, *Melbourne Beach and Indialantic* (Charleston, S.C.: Arcadia, 1999), p. 21. By the year 2000, owing to the negative ecological effects from the opening of the ocean inlets in the late 1940s, the Indian River Lagoon had become very brackish, causing a host of new environmental problems. The increased salinity of the lagoon, for example, killed off the oyster beds. Another problem has been pollution and pesticides being flushed into the Indian River from man-made canals.

2. Arthur C. Bent, *Life Histories of North American Petrels and Pelicans and Their Allies* (Washington, D.C.: United States Government Printing Office, 1922, No. 121).

3. Ibid.

4. James Alexander Henshall, *Camping and Cruising in Florida* (Cincinnati, Ohio: Robert Clarke & Co., 1884), p. 57 and Robert R. Cointepoix, "Early Ornithologists," *Tales of Sebastian* (Vero Beach, Fla.: Sebastian River Area Historical Society, 1990), p. 127.

5. "The Stork Facts," *Kingdom* (December 17, 2002).

6. *White Stork File* (Washington, D.C.: National Zoological Park, Smithsonian

Institution). See also J. A. Hancock, J. A. Kushlan, and M. P. Kahl, *Storks, Ibises, and Spoonbills of the World* (London: Academic, 1992).

7. Benjamin Thorpe, *Northern Mythology*, English Edition, Vol. II (London: Edward Cumley, 1941), pp. 271–274.

8. Jackie Wullschlager, *Hans Christian Andersen: The Life of a Storyteller* (New York: Knopf, 2001), p. 194.

9. "The Kroegel Family Stories," in *The Original Tales from Sebastian* (Sebastian, Fla.: Sebastian River Area Historical Society, 1992), pp. 45–48.

10. Arline Westfahl and George Keyes, *One Person Can Make a Difference: A Story of Paul Kroegel and Pelican Island* (Vero Beach, Fla.: Sebastian River Area Historical Society, 2003).

11. Wallace Stegner, *The American West as Living Space* (Ann Arbor: University of Michigan Press, 1987), p. v.

12. Pelican Island National Wildlife Refuge Archives, Vero Beach, Fla. More than thirty bird species used Pelican Island as a rookery, feeding ground, or loafing area. Among the most common besides brown pelicans were the wood stork, great egret, snowy egret, reddish egret, great blue heron, little blue heron, double-crested cormorant, anhinga, white ibis, American oystercatcher, and common moorhen.

13. Thomas Gilbert Pearson, *Adventures in Bird Protection* (New York: Appleton-Century, 1937), p. 41.

14. Ramona Vickers, "The Kroegel Family Story," in *The Original Tales from Sebastian* (Vero Beach, Fla.: Sebastian River Area Historical Society, 1992), p. 45.

15. Edmund Berkeley and Dorothy Smith Berkeley, *The Correspondence of John Bartram 1734–1777* (Gainesville: University Press of Florida, 1992), p. 685.

16. Author interview with Douglas Kroegel.

17. Author interview with Janice Kroegel Timinsky (June 20, 2007), Sebastian, Fla.

18. "Paul Kroegel (1864–1948)," U.S. Fish and Wildlife Service, Conservation Files, Pelican Island, Fla.

19. Westfahl and Keyes, *One Person Can Make a Difference*, p. 6.

20. Ted Williams, "The Second Century," *Audubon* (June 2003), p. 73.

21. George Laycock, *Wild Refuges* (Garden City, N.Y.: Natural History Press, 1969), pp. 12–20.

22. Frank M. Chapman, "Introduction," in *Adventures in Bird Protection* (New York: Appleton-Century, 1937), p. xiv.

23. John Muir, *A Thousand Mile Walk to the Gulf* (Boston and New York: Houghton Mifflin, 1916), p. 101.

24. Frank Chapman, *Autobiography of a Bird-Lover* (New York: Appleton-Century, 1933), pp. 45–46.

25. Frank Chapman, *Bird Studies with a Camera* (New York: Appleton, 1900), p. 1.

26. Ibid., p. 3.

27. Ibid.

28. Ibid., pp. 196–199.

29. Ibid., p. 207.

30. U.S. Fish and Wildlife Service (content source), J. Emmett Duffy (topic ed.), "History of Pelican Island National Wildlife Refuge," in Cutler J. Cleveland (ed.), *Encyclopedia of Earth* (Washington, D.C.: Environmental Information Coalition, National Council for Science and the Environment). (First published October 16, 2006; last revised January 31, 2007; retrieved September 13, 2007.)

31. Robert E. Kohler, *All Creatures: Naturalists, Collectors, and Biodiversity 1850–1950* (Princeton, N.J.: Princeton University Press, 2006), p. 170.

32. Chapman, *Autobiography of a Bird-Lover*, pp. 88–90.

33. Elizabeth S. Austin (ed.), *Frank M. Chapman in Florida: His Journals and Letters* (Gainesville: University of Florida Press, 1967).

34. Chapman quoted in Frank Graham, Jr., "Where Wildlife Rules," *Audubon* (June 2003), p. 47.

35. "History of Pelican Island National Wildlife Refuge," *Encyclopedia of Earth* (September 2007). Also special thanks to William Reffalt.

36. William Reffalt, "A Prologue to Pelican Island" (February 2003). (Unpublished. Reffalt, the original author, is a retiree of the U.S. Fish and Wildlife Service, a former chief of the Division of Refuges, and a current volunteer.)

37. "History of Pelican Island National Wildlife Refuge," *Encyclopedia of Earth* (September 2007).

38. T. S. Palmer, "In Memoriam: William Dutcher," *The Auk: A Quarterly Journal of Ornithology*, Vol. 38 (October 1921).

39. Ibid.

40. *Yearbook of the United States Department of Agriculture: 1903* (Washington, D.C.: Government Printing Office, 1904), p. 569.

41. William Dutcher to Paul Kroegel (March 24, 1903), Personal Papers of Janice Kroegel Timinsky, Vero Beach, Fla.

42. Author interview with Janice Kroegel Timinsky, May 15, 2007.

43. Weona Cleveland, "Pelican Island Was First Wildlife Refuge," *Evening Times* (June 7, 1978).

44. William Dutcher to Paul Kroegel (April 28, 1902), Personal Papers of Janice Kroegel Timinsky, Vero Beach, Fla.

45. Clara Barrus, *The Heart of Burroughs's Journals* (Boston, Mass.: Houghton Mifflin, 1928), p. 320.

46. Author interview with Janice Kroegel Timinsky, May 15, 2007.

47. William Reffalt, "Pelican Island, Florida—Chronology of Early Events and Pelican Nesting Data" (May 2006). (Unpublished.)

48. McIver, *Death in the Everglades*, pp. 147–169.

49. Ibid.

50. Charles W. Tebeau, *Man in the Everglades* (Coral Gables, Fla.: University of Miami Press, 1968).

51. Stuart B. McIver, *Death in the Everglades: The Murder of Guy Bradley, America's First Martyr to Environmentalism* (Gainesville: University of Florida, 2003), p. 136.

52. Frank M. Chapman, *Camps and Cruises of an Ornithologist* (New York: Appleton, 1908), p. 136.

53. Jack E. Davis, *An Everglades Providence* (Athens, GA: University of Georgia Press, 2009), p. 189.

54. T.R., *A Book-Lover's Holidays in the Open* (New York: Scribner, 1916), p. 286.

55. McIver, *Death in the Everglades*, p. 153.

56. Michael Grunwald, *The Swamp* (New York: Simon & Schuster, 2006), pp. 1–80 and Frank Graham, Jr., *The Audubon Ark: A History of the National Audubon Society* (New York: Alfred A. Knopf, 1990), pp. 50–68.

57. Frank Graham, Jr., *The Audubon Ark: A History of the National Audubon Society* (New York: Knopf, 1990), pp. 58–59.

58. Dutcher quoted in "History of Pelican Island National Wildlife Refuge."

59. Robert R. Cointepoix, "Early Ornithologists," p. 128.

60. Paul Tritaik to Douglas Brinkley, March 25, 2009. Spoke to Tritaik around a dozen times.

19: Passports to the Parks

1. T.R. to John Pitcher (February 18, 1903).

2. T.R. to John Pitcher (March 2, 1900).

3. Aubrey L. Haines, *The Yellowstone Stories*, Vol. 2 (Yellowstone National Park, Wyo.: Yellowstone Library and Museum Association, 1977), p. 81.

4. T.R. to Frederick Weyerhaeuser (March 5, 1903).

5. T.R. to John Burroughs (March 7, 1903).

6. Edward J. Renehan, Jr., *John Burroughs* (Post Mills, Vt.: Chelsea Green, 1992), pp. 227–228.

7. John Burroughs, "Real and Sham Natural History," *Atlantic Monthly*, Vol. 91, (March 1903).

8. T.R. to Ernest Thompson Seton, quoted in Paul Russell Cutright, *Theodore Roosevelt: The Naturalist* (New York: Harper, 1956), p. 131.

9. Renehan, *John Burroughs*, pp. 232–233.

10. Ernest Thompson Seton, *Life Histories of Northern Animals: An Account of the Mammals of Manitoba* (New York: Scribner, 1909). Ernest Thompson Seton, *Lives of Game Animals* (New York: Doubleday, 1929).

11. John Burroughs to Julian Burroughs (March 31, 1903), Vassar Library Collection, Poughkeepsie, N.Y.

12. Burroughs, "Real and Sham Natural History," *Atlantic Monthly*, 91 (March 1903).

13. T.R. to John Burroughs (March 7, 1903).

14. Paul Schullery, "Theodore Roosevelt: The Scandal of the Hunter as Nature Lover," in Natalie A. Naylor, Douglas Brinkley, and John Allen Gable (eds.), *Theodore Roosevelt: Many Sided American* (Interlaken, N.Y.: Heart of the Lakes, 1992), p. 229.

15. T.R. quoted in Aubrey L. Haines, *The Yellowstone Story: A History of Our First National Park* (Boulder: Colorado Associated University Press, 1977), p. 81.

16. Donald Worster, *Nature's Economy* (Cambridge: Cambridge University Press, 1985), p. 260. Also Eric Busch of University of Texas at Austin (a PhD student in history) helped me formulate this idea. He is a cutting-edge new environmental historian whose expertise pertains to the Rocky Mountains.

17. "President's Train Ready," *New York Times* (April 1, 1903), p. 8.

18. John Burroughs, *Camping and Tramping with Roosevelt* (Boston, Mass.: Houghton Mifflin, 1906), p. viii.

19. "Snow in Yellowstone Park," *New York Times* (March 24, 1903), p. 5.

20. "Invites John Burroughs," *New York Times* (March 16, 1903), p. 1.

21. *New York Times* (April 29, 1903). (Obituary.)

22. "Wyoming for Roosevelt," *New York Times* (March 18, 1903), p. 3.

23. "Rival Towns Upset by the President's Trip," *New York Times* (March 24, 1903), p. 5.

24. "A Bear for the President," *New York Times* (March 25, 1903), p. 1.

25. "Cowboys to Greet President," *New York Times* (April 2, 1903), p. 1.

26. "Dynamite Salute Planned," *New York Times* (April 6, 1903), p. 1.

27. "The President's Progress," *New York Times* (April 2, 1903), p. 8.

28. T.R. to John Burroughs (March 14, 1903).

29. Burroughs, *Camping and Tramping with Roosevelt*, pp. 5–6.

30. T.R. to Dr. C. Hart Merriam (March 31, 1903).

31. Howells quoted in William M. Gibson, *Theodore Roosevelt among Humorists* (Knoxville: University of Tennessee Press, 1980), p. 21.

32. John Burroughs, "Camping with President Roosevelt," *The Atlantic Monthly*, May 1906 (Vol. 97, No. 5).

33. "President Discusses the Monroe Doctrine," *New York Times* (April 3, 1903), p. 1.

34. H. Paul Jeffers, *Roosevelt the Explorer: Teddy Roosevelt's Amazing Adventures as a Naturalist, Conservationist, and Explorer* (New York: Taylor Trade, 2003).

35. " 'Dooleyized' the President: University of Chicago Students Adopt a Popular Song in Welcoming Mr. Roosevelt," *New York Times* (April 3, 1903), p. 1.

36. Paul Schullery, "Buffalo Jones and the Bison Herd in Yellowstone," *Montana: The Magazine of Western History*, Vol. 26, No. 3 (July 1986), pp. 40–51.

37. T.R. to Clinton Hart Merriam, April 16, 1903.

38. T.R. to Lieutenant General S. B. M. Young (January 22, 1908), Yellowstone Reference Library, Yellowstone National Park.

39. Brodie Farquehar, "Centennial Anniversary of Visit This Month," *Caspar Star Tribune* (April 1, 2003), p. C1.

40. T.R. to C. Hart Merriam (April 22, 1903).

41. Burroughs, *Camping and Tramping with Roosevelt*, p. 8.

42. Ibid., pp. 26, 29.

43. Ibid., pp. 111, 60.

44. "President Kills Lion in Yellowstone Park," *New York Times* (April 12, 1903), p. 1.

45. "President on the Move," *New York Times* (April 15, 1903), p. 1.

46. Burroughs, *Camping and Tramping with Roosevelt*, p. 66.

47. T.R. to Dr. C. Hart Merriam (April 22, 1903).

48. Haines, *The Yellowstone Story*, Vol. 2, pp. 229–237.

49. "Roosevelt Delights in Yellowstone," *Caspar Star Tribune* (April 11, 2003), p. C1.

50. "The President in the Park," *Forest and Stream*, Vol. 60, No. 18 (May 2, 1903).

51. Erin H. Turner, *It Happened in Yellowstone* (Guilford, Conn.: Morris, 2001), p. 47.

52. "The President in the Park."

53. "Resumes His Tour," *Washington Post* (April 25, 1903), p. 1.

54. Liz Nelson, "The Hermit of Ravenswood," *Special Places*, Vol. 14, No. 3 (Fall 2006), pp. 8–10.

55. T.R. to George Bird Grinnell (April 24, 1903).

56. Morison (ed.), T.R. in *The Letters of Theodore Roosevelt*, Vol. 3.

57. Burroughs, *Camping and Tramping with Roosevelt*, pp. 59–60.

58. T.R. to Gifford Pinchot (April 9, 1906), Series 2, Vol. 62, Reel 241, p. 444.

59. R. Douglas Hurt, "Forestry on the Great Plains, 1902–1904," Lecture for Kansas State University's People, Prairies, and Plains, N.E.H. Summer Teachers' Institute on Environmental History (July–August 1996).

60. "The President in Iowa," *New York Times* (April 29, 1903), p. 1.

61. Address by John F. Lacey before Iowa Federation of Women's Clubs, Waterloo (May 12, 1905). (Transcript.)

62. David Dary, *The Buffalo Book: The Full Saga of the American Animal* (Chicago, Ill.: Swallow, 1974), pp. 233–236.

63. "President Roosevelt Reaches St. Louis," *New York Times* (April 30, 1903), p. 3.

64. "Secretary Hitchcock Now Faces Charges," *New York Times* (August 20, 1903), p. 1.

65. Henry S. Brown, "Punishing the Landlooters," *Outlook* (February 23, 1907).

66. "President's Train Ready," *New York Times* (April 1, 1903), p. 8.

67. Barbara Kerley, "Josiah, the White House Badger," *Highlights* (April 2006), pp. 32–35.

68. "The President's Sunday at Sharon Springs, Kansas," *New York Times* (May 4, 1903), p. 2. Also Sagamore Hill National Historic Site Archive (pet file).

69. T.R. to children (May 10, 1903), Sagamore Hill Archives, Oyster Bay, N.Y. (Group letter from Del Monte, California.)

70. "President in Colorado," *New York Times* (May 5, 1903), p. 9.

71. "Denver in Readiness," *Washington Post* (May 4, 1903), p. 1.

72. Burroughs, *Camping and Tramping With Roosevelt*, p. 53.

73. "Mr. Roosevelt Tells New Mexico to Grow," *New York Times* (May 6, 1903), p. 3.

74. C. G. Turner II, *Petroglyphs of the Glen Canyon Region* (Flagstaff: Museum of Northern Arizona Bulletin, No. 38). Also see David S. Whitney, *Handbook of Rock Art Research* (Walnut Creek, Calif.: Alta Mira, 2001), pp. 385–386.

75. Max Frost and Paul A. F. Walter, *The Land of Sunshine: A Handbook of the Resources, Products, Industries, and Climate of New Mexico* (Santa Fe: New Mexican Printing, 1904).

76. T.R., *A Book-Lover's Holidays in the Open* (New York: Scribner, 1916), p. 1.

77. "Mr. Roosevelt Sees the Grand Canyon," *New York Times* (May 7, 1903), p. 2.

78. Stephen R. Whitney, *A Field Guide to the Grand Canyon* (Seattle, Wash.: Mountains, 1996), p. 1.

79. T.R., *A Book-Lover's Holidays*, p. 8.

80. "Mr. Roosevelt Sees the Grand Canyon," *New York Times* (May 7, 1903), p. 2.

81. John Burroughs, *Locusts and Wild Honey* (Boston, Mass.: Houghton, Osgood, 1879), p. 200.

82. T.R. quoted in Stephen J. Pyne, *How The Canyon Became Grand: A Short History* (New York: Viking, 1998), p. 38.

83. David S. Whitney, *A Field Guide to the Grand Canyon* (Seattle, Wash.: Mountaineers, 1996), pp. 13–19.

84. Rose Houk, *An Introduction to Grand Canyon Ecology* (Grand Canyon, AZ: Grand Canyon Association, 1996), pp. 4–45.

85. Richard G. Beidleman, *California's Frontier Naturalists* (Berkeley: University of California Press, 2006), p. 369.

86. "President Roosevelt in California," *New York Times* (May 8, 1903), p. 1.

87. "The Pope to Mr. Roosevelt," *New York Times* (May 10, 1903), p. 5.

88. "President Talks of California's Big Trees," *New York Times* (May 12, 1903), p. 2.

89. "Degree Conferred upon President Roosevelt," *New York Times* (May 15, 1903), p. 1.

90. "President Talks of California's Big Trees."

91. Ibid.

92. Burroughs quoted in Worster, *Nature's Economy*, p. 17.

93. "America's Destiny on the Pacific," *New York Times* (May 14, 1903), p. 1.

94. Lynn Readicker-Henderson and Ed Readicker-Henderson, *Adventure Guide: Inside Passage and Coastal Alaska*, 4th ed.(Edison, N.J.: Hunter, 2002), pp. 55–57.

95. A. Lincoln, "Roosevelt and Muir at Yosemite," *Pacific Discovery* Vol. 16 (January–February 1963), pp. 18–22.

96. Shirley Sargent, *Yosemite's Famous Guests* (Yosemite, Calif.: Flying Spur, 1970), pp. 18–21.

97. "How Big Are Big Trees?" *California State Parks* (2008). (Pamphlet produced by the state of California.)

98. Ted Kerasote, "Roosevelt and Muir," *Bugle* (Winter 1997), p. 78.

99. Osborn quoted in Edwin Way Teale (ed.), *The Wilderness World of John Muir* (Boston, Mass.: Houghton Mifflin, 1954), p. 181.

100. Ted Kerasote, "Roosevelt and Muir," *Bugle* (Winter 1997), p. 78.

101. Lincoln, "Roosevelt and Muir at Yosemite."

102. Charlie Leidig, "Report of President Roosevelt's Visit in May, 1903," Yosemite National Park Archive, Yosemite, Calif.

103. T.R., "John Muir: An Appreciation," *Outlook*, Vol. 109 (January 6, 1915), pp. 27–28.

104. Kerasote, "Roosevelt and Muir."

105. Teale (ed.), *The Wilderness World of John Muir*, p. xvii.

106. T.R., *An Autobiography* (New York: Macmillan, 1913), p. 332–333.

107. Donald Worster, *A Passion for Nature: The Life of John Muir* (New York: Oxford University Press, 2008), p. 346.

108. Ibid.

109. Lincoln, "Roosevelt and Muir at Yosemite."

110. Worster, *A Passion for Nature*, pp. 208–211.

111. Clara Barrus, "In the Yosemite with John Muir," *The Craftsman*, Vol. 23, No. 3 (December 1912), pp. 324–335.

112. Worster, *A Passion for Nature*, p. 509.

113. T.R., "John Muir: An Appreciation."

114. Linnie Marsh Wolfe, *Son of the Wilderness: The Life of John Muir* (New York: Knopf, 1945), pp. 288–289.

115. Worster, *A Passion for Nature*, p. 366.

116. Wolfe, *Son of the Wilderness*, p. 290.

117. Robert Underwood Johnson, *Remembered Yesterdays* (Boston, Mass.: Little, Brown, 1923), p. 388.

118. Worster, *A Passion for Nature*, p. 369.

119. Alfred Runte, *Yosemite: The Embattled Wilderness* (Lincoln: University of Nebraska Press, 1990), p. 87.

120. T.R. quoted in Sargent, *Yosemite's Famous Guests*, p. 19.

121. John Muir, letter to Robert Underwood Johnson (1889), cited in Frank Bergon, *The Wilderness Reader* (Lincoln: University of Nevada Press, 1994), p. 251.

122. John Muir, Linnie Marsh Wolfe (ed.), *John of the Mountains: The Unpublished Journals of John Muir.* (Boston: Houghton Mifflin, 1938), p. 277.

123. John Muir to Louis Muir (May 19, 1903), John Muir Papers, University of the Pacific.

124. John Muir to Dr. and Mrs. C. Hart Merriam and the Baileys (January 1, 1904), Muir Papers, University of the Pacific.

125. Runte, *Yosemite: The Embattled Wilderness*, p. 87.

126. Leidig, "Report of President Roosevelt's Visit in May, 1903."

127. Photograph included in *Our National Parks* (Pleasantville, N.Y.: Reader's Digest Association, 1985), p. 15.

128. T.R., "John Muir: An Appreciation."

129. Ibid.

130. T.R. to Ethan Allen Hitchcock (May 19, 1903).

131. T.R. to John Muir (May 19, 1903).

132. Rod Miller, *John Muir's Magnificent Tramp* (New York: A Tom Doherty Associates Book, 2005), p. 146.

133. T.R., *California Addresses* (San Francisco: The California Promotion Committee, 1903), printed by the Tomoyé Press, San Francisco, p. 140.

134. George Wharton James, "Harry Cassie Best: Painter of the Yosemite Valley and the California Mountains," *Out West*, New Series, Vol. 7, No. 1 (January 1914), p. 11.

135. "President Quits California," *New York Times* (May 21, 1903), p. 8.

136. T.R. to Harry Cassie Best (November 12, 1908). The actual title of the painting was *Evening at Mt. Shasta.* According to *Out West*, T.R. told Best, "That afterglow on Mt. Shasta is the grandest sight in Nature I have ever witnessed, and I never expected to see such a good reproduction of it on canvas."

137. James, "Harry Cassie Best: Painter of the Yosemite Valley and the California Mountains."

138. "President's Oregon Tour," *New York Times* (May 22, 1903), p. 7.

139. Kohler, *All Creatures*, p. 84. Burroughs thought using Linnaean binomials made "readers feel ignorant and mystified."

140. William L. Finley, "Birds about an Oregon Pond," *Sunset Magazine* (December 1907).

141. Worth Mathewson, *William L. Finley's Pioneer Wildlife Photography* (Corvallis: Oregon State University, 1986), p. 38.

142. Russell D. Butcher, *America's National Wildlife Refuges* (Lanham, Md.: Rowman and Littlefield, 2003), pp. 531–532.

143. Ibid.

144. Mathewson, *William L. Finley's Pioneer Wildlife Photography*, pp. 17–18.

145. Tom McAllister, *Audubon-Warbler* (April 1959).

146. Mathewson, *William L. Finley's Pioneer Wildlife Photography*, pp. 6–7.

147. U.S. Senate History, Expulsion and Censure, Senate Historical Archive, Washington, D.C.

148. T.R. to Alice Lee Roosevelt (May 27, 1903).

149. Ethan Trex, "White House Pets: Hippo, Gator, and 'Satan'" (transcript), CNN Archives (November 7, 2008).

150. T.R., letter from the White House (June 6, 1903), Sagamore Hill National Historic Site, Oyster Bay, N.Y.

20: Beauty Unmarred

1. T.R. and Henry Cabot Lodge, *Hero Tales from American History* (New York: The Century Company, 1895), p. 169.

2. John Milton Cooper, "Theodore Roosevelt: On Clio's Active Service," *The Virginia Quarterly Review* (Winter 1986), pp. 21–37.

3. T.R. to George Otto Trevelyan (January 25, 1904).

4. George Bird Grinnell, "Preface" in George Bird Grinnell (ed.), *American Big Game in Its Haunts* (New York: Forest and Stream, 1904), pp. 19–20.

5. George Bird Grinnell, "Theodore Roosevelt," ibid., pp. 19–21.

6. T.R., "Wilderness Reserves," ibid., pp. 20–51. (T.R. included this essay a year later in his *Outdoor Pastimes of an American Hunter*, giving it a more widespread readership.)

7. Ibid., p. 51.

8. T.R. to James Wilson (March 12, 1904).

9. T.R. to Lawrence Fraser Abbot (March 14, 1904).

10. Lewis L. Gould, *The Presidency of Theodore Roosevelt* (Lawrence: University Press of Kansas, 1991), pp. 101–104.

11. Barbara Kerley, "Josiah, the White House Badger," *Highlights* (April 2006), pp. 32–33.

12. Jacob Riis, *Theodore Roosevelt: The Citizen* (New York: Macmillan, 1904), pp. 318–319.

13. T.R., *An Autobiography* (New York: Macmillan, 1913), p. 357.

14. Ibid., pp. 355–357.

15. "Pets at White House," *Washington Post* (January 22, 1907), p. 18.

16. "Pet Lamb for Theodore Roosevelt Jr.," *New York Times* (October 9, 1902), p. 9.

17. T.R. to Kermit Roosevelt (January 8, 1903).

18. Ibid.

19. T.R. to Kermit Roosevelt (January 18, 1904).

20. *Washington Evening Star* (January 22, 1908). White House Historical Assocation archives (2009 updated).

21. Irwin Hood Hoover, *Forty-Two Years in the White House* (Boston, Mass.: Houghton Mifflin, 1934), p. 28.

22. "Archie Roosevelt Is Ill," *New York Times* (April 14, 1903), p. 1.

23. "Pony in the White House," *New York Times* (April 27, 1903), p. 1. Also Sagamore Hill Pet Archive, Oyster Bay, N.Y.

24. "Pets at White House," *Washington Post* (January 22, 1907), p. 18.

25. Ibid.

26. "Menelik to Roosevelt," *New York Times* (March 5, 1904), p. 2.

27. Steve Kemper, "Who's Laughing Now?," *Smithsonian* (May 2008).

28. "Roosevelt Dog Is Found," *New York Times* (October 1, 1909), p. 20.

29. "Cultural Landscape Report by Sagamore Hill National Historic Site" (prepared by Regina M. Bellavia and George W. Curry), Sagamore Hill National Historic Site Archive, Oyster Bay, N.Y. (2003 reprint.)

30. Jenks Cameron, *The Bureau of Biological Survey* (New York: Arno, 1974), pp. 110–111.

31. Minutes of Executive Committee of Boone and Crockett Club (October 27, 1913). (Transcript.)

32. T.R. to John F. Lacey (April 21, 1904).

33. Cameron, *The Bureau of Biological Survey*, pp. 113–116.

34. "Ranger Boats," Tongass National Forest Facts, Tongass National Forest (history file).

35. T.R. to Kentaro Kaneko (April 23, 1904).

36. T.R. quoted in *The Russo-Japanese War Research Society* (February 1904–September 1905) time line. Online study group.

37. Timothy Foote, "Where the Gooney Birds Are," *Smithsonian Magazine* (September 2001), p. 95.

38. Donald J. Pisani, *Water, Land, Law in the West: The Limits of Public Policy, 1850–1920* (Lawrence: University Press of Kansas, 1996), p. 116.

39. "President Opens Fair with Golden Button," *New York Times* (May 1, 1904), p. 1.

40. Enos A. Mills, *Your National Parks* (Boston, Mass.: Houghton Mifflin, 1917), pp. 244–245.

41. *North Dakota: A Guide for the Northern Prairie State* (New York: Oxford University Press, 1950), p. 256. (This is a revised edition of the first printing in 1938.)

42. George Bird Grinnell, "Forest Reserves of North Dakota," in *American Big Game in Its Haunts*, pp. 458–466.

43. Joseph Maxwell, "Sullys Hill National Game Preserve," *North Dakota Outdoors* (March 2003), p. 22. The USDA used *Sully's* during the T.R. years but today its *Sullys*. The apostrophe has been deleted.

44. T.R. to Edward Howe Forbush (July 21, 1904).

45. "Early State Forestry Efforts" (Washington, D.C.: Forest Service, U.S. Department of Agriculture Brochure No. 9, 2008). Part of the Mini-histories of the Forest Service Series.

46. T.R. to James Rudolph Garfield (July 13, 1904).

47. Buffalo Jones to T.R. (July 27, 1903). Yellowstone National Park, Jackson Hole, Wyo.

48. T.R. to John Burroughs (August 12, 1904).

49. "Breton Island," National Wildlife Refuge, U.S. Fish and Wildlife, Shepherdstown, W. Va.

50. T.R., *A Book-Lover's Holidays in the Open*, pp. 285–289.

51. Ibid., pp. 286–287.

52. *Official Report of the Proceedings of the Sixteenth Republican National Convention* (New York: LaFayette B. Gleason, 1916), p. 264. Also see Lewis Gould, ed., "Charles Warren Fairbanks and the Republican National Convention of 1900: A Memoir," *Indiana Magazine of History*, Vol. 77 (December 1981), p. 370.

53. T.R. to Kermit Roosevelt (October 15, 1904).

54. T.R. to George Bruce Cortelyou (October 26, 1904).

55. T.R. to Owen Wister (November 19, 1904).

56. "Menelik's Gifts Here," *New York Times* (November 8, 1904), p. 1.

57. "Gifts to the President," *New York Times* (November 23, 1905), p. 1.

58. *Carnegie Institution Yearbook, 1906*, quoted in "A Year's Work of the Carnegie Institution," *Nature*, Vol. 75, No. 1956 (April 25, 1907).

59. T.R. to Andrew Carnegie (November 10, 1904).

60. Mike Thompson, *The Travels and Tribulations of Theodore Roosevelt's Cabin* (San Angelo, Tex.: Laughing Horse Enterprises, 2004), pp. 30–34.

61. T.R. to Kermit Roosevelt (November 29, 1904) and T.R. to Oliver Wendell Holmes (December 5, 1904).

62. T.R. to Robert Underwood Johnson (January 17, 1905).

63. T.R. to Grant LaFarge (January 27, 1905).

64. Tyler Dennett, *John Hay: From Poetry to Politics* (New York: Dodd, Mead & Co., 1933), pp. 184–186; and T.R., *State Papers*, National Edition, Vol. 15, pp. 318–401.

65. T.R. to Orville Hitchcock Platt (February 23, 1905).

66. "Forest Transfer Act of 1905," Issue 15, *Celebrating a Century of Service, A Glance at the Agency's History U.S. Forestry Service*, Bi-Weekly Postings, U.S. Forest Service, International Programs Archives, Washington, D.C.

67. Dennis M. Roth, *A History of Wildlife Management in the Forest Service* (Washington, D.C.: USDA Forest Service History Unit, 1989). This is an unpublished manuscript. See also Jack Ward Thomas, *Wildlife Habitats in Managed Forests*, Agricultural Handbook 553 (Washington D.C.: USDA Forest Service, 1979).

68. Ted Kerasote, "Roosevelt and Muir," *Bugle* (Winter 1997), p. 85.

69. Thomas Mallon, "Set in Stone," *New Yorker* (October 13, 2008).

70. "Washington Snow-Clad on Inauguration's Eve," *New York Times* (March 2, 1905), p. 1.

71. "Washington, Aflutter Donning Gala Attire," *New York Times* (March 3, 1905), p. 1.

72. "Indians At the Inaugural," *New York Times* (February 3, 1905), p. 8.

73. David Dary, *True Tales of the Prairies and Plains* (Lawrence: University Press of Kansas, 2007), p. 119.

74. T.R. to Theodore Roosevelt Jr. (March 1, 1903).

75. "Big Sticks for Souvenirs," *New York Times* (March 5, 1905), p. 6.

76. "Nation Mirrored in Marching Host," *New York Times* (March 5, 1905), p. 2.

77. "President Chooses Bible," *New York Times* (March 4, 1905), p. 2.

78. T.R. to Robert Barnwell Roosevelt (March 6, 1905).

79. "Devil's Lake Basin in North Dakota," North Dakota Science Society (July 2008).

80. Stan Tekiela, *Birds of the Dakotas* (Cambridge, Minn.: Adventure, 2003), p. 275.

81. Craig Bihrle, "100 Years of Refuges in North Dakota Is Centerpiece for National Event," *North Dakota Outdoors* (March 2003), p. 3.

82. *Annual Reports of the Department of Agriculture, Fiscal Year Ended June 30* (Washington, D.C.: Government Printing Office, 1905), p. 310.

21: The Oklahoma Hills

1. Lewis L. Gould, "Theodore Roosevelt, William Howard Taft, and the Disputed Delegates in 1912," *Southwestern Historical Quarterly*, Vol. 80 (July 1976); and Paul D. Casdorph, *A History of the Republican Party in Texas, 1865–1965* (Austin: Pemberton, 1965).

2. T.R. to Cecil Andrew Lyon (March 16, 1905).

3. "Negro Mob Killed Sheriff," *New York Times* (March 17, 1905), p. 6.

4. William Caire, Jack D. Tyler, Bryan P. Glass, and Michael A. Mares, *Mammals of Oklahoma* (Norman: University of Oklahoma Press, 1989), p. xi.

5. Bill Neeley, *The Last Comanche Chief: The Life and Times of Quanah Parker* (Hoboken, N.J.: Wiley, 1995), p. 221. Also Edward Charles Ellenbrook, *Outdoor and Trail Guide to the Wichita Mountains of Southwest Oklahoma*, 8th rev. ed. (Lawton, Okla.: In the Valley of the Wichitas, 2008), pp. 6–9.

6. George Bird Grinnell, *When Buffalo Ran* (New Haven, Conn.: Yale University Press, 1920), p. 22. Also Richard C. Rattenbury, *Hunting the American West: The Pursuit of Big Game for Life, Profit, and Sport, 1800–1900* (Missoula, Mont.: Boone and Crockett Club, 2008), p. 207.

7. "The Wichita National Forest and Game Preserve," Miscellaneous Circular No. 36, USDA (May 1925).

8. Alfred Runte, *Trains of Discovery:*

Western Railroads and the National Parks (Niwot, Colo.: Roberts Rinehart, 1990), pp. 19–21. Reprint.

9. Andrew C. Isenberg, *The Destruction of Bison: An Environmental History* (New York: Cambridge University Press, 2000), p. 177.

10. Raymond Gorges, *Ernest Harold Baynes: Naturalist and Crusader* (New York: Houghton Mifflin, 1928), pp. 74–75. Also Joel Berger and Carol Cunningham, *Bison: Mating and Conservation in Small Populations* (New York: Columbia University Press, 1994), p. 29.

11. *Congressional Record*, 59 Cong. 1 Sess; Pt. I, p. 103.

12. Officially the bison were protected by proclamation (June 2, 1905, 34 Stat. 3062) by President Theodore Roosevelt, in Otis H. Gates (comp.), *Laws Applicable to the United States Department of Agriculture* (Washington, D.C.: Government Printing Office, 1913, rev. 1912), p. 111.

13. David Dary, *The Buffalo Book: The Full Saga of the American Animal* (Chicago, Ill.: Swallow, 1974), pp. 233–236.

14. Jack Dan Haley, "A History of the Establishment of the Wichita National Forest and Game Preserve, 1901–1908," unpublished master's thesis, University of Oklahoma, 1973.

15. T.R., *Outdoor Pastimes of an American Hunter* (New York: Macmillan, 1902), p. 102.

16. Neeley, *The Last Comanche Chief*, p. 143.

17. "History Files," Wichita Mountains National Wildlife Refuge, Refuge Headquarters, Indiahoma, Okla.

18. Wichita Mountains (Albuquerque: Southwest Natural and Cultural Heritage Association, 1992). This monograph was compiled by the staff at the Wichita Mountains Wildlife Reserve.

19. John R. (Jack) Abernathy, *In Camp with Theodore Roosevelt, or the Life of John R. (Jack) Abernathy* (Oklahoma City: Times-Journal, 1933).

20. Jon T. Coleman, "Foreword," in John R. Abernathy, *Catch 'Em Alive Jack* (Lincoln: University of Nebraska Press, 2006), p. v.

21. Matthew Rex Cox, "Roosevelt's Wolf Hunt." (Advance article from the *Oklahoma Encyclopedia.*)

22. T.R., *Outdoor Pastimes of an American Hunter*, p. 111.

23. "President Off to Hunt; Taft Sits on Lid," *New York Times* (April 4, 1905), p. 1.

24. W. LaBarre, *The Peyote Cult* (Norman: University of Oklahoma Press, 1989).

25. William T. Hagan, *Quanah Parker, Comanche Chief* (Norman: University of Oklahoma Press, 1993), p. 57.

26. "Star House," *Prairie Lore*, Vol. 41, No. 2, Book 15.

27. Neeley, *The Last Comanche Chief*, p. 199.

28. "Buffalo Hunt Is Held: Game Shot from Auto," *New York Times* (June 11, 1903), p. 5.

29. "Killed by Roosevelt's Train," *New York Times* (April 5, 1905), p. 2.

30. "Roosevelt Says He's a Typical President," *New York Times* (April 6, 1905), p. 2.

31. "Col. Roosevelt Greets His Old Rough Riders," *New York Times* (April 8, 1905), p. 1.

32. T.R., *Outdoor Pastimes of an American Hunter*, p. 100.

33. David Minor, "Samuel Burk Burnett," *The Handbook of Texas* (online; January 9, 2008, update).

34. *Time* (May 22, 1939).

35. "Dr. Lambert Dies; Narcotics Expert," *New York Times* (May 10, 1939), p. 23.

36. *Frederick Enterprise* (April 15, 1905). (Summary story.)

37. Abernathy, *Catch 'Em Alive Jack*, p. 100.

38. W. M. Draper Lewis, *The Life of Theodore Roosevelt* (Philadelphia and Chicago: John C. Winston, 1919), p. 177.

39. *Frederick Enterprise* (April 15, 1905).

40. "President in Wild," *Washington Post* (April 10, 1905), p. 1.

41. Abernathy, *Catch 'Em Alive Jack*, pp. 103–104.

42. Coleman, "Foreward," in Abernathy, *Catch 'Em Alive Jack*, p. ix.

43. Caire et al., *Mammals of Oklahoma*, pp. 281–285.

44. "Why a Refuge," Wichita Mountains Wildlife Refuge Archive, Indiahoma, Okla.

45. T.R., *Outdoor Pastimes of an American Hunter*, p. 101.

46. Abernathy, *Catch 'Em Alive Jack*, p. 102.

47. Francis Haines, *The Buffalo* (Norman: University of Oklahoma Press, 2001), pp. 200–201.

48. T.R., *Outdoor Pastimes of an American Hunter*, p. 103.

49. "President in Foot Races," *New York Times* (April 13, 1905), p. 1.

50. George Bird Grinnell, *When the Buffalo Ran* (New Haven: Yale University Press, 1920), p. 82.

51. T.R., *Outdoor Pastimes of an American Hunter*, p. 104.

52. David E. Lantz, *The Relation of Coyotes to Stock Raising in the West* (Washington, D.C.: Government Printing Office, 1905).

53. Abernathy, *Catch 'Em Alive Jack*, p. 115.

54. T.R., *Outdoor Pastimes of an American Hunter*, pp. 113–114.

55. Abernathy, *Catch 'Em Alive Jack*, p. 127.

56. Abernathy, *Catch 'Em Alive Jack*, p. 115.

57 *Frederick Enterprise* (April 15, 1905). (Clipping at the Wichita Wildlife Refuge in Oklahoma.)

58. T.R., *Outdoor Pastimes of an American Hunter*, p. 116.

59. Ibid., p. 106.

60. Haines, *The Buffalo*, p. 6.

61. Neeley, *The Last Comanche*, pp. 220–221.

62. Clyde L. Jackson and Grace Jackson, *Quanah Parker: The Last Chief of the Comanches—A Study in Frontier History* (New York: Exposition, 1963), p. 129.

63. Ibid., p. 128.

64. Alice Marriot and Carol K. Rachlin, *American Indian Mythology* (New York: Mentor, 1972), p. 170.

65. Wichita Mountain Wildlife Refuge (Albuquerque, N.M.: Southwest Natural and Cultural Heritage Association, 1997).

66. Ernest Wallace and E. Adamson Hoebel, *The Comanches: Lords of the South Plains* (Norman: University of Oklahoma Press, 1952), p. 206.

67. T.R., *Hunting Trips of a Ranchman* (New York: G. P. Putnam's Sons, 1886), p. 260.

68. *Tenth Annual Report of the Bison Society, 1915–1916* (New York: American Bison Society, 1916), pp. 20–22. Also Robert Dorman, *It Happened in Oklahoma* (New York: Morris Book Publishing, 2006), pp. 53–56.

69. Abernathy, *Catch 'Em Alive Jack*, p. 126.

70. "Speeding to the Rockies," *Washington Post* (April 14, 1905), p. 3.

71. "President Appeals to Press," *New York Times* (April 15, 1905), p. 1.

72. "Orville H. Platt Dies," *New York Times* (April 22, 1905), p. 1.

73. Douglas C. McChristian, "The Great Health Mecca and Summer Resort," Historical Resources Study (June 2003), Chickasaw National Recreation Area, Sulphur, Okla. (Unpublished.)

74. *Reports of the Department of Interior 1919*, Vol. 1 (Washington, D.C.: Government Printing Office, 1919), p. 1025.

75. Louis A. Coolidge, *An Old Fashioned Senator: Orville H. Platt of Connecticut* (New York: Putnam, 1910), p. 623.

76. Edward E. Dale, Jr., "The Grasslands of Platt National Park, Oklahoma," *Southwestern Naturalist*, Vol. 4, No. 2 (September 15, 1959), pp. 45–60.

77. Platt Historical District File, Chickasaw National Recreation Area, Sulphur, Okla.

78. "The President's Return," *New York Times* (April 24, 1905), p. 10.

79. "President Cheered at Open-Air Church," *New York Times* (May 1, 1905), p. 1.

80. "Skip," *Washington Post* (April 11, 1907), p. 12.

81. T.R. to Kermit Roosevelt (May 25, 1905).

82. William H. Harbaugh, *The Theodore Roosevelts' Retreat in Southern Albemarle: Pine Knot 1905–1908* (Charlottesville, Va.: Albemarle County Historical Society, 1993).

83. T.R., *A Book-Lover's Holidays in the Open* (New York: Scribner, 1916), app. B, p. 366.

84. Ibid., pp. 96–97.

85. Harbaugh, *The Theodore Roosevelts' Retreat in Southern Albemarle*, p. 4.

86. T.R. to Kermit Roosevelt, June 11, 1905.

87. Ibid.

88. Sylvia Jukes Morris, *Edith Kermit Roosevelt: Portrait of a First Lady* (New York: Coward, McCann, and Geoghegan, 1980), p. 3.

89. T.R. to George Herbert Locke (September 27, 1905).

90. Abernathy, *Catch 'Em Alive Jack*, p. 149.

91. "Sat in President's Chair," *New York Times* (February 10, 1906), p. 1. (Special to the *Times*.)

92. T.R., *Outdoors Pastimes of an American Hunter*, p. 124.

93. T.R., *Outdoor Pastimes of an American Hunter*, p. 287.

94. T.R. to John Burroughs (October 2, 1905).

95. "Strenuous Sport," *New York Times Book Review* (November 4, 1905).

96. Foster Harris, "T.R. and the Great Wolf Hunt," *Oklahoma Today* (Fall 1958), p. 31.

97. Abernathy, *Catch 'Em Alive Jack*, p. 168.

98. Ibid., p. 172.

99. T.R. to John Abernathy (June 4, 1906).

100. Abernathy, *Catch 'Em Alive Jack*, p. 173.

101. T.R. to Clarence Don Clarke (December 8, 1905).

102. T.R. "Wichita Mountains," presidential proclamation (June 2, 1905). See John T. Wolley and Bernard Peters, *The American Presidency Project*. (Online: University of Santa Barbara–California, host.)

103. Caspar Whitney, "The View-Point," *Outing Magazine* (April 1907), p. 102.

104. "American Bison Society," *Saving Wildlife* (September 2007).

105. J. Alden Loring, "The Wichita Buffalo Range" *in Tenth Annual Report of the New York Zoological Society for the Year 1905*, pp. 180–200.

106. "Roosevelt to Pay His Hunt Expenses," *New York Times* (December 6, 1908), p. 1.

107. Betsy Rosenbaum, "Buffalo, or Is It Bison?" Courtesy of Outdoor Recreation Planner, Wichita Mountains Wildlife Refuge Archive (Courtesy of Jeff Rupert.)

108. *Tenth Annual Report of the Bison Society, 1915–1916* (New York: American Bison Society, 1916), pp. 20–22.

109. Sanborn quoted in John G. Mitchell, "The Way We Shipped Off the Buffalo," *Wildlife Conservation* (January–February 1993), pp. 46–50.

110. Elwin R. Sanborn, "An Object Lesson in Bison Preservation: the Wichita National Bison Herd after Five Years," *Zoological Society Bulletin* (Wildlife Protection Number), Vol. 16, No. 57 (May 1913), pp. 990–993. R. B. Thomas, "The Wichita National Forest and Game Preserve" (1936), in Miscellaneous Papers of the W.P.A. Project File, Oklahoma Historical Society Library. Clara Ruth, "Preserves and Ranges Maintained for Buffalo and Other Big Game" (Washington, D.C.: United States Department of Agriculture, Bureau of Biological Survey, Wildlife Research and Management Leaflet BS-95, September 1937), pp. 1–21.

111. Harry B. Candell, "History of the Bison Herd," Wichita Mountain Wildlife Reserve, U.S. Fish and Wildlife Service Archives, Indiahoma, Okla. (March 19, 2009).

112. "Traditional Uses of Bison" (Rapid City, S. Dak.: Intertribal Bison Cooperative and Administration for Native Americans, 2008).

113. Author interview with Jeff Rupert, Wichita Mountains Wildlife Refuge, Cache, Oklahoma.

114. Rush quoted in Tom McHugh, *The Time of the Buffalo* (New York: Knopf, 1972), p. 303.

115. McHugh, *The Time of the Buffalo*, p. 303.

116. James B. Trefethen, *An American Crusade for Wildlife* (New York: Winchester Press, 1975), pp. 95–96.

117. Isenberg, *The Destruction of the Bison*, p. 165.

118. Frank Graham, Jr., "Where Wildlife Rules," *Audubon* (June 2003).

119. Jim Pisarowicz, "Wildlife Management" (April 29, 2006), Wind Cave National Park Archives, Hot Springs, South Dakota.

120. William Temple Hornaday, *Annual Report of the American Bison Society* (1911), p. 32.

121. Shannon Peterson, *Acting for Endangered Species* (Lawrence: University Press of Kansas, 2002), p. 10.

122. Ellenbrook, *Outdoor and Trail Guide to the Wichita Mountains of Southwest Oklahoma*, pp. 20–21.

123. Betsy Rosenbaum, "Buffalo, or Is It Bison?"

124. Caire et al., *Mammals of Oklahoma*, p. 370.

125. "President and Mrs. Bush Host Celebration in Honor of Theodore Roosevelt's 150th Birthday" (October 27, 2008). Transcript. Laura Bush told the story in the East Room, Office of the Press Secretary, Washington, D.C.

126. Stacy A. Cordery, *Alice: Alice Roosevelt Longworth, from White House Princess to Washington Power Broker* (New York: Viking, 2007), p. 456.

22: The National Monuments of 1906

1. T.R. to Kermit Roosevelt, March 11, 1906, quoted in Joseph Bucklin Bishop, *Theodore Roosevelt's Letters to His Children* (New York: Scribner, 1919), pp. 152–153.

2. Ray H. Mattison, "Devils Tower" (National Park Service, 1955), Devils Tower Wyoming Archive. George L. San Miguel, "How Is Devils Tower a Sacred Site to American Indians" (U.S. National Park Service, August 1994).

3. N. Scott Momaday, *The Way to Rainy Mountain* (Albuquerque: University of New Mexico Press, 1976), p. 8.

4. Richard I. Dodge, *The Black Hills* (New York: James Miller, 1876), p. 95.

5. Newton quoted in Raymond J. Demallie, "Introduction," in Mary Alice Gunderson, *Devils Tower: Stories in Stone* (Glendo, Wyo.: High Plains Press, 1988), p. x.

6. Gunderson, *Devils Tower.*

7. Mattison, "Devils Tower."

8. Rebecca Conrad, "John F. Lacey: Conservation's Public Servant" in David Harman, Francis P. McManamon, and Dwight T. Pitcaithley, *The Antiquities Act: A Century of American Archaeology, Historic Preservation, and Nature Conservation* (Tucson: University of Arizona Press, 2006), p. 57.

9. T.R. quoted in Edmund Morris, *Theodore Rex* (New York: Random House, 2001), p. 507.

10. T.R. to John Pitcher (January 8, 1906).

11. John P. Avlon, "TR's Enduring Lessons," *Theodore Roosevelt Association Journal*, Vol. 26, No. 1 (2004), pp. 16–17.

12. Edward Wagenknecht, *The Seven Worlds of Theodore Roosevelt* (New York: Longmans, Green, 1958), p. 17.

13. Simon Winchester, *A Crack in the Edge of the World* (New York: Harper Collins, 2006), p. 16.

14. Suzanne Herel, "San Francisco 1906 Quake Toll Disputed, Supervisors Asked to Recognize Higher Number Who Perished," *San Francisco Chronicle* (January 15, 2005).

15. "Roosevelt Offers Aid," *New York Times* (April 19, 1906), p. 8.

16. "All San Francisco May Burn," *New York Times* (April 19, 1906), p. 1.

17. Elting Morison (ed.), *The Letters of Theodore Roosevelt*, Vol. 5 (Cambridge, Mass.: Harvard University Press, 1952), p. 154.

18. "Remington's Novel," *New York Times* (October 25, 1901), p. BR2.

19. Allen P. Splete and Marilyn D. Splete, *Frederic Remington: Selected Letters* (New York: Abbeville, 1988), p. 359.

20. Frederic Remington to T.R. (Summer 1906).

21. T.R. to Frederic Remington (August 6, 1906).

22. T.R. to John Burroughs (May 5, 1906).

23. T.R. to Owen Wister (April 27, 1906).

24. "President's Threat with Meat Report," *New York Times* (June 5, 1906), p. 1.

25. T.R. to Henry Bryant Bigelow (May 29, 1906).

26. T.R. to George Clement Perkins (June 5, 1906).

27. Hal Rothman, "The Antiquities Act and the National Monuments: A Progressive Conservation Legacy," Culture Resource Management, National Park Service, No. 4 (1999), pp. 16–18.

28. Harmon, McManamon, and Pitcaithley, *The Antiquities Act*, p. 3.

29. Robert W. Righter, "National Monuments to National Parks: The Use of the Antiquities Act of 1906," *Western Historical Quarterly*, Vol. 20, No. 3 (August 1989), pp. 281–301.

30. Samuel P. Hays, *Conservation and the Gospel of Efficiency* (Cambridge, Mass.: Harvard University Press, 1959), p. 3.

31. Harvey Leake, "John Wetherill," http://wetherillfamily.com/john_wetherill.htm.

32. John F. Lacey, "The Petrified Forest National Park of Arizona," *Shield's Magazine*, Vol. I, No. 5 (July 1905).

33. Ibid.

34. Ibid.

35. Conrad, "John F. Lacey: Conservation's Public Servant," p. 61.

36. John F. Lacey, "The Petrified Forest National Park of Arizona."

37. "Elephant Routs G.O.P.," *New York Times* (June 10, 1906), p. 1.

38. "R. B. Roosevelt No Better," *New York Times* (July 12, 1906), p. 1.

39. "Robert B. Roosevelt Ill," *New York Times* (June 11, 1906), p. 1.

40. Ibid.

41. Eric Jay Dolin, *Smithsonian Book of National Wildlife Refuges* (Washington, D.C.: Smithsonian, 2003), p. 58.

42. T.R. to Mark A. Rodgers (June 27, 1906).

43. John Burroughs, *Time and Change* (Boston: Houghton Mifflin, 1912), p. 246.

44. Raymond Esthus, *Theodore Roosevelt and Japan* (Seattle: University of Washington Press, 1967), pp. 132–135.

45. William C. Dennis Memorandum to President Roosevelt, September 10, 1907.

46. Duane A. Smith, *Women to the Rescue* (Durango, Colo.: Durango Herald Small Press, 2005), p. iv.

47. Char Miller, "Landmark Decision: The Antiquities Act, Big Stick Conservation, and the Modern State," in David Harmon, Francis P. McManamon, and Dwight T. Pitcaithley (eds.), *The Antiquities Act* (Tucson: The University of Arizona Press, 2006), pp. 64–78.

48. Smith, *Women to the Rescue*, pp. 54–55.

49. Ibid., p. 56.

50. "Two Roosevelt Bears for the Bronx Zoo; Cubs Caught in Colorado Brought to the Park. Named Teddy B and Teddy G. Presented to the Society by an Admirer of Their Namesake in the Sunday Times," *New York Times* (June 1, 1906), p. 9.

51. Ibid.

52. Phillips Verner Bradford and Harvey Blume, *Ota Benga: The Pygmy in the Zoo* (New York: St. Martin's Press, 1992), p. 177.

53. Jay Maeder, "The Little Man in the Zoo," in *Big Town, Big Time: A New York Epic, 1898–1998* (New York: *New York Daily News*, 1999), p. 23.

54. Ibid.

55. "Benga Tried to Kill; Pygmy Slashes at Keeper Who Objected to His Garb," *New York Times* (September 25, 1906), p. 1.

56. Hornaday quoted in Maeder, "The Little Man in the Zoo."

57. "Negro Ministers Act to Free the Pygmy," *New York Times* (September 11, 1906), p. 2.

58. "African Pygmy's Fate Is Still Undecided," *New York Times* (September 18, 1906), p. 9.

59. Ibid.

60. Bradford and Blume, *Ota Benga*, p. 192.

61. Hornaday quoted ibid., p. 220.

62. Karl W. Gibson, *Saving Darwin: How to Be a Christian and Believe in Evolution* (New York: HarperOne, 2008), p. 73.

63. Mattison, "Devils Tower."

64. Richard West Sellars, *Preserving Nature in the National Parks: A History* (New Haven: Yale University Press, 1997), p. 13.

65. Roy M. Robbins, *Our Landed Heritage: The Public Domain 1776–1936* (Lincoln: University of Nebraska Press, 1962), p. 333.

66. *Ellensburg* (Washington) *Dawn* (October 18, 1902), reprinted ibid.

67. T.R. to Gifford Pinchot (August 24, 1906).

68. Ibid.

69. Ibid.

70. Recommendations reported by W. A. Richards, F. H. Newell and Gifford Pinchot, *Annual Reports of the Interior*

(Washington, D.C.: Government Printing Office, 1905).

71. Deanne Stillman, *Mustang: The Saga of the Wild Horse in the American West* (Boston, Mass.: Houghton Mifflin, 2008), p. 239.

72. *New York Times* (November 5, 1906), and *Washington Post* (November 5, 1906).

73. T.R. to Kermit Roosevelt (November 4, 1906), and T.R. to William Sewall (January 2, 1907).

74. T.R., "Small Country Neighbors," *Scribner's Magazine*, Vol. 42, No. 4 (Ocober 1907).

75. T.R. to William S. Harvey (September 16, 1906).

76. Robbins, *Our Landed Heritage*, p. 336.

77. William T. Hornaday, "John F. Lacey," *Annals of Iowa*, XI, No. 1 (Des Moines, Iowa, April 1913, 3D series), pp. 582–584.

78. T.R. to Kermit Roosevelt (November 20, 1906).

79. H. E. Anthony, "Panama Mammals Collected in 1914–1915," *Bulletin of American Museum of Natural History*, Vol. 35 (New York: American Museum of Natural History, 1916), pp. 357–376.

80. T.R. to Kermit Roosevelt (November 23, 1906).

81. T.R. to Kermit Roosevelt (December 5, 1906).

82. Charles F. Lummis, "Strange Corners of our Country," *St. Nicholas* (1891).

83. Josh Protas, *A Past Preserved in Stone: A History of Montezuma Castle National Monument* (Tucson, Ariz.: Western National Parks Association, 2002).

84. John B. Jackson, *A Sense of Place, A Sense of Time* (New Haven: Yale University Press, 1994), pp. 24–25.

85. John F. Lacey, "The Petrified Forest National Park of Arizona."

23: The Prehistoric Sites of 1907

1. "James Rudolph Garfield," *Washington Post* (April 1, 1907), p. E4 from the *Saturday Evening Post.*

2. Ibid.

3. George Bird Grinnell, *Blackfoot Lodge Tales: The Story of a Prairie People* (New York: Charles Scribner's Sons, 1892); *By Cheyenne Campfires* (New Haven: Yale University Press, 1926); *Pawnee Hero Stories and Folk Tales: With Notes on the Origin, Customs, and Character of the Pawnee People* (New York: Forest and Stream, 1889).

4. T.R. to James Garfield (February 1, 1908).

5. Roy M. Robbins, *Our Landed Heritage: The Public Domain, 1776–1970* (second edition) (Lincoln: University of Nebraska Press, 1976), pp. 346–347.

6. Ibid., p. 347.

7. T.R. Memorandum (March 2, 1907).

8. Terry Richard, "Teddy Roosevelt and North Dakota," *The Oregonian*, August 10, 2003 (put online March 11, 2008).

9. William H. Harbaugh, "The Theodore Roosevelts' Retreat in Southern Albemarle: Pine Knot, 1905–1908," *Magazine of Albemarle County History*, Vol. 51 (1933), p. 29.

10. Charles F. Clarke, *Theodore Roosevelt and the Great Adventure* (Des Moines: Garner Publishing Co., 1959), p. 109.

11. T.R., *An Autobiography* (New York: Macmillan, 1913), pp. 419–420.

12. T.R. to John Burroughs (March 12, 1907).

13. *Congressional Record* (March 10, 1914), pp. 4, 633.

14. T.R., "The People of the Pacific Coast," *Outlook* Vol. 99, No. 4 (September 23, 1911).

15. T.R. to James Wilson (June 7, 1907).

16. *Centennial* (Wyoming) *Post* (March 30, 1908).

17. T.R. to Kermit Roosevelt (March 31, 1907).

18. Ellen Glasgow, *The Woman Within* (New York: Harcourt Brace, 1954), pp. 208–209.

19. Ron Chernow, *The Titan* (Random House: New York, 1998), p. 435.

20. David Nasaw, *The Chief: The Life of William Randolph Hearst* (Boston, Mass.: Houghton Mifflin, 2000), pp. 104–149.

21. "The History of Arbor Day," Arbor Day Foundation, www.arborday.org/arborday/history.cfm (Archive in Nebraska City, Nebraska).

22. "President for Trees," *Washington Post* (April 15, 1907), p. 1.

23. Ibid.

24. "Roosevelt to Children," *New York Times* (April 15, 1907), p. 5.

25. T.R. to James Wilson (June 7, 1907).

26. T.R. quoted in Edmund Morris, *Theodore Rex* (New York: Random House, 2001), p. 507.

27. T.R. to Kermit Roosevelt (June 5, 1907) and T.R., "Small Country Neighbors," *Scribner's Magazine*, Vol. 42, No. 4 (October 1907).

28. Andrew D. Blechman, *Pigeons: The Fascinating Saga of the World's Most Revered and Reviled Bird* (New York: Grove, 2006), pp. 52–53.

29. Morris, *Theodore Rex*, pp. 490–491.

30. Harbaugh, "The Theodore Roosevelts' Retreat in Southern Albemarle," pp. 31–32.

31. W. B. Mershon, *The Passenger Pigeon* (New York: Outing, 1907).

32. Gary Scharnhorst, "Introduction," in Frederic Remington, *John Ermine of the Yellowstone* (Lincoln: University of Nebraska Press, 2008), pp. v–xii.

33. *Independent* (February 26, 1903), p. 506.

34. T.R. to Frederic Remington (July 17, 1907).

35. David Starr Jordan, "Personal Glimpses of Theodore Roosevelt," *Natural History*, Vol. 19 (January 19, 1919), p. 16.

36. T.R., *A Book-Lover's Holidays in the Open* (New York: Charles Scribner's Sons, 1916), p. 42.

37. "How Old Is Cinder Cone?—Solving a Mystery in Lassen Volcanic National Park" (Washington D.C.: National Park Service, U.S. Department of the Interior, 2009).

38. Ibid.

39. "Nature and Science" (Lassen Volcanic National Park, National Park Service).

40. "Theodore Roosevelt before National Editorial Association, Jamestown, Virginia (June 10, 1907)," *Presidential Addresses and State Papers*, Vol. 6 (New York: The Review of Reviews Company, 1910), pp. 1310–1311.

41. T.R. to Archie Roosevelt (September 21, 1907).

42. Mark Twain, *Life on the Mississippi* (New York: Harper, 1917), p. 236.

43. T.R. to John Parker (March 5, 1913).

44. T.R., "In the Louisiana Canebrakes," *Scribner's Magazine*, Vol. 43 (January–June 1908).

45. Stephen E. Ambrose and Douglas G. Brinkley, *The Mississippi and the Making of a Nation* (Washington D.C.: National Geographic Books, 2003), pp. 132–149.

46. *Louisiana Federal Writers' Project* (Louisiana State University, 1941), p. 593.

47. T.R., "Our Vanishing Wild Life," *Outlook* (January 25, 1913).

48. Minor Ferris Buchanan, *Holt Collier* (Jackson, Mississippi: Centennial Press, 2002), p. 189.

49. Buchanan, *Holt Collier*, p. 189.

50. T.R., "In the Louisiana Canebrakes."

51. Dutch Salmon, "Mountain Men of the Gila," www.southernnewmexico.com (January 11, 2003).

52. T.R., "In the Louisiana Canebrakes."

53. Ibid.

54. Ibid.

55. Harris Dickson, "When the President Hunts," *Saturday Evening Post* (August 8, 1908), p. 24.

56. T.R., "In the Louisiana Canebrakes."

57. Ibid.

58. Ibid.

59. Ibid.

60. R. L. Wilson, *Theodore Roosevelt, Outdoorsman* rev. ed. (Agoura, Calif.: Trophy Room, 1994), p. 210.

61. William M. Gibson, *Theodore Roosevelt among the Humorists: W. D. Howells, Mark Twain, and Mr. Dooley* (Knoxville: University of Tennessee Press, 1980), p. 24.

62. Bernard DeVoto (ed.), *Mark Twain in Eruption* (New York: Harper, 1940), pp. 10–18.

63. T.R. to George Otto Trevelyan (January 22, 1906).

64. Chris Darimont quoted in Anne Minard, "Hunters Speeding Up Evolution of Trophy Prey," *National Geographic News* (January 12, 2009).

65. Karyl Whitman, Anthony M. Starfield, Henley S. Quadling and Craig Parker, "Sustainable Trophy Hunting of African Lions," *Nature*, 428 (March 11, 2004), pp. 175–178. Also see Lily Huang, "It's Survival of the Weak and Scrawny," *Newsweek* (January 12, 2009).

66. Cornelia Dean, "Research Ties Human Acts to Harmful Rates of Species Evolution," *New York Times* (January 13, 2009), p. D3.

67. Interview with Chris Darimont (January 15, 2009).

68. DeVoto (ed.), *Mark Twain in Eruption*, p. 49.

69. "Gila Cliff Dwellings: An Administrative History" (Washington, D.C.: Gila Cliff Dwellings National Monument, National Park Service, Department of the Interior, 2009).

70. Salmon, "Mountain Men of the Gila."

71. Aldo Leopold, *A Sand County Almanac* (New York: Oxford University Press, 1949), p. 130.

24: Mighty Birds

1. Herbert Keightley Job Collection, Watkinson Library, Trinity College, Hartford, Connecticut. The collection includes more than 400 letters, 326 glass plate slides, and fourteen notebooks.

2. Herbert K. Job, *Wild Wings: Adventures of a Camera Hunter among the Larger Wild Birds of North America on Sea and Land* (Boston, Mass.: Houghton Mifflin, 1905), p. 44.

3. "Mr. H. K. Job on Bird Photography," *Harvard Crimson* (April 27, 1906).

4. Herbert K. Job, "Bird Castles in the Rocks," *Outing Magazine* (June 1909), Vol. 54, No. 3.

5. T.R. to Herbert K. Job, quoted in "The Viewpoint: A Humane Sportsman and a Gentle Naturalist," *Outing Magazine*, Vol. 54, No. 3 (June 1909).

6. Job, *Wild Wings*, pp. 54–55.

7. "In and Around Boston," *Congregationalist and Christian World*, Vol. 90, No. 39 (September 30, 1905).

8. Ibid. Also "Wild Wings at the Minister's Meeting," *Congregationalist and Christian World*, Vol. 90, No. 30 (September 30, 1905), p. 442.

9. T.R. to Herbert K. Job (introductory letter to *Wild Wings*). Also see Herbert K. Job, *Among the Water-Fowl* (New York: Doubleday, Page, 1905).

10. T.R., "The People of the Pacific Coast," *Outlook*, Vol. 99, No. 4 (September 23, 1911).

11. "Where Birds Are Safe from Guns," *Friends' Intelligencer*, Vol. 66, No. 27 (July 3, 1909).

12. John Muir, *Steep Trails* (Boston, Mass., and New York: Houghton Mifflin, 1918), p. 146.

13. W. C. Henderson, "1885—Fiftieth Anniversary Notes—1935," Vol. 16, Nos. 4–6, *Survey* (April–June 1935), p. 65.

14. Ernest Harold Baynes, *Wild Bird Guests: How to Entertain Them* (New York: Dutton, 1915), pp. 40–41.

15. Ira N. Gabrielson, *Wildlife Refuges* (New York: Macmillan, 1943), pp. 3–8.

16. Aldo Leopold, *Game Management* (New York: Scribner, 1933), p. 17.

17. "Annual Report for 1907," quoted in Ira N. Gabrielson, *Wildlife Refuges*, p. 10.

18. *United States Coast Pilot: Pacific Coast: California, Oregon, and Washington* (Washington, D.C.: Government Printing Office, 1917), p. 145.

19. T.R., *A Book-Lover's Holidays in the Open* (New York: Scribner, 1916), pp. 365–368.

20. Roger Tory Peterson, "Foreword," in Worth Mathewson, *William L. Finley: Pioneer Wildlife Photographer* (Corvallis: Oregon State University Press, 1986), pp. 1–2.

21. "Three Arch Rocks Refuge Celebrates Centennial," Oregon Coast National Wildlife Refuge, Complex Archive news release (September 27, 2007).

22. Robin W. Doughty, *Feather Fashions and Bird Preservation*, pp. 19–20.

23. Robert L. Fischman, *The National Wildlife Refuges: Coordinating a Conservation System through Law* (Washington, D.C.: Island, 2003), p. 212.

24. Dallas Lore Sharp, *Sanctuary! Sanctuary!* (New York: Harper, 1971), pp. 19–20.

(Originally printed by Dallas Lore Sharp in 1926.)

25. "Flattery Rocks National Wildlife Refuge," U.S. Fish and Wildlife Information Sheet.

26. Berthold Seemann, F.L.S., "*Narrative of the Voyage of H.M.S. Herald: During the Years 1845–1851, under the Command of Captain Henry Kellett, in Two Volumes*, Vol. 1 (London: Beeve, Henrietta Street, Covent Garden, 1853).

27. Author interview, Kevin Ryan, Clallam County, Washington (September 30, 2009).

28. Roy Crandall, "To Give the Birds a Refuge," *Technical World Magazine*, Vol. 11 (April 1909).

29. "What's in a Refuge Name?" *Fish and Wildlife News*, Special Edition (December 1978–January 1979), p. 8.

30. William Burt, *The Disappearing Eden* (New Haven: Yale University Press, 2001), pp. 102–107.

31. William L. Finley, "Federal Bird Reservations," *Nature Magazine* (May 1926).

32. E.W.S., "Save the Birds," *Outlook* (February 27, 1909).

33. C. Hart Merriam, *Report of the Chief of the Bureau of Biological Survey 1909* (Washington, D.C.: Government Printing Office, 1909), p. 16.

34. Warren Zeiller, *Introducing the Manatee* (Gainesville: University Press of Florida, 1992), p. 116.

35. T.R., "The Conservation of Wildlife," *Outlook* (January 20, 1915).

36. Clara Barrus, *The Life and Letters of John Burroughs*, Vol. 2 (Boston: Mass.: Houghton Mifflin, 1925), p. 42.

37. Alfred Henry Garrod, "Notes on the Manatee (Manatus Americanous) Recently Living in Society's Gardens," in *The Collected Scientific Papers of the Late Alfred Henry Garrod*, W. A. Forbes (ed.) (London: R. H. Porter, 1881) pp. 303–313. Outram Bangs, "The Present Standing of the Florida Manatee, Trichechus Latirostris (Harlan) in the Indian River Waters," *American Naturalist*, Vol. 29 (September 1895), p. 345.

38. Robert Barnwell Roosevelt, *Florida and the Game Water-Birds of the Atlantic Coast and the Lakes of the United States* (New York: Orange Judd, 1884), p. 10.

39. Zeiller, *Introducing the Manatee*, p. 98; and "Along the Florida Reef," *Harper's New Monthly Magazine* (1876).

40. Marston Bates, *The Forest and the Sea* (New York: Knopf, 1960), p. 31.

41. Carl Safina, *Voyage of the Turtle: In Pursuit of the Earth's Last Dinosaur* (New York: Holt, 2006), p. 61.

42. Kathryn Hall Proby, *Audubon in Florida* (Coral Gables: University of Miami Press, 1974), p. 75.

43. "Dry Tortugas National Park: Superintendent Annual Narrative Report (fiscal year 2004)" and "Park Vision" (Archives) 2007 Timeline, Dry Tortugas, Fla.

44. Rachel Carson, *The Edge of the Sea* (Boston, Mass.: Houghton Mifflin, 1955), p. 204.

45. Mark H. Lytle, *The Gentle Subversive* (New York: Oxford University Press, 2007), p. 79.

46. Job, *Wild Wings*, p. 87.

47. Alexander Sprunt, Jr., "The Tern Colonies of the Dry Tortugas Keys," *Auk*, Vol. 65, No. 1 (January 1948), pp. 5–6.

48. James W. Porter and Karen G. Porter, *The Everglades, Florida Bay, and Coral Reefs of the Florida Keys* (Boca Raton: CRC, 2002), pp. 829–830. See also Florida Department of Environmental Protection Report (2007).

49. Elizabeth Weise, "Scientists: Global Warming Could Kill Off Reefs by 2050," *USA Today* (December 13, 2007), p. A1.

50. *Bird-Lore*, October 1, 1907, Vol. IX, No. 5.

51. Pine Island Files, Mission Statements, Pine Island National Wildlife Refuge, Sanibel, Fla. Today Pine Island is part of the J. N. "Ding" Darling National Wildlife Refuge Complex, named after a famous political cartoonist who had a penchant for conservation.

52. Dr. William Wilbanks, *Forgotten Heroes: Police Officers Killed in Early Florida, 1840–1925* (Paducah, Ky.: Turner, 1998), p. 98.

53. Michael Wisenbaker, "The Hermit Warden of Cayo Pelau," *Florida Monthly* (April 2005), pp. 26–27.

54. B. S. Bowdish, "Ornithological Miscellany from Audubon Wardens," *Auk*, Vol. 26, No. 1 (January 9, 1909).

55. Columbus McLeod, "White Pelicans," *Auk*, Vol. 26 (January 1909).

56. "Public Opinion," E.W.S., "Save the Birds," *Outlook* (February 27, 1909).

57. Frank M. Chapman, "A Case in Point," *Bird-Lore*, Vol. 18 (1916).

58. Wisenbaker, "The Hermit Warden of Cayo Pelau," p. 26.

59. Lindsey Williams, "Audubon Warden McLeod Murdered for Feathers on Ladies Hats," *Charlotte Sun Herald*, April 26, 1992 and Joe Crankshaw, "Warden's Valor Saved Egrets From Extinction," *Miami Herald Tropic Magazine*, June 8, 1992.

60. *St. Augustine Record*, quoted in Lindsey Williams and U. S. Cleveland, *Our Fascinating Past* (Punta Gorda, Fla.: Charlotte Harbor Area Historical Society, 1993), p. 211.

61. E.W.S., "Save the Birds," *Outlook* (February 27, 1909).

62. Raymond Ditmars, *The Reptile Book* (New York: Doubleday Page, 1908).

63. L. N. Wood, *Raymond L. Ditmars: His Exciting Career with Reptiles, Animals, and Insects* (New York: Julian Messner, 1944), p. 132.

64. T.R. quoted in *University of State of New York Bulletin* (March 1, 1914), pp. 39–44.

65. T.R. to Raymond Ditmars ([n.d.] 1907).

66. Ibid. This letter is quoted in Paul Russell Cutright, *Theodore Roosevelt: The Naturalist* (New York: Harper, 1956), p. 143.

67. T.R., "Notes on Florida Turtles," *American Museum Journal*, Vol. 17, No. 5 (May 1917).

68. Don Arp, Jr., "Hunting the Dragons: TR and the World's Crocodilians," *Theodore Roosevelt Association Journal*, Vol. 24, No. 4 (2001), pp. 5–9.

69. John Mortimer Murphy, "Alligator Shooting in Florida," *Outing Magazine* (1899).

70. T.R., *Outdoor Pastimes of an American Hunter* (New York: Macmillan, 1902), pp. 362–363.

71. Joseph Bucklin Bishop (ed.), *Theodore Roosevelt's Letters to His Children* (New York: Scribner, 1914), p. 184.

72. T.R., *African Game Trails* (New York: Charles Scribner's Sons, 1910), p. 341.

73. Michael Grunwald, *The Swamp* (New York: Simon & Schuster, 2006), p. 128.

74. "Dr. John C. Gifford, Forestry Authority," *New York Times* (June 27, 1949), p. 27.

75. Nathaniel Southgate Shaler, *The United States of America* (New York: Appleton, 1894), p. 278.

76. "John Clayton Gifford," in *Reclaiming the Everglades: South Florida's Natural History, 1884–1934*, Everglades Archival Library and Museum, Fla.

77. Bureau of Reclamation, *Reclamation Project Data* (Washington, D.C.: U.S. Department of the Interior, 1948).

78. I. F. Eldredge, "Fire Problem on the Florida Native Forest," *Proceedings of the Society of American Foresters* (Washington, D.C.: Judd and Detweiller, 1911), pp. 166–168.

79. Thomas Barbour, *The Vanishing Eden: A Naturalist's Florida* (Boston, Mass.: Little, Brown, 1944).

80. T.R., "Notes on Florida Turtles," *American Museum Journal*, Vol. 17 (1917).

81. Oliver P. Pearson, "The Metabolism of Hummingbirds," *The Condor*, Vol. 52, No. 4 (July–August, 1950), pp. 145–152.

82. "U.S. Fish and Wildlife Overview/History," Island Bay National Wildlife Refuge Archive, Sanibel, Fla. (April 9, 2009).

83. On October 23, 1970, President Richard Nixon, recognizing how exceptional the islands were, declared the refuge a Wilderness Area. Thus no tourists are allowed to visit them.

84. "U.S. Fish and Wildlife Overview/History."

85. Mark V. Barrow, Jr., *A Passion for Birds: American Ornithology after Audubon* (Princeton, N.J.: Princeton University Press, 1997), p. 142.

86. William Dutcher to William Moody (July 2, 1903), in *Auk*, Vol. 21 (January 1904).

87. Herbert K. Job, Report to William Dutcher (1904).

88. Hermann Hagedorn and Sidney Wallach, *A Theodore Roosevelt Round-Up*

(New York: Theodore Roosevelt Association, 1958), p. 64.

89. For Darling's childhood experiences I consulted the Darling Papers, Special Collections, University of Iowa, Iowa City.

90. David L. Lendt, *Ding: The Life of Jay Norwood Darling* (Ames: Iowa State University Press, 1979), pp. 3–17.

91. Joseph P. Dudley, "Jay Norwood 'Ding' Darling: A Retrospect," *Conservation Biology*, Vol. 7, No. 1 (March 1993), pp. 200–203. (This article includes two of Darling's cartoons.)

92. Lendt, *Ding*, p. 21.

93. Worth Mathewson, *William L. Finley: Pioneer Wildlife Photographer* (Corvallis: Oregon State University Press, 1986), p. 9.

94. Eric Jay Dolan and Bob Pumaine, *The Duck Stamp Story* (privately published), pp. 34–77.

95. Ding Darling U.S. Fish and Wildlife Files. Sanibel, Florida U.S. Fish and Wildlife, December 18, 2008.

96. Ibid.

97. "Ding Darling National Wildlife Center," *Duck Report*, No. 32 (2008).

98. Author interview with Ding Darling's grandson, Christopher D. Koss, Key Biscayne, Fla. (November 4, 2007).

99. Elting Morison, "Introduction," *The Letters of Theodore Roosevelt: The Big Stick: 1905–1907*, Vol. V (Cambridge, Mass.: Harvard University Press, 1954), p. xiv.

100. Wendell Berry, "The Peace of Wild Things," in *Collected Poems: 1957–1982* (San Francisco: North Point, 1985). Also see Rodger Shlickeisen, "Finding Solace with the Wood Drake," *Fish and Wildlife News* (Spring 2008), p. 45.

101. Mathewson, *William L. Finley*, p. 57.

102. "Klamath Basin National Wildlife Refuges," Mission Statement, Klamath Basin National Wildlife Reservation Archive, Tulelake, Calif.

103. William L. Finley, *American Birds* (New York: Scribner, 1907).

104. Mathewson, *William L. Finley*, pp. 57–84.

105. William L. Finley, "Among the Pelicans," *The Condor*, Vol. 9, No. 2 (March–April 1907); William L. Finley, "The Grebes of Southern Oregon," *The Condor*, Vol. 9, No. 4 (July–August 1907). William L. Finley, "Among the Gulls on Klamath Lake," *The Condor*, Vol. 9, No. 1 (January–February 1907).

106. Finley, "The Grebes of Southern Oregon."

107. William Kittredge, *Balancing Water: Restoring the Klamath Basin* (Berkeley: University of California Press, 2000), pp. 76–79.

108. Finley, "Among the Pelicans," p. 40.

109. Mathewson, *William L. Finley*, p. 9.

110. Finley quoted in *National Geographic* (August 1923).

111. "Oregon Governor Oswald West," National Governors Association, Biography File, Washington, D.C.

112. T.R., "The People of the Pacific Coast," *Outlook*, Vol. 99, No. 4 (September 23, 1911).

113. Ibid.

114. "Rogue Goes to the Birds," *Rogue Wire Service Report* (March 28, 2008).

115. Butcher, *America's National Wildlife Refuges*, pp. 531–532.

25: The Preservationist Revolution of 1908

1. Donald Worster, *A Passion for Nature* (Oxford: Oxford University Press, 2008), p. 421.

2. Terry Gifford, *Reconnecting with John Muir: Essays in Post-Pastoral Practice* (Athens, Ga.: University of Georgia Press, 2006) p. 42.

3. Roderick Nash, *Wilderness and the American Mind* (New Haven, Conn.: Yale University Press, 1967), pp. 172–173.

4. William Kent to John Muir (January 16–17, 1908). *John Muir Papers* (Microfilm Edition of Ronald H. Limbaugh and Kristen E. Lewis (eds.), *John Muir Papers*, Reel 17, Frame 9495–9500).

5. Galen Clark, "The Big Trees of California" (1907), Yosemite National Park Archive, Calif.

6. John Muir to William Kent (January 14, 1908). *John Muir Papers*, (Reel 17, Frame 9487).

7. T.R. to William Kent (January 22, 1908), Muir Woods National Monument Archive, Mill Valley, California.

8. William Kent to T.R. (January 30, 1908), Muir Woods National Monument Archive.

9. T.R. to Douglas Robinson (January 10, 1908).

10. Sandburg quoted in Stephen J. Pyne, *How the Canyon Became Grand* (New York: Viking, 1998), p. 159.

11. Pyne, p. 158.

12. Robert H. Webb, *Grand Canyon: A Century of Change—Rephotography of the 1889–1890 Stanton Expedition* (Tucson: University of Arizona Press, 1996), p. 208.

13. Hal Rothman, *Preserving Different Pasts: The American National Monuments* (Urbana: University of Illinois Press, 1989), pp. 16–18.

14. Hal Rothman, "The Antiquities Act and National Monuments: A Progressive Conservation Legacy," *CRM Bulletin*, Vol. 22, No. 4 (1999), pp. 16–18.

15. Pyne, *How the Canyon Became Grand*, p. 111.

16. T.R., *A Book-Lover's Holidays in the Open* (New York: Scribner, 1916), pp. 96–97.

17. William M. Gibson, *Theodore Roosevelt among the Humorists.* (Knoxville: University of Tennessee Press, 1989), p. 34.

18. Webb, *Grand Canyon: A Century of Change*, p. 208.

19. Ibid.

20. T.R., *A Book-Lover's Holidays in the Open*, p. 28.

21. Address by Robert Glenn to the National Conference of Governors (May 13–15, 1908), published in *Proceedings of a Conference of Governors* (Washington, D.C.: Government Printing Office, 1909), p. 121.

22. Patricia O'Toole, *When Trumpets Call: Theodore Roosevelt after the White House* (New York: Simon and Schuster, 2005), p. 228.

23. David H. Dickason, "David Starr Jordan as a Literary Man," *Indiana Magazine of History*, Vol. 38 (1941), pp. 343–358; and David Starr Jordan, *Evolution and Animal Life: An Elementary Discussion of Facts, Processes, Laws, and Theories Relating to the Life and Evolution of Animals* (New York: Appleton, 1907).

24. Char Miller, "Landmark Decision: The Antiquities Act, Big Stick Conservation, and the Modern State," in David Harmon, Francis P. McManamon, and Dwight T. Pitcaithley (eds.), *The Antiquities Act: A Century of American Archeology, Historic Preservation, and Nature Conservation* (Tucson: University of Arizona Press, 2006), pp. 64–78.

25. Ron Chernow, *Titan: The Life of John D. Rockefeller, Sr.* (New York: Random House, 1998), p. 435.

26. T.R. to Kermit Roosevelt (February 23, 1908).

27. "Jewel Cave National Monument," *National Park Service Archive*, Jewel Cave, South Dakota.

28. Gail Evans-Hatch and Michael Evans-Hatch, *Place of Passages: Jewel Cave National Monument Historic Resource Study* (Omaha, Neb.: Midwestern Regional National Park Service, 2006), pp. 173–177.

29. Ibid., p. 4.

30. "Jewel Cave National Monument, South Dakota, by the President of the United States, A Proclamation," Box 1, Jewel Cave National Monument Archives, Mount Rushmore National Monument.

31. Ibid.

32. In fact, the Biological Survey called for the wholesale extermination of English sparrows, as they had become a menace to fruit trees and other crops from coast to coast. *Report of the Chief of the Bureau of the Biological Survey for 1908* (Washington, D.C.: Government Printing Office, 1908), p. 9.

33. T.R. to Dr. C. Hart Merriam (March 15, 1908).

34. T.R. to Theodore Roosevelt, Jr. (May 23, 1908).

35. J. C. Kerbis Peterhans and T. P. Gnoske, "Man-Eaters of Tsavo," *Journal of East African Natural History*, Vol. 90 (2001), pp. 1–40.

36. T.R. to Lieutenant Colonel John Henry Patterson (March 20, 1908).

37. T.R., *African Game Trails* (New York: Scribner, 1909), p. ix.

38. "The Colossal Natural Bridges of

Utah," *National Geographic*, Vol. 15 (1904), pp. 367–369. (Author unknown.)

39. There is some controversy over the use of the word "Anasazi." While recognizing its limitations, I have chosen it both for its brevity and because there is no agreed-on alternative.

40. T.R., *A Book-Lover's Holidays in the Open*, pp. 39–62.

41. Ibid., pp. 53–54.

42. T.R. at the creation of Natural Bridges National Monument (April 16, 1908), Natural Bridges National Monument Archive, Lake Powell, Utah.

43. G. Michael McCarthy, *Hour of Trial: The Conservation Conflict in Colorado and the West, 1891–1907* (Norman: University of Oklahoma Press, 1977).

44. George F. Kunz, "The Preservation of Scenic Beauty" in *Proceedings of the Conference of Governors*, pp. 408–419.

45. T.R. to Cecil Arthur Spring-Rice (April 11, 1908).

46. Edward J. Renehan, Jr., *John Burroughs: An American Naturalist* (Post Mills, Vt.: Chelsea Green, 1992), p. 250.

47. T.R. to John Burroughs (June 29, 1907), *Theodore Roosevelt Papers*, Reel 346.

48. John Burroughs, "With Roosevelt at Pine Knot," *Outlook* (May 25, 1921).

49. John Burroughs, *Camping and Tramping with Roosevelt* (Boston, Mass., and New York: Houghton Mifflin, 1906), pp. 102–103.

50. Renehan, *John Burroughs*, p. 250; Lifton Johnson, *John Burroughs Talks* (Boston and New York: Houghton Mifflin, 1922), pp. 237–241; and Clara Barrus (ed.), Burroughs, *The Life and Letters of John Burroughs*, Vol. 2 (Boston: Houghton Mifflin Company, 1925), p. 363.

51. William Harbaugh, "The Theodore Roosevelts' Retreat in Southern Albemarle, Pine Knot 1905–1908," *Magazine of Albermarle Country History*, Vol. 51, 1993, pp. 37–41.

52. Johnson, *John Burroughs Talks*, p. 290.

53. T.R. to Archie Roosevelt (May 10, 1908) in Joseph Bucklin Bishop (ed.), *Theodore Roosevelt's Letters to His Children* (New York: Charles Scribner's Sons, 1919), pp. 226–227.

54. Ibid.

55. Paul Russell Cutright, *Theodore Roosevelt: The Naturalist* (New York: Harper, 1956), p. 180.

56. Charles F. Clark, *Theodore Roosevelt and the Great Adventure* (Des Moines, Iowa: Garner, 1959), p. 111.

57. T.R. quoted in ibid., p. 112.

58. T.R., *Address to the National Governors' Conference, May 13–15* (Washington, D.C.: Government Printing Office, 1909), p. 8.

59. Address of Edwin L. Norris in *Proceedings of the Conference of Governors* (Washington, D.C.: Goverment Printing Office, 1909), pp. 172–173.

60. L. O. Howard, *Fighting the Insects* (New York: Macmillan, 1933), pp. 239–240.

61. T.R. to Theodore Elijah Burton (June 8, 1908).

62. T.R. to Archie Roosevelt (May 17, 1908), in Bishop (ed.), *Theodore Roosevelt's Letters to His Children*, p. 228.

63. T.R. to Frank M. Chapman (May 10, 1908).

64. *Report to the Chief of the Biological Survey for 1907* (Washington, D.C.: Government Printing Office, 1908), p. 9.

65. T.R. to Frank M. Chapman (June 7, 1908).

66. Charles Herner, *The Arizona Rough Riders* (Prescott, Ariz.: Scharlot Hall Museum, 1998), p. 222.

67. Johnson, *John Burroughs Talks*, p. 291.

68. Michael F. Anderson, *Polishing the Jewel: An Administrative History of the Grand Canyon* (Grand Canyon, Ariz,: Grand Canyon Association, 2000), pp. 15–108; and Stephen R. Whitney, *A Field Guide to the Grand Canyon* (Seattle: The Mountaineers, 1996), pp. 53–65.

69. T.R., *A Book-Lover's Holidays in the Open*, p. 5.

70. "Clinton G. Smith," in *Biographical Record of the Graduates and Former Students in the Yale Forestry School* (New Haven, Conn.: Yale Forestry School, 1913), p. 86.

71. T.R., "Forestry and Foresters," speech before the Society of American Foresters, March 26, 1903 (U.S. Department of

Agriculture, Bureau of Forestry, Circular No. 25, June 11, 1903).

72. T.R. to Henry Cabot Lodge (June 24, 1908).

73. Elting Morrison (ed.), *The Letters of Theodore Roosevelt*, Vol. 7 (Cambridge, Mass.: Harvard University Press, 1954), p. 4. (Editorial footnote.)

74. T.R. to Henry Fairfield Osborn (August 5, 1908).

75. T.R. to Henry Cabot Lodge (August 18, 1908).

76. Jack London, "The Other Animals," *Collier's* (September 5, 1908); and "London Answers Roosevelt," *New York Times* (August 31, 1908), p. 7.

77. T.R. to Mark Sullivan (September 9, 1908).

78. London, "The Other Animals."

79. Jonathan Swift, *Gulliver's Travels: The Voyages to Lilliput and Brobdingnag* (New York: American Book Company, 1914), p. 129.

80. T.R. to Professor L. H. Bailey (August 4, 1908).

81. T.R., "Country Life Commission," *Century Magazine* (October 1913).

82. T.R. to Herbert Myrick (September 10, 1908).

83. Ibid.

84. T.R. to Herbert Mynick (September 10, 1908).

85. T.R. to William Jennings Bryan (September 27, 1908).

86. T.R. to William Kent (September 28, 1908).

87. "Attacks Gifford Pinchot," *New York Times* (October 1, 1908), p. D3.

88. Gould, *The Presidency of Theodore Roosevelt*, p. 289.

89. T.R. to John Raleigh Mott (October 12, 1908).

90. T.R. to Theodore Roosevelt, Jr. (November 27, 1908).

91. T.R. to George Otto Trevelyan (December 1, 1908).

92. T.R. to John St. Lee Strachey (February 22, 1907).

93. T.R. to John Hay (May 22, 1903).

94. T.R. to Whitelaw Reid (December 4, 1908).

95. Dave Cooper, "Wild Hike Reveals Right Tuff in Wheeler Geologic Area," *Denver Post* (September 24, 2006).

96. T.R. to Robert Underwood Johnson (December 17, 1908).

97. T.R., *An Autobiography*, p. 424.

98. Eric Jay Dolin, *Smithsonian Book of Natural Wildlife Refuges* (Washington, D.C.: Smithsonian Institution Press, 2003), pp. 60–61.

99. T.R. to Sydney Brooks (November 20, 1908).

26: Dangerous Antagonist

1. T.R. to William Howard Taft (December 31, 1908).

2. Nathan Miller, *Theodore Roosevelt: A Life* (New York: Morrow, 1992), p. 483.

3. Ibid., p. 485.

4. M. Nelson McGeary, *Gifford Pinchot: Forester-Politician* (Princeton, N.J.: Princeton University Press, 1960), p. 115.

5. Miller, *Theodore Roosevelt: A Life*, p. 490.

6. T.R., "A Hunter-Naturalist in Europe and Africa," *Outlook*, Vol. 99, No. 3 (September 16, 1911).

7. T.R. to Winston Churchill (January 6, 1909).

8. T.R. to Kermit Roosevelt (January 14, 1909).

9. T.R. to Kermit Roosevelt (January 9, 1909).

10. Henry Litchfield West, "The Incoming Taft Administration," *Forum* (March 1909).

11. T.R. to Theodore Roosevelt, Jr. (January 31, 1909).

12. "The Northwestern Hawaiian Islands: 100 Years of Presidential Protection," Marine Conservation Biology Institute (Washington, D.C.). (Pamphlet, 2007.)

13. William Alanson Bryan, *Natural History of Hawaii: Being an Account of the Hawaiian People, the Geology and Geography of the Islands, and the Native and Introduced Plants and the Animals of the Group* (Honolulu: Hawaiian Gazette, 1915), p. 93.

14. Mark Twain, *Roughing It*, Vol. 1 (Hartford, Conn.: American, 1871), pp. 265–266.

15. Nona Beamer, *Nā Mele Hula: A*

Collection of Hawaiian Hula Chants (Lã' cie, Hawaii: Pacific Institute Press, 1987), p. 46.

16. Bryan, *Natural History of Hawaii*, p. 93.

17. Turner Morton, "Laysan—A Bird Paradise," *Pearson's Magazine* (May 1901).

18. "An Island Owned by Birds," *Friend: A Religious and Literary Journal*, Vol. 75, No. 11 (September 28, 1909).

19. Bryan, *Natural History of Hawaii*, p. 97.

20. Ibid.

21. "Laysan Island," *Youth's Companion* (March 9, 1905), p. iii.

22. "Laysan Island," Northwestern Hawaiian Islands Multi-Agency Education Project File, Laboratory for Interactive Learning Technologies, University of Hawaii (July 2008).

23. Bryan, *Natural History of Hawaii*, p. 95.

24. Morton, "Laysan—A Bird Paradise."

25. A. Binion Amerson, Jr., *The Natural History of French Frigate Shoals: Northwestern Hawaiian Islands* (Washington, D.C.: Paper No. 79, Pacific Ocean Biological Survey Program, Smithsonian Institution, 1971).

26. Northwestern Hawaiian Islands Multi-Agency Education Project Archive, Honolulu.

27. T.R., *A Book-Lover's Holidays in the Open* (New York: Scribner, 1916), p. 368.

28. "Birds in Millions Inhabit Laysan Island in Pacific," *Christian Science Monitor* (August 8, 1911), p. 7.

29. *Annual Reports of the Department of Agriculture* (Washington, D.C.: Government Printing Office, 1916), p. 241.

30. "Farallon National Wildlife Refuge Brochure," Farallon National Wildlife Refuge, San Francisco Bay National Wildlife Refuge Complex File, Newark, Calif. (September 2002).

31. Michael Grunwald, *The Swamp: The Everglades, Florida, and the Politics of Paradise* (New York: Simon and Schuster, 2006), pp. 125–126.

32. T.R. to Theodore Roosevelt Jr. (February 6, 1909).

33. Flagler, quoted ibid., p. 113. See also David Chandler, *Henry Flagler: The Astonishing Life and Times of the Visionary Robber Baron Who Founded Florida* (New York: Macmillan, 1986), p. 236.

34. Hermann Hagedorn and Sidney Wallach, *A Theodore Roosevelt Round-Up* (New York: Theodore Roosevelt Association, 1958), pp. 154–155.

35. Clara Barrus (ed.), *The Heart of Burroughs's Journals* (Boston, Mass.: Houghton Mifflin, 1928), p. 320.

36. Three-volume report of the National Conservation Commission (Washington, D.C.: Goverment Printing Office, 1909).

37. T.R., "Devilfish Harpooning," *Scribner's Magazine*, Vol. 62 (July–December 1917).

38. T.R., *An Autobiography* (New York: Macmillan, 1913), p.p 394–410.

39. "The All-American Interest in Conservation," *American Review of Reviews*, Vol. 34 (January–June 1909), p. 405.

40. "Saving of America," *Washington Post*, February 19, 1909, p. 1.

41. Barry Walden Walsh, "Gifford Pinchot, Conservationist," *Theodore Roosevelt Association Journal*, Vol. 24, No. 3 (2001), pp. 3–7.

42. "Forty-Five Nations to Hold Council on World Resources," *Christian Science Monitor* (February 20, 1909), p. 1.

43. Paul Russell Cutright, *Theodore Roosevelt: The Naturalist* (New York: Harper, 1956), p. 182.

44. McGeary, *Gifford Pinchot*, p. 108.

45. "Roosevelt and Taft Address a Meeting," *New York Times* (December 9, 1908), p. 5.

46. T.R. to James Rudolph Garfield (February 16, 1909).

47. Verlyn Klinkenborg, "Walking with Henry," *New York Times* (February 22, 2009).

48. John Allen Gable, "National Forests Created by Theodore Roosevelt," *T.R.A. Journal* (November 2005).

49. A. W. Greely, *Handbook of Alaska* (New York: Scribner, 1909), p. 26.

50. "T.R.'s Alaskan Views," *Washington Post* (August 11, 1911), p. 3.

51. Joshua David Hawley, *Theodore Roosevelt: Preacher of Righteousness* (New

Haven, Conn.: Yale University Press, 2008), p. 177.

52. A. W. Greely, *Handbook of Alaska: Its Resources, Products, and Attractions* (New York: Charles Scribner's Sons, 1909), pp. 27–203.

53. C. Hart Merriam, "Roosevelt, the Naturalist," *Science*, New Series, Vol. 75, No. 1937 (February 12, 1932). Also see C. Hart Merriam in Hagedorn and Sidney Wallach, *A Theodore Roosevelt Round-Up*, p. 137.

54. T.R. to Caspar Whitney (January 31, 1909).

55. T.R., "The Pioneer Spirit and American Problems," *Outlook*, Vol. 96, No. 2 (September 10, 1910), p. 56.

56. T.R. to the United States Army War College (February 8, 1909).

57. "Mount Olympus Park," *Time* (July 11, 1938).

58. Ronald F. Lee, *Family Tree of the National Park Service: A Chart with Accompanying Text Designed to Illustrate the Growth of the National Park System 1872–1972* (Philadelphia, Pa.: Eastern National Park and Monument Association, 1972), part 3. (This book is available online from the National Park Service.)

59. "Mount Olympus National Monument" (Washington, D.C.: National Park Service Archives, 1909).

60. "A Norwegian Explanation," *New York Times* (May 7, 1910), p. 8.

61. *Our National Parks* (Pleasantville, N.Y., 1985), pp. 222–232.

62. T.R. to James Joseph Walsh (February 23, 1909).

63. T.R., "The Pigskin Library," *Outlook*, Vol. 94, No. 18 (April 30, 1910).

64. Hans Driesch, *The Science and Philosophy of the Organism: Gifford Lectures Delivered at Aberdeen University, 1907* (Aberdeen, Scotland: Printed for the University, 1908), pp. 261–263.

65. T.R. to Robert Simpson Woodward (January 22, 1909).

66. T.R. to Jean-Jules Jusserand (February 25, 1909).

67. John A. Lomax, *Cowboy Songs and Other Frontier Ballads* (New York: Sturgis and Walton, 1910).

68. Timothy Egan, "This Land Was My Land," *New York Times* (June 23, 2007).

69. T.R. to Gifford Pinchot (March 2, 1909).

70. Lucy Maynard, "President Roosevelt's List of Birds," *Bird-Lore*, Vol. 12, No. 2 (March–April 1910), pp. 53–54.

71. T.R., "White House Bird List," in Lucy Maynard, *Birds of Washington and Vicinity*, 3rd ed. (Washington, D.C.: Woodward & Lothrop, 1909).

72. Jim Bendat, *Democracy's Big Day: The Inauguration of Our President, 1789–2009* (Lincoln, Neb.: iUniverse Star, 2008), p. 40.

73. Edmund Morris, *Theodore Rex* (New York: Random House, 2001), p. 544.

74. Nathan Miller, *Theodore Roosevelt: A Life* (New York: Morrow, 1992), pp. 494–495.

75. "Wife to Ride with Taft," *New York Times* (March 1, 1909), p. 1.

76. "Bible for Taft Inaugural," *New York Times* (February 14, 1909), p. 10.

77. Morris, *Theodore Rex*, pp. 550–555.

78. Ibid., p. 554.

79. "Roosevelt Says Good-Bye," *New York Times* (March 5, 1909), p. 3.

80. T.R. to William Allen White (February 19, 1909).

81. T.R., "Our Vanishing Wild Life," *Outlook* (January 25, 1913). This was a book review of William T. Hornaday's *Our Vanishing Wild Life* (New York: Charles Scribner's Sons, 1913).

ACKNOWLEDGMENTS

Naturalist Edward O. Wilson of Harvard University has written about a human condition he calls biophilia, the desire to affiliate with other forms of life. If ever there was somebody possessed with biophilia, it was Theodore Roosevelt. Wilson surmises that *Homo sapiens*, as a rule, has a genuine love of nature which is biological. Almost everybody responds intuitively with oohs and ahs when viewing a gorgeous valley, hiking a red-rock canyon, or hearing a loon call from a mud bog. Wilson's biophilia theory suggests that, at heart, humans *want* to be touched by nature in their daily lives. Wilson's hypothesis is the key to understanding why Roosevelt helped add over 234 million acres to the public domain between 1901 and 1909; it's his most enduring national legacy. Roosevelt responded both scientifically and emotively to wilderness. Therefore, I've purposefully avoided the fairly shopworn debate over whether Roosevelt was a nature preservationist or a utilitarian conservationist. He was both. Roosevelt was too many-sided and paradoxical to be pigeonholed. If forced to attach a single label to Roosevelt, I'd go with "Darwinian naturalist" (albeit one imbued with excessive biophilic needs).

Roosevelt's voluminous correspondence, books, articles, and diaries about his so-called outdoors life proved invaluable in writing this book. My institutional partner in tracking down all of Roosevelt's leavings was Dickinson State University. Under the leadership of Professor Clay Jenkinson, this fine North Dakota higher-learning institution created the Theodore Roosevelt Center in 2007. The center has undertaken a complete digitization of the Library of Congress's holdings of our twenty-sixth president. The center is also digitizing all T.R. photographs, films, audio, and ancillary papers.

The Library of Congress has also put T.R.'s papers on microfilm, an admirable move that made the collection user-friendly at the library of the University of Texas–Austin.

Another fine resource is the eight-volume *The Letters of Theodore Roosevelt* (Cambridge, Mass.: Harvard University Press, 1951–1954), selected and edited by Elting E. Morison. When convenient, I have quoted from the hundreds of missives in this outstanding primary source set. Morison and his associate editors, however, only partially tapped the reservoir of Roosevelt's brilliant correspondence. Bird watching, big game hunting, Interior and Agriculture department reforms, and Marshian conservation got short shrift in the *Letters* compared to political campaigns, foreign affairs, and trust busting. This gave me quite an opening: many of the excerpts of Roosevelt's letters and diary entries in this volume are appearing in print for the first time. Nobody before had systematically gone through Roosevelt's complete correspondence with an eye trained on his observations of the natural world. The result is that the intellectual influence of Charles Darwin on Roosevelt looms larger than previously discerned.

Special thanks to Aunna Carlton of Austin for helping me go through rolls and rolls of T.R. microfilm. When I taught at Tulane University, Andrew Travers did the same with reels of the Gifford Pinchot papers (courtesy of the Library of Congress).

Wallace Dailey, curator of the Theodore Roosevelt Papers at Harvard University, was a marvelous facilitator. An old-school archivist extraordinaire, Wallace helped me track down obscure photographs and boyhood diaries. The bulk of T.R.'s correspondence is held at the Library of Congress in letterbooks (which he started in 1897). Originals of these typed letters, however, are scattered about various institutional collections or held in private hands. A significant number of letters to family members are housed at Harvard, where I conducted research in the Houghton Library's reading room. All told, Roosevelt wrote more than 150,000 letters. I've read most of them with a keen eye for information pertaining to conservation. Meanwhile, the reader should be aware of my occasional use of Latin binomials to designate wildlife in taxonomic terms, which is limited to instances when Roosevelt used the Linnaean classification himself (or when it was absolutely pertinent to the narrative flow).

Since 1992, I've spent summers in the Badlands of North Dakota (first started as a Civilian Conservation Corps project in 1934, it became Theodore Roosevelt National Park in 1978). My family considers Medora—the park's gateway hamlet—our second home. When Roosevelt first arrived in the Badlands in September 1883, the area was still a frontier wilderness. Today, it's the easiest place in the Great Plains to encounter wild horses, buffalo, antelope, and prairie dogs. Numerous Roosevelt conservation sites in the Badlands have deeply inspired me (such as his Elkhorn Ranch, located thirty-five miles north of Medora). Everybody needs to discover a landscape that speaks to them, and mine is the washed-out prairie, rock-strewn slopes, and thick woody draws along the Little Missouri River. My true-blue friendships in North Dakota run extremely deep. I'd like to thank and send love to my North Dakotan friends, especially Sheila Schafer (an angel); Randy and Laurie Hatzenbuhler (T.R. Medora Foundation); Ed Schafer (former U.S. secretary of agriculture); Byron Dorgan (U.S. senator); Kent Conrad (U.S. senator); and Douglas and Mary Ellison (Western Edge Books). Valerie Naylor, the superintendent of Theodore Roosevelt National Park, dutifully read chapters pertaining to the American West. She is an amazing public servant.

Dr. John Allen Gable of the Theodore Roosevelt Association (TRA) of Oyster Bay, New York, was largely responsible for my writing this book. As longtime executive director of the TRA, John knew more minutiae about our twenty-sixth president than all the award-winning biographers combined. Back in April 1990, John and I cohosted the Theodore Roosevelt Conference at Hofstra University. (We later coedited a conference volume of the academic papers with Natalie Naylor.) In October 2002, John and I, along with Barbara Berryman Brandt, cochaired another T.R. conference ("The Big Stick and the Square Deal") at Canisius College in Buffalo. At both of these meetings, I lectured on T.R. and the environment. Following the Buffalo event—which featured biographers Edmund Morris, Kathleen Dalton, Candice Millard, Patricia O'Toole, and H. W. Brands—John encouraged me to write a definitive book on T.R. as our naturalist president. As an incentive, John put together the first comprehensive list of all the parks, forests, monuments, and bird reservations Roosevelt had saved. Gilding the lily, he opened up previously closed T.R. papers for me to use, including illuminating new material on Robert B. Roosevelt. Unfortunately, John died of cancer in 2005. He was only sixty-two years old. This book was written—in part—for him.

Tweed Roosevelt likewise encouraged me to write about his great-grandfather's conservationist legacy. Tweed's article "Theodore Roosevelt: The Mystery of the Unrecorded Environmentalist" (published in a 2002 issue of *Theodore Roosevelt Association Journal*) considered the notion that the creation of the U.S. Fish and Wildlife Service was Roosevelt's great institutional accomplishment. Often misunderstood by a public enamored with "scenic wonders" such as the Grand Canyon or Yellowstone, the U.S. Fish and Wildlife rangers and biologists (Rachel Carson was one) protect more than 280 different endangered species and their habitats in 550 national wildlife refuges. Every day, these men and women serve as Theodore Roosevelt's environmental foot soldiers. The U.S. Fish and Wildlife archive in Shepherdstown, West Virginia, is a treasure trove. If I ever execute my planned America in the Age of Conservation quartet—with *The Wilderness Warrior* serving as the first volume—Shepherdstown will surely become my new Medora.

Throughout the writing process I had five guardian angels: Paul Tritaik of Ding Darling National Wildlife Refuge; Mark Madison, historian of U.S. Fish and Wildlife in Shepherdstown; Lowell Baier, president of the Boone and Crockett Club; Professor John

Reiger of Ohio University–Chillicothe, the leading scholar on George Bird Grinnell; and Robert M. Utley, former chief historian of the National Park Service. They carefully read chapters and offered their expertise.

After I finished the first draft of the manuscript, I asked numerous specialists and friends to comment on chapters. My honor roll includes the following experts: Donald Worster of the University of Kansas; Doris Kearns Goodwin of Concord, Massachusetts; David Dary of the University of Oklahoma; Paul Schullery of Yellowstone National Park; Chris Darimont of the University of California–Santa Cruz; Mike Grunwald of *TIME;* Wendell Swank of the Boone and Crockett Club; Stephen Mark of Crater Lake National Park; Dorothy FireCloud and Hugh Hawthorne of Devils Tower National Monument; Harvey Leake, historian to the Wetherill family; Tom Farrell of Wind Cave National Park; Jeff Rupert of Wichita Mountains National Wildlife Refuge; Bruce Noble of the Chickasaw National Recreation Area; Debbie Baroff at the Museum of the Great Plains, Lawton, Oklahoma; Elizabeth Sims and Leslie Klinger of the Biltmore Estate, Asheville, North Carolina; John Flicker, president of the National Audubon Society; Edward Renehan, Jr., author of *John Burroughs: An American Naturalist*; Joan Burroughs, granddaughter of the legendary naturalist; Bob Edwards of the Jim Gatchell Memorial Museum, Buffalo, Wyoming; Martha Resk of the Audubon House, Key West, Florida; Ann Hornaday of the *Washington Post*; Michael Tuerkay of the Senckenberg Research Institute, Germany; Denise Pope of Trinity University in San Antonio, Texas; Stephen L. Zawistowski and Alison Zaccone of the American Society for the Prevention of Cruelty to Animals; Susan Blair of the National Geographic Society; Mike Gipple of the Mahaska County Conservation Board in Iowa; Patrick Sharp of California State University–Los Angeles; Minor Ferris Buchanan of Jackson, Mississippi; Jeff Johns of the Brooks Institute; Ellen Allers of the Smithsonian Institution; John Coleman of the University of Notre Dame; Pat Romero of the Rough Riders Museum, Las Vegas, New Mexico; Ryan Hathaway, David Bennett, and Chris Murray of the University of Delaware; Bill Kight of the White River National Forest; Richard Paterson of Grey Towers National Monument; Jennifer Capps of the Benjamin Harrison Home, Indianapolis, Indiana; Jamie Fowler of Ohio University; Cathy Engstrom and Greg Beisker of the Iowa National Heritage Foundation; and Nancy Freeman of NWR Center.

Rice University, where I teach history, is a fantastic, top-tier place of higher education. I would like to thank Professor Allen Matusow and Ambassador Edward Djerejian at the James A. Baker III Institute for Public Policy for allowing me to host a conservation public policy seminar. The staff of Fondren Library at Rice tolerated many demands. Thanks to Sara Lowman, Randy Tibbits, Suellen Denton, Cheryl Cormier, Barbara Hansel, Karol Comie, Ginny Martin, and Lea Martinello. Also, the Rice University biology department kept me honest about Darwin, especially Lesley Campbell, David Queller, and Jim Coleman. My colleagues in the history department at Rice allowed me to teach a graduate course on Theodore Roosevelt and conservation, which proved very beneficial to me.

The Smithsonian Institution unfailingly helped me better understand Spencer Fullerton Baird, sharing his correspondence with Robert B. Roosevelt and William T. Hornaday. Members of the World Conservation Society opened papers up to me, read chapters, and encouraged the writing process in numerous ways. Special thanks to the Bronx Zoo's director, Jim Breheny, and its vice president of communications, Mary Dixon. There is no finer way to spend a day than wandering around the 265-acre zoo, whose educational displays would make Charles Darwin proud. The staff of the American Museum of Natural History in New York City, especially Ellen V. Futter, did more than any other institution to encourage me in writing this book. Likewise, the Boone and Crockett Club Archive in Missoula, Montana—all 130 boxes—proved indispensable. The SPCA in New York enhanced my understanding of Henry Bergh mightily.

In Austin, Eric Busch, a history PhD student at the University of Texas, aided me on a couple of chapters. He's on the fast track to becoming a premier environmental historian. A knot of local historian friends graciously listened to my T.R. yarns at dinner parties with good cheer, including H. W. Brands, David and Jane Oshinsky, Robert Utley and Melody Webb, Lawrence and Roberta Wright, Don and Suzanne Carleton, Evan and Julia Smith, and Tom and Muffy Staley. They make living in Austin special.

I've spent so many days in Oyster Bay, New York (a town since 1653), that the mayor should issue me an honorary residence certificate. The drive to Sagamore Hill along Cove Neck Road remains the most surefire way to transport me back one hundred years to when T.R. was the reigning squire. At Sagamore Hill National Historic Site—run by the U.S. Department of the Interior—a special thanks is due to Thomas Ross, Charles Markis, Amy Verone, Eric Witzke, and Julie Abbate. When visiting Sagamore Hill, one must be sure to stop at the Theodore Roosevelt Sanctuary and Audubon Center, where there are nature programs featuring live birds of prey and ecosystem rehabilitation workshops. Adjacent to the sanctuary is Youngs Memorial Cemetery, which contains the graves of Theodore and Edith Roosevelt.

I've been a member of the Theodore Roosevelt Association since 1990. Besides attending annual meetings, I've participated in TRA excursions to the Netherlands, North Dakota, San Antonio, and Tampa Bay. A few stalwart friends I've made in the TRA include Theodore Roosevelt IV, Edmund and Sylvia Morris, Admiral C. S. Abbot, Mark Ames, Dr. William N. Tilchin, Robert D. Dalziel, Norman Parsons, Lucky Roosevelt, Stephen and Regina Jefferies, Simon Roosevelt, Elizabeth Moore, and Dr. Cornelius A. van Minnen. Two ardent Rooseveltian conservationist friends—Nate Brostrom of the University of California–Berkeley and William J. vanden Heuvel of the Franklin and Eleanor Roosevelt Institute—have helped me every step of the way.

At HarperCollins, my editor, Tim Duggan, was his trademark self—smart, proactive, and devoted. Anybody who believes the days of quality editing are over has never worked with Tim, whose standards are the best in the business. Jonathan Burnham epitomizes an excellent publisher in 2009—a lucid thinker and marketplace-savvy friend. Brian Murray, the president and CEO of HarperCollins worldwide, was typically helpful at every stage. I'll always be grateful for his wise counsel. Others at HarperCollins who deserve thanks are Allison Lorentzen (a genius facilitator), Susan Gamer (copyeditor), Katharine Baker (production editor), and Leah Carlson-Stanisic (designer). The indomitable Trent Duffy of New York assisted me in editing the first half of the manuscript. Likewise, the ethereal Emma Juniper assisted me during my final stages. Lisa Bankoff of ICM—my agent since 1992—offered her trademark sound counsel.

My indispensable research associate for the past two years has been Stone Weeks. Stone's father, Linton Weeks of National Public Radio, has been a longtime acquaintance of mine. One afternoon, he telephoned me about grabbing a cup of coffee with his twenty-two-year-old son Stone, who was poised to graduate from the University of Delaware. I said, "Sure." After meeting Stone in New Orleans, I was deeply impressed by his intellect, drive, and computer skills. Rice University president David Leebron hired him to be my personal assistant on this book, and we've worked closely together over the last two years. An Internet wizard, he helped me track down obscure books, academic articles, and interlibrary loan titles. During the manuscript preparation phase he was exemplary. He became like family. The world can expect great things from him in the coming decades.

As always, my most profound gratitude is reserved for my wife, Anne, who shared this adventure into T.R.'s America every step of the way. When other families were visiting Disneyland, the Brinkleys spent their summers at Roosevelt sites such as Muir Woods, Mesa Verde, Crater Lake, and Ocala National Forest. Together, Anne and

I have tried to bring nature into our children's lives. Our backyard in Austin is full of wildlife—a family of deer lives in our mesquite forest, and a raccoon mother recently gave birth to a kit under my daughter Cassady's bedroom. Weekly we jog around Lady Bird Lake in Zilker Park, looking for great blue heron and wood duck. Like pied pipers, whenever we have free time, we march Benton, Johnny, and Cassady to the Wild Basin just down Highway 360 to hunt for pinecones and smooth rocks near the magical waterfall. These regular family outings have convinced me that Theodore Roosevelt was right: we all need to bring nature into our daily lives.

Austin, Texas
April 12, 2009

IMAGE CREDITS

Grateful acknowledgment is made for permission to reproduce the images on the following pages:

Page xviii: T.R. visiting the Arizona Territory in 1913. (*Courtesy of Theodore Roosevelt Collection, Harvard College Library*)
Page 17: Illustration of T.R. petting a brown pelican. (*Courtesy of Michael McCurdy*)
Page 21: T.R. at Yosemite National Park. (*Courtesy of the National Park Service*)
Page 23: A precocious young T.R. (*Courtesy of Theodore Roosevelt Collection, Harvard College Library*)
Page 34: T.R. sketches of mammals. (*Courtesy of Theodore Roosevelt Collection, Harvard College Library*)
Page 42: Theodore Roosevelt, Sr. (*Courtesy of American Museum of Natural History*)
Page 49: Henry Bergh cartoon in *Puck*. (*Courtesy of the ASPCA*)
Page 56: Roosevelt family group shot. (*Courtesy of Theodore Roosevelt Collection, Harvard College Library*)
Page 62: T.R.'s Darwin evolution drawings. (*Courtesy of Theodore Roosevelt Collection, Harvard College Library*)
Page 78: Robert B. Roosevelt. (*Courtesy of the Theodore Roosevelt Association*)
Page 99: T.R. at Harvard. (*Courtesy of the Theodore Roosevelt Collection, Harvard College Library*)
Page 114: T.R. on North Woods of Maine hike. (*Courtesy of Theodore Roosevelt Collection, Harvard College Library*)
Page 127: T.R. and Elliott with big-game hunting coach. (*Courtesy of Theodore Roosevelt Collection, Harvard College Library*)
Page 159: Map of the Little Missouri River. (*Courtesy of T.R. Medora Foundation*)
Page 178: Roosevelt in customized Badlands costume (*Courtesy of Theodore Roosevelt Collection, Harvard College Library*)
Page 185: George Bird Grinnell. (*Courtesy of John F. Reiger*)
Page 192: T.R. guarding the boat thieves. (*Courtesy of Theodore Roosevelt Collection, Harvard College Library*)
Page 203: T.R. with Boone and Crockett Club antlers. (*Courtesy of the Boone and Crockett Club*)
Page 208: Frederic Remington's *Buffalo Hunter Spitting a Bullet into a Gun.* (*Courtesy of the Frederic Remington Art Museum, Ogdensburg, New York*)
Page 218: T.R. and John Burroughs at Yellowstone camp. (*Courtesy of Theodore Roosevelt Collection, Harvard College Library*)
Page 277: *The Great American Buffalo.* (*Courtesy of the Boone and Crockett Club*)
Page 291: T.R. with Grover Cleveland. (*Courtesy of the Library of Congress*)
Page 316: T.R. in Rough Riders uniform. (*Courtesy of Theodore Roosevelt Collection, Harvard College Library*)
Page 317: Colonel Roosevelt with mascots. (*Courtesy of Theodore Roosevelt Collection, Harvard College Library*)
Page 338: Roosevelt as governor. (*Courtesy of the Library of Congress*)
Page 343: Gifford Pinchot and forestry team. (*Courtesy of Gray Towers National Historic Site, Milford, Pennsylvania*)
Page 366: Congressman John F. Lacey. (*Courtesy of the University of Iowa Special Collections*)
Page 375: McKinley and T.R. (*Courtesy of the Theodore Roosevelt Association*)

Page 402: T.R. with Seth Bullock. (*Courtesy of David Dary*)
Page 405: T.R. with Booker T. Washington. (*Courtesy of the Theodore Roosevelt Association*)
Page 436: Holt Collier. (*Courtesy of Minor Ferris Buchanan*)
Page 441: Clifford Berryman's "Drawing the Line in Mississippi." (*Courtesy of the Cartoon Museum*)
Page 448: Ted Roosevelt Jr. with parrot. (*Courtesy of Theodore Roosevelt Collection, Harvard College Library*)
Page 451: "Use Forest Reserve Tonic." (*Courtesy of Ding Darling Estate*)
Page 453: Crater Lake National Park. (*Courtesy of the National Park Service*)
Page 475: T.R. as a rower. (*Courtesy of Theodore Roosevelt Collection, Harvard College Library*)
Page 480: Paul Kroegel. (*Courtesy of U.S. Fish and Wildlife Service*)
Page 486: Frank M. Chapman. (*Courtesy of the American Museum of National History*)
Page 498: Warden Guy Bradley. (*Courtesy of U.S. Fish and Wildlife Service*)
Page 508: T.R. and Burroughs near geyser. (*Courtesy of Theodore Roosevelt Collection, Harvard College Library*)
Page 526: T.R. standing at the Grand Canyon. (*Courtesy of the National Park Service*)
Page 539: T.R. and Muir at Mariposa Grove. (*Courtesy of the National Park Service*)
Page 542: T.R. and Muir at Yosemite National Park. (*Courtesy of the Sierra Club*)
Page 549: Finley and Bohlman. (*Courtesy of U.S. Fish and Wildlife Service*)
Page 559: First Lady Edith Roosevelt. (*Courtesy of Theodore Roosevelt Collection, Harvard College Library*)
Page 562: T.R. with pet dog. (*Courtesy of Theodore Roosevelt Collection, Harvard College Library*)
Page 573: T.R. with Charles W. Fairbanks. (*Courtesy of Theodore Roosevelt Collection, Harvard College Library*)
Page 596: Quanah Parker. (*Courtesy of U.S. Fish and Wildlife Service*)
Page 599: The great pasture hunt. (*Courtesy of Theodore Roosevelt Collection, Harvard College Library*)
Page 608: T.R. with a roped wolf. (*Courtesy of Theodore Roosevelt Collection, Harvard College Library*)
Page 616: T.R. in Colorado cabin with Skip. (*Courtesy of Theodore Roosevelt Collection, Harvard College Library*)
Page 632: Devils Tower National Monument. (*Courtesy of the National Park Service*)
Page 659: Ota Benga. (*Courtesy of the American Museum of Natural History*)
Page 666: T.R. races off to Pine Knot. (*Courtesy of the Edith and Theodore Roosevelt Pine Knot Foundation*)
Page 674: "Presidential Timberland." (*Courtesy of the Theodore Roosevelt Association*)
Page 685: T.R. at Arbor Day tree planting. (*Courtesy of Theodore Roosevelt Collection, Harvard College Library*)
Page 694: T.R. and Pinchot. (*Courtesy of Theodore Roosevelt Collection, Harvard College Library*)
Page 701: T.R. at the Louisiana canebrakes hunt. (*Courtesy of Theodore Roosevelt Collection, Harvard College Library*)
Page 711: Herbert K. Job photos. (*Courtesy of Trinity College, Hartford, Connecticut*)
Page 744: William L. Finley. (*Courtesy of U.S. Fish and Wildlife Service*)
Page 791: T.R. hiking the canyonlands of Utah. (*Courtesy of Theodore Roosevelt Collection, Harvard College Library*)
Page 793: T.R. chopping firewood. (*Courtesy of Theodore Roosevelt Collection, Harvard College Library*)
Page 817: T.R. inspired children to join the conservation movement. (*Courtesy of Theodore Roosevelt Collection, Harvard College Library*)

INDEX

Page numbers in *italics* refer to illustrations.

BOOKS BY DOUGLAS BRINKLEY

THE WILDERNESS WARRIOR
Theodore Roosevelt and the Crusade for America

ISBN 978-0-06-056531-2 (paperback)

A sweeping historical narrative and eye-opening look at the pioneering environmental policies of President Theodore Roosevelt, avid bird-watcher, naturalist, and the founding father of America's conservation movement.

THE GREAT DELUGE
Hurricane Katrina, New Orleans, and the Mississippi Gulf Coast

ISBN 978-0-06-114849-1 (paperback) • 978-0-06-112894-3 (audio CD)

The complete tale of the terrible storm, offering a unique, piercing analysis of the ongoing crisis, its historical roots, and its repercussions for America.

THE BOYS OF POINTE DU HOC
Ronald Reagan, D-Day, and the U.S. Army 2nd Ranger Battalion

ISBN 978-0-06-056530-5 (paperback) • 978-0-06-075959-9 (audio CD)
A chronicle of the brave men who conquered Pointe du Hoc on June 6, 1944, as well as the Presidential speech made forty years later in tribute to their duty, honor, and courage.

PARISH PRIEST
Father Michael McGivney and American Catholicism

ISBN 978-0-06-077685-5 (paperback) • 978-0-06-085340-2 (audio CD)
978-0-06-085348-8 (large print)

An in-depth biography of Father Michael J. McGivney, the Roman Catholic priest who stood up to anti-Papal prejudice in America and founded the Knights of Columbus.

TOUR OF DUTY
John Kerry and the Vietnam War

ISBN 978-0-06-056529-9 (paperback)

With exclusive access to personal diaries and letters, Douglas Brinkley explores Senator John Kerry's odyssey from highly-decorated war veteran to outspoken antiwar activist.